Petroleum Geology of the Irish Sea and Adjacent Areas

Petroleum Geology of the Irish Sea and Adjacent Areas

EDITED BY

N. S. MEADOWS
Redrock Associates Limited, Prenton, Wirral, UK

S. P. TRUEBLOOD
BHP Petroleum, London, UK

M. HARDMAN
BG Exploration and Production, Reading, UK

&

G. COWAN
BG plc, Reading, UK

1997
Published by
The Geological Society
London

THE GEOLOGICAL SOCIETY

The Society was founded in 1807 as The Geological Society of London and is the oldest geological society in the world. It received its Royal Charter in 1825 for the purpose of 'investigating the mineral structure of the Earth'. The Society is Britain's national society for geology with a membership of around 8000. It has countrywide coverage and approximately 1000 members reside overseas. The Society is responsible for all aspects of the geological sciences including professional matters. The Society has its own publishing house, which produces the Society's international journals, books and maps, and which acts as the European distributor for publications of the American Association of Petroleum Geologists, SEPM and the Geological Society of America.

Fellowship is open to those holding a recognized honours degree in geology or cognate subject and who have at least two years' relevant postgraduate experience, or who have not less than six years' relevant experience in geology or a cognate subject. A Fellow who has not less than five years' relevant postgraduate experience in the practice of geology may apply for validation and, subject to approval, may be able to use the designatory letters C Geol (Chartered Geologist).

Further information about the Society is available from the Membership Manager, The Geological Society, Burlington House, Piccadilly, London W1V 0JU, UK. The Society is a Registered Charity, No. 210161.

Published by the Geological Society from:
The Geological Society Publishing House
Unit 7
Brassmill Enterprise Centre
Brassmill Lane
Bath BA1 3JN
UK
(*Orders*: Tel 01225 445046
            Fax 01225 442836)

First published 1997

The publishers make no representation, express or implied, with regard to the accuracy of the information contained in this book and cannot accept any legal responsibility for any errors or omissions that may be made.

British Library Cataloguing in Publication Data
A catalogue record for this book is available from the British Library.

ISBN 1-897799-84-5
ISSN 0305-8719

Typeset by E & M Graphics,
Midsomer Norton, Bath BA3 2NS, UK

Printed by The Alden Press, Osney Mead, Oxford, UK

Distributors

USA
  AAPG Bookstore
  PO Box 979
  Tulsa
  OK 74101-0979
  USA

(*Orders*: Tel. (918) 584-2555
            Fax (918) 560-2652)

Australia
  Australian Mineral Foundation
  63 Conyngham Street
  Glenside
  South Australia 5065
  Australia

(*Orders*: Tel. (08) 379-0444
            Fax (08) 379-4634)

India
  Affiliated East-West Press PVT Ltd
  G-1/16 Ansari Road
  New Delhi 110 002
  India

(*Orders*: Tel. (11) 327-9113
            Fax (11) 326-0538)

Japan
  Kanda Book Trading Co.
  Tanikawa Building
  3-2 Kanda Surugadai
  Chiyoda-Ku
  Tokyo 101
  Japan

(*Orders*: Tel. (03) 3255-3497
            Fax (03) 3255-3495)

Contents

Sedimentology

Fault analysis and diagenesis

Field studies

Preface

Historically the sedimentary basins to the west of mainland Britain have taken a backseat role in the exploration of the UK Continental Shelf. Throughout the 1980s the discovery and development of fields in the prolific basins of the North Sea retained both the technical focus and, more importantly, the budget funds of major operators. Consequently, the history of exploration in western areas, and particularly in the East Irish Sea Basin, is one of alternating feasts and famines (Colter, this volume). In this context the recent years of 1990–1995 represent a major feast, the second such feast in the exploration history of the region. More importantly, a veritable feast both in the volume of hydrocarbons discovered and the apparent advancement of geological understanding.

The initial leap forward, made during the mid 1970s in the East Irish Sea Basin, is, however, likely to remain the volumetrically most important. Current workers in these basins owe a great debt to the 'early pioneers' who carried out the initial exploration and simultaneously had to break the mould of traditional North Sea thinking.

Ultimately it was the discovery of the giant Morecambe Field in 1974 which proved the value of exploration in the west but the full potential of the area has, we believe, yet to be realized. Witness the efforts of the early 1990s which overturned the 'one field gas basin' paradigm and transformed the East Irish Sea Basin into a mature oil and gas province. This gave impetus to exploration, drew attention to the geological complexity and, in so doing, added more than a dozen fields to the maps of the area, including the oil fields of Douglas and Lennox.

This volume is not intended to comprise the definitive statement on the petroleum geology of the East Irish Sea and adjacent areas, but rather to provide a snap-shot of current understanding which will hopefully prove a platform for further exploration in basins west of Britain. We hope that our joint efforts will encourage the continuing exploration of the region, particularly in the lesser explored basins. We wish success to those with the belief, imagination and funds to give the west a try.

In producing this volume we are indebted to a number of individuals and organizations: the authors and referees who have given freely of their time and expertise, the companies and universities who have permitted publication and also our employers for giving us the time to devote to the task. Mention must also be made of the Geological Society staff and officers who have guided us through the organization of both the conference from which the volume stemmed and the volume itself.

Neil Meadows
Steve Trueblood
Martin Hardman
Greig Cowan

The East Irish Sea Basin – from caterpillar to butterfly, a thirty-year metamorphosis

V. S. COLTER

Pammenters House, The Old School, High Street, Wendover,

Bucks HP22 6DK, UK

Abstract: In the thirty years since the award of the first licenses, the East Irish Sea Basin has emerged as a significant hydrocarbon province. This paper first lists some of the occasionally almost arbitrary events that led to the first success in the basin, the discovery of the Morecambe Field in 1974. An attempt is made to review progress over those thirty years in certain topics, namely (1) stratigraphy, (2) structure, (3) sedimentology, (4) diagenesis – 'the Platy Illite problem' and others, (5) uplift and inversion, (6) hydrocarbon sources and types, (7) East Irish Sea Basin analogues. The paper concludes by summarizing the current state of knowledge.

1995 represented the thirtieth anniversary of the first award of licences in the East Irish Sea Basin (EISB), the twenty-sixth anniversary of the drilling of the first wells and the twenty-first anniversary of the discovery of the South Morecambe Field. It is, perhaps appropriate to look back on some of the crucial events of the last thirty years, and on some of the earlier concepts and interpretations, with a view to seeing how these last have progressed or been found wanting. Undoubtedly, the biggest spur to exploration was the discovery in 1974 by the British Gas Corporation subsidiary HGB Ltd of the South Morecambe Field. How a fledgling organization came to discover this 5 TCF gas accumulation in its first venture, is itself noteworthy.

Key events that led to this discovery might be summarized as follows:

(1) the wish of the government in power in 1969 to foster a national presence in the exploration of British waters by allowing the then Gas Council to take 100% interests in EISB blocks;

(2) the readiness of the Gas Council to exploit this opportunity;

(3) the unwillingness of the Council's main North Sea partner to participate in the proposed application, with its offer of a firm well;

(4) the unwillingness of other majors to join the application;

(5) the informal gift by Gulf in 1973 of the logs of wells 110/8-1 and -2, on the basis of a 'gentleman's agreement' that HGB would likewise make data available from any wells it might drill;

(6) the brilliant re-interpretation of the logs of well 110/8-2 by John Bains in 1974, to reveal approximately 600 feet of untested pay in the Sherwood Sandstone;

(7) the persistence of the Exploration Manager Peter Hinde in spending three months to persuade the then General Manager that a well on trend with 110/8-2, in Block 110/2, was preferable to one on the seemingly simple egg-shaped structure on Block 110/7 that had been the preferred location. Had Bains' reinterpretations not been available, or had the dry hole 110/7-1 been drilled first anyway, it is easy to imagine that a subsequent well on trend with the seemingly dry 110/8-2 would have met with extreme resistance.

Undoubtedly, Bains and Hinde must be regarded as the first heroes in the EISB. Many have followed in their wake and the EISB now boasts some twenty hydrocarbon accumulations, significantly including the oil fields of Douglas (Yaliz, this volume) and Lennox (Haig *et al.*, this volume) discovered by BHP (then Hamilton Oil) in 1990 and 1992 respectively.

Stratigraphy

The acquisition of the logs of 110/8-2 showed at once that the initial concept of the East Irish Sea being a direct analogue of the Southern North Sea was false, at least as far as the licensed areas were concerned, inasmuch as the hoped for Collyhurst Sandstone reservoir was lacking in that well. Increased drilling over the years, and more recent thinking have revealed some of the shortcomings of the early attempts at Permo-Triassic correlation in the EISB (Colter & Barr 1975; Colter 1978). The Lower Permian correlations in these wells implied the presence at the centre of the EISB of an area of shale and mudstone, equivalent to the coarser Collyhurst sandstone (Colter & Barr 1975, fig. 5A) The analogy was thought to be the Silver Pit Formation of the North Sea. This interpretation, based as it was on the presence in 110/8-2 of a unit interpreted to be Permian in age, with the benefit of hindsight glossed over an ever-present problem in defining the base of the Permo-Trias, namely that of distinguishing between true Permo-Trias and reddened Carboniferous. Although Jackson *et al.* (1987, fig. 7a) refined Colter & Barr's interpretation, Jackson *et al.* (this volume) now interpret the supposed playa lake sequences in wells 110/8-1, 110/8-2 and 113/26-1 to be largely or wholly Carboniferous. The fact that any equivalent to the Collyhurst Sandstone appears to be absent from the interpreted Quadrant 109 Arch should come as no surprise, given the thickness variations in the Formby wells and at the outcrop (Colter & Barr 1975, p.66).

Although at the time of Colter's & Barr's 1975 paper Zechstein correlations were being carried out in the North Sea (Taylor & Colter 1975), it was not thought possible to extend this work to the Irish Sea. Recent work (Holliday 1993*a* and subsequent discussions) shows that this problem and the palaeographic consequences remain. The lithostratigraphical subdivision of the Carboniferous by Jackson *et al.* (this volume) provides a basis for correlating these older sequences although there is still very little known about them in a regional sense. The few wells that have penetrated the Carboniferous provide only a glimpse of early palaeogeographies although the significant thickness of Tournasian to early Visean evaporites recorded in the Edinburgh Oil and Gas well, Easton-1, by Ward (this volume) provides a valuable insight.

The overlying 'silicified' zone of the Sherwood Sandstone, so called by Colter & Barr, (Rottington Sandstone Member of Jackson *et al.*, this volume), has been since shown to be represented in the Larne No 1 Borehole (Penn *et al.* 1993), proving its presence in three basins. The long recognized unconformity at the base of the old Keuper Sandstone (Helsby and Ormskirk Sandstone formations of the Cheshire and East Irish Sea basins respectively) was interpreted in 1975 to be marked by an abrupt decrease in gamma-ray activity near the top of the Sherwood Sandstone in the onshore and offshore wells studied. Above this, in Knutsford No.1 a threefold subdivision of the remaining section was correlated with David Thompson's three outcrop members (Thompson 1970). Above the Sherwood Sandstone was a unit thought to represent the Keuper Waterstones (Tarporley Siltstone Formation). These four units were correlated with apparently similar units in the Formby wells and thence to 110/8-2. One problem with this correlation was the palaeogeographic consequence of having a more or less fluvial sandstone at the top of the Sherwood Sandstone of the EISB equivalent to the apparently intertidal Waterstones onshore. There were also some

mutterings about the impropriety of extending onshore member names into another basin, whose supposed equivalents were of different facies. The surviving author must take the blame for this breach of procedure, although in the light of facies interpretations presented in this volume, the exact environment of deposition seems to be a moving target (see 'Sedimentology', below).

Recent work has, however, called this whole correlation into question. Evans *et al.* (1993) put Thompson's lowest, onshore Keuper Sandstone member (the Thurstaston Sandstone Member) in the Wilmslow Sandstone (i.e. below the unconformity). Offshore, Jackson *et al.* (this volume) believe that the offshore 'Waterstones' of Colter & Barr are equivalent to the upper part of the onshore Frodsham Member. Contrasting with this interpretation, however, Thompson & Meadows (this volume), regard the onshore and nearshore Frodsham Member, as the age-equivalent of the lowest Mercia Mudstone in the central parts of the EISB. Permo-Triassic correlation problems have been with us for more than a century, and it seems that they will be providing geologists with harmless amusement for many years yet.

Structure

The huge increase in the amount of structural information entering the public domain can be guessed at by comparing the simple maps of the EISB published in 1975 by Colter & Barr (figs 1 & 2) with that of Jackson & Mulholland published in 1993 (fig. 2). Until at least the early 1980s, structural interpretations were hampered by a paucity of continuous reflections. In general, two events were mapped; these being the base Mercia Mudstone and 'base' Permo-Triassic. Variations in Mercia Mudstone salt sections across faults led to incorrect picks of the first of these, from one side of some faults to the other. Absence of reflectors also contributed to uncertainty in defining fault geometry; i.e. should they be drawn automatically as listric, as was preferred by some during discussions of the South Morecambe Annex B, or should caution prevail and they be represented as planes. This difference of opinion was not without its broader implications, as the listric interpretation reduced estimated reserves somewhat, at a time when public sector expenditure was under pressure. Since then, horizon discrimination by seismic has advanced beyond recognition, and Jackson *et al.* (1987) were able to show a top Ormskirk Sandstone, top St Bees Sandstone, 'Yellow' (top silicified zone) and 'Brown' (top Carboniferous) reflectors, while deep reflection data has begun to reveal the lower crustal structure (England & Soper, this volume).

Further, analysis of onshore faults (Beach *et al.*, this volume) has provided much more detailed models of fault relationships and transmissibility that can now be used in assessments of seal potential and, hence, reservoir production that, in turn, impact directly on offshore field development models. Similarily, detailed study of the Isle of Man exposures has allowed Quirk (this volume) to shed more light on the Devonian and early Carboniferous evolution of the area. The interpretation of these older sequences, however, requires some care, particularly on the basis of seismic data alone, as amply illustrated by the extraordinary, and highly seismically reflective, Lower Carboniferous evaporite sequence discussed by Ward (this volume) in the Easton-1 well.

Sedimentology

In the early days of the discovery and delineation of the Morecambe Field, little purely depositional sedimentological work was carried out. This was in part the result of a rather parsimonious approach to expenditure on coring, and also to a perceived need to apply limited manpower to what were seen as more pressing matters, i.c. exploration and delineation drilling and diagenetic studies of the reservoir. It was thought that the succession was largely fluvial with some probable aeolian content. Given the wide spacing of the available wells and the, by and large, sporadic cores, it was not clear what a more refined interpretation could contribute at that time. In the light of this meeting, it would probably have been wrong anyway!

In the 1980s, with more core data from platform wells as well as new ideas, Bushell (1986) interpreted the Morecambe reservoir succession in terms of a fluvial model (figs 5 & 6). Later, Woodward & Curtis (1987, pp. 211–214 and figs 11 & 12), following the work of Macchi (1987), recognized the existence of aeolian intervals in the platy illite-affected parts of the Morecambe reservoir, and their special contribution to the production from this zone. Cowan *et al.* (1993) reported on the use of dipmeter logs in palaeocurrent analysis in the Morecambe Field. These authors noted the existence of strong westsouthwest depositional dips throughout the reservoir, which were interpreted to represent dips in sands deposited in the west flowing channels. This interpretation differs markedly from the more usual palaeogeographic interpretation of drainage from the south, passing through the Worcester Graben and the Cheshire basin.

This same phenomenon had been noted in the 1970s by the late Dr K. W. Barr, in one of the early wells, but at the time the dips seemed too regular in azimuth to represent depositional dips, and the

observation was filed away for future consideration. The Cowan *et al.* (1993) interpretation of west flowing channels seems to conflict with that of Meadows & Beach (1993, p.831, fig. 8) who suggested that drainage systems entered the EISB from the south and exited from the north. The apparent contradiction between these two views may be a question of scale, in that the Cowan *et al.* (1993) interpretation may represent a more local system that was structurally controlled and drained towards the Keys Fault as these authors note (1993, p.881), whereas Meadows & Beach address the basin as a whole. Cowan *et al.* (1993) also drew attention to aeolian deposits in the dominantly fluvial Sherwood Sandstone, especially in connection with their economic significance as thin high productivity layers. The aeolian interpretation seems to be gaining ground, however, with Herries & Cowan (this volume) interpreting the 'sheetflood' sands of Stuart (1993, fig. 8) in the North Morecambe Field as aeolian sabkha deposits. This change of interpretation affects close to half the interval penetrated and one can only speculate on what the long-term effects may be on development decisions and reservoir models based on long held, but now discarded, views of depositional environments. Unless, that is, these facies interpretations are more theoretical than practical.

Diagenesis

In the mid 1970s, it was thought useful to try to relate porosity/permeability measurements to the depositional and diagenetic fabrics seen in the plugs from the Morecambe reservoir. The most significant data were very low permeability values associated with high porosities in some samples from the lower part of the Ormskirk Sandstone (Colter & Ebbern 1978, plate 4). Diagenetic clay minerals were suspected and the presence of platy illite was confirmed by John Ebbern and the SEM operator Ramendra Sinha, at BG's London Research Station (Colter & Ebbern, 1978, plate 2e and f). As a result of this discovery, the frugal coring policy was relaxed, and the reservoir in well 110/2-6 was fully cored, confirming the existence of a 'Platy Illite Zone' in its lower part (Colter & Ebbern 1978). At the time, a fossil gas–water contact was suspected, it being surmised that the gas had expanded during basin inversion, pushing down the gas–water contact (GWC) to its present day level. Lack of firm data prevented any volumetric check on this view and apatite fission track data now show that since cooling has also occurred gas volumes could in fact shrink (G. Cowan, pers. comm.). In the 1980s, however, Bushell (1986, p.204 and figs 12 & 13) confirmed the likely hydrocarbon–water contact nature of the Platy Illite Zone, and its more or less field-wide distribution. This author then had K/Ar dating carried out on the illites, revealing a Jurassic age (Bushell 1986, p.202 & fig. 10).

It was proposed that the Platy Illite Zone represented a hydrocarbon–water interface beneath an earlier, possibly oil, accumulation, which had escaped during Cimmerian breaching. At first sight this theory suffered from the fact that it was necessary to postulate the existence of something no longer present. Woodward & Curtis (1987, p.208), however, revealed that the non-illite zone in Morecambe had suffered common post-illite quartz cement, demonstrating a later flushing of the reservoir. Hardman *et al.* (1993, pp. 814–815 and fig.8) have since shown that either this cementation was in two phases or that a subsequent heating event affected fluid inclusions present in a single cement phase. In this volume Cowan & Bradney present data that indicate solutions were funnelled from the Keys Basin in the northwest, between the Keys and Tynwald faults, onto the crest of the Morecambe structure where they escaped and dissipated themselves, leaving the east flank of the field largely illite-free.

The relationships between diagenesis and hydrocarbon emplacement have also shed further light on various aspects of basin evolution. Greenwood & Habesch (this volume) use a range of stable isotope, fluid inclusion and cathodoluminescence data to reveal some of the complexity of the diagenetic history. Early calcites, predictably, precipitated from meteoric fluids but, more significantly, early sulphates in the upper parts of the Ormskirk Sandstone are shown to have formed from Triassic seawater. These authors also demonstrate several episodes of oil and gas migration in the reservoir, each of which has been recorded in diagenetic phases and has implications for any models of burial and thermal history. A broadly analagous situation onshore is discussed by Rowe & Burley (this volume) within the Cheshire Basin at Alderley Edge. Here significant permeability contrasts are related to the combination of sedimentary characteristics, structural deformation and highly variable cement precipitation, the latter involving locally intense mineralization, related to migration pathways and fault patterns, that again represent multiple episodes of fluid flow. On a broader scale, Millwood-Hargrave & Francis (this volume) have demonstrated that the diagenetic effects associated with fossil hydrocarbon–water contacts may be seen on seismic. It will be up to individuals to assess the significance of all these aspects in terms of their relevance to hydrocarbon migration, structural evolution and the prospect of present day undiscovered accumulations.

Uplift and inversion

Although the general lack of post-Triassic sediments in the EISB had long suggested uplift and erosion of a considerable section of younger Mesozoic beds, it was difficult in the early days to demonstrate either that this was the case or, if so, the amount of that uplift. The drilling of well 110/3-2 in 1976, however, suggested a means of estimating the amount of uplift that had taken place. This well showed not only unusually high sonic velocities in the Mercia Mudstone, but a distinct gradient from top to bottom. These values were plotted on the then recently published North Sea Bröckleschiefer burial curve of Marie (1975, fig. 1D), and suggested about 2000 m of uplift and erosion might have taken place, subject to all kinds of reservations (Colter 1978, fig. 6). Although no timing for the uplift was implicit in this interpretation it was thought most likely to be early Tertiary in age on regional grounds. Bushell (1986) published burial history curves for the Morecambe Field showing an interrupted 4500 feet of Cimmerian uplift, followed by reburial and gradual uplift, starting about 50 million years ago and persisting to the present day. The question of the amount of Cimmerian uplift and erosion, if present, was addressed by Hardman *et al.* (1993), who pronounced its existence to be 'an enigma' (p.820). This question appears to have remained unresolved despite further discussions during the meeting that prompted this volume.

Over the years since 1978, many workers have given estimates of Irish Sea Basin uplift and /or inversion, using a variety of indirect evidence. Jackson *et al.* (1987, p.202) confirmed the figure of 2000 m of uplift, using the same sonic log analysis. Hardman *et al.* (1993) used the same method to propose 5100–5500 ft (1555–1676 m) of uplift at the site of an unspecified well. In the same paper, however, these authors quoted figures of up to 10 000 ft (3550 m), based on vitrinite reflectance (VR) measurements. This estimate is not dissimilar to that quoted by Jenner (1981, p.431) for the Kish Bank Basin, based on spore coloration. Fault analysis, however, led Knipe *et al.* (1993, p.864) to the conclusion that about 2000 m of uplift had taken place in the EISB.

Meanwhile, work on fission track analysis had led its users to propose considerable (1–2 km) uplift away from known inversion axes. Green *et al.* (1993*a*, p.1069) pointed out that the sonic log method used by Marie (1975) and Colter (1978) only works if the reference section itself has suffered no uplift. By implication, 1–2 km of additional uplift might have to be added to any value arrived at using this method. It began to appear that methods relying on heating (VR, spore coloration and fission track analysis) were coming up with interpreted uplifts considerably in excess of those based on shale velocity and structural reconstruction. Cope (1986) pointed out that VR measurements from the East Midlands Shelf were incompatible with the known burial history. Nevertheless, Green *et al.* (1993*a* p.1069) reconciled the VR measurements with burial histories based on fission track analysis, and Holliday (1993*b*) raised some objections to the assumptions used in converting fission track temperatures to depths of burial. In a later paper, however, Green *et al.* (1993*b*) apparently rather suddenly also questioned the belief that fission track analysis results can be interpreted simply in terms of depth burial in the northwest of Great Britain, and noted that in the EISB the results are best explained by about 2000 m of uplift, accompanied by additional heating caused by fluid flow.

These figures are repeated by both Green *et al.* (this volume) and by Quirk for the Isle of Man area (this volume), while the stratigraphic approach to the post-Triassic sequences of Warrington (also in this volume) indicate a pre-Cretaceous-Chalk inversion event as well as the more familiar Early Tertiary episode. The regional significance of the earlier event is indicated by evidence of its effects in Northern Ireland provided by Shelton (this volume) in his discussion of the Larne Basin. In a quantitative stratigraphic approach to the question of lost section, Cope (this volume) extended his 1994 work to suggest that perhaps 2130 m of lost Mesozoic section could be postulated. This author proposes a mantle plume as the cause of the uplift. A consensus seems to be emerging, then, that around 2000 m of uplift and erosion in early Tertiary times, is compatible with most evidence from the EISB, a figure remarkably like that proposed on scanty evidence in 1978. The last word, however, is unlikely to have been written on this subject. An implication of a late heating event resulting from fluid flow is that early Tertiary inversion structures may be able to reservoir any hydrocarbon so produced, a possibility that has embarrassed the writer, who was, until recently, still living in a post-Wytch Farm world, in which 'alpine' inversion structures post-dated migration.

Hydrocarbon sources and types

When the Gas Council first went into the EISB in 1969, the thoughts foremost in the mind were of the gassy Point of Ayr and Whitehaven collieries, both of which had fed gas into the local gas works. At the back of the mind were the Formby oil seep and a compendium of oil shows in Lancashire coal mines. The Irish Sea was seen, however, mainly as a Southern North Sea analogue. Although little was known about the distribution of Carboniferous rocks in the EISB, and less of their source potential, the discovery of the Morecambe Field, and even of sporadic oil traces did not seem to conflict with the original notion that the Coal Measures might be the principal source.

More recently, improved knowledge of the Carboniferous succession and its source potential, plus the discovery of oil fields, has focused attention on the Namurian in a basinal facies, as the principal source. Hardman *et al.* (1993 pp. 811 & 812) noted the largely inertinitic kerogens of the Coal Measures, but the presence of oil-prone kerogen in the equivalents of the Namurian Sabden Shale. Jackson *et al.* (this volume) showed a variety of kerogen types in the basinal Namurian, from gas prone in the northern, proximal area to oil prone in the southern, distal area. Approaching from the south side of the basin, Armstrong *et al.* (this volume) propose that the most likely candidate for the source of oil in the 110/13-10 and 110/15-6 wells is in the lower part of the Namurian Hollywell Shale, with some possible input from higher, more proximal source rocks in the upper part of this unit.

At the time of the Morecambe discovery, the timing of gas generation was thought, vaguely, to be some time before the assumed early Tertiary inversion. Later, Bushell (1986) produced evidence of an early accumulation in the Morecambe structure, of Jurassic age, which escaped during Cimmerian breaching. The present accumulation was thought to have formed in Cretaceous to Early Tertiary times. Parnell & Veale (this volume) have presented mid-Mesozoic dates for uraniferous and thoriferous bitumens in the Solway Basin, which may not be inconsistent with the Bushell (1986) Jurassic generation episode. On the basis of fission track analysis, Green *et al.* (this volume) present evidence of heating events of 290 ma, 260–220 Ma and 140–110 Ma. Green *et al.* also refer to a possible Jurassic event, but suggest that peak generation in the southern part of the EISB was just before the early Tertiary cooling event, which followed the suggested hot fluid flush.

Combinations of various periods of burial and uplift have obviously produced a wide variety of thermal histories, and earlier-generated hydrocarbons would have had to survive the gamut of tectonics and heating in order to persist to the present day. As Green *et al.* (this volume) state, it appears that the southern part of the EISB had the most favourable tectonic and thermal history for the survival of reservoired hydrocarbons. The wide variety of hydrocarbon types and non-hydrocarbon gasses no doubt reflects the vicissitudes of the thermal history, but the sporadic occurrence of H_2S is perhaps of special interest. The H_2S and nitrogen in the Morecambe gas were, for want of a better theory, assumed in the 1970s to be the products of higher levels of organic metamorphism before inversion. The much higher concentrations of H_2S in the southern areas of the basin were first noted with the drilling of 110/7a-3 in 1982 (the discovery well for the Calder Field discussed by Blow & Hardman, this volume). These abundancies were confirmed with the later discovery of high H_2S contents in the Hamilton, Douglas (Yaliz, this volume) and Lennox fields (Haig *et al.*, this volume). The relatively newly identified hot fluid pulse offers a non-depth related mechanism. It might also be speculated that the presence of anhydrite in the Ormskirk Sandstone (Thompson & Meadows, this volume) could have some bearing on H_2S distribution (Hardman, pers. comm.).

Application of heat is, however, not the only possible mechanism for H_2S generation. McGovern-Traa & Spark (this volume) suggest that sulphate-reducing bacteria may be alive and well in reservoir waters, and could be responsible for the H_2S generation. This view echoes work done by Dr Eileen Pankhurst at the Gas Council's London Research Station in the 1960s. In those pre-North Sea days, the Council was interested in storing town gas in shallow structures onshore, and was anxious that it did not put sweet gas into the aquifer and get sour gas out. In order to investigate the value of bactericides in drilling fluids, the presence of live, indigenous bacteria in the subsurface was investigated. The presence of indigenous bacteria in the Lockton Zechstein sour gas accumulation in Yorkshire was suspected, but this work folded with the advent of the North Sea Gas and different storage requirements.

East Irish Sea Basin analogues

It might be expected to be unlikely that the EISB would be the only one of a number of superficially similar basins to contain commercial hydrocarbons. The experience to date is, however, disappointing. Green *et al.* (this volume) mention the suitability of the thermal history of the southern part of the EISB for hydrocarbon generation and preservation. Adjacent regions, presumably including the rest of EISB, are said to have more complex and, in some cases, less suitable histories.

Before the discovery of the South Morecambe Field, and the realization that the Sherwood Sandstone was the principal target, Gas Council (Exploration) Ltd drilled the well Knutsford No 1 on behalf of itself and BP. The target was the Permian Collyhurst Sandstone capped by the Manchester Marls. The well was off-structure at top Sherwood Sandstone level. In the event, whatever the structural integrity of the 'Base Permian'

structure drilled, or the thermal history, the Manchester Marl was poorly developed and no seal existed. Since then, a number of structures at Top Sherwood Sandstone level have been drilled, without success. Mikkelsen & Floodpage (this volume) suggest that the nature of the Carboniferous subcrop may be critical, most dry holes having entered pre-Permian sections with little or no source potential. This interpretation presumably discounts long range migration in the Sherwood Sandstone. Chadwick & Rowley (this volume), however, are more optimistic and point out the many similarities between the Cheshire Basin and the offshore EISB as illustrated by their structural modelling.

In the Cardigan Bay Basin, GC(E) participated in the drilling of two Sherwood Sandstone wells, one (106/28-1) as an operator with Deminex, and the other (103/2-1) with Texaco as operator. Both penetrated Sherwood Sandstone sections, and 103/2-1 bottomed in interpreted Westphalian D Barren Red Measures (Barr *et al.* 1981, p.438). No information was available at the time of that paper on the maturity of the Late Carboniferous in 103/2-1, but the authors noted the immaturity of Carboniferous in the HGB well 107/21-1 (VR 0.5% at TD). Minor gas shows were recorded from the Lias below 1960 m, and the down-dip equivalents are presumably more mature. Nevertheless, Green (this volume) note that the palaeotemperatures based on fission track analysis drop off into the Welsh Massif, suggesting that this acted as a buttress against fluid transport. The Cardigan Bay Basin may, therefore, have escaped the heat phase critical to hydrocarbon generation in the EISB. The Kish Bank Basin, on the other hand, may not have escaped this heating event, if the Jenner (1981) estimate of 3000 m of uplift, based on spore coloration is viewed in the same light as similar heat-based observations in the EISB. Naylor *et al.* (1993), however, propose that lack of Carboniferous Coal Measures in the deeper part of the basin may be a reason for failure to find hydrocarbons in this basin.

The onshore Worcester Graben has been explored by the drill without success, due mainly it seems to the fact that the Permo-Triassic section overlies Precambrian and Lower Palaeozoic strata (Scott & Colter 1987, p.102). A measure, however, of enthusiasm for still virgin basins is presented by the British Gas exploration of the North Channel Basin (Maddox & Blow, this volume) and by the drilling, carried out or proposed, of the Central Irish Sea, Peel and Solway basins by BHP, Elf Enterprise Caledonia and Esso respectively. At the time that the ancestor of BG E&P, the Gas Council, first took the plunge and took out 100% licences in the EISB, at least one geologist (the author) had some uneasiness that this small basin could, in an area a fraction of the size of the North Sea Basin, have all of the essentials for commercial hydrocarbons, and have them in the right order. The present organization, a quarter of a century later, is happy to explore a basin covering half a dozen blocks, overlain by 300 m of water containing unexploded ordnance and the playground of nuclear submarines. Such is either the romance of these basins or the shortage of good exploration acreage 25 years on.

The only other comparable Triassic basin that has had success is the Wessex Basin, but this depended for its romance on a favourable juxtaposition of Sherwood Sandstone and a Liassic source rock. This possibility is, alas, lacking from most such basins.

Conclusions

In the thirty years since the award of the first licences in the EISB, this basin has passed from exciting little interest in some quarters, and none at all in others, through the stage of having been a 'one field basin' to demonstrating the existence of what Chris Cornford (Irish Sea meeting in February, 1995) called 'a spectacular diversity of hydrocarbon types'. Perhaps even more spectacular is the explosion of geological information that has occurred, both before and in the preparation of this volume. The sheer weight of science that has been brought to bear makes the early interpretations seem embarrassingly unsophisticated. Those of us who were in a modest way pioneers in this basin can take real satisfaction from the success, both economic and scientific, that has resulted from the application of so much original thought to the problems of this basin. With the continuing application of expertise of the sort represented here, it seems inescapable that more success will be achieved, and that more geological problems will yield to painstaking analysis.

I wish to thank Greig Cowan and Martin Hardman of BG and Ian Jackson for their help and patience in bringing my ideas on the EISB into the 1990s, and Heidi Vietz for typing this manuscript. Greig, Martin and Neil Meadows also knocked my original contribution into its present published form.

References

ARMSTRONG, J. P., SMITH, J., D'ELIA, V. A. A. & TRUEBLOOD, S. P. 1997. The occurrence and correlation of oils and Namurian source rocks in the Liverpool Bay–North Wales area. *This volume.*

BARR, K. W., COLTER, V. S. & YOUNG, R. 1981. The geology of the Cardigan Bay–St George's Channel Basin; *In*: ILLING, L. V. & HOBSON, G. D. (eds) *Petroleum Geology of the Continental Shelf of N.W.Europe.* Institute of Petroleum, 432–443.

BEACH, A., BROWN, J. L., WELBON, A. I., MCCALLUM, J. E., BROCKBANK, P. & KNOTT, S. Characteristics of fault zones in sandstones from NW England: application to fault transmissibility. *This volume.*

BLOW, R. A. & HARDMAN, M. 1997. Calder Field appraisal well 110/7a-8, East Irish Sea Basin. *This volume.*

BUSHELL, T. P. 1986. Reservoir geology of the Morecambe Field. *In*: BROOKS, J., GOFF, J. C. & VAN HORN, B. (eds) *Habitat of Palaeozoic Gas in N.W.Europe* Geological Society, London, Special Publication **23**, 189–208.

CHADWICK, R. A. 1997. Fault analysis of the Cheshire Basin, NW England. *This volume.*

COLTER, V. S. 1978. Exploration for gas in the Irish Sea. *In*: VAN LOON, A. J. (ed.) Key notes of the MEGS-II (Amsterdam 1978). *Geologie en Mijnbouw*, **57**, 503–516.

—— & BARR, K. W. 1975. Recent developments in the geology of the Irish Sea and Cheshire Basins. *In*: WOODLAND, A. W. (ed.) *Petroleum and the Continental Shelf of North West Europe, Vol. 1: Geology.* Applied Science, London, 61–75.

—— & EBBERN, J. 1978. The petrography and reservoir properties of some Triassic sandstones from the Morecambe Field, Irish Sea, UK. *Journal of the Geological Society, London*, **135**, 57–62.

—— & —— 1979. SEM studies of Triassic reservoir sandstones from the Morecambe Field, Irish Sea, U.K. *Scanning Electron Microscopy*, **1**, 531–538.

COPE, J. C. W. 1994. A latest Cretaceous hotspot and the south easterly tilt of Britain. *Journal of the Geological Society, London*, **151**, 905–908.

—— 1997. The Mesozoic and Tertiary history of the Irish Sea. *This volume.*

COPE, M. J. 1986. An interpretation of vitrinite reflectance data from the Southern North Sea Basin. *In*: BROOKS, J., GOFF, J. C. & VAN HOORN, B. (eds) *Habitat of Palaeozoic Gas in NW Europe.* Geological Society, London, Special Publication, **23**, 85–98.

COWAN, G. 1993. The identification and significance of aeolian deposits in the dominantly fluvial Sherwood Sandstone Group of the EISB, UK. *In*: NORTH, C. P. & PROSSER, D. J. (eds) *Characterization of Fluvial and Aeolian Reservoirs.* Geological Society, London, Special Publication, **73**, 231–245.

—— & BRADNEY, J. 1997. Regional diagenetic controls on reservoir properties in the Millom accumulation: implications for field development. *This volume.*

——, OTTESEN, C. & STUART, I. A. 1993. The use of dipmeter logs in the structural interpretation and palaeocurrent analysis of Morecambe Fields, EISB. *In*: PARKER, J. R. (ed.) *Petroleum Geology of Northwest Europe: Proceedings of the 4th Conference.* Geological Society, London, 867–882.

ENGLAND, R. W. & SOPER, N. J. 1997. Lower crustal structure of the East Irish Sea from deep seismic reflection data. *This volume.*

EVANS, D. J., REES, J. G. & HOLLAWAY, S. 1993. The Permian to Jurassic stratigraphic and structural evolution of the central Cheshire Basin. *Journal of the Geological Society, London,* **150**, 857–870.

FRANCIS, A., MILLWOOD HARGRAVE, M., MULHOLLAND, P. & WILLIAMS, D. 1997. Real and relict direct hydrocarbon indicators in the East Irish Sea Basin. *This volume.*

GREEN, P. F., DUDDY, I. R. & BRAY, R. J. 1997. Variation in thermal history styles around the Irish Sea and adjacent areas: implications for hydrocarbon occurrence and tectonic evolution. *This volume.*

——, ——, —— & LEWIS, C. L. E. 1993a. Elevated palaeotemperatures prior to Early Tertiary cooling throughout the U.K region: implications for hydrocarbon generation. *In:* PARKER, J. R. (ed.) *Petroleum Geology of Northwest Europe: Proceedings of the 4th Conference.* Geological Society, London, 1067–1074.

——, ——, —— & —— 1993b. Early Tertiary heating in Northwest England: fluids & burial (or both?), *Geofluids '93, Extended abstracts,* 119–123.

GREENWOOD, J. & HABESCH, S. 1997. Diagenesis of the Sherwood Sandstone Group in the southern East Irish Sea Basin (Blocks 110/13, 110/14 and 110/15): Constraints from preliminary isotopic and fluid inclusion studies. *This volume.*

HAIG, B., PICKERING, S. C. & PROBERT, R. 1997. The Lennox oil and gas field. *This volume.*

HARDMAN, M., BUCHANAN, J., HERRINGTON, P. & CARR, A. 1993. Geochemical modelling of the EISB: its influence on predicting hydrocarbon type and quality. *In:* PARKER, J. R. (ed.) *Petroleum Geology of Northwest Europe: Proceedings of the 4th Conference,* Geological Society, London, 809–821.

HERRIES, R. & COWAN, G. 1997. Challenging the 'sheetflood' myth: the role of water-table-controlled sabkha deposits in redefining the depositional model for the Ormskirk Sandstone formation (Lower Triassic), East Irish Sea Basin. *This volume.*

HOLLIDAY, D. W. 1993a. Geophysical log signatures in the Eden Shales (Permo Triassic) of Cumbria and their regional significance, *Proceedings of the Yorkshire Geological Society,* **49**, 345–354.

—— 1993b. Mesozoic cover over northern England: interpretation of apatite fission track data, *Journal of the Geological Society, London,* **150**, 657–660.

JACKSON, D. J. & MULHOLLAND, P. 1993. Tectonic and stratigraphic aspects of the EISB and adjacent areas: contrasts in their post-Carboniferous structural styles. *In:* PARKER, J. R. (ed.) *Petroleum Geology of Northwest Europe: proceedings of the 4th Conference.* Geological Society, London, 791–808.

——, JOHNSON, H. & SMITH, N. J. P. 1997. Stratigraphical relationships and a revised lithostratigraphical nomenclature for the Carboniferous, Permian and Triassic rocks of the offshore East Irish Sea Basin. *This volume.*

——, MULHOLLAND, P., JONES, S. M. & WARRINGTON, G. 1987. The geological framework of the EISB. *In:* BROOKS, J. & GLENNIE, K. (eds) *Petroleum Geology*

of Northwest Europe, Graham & Trotman, London, 191–203.

JENNER, J. K. 1981. The structure and stratigraphy of the Kish Bank basin. *In:* ILLING, L. V. & HOBSON, G. D. (eds) *Petroleum Geology of the Continental Shelf of NW Europe.* Institute of Petrolum, London, 426–431.

KNIPE, R., COWAN, G. & BALENDRAN, V. S. 1993. The tectonic history of the EISB with reference to the Morecambe Fields. *In:* PARKER, J. R. (ed.) *Petroleum Geology of Northwest Europe: Proceedings of the 4th Conference.* Geological Society, London, 857–866.

MACCHI, L. 1987. A review of sandstone illite cements and aspects of their significance to hydrocarbon exploration and development. *Geological Journal,* **22**, 333–345.

McGOVERN-TRAA, C., YIU-LEU, J., SPARK, I. S. C. & HAMILTON, W. A. 1997. The presence of sulphate reducing bacteria in live drilling muds, core materials and reservoir formation brine from new oilfields. *This volume.*

MADDOX, S. J., BLOW, R. A. & O'BRIEN, S. R. 1997. The geology and hydrocarbon prospectivity of the North Channel Basin. *This volume.*

MARIE, J. P. P. 1975. Rotliegendes stratigraphy and diagenesis. *In:* WOODLAND, A. W. (ed.) *Petroleum and the Continental Shelf of NW Europe.* Institute of Petroleum, London, 205–211.

MEADOWS, N. S. & BEACH, A. 1993. Controls on reservoir quality in the Triassic Sherwood Sandstone of the Irish Sea. *In:* PARKER, J. R. (ed.) *Petroleum Geology of Northwest Europe: Proceedings of the 4th Conference.* Geological Society, London, 823–833.

MIKKELSEN, P. W. & FLOODPAGE, J. B. 1997. The hydrocarbon potential of the Cheshire Basin. *This volume.*

NAYLOR, D., HAUGHEY, N., CLAYTON, G. & GRAHAM, J. R. 1993. The Irish Bank Basin, Offshore Ireland. *In:* PARKER, J. R. (ed.) *Petroleum Geology of Northwest Europe: Proceedings of the 4th Conference.* Geological Society, London, 845–855.

PARNELL, J. 1997. Fluid migration history in the North Irish Sea–North Channel region. *This volume.*

PENN, I. E., HOLLIDAY, D. W., KIRBY, G. A., KUBALA, M., SOKEY, R. A. *ET AL.* 1993. The Larne No.2 borehole; discovery of a new Permian volcanic centre. *Scottish Journal of Geology,* **19**(3), 333–346.

QUIRK, D. G. & KIMBELL, G.S. 1997. Structural evolution of the Isle of Man and central part of the Irish Sea. *This volume.*

ROWE, J. & BURLEY, S. 1997. Faulting and porosity modification in the Sherwood Sandstone at Alderley Edge, northeastern Cheshire: an exhumed example of fault-related diagenesis. *This volume.*

SCOTT, J. & COLTER, V. S. 1987. Geological aspects of current onshore G.B. exploration plays. *In:* BROOKS, J. & GLENNIE, K. (eds) *Petroleum Geology of Northwest Europe.* Graham & Trotman, London, 95–107.

SHELTON, R. 1997. Tectonic evolution of the Larne Basin. *This volume.*

STUART, I. A. 1993. The geology of the North Morecambe Gas Field, EISB. *In:* PARKER, J. R. (ed.) *Petroleum*

Geology of Northwest Europe: Proceedings of the 4th Conference. Geological Society, London, 883–895.

TAYLOR, J. C. M. & COLTER, V. S. 1975. Zechstein of the English sector of the Southern North Sea. *In:* WOODLAND, A. W. (ed.) *Petroleum Geology and the Continental Shelf of NW Europe.* Institute of Petroleum, London, 249–263.

THOMPSON, D. B. 1970. The stratigraphy of the so-called 'Keuper' Sandstone Formation. *Quarterly Journal of the Geological Society, London,* **126**, 151–179.

THOMPSON, J. & MEADOWS, N. S. 1997. Clastic sabkhas and diachroneity at the top of the Sherwood Sandstone Group: East Irish Sea. *This volume.*

WARD, J. 1997. Early Dinantian evaporites from the Easton-1 well in the Solway Basin, Cumbria. *This volume.*

WARRINGTON, G. 1997. The Penarth Group–Lias Group succession (Late Triassic–Early Jurassic) in the East Irish Sea Basin and neighbouring areas: a stratigraphical review. *This volume.*

WOODWARD, K. & CURTIS, C. D. 1987. Predictive modelling for the distribution of production constraining illites – Morecambe Gas Field, Irish Sea, Offshore U.K. *In:* BROOKS, J. & GLENNIE, K. (eds) *Petroleum Geology of Northwest Europe.* Graham & Trotman, London, 205–215.

YALIZ, A. 1997. The Douglas oil field. *This volume.*

Stratigraphical relationships and a revised lithostratigraphical nomenclature for the Carboniferous, Permian and Triassic rocks of the offshore East Irish Sea Basin

D. I. JACKSON[1], H. JOHNSON[2] & N. J. P. SMITH[3]

[1]*British Geological Survey, Murchison House, West Mains Road, Edinburgh EH9 3LA, UK*
(*Present address: 19 Swanston Crescent, Fairmilehead, Edinburgh, EH10 7EL, UK*)
[2]*British Geological Survey, Murchison House, West Mains Road, Edinburgh EH9 3LA, UK*
[3]*British Geological Survey, Kingsley Dunham Centre, Keyworth, Nottingham NG12 5GG, UK*

Abstract: The Lower Permian and Triassic successions of the offshore East Irish Sea Basin are probably the thickest of the United Kingdom Continental Shelf or UK mainland, but currently, no hierarchical lithostratigraphy exists for this offshore area. A summary is presented of the first comprehensive hierarchical lithostratigraphical nomenclature scheme to be established for the Carboniferous, Permian and Triassic rock successions of the offshore East Irish Sea Basin. This scheme, which has been sponsored by interested exploration companies, identifies 25 formal lithostratigraphical divisions following the methodology already established in BGS–UKOOA (British Geological Survey–United Kingdom Offshore Operators' Association) revisions of North Sea lithostratigraphy.

Sixteen new lithostratigraphical names are introduced (nine in the Mercia Mudstone Group, one in the Sherwood Sandstone Group, three each within the Permian and Carboniferous successions). The lithology, wireline log characteristics and nature of the lower boundary are discussed briefly for each lithostratigraphical unit, together with correlations to the adjacent onshore successions where appropriate. Formal definitions of lithostratigraphic units, including type and reference sections are fully documented elsewhere. The aims of this study are to rationalize the existing terminology and to provide a practical scheme that will have the widest acceptance within the oil industry.

Following deposition of a thick and relatively continuous marine to paralic Carboniferous succession that was uplifted during the Variscan movements, the geological record of the Permo-Triassic East Irish Sea Basin consists of five depositional sequences (Lower Permian, Upper Permian, Sherwood Sandstone Group, Mercia Mudstone Group, Penarth Group) separated by major sequence boundaries or disconformities (Fig. 1; cf. Jackson *et al.* 1987). After the Variscan movements, subsidence recommenced in the early Permian in a series of discrete intermontane basins, probably partly below sea-level (Smith 1972). These basins were later united by the first Bakevellia transgression of the late Permian (Smith & Taylor 1992) to create a pocket-basin (the Bakevellia Sea Basin) in which, at maximum development, a central brine pan was surrounded by a muddy coastal plain passing to basin margin dune sands and alluvial fan breccias (Smith & Taylor 1992). Breaching of distant interfluves at the start of the Triassic led to the deposition of thick fluvial sands along a northward-sloping rift valley whose sediments overwhelmed the Bakevellia Sea Basin. The rift structure was flooded subsequently by the Röt transgression or an early precursor, and the area evolved into a northward-sloping mudstone and halite coastal salina complex during the Mid-Late Triassic (cf. Fig. 2).

Although the offshore East Irish Sea Basin (hereafter EISB) contains the thickest aggregate Lower Permian (Appleby Group), Lower–Middle Triassic (Sherwood Sandstone Group) and Middle–Upper Triassic (Mercia Mudstone Group) succession of the UKCS, a formal lithostratigraphy has not been erected previously and is long overdue. Offshore, lithostratigraphical nomenclature has evolved piecemeal in the 20 years since the first study on the deep geology of the basin by Colter & Barr (1975), on occasions by modifying the stratigraphical definition and 'sense' of onshore basin margin terms. In general, the earlier offshore terminology has been applied basinwide with little regard to lateral differences in facies or log character.

The British Geological Survey (hereafter BGS) has embarked upon a revision series of litho-stratigraphical nomenclature schemes for the western United Kingdom Continental Shelf (hereafter UKCS), sponsored by the oil industry, to

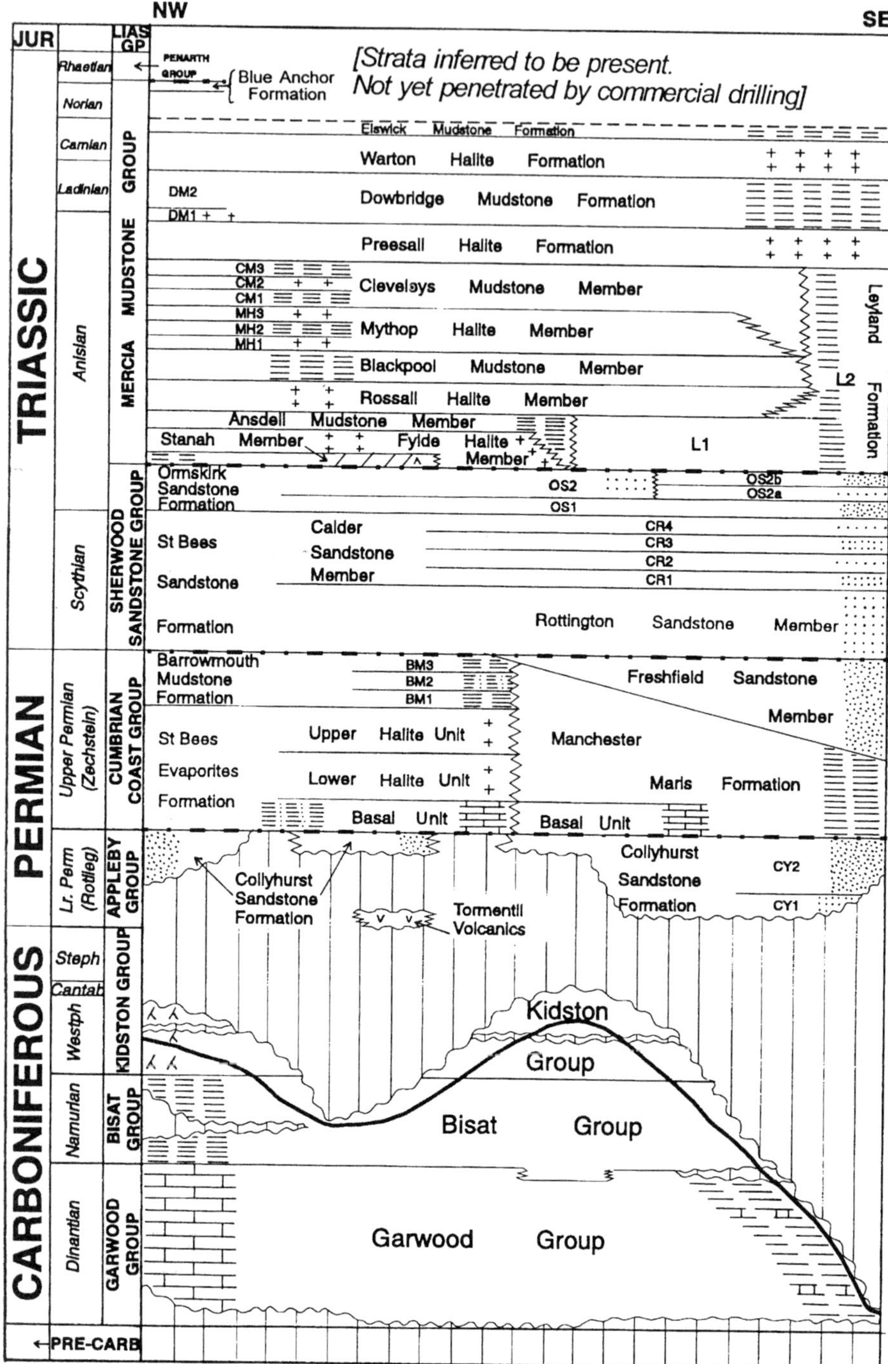

Notes. 1. Tertiary intrusions penetrate the entire succession.

      2. Suggested non-sequence   · · — — · · — —

      3. Lower limit of secondarily reddened Carboniferous   ————

      4. For key to ornaments, see Fig. 5

complement those funded jointly with the United Kingdom Offshore Operators' Association (hereafter UKOOA) for the North Sea (Knox & Cordey 1992–4); this volume provides a timely introduction to the full EISB scheme (Fig. 1). This contribution discusses and summarizes the stratigraphical relationships, of the newly proposed lithostratigraphical units in the EISB (Fig. 1), formally set out in much greater detail in Jackson & Johnson (1996). Following the philosophy in the companion BGS–UKOOA volumes, a conservative approach has been adopted with regard to the introduction of a revised nomenclature; despite this, 16 new names are proposed (of which nine are in the Mercia Mudstone Group). Although intended primarily for the offshore EISB where the chief means of correlation is by wireline logs, the scheme can be dovetailed into the coastal stratigraphy of Lancashire, Cumbria, Cheshire and Clwyd.

The study is based on 32 released offshore exploration and appraisal wells supplemented by 5 offshore wells for which pre-release consent has kindly been given by participating companies. The well database is uneven and biased towards wells in and around the Morecambe Field. In this account, the scheme is described in inverse stratigraphical order, as in the drilling of an imaginary stratigraphic test well from the Tertiary to the Carboniferous. Terms prefaced by * are new names introduced in Jackson & Johnson (1996). As in the BGS–UKOOA volumes, a system of numbered abbreviations has been employed for the informal subdivision of a formation or member. The maximum thicknesses quoted for particular units are usually from individual wells (rarely the type section) rather than from seismically-determined basin maxima. Structural and place names, and well locations mentioned in the text are included on the inset map of Fig. 3. Most type sections have not been figured here, because of lack of space and the need to compromise on well selection in Figs 3–6; full details of the type and reference sections, and the etymology of the newly introduced terms are contained in Jackson & Johnson (1996).

Tertiary

The non-specific description 'Tertiary intrusion' is the preferred label for dykes and sills of Tertiary age in offshore areas, in view of their widespread occurrence (e.g. in the Isle of Man, North Wales, and the Northwest Anglesey Dykes) outside the restricted area of the Fleetwood Dyke Group (Kirton & Donato 1985).

Lias Group (Lower Jurassic) and Penarth Group (Rhaetian)

Palaeogeographical reconstructions (Warrington & Ivimey-Cook 1992) and the presence of scattered Liassic outliers encircling the EISB (near Carlisle, at Mochras, in Northern Ireland, the Kish Bank Basin and the Cheshire Basin) suggest that the Lias Group sediments proved in the Keys Basin and postulated in the Berw and Solway Firth basins (Jackson & Mulholland 1993) overlie the undrilled Penarth Group and Blue Anchor Formation at a disconformity (cf. Warrington et al. 1980). Use of the terms Penarth Group and Blue Anchor Formation offshore, together with their formally defined subdivisions, is appropriate, since their lithologies and log characteristics are relatively consistent throughout Britain (see Whittaker et al. 1985; Ivimey-Cook et al. 1995). As in the Southern North Sea (Johnson et al. 1994, p.7), the base of the Lias Group is placed on log and lithological character rather than biostratigraphical determination (cf. Whittaker et al. 1985).

Mercia Mudstone Group (Middle and Upper Triassic)

General

The Mercia Mudstone Group (MMG) succession of the EISB (c. 3025 m thick) is the thickest Middle and Upper Triassic succession known to date from the UK mainland or continental shelf, and comprises a two-component system of thick alternating silty mudstone and halite units deposited as a northward-thickening onlapping evaporite wedge (Figs 1, 2). Each halite and mudstone unit can be identified by its intrinsic internal features and associated log characteristics. Bed-by-bed correlations (e.g. mudstones as thin as 1 m in the Preesall Halite Formation) can be made basinwide within both mudstone and halite units, and will

---

Fig. 1. Lithostratigraphical nomenclature scheme for the Carboniferous, Permian and Triassic rocks of the offshore East Irish Sea Basin. Chronostratigraphical scale and framework based on George et al. (1976), Ramsbottom et al. (1978), Smith et al. (1974) and Warrington et al. (1980), but scale expanded for sake of clarity especially in the Zechstein and Anisian intervals. Stage boundaries in the Triassic taken from Benton et al. (1994). Lateral equivalents in Permo-Triassic of adjacent basins shown on fig. 3 of Jackson & Mulholland (1993). See Fig. 5 for key to lithology ornaments.

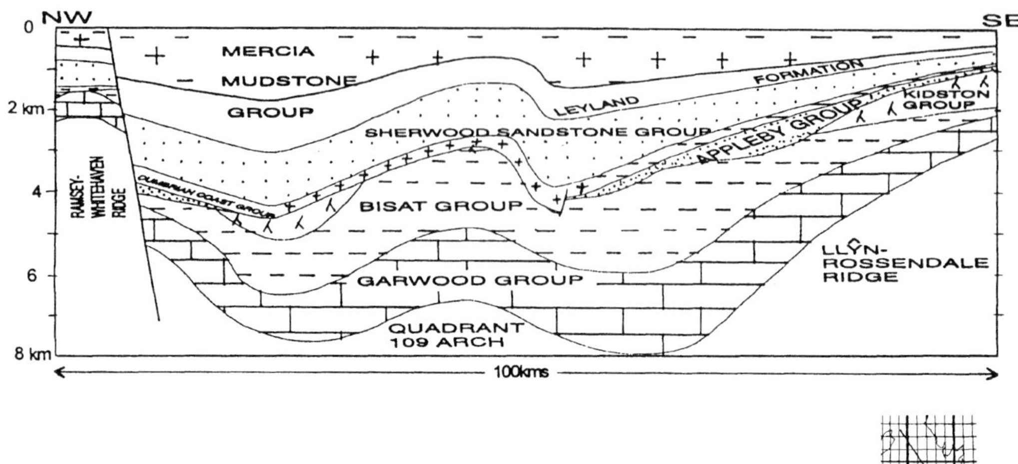

Fig. 2. NW–SE transect across the offshore East Irish Sea Basin showing geometry of fill for the various lithostratigraphical groups.

provide the basis for a very detailed sequence stratigraphy in future. Massive ('layered') halites deposited in brine pans pass outwards laterally into haselgebirge (a mudstone host rock with variable content of displacive crystalline halite, cf. Wilson 1990), which in turn passes laterally into mudstone or a dolomitic siltstone ('skerry' of onshore usage) (Wilson 1990; Arthurton 1980). A transition zone of thin interbedded halites and mudstones occurs at the top and bottom of most halite units. The transition zones are halite-dominated in the north and centre of the basin, and mudstone-dominated in the south and on the basin margins. Following guidance from the oil industry, supplemented by cuttings descriptions, the boundaries of the halite units are taken here at the lowest and highest occurrences of significant halite beds which generally display gamma ray values less than *c.*35 API units. Stage boundaries in Fig. 1 are taken from Benton *et al.* (1994).

In the scheme (Fig. 1), halites and mudstones of basinwide occurrence have been accorded formation status (the Preesall Halite and overlying formations); locally-developed halites and mudstones below the Preesall Halite Formation have been given member status and are assigned to the *Leyland Formation. Halite members in the Leyland Formation (Fig. 1) comprise halite leaves (3–20 m thick) alternating with mudstone intercalations; they contrast with the slightly cleaner and much thicker Preesall and Warton halites which contain fewer mudstones, generally as partings (Fig. 3). Lateral equivalents of the Leyland Formation halites in west Lancashire are mostly haselgebirge (Wilson 1990). The halite

members thin markedly towards the basin margins and were probably basin-filling types eliminating the contemporary basin topography and confined to the areas of greatest subsidence. The mudstone members of the Leyland Formation can be differentiated only where the halite members are recognized; beyond the halite feather-edge they pass insensibly into the undivided Leyland Formation (Fig. 1). Mudstones tend to be anhydritic in the north, and more silty, dolomitic and with a higher average velocity in the south. The depositional area of successive MMG halites expanded with time (limits shown in Jackson & Mulholland 1993, fig. 5), in contrast to the pattern of coeval halites in the Southern North Sea. Lateral continuity of halite deposition with the Cheshire Basin over the intervening Llŷn–Rossendale Ridge was first established during the precipitation of the Preesall/Northwich Halite Formation.

The Mercia Mudstone Group shows a regional northwards thickening in all stratigraphical units away from the Llŷn–Rossendale Ridge towards a depocentre in the Keys Basin possibly associated with the Sigurd Fault (Jackson & Mulholland 1993). The halite units show a greater expansion in thickness across the basin than the intervening mudstone units. However, during and after deposition of the Cleveleys Mudstone Member, a large periclinal structure antecedent to the present-day Morecambe Field (cf. Fig. 2) is thought to have been rejuvenated astride the Caledonoid Quadrant 109 Arch-High Haume Anticline (Jackson & Mulholland 1993), as a result of reduced subsidence over the high (cf. Stuart 1993) combined with a NW–SE rollover to the west.

Anomalies departing from these regional patterns are provided by local expansion in the hangingwall of growth faults and in Mercia Mudstone Group collapse graben (e.g. 110/3-2, Fig. 3, within the Crosh Vusta Fault Complex). Expanded sections are counterbalanced by condensed successions, especially of the higher halites, over the rejuvenated Quadrant 109 Arch (Figs 1, 3; Jackson & Mulholland 1993). Wells drilled to Ormskirk Sandstone Formation targets commonly also intersect the bounding tilt-block normal fault within the Mercia Mudstone Group succession (cf. Fig. 3; Jackson *et al.* 1987, fig. 3b; Jackson & Mulholland 1993, fig. 10), and prove expanded hangingwall successions overlying relatively condensed footwall successions, which can distort isopach maps and palaeogeographical reconstructions. This local thickness variation (e.g. between 110/3-2 and the nearby 110/3b-4) is conveniently ascribed to syndepositional faulting, possibly accentuated by sliding on halite detachments, and appears to date the creation of the Mercia Mudstone Group collapse graben (Jackson & Mulholland 1993) as late Triassic. Most drilled successions are disrupted by minor internal faulting (e.g. 110/2-5, Fig. 3), and ductile halite layers provide a focus for listric fault fans, including a basal detachment on or directly above the top of the Sherwood Sandstone Group (e.g. 110/3-2, 110/2a-8, Fig. 3). At shallow depths the Preesall and Warton halite formations appear in several cases to have been subjected to 'wet' rockhead conditions offshore (cf. Earp & Taylor 1986; Wilson & Evans 1990); only the lower portions of the halite are preserved and/or are overlain by abnormal thicknesses of mudstone (and ?preferentially thickened Quaternary deposits).

As *aide-mémoires*, the geographical names Ansdell (a district of Lytham St Anne's), Blackpool, Cleveleys (a suburb of Thornton), Dowbridge (near Kirkham) and Elswick (east of Blackpool) were chosen from west Lancashire as replacement geographical qualifiers for the Unit A, B, C, D, and E mudstones of Jackson *et al.* (1987); similarly the initial W of Warton (near Lytham St Anne's) was chosen for the lateral equivalent of the Wilkesley Halite, and finally, 'Stanah' (near Thornton), replacing the former 'anhydrite-dolomite marker', has a passing onomatopoetic resemblance to 'anhydrite'.

Lithostratigraphy

The chocolate to red-brown mudstones of the ***Elswick Mudstone Formation** (assessed as latest Carnian to Norian in age) have been proved by drilling to be up to 69.5 m thick (110/13-8, Fig. 3) but, from comparisons with the laterally equivalent

Brooks Mill Mudstone Formation of the Cheshire Basin (Wilson 1993; the Prees-1 and Wilkesley bores provide a typical log response, Evans *et al.* 1993), are thought to exceed 800 m in the north of the EISB. Lateral equivalents in Northern Ireland, the Southern North Sea and the English Midlands contain bedded and nodular gypsum (Warrington *et al.* 1980; Wilson 1993). The upper boundary has not been drilled in the EISB, but by analogy with adjacent areas is placed at the base of the grey-green mudstones of the Blue Anchor Formation (Warrington *et al.* 1980; Ivimey-Cook *et al.* 1995).

The clean 'massive' halite of the ***Warton Halite Formation** (late Ladinian to Carnian) has a maximum drilled thickness of 269 m (110/13-8, Fig. 3), but seismic data and inter-basinal thickness comparisons suggest that the unit may be the thickest halite (1000 m) in the Mercia Mudstone Group of the EISB. Detailed correlations of the mudstone partings with those in the correlative Wilkesley Halite Formation of the Cheshire Basin (Warrington *et al.* 1980; Wilson 1993), demonstrate that all available sections of the Warton Halite are incomplete and preserved beneath 'wet' rockhead; similarly, a 65 m thick collapse-breccia proved in west Lancashire is thought here to represent merely the lower *c.*40% of the Warton Halite Formation equivalent (Wilson 1990; Wilson & Evans 1990). Offshore, the mudstone partings are more numerous than in the Preesall Halite Formation and of greater lateral persistence.

The lithologically varied ***Dowbridge Mudstone Formation** (up to 573.5 m drilled, but ?exceeding 1000 m elsewhere in the basin; latest Anisian to late Ladinian) consists dominantly of structureless, red silty mudstones with siltstone packets. In most wells the lower boundary has been positioned at the sharp log breaks with the clean Preesall Halite Formation (Fig. 3). However, a minority of expanded sequences record a thick basal transition zone (unit DM1, Fig. 1, and 105 m thick in 110/7-1) of interbedded halite and mudstone; thick mudstones in the overlying strata in which halites are largely absent (unit DM2), are probably loessic in origin (cf. Wilson 1990). The formation is equivalent to the combined Coat Walls Mudstone Member and the lower part of the Breckells Mudstone Formation of west Lancashire (Wilson 1990), and to the combined Byley and Wych mudstone formations of the Cheshire Basin (Wilson 1993); unit DM1 is tentatively equated with the green/red colour-banded lacustrine laminites (cf. Arthurton 1980) of the lower Coat Walls Mudstones.

The **Preesall Halite Formation** (Warrington *et al.* 1980, table 4; cf. Wilson 1990; Rose & Dunham 1977; 95–620 m offshore, late Anisian) is the cleanest and the thickest halite drilled to date in the

D. I. JACKSON *ET AL.*

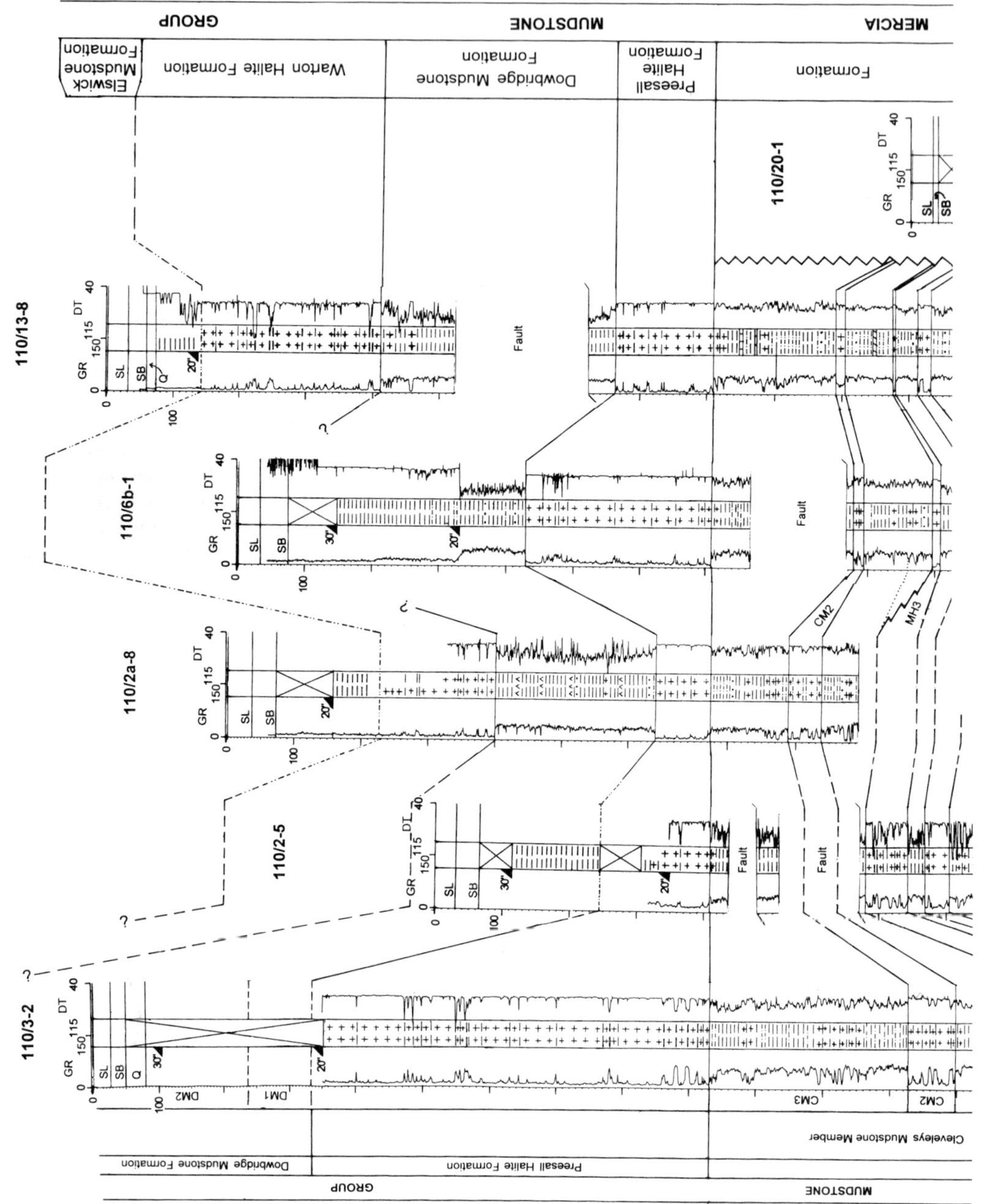

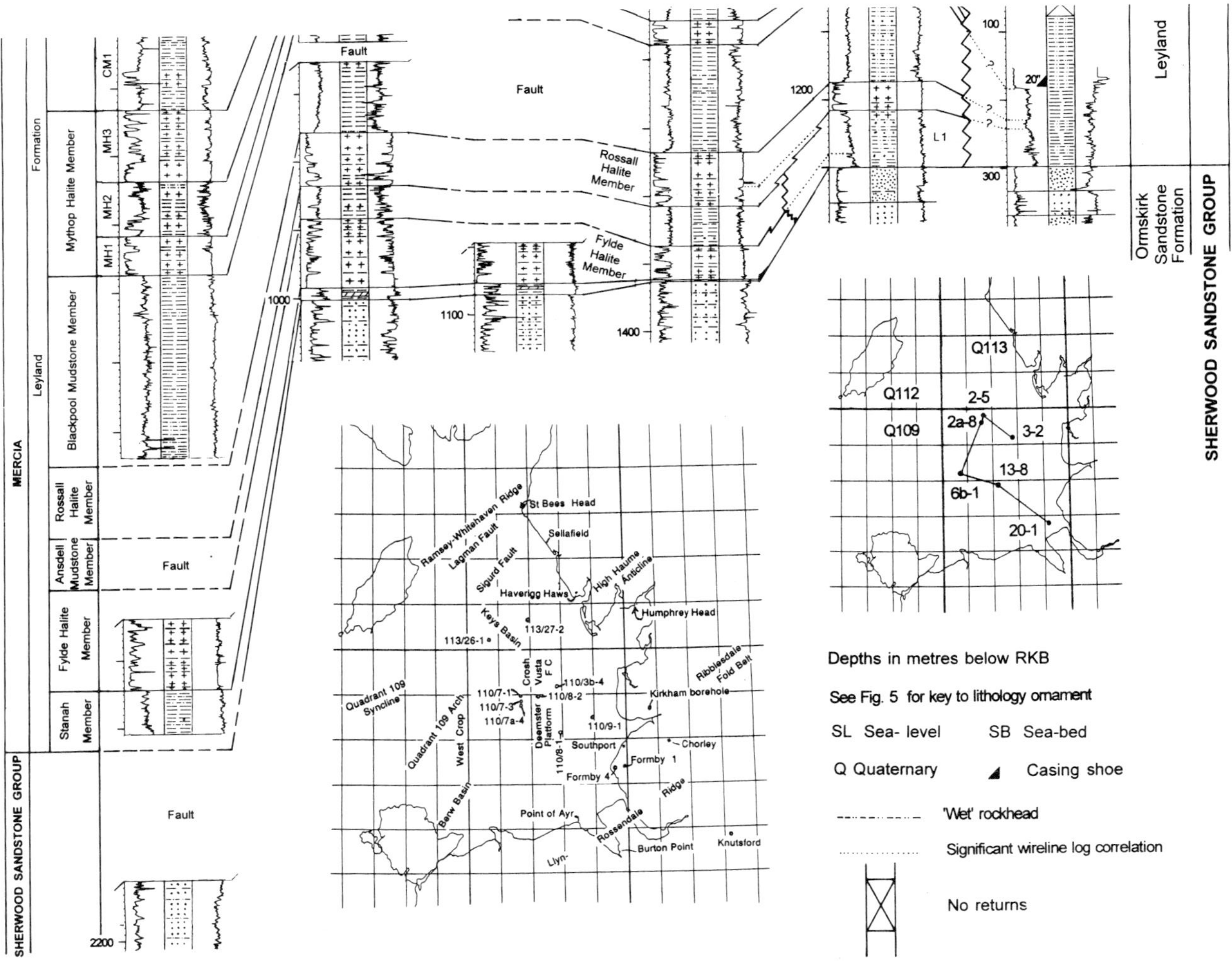

Fig. 3. Lithostratigraphy and correlation of the Mercia Mudstone Group in selected wells in the offshore East Irish Sea Basin. Inset map shows names of main structural elements, localities and well locations quoted in the text.

EISB. Mudstone partings (generally 1–3 m) constitute the only other significant lithology and provide the basis for achieving detailed correlations across the basin, and also with the correlative Northwich Halite Formation. The lower boundary is taken at the base of the lowest significant halite (upper cut-off *c.*35 API units) rather than at the base of the lowest halite stringer of the transition zone (Fig 3). Abnormally thin successions of the formation have been identified in two settings – firstly, beneath a 'wet' rockhead, down to depths exceeding 220 m BOD (e.g. 110/9-1, cf. Jackson *et al.* 1987, fig. 9) and secondly, as sections in the Calder and Morecambe fields (e.g. 110/2a-8, Fig. 3) regarded here as unfaulted but condensed, and which provide firm evidence for a reactivated Quadrant 109 Arch-High Haume Anticline.

Over most of the basin, the *Leyland Formation (250–1200 m, Anisian) can be subdivided fully into constituent evaporite (mostly halite) and mudstone members (Fig. 1). In the south and especially over the Llŷn–Rossendale Ridge where the halites fail, the formation is not subdivided formally, but a thin basal unit (L1) of grey mudstones can be differentiated from the overlying red-brown mudstones (unit L2; e.g. 110/20-1, Fig. 3). Successions lacking halites also occur as lateral equivalents on the eastern margin (e.g. the Kirkham borehole, and probably near Southport, Wilson 1990) and are also predicted on the West Crop of the EISB (Jackson & Mulholland 1993).

The *Cleveleys Mudstone Member (117–503 m, mid to late Anisian) is a thick, lithologically varied, mudstone and siltstone unit which contains the highest percentage of halite in any mudstone unit of the Mercia Mudstone Group. The base is taken at the first downward occurrence of significant halite in the upper group of Mythop halites, some 45–50 m lower than the base of the distinctive green and red colour-banded Thornton Mudstone Member (113 m maximum thickness) of west Lancashire (Wilson 1990, fig. 4). Offshore, where greater reliance on unit recognition is placed on wireline log response than colour change, a new name was deemed preferable to re-defining the boundaries of the Thornton Mudstone.

An informal tripartite subdivision of the Cleveleys Mudstone Member is manifest throughout much of the basin (e.g. 110/3-2, Fig. 3), though recognition of this pattern, especially in the Morecambe Field area, is commonly obscured by faulting. Unit CM3 consists of mudstones with subordinate upward-fining siltstones, sandstones, thin individual dolomites and anhydrites; by analogy with onshore sections the mudstones are interpreted as lacustrine laminites (Wilson 1990). Excluding the transition zone halites at the top, two variable packets of halite are present in the unit,

generally confined to the north of the basin, and passing into dolomitic siltstones elsewhere. Unit CM2 (80 m maximum), consists mostly of halite leaves with mudstone partings; thin (1 m) very high velocity dolomites directly overlie some of the higher halites and generate distinctive log marker peaks (e.g. 110/3-2, Fig. 3). Unit CM1, contains similar lithologies to unit CM3, together with a persistent paired halite (20 m thick), *c.*30 m above the base in the north and centre of the basin (e.g. well 110/3-2); the mudstone parting of this halite marks the base of the Thornton Mudstone Member onshore (Wilson & Evans 1990). The Cleveleys Mudstone Member probably denotes the start of significant differential subsidence over a nascent Morecambe Field high (Fig. 2; cf. Stuart 1993).

As used here, the term **Mythop Halite Member** (0–241.5 m, mid-Anisian) is conterminous with 'the main group of Mythop Salts' in west Lancashire (Wilson & Evans 1990, p.13, fig. 6). Offshore, the member comprises up to seven halite leaves, separated by laterally persistent silty mudstone intercalations whose total proportion exceeds that of any other Mercia Mudstone Group halite member. The lower boundary is taken at the base of the thick lowest halite (Fig. 3). Offshore, the Mythop Halite Member is bifid, as in the main group of salts onshore (Wilson & Evans 1990, fig. 6), with an upper group of up to five thinner halites (unit MH3) separated by a thick near-median mudstone (unit MH2, 55 m maximum thickness) from a lower group of two persistent halites (unit MH1; e.g. 110/3-2). This tripartite subdivision is traceable throughout the basin and the central mudstone 'waist' is a characteristic feature on wireline logs (Fig. 3). Although most of the upper halites pass laterally into haselgebirge and/or dolomitic siltstones south of the Morecambe Field, the lowermost halite of unit MH3 persists, in available wells, with the two halites of unit MH1 to the southern feather-edge of the member (e.g. 110/13-8, Fig. 3).

The uniform red structureless mudstones of the *Blackpool Mudstone Member** (99–269 m+; early to mid Anisian) contain the lowest proportion of halite of any formal unit in the entire Mercia Mudstone Group of the offshore area. The beds, which are identified by the distinctive linear gamma trace and, commonly in the north, by the widespread low velocity zone at the top (Fig. 3), are believed to comprise aeolian loessic deposits (cf. Wilson 1990). The abrupt lower contact marks a prominent log break with the highest and cleanest halite of the Rossall Halite Member. The member is differentiated only where the Rossall and Mythop halites are present (Fig. 1; Jackson & Mulholland 1993, fig. 5), and beyond these limits passes southwards into equivalents within the undivided

Leyland Formation and eastwards into equivalents within the Singleton Mudstone Formation (Wilson 1990).

In the north and centre of the basin, the **Rossall Halite Member** (0–148 m, early Anisian) consists of four leaves of clean halite, separated by widespread mudstone intercalations that become progressively thinner upwards (Fig. 3). The successive leaves of halite were precipitated over progressively larger areas, repeatedly infilling the local topography, as subsidence attempted to outpace sedimentation. Towards the south, the three lower leaves of halite pass laterally to dolomitic siltstones (e.g. 110/13-8, Fig. 3), but the uppermost leaf fuses into a single clean halite, which is the direct lateral equivalent of the 11.50 m thick onshore Rossall Salt Member of haselgebirge lithology (Wilson 1993, 1990).

The *Ansdell Mudstone Member (41.5–82 m, early Anisian), comprises red with subordinate green silty mudstones (largely lacustrine laminites with some aeolian loessic intervals) and is characterized by a single, thin (3 m) but widespread, near-median halite over and to the north of the Morecambe Field. The basal contact is positioned at the top of the highest developed halite (<c.35 API units) of the Fylde Halite Member (e.g. 110/2-5; Fig. 3), which, because of facies changes, is taken at stratigraphically lower halites towards the south. Beyond the feather edge of the Fylde Halite Member, the red Ansdell Mudstone Member cannot be differentiated and apparently passes into dominantly grey mudstones of the higher part of unit L1 of the Leyland Formation; the Ansdell Mudstone Member is taken here to be the correlative of the red basal Singleton Mudstone Formation of west Lancashire (unit 1 of Wilson 1990; Wilson & Evans 1990; Jackson & Johnson 1996).

The *Fylde Halite Member (0–182.5 m, early Anisian) comprises up to six halite leaves. The proportions and thicknesses of halite and mudstone are similar to those of the Rossall Halite Member, but the partings and laterally persistent intercalations of mudstone become individually thicker and more numerous towards the top, which is the reverse of the pattern exhibited by the Rossall Halite (Fig. 3). The lower boundary is taken at the sharp base of halite overlying dolomite, anhydrite or siltstone of the Stanah Member, or, further to the south, overlying sandstone of the Ormskirk Sandstone Formation. The member shows a marked internal thinning towards the south. In addition, the highest halites fail by lateral passage into dolomitic siltstones and grey mudstones (e.g. 110/6b-1, Fig. 3); concomitantly, much of the thick basal halite leaf is lost, probably by onlap, at a non-sequence (cf. 110/2a-8, 110/6b-1) against the

Stanah Member or further to the south, against the Ormskirk Sandstone Formation. The Fylde Halite Member dies out abruptly south of 110/6b-1 (Fig. 3) and is absent onshore where the Hambleton Mudstone Formation, the approximate lateral equivalent, contains numerous horizons of pseudomorphs after halite (Wilson 1990). The Fylde Halite Member is a basin-filling halite like the Rossall Halite Member, but in contrast, the brine pan contracted in area with time – a trait shared with the Mythop Halite Member.

In the north and centre of the basin, a distinctive 'hourglass'-shaped log response, converging on the 'neck' of the Ansdell Mudstone Member, is discernable from the disposition of halite and mudstone intercalations and partings in the Fylde and Rossall halites (e.g. 110/2-5, Fig. 3).

The lithologically variable sequence of the *Stanah Member (0–67.5 m, early Anisian) consists generally of a condensed succession of thin very high gamma, grey mudstone beds alternating with thin anhydrite, dolomite, sandstone, and locally halite, beds. This condensed facies was formerly identified as the 'anhydrite and/or dolomite bed' (Colter & Barr 1975) or the 'anhydrite-dolomite marker' (Jackson et al. 1987), and constitutes a c.15 m thick, high velocity, marker unit. The basal high gamma mudstone represents the transgressive surface beneath a systems tract, and the overlying evaporites are thought to represent a coastal sabkha of the Röt transgression or an early precursor, though marine biota have not been recovered offshore to date. The member thins southwards, probably both internally and by onlap against the Ormskirk Sandstone Formation, and it is absent from onshore areas. The lower boundary is taken at the sharp junction with sandstone of the Ormskirk Sandstone and is believed here to be a major sequence boundary, though the nearest demonstrable unconformable relationship between the Mercia Mudstone Group and Sherwood Sandstone Group lies at the southern end of the Pennines (IGS 1972; cf. Thompson 1991).

Sherwood Sandstone Group (Lower and basal Middle Triassic)

The Sherwood Sandstone Group was deposited within an active NNW–SSE rift by the major northward-flowing Budleighensis River (Wills 1970), during a long terrestrial interlude between two quasi-marine argillaceous–evaporitic episodes. In this study the group is divided unequally into two formations – the Ormskirk Sandstone Formation and the St Bees Sandstone Formation (Fig. 1). The early fluvial palaeocurrents were unidirectional towards the NNW, but the later palaeocurrent vectors are at variance with the regional stratigraphical considerations and indicate a burgeoning easterly supply from the Lake District that apparently dominates during deposition of the Ormskirk Sandstone Formation (Cowan *et al.* 1993; Meadows & Beach 1993). Deposition of aeolian sandstones under easterly palaeowinds increased with time and also became proportionally more important towards the eastern basin margin, and over interfluves and structural divides (Meadows & Beach 1993; Cowan 1993). The Sherwood Sandstone Group overlaps the Upper Permian at a sequence boundary on the basin margin, both at outcrop (e.g. near Chorley, and on the flanks of the High Haume Anticline, Dunham & Rose 1949) and in the subsurface (e.g. in the coastal fringe around Sellafield, Smith 1924; Trotter *et al.* 1937).

The **Ormskirk Sandstone Formation** (128–333 m, earliest Anisian; Warrington *et al.* 1980; Jackson *et al.* 1987; Benton *et al.* 1994) comprises sandstones deposited in braided streams interbedded with locally dominant aeolian beds (Stuart & Cowan 1991; Cowan 1993; Meadows & Beach 1993; Stuart 1993; Cowan *et al.* 1993; Jones & Ambrose 1994). The formation has never been defined satisfactorily offshore or onshore, and 110/13-1 is therefore proposed as the type section (765–930 m RKB, Fig. 4; Jackson & Johnson 1996). The lower boundary is taken at the basinwide downward increase in velocity from aeolian sandstone to fluvial St Bees Sandstone Formation, and is locally coincident with the uppermost high gamma mudstone of the Calder Sandstone Member (Fig. 4).

Over large parts of the basin including the Morecambe Field, an informal and unequal bipartite subdivision of the Ormskirk Sandstone Formation into upper, dominantly fluvial sandstones (unit OS2) and lower, dominantly aeolian sandstones (unit OS1) is proposed on the basis of lithological criteria and subtle wireline log characteristics (cf. Cowan 1993; Meadows & Beach 1993). In the north, the highest 45 m of unit OS2 contain a higher frequency of thin high gamma mudstones and siltstones, and correspond to the 'Keuper Waterstones' of Colter & Barr (1975) and to the Upper Ormskirk Sandstone Formation of Cowan *et al.* (1993). Despite detailed re-examination of wireline logs, an unequivocal and consistent lower boundary for these 45 m beds could not be identified areally, and no informal subdivision is proposed; further subdivision of unit OS2 in the north and centre of the basin is best incorporated into detailed reservoir classifications of individual fields (cf. Stuart & Cowan 1991; Stuart 1993).

Unit OS1 (35–75 m) essentially comprises a single major upward-drying cycle of dominantly sandflat facies (Cowan 1993; Meadows & Beach 1993, fig. 4), with aeolian sandsheet deposits developed preferentially in the north and centre of the basin at the expense of aeolian dune systems towards the south and basin margins (Thompson 1970; Cowan 1993; Meadows & Beach 1993; Barnes *et al.* 1994; Jones & Ambrose 1994). The top of unit OS1 is taken at the sharp downward decrease in velocity; the gamma response at the boundary is variable (Fig. 4). This boundary represents the basinwide replacement of a low velocity, locally thin, aeolian ?dune phase, comprising the uppermost bed of unit OS1, by dominantly fluvial deposition (Cowan 1993; Meadows & Beach 1993); the boundary lies some 6–25 m below the top of the former so-called 'Thurstaston Member' of Colter & Ebbern (1979) and Stuart & Cowan (1991).

In the southern part of the basin (e.g. 110/13-1, Fig. 4; Haig *et al.* 1997; Yaliz 1997), an informal threefold subdivision into units OS1, OS2a and OS2b is recognized. The sandstones of unit OS2a are assessed as dominantly fluvial in origin whereas the sandstones in unit OS2b, like those of unit OS1, exhibit a lower average velocity, contain fewer mudstone partings and are interpreted as dominantly aeolian (110/13-1, Fig. 4; Cowan 1993; Meadows & Beach 1993). In four available wells (110/7-3, 110/7a-4, 110/8-2 and 110/9-1), the top of the aeolian unit OS1 is not recognizable with certainty and it is suggested that the Ormskirk Sandstone Formation should not be subdivided in those cases. Correspondingly, at sites on the basin margin, the formation may comprise largely aeolian sandstones (e.g. at Sellafield, Barnes *et al.* 1994; Jones & Ambrose 1994), and again an undivided Ormskirk Sandstone Formation is advocated.

The **St Bees Sandstone Formation** (371.5–1343.5 m, Scythian) is a thick monotonous sandstone subdivided into two units of approximately equal thickness – an upper mixed fluvial–aeolian subdivision (here the Calder Sandstone Member) and a lower fluvial subdivision (here the *Rottington Sandstone Member). With oil industry agreement, the term St Bees Sandstone Formation has been applied in the scheme (Fig. 1; Jackson & Johnson 1996) according to the traditional usage, both onshore (Warrington *et al.* 1980; cf. Rose & Dunham 1977) and offshore (Colter 1978; Jackson *et al.* 1987) rather than adopting the re-definition by Barnes *et al.* (1994).

The **Calder Sandstone Member** (197–730 m; the Calder Sandstone Formation of Barnes *et al.* 1994) generally comprises non-silicified sandstones with scattered argillaceous partings. Offshore, the lower boundary (the Top Silicified Zone regional marker of Colter & Barr 1975) is taken normally at the slight but sharp downward change

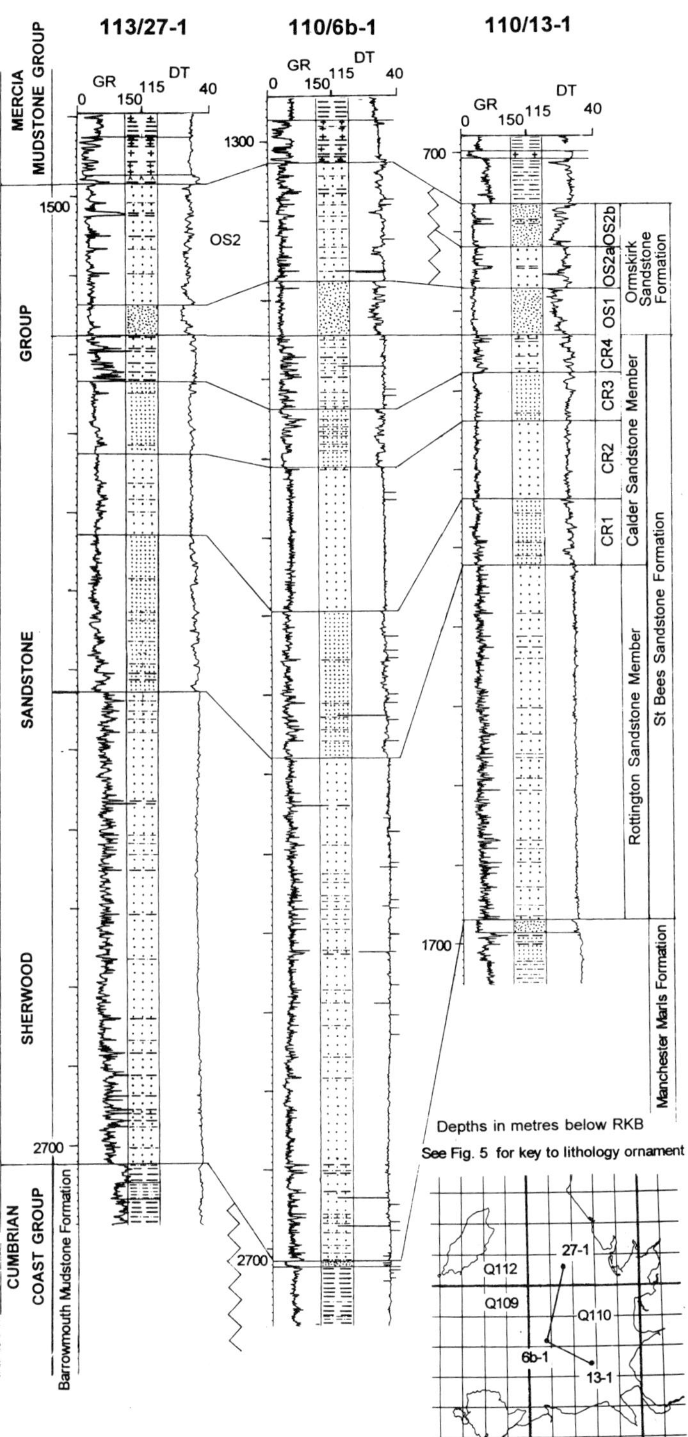

Fig. 4. Lithostratigraphy and correlation of the Sherwood Sandstone Group in selected wells in the offshore East Irish Sea Basin.

to the linear, very high velocity response of the Rottington Sandstone Member; this position is also coincident with the downward increase in overbank mudstone partings and in the higher background gamma value of the sandstones (e.g. 113/27-1, Fig. 4; cf. Barnes *et al.* 1994; Jones & Ambrose 1994). However, at shallow burial depths where the velocity contrast between the two members is less marked (e.g. 110/8-1; 110/20-1, Jackson & Johnson 1996), the gamma log may provide a firmer boundary pick (cf. Chadwick *et al.* 1994).

By analogy with similar log responses in the Ormskirk Sandstone Formation (Cowan 1993; Meadows & Beach 1993), four informal sub-divisions (units CR1, CR2, CR3 and CR4) of fluvial alternating with interbedded aeolian/fluvial sandstones can be recognized widely in the Calder Sandstone Member (cf. Jones & Ambrose 1994). The largely fluvial unit CR4, contains the highest proportion of thin high gamma mudstones, and overlies a mixed aeolian and fluvial sandstone interval (unit CR3; e.g. 113/27-1, Fig. 4). The largely waterlain unit CR2 with relatively linear gamma and sonic traces, constitutes a higher 'silicified zone' and overlies unit CR1, which has a similar log response to unit CR3 (e.g. 110/6b-1, Fig. 4). The Calder Sandstone Member equates with the St Bees Sandstone Formation (upper unit) of Jackson *et al.* (1987), and with the Wilmslow Sandstone Formation of the Cheshire Basin (*sensu* Warrington *et al.* 1980, p.63, and Jackson *et al.* 1987, but *contra* Colter 1978 and Evans *et al.* 1993).

The ***Rottington Sandstone Member** (175–666 m) is a thick monotonous 'orthoquartzitic' sandstone deposited in stacked low-sinuosity fluvial channels, and characterized by a distinctive, linear, near-matrix sonic velocity (58–65 μs ft^{-1}) reflecting the siliceous and dolomitic cement (cf. Barnes *et al.* 1994). The spiky gamma log trace arises from numerous thin high gamma, overbank silty mudstone and siltstone partings, mudstone-rich channel lags and micaceous sandstones (cf. Macchi 1991). In the north of the EISB, the lower boundary is taken at the sharp downward change to high gamma mudstone (or locally siltstone) of the Barrowmouth Mudstone Formation (e.g. 113/27-1, Fig. 4). In the centre and south, it is taken at the downward change to the Freshfield Sandstone Member – a thin dominantly aeolian sandstone with lower gamma values (e.g. 110/6b-1, Figs 4, 5) and generally with a slightly lower velocity and fewer mudstone spikes (e.g. 110/13-1, Fig. 4; cf. also Macchi 1991, p.71). The Rottington Sandstone Member may not have been deposited on the higher trailing margins of certain half graben (cf. Jackson *et al.* 1987, fig. 3b).

The Rottington Sandstone Member equates with the lower unit of the St Bees Sandstone Formation of Jackson *et al.* (1987), and with the Chester Pebble Beds Formation (*sensu* Warrington *et al.* 1980, but *contra* Evans *et al.* 1993 and Colter 1978). In the EISB, equivalent sandstones locally contain scattered pebbles (e.g. Formby 1, Kent 1948), but until the regional log motif of the pebbly units is better known, the Llŷn–Rossendale Ridge provides an appropriate northern limit for the Chester Pebble Beds Formation. Rottington is a hamlet near St Bees Head.

Cumbrian Coast Group (Upper Permian)

The term Cumbrian Coast Group (Fig. 5) has been introduced recently for the evaporitic–argillaceous Upper Permian strata of NW England and Southern Scotland (Brandon *et al.* in press). Constituent formations, recently formalized by BGS, include the St Bees Shales Formation, the St Bees Evaporites Formation and the Manchester Marls Formation (Smith *et al.* 1974; Arthurton *et al.* 1978). Uniform red-brown barren laminated mudstones comprise the highest strata over wide areas, both as individual formations (e.g. the St Bees Shales at St Bees Head, Arthurton & Hemingway 1972) and as lateral equivalents incorporated within more broad-ranging dominantly argillaceous successions (e.g. Manchester Marls Formation, Eden Shales Formation). In contrast, the remainder of the group offshore is characterized by marked lateral variability, with halite-dominated successions (St Bees Evaporites Formation) in the EISB depocentre passing abruptly outwards to marine and coastal plain mudstones (Manchester Marls Formation). Further south and over the Llŷn–Rossendale Ridge, the Manchester Marls Formation passes with increasing amounts of sandstone, initially water-laid but becoming progressively wind-derived to the south, into an aeolian-dominated sandstone succession (namely the Upper Permian component of the Kinnerton Sandstone Formation, Warrington *et al.* 1980); in the east, fringing the Lake District Massif (and locally the Pennines), the evaporites and coastal plain mudstones pass into alluvial and piedmont-fan breccias (the Upper Permian component of the Brockram, Arthurton *et al.* 1978).

The Upper Permian succession thickens as a sag (Fig. 2) centripetally towards the evaporite depocentre but reveals local maxima in the hangingwall of growth faults, both interiorly (e.g. 110/8-2, adjacent to the Deemster Platform) and at the basin margin (Humphrey Head, Rose & Dunham 1977). The base of the Cumbrian Coast Group represents an important sequence boundary, and overstep across Lower Permian strata is reported from the basin margins, e.g. the West Crop of the EISB (Jackson & Mulholland 1993, fig. 9a), in Formby 4

(Falcon & Kent 1960), and in Furness (Dunham & Rose 1949; Jackson & Mulholland 1993).

The term *Barrowmouth Mudstone Formation** (83–135 m) is introduced as the offshore equivalent to the upper part (laminated facies) of the St Bees Shales of St Bees Head (Arthurton & Hemingway 1972), in order that the present nomenclative ambiguities in the use of 'St Bees Shales' can be avoided. The term 'St Bees Shales' is currently employed in three senses – (a) in west Cumbria, as redefined by Arthurton & Hemingway (1972) for the mudstones overlying the St Bees Evaporites Formation at St Bees Head , (b) in south Cumbria, for the entire Upper Permian succession excluding the Grey Beds and Gleaston Dolomite, but including equivalents of the higher St Bees Evaporites (Arthurton *et al.* 1978; Rose & Dunham 1977), and (c) in the Carlisle Plain, for the entire Upper Permian succession as an equivalent of the Eden Shales Formation (Eastwood *et al.* 1968; Arthurton *et al.* 1978, p.193).

The lower boundary of the Barrowmouth Mudstone Formation is abrupt and is positioned, in undisturbed sequences, either at the first downward occurrence of halite in the St Bees Evaporites Formation (e.g. 112/25a-1, Fig. 5), or, where present, at a thin anhydrite capping the halite. Several wells demonstrate tectonic and halokinetic disturbance where the beds rest on horizons low in the St Bees Evaporites Formation, e.g. at a halite detachment (110/8-2) or halite dissolution front (110/2b-10, Fig. 5; *q.v.* Jackson & Johnson 1996). The sigmoidal gamma log trace of the Barrowmouth Mudstone Formation is distinctive (e.g. 112/25a-1, Fig. 5) and is correlated here with the Bröckelschiefer Member (cf. Johnson *et al.* 1994), and with the highest Eden Shales Formation in the Hilton borehole (Holliday 1993). Offshore, three subdivisions of approximately equal thickness can be recognized mainly from the gamma log. Units BM1 and BM3 consist of upward-increasing gamma values and are separated by a more silty interval (unit BM2), displaying a relatively linear, lower gamma response (e.g. 112/25a-1, 113/27-3, Fig. 5) and believed here to be equivalent to the Hewett Sandstone of the Southern North Sea (Johnson *et al.* 1994). Barrowmouth is near St Bees Head.

In the EISB, recognition on wireline logs of the Barrowmouth Mudstone Formation as a correlative of the Bröckelschiefer Member of the Southern North Sea constrains, for the first time, the discrepancy between the base of the Triassic as adopted in the two areas (cf. Warrington *et al.* 1980). In the absence of biostratigraphical control, the base of the Triassic in western Britain (and this study) has been taken, conventionally, as the base of the Budleigh Salterton Pebble Beds (=Chester Pebble Beds Formation and Rottington Sandstone Member and equivalents, i.e. the **top** of the Barrowmouth Mudstone Formation). This contrasts with a lower stratigraphic position for the boundary in the UK Southern North Sea where the base Triassic is defined, by convention, at the base of the Bröckelschiefer Member (Rhys 1974; Johnson *et al.* 1994), equivalent to the **base** of the Barrowmouth Mudstone Formation.

Offshore, the **St Bees Evaporites Formation** (145.5–220 m aggregate, but possibly *c.*350 m thick near 110/8-2) comprises a Basal Siltstone/Carbonate Unit (21.5 m) overlain by a clean Lower Halite Unit (87 m) and a more argillaceous Upper Halite Unit (110m; Figs 1, 5; cf. Jackson 1994). Lateral variation in the Basal Unit is reflected in the preferential development of carbonate near the basin margin and over oxygenated highs (e.g. 113/27-2), whereas grey calcareous siltstone is widely developed in the dysaerobic basinal areas. The lithological elements of the evaporite succession are laterally persistent and correlations with the Zechstein Basin and the basin margin sections around the Bakevellia Sea Basin are discussed in Jackson (1994, fig. 2), updated in Jackson & Johnson (1996) and are not repeated here. Offshore, the lower boundary is taken at the abrupt downward change to Collyhurst Sandstone Formation (usually thin) or, over the Quadrant 109 Arch, at the contact with secondarily reddened Carboniferous (e.g. 110/3b-4). Halite dissolution (e.g. 110/2b-10, Fig. 5), halite detachments (e.g. 112/25a-1, Fig. 5) and halokinesis (e.g. 110/8-2, Jackson *et al.* 1987, fig. 6a) occur locally and give rise to anomalous sequences.

The St Bees Evaporites Formation passes abruptly southwards into the Manchester Marls Formation. The basin-filling halites thin substantially and pass into mudstones and dolomitic siltstones across a hinge line or a fault-controlled depositional slope. Only two evaporite-bearing sequences are known from the narrow transition zone (110/6b-1, Fig. 5, and 110/8-1, both in the south of the zone), and both are assigned here to the Manchester Marls Formation. Based on this limited evidence, it is suggested that where the evaporite proportion in unfaulted successions exceeds *c.*40%, then the beds are referred to the St Bees Evaporites Formation, but where less than *c.*40%, then the beds are placed in the Manchester Marls Formation. Applied onshore, this definition ensures compatability with the type section of the St Bees Evaporites Formation at St Bees Head (Arthurton & Hemingway 1972), whereas the coeval, clastic-dominated, basin margin successions of Haverigg Haws and south Cumbria are excluded, as at present (Rose & Dunham 1977; Arthurton *et al.* 1978).

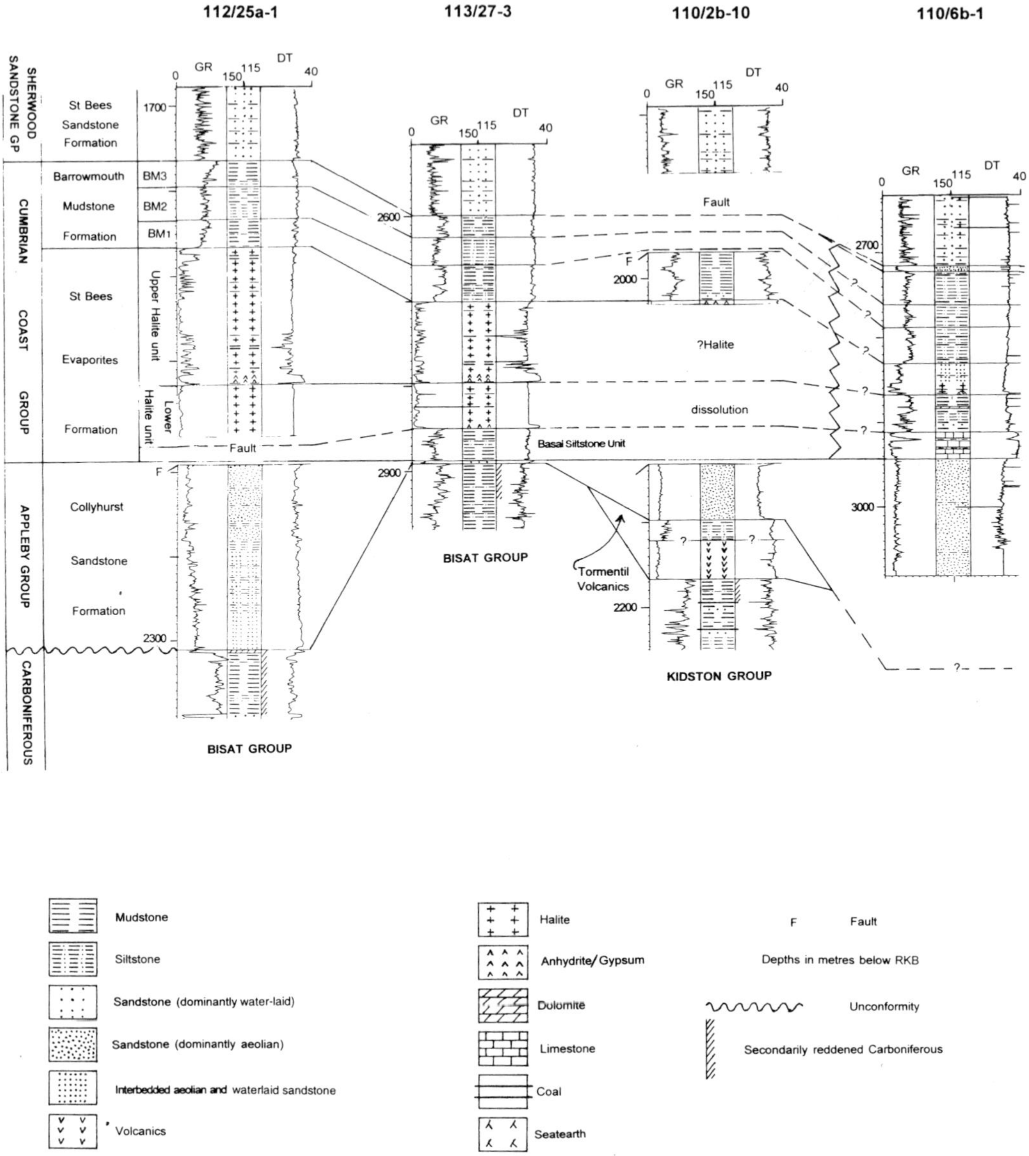

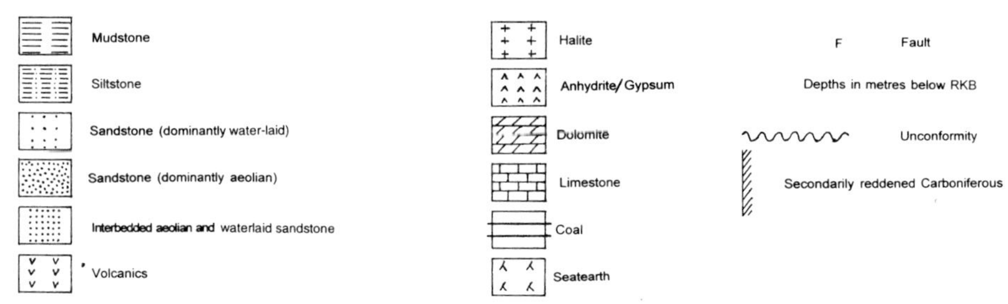

The **Manchester Marls Formation** (69–227.5 m) is laterally equivalent to the combined St Bees Evaporites and Barrowmouth Mudstone formations and dominantly comprises red mudstones, marine in origin at the base where, however, they are commonly grey. Marked lateral facies changes occur. The three subdivisions of the Barrowmouth Mudstone Formation can be recognized on log character only in the northernmost developments of the Manchester Marls Formation (e.g. 110/6b-1, Fig. 5). Sandstones and siltstones become increasingly common and thicker upwards, and towards all the basin margins; onshore, a laterally equivalent facies of interbedded mudstones, fluvial and aeolian sandstones forms a wide transition zone (the Bold Formation,

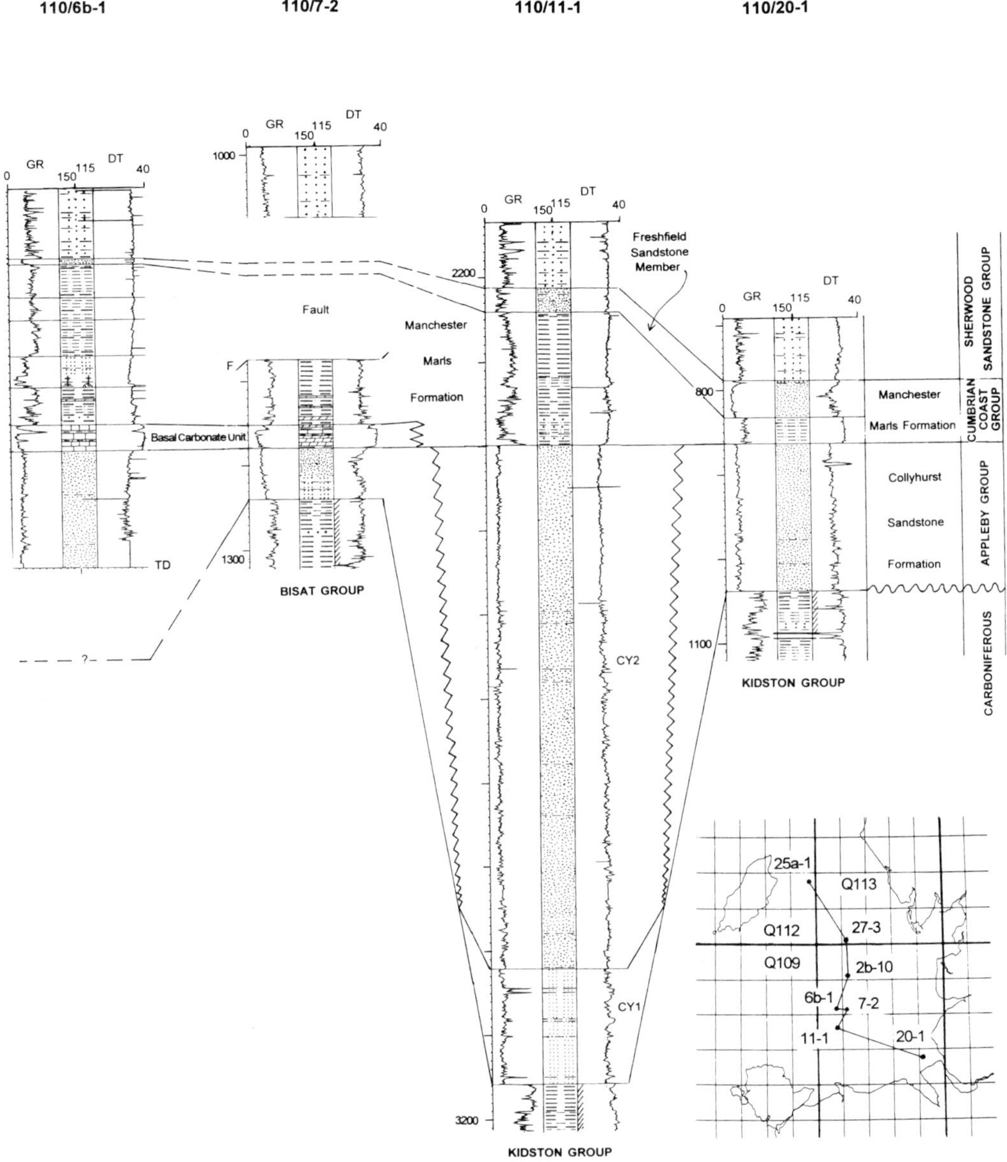

Fig. 5. Lithostratigraphy and correlation of the Cumbrian Coast Group and Appleby Group in selected wells in the offshore East Irish Sea Basin. Also includes key to lithology ornaments used on figures.

Thompson 1989; Evans *et al.* 1993) developed over and around the Llŷn–Rossendale Ridge. In most wells sited in areas of slow Permian subsidence, a Basal Carbonate Unit occurs (e.g. 110/6b-1, 110/7-2, Fig. 5; Jackson *et al.* 1987, fig. 6a) and is believed to be laterally equivalent to the Basal Carbonate Unit of the St Bees Evaporites Formation; elsewhere the basal beds comprise a thin high gamma mudstone ('hot shale') or siltstone representing the transgressive surface of the first Bakevellia transgression (e.g. 110/11-1, Fig. 5). The lower boundary of the Manchester Marls

Formation marks the abrupt downward change to aeolian Collyhurst Sandstone. The term Manchester Marls (cf. Smith *et al.* 1974) has precedence over the Manchester Marl of Colter & Barr (1975).

The term (Bunter) Lower Mottled Sandstone has been used in two different stratigraphical senses in the past. Firstly, the name was applied to a relatively thin (40 m) unit of largely aeolian sandstone separating the mudstones of the Manchester Marls Formation from the fluvial sandstones of the Sherwood Sandstone Group (e.g. Formby 1, Kent 1948; Knutsford 1, Colter 1978); the term *Freshfield Sandstone Member is now introduced for the offshore equivalent of these thin aeolian sandstones. Secondly, the name was given to a thick (150–380 m) aeolian sandstone of combined early and late Permian age (which was also overlain by the fluvial Sherwood Sandstone Group), in areas where the Manchester Marls Formation could not be recognized (e.g. over the Llŷn-Rossendale Ridge; Burton Point, Macchi 1991). Warrington *et al.* (1980) introduced the term Kinnerton Sandstone Formation (see below) for these thick aeolian sandstones, but included them in the Sherwood Sandstone Group, as a unit of partly Triassic and partly Permian age.

The ***Freshfield Sandstone Member** (0–44 m) is a low gamma marker sandstone, generally of moderately low velocity, comprising the highest Manchester Marls Formation; it thickens progressively southwards both in the EISB (Fig. 5) and onshore as an equivalent in the Cheshire Basin (e.g. Knutsford 1, Colter 1978). Exceptionally, on its northernmost feather-edge in 110/3b-4, the sandstone constitutes the uppermost Barrowmouth Mudstone Formation. The sandstone was specifically excluded from the Kinnerton Sandstone Formation by Warrington *et al.* (1980, pp. 31, 59), and regarded as a local pebble-free unit of the Chester Pebble Beds Formation (i.e. Triassic) by those authors, whereas Jackson *et al.* (1987) and Evans *et al.* (1993) preferred to view the sandstone as an upper leaf of the largely aeolian Kinnerton Sandstone Formation. In the Bakevellia Sea Basin the wireline log response (Fig. 5), together with sedimentological analysis of presumed equivalent beds at Burton Point (Macchi 1991) and in Formby 1 (Kent 1948), strongly suggest that the unit is largely aeolian in origin and directly precedes the first fluvial input of the Budleighensis River.

Appleby Group (Lower Permian)

The Appleby Group is a newly introduced term for the continental Lower Permian rocks of NW England and Southern Scotland (Brandon *et al.* in press) that includes the recently formalized Penrith,

Collyhurst, and Locharbriggs sandstone formations (cf. Smith *et al.* 1974). The beds are unconformable everywhere on the beds beneath and abrupt thickness changes are common, especially at basin margins (e.g. Poole & Whiteman 1955; Falcon & Kent 1960). The group is characterized by sequences of monotonous aeolian sandstone deposited in discrete intermontane basins many of which probably lay below sea-level (Smith 1972; Smith & Taylor 1992).

Offshore, the largely aeolian **Collyhurst Sandstone Formation** (0–763 m) generally rests on Carboniferous rocks unconformably, but locally is presumed to lie on ?early Permian volcanics at a disconformity. The red sandstone, which generates a low gamma, box-car log response, is locally grey in colour probably as a result of reduction by hydrocarbons. In some expanded successions (e.g. 110/11-1, Fig. 5; 110/13-1), a thin basal waterlaid unit of higher or more variable velocity has been differentiated locally as unit CY1 (e.g. 110/11-1); it may be analogous to the basal, locally calcareous, breccia and pebbly sandstone described from thick Collyhurst Sandstone successions in the Formby and Stockport areas (Kent 1948; Taylor *et al.* 1963; cf. Poole & Whiteman 1955) and to the Penrith Brockram (e.g. below 1270 m in Silloth 1A, Holliday 1993; cf. Arthurton *et al.* 1978) (see also Smith 1972; Brookfield 1978). The overlying unit CY2 is largely a monotonous aeolian sandstone, assumed to be mostly dune-bedded but probably containing deposits from inter-dune sites (identified by anhydrite cements, e.g. 110/12a-1) and ephemeral playas (represented by very scarce thin mudstones, cf. Taylor *et al.* 1963, plate 4; Kent 1948). In most cases, the base of the formation is easily identified where the monotonous sandstones abruptly overlie high gamma secondarily reddened Silesian measures (Fig. 5), or less commonly, Dinantian carbonates; the contact may be difficult to identify from log character where Permian sandstone directly overlies a Silesian sandstone (?cf. 112/25a-1, 110/7-2, Fig. 5), especially if the high velocity unit CY1 is present.

During the early Permian, sediment accumulation over the Quadrant 109 Arch-High Haume Anticline (Jackson & Mulholland 1993) occurred in small pockets only or was entirely absent (Fig. 2); the structure separated a smaller northern half-graben (e.g. 112/25a-1, Fig. 5), bounded on the north by the Lagman Fault, from a larger southern sub-basin in which much thicker dune sandstones were deposited in a small erg (e.g. 110/11-1, Fig. 5). The successions in the south are believed here to have infilled, passively, a subsea tectonic depression, coincident in part with a Caledonoid-trending, Westphalian-cored syncline (Smith 1972; Jackson & Mulholland 1993). The aeolian sediments were banked in some cases against pre-existing fault scarps (cf. Poole & Whiteman 1955), though other faults remained active, especially at the basin margins. Along the northeast margin of the EISB, both onshore and offshore, the Collyhurst Sandstone Formation lies basinward of a clast-supported, breccio-conglomerate (the **Basal Breccia**, 1–26 m thick, Arthurton & Hemingway 1972; Arthurton *et al.* 1978; log response (incomplete) at Sellafield shown in Barclay *et al.* 1994).

A unit of thinly bedded red sandstones, siltstones and mudstones believed to comprise a Lower Permian playa facies analogous to the Silverpit Formation of the Southern North Sea has been identified in the centre of the EISB by Smith *et al.* (1974), Colter & Barr (1975) and Jackson *et al.* (1987). However, recognition of the Quadrant 109 Arch-High Haume Anticline as a positive feature during the early Permian has prompted a reappraisal of these supposed playa sequences (e.g. wells 110/8-2 and 113/26-1). They are regarded here, largely or wholly, as secondarily reddened thinly bedded Silesian measures (Fig. 1).

The *Tormentil Volcanics comprise low gamma, high velocity, altered basalts (45.5 m thick) and possible overlying volcaniclastic rocks (22 m) in 110/2b-10 (Fig. 5; Hardman 1992); they are regarded here as early Permian (possibly to ?latest Carboniferous) in age, approximately contemporaneous with the volcanic activity known from Larne (Penn *et al.* 1983), Thornhill and elsewhere in Southern Scotland (Brookfield 1978), and from further afield in the Southern North Sea (Cameron 1993) and SW England. In 110/2b-10, the upper and lower contacts (2121.5–2167 m) with the low velocity Collyhurst Sandstone Formation and high gamma Silesian mudstones respectively, are both sharp. The localized volcanic activity occurred at a tectonic contrast between areas of basin extension and the more rigid structure of the positive Quadrant 109 Arch (cf. Dixon *et al.* 1981). Tormentil is a small yellow grassland flower.

Undivided Permian successions at basin margins

Over the higher flanks and the crest of the Llŷn–Rossendale Ridge, Upper and Lower Permian rocks are reported to be indistinguishable on lithological criteria, and this undifferentiated, largely aeolian, sandstone (formerly the Bunter Lower Mottled Sandstone) is now termed the **Kinnerton Sandstone Formation** (Warrington *et al.* 1980). The sandstone has not been proved in offshore wells to date, though evidence from the Cheshire Basin (e.g. Prees-1, Knutsford-1, Evans *et*

al. 1993) suggests that the Upper and Lower Permian intervals **can** be differentiated on wireline log character.

Adjacent to the Lake District, particularly near active faults (e.g. Humphrey Head, Rose & Dunham 1977), the Upper Permian rocks unite with the Basal Breccia and pass laterally into the canyon breccia fills of the **Brockram** (120–257 m+) of undivided Permian age (Smith 1924; Arthurton *et al.* 1978). Although breccia tongues occur offshore in Upper Permian sequences (cf. Haverigg Haws, Rose & Dunham 1977; Jackson *et al.* 1987), it is recommended that use of the *lithostratigraphical* term Brockram should be reserved for successions where the arenaceous and rudaceous content exceeds *c.*60% (e.g. Humphrey Head; see Barclay *et al.* 1994 for log response (incomplete) in the Sellafield area).

Carboniferous

Bearing in mind the complex and varied Carboniferous lithostratigraphical classifications onshore (George *et al.* 1976; Ramsbottom *et al.* 1978), it is considered premature at this stage to erect a hierarchical lithostratigraphy for Carboniferous successions of the offshore EISB, from merely the limited stratigraphical range and minimal duplication of sequences in available wells and the scanty biostratigraphical determinations. Information on stage boundaries presented on Fig. 6 has been simplified from composite logs and/or commercial biostratigraphical reports. In many cases, suitable onshore time-equivalent names do not exist, are applicable only locally or, the log response of onshore correlative units is poorly known. Three new groups, the Kidston, Bisat and Garwood groups, corresponding as far as practicable to beds of Westphalian, Namurian and Dinantian age, are therefore introduced for the Carboniferous rocks of the EISB (Figs 1, 2; cf. Ramsbottom *et al.* 1978, fig. 1). Their status may require raising to supergroup when additional wireline log data enable subdivisions to be erected. The identification of a widespread sandstone unit in the Bisat Group shows the potential of Carboniferous wireline log correlation in the EISB (Fig. 6) and strongly suggests that future correlations will match the precision of those in Permo-Triassic successions and thus provide vital information on the stratigraphical distribution of the various source rocks.

The ***Kidston Group** (500–?4000 m in the Quadrant 109 Syncline) comprises rocks of mainly Westphalian and Stephanian age, but also incorporates G_1 strata in areas where the Subcrenatum Marine Band cannot be recognized. The group

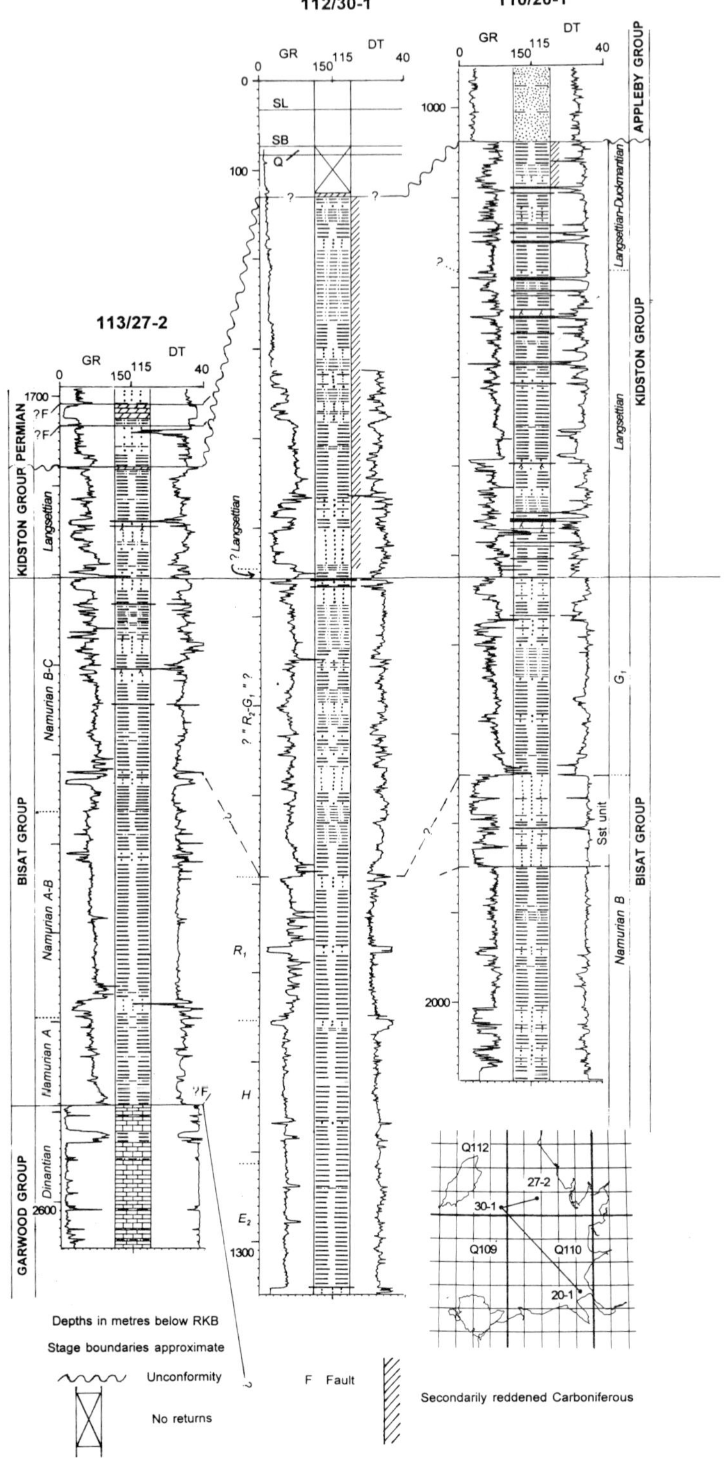

Fig. 6. Lithostratigraphy of the Carboniferous succession (Kidston, Bisat and Garwood groups) in selected wells in the offshore East Irish Sea Basin; the figure is hung on the Top Bisat Group.

consists of grey, coal-bearing, upward-coarsening cyclothems of Langsettian to early Bolsovian age of upper delta plain and alluvial plain origin (Fig. 6), together with the primary red (and subordinate grey) upward-fining, more arenaceous, alluvial channel and floodplain cycles of late Bolsovian to Westphalian D age (Ramsbottom *et al.* 1978; Fraser & Gawthorpe 1990; Cope *et al.* 1992); Cantabrian and Stephanian strata may occur in the core of the Quadrant 109 Syncline (Jackson & Mulholland 1993). Precise recognition of the lower boundary is unsatisfactory unless core and/or macrofauna are available, and offshore is dependent largely on identification of the log signature of the Sub-crenatum Marine Band, generally a 'hot shale' (Whittaker *et al.* 1985; *q.v.* Jackson & Johnson 1996). Future subdivision into formations is envisaged. The group name is derived from R. Kidston, who investigated the taxonomy and stratigraphical range of Westphalian plants.

The ***Bisat Group**** (150–2900 m) comprises strata chiefly of Namurian age, but might also include Langsettian measures where the base of the Kidston Group is positioned provisionally above the presumed or unrecognized Subcrenatum Marine Band. The group also includes rocks of Dinantian age in locations where the Cravenoceras leion Marine Band cannot be identified (cf. Ramsbottom *et al.* 1978), in which case the base of the Lower Bowland Shale Formation equivalent is taken provisionally as the junction, e.g. in the E–W axial zone along-strike from the Ribblesdale Fold Belt (Earp *et al.* 1961; Aitkenhead *et al.* 1992). The Bisat Group as proved in available wells (Fig. 6), is more argillaceous than the laterally equivalent Namurian successions of Lancashire, Cumbria and North Wales (excluding the area around the Point of Ayr), where sandstones are generally well developed (Ramsbottom 1969*a*; Ramsbottom *et al.* 1978). The mudstones show considerable variation, both laterally and vertically (Jackson *et al.* 1987). The EISB appears to have lain beyond the SSW limit of the feldspathic, northerly derived, turbidite-fronted Millstone Grit deltas, and north of the main occurrence of thin orthoquartzitic sandstones erived from the Wales–Brabant Massif to the south (Ramsbottom 1969*a*; Collinson 1988). Strati-graphically, the sequence shows the ubiquitous overall upward-shallowing and upward-coarsening from the E_1 deep-water marine mudstones of Namurian A, through the delta slope sediments of Namurian B, to the shallow-water sheet-delta and lower delta plain sequences of Namurian C with Westphalian-type cyclothems and siltstones (Fraser & Gawthorpe 1990; Cope *et al.* 1992).

The lower boundary of the Bisat Group is taken at the base of the 'hot shale' of the Cravenoceras leion Marine Band (Whittaker *et al.* 1985). The log signature may be difficult to identify unequivocally and, as with the Kidston Group, the lower boundary choice should then be guided by other faunal and geographical constraints (*q.v.* Jackson & Johnson 1996). As more wireline logs become available, formal subdivision of the Bisat Group will be possible; for example, the gamma signature of a sandstone unit, dated as Namurian C/Langsettian in 110/20-1 but Namurian B/C in 110/11-1, shows close similarities in several adjacent wells. The group name derives from W.S. Bisat who pioneered the detailed use of goniatites in Namurian strati-graphy.

Rocks of the *****Garwood Group**** (300–2050 m) are largely of Dinantian age, but provisionally, in the extreme north might include early E_1 cyclothems of limestone-rich Yoredale facies, where the Cravenoceras leion Marine Band cannot be identified (Rose & Dunham 1977; Mitchell *et al.* 1978). The carbonate (chiefly limestone) lithologies of the group were deposited in environments ranging from shallow-water evaporitic sabkhas and oolite shoals to thinly bedded deep-water limestones (Fig. 6; Gawthorpe *et al.* 1989); thinly bedded basinal limestones and calcareous mudstones are predicted along the line of the Ribblesdale Fold Belt/Quadrant 109 Syncline/Dublin Basin (Ramsbottom 1969*b*; Fraser & Gawthorpe 1990; Cope *et al.* 1992; Jackson & Mulholland 1993). The lower boundary has not been proved in offshore wells, but from onshore analogies is taken variously at the downward change from limestone or basal conglomerate resting unconformably upon red Devonian sand-stones or mudstones, Lower Palaeozoic grey-wackes or cleaved mudstones, or Precambrian metasediments (BGS 1994). The group name is derived from E.J. Garwood who worked exten-sively on the Dinantian of NW England.

Secondarily reddened Carboniferous. Except where the Top Carboniferous is faulted, a zone (50–500 m thick) of secondarily reddened Carbon-iferous rocks occurs over large areas of the EISB directly beneath the Variscan Unconformity (Fig. 6; Jackson *et al.* 1987; Trotter 1953). The reddening affects Carboniferous rocks of all ages but the effects were more penetrative in the thinly bedded Silesian measures and in some deep-water Dinantian mudstone successions, rather than in the thick-bedded Dinantian limestones (Fig. 1). The zone is characterized by drab purple, maroon, dull red and variegated hues, rather than the slightly brighter red colours of primary red beds. Mudstones are commonly affected to greater depths than sandstones; carbonaceous organic matter within the zone is destroyed, including the conversion in extreme circumstances, of coal into dolomite (Strahan 1901). The lower boundary is transitional and is taken at the downward change to grey strata, though the base probably cannot be positioned accurately within primary red bed sequences of post intra-Bolsovian age.

The work upon which this contribution is based was funded partially by an industry consortium (in alphabetical order, BHP Petroleum Ltd [formerly Hamilton Brothers Oil and Gas Ltd], British Gas Exploration and Production Ltd, Lasmo North Sea PLC and Monument Oil and Gas plc), and we are most grateful to those companies and their partners for permission to present a summary of the results. We wish to thank the following companies for consent to pre-release and figure well intervals: BHP for parts of 110/13-1 and 110/13-8, British Gas for part of 110/2b-10, and Shell for parts of 110/20-1 (Liverpool Bay 1). Comments at various stages in this work have been welcomed from Greig Cowan, Brian Haig, Robert Knox and Ian Longley. Thanks also to Sheila Jones, Sandy Henderson, John McInnes and Heather Baily for help with computing, and to Fred Broadhurst and an anonymous referee for helpful comments. This paper is published with the permission of the Director of the British Geological Survey (NERC).

References

AITKENHEAD, N., BRIDGE, D. McC., RILEY, N. J. & KIMBELL, S. F. 1992. *Geology of the country around Garstang.* Memoir of the British Geological Survey, Sheet 67 (England and Wales), HMSO.

ARTHURTON, R. S. 1980. Rhythmic sedimentary sequences in the Triassic Keuper Marl (Mercia Mudstone Group) of Cheshire, northwest England. *Geological Journal,* **15**, 43–58.

—— & HEMINGWAY, J. E. 1972. The St Bees Evaporites – a carbonate-evaporite formation of Upper Permian age in West Cumberland, England. *Proceedings of the Yorkshire Geological Society,* **38**, 565–592.

——, BURGESS, I. C. & HOLLIDAY, D. W. 1978. Permian and Triassic. *In:* MOSELEY, F. (ed.) *The Geology of the Lake District.* Yorkshire Geological Society Occasional Publication, No. 3, 189–206.

BARCLAY, W. J., RILEY, N. J. & STRONG, G. E. 1994. The Dinantian rocks of the Sellafield area, west Cumbria. *Proceedings of the Yorkshire Geological Society,* **50**, 37–49.

BARNES, R. P., AMBROSE, K., HOLLIDAY, D. W. & JONES, N. S. 1994. Lithostratigraphical subdivision of the Triassic Sherwood Sandstone Group in west Cumbria. *Proceedings of the Yorkshire Geological Society,* **50**, 51–60.

BENTON, M. J., WARRINGTON, G., NEWELL, A. J. & SPENCER, P. S. 1994. A review of the British Middle Triassic tetrapod assemblages. *In:* FRASER, N. C. & SUES, H-D. (eds) *In the shadow of the dinosaurs: early Mesozoic tetrapods.* Cambridge University Press, New York, 131–160.

BGS. 1994. *East Irish Sea* (Special Sheet Edition, Solid Geology). 1:250 000, British Geological Survey, Edinburgh.

BRANDON, A., AITKENHEAD, N., CROFTS, R. G., ELLISON, R. A., EVANS, D. J. & RILEY, N. J. In press. *Geology of the country around Lancaster.* Memoir of the British Geological Survey, Sheet 59 (England and Wales), HMSO.

BROOKFIELD, M. E. 1978. Revision of the stratigraphy of Permian and supposed Permian Rocks of Southern Scotland. *Geologische Rundschau,* **67**, 110–149.

CAMERON, T. D. J. 1993. Volume 4. Triassic, Permian and pre-Permian of the Central and Northern North Sea. *In:* KNOX, R. W. O'B. & CORDEY, W. G. (eds) *Lithostratigraphic nomenclature of the UK North Sea.* British Geological Survey, Nottingham.

CHADWICK, R. A., KIRBY, G. A. & BAILY, H. E. 1994. The post-Triassic structural evolution of north-west England and adjacent parts of the East Irish Sea. *Proceedings of the Yorkshire Geological Society,* **50**, 91–102.

COLLINSON, J. D. 1988. Controls on Namurian sedimentation in the Central Province basins of northern England. *In:* BESLY, B. M. & KELLING, G. (eds) *Sedimentation in a Synorogenic Basin Complex. The Upper Carboniferous of Northwest Europe.* Blackie, Glasgow, 85–101.

COLTER, V. S. 1978. Exploration for gas in the Irish Sea. *In:* VAN LOON, A. J. (ed.) *Key-notes of the MEGS-II (Amsterdam, 1978). Geologie en Mijnbouw,* **57**, 503–516.

—— & BARR, K. W. 1975. Recent Developments in the Geology of the Irish Sea and Cheshire Basins. *In:* WOODLAND, A. W. (ed.) *Petroleum and the Continental Shelf of North-West Europe. Volume 1: Geology.* Applied Science Publishers, London, 61–75.

—— & EBBERN, J. 1979. SEM studies of Triassic reservoir sandstones from the Morecambe Field, Irish Sea, U.K. *Scanning Electron Microscopy,* **1**, 531–538.

COPE, J. C. W., GUION, P. D., SEVASTOPULO, G. D. & SWAN, A. R. H. 1992. Carboniferous. *In:* COPE, J. C. W., INGHAM, J. K. & RAWSON, P. F. (eds) *Atlas of Palaeogeography and Lithofacies.* Geological Society, London, Memoir **13**, 67–86.

COWAN, G. 1993. Identification and significance of aeolian deposits within the dominantly fluvial Sherwood Sandstone Group of the East Irish Sea Basin, UK. *In:* NORTH, C. P. & PROSSER, D. J. (eds) *Characterization of Fluvial and Aeolian Reservoirs.* Geological Society, London, Special Publications, **73**, 231–245.

——, OTTESEN, C. & STUART, I. A. 1993. The use of dipmeter logs in the structural interpretation and palaeocurrent analysis of Morecambe Fields, East Irish Sea Basin. *In:* PARKER, J. R. (ed.) *Petroleum Geology of Northwest Europe: Proceedings of the 4th Conference.* The Geological Society, London, 867–882.

DIXON, J. E., FITTON, J. G. & FROST, R. T. C. 1981. The Tectonic Significance of Post-Carboniferous Igneous Activity in the North Sea Basin. *In:* ILLING, L. V. & HOBSON, G. D. (eds) *Petroleum geology of the continental shelf of North-West Europe.* Heyden, London, 121–137.

DUNHAM, K. C. & ROSE, W. C. C. 1949. Permo-Triassic geology of south Cumberland and Furness. *Proceedings of the Geologists' Association,* **60**, 11–40.

EARP, J. R. & TAYLOR, B. J. 1986. *Geology of the Country Around Chester and Winsford.* Memoir of the Geological Survey of Great Britain, Sheet 109 (England and Wales), HMSO.

——, MAGRAW, D., POOLE, E. G., LAND, D. H. & WHITEMAN, A. J. 1961. *Geology of the Country around Clitheroe and Nelson.* Memoir of the Geological Survey of Great Britain, Sheet 68 (England and Wales).

EASTWOOD, T., HOLLINGWORTH, S. E., ROSE, W. C. C. & TROTTER, F. M. 1968. *Geology of the country around Cockermouth and Caldbeck.* Memoir of the Geological Survey of Great Britain, Sheet 23 (England and Wales).

EVANS, D. J., REES, J. G. & HOLLOWAY, S. 1993. The Permian to Jurassic stratigraphy and structural evolution of the central Cheshire Basin. *Journal of the Geological Society, London,* **150**, 857–870.

FALCON, N. L. & KENT, P. E. 1960. *Geological results of petroleum exploration in Britain, 1945–1957.* Geological Society, London, Memoir No. 2.

FRASER, A. J. & GAWTHORPE, R. L. 1990. Tectono-stratigraphic development and hydrocarbon habitat of the Carboniferous in northern England. *In:* HARDMAN, R. F. P. & BROOKS, J. (eds) *Tectonic Events Responsible for Britain's Oil and Gas Reserves.* Geological Society, London, Special Publications, **55**, 49–86.

GAWTHORPE, R. L., GUTTERIDGE, P. & LEEDER, M. R. 1989. Late Devonian and Dinantian basin evolution in northern England and North Wales. *In:* ARTHURTON, R. S., GUTTERIDGE, P. & NOLAN, S. C. (eds) *The role of tectonics in Devonian and Carboniferous sedimentation in the British Isles.* Yorkshire Geological Society Occasional Publication, No. 6, 1–23.

GEORGE, T. N., JOHNSON, G. A. L., MITCHELL, M., PRENTICE, J. E., RAMSBOTTOM, W. H. C., SEVASTOPULO, G. D. & WILSON, R. B. 1976. *A correlation of Dinantian rocks in the British Isles.* Geological Society, London, Special Reports, **7**.

HAIG, D. B., PICKERING, S. C. & PROBERT, R. 1997. The Lennox Oil and Gas Field. *This volume.*

HARDMAN, M. 1992. The East Irish Sea Basin. ?Direct evidence for Permo-Carboniferous rifting. *In:* Abstracts with Programme. *Conference on Permo-Triassic rifting, Geological Society of London, October 1992.*

HOLLIDAY, D. W. 1993. Geophysical log signatures in the Eden Shales (Permo-Triassic) of Cumbria and their regional significance. *Proceedings of the Yorkshire Geological Society,* **49**, 345–354.

Institute of Geological Sciences. 1972. 1:50, 000 Sheet 126 (Nottingham).

IVIMEY-COOK, H. C., WARRINGTON, G., WORLEY, N. E., HOLLOWAY, S. & YOUNG, B. 1995. Rocks of Late Triassic and Early Jurassic age in the Carlisle Basin, Cumbria (north-west England). *Proceedings of the Yorkshire Geological Society,* **50**, 305–316.

JACKSON, D. I. 1994. *Discussion of* Geophysical log signatures in the Eden Shales (Permo-Triassic) of Cumbria and their regional significance. *Proceedings of the Yorkshire Geological Society,* **50**, 175–184.

—— & MULHOLLAND, P. 1993. Tectonic and stratigraphic aspects of the East Irish Sea Basin and adjacent areas: contrasts in their post-Carboniferous structural styles. *In:* PARKER, J. R. (ed.) *Petroleum Geology of Northwest Europe: Proceedings of the 4th Conference.* The Geological Society, London, 791–808.

—— & JOHNSON, H. 1996. *Lithostratigraphic nomenclature of the Triassic, Permian and Carboniferous of the UK offshore East Irish Sea Basin.* British Geological Survey, Nottingham.

——, MULHOLLAND, P., JONES, S. M. & WARRINGTON, G. 1987. The geological framework of the East Irish Sea Basin. *In:* BROOKS, J. & GLENNIE, K. W. (eds) *Petroleum Geology of North West Europe.* Graham and Trotman, London, 191–203.

JOHNSON, H., WARRINGTON, G. & STOKER, S. J. 1994. Volume 6. Permian and Triassic of the Southern North Sea. *In:* KNOX, R. W. O'B. & CORDEY, W. G. (eds) *Lithostratigraphic nomenclature of the UK North Sea.* British Geological Survey, Nottingham.

JONES, N. S. & AMBROSE, K. 1994. Triassic sandy braidplain and aeolian sedimentation in the Sherwood Sandstone Group of the Sellafield area, West Cumbria. *Proceedings of the Yorkshire Geological Society,* **50**, 61–76.

KENT, P. E. 1948. A Deep Borehole at Formby, Lancashire. *Geological Magazine,* **85**, 253–264.

KIRTON, S. R. & DONATO, J. A. 1985. Some buried Tertiary dykes of Britain and surrounding waters deduced by magnetic modelling and seismic reflection methods. *Journal of the Geological Society, London,* **142**, 1047–1058.

KNOX, R. W. O'B. & CORDEY, W. C. (eds) 1992–1994. *Lithostratigraphic nomenclature of the UK North Sea.* British Geological Survey, Nottingham. Volumes 1–7.

MACCHI, L. 1991. *A Field Guide to the continental Permo-Triassic rocks of Cumbria and northwest Cheshire.* Liverpool Geological Society.

MEADOWS, N. S. & BEACH, A. 1993. Structural and climatic controls on facies distribution in a mixed fluvial and aeolian reservoir: the Triassic Sherwood Sandstone Group in the Irish Sea. *In:* NORTH, C. P. & PROSSER, D. J. (eds) *Characterization of Fluvial and Aeolian Reservoirs.* Geological Society, London, Special Publications, **73**, 247–264.

MITCHELL, M., TAYLOR, B. J. & RAMSBOTTOM, W. H. C. 1978. Carboniferous. *In:* MOSELEY, F. (ed.) *The Geology of the Lake District.* Yorkshire Geological Society Occasional Publication, No. 3, 168–188.

PENN, I. E., HOLLIDAY, D. W., KIRBY, G. A. *ET AL.* 1983. The Larne No. 2 Borehole: discovery of a new Permian volcanic centre. *Scottish Journal of Geology,* **19**, 333–346.

POOLE, E. G. & WHITEMAN, A. J. 1955. Variations in the thickness of the Collyhurst Sandstone in the Manchester area. *Bulletin of the Geological Survey of Great Britain,* **9**, 33–41.

RAMSBOTTOM, W. H. C. 1969*a*. The Namurian of Britain. *Compte Rendu du Sixième (6me) Congrès International de Stratigraphie et de Géologie du Carbonifère, Sheffield, 1967*, **Vol. 1**, 219–232.

—— 1969*b*. Reef distribution in the British Lower Carboniferous. *Nature, London*, **222**, 765–766.

——, CALVER, M. A., EAGAR, R. M. C., HODSON, F., HOLLIDAY, D. W., STUBBLEFIELD, C. J. & WILSON, R. B. 1978. *A correlation of Silesian rocks in the British Isles*. Geological Society, London, Special Reports, **10**.

RHYS, G. H. 1974. *A proposed standard lithostratigraphic nomenclature for the southern North Sea and an outline structural nomenclature for the whole of the (UK) North Sea*. Institute of Geological Sciences Report, 74/8.

ROSE, W. C. C. & DUNHAM, K. C. 1977. *Geology and hematite deposits of South Cumbria*. Geological Survey of Great Britain Economic Memoir, Sheets 58, part 48 (England and Wales).

SMITH, B. 1924. On the West Cumberland Brockram and its associated Rocks. *Geological Magazine*, **61**, 289–308.

SMITH, D. B. 1972. The British Isles. *In*: FALKE, H. (ed.) *Rotliegend: essays on European Lower Permian*. Brill, Leiden, 1–33.

—— & TAYLOR, J. C. M. 1992. Permian. *In*: COPE, J. C. W., INGHAM, J. K. & RAWSON, P. F. (eds) *Atlas of Palaeogeography and Lithofacies*. Geological Society, London, Memoir **13**, 87–96.

——, BRUNSTROM, R. G. W., MANNING, P. I., SIMPSON, S. & SHOTTON, F. W. 1974. *A correlation of Permian rocks in the British Isles*. Geological Society, London, Special Reports, **5**.

STRAHAN, A. 1901. On the Passage of a Seam of Coal into a Seam of Dolomite. *Quarterly Journal of the Geological Society of London*, **57**, 297–306.

STUART, I. A. 1993. The geology of the North Morecambe Gas Field, East Irish Sea Basin. *In*: PARKER, J. R. (ed.) *Petroleum Geology of Northwest Europe: Proceedings of the 4th Conference*. The Geological Society, London, 883–895.

—— & COWAN, G. 1991. The South Morecambe Field, Blocks 110/2a, 110/3a, 110/8a, UK East Irish Sea. *In*: ABBOTTS, I. L. (ed.) *United Kingdom Oil and Gas Fields, 25 Years Commemorative Volume*. Geological Society, London, Memoir **14**, 527–541.

TAYLOR, B. J., PRICE, R. H. & TROTTER, F. M. 1963. *Geology of the country around Stockport and Knutsford*. Memoir of the Geological Survey of Great Britain, Sheet 98 (England and Wales).

THOMPSON, D. B. 1970. The stratigraphy of the so-called Keuper Sandstone Formation (Scythian–?Anisian) in the Permo-Triassic Cheshire Basin. *Quarterly Journal of the Geological Society of London*, **126**, 151–181.

—— 1989. The geology of the neighbourhood of Chester – an essay review. *Amateur Geologist*, **XIII** (1), 45–54.

—— 1991. Itineraries VIII to XI, Triassic rocks of the Cheshire Basin. *In*: EAGAR, R. M. C. & BROADHURST, F. M. *Geology of the Manchester Area*. Geologists' Association Guide No. 7, The Geologists' Association.

TROTTER, F. M. 1953. Reddened Beds of Carboniferous age in north-west England and their origin. *Proceedings of the Yorkshire Geological Society*, **29**, 1–20.

——, HOLLINGWORTH, S. E., EASTWOOD, T. & ROSE, W. C. C. 1937. *Gosforth District*. Memoir of the Geological Survey of Great Britain, Sheet 37 (England and Wales).

WARRINGTON, G. & IVIMEY-COOK, H. C. 1992. Triassic. *In*: COPE, J. C. W., INGHAM, J. K. & RAWSON, P. F. (eds) *Atlas of Palaeogeography and Lithofacies*. Geological Society, London, Memoir **13**, 97–106.

——, AUDLEY-CHARLES, M. G., ELLIOTT, R. E. *ET AL.* 1980. *A correlation of Triassic rocks in the British Isles*. Geological Society, London, Special Reports, **13**.

WHITTAKER, A., HOLLIDAY, D. W. & PENN, I. E. 1985. *Geophysical logs in British stratigraphy*. Geological Society, London, Special Reports, **18**.

WILLS, L. J. 1970. The Triassic succession in the central Midlands in its regional setting. *Quarterly Journal of the Geological Society of London*, **126**, 225–285.

WILSON, A. A. 1990. The Mercia Mudstone Group (Trias) of the East Irish Sea Basin. *Proceedings of the Yorkshire Geological Society*, **48**, 1–22.

—— 1993. The Mercia Mudstone Group (Trias) of the Cheshire Basin. *Proceedings of the Yorkshire Geological Society*, **49**, 171–188.

—— & EVANS, W. B. 1990. *Geology of the country around Blackpool*. Memoir of the British Geological Survey, Sheet 66 (England and Wales).

YALIZ, A. 1997. The Douglas Oil Field. *This volume*.

The Penarth Group–Lias Group succession (Late Triassic–Early Jurassic) in the East Irish Sea Basin and neighbouring areas: a stratigraphical review

GEOFFREY WARRINGTON

*Biostratigraphy & Sedimentology Group, British Geological Survey,
Kingsley Dunham Centre, Keyworth, Nottingham NG12 5GG, UK*

Abstract: Small, widely separated outliers of latest Triassic and Early Jurassic rocks in and around the East Irish Sea are preserved in the hangingwall blocks of faults that were active during post-Pliensbachian–pre-Cenomanian times. The successions preserved in the outliers comprise the Penarth and Lias groups, in typical facies and with characteristic microfloras and macrofaunas, and are vestiges of a formerly extensive cover of those rocks to the northwest of their main outcrops in England. The Penarth Group, of Rhaetian (Late Triassic) age, comprises the Westbury Formation and the succeeding Lilstock Formation; the latter is represented largely, or entirely, by the Cotham Member. Lias Group successions preserved in the outliers include, above the basal ('pre-planorbis') beds, of latest Triassic age, Early Jurassic beds ranging in age from Hettangian (*planorbis* Zone) to Pliensbachian (*spinatum* Zone) and are considerably thicker than their correlatives in eastern England, indicating that subsidence was slower there than in western Britain in Hettangian, Sinemurian and Pliensbachian times.

The area covered by this account is defined, to the southeast and northeast, by Lias Group outliers in the Cheshire and Carlisle basins respectively, and to the southwest and northwest by those in the offshore Kish Bank and Loch Indaal basins (Fig. 1). Within this area of some 50 000 km² the Lias Group (Powell 1984) also occurs in the Keys Basin, in the East Irish Sea, and in the Larne–Lough Neagh and Rathlin basins in Northern Ireland. The Penarth Group (Warrington *et al.* 1980) is present beneath the Lias Group in Northern Ireland and in the Loch Indaal, Carlisle and Cheshire basins, but is not yet proven in the Keys and Kish Bank basins. In Northern Ireland the Penarth and Lias groups are largely concealed by an unconformable cover of Upper Cretaceous (Cenomanian to Maastrichtian) sediments overlain by Palaeogene volcanics; Palaeogene to Neogene deposits partially mask Triassic and Jurassic rocks in the Kish Bank Basin but elsewhere those rocks are overlain only by Quaternary deposits. Study of the latter occurrences, especially onshore in the Carlisle and Cheshire basins, aids interpretation of their largely concealed analogues in Northern Ireland; conversely, the relationships seen at outcrop and proved by boreholes in Northern Ireland may be inferred for the areas where a cover of Cretaceous rocks no longer exists. The extent of the concealed development in Northern Ireland is not known with certainty but the combined area of the outcrops of the Penarth and Lias groups there and in the other outliers is less than 2% of that covered by this review.

Information on the Lias Group and, where proven, Penarth Group successions in the outliers mentioned above is reviewed and, where necessary, revised; some previously unpublished biostratigraphical information is, pending fuller documentation, summarized in an Appendix. Jurassic ammonite 'zones' and 'subzones' used in this account are biozones and sub-biozones, the boundaries of which are based upon the first stratigraphic occurrences of index ammonite taxa; none are formally defined chronozones, though they are commonly misconstrued as such (*vide* Cox 1990).

Offshore occurrences

Keys Basin

The Keys Basin, situated in the northern part of the East Irish Sea Basin, is approximately equidistant from the Isle of Man and the coast of Cumbria (Fig. 1). Jackson *et al.* (1987) suggested that beds seen on seismic sections to rest disconformably upon Triassic (Mercia Mudstone Group) rocks could be Jurassic. A shallow borehole (89/11A; Fig. 2a), drilled by the British Geological Survey, confirmed the presence of the Lias Group (Jackson & Mulholland 1993; Jackson *et al.*, 1995). Between the base of Quaternary sediments, at 85.81 m, and TD at 95.00 m, the borehole encountered dark grey and black fossiliferous mudstones, commonly silty and micaceous, and with some thin shelly limestones; these beds yielded spores, pollen, acrit-

From Meadows, N. S., Trueblood, S. P., Hardman, M. & Cowan, G. (eds), 1997, *Petroleum Geology of the Irish Sea and Adjacent Areas,* Geological Society Special Publication No. 124, pp. 33–46.

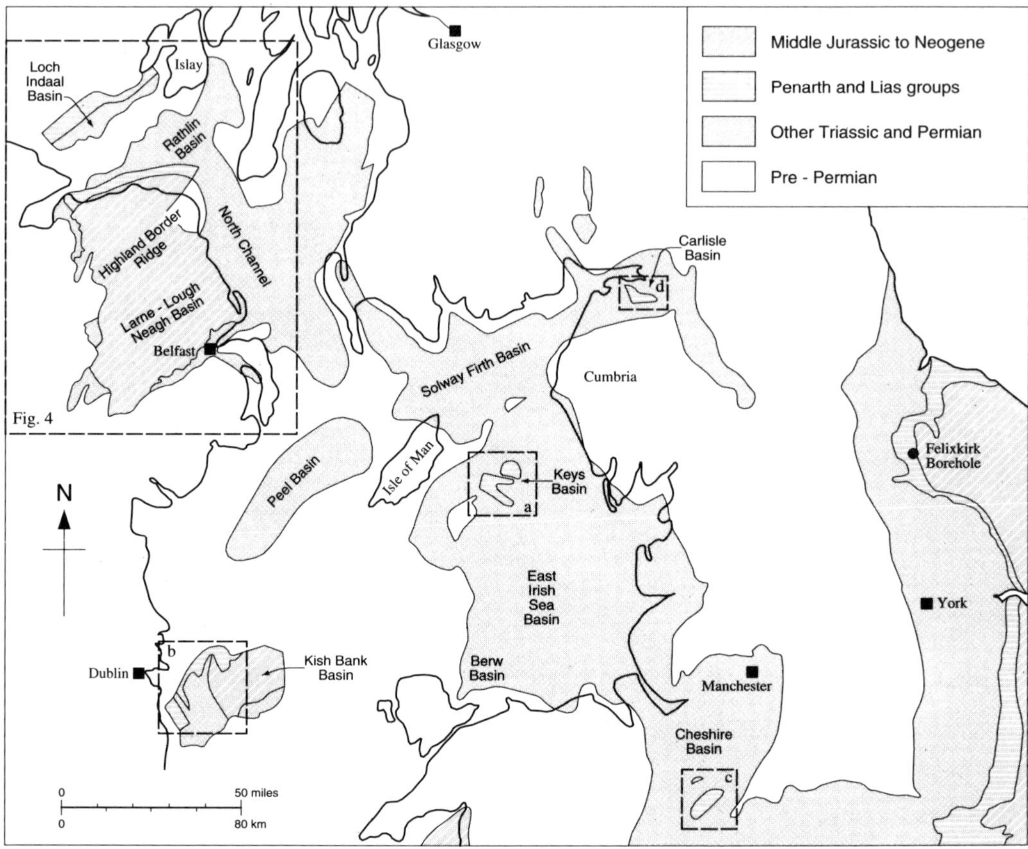

Fig. 1. General location map with areas shown in Figs 2(a–d) and 4. Geology based upon British Geological Survey 1:1 000 000 scale map *Geology of the United Kingdom, Ireland and the adjacent continental shelf* (north and south sheets, 1991), Jenner (1981), Fyfe *et al.* (1993) and Jackson & Mulholland (1993); topography based upon Ordnance Survey mapping.

archs, gastropods, bivalves, ammonites, ostracods and echinoderms (see Appendix). The ammonites are mostly *Arnioceras*, including forms referable to the *A. semicostatum* group which is present in the Sinemurian *semicostatum*, *turneri* and *obtusum* zones. A fragment of *Caenisites?* may indicate the *turneri* Zone (top Lower Sinemurian) though the crinoid columnal cf. *Hispidocrinus*, which Simms (1989) regards as occurring in the Upper Sinemurian, could indicate the younger *obtusum* Zone (H. C. Ivimey-Cook, pers. comm.) (Fig. 3). The miospore assemblage is, by comparison with records from ammonite-controlled successions in southern England (unpublished BGS records), also indicative of a position within the *bucklandi* to *obtusum* Zone sequence.

Jackson & Mulholland (1993) estimated that the beds proved in borehole 89/11A lie 600 m above the 'Near Top Triassic disconformity' of Jackson *et*

al. (1987). Those beds dip northwestwards and are thrown down, in the hangingwall block of the Sigurd Fault, against Triassic (Mercia Mudstone Group) rocks of the Lagman Basin succession to the northwest (Fig. 2a); they were mapped, on seismic evidence, over an area of about 135 km² (British Geological Survey 1994).

The Penarth Group has not been proved in the Keys Basin but Jackson & Mulholland (1993) considered that the 'Near Top Triassic disconformity' may mark its base in that basin. They suggested that the same feature is recognizable on seismic sections of the Berw Basin, some 80 km to the south, and of the Solway Firth Basin to the north (Fig. 1), and thus implied that the Penarth and Lias groups may occur in those basins also. The lower and middle parts of a 650 m thick sequence thought, on seismic evidence, to rest disconformably upon the Mercia Mudstone Group in the

Solway Firth Basin may, by comparison with the Keys Basin, comprise the Lias Group (Jackson *et al.* 1995) but biostratigraphic evidence is lacking and the sparse lithological evidence is ambiguous. Some 200 m of strata thought, on seismic evidence, to comprise an outlier in the Berw Basin may consist largely of the Lias Group, small outliers of which may also be preserved further northeast, in the Tynwald and Crosh Vusta fault complexes (Jackson *et al.* 1995).

Kish Bank Basin

The Kish Bank Basin, about 150 km southwest of the Keys Basin (Fig. 1), was first detected by gravity studies (Bott 1968; Bott & Young 1971). It was subsequently defined on seismic evidence and interpreted as a half-graben occupied by a north-westerly-dipping sequence of Carboniferous to Triassic rocks overlain by the Lias Group and post-Mesozoic sediments. The basin covers an area of about 1200 km^2 and is largely bounded to the northwest by the Lambay and Dalkey faults (Fig. 2b), and to the southwest by the Bray Fault (Dobson & Whittington 1979; Jenner 1981).

A shallow core from a site 21 km northeast of the Kish Bank lighthouse (Fig. 2b) yielded micro-faunas interpreted, by B. Johnson, as 'Lower Liassic' in age (Dobson & Whittington 1979, p. 247). This remains the only direct evidence of the presence of Jurassic rocks in the basin; wells drilled subsequently have proved, beneath Quaternary

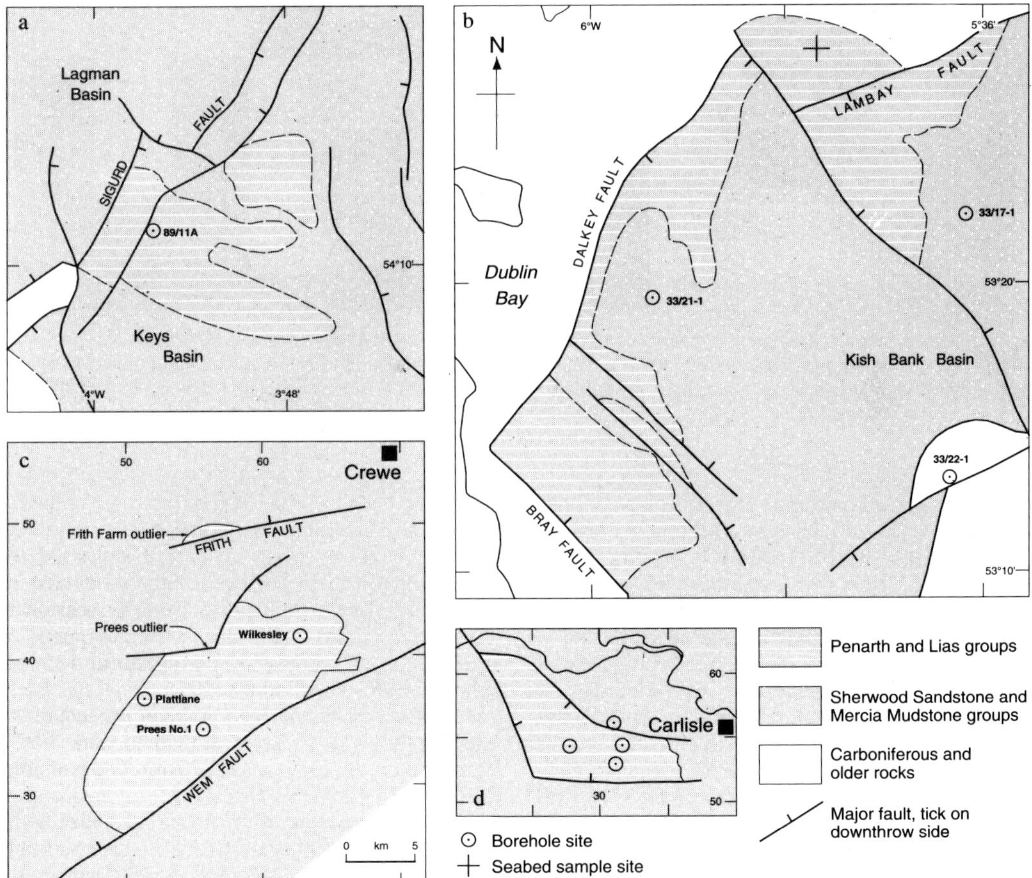

Fig. 2. Occurrences of Penarth and Lias group rocks in areas a to d (Fig. 1): (**a**) Keys Basin (based upon British Geological Survey 1:1 000 000 scale map *Geology of the United Kingdom, Ireland and the adjacent continental shelf* (south sheet, 1991); (**b**) western Kish Bank Basin (based upon Jenner 1981); (**c**) Cheshire Basin (based upon Wilson 1993); (**d**) Carlisle Basin (based upon Ivimey-Cook *et al.* 1995). Topography in (c) and (d) based upon Ordnance Survey mapping.

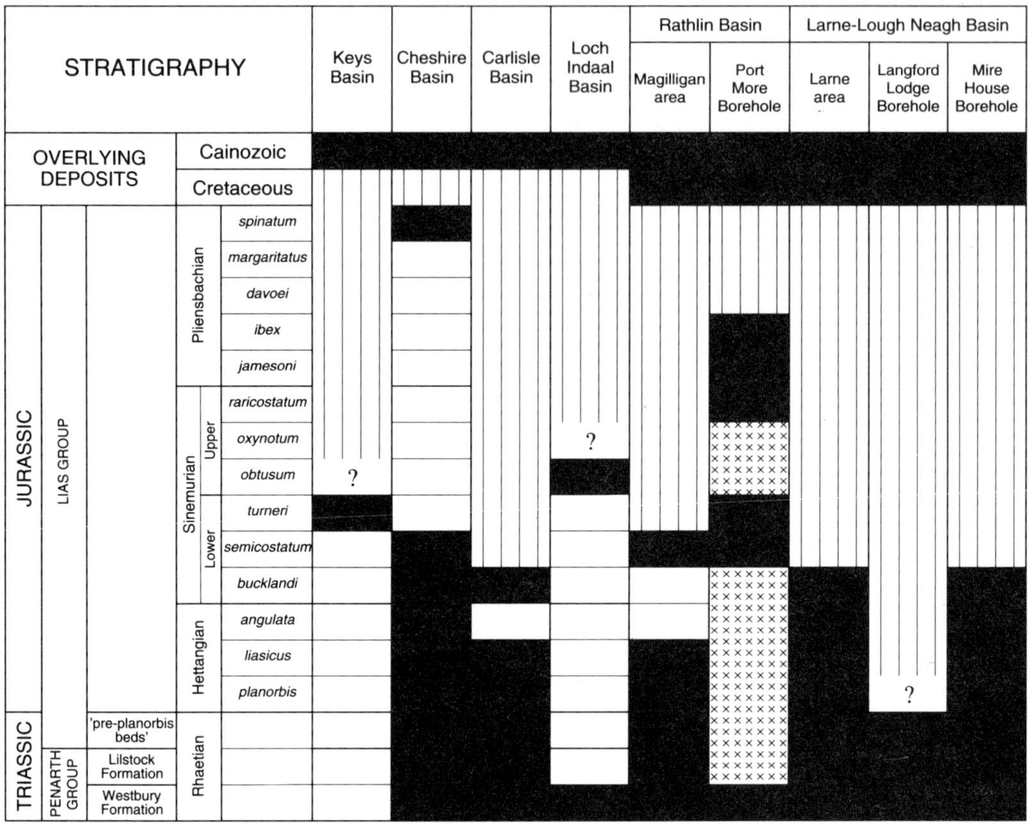

Fig. 3. The Penarth and Lias group successions and overlying beds in the East Irish Sea region. Black – unit proved; white – not proved but presumed present below or between proven units; ? -not proved but may occur above a proven unit; crosses – missing (sequence interrupted by igneous intrusion); vertical ruling – missing below unconformable base of overlying beds.

deposits, a Carboniferous succession (33/22-1; Jenner 1981) and Triassic (Mercia Mudstone Group) rocks (33/17-1, 33/21-1; Naylor *et al.* 1993).

Jenner (1981) depicted an outcrop of the Lias Group covering *c.* 35 km² on the upthrow side of the Lambay Fault, around the site of the 'Lower Liassic' samples reported by Dobson & Whittington (1979). Jenner (1981, fig. 2: key) also suggested that Jurassic rocks are probably present elsewhere in the basin, where the 'time to Base Keuper exceeds one second'. Following this proposal, Broughan *et al.* (1989) inferred the presence of additional outcrops of Jurassic rocks totalling some 285 km² in the hangingwall blocks of the Lambay, Dalkey and Bray faults (Fig. 2b). They also suggested that 2700 m of beds, interpreted as dominantly evenly bedded mudstones and thin limestones and 'in the main of Liassic age', occur beneath Quaternary deposits at the western side of the basin, within 2 km of the coast immediately south of Dublin.

Lower Jurassic fossils recovered from glacial deposits at Kill-o'-the-Grange, on the south side of Dublin Bay, were thought to have a Hebridean provenance (Sollas & Praeger 1895). Lamplugh *et al.* (1903, p.114), following a suggestion made by Kilroe in 1896, proposed that this material had been transported first to the Irish Sea and subsequently 'pushed upward ... from the Irish Sea bed'. Broughan *et al.* (1989) also considered it probable that these fossils were transported from the floor of the Irish Sea, but that they originated there, from outcrops in the Kish Bank Basin. If this is so, the fauna reported by Sollas & Praeger (1895) indicates that the succession in that basin will include the Hettangian *angulata* Zone, with *Schlotheimia angulata*, and extend up to the Toarcian, to the level of beds yielding 'Harpoceras bifrons *Brug.*' [*sic*.].

Loch Indaal Basin

The basin extends from Loch Indaal, offshore southern Islay, southwestwards for *c.* 50 km towards the coast of Donegal, Ireland (Figs 1, 4). First detected by a gravity survey (Riddihough & Young 1970), it was defined by further geophysical studies and interpreted as bounded to the northwest by the Leannan–Loch Gruinart fault system and occupied by northwesterly-dipping Permo-Triassic rocks (Dobson & Evans 1974; Dobson *et al.* 1975). Shallow-cored boreholes drilled in 1975 at the northeastern end of the basin proved the presence of Triassic (Mercia Mudstone and Penarth groups) and Jurassic (Lias Group) deposits (IGS 1976; Owens & Marshall 1978; Evans *et al.* 1980, 1982).

Borehole 75/41 encountered, beneath 17.80 m of Quaternary deposits, dark grey and black mudstones and shales (12.57 m) which were cored from 22.00 m and which rested, with a sharp, irregular contact at 30.37 m, upon grey-green mudstones. In the core the grey and black mudstones included thin siltstones, sandstones and limestones and proved fossiliferous. They were identified, on lithological character, as a representative of the Westbury Formation (Penarth Group); palynomorphs and other fossils recovered from these beds (see Appendix) confirmed this attribution and indicated a latest Triassic (Rhaetian) age (Fig. 3). The underlying grey-green mudstones are unfossiliferous and represent the Blue Anchor Formation (Mercia Mudstone Group). The thickness given here for the Westbury Formation is 0.70 m greater than that (11.87 m) cited in most previous accounts (IGS 1976; Owens & Marshall 1978; Evans *et al.* 1980, 1982) and follows a reinterpretation of the depth to rockhead, based upon original drilling records; the thickness of 18.50 m given by Fyfe *et al.* (1993) is incorrect.

Borehole 75/44 encountered, beneath Quaternary deposits (35.31 m), very dark grey, laminated, silty, calcareous, fossiliferous mudstones which were proved for 15.14 m to TD at 50.45 m. These beds yielded tasmanitid algae, dinoflagellate cysts, spores, pollen, acritarchs, foraminifers, brachiopods, scaphopods, gastropods, bivalves, ammonites, belemnites and ostracods (see Appendix). With the exception of a possible specimen of *Laevitrigonia troedssoni*, a bivalve which, in England, is usually found in the Lower Pliensbachian, the macrofauna indicates the Upper Sinemurian *obtusum* Zone (Fig. 3), with the ammonite *Eparietites?* suggesting the *denotatus* Subzone (H. C. Ivimey-Cook, pers. comm.). Microfaunas indicate a similar (*turneri* or *obtusum* biozone) age (BGS unpublished records). The palynomorphs recovered (see Appendix) are consistent with a Sinemurian age; Fyfe *et al.* (1993) reported a dinoflagellate cyst (*Liasidium variabile*) which is indicative of an Upper Sinemurian (*obtusum* to *raricostatum* Zone) horizon (Woollam & Riding 1983; Riding 1984).

The sequence infilling the Loch Indaal Basin lies in the hangingwall block of the Leannan–Loch Gruinart fault system (Fig. 4) and may be up to 2400 m thick (Evans *et al.* 1980; Fyfe *et al.* 1993). Though the thickness of the Penarth Group–Lias Group succession there is not known, it is likely to comprise a relatively small part of that total (Fyfe *et al.* 1993, fig. 30); those beds may, however, crop out over as much as 190 km^2, nearly half the area occupied by the basin (Fyfe *et al.* 1993, fig. 32). The age of the youngest Jurassic rocks preserved in the basin is not known but, because of the northwesterly dip, is likely to be slightly younger than that proved at the Borehole 75/44 site.

Onshore occurrences

Cheshire Basin

The Cheshire Basin lies between the East Irish Sea Basin and outcrops of the Penarth and Lias groups in eastern and central England (Fig. 1). The groups form the Frith Farm (*c.* 3 km^2) and Prees (*c.* 140 km^2) outliers in the southern part of the basin where the Lias Group was first recognized by Murchison (1835) and the Penarth Group by Maw (1870). The outliers were described by Pocock & Wray (1925) and Poole & Whiteman (1966); the interpreted form of the Prees outlier has been modified by Wilson (1993). The Frith Farm outlier is bounded to the south by the west-southwest-trending Frith Fault and the Prees outlier is bounded to the southeast by the southwesterly-trending Wem Fault (Fig. 2c). The southeasterly-dipping beds in the outliers occur in the hangingwall blocks of those faults and are downthrown against Triassic (Mercia Mudstone Group) rocks to the southeast. Beds in the Prees outlier are seen in small exposures but the succession there is known mainly from the cored Geological Survey Plattlane [SJ 5140 3645] and Wilkesley [SJ 6286 4144] boreholes (Poole & Whiteman 1966) and the Trend Petroleum Prees No.1 Borehole [SJ 5572 3447] (Colter & Barr 1975; Colter 1978; Penn 1987; Evans *et al.* 1993) which form the basis for the following account. The succession in the Frith Farm outlier is, in contrast, known only from small exposures in the banks of the River Weaver (Poole & Whiteman 1966).

The Penarth Group overlies a minor unconformity, represented by an erosion surface at the top of the Blue Anchor Formation (Mercia Mudstone Group). The interpretation of the Penarth Group ('Rhaetic') succession in the Plattlane and

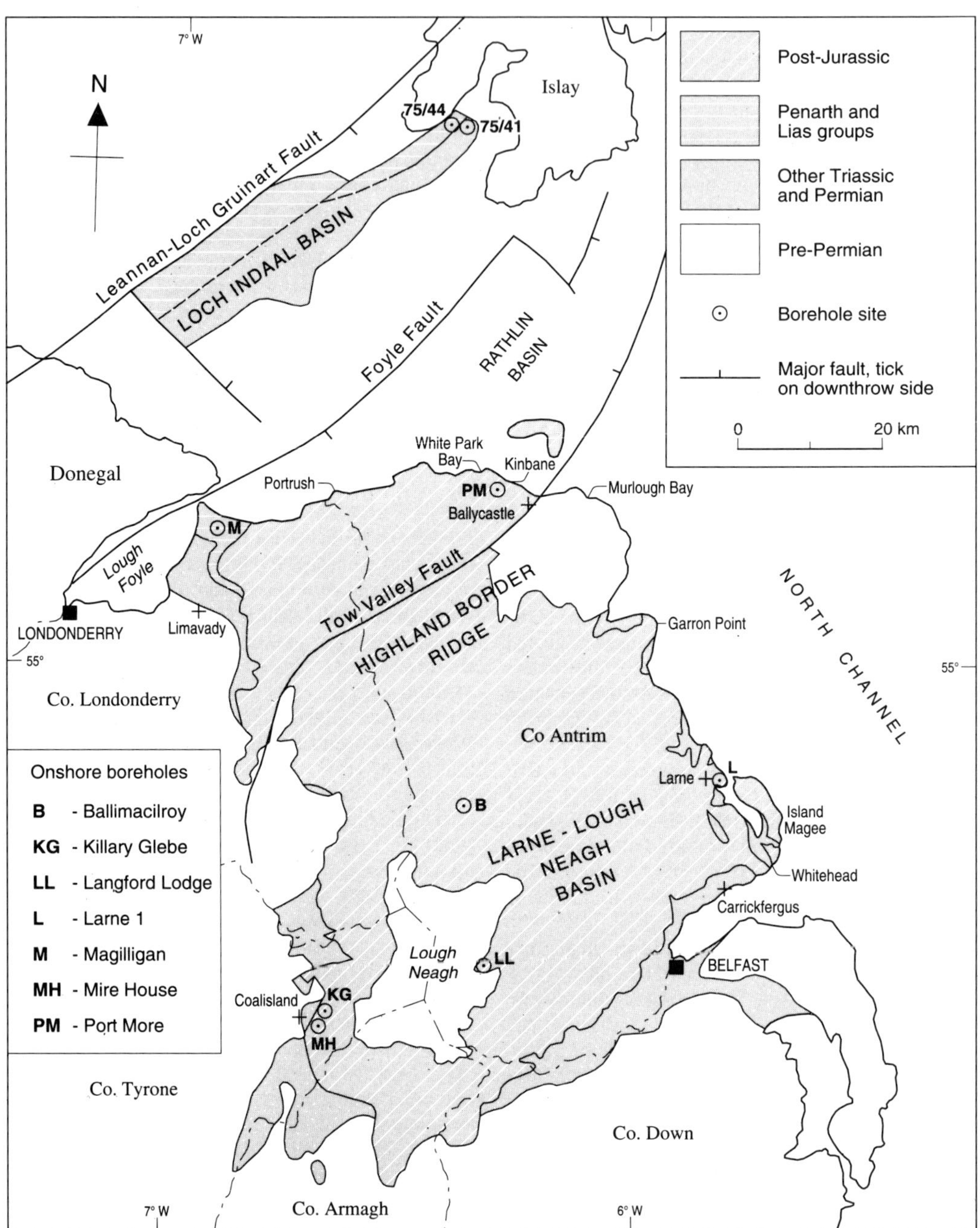

Fig. 4. Distribution of Penarth and Lias groups onshore in Northern Ireland and in the offshore Loch Indaal Basin (for other offshore occurrences see Fig. 1). Based upon British Geological Survey 1:1 000 000 scale map *Geology of the United Kingdom, Ireland and the adjacent continental shelf* (north and south sheets, 1991) and Fyfe *et al.* (1993). Topography based upon Ordnance Survey mapping.

Wilkesley boreholes (Poole & Whiteman 1966) has been revised by Ivimey Cook and Warrington (*in* Plant *et al.*, 1996). The group is 13.55 m thick; the lower part comprises dark grey and black, fissile, micaceous, fossiliferous mudstones, with thin beds of siltstone, fine sandstone and limestone, of the Westbury Formation (7.8 m). Overlying pale grey-green, calcareous mudstones and siltstones are assigned, with a succeeding thin (<0.15 m), fine-grained, argillaceous limestone, to the Lilstock Formation (5.75 m). The latter consists largely of the Cotham Member. The thin limestone, which has an irregular base and, in Wilkesley Borehole, an eroded upper surface, may represent the Langport Member; it was formerly regarded as the basal bed of the 'pre-planorbis beds' of the Lias Group (Poole & Whiteman 1966).

Anderson (1964) recorded ostracods from the Penarth Group in the Plattlane Borehole and the macrofaunas there and in the Wilkesley Borehole were listed by Ivimey-Cook (*in* Poole & Whiteman 1966). The micro- and macrofaunas are indicative of a latest Triassic (Rhaetian) age.

The thickness of beds assigned to the Penarth Group in Prees No.1 Borehole (*c.* 19 m), on the basis of borehole geophysical log interpretation (Penn 1987), was greater than that recognized in the cored Plattlane and Wilkesley boreholes. Comparison of the gamma-ray log of Prees No.1 Borehole with that calibrated against the Wilkesley core indicates that the top of the Penarth Group was placed about 5 m too high in Prees No.1 Borehole; the revised thickness of the group there (*c.* 14 m) is comparable with that recorded at Wilkesley.

The Triassic–Jurassic boundary is placed at the level of the appearance of ammonites of the genus *Psiloceras* (Cope *et al.* 1980; Warrington *et al.* 1980). In Wilkesley Borehole this level is 9.37 m above the base of the Lias Group, the lowest beds of which (the 'pre-planorbis beds', *auctt.*) are, therefore, assigned to the latest Triassic (Rhaetian). The interpretation of the Lias Group succession in Wilkesley Borehole (Poole & Whiteman 1966) has been revised by Ivimey-Cook (*in* Plant *et al.* 1996). About 130 m of the group in that borehole is of Jurassic age, and ammonite faunas indicate the presence of the *planorbis* (25.43 m), *liasicus* (32.59 m) and *angulata* (45.16 m) zones (Hettangian), and the *bucklandi* Zone (27.28 m) and 0.3 m of the succeeding *semicostatum* Zone (Sinemurian) (Fig. 3). In Plattlane Borehole the Jurassic is represented only by about 6.9 m of beds of the *planorbis* Zone which are overlain by Quaternary deposits. The foraminifer biostratigraphy of the Lias Group in the Plattlane and Wilkesley boreholes was reported by Copestake (1989) and Copestake & Johnson (1989).

Following the revision of the position of the Penarth Group–Lias Group boundary in Prees No.1 Borehole (see above) the thickness of the Lower Lias there is 597 m. No biostratigraphy is available but comparison with the nearby Wilkesley Borehole, where the 'pre-planorbis beds' to basal Sinemurian sequence is *c.* 130 m thick, indicates that the upper c. 457 m at Prees comprises Lower Sinemurian (*semicostatum* Zone) and younger beds. The beds proved in Prees No.1 Borehole are succeeded, at outcrop, by 30 m of Middle Lias in which the youngest bed recognized, the Marlstone Rock, is of Pliensbachian (*spinatum* Zone) age (Pocock & Wray 1925; Poole & Whiteman 1966; Cope *et al.* 1980). The Jurassic succession preserved in the Prees outlier therefore comprises at least 617 m of beds of earliest Hettangian (*planorbis* Zone) to late Pliensbachian (*spinatum* Zone) age (Fig. 3).

Carlisle Basin

Binney (1859) first recognized the presence of the Lias Group at a number of localities west of Carlisle. Holmes (1881) considered that those beds formed an outlier covering some 50 km^2 and extending to about 10 km west of Carlisle. The outlier lies c.100 km northeast of the Keys Basin (Fig. 1). Its form has been completely reinterpreted by Ivimey-Cook *et al.* (1995) on the basis of boreholes drilled in 1989, two of which provided the first proof of the presence of the Penarth Group in this area, and from seismic sections.

The outlier is now interpreted as broadly triangular in outline and covering an area of up to 65 km^2. The Penarth and Lias groups dip southwest at about 2° and are downfaulted against Triassic (Mercia Mudstone Group) rocks to the south, west and northeast (Fig. 2d). The Penarth Group is not exposed but may have a small outcrop, masked by glacial deposits, at the eastern end of the outlier. The Lias Group is exposed in small stream bank sections, mostly near the south side of the outlier. The following account is based upon the new borehole records and the reinterpretation of older records by Ivimey-Cook *et al.* (1995).

The Penarth Group rests disconformably upon the Blue Anchor Formation (Mercia Mudstone Group) and is succeeded conformably by the Lias Group. The Penarth Group (13.14 m thick) comprises dark grey, fossiliferous mudstones and silty mudstones, with thin sandstones and limestones, of the Westbury Formation (8.11 m), overlain by the Lilstock Formation (5.03 m) which comprises pale grey, calcareous mudstones, with silty laminae, capped by a thin (0.8 m), grey, micritic limestone. The Lilstock Formation consists largely of the Cotham Member but the thin limestone at the top of the formation may represent the Langport Member. Macrofaunas from the group are indicative of a latest Triassic (Rhaetian) age (Ivimey-Cook *et al.* 1995).

The Triassic–Jurassic boundary, at the level of the appearance of ammonites of the genus *Psiloceras*, is 18.11 m above the base of the Lias Group, the lowest ('pre-planorbis') beds of which are assigned a latest Triassic (Rhaetian) age. About 92 m of the 110 m of Lias Group deposits which may be preserved in the outlier are, therefore, of Jurassic age. Within the Jurassic section, the basal Hettangian *planorbis* Subzone (lower *planorbis* Zone) is proved in cored boreholes and is at least 6.4 m thick (Ivimey-Cook *et al.* 1995). Biostratigraphical evidence from younger Jurassic beds is discontinuous and comes from three outcrop localities. At one site a *johnstoni* Subzone (upper *planorbis* Zone) fauna is probably present, though the '*Ammonites johnstoni*' recorded by Holmes (1899) has not been located and confirmed. A second site yielded a *Waehneroceras*, indicative of the *liasicus* Zone, and lowest Sinemurian (*bucklandi* Zone) faunas are known from the area of the thickest preserved Lias Group (Ivimey-Cook *et al.* 1995). The *angulata* Zone is not proved but is thought to be well-developed in the c. 92 m thick succession which includes proven older Hettangian beds and the basal Sinemurian (Fig. 3).

Northern Ireland

Rocks attributable to the Lias Group were apparently first noted at Portrush, north Co. Antrim, by Richardson 'in or before the year 1799' (Portlock 1843, p. 38); the term 'lias' appears to have been first used by Sampson in 1802 (Portlock 1843, pp.47–49). Other early accounts of these beds are by Berger & Conybeare (1816), Portlock (1843) and Tate (1864, 1867, 1870a). Rocks of the Penarth Group were apparently first recognized, as the '*Avicula contorta* beds' and the 'White Lias', by Tate (1864) though some of these beds were apparently noticed by Sampson as early as 1802 (Portlock 1843, pp.48–49) and, with their fauna, including the characteristic bivalve '*Avicula contorta*' (now *Rhaetavicula contorta* (Portlock)), were recorded by Portlock (1843).

Penarth and Lias group rocks occur in the Rathlin and Larne–Lough Neagh basins, to the north and south, respectively, of the Highland Border Ridge (Figs 1, 4). The Rathlin Basin parallels the Loch Indaal Basin, to the northwest, and extends from northern Co. Antrim and Co. Londonderry, northeastwards for *c.* 50 km, into the offshore area south and east of Islay (Figs 1, 4). It is bounded to the northwest and southeast, respectively, by the Foyle and Tow Valley faults (Evans *et al.* 1980; Fyfe *et al.* 1993). The Larne–Lough

Neagh Basin extends eastwards from eastern Co. Tyrone, through south Co. Antrim and into the North Channel offshore area (Figs 1, 4). There is little direct information on the successions in the offshore parts of these basins and the following account is based upon the onshore occurrences. In both basins Upper Cretaceous rocks rest unconformably upon the Lias Group and overstep onto the Penarth Group and older Triassic beds (George 1967; Fletcher 1977, fig. 8).

In the Rathlin Basin the Penarth and Lias groups crop out over c. 25 km² in eastern Co. Londonderry, from Lough Foyle southwards to *c.* 5 km northeast of Limavady where they are overstepped by Cretaceous rocks (IGS 1981). The Penarth Group and succeeding beds in the Lias Group were proved in Magilligan Borehole [C 6830 3534], Co. Londonderry (Bullerwell 1964; Geological Survey 1964, p.60), and are exposed further to the south and southeast (Wilson 1972, 1981). To the east, small outcrops of the Lias Group occur on the north Co. Antrim coast at Portrush, White Park Bay and near Kinbane (Wilson & Robbie 1966; Wilson & Manning 1978; Ivimey-Cook *in* Cope *et al.* 1980, pp.45, 46, fig.6).

The thickness of the Penarth and Lias group succession in Magilligan Borehole is 98.1 m, of which 7.93 m were originally assigned to the 'Rhaetic' and 90.17 m to the 'Lower Lias' (Bullerwell 1964; Geological Survey 1964, p.60). Wilson (1972) revised the thickness of the Penarth Group to 22 m but did not make a corresponding adjustment to that of the Lias Group. The Penarth and Lias groups are now regarded as *c.* 22 m and *c.* 76 m thick, respectively, in Magilligan Borehole (IGS 1981) but other accounts have given alternative, and confusing, interpretations; Wilson (1981) gave thicknesses of 7.3 m and 90 m respectively for the 'Rhaetic' and 'Liassic' there and Fyfe *et al.* (1993) cited the thicknesses proposed by Wilson (1972). The Lias Group may succeed the Penarth Group non-sequentially in Magilligan Borehole (Warrington *et al.* 1980, p. 29) though the *planorbis* Zone is recognized in that borehole (Wilson 1972, 1981). Details of that section and nearby exposures, in which *liasicus* Zone and Lower Sinemurian *semicostatum* Zone beds (Fig. 3) have also been proved (Ivimey-Cook *in* Cope *et al.* 1980, p.46, fig.6; Wilson 1981; Fyfe *et al.* 1993), will appear in an account of the geology of the country around Limavady and Londonderry (Bazley *et al.*, in press).

In north Co. Antrim, the Penarth Group is known only in Port More Borehole [D 0690 4347] where it is represented by 4.07 m of dark grey and black, shaly, pyritous, fossiliferous mudstones of the Westbury Formation. These rest upon the Collin Glen Formation, the Northern Ireland equivalent of the Blue Anchor Formation (Mercia Mudstone Group); the nature of the junction is not known, because of core loss. The Westbury Formation is truncated, upwards, by an igneous intrusion (Fig. 3); higher Penarth Group beds and the lower part of the Lias Group are missing (Wilson & Manning 1978). The oldest beds present above the igneous intrusion are of Early Sinemurian (*semicostatum* Zone, *scipionianum* Subzone) age (Wilson & Manning 1978). The Jurassic succession preserved above that level comprises 76.5 m of Lower Sinemurian (*semicostatum* and *turneri* Zone) beds separated, by another igneous intrusion, from 172.5 m of Upper Sinemurian (*raricostatum* Zone) to Lower Pliensbachian (*jamesoni* and *ibex* Zone) beds (Fig. 3); the intervening *obtusum* and *oxynotum* zones are missing (Wilson & Manning 1978). The Pliensbachian sequence is overlain unconformably by Cretaceous rocks. Debris containing Lower Pliensbachian, possibly *jamesoni* Zone, faunas occurs in Palaeogene volcanic ash near the borehole site (Manning 1962; Wilson & Manning 1978, pp.14, 78).

Faunas from Lias Group outcrops at Portrush indicate the presence of the *raricostatum* Zone (Wilson & Manning 1978). The *ibex* Zone crops out at Kinbane (Wilson & Robbie 1966) and, with the *raricostatum* Zone, at White Park Bay (Wilson & Manning 1978). Foraminifers regarded as indicative of the *angulata* Zone were recorded from the last locality (McGugan 1965) and beds as low as the *planorbis* Zone and as high as the *davoei* Zone have also been reported to occur there (Charlesworth 1935, 1963). However, the presence, at White Park Bay, of beds other than those of *raricostatum* and *ibex* Zone age has not been confirmed (Wilson 1972; Wilson & Manning 1978, p.78).

Faunas indicative of horizons in the Middle Lias occur in glacial drift in the Ballycastle area (Gray 1870; Tate 1870b; Langtry 1875; Wilson 1972; Wilson & Robbie 1966; Wilson & Manning 1978). Faunas reworked from Middle and Upper Lias beds occur in conglomerates at the base of the Cretaceous further east, in Murlough Bay (Hartley 1933; Versey 1958; Savage 1963; Wilson & Robbie 1966; Wilson 1972, 1981; Wilson & Manning 1978). These derived faunas may reflect the former presence of such deposits in Northern Ireland or may have originated from nearby offshore outcrops (Gray 1870; Hartley 1933; Wilson & Robbie 1966; Wilson 1972; Wilson & Manning 1978); alternatively, they may have a more northerly provenance, in the known outcrops in the Hebrides (Tate 1870b; Versey 1958).

The Penarth and Lias group succession in the Larne–Lough Neagh Basin is exposed mainly in east Co. Antrim, from Garron Point southwards to Larne, Island Magee and the Whitehead–Carrickfergus area (Ivimey-Cook 1975; Griffith & Wilson 1982); the outcrops here are mostly small, discontinuous and faulted and are commonly obscured by landslips. The succession in this area was partly cored in the Larne No.1 Borehole [D 4010 0239] (Ivimey-Cook 1975; Manning & Wilson 1975). Westwards from the Carrickfergus area the succession is overstepped by Cretaceous rocks though small isolated outcrops occur further west, to the north and west of Belfast (Manning *et al.* 1970). Concealed occurrences beneath the Cretaceous and younger rocks elsewhere in south Co. Antrim and east Co. Tyrone are similarly discontinuous, the Lias Group being present below Cretaceous rocks in boreholes such as Ballymacilroy [J 0574 9761] (Thompson 1979), Langford Lodge [J 090 748] (Manning *et al.* 1970) and Mire House [H 861 663] (Fowler & Robbie 1961; IGS 1969, p.98), but not in others, such as Killary Glebe [H 8694 6788] (Johnston 1979) where the Cretaceous rests upon Triassic (Mercia Mudstone Group) rocks. Thus, the relationships apparent at outcrop persist beneath the cover of Cretaceous and younger rocks.

The Penarth Group in the Larne–Lough Neagh Basin contains palynomorphs and a fauna indicative of a latest Triassic (Rhaetian) age (Manning *et al.* 1970; Ivimey-Cook 1975; Manning & Wilson 1975; Warrington & Harland 1975). The thickness of the unit varies considerably within the basin. In Larne No.1 Borehole it is 20.55 m thick (Manning & Wilson 1975); reinterpretation of the records of that borehole suggests, however, that those authors placed the Westbury Formation–Lilstock Formation boundary too high and that the thickness of the formations should be revised to 8.59 and 11.96 m, respectively. The Penarth Group is *c.* 17 m thick in Langford Lodge Borehole (Manning *et al.* 1970). Further west, near Coalisland, east Co. Tyrone, the thickness in Mire House Borehole [H 861 663] was considered to be only 9.75 m (Fowler & Robbie 1961). Reinterpretation of that section suggests that the Collin Glen Formation (Mercia Mudstone Group) was included in that thickness and that the thickness of the Penarth Group there should be revised to 4.72 m.

The Jurassic succession preserved beneath Cretaceous rocks in the Larne–Lough Neagh Basin is less complete than that in the Rathlin Basin, where the youngest beds are Pliensbachian (*ibex* Zone) (Fig. 3). In the Larne–Lough Neagh Basin the youngest Jurassic rocks are Early Sinemurian (Fig. 3). These are present below the Cretaceous in Mire House Borehole where the Lias Group is *c.* 125 m thick, of which 118 m were dated, by macrofaunas, as Hettangian (Fowler & Robbie

1961) but have been reinterpreted as comprising Hettangian and Early Sinemurian deposits (IGS 1969, p.98; Ivimey-Cook *in* Cope *et al.* 1980, p.46, fig.6). The interpretations of this borehole given by Wilson (1972, 1981) are incorrect. Lower Sinemurian (*bucklandi* Zone) beds occur below the Cretaceous in Ballymacilroy Borehole (Thompson 1979), at outcrop in the Larne area (Fig. 3) and, further north, at Garron Point (Ivimey-Cook 1975). In Langford Lodge Borehole the 9.30 m of Lias Group beds present between the Penarth Group and the Cretaceous may comprise the youngest Triassic (Rhaetian) 'pre-planorbis' beds (Fig. 3) as there is no faunal evidence of Jurassic (Hettangian) beds in that section. The variations in the completeness of the Lower Jurassic succession preserved in different parts of Northern Ireland reflect faulting, followed by erosion, prior to overstep by the unconformable Cretaceous deposits (George 1967; Fletcher 1977).

Summary and conclusions

The outliers described provide scattered data points for the interpretation of the history and palaeo-geography of a large area of northwestern Britain (Fig. 1) in Rhaetian to Pliensbachian times (*c.* 210–187 Ma, Forster & Warrington 1985; Hallam *et al.* 1985). The cover of Cretaceous rocks in Northern Ireland was probably continuous with that in England by Campanian times (Hancock & Rawson 1992). The overstepping relationship of Cretaceous beds to the faulted occurrences of Pliensbachian to Triassic rocks in Northern Ireland therefore provides an indication of the relationships that are likely to have existed elsewhere in the region prior to the removal of a cover of Cretaceous rocks.

The Penarth Group is proved in most of the outliers described and is likely to be present in the Keys and Kish Bank basins also. Throughout the area considered it comprises the Westbury Formation, in typical facies and with characteristic microfloras and macrofaunas, succeeded by the Lilstock Formation which is represented largely, or entirely, by the Cotham Member; the Langport Member is either not developed or is very thin. The occurrence of the Westbury Formation overlying the Blue Anchor Formation (Mercia Mudstone Group) in the Loch Indaal Basin is the most northerly example of that sequence in the British Isles. Full sections of the Penarth Group are, with the exception of that in Mire House Borehole, Co. Tyrone, thicker in the Rathlin and Larne–Lough Neagh basins than elsewhere in the area considered.

The Lias Group is present in all the outliers described, though the sequences in the Keys, Kish Bank, Carlisle and Loch Indaal basins are imperfectly known and those elsewhere are only partly proved because of lack of continuous core or exposure or, in the case of the Rathlin Basin, because of the presence of igneous intrusions (Fig. 3). The youngest beds proved beneath the Upper Cretaceous in Northern Ireland are Pliensbachian (*ibex* Zone). In the other outliers no cover of Cretaceous rocks remains but in the Cheshire Basin the youngest preserved beds are slightly higher in the Pliensbachian (*spinatum* Zone) (Fig. 3). There is indirect evidence, from reworked material, that successions in the offshore part of the Rathlin Basin and in the Kish Bank Basin may include beds as young as Toarcian. Some Mid and Late Jurassic sediments may also have been deposited but the region is thought to have been emergent between Kimmeridgian and Cenomanian times (Bradshaw *et al.* 1992; Hancock & Rawson 1992).

Where biostratigraphical evidence permits objective comparison the Jurassic successions in the outliers are considerably thicker than their equivalents in eastern England. For example, the Hettangian succession is *c.* 80 m thick in the Carlisle Basin and 103 m in the Cheshire Basin, in contrast to 30 m in the Felixkirk Borehole (Fig. 1), North Yorkshire (Ivimey-Cook & Powell 1991). The Hettangian (*planorbis* Zone) to Pliensbachian (*spinatum* Zone) sequence in the Cheshire Basin is at least 617 m thick, compared with 217 m in the Felixkirk Borehole. The more rapid subsidence in western parts of Britain, which is reflected by these differences, resulted from movement on basement-controlled faults at the margins of basins forming the 'Clyde Belt' (McLean 1978; Jackson & Mulholland 1993). The occurrences of the Penarth and Lias groups described above are preserved in the hangingwall blocks of faults **within** some of those basins.

Where sufficient evidence is available, for example, in Northern Ireland and the Cheshire Basin, the Penarth and Lias groups do not appear to show lateral variations that would indicate syn-depositional movement on the faults bounding the outliers. Those structures therefore reflect post-Pliensbachian movement which, on certain faults in the region, may account for up to one-third of the total post-Carboniferous displacement (Jackson & Mulholland 1993). The outliers of the Penarth and Lias groups are interpreted as the vestiges of a cover of those beds that was probably originally continuous throughout the area described. Their separation, and the variations in the completeness of the Rhaetian to Pliensbachian sequence preserved in them, are consequences of the structural character imposed on the region during the 90 Ma between the Pliensbachian and the Cenomanian.

Faulting during the earlier part of this interval produced tilted fault blocks which, during a subsequent, possibly post-Kimmeridgian, uplift phase, were eroded down to the levels preserved in the outliers before overstep by Upper Cretaceous deposits occurred in Cenomanian to Campanian times.

The author is grateful to Dr H. C. Ivimey-Cook, formerly of the British Geological Survey, for advice and for permission to include his determinations and interpretations of the macrofaunas from boreholes in the Keys and Loch Indaal basins in this account. Drs R. A. B. Bazley (Geological Survey of Northern Ireland, Belfast), B. M. Cox (British Geological Survey, Keyworth) and D. Evans (British Geological Survey, Edinburgh) kindly read and made helpful comments upon a draft version of this contribution which also benefited from the careful and constructive attention of an anonymous reviewer. The illustrations were prepared by staff of the Publications Services Group at the British Geological Survey, Keyworth. The account is published with the approval of the Director, British Geological Survey (NERC).

References

ANDERSON, F. W. 1964. Rhaetic Ostracoda. *Bulletin of the Geological Survey of Great Britain*, **21**, 133–174.

BAZLEY, R. A. B., BRANDON, A. & ARTHURS, J. W. In press. Geology of the country around Limavady and Londonderry. *Geological Survey of Northern Ireland, Technical Report*, **GSNI/97/1**.

BERGER, J. F. & CONYBEARE, W. 1816. On the geological features of the north-eastern counties of Ireland. *Transactions of the Geological Society of London*, **3**, 121–195.

BINNEY, E. W. 1859. Notice of Lias deposits at Quarry-Gill and other places near Carlisle. *Quarterly Journal of the Geological Society of London*, **15**, 549–551.

BOTT, M. H. P. 1968. The geological structure of the Irish Sea basin. *In:* DONOVAN, D. T. (ed.) *Geology of shelf seas*. Oliver & Boyd, Edinburgh, 93–115.

—— & YOUNG, D. G. G. 1971. Gravity measurements in the north Irish Sea. *Quarterly Journal of the Geological Society of London*, **126**, 413–434.

BRADSHAW, M. J., COPE, J. C. W., CRIPPS, D. W. ET AL. 1992. Jurassic. *In:* COPE, J. C. W., INGHAM, J. K. & RAWSON, P. F. (eds) *Atlas of Palaeogeography and Lithofacies*. Geological Society, London, Memoir, **13**, 107–129.

BRITISH GEOLOGICAL SURVEY. 1994. *East Irish Sea* (Special Sheet Edition). 1:250 000. (Edinburgh, Scotland: British Geological Survey).

BROUGHAN, F. M., NAYLOR, D. & ANSTEY, N. A. 1989. Jurassic rocks in the Kish Bank Basin. *Irish Journal of Earth Sciences*, **10**, 99–106.

BULLERWELL, W. 1964. *Geophysical and drilling exploration in Northern Ireland: progress report*. HMSO, Belfast.

CHARLESWORTH, J. K. 1935. The geology of north-east Ireland. X. Rhaetic and Lias. *Proceedings of the Geologists' Association*, **44**, 460–461.

—— 1963. *Historical geology of Ireland*. Oliver & Boyd, Edinburgh & London.

COLTER, V. S. 1978. Exploration for gas in the Irish Sea. *Geologie en Mijnbouw*, **57**, 503–516.

—— & BARR, K. W. 1975. Recent developments in the geology of the Irish Sea and Cheshire basins. *In:* WOODLAND, A. W. (ed.) *Petroleum and the continental shelf*, **1**. Applied Science Publishers, London, 61–73.

COPE, J. C. W., GETTY, T. A., HOWARTH, M. K., MORTON, N. & TORRENS, H. S. 1980. *A correlation of Jurassic rocks in the British Isles*. Geological Society, London, Special Report, **14**.

COPESTAKE, P. 1989. Triassic. *In:* JENKINS, D. G. & MURRAY, J. W. (eds) *Stratigraphical atlas of fossil foraminifera* (2nd edn). Ellis Horwood, Chichester, 97–124.

—— & JOHNSON, B. 1989. The Hettangian to Toarcian (Lower Jurassic). *In:* JENKINS, D. G. & MURRAY, J. W. (eds) *Stratigraphical atlas of fossil foraminifera* (2nd edn). Ellis Horwood, Chichester, 129–188.

COX, B. M. 1990. A review of Jurassic chronostratigraphy and age indicators for the UK. *In:* HARDMAN, R. F. P. & BROOKS, J. (eds) *Tectonic Events Responsible for Britain's Oil and Gas Reserves*. Geological Society, London, Special Publication, **55**, 169–190.

DOBSON, M. R. & EVANS, D. 1974. Geological structure of the Malin Sea. *Journal of the Geological Society, London*, **130**, 475–478.

—— & WHITTINGTON, R. J. 1979. The geology of the Kish Bank Basin. *Journal of the Geological Society, London*, **136**, 243–249.

——, EVANS, D. & WHITTINGTON, R. 1975. The offshore extension of the Loch Gruinart Fault, Islay. *Scottish Journal of Geology*, **11**, 23–35.

EVANS, D., KENOLTY, N., DOBSON, M. R. & WHITTINGTON, R. J. 1980. The geology of the Malin Sea. *Report of the Institute of Geological Sciences*, **79/15**.

——, CHESHER, J. A., DEEGAN, C. E. & FANNIN, N. G. T. 1982. The offshore geology of Scotland in relation to the IGS shallow drilling programme, 1970–1978. *Report of the Institute of Geological Sciences*, **81/12**.

EVANS, D. J., REES, J. G. & HOLLOWAY, S. 1993. The Permian to Jurassic stratigraphy and structural evolution of the central Cheshire Basin. *Journal of the Geological Society, London*, **150**, 857–870.

FLETCHER, T. P. 1977. Lithostratigraphy of the Chalk (Ulster White Limestone Formation) in Northern Ireland. *Report of the Institute of Geological Sciences*, **77/24**.

FORSTER, S. C. & WARRINGTON, G. 1985. Geochronology of the Carboniferous, Permian and Triassic. *In:* SNELLING, N. J. (ed.) *The Chronology of the Geological Record*. Geological Society, London, Memoir, **10**, 99–113.

FOWLER, A. & ROBBIE, J. A. 1961. *Geology of the country around Dungannon*. Memoirs of the Geological Survey of Northern Ireland, sheet 35. HMSO, Belfast.

FYFE, J. A., LONG, D. & EVANS, D. 1993. *United Kingdom offshore regional report: the geology of the Malin–Hebrides sea area.* British Geological Survey, HMSO, London.

GEOLOGICAL SURVEY. 1964. *Summary of Progress of the Geological Survey of Great Britain and the Museum of Practical Geology for the year 1963.* HMSO, London.

GEORGE, T. N. 1967. Landform and structure in Ulster. *Scottish Journal of Geology,* **3**, 413–448.

GRAY, W. 1870. *Seventh Annual Report of the Belfast Naturalists' Field Club,* 49–50.

GRIFFITH, A. E. & WILSON, H. E. 1982. *Geology of the country around Carrickfergus and Bangor.* Memoirs of the Geological Survey of Northern Ireland, sheet 29. HMSO, Belfast.

HALLAM, A., HANCOCK, J. M., LaBREQUE, J. L., LOWRIE, W. & CHANNELL, J. E. T. 1985. Jurassic to Paleogene: Part I. Jurassic and Cretaceous geochronology and Jurassic to Paleogene magnetostratigraphy. *In:* SNELLING, N. J. (ed.) *The Chronology of the Geological Record.* Geological Society, London, Memoir, **10**, 118–140.

HANCOCK, J. M. & RAWSON, P. F. 1992. Cretaceous. *In:* COPE, J. C. W., INGHAM, J. K. & RAWSON, P. F. (eds) *Atlas of Palaeogeography and Lithofacies.* Geological Society, London, Memoir, **13**, 131–139.

HARTLEY, J. J. 1933. Notes on fossils recently obtained from the "chloritic" conglomerate of Murlough Bay, Co. Antrim. *Irish Naturalists' Journal,* **4**, 238–240.

HOLMES, T. V. 1881. The Permian, Triassic and Liassic rocks of the Carlisle Basin. *Quarterly Journal of the Geological Society of London,* **37**, 286–298.

—— 1899. *The geology of the country around Carlisle.* Memoirs of the Geological Survey of Great Britain, sheets 16 and 17, with parts of 12, 18, 22 and 23. HMSO, London.

IGS. 1969. *Institute of Geological Sciences, Annual Report for 1968.* HMSO, London.

—— 1976. IGS boreholes 1975. *Report of the Institute of Geological Sciences,* **76/10**.

—— 1981. *Northern Ireland 1:50 000-scale geological sheet (solid), 12 and part of 6 (Limavady).* Institute of Geological Sciences.

IVIMEY-COOK, H. C. 1975. The stratigraphy of the Rhaetic and Lower Jurassic in East Antrim. *Bulletin of the Geological Survey of Great Britain,* **50**, 51–69.

—— & POWELL, J. H. 1991. Late Triassic and early Jurassic biostratigraphy of the Felixkirk Borehole, North Yorkshire. *Proceedings of the Yorkshire Geological Society,* **48**, 367–374.

——, WARRINGTON, G., WORLEY, N. E., HOLLOWAY, S. & YOUNG, B. 1995. Rocks of Late Triassic and Early Jurassic age in the Carlisle Basin, Cumbria (north-west England). *Proceedings of the Yorkshire Geological Society,* **50**, 305–316.

JACKSON, D. I. & MULHOLLAND, P. 1993. Tectonic and stratigraphic aspects of the East Irish Sea Basin and adjacent areas: contrasts in their post–Carboniferous structural styles. *In:* PARKER, J. R. (ed.) *Petroleum Geology of Northwest Europe: Proceedings of the 4th Conference.* The Geological Society, London, 791–808.

——, JACKSON, A. A., EVANS, D., WINGFIELD, R. T. R., BARNES, R. P. & ARTHUR, M. J. 1995. *United Kingdom offshore regional report: the geology of the Irish Sea.* British Geological Survey, HMSO, London.

——, MULHOLLAND, P., JONES, S. M. & WARRINGTON, G. 1987. The geological framework of the East Irish Sea Basin. *In:* BROOKS, J. & GLENNIE, K. (eds) *Petroleum geology of north west Europe.* Graham & Trotman, London, 191–203.

JENNER, J. K. 1981. The structure and stratigraphy of the Kish Bank Basin. *In:* IILLING, L. V. & HOBSON, G. D. (eds) *Petroleum geology of the continental shelf of north-west Europe.* Institute of Petroleum, London, 426–431.

JOHNSTON, T. P. 1979. Preliminary report on the Killary Glebe No.1 Borehole, Coalisland, Co. Tyrone. *Geological Survey of Northern Ireland, Open File Report,* **62**.

LAMPLUGH, G. W., KILROE, J. R., M'HENRY, A., SEYMOUR, H. J. & WRIGHT, W. B. 1903. *The geology of the country around Dublin.* Memoirs of the Geological Survey, Ireland, sheet 112. HMSO, Dublin.

LANGTRY, G. 1875. On the occurrence of the Middle Lias at Ballycastle. *Report of the forty-fourth meeting of the British Association for the Advancement of Science; Belfast, 1874. Transactions of the Sections,* 88.

MANNING, P. I. 1962. *Proceedings of the Geological Society of London,* No. 1597, 74.

—— & WILSON, H. E. 1975. The stratigraphy of the Larne Borehole, County Antrim. *Bulletin of the Geological Survey of Great Britain,* **50**, 1–50.

——, ROBBIE, J. A. & WILSON, H. E. 1970. *Geology of Belfast and the Lagan Valley.* Memoirs of the Geological Survey of Northern Ireland, sheet 36. HMSO, Belfast.

MAW, G. 1870. On the occurrence of the Rhaetic beds in north Shropshire and Cheshire. *Geological Magazine,* **7**, 203–204.

McGUGAN, A. 1965. Liassic Foraminifera from Whitepark Bay, County Antrim. *The Irish Naturalists' Journal,* **15**, 85–87.

McLEAN, A. C. 1978. Evolution of fault-controlled ensialic basins in northwestern Britain. *In:* BOWES, D. R. & LEAKE, B. E. (eds) *Crustal evolution in northwestern Britain and adjacent regions.* Geological Journal Special Issue **10**, 325–346.

MURCHISON, R. I. 1835. On an outlying basin of Lias on the borders of Salop and Cheshire, with a short account of the lower Lias between Gloucester and Worcester. *Proceedings of the Geological Society of London,* **2**, 114–115.

NAYLOR, D., HAUGHEY, N., CLAYTON, G. & GRAHAM, J. R. 1993. The Kish Bank Basin, offshore Ireland. *In:* PARKER, J. R. (ed.) *Petroleum Geology of Northwest Europe: Proceedings of the 4th Conference.* The Geological Society, London, 845–855.

OWENS, B. & MARSHALL, J. (Compilers). 1978. Micropalaeontological biostratigraphy of samples from around the coasts of Scotland. *Report of the Institute of Geological Sciences,* **78/20**.

PENN, I. E. 1987. Geophysical logs in the stratigraphy of Wales and adjacent offshore and onshore areas.

Proceedings of the Geologists' Association, **98**, 275–314.

PLANT, J. A., JONES, D. G. & HASLAM, H. W. (Eds). 1996. Basin evolution, fluid movement and mineral resources in a Permo-Triassic rift setting: the Cheshire Basin. *Technical Report of the British Geological Survey*, WP/96/14R.

POCOCK, R. W. & WRAY, D. A. 1925. *The geology of the country around Wem*. Memoirs of the Geological Survey of Great Britain (England and Wales), sheet 138. HMSO, London.

POOLE, E. G. & WHITEMAN, A. J. 1966. *Geology of the country around Nantwich and Whitchurch*. Memoirs of the Geological Survey of Great Britain (England and Wales), sheet 122. HMSO, London.

PORTLOCK, J. E. 1843. *Report on the geology of the County of Londonderry, and of parts of Tyrone and Fermanagh*. HMSO, Dublin.

POWELL, J. H. 1984. Lithostratigraphical nomenclature of the Lias Group in the Yorkshire Basin. *Proceedings of the Yorkshire Geological Society*, **45**, 51–57.

RIDDIHOUGH, R. P. & YOUNG, D. G. G. 1970. Gravity and magnetic surveys of Inishowen and adjoining sea areas off the north coast of Ireland. *Proceedings of the Geological Society of London*, **1664**, 215–220.

RIDING, J. B. 1984. Dinoflagellate cyst range-top biostratigraphy of the uppermost Triassic to lowermost Cretaceous of northwest Europe. *Palynology*, **8**, 195–210.

SAVAGE, R. J. G. 1963. Upper Lias ammonite from Cretaceous conglomerate of Murlough Bay. *The Irish Naturalists' Journal*, **14**, 179–180.

SIMMS, M. J. 1989. British Lower Jurassic crinoids. *Monograph of the Palaeontographical Society, London*, **142**, 1–103.

SOLLAS, W. J. & PRAEGER, R. L. 1895. Notes on glacial deposits in Ireland. II.-Kill-o'-the Grange. *The Irish Naturalist*, **4**, 321–329.

TATE, R. 1864. On the Liassic strata of the neighbourhood of Belfast. *Quarterly Journal of the Geological Society of London*, **20**, 103–114.

—— 1867. On the Lower Lias of the north-east of Ireland. *Quarterly Journal of the Geological Society of London*, **23**, 297–305.

—— 1870a. A list of Irish Liassic fossils, with notes on new and critical species. *Seventh Annual Report of the Belfast Naturalists' Field Club*, Appendix I.

—— 1870b. Note on the Middle Lias in the north-east of Ireland. *Quarterly Journal of the Geological Society of London*, **26**, 324–325

THOMPSON, S. J. 1979. Preliminary report on the Ballymacilroy No. 1 Borehole, Ahoghill, Co. Antrim. *Geological Survey of Northern Ireland, Open File Report*, **63**.

VERSEY, H. C. 1958. Derived ammonites in basal Cretaceous conglomerate. *Geological Magazine*, **95**, 440.

WARRINGTON, G. & HARLAND, R. 1975. Palynology of the Trias and Lower Lias of the Larne Borehole. *Bulletin of the Geological Survey of Great Britain*, **50**, 37–50.

——, AUDLEY-CHARLES, M. G., ELLIOTT, R. E. ET AL. 1980. *A correlation of Triassic rocks in the British Isles*. Geological Society of London, Special Report, **13**.

WILSON, A. A. 1993. The Mercia Mudstone Group (Trias) of the Cheshire Basin. *Proceedings of the Yorkshire Geological Society*, **49**, 171–188.

WILSON, H. E. 1972. *Regional geology of Northern Ireland*. HMSO, Belfast.

—— 1981. Permian and Mesozoic. *In*: HOLLAND, C. H. (ed.) *A geology of Ireland*. Scottish Academic Press, Edinburgh, 201–212.

—— & ROBBIE, J. A. 1966. *Geology of the country around Ballycastle*. Memoirs of the Geological Survey of Northern Ireland, sheet 8. HMSO, Belfast.

—— & MANNING, P. I. 1978. *Geology of the Causeway Coast* (2 vols). Memoirs of the Geological Survey of Northern Ireland, sheet 7. HMSO, Belfast.

WOOLLAM, R. & RIDING, J. B. 1983. Dinoflagellate cyst zonation of the English Jurassic. *Report of the Institute of Geological Sciences*, **83/2**.

Appendix

Faunas (determinations by Dr H. C. Ivimey-Cook) and palynomorphs (determinations by G. Warrington) from shallow boreholes in the Keys and Loch Indaal basins.

Borehole 89/11A, Keys Basin

Fauna (85.81–95.00 m)

- Gastropod – *Procerithium* sp.
- Bivalves – *Bakevellia* cf. *laevis* (J. Buckman), *Camptonectes* sp., *Cardinia* sp., *Dacryomya* sp., *Lucina limbata* (Terquem & Piette), *Oxytoma?*, *Palaeoneilo galatea* (d'Orbigny), *Plagiostoma* sp., *Plicatula?*, *Protocardia* sp., *Ryderia doris* (d'Orbigny), *Tutcheria* sp.
- Ammonites – *Arnioceras* cf. *semicostatum* (Young & Bird), *Caenisites?*
- Ostracods
- Echinoderms – crinoid fragments, including columnals of cf. *Hispidocrinus* sp.; diademopsid spines.

Palynomorphs (94.50–94.60 m)

- Miospores – *Abietineaepollenites dunrobinensis* Couper 1958, *Alisporites* sp., *Cerebropollenites thiergartii* Schulz 1967, *Chasmatosporites magnolioides* (Erdtman) Nilsson 1958, *Classopollis torosus* (Reissinger) Balme 1957, *Cyathidites minor* Couper 1953, *Gliscopollis meyeriana* (Klaus) Venkatachala 1966, *Kraeuselisporites reissingeri* (Harris) Morbey 1975, *Osmundacidites wellmanii* Couper 1953, *Perinopollenites elatoides* Couper 1958, *Quadraeculina anellaeformis* Maljavkina 1949,

Retitriletes austroclavatidites (Cookson) Döring, Krutzsch, Mai & Schulz 1963, *R. clavatoides* (Couper) Döring, Krutzsch, Mai & Schulz 1963, *Todisporites minor* Couper 1958, *Tsugaepollenites mesozoicus* Couper 1958,
- Acritarchs – *Baltisphaeridium delicatum* Wall 1965, *Micrhystridium fragile* Deflandre 1947, *M. intromittum* var. *intromittum* Wall 1965, *M. lymense* var. *gliscum* Wall 1965, *M.* cf. *stellatum* Deflandre 1945.

Borehole 75/41, Loch Indaal Basin

Fauna (22.70–30.30 m)
- Gastropod – *'Natica' oppelii* Moore
- Bivalves – *Arcomya?*, *Cardinia?*, *Cercomya?*. *Eotrapezium concentricum* (Moore), *E. germari* (Dunker), *'Gervillia' praecursor* (Quenstedt), *Lyriomyophoria postera* (Quenstedt), *Modiolus hillanoides* (Chapuis & Dewalque), *Pleuromya?*, *Plicatula* sp., *Protocardia rhaetica* (Merian), *Pteromya* sp., *Rhaetavicula contorta* (Portlock).
- Ostracods
- Fish remains

The bivalve *R. contorta*, a characteristic component of Rhaetian shallow water benthic faunas, was recorded throughout the section examined, which comprises almost the full cored section of the black shales; other taxa and groups were recorded less regularly.

Palynomorphs (24.10–30.30 m)
- Miospores – *Acanthotriletes ovalis* Nilsson 1958, *?A. varius* Nilsson 1958, *Alisporites* cf. *thomasii* (Couper) Nilsson 1958, *Carnisporites lecythus* Morbey 1975, *C. spiniger* (Leschik) Morbey 1975, *Chasmatosporites* cf. *apertus* (Rogalska) Nilsson 1958, *C. magnolioides* (Erdtman) Nilsson 1958, *Classopollis torosus* (Reissinger) Balme 1957, *?Convolutispora microrugulata* Schulz 1967, *Deltoidospora* sp., *Gliscopollis meyeriana* (Klaus) Venkatachala 1966, *Kraeuselisporites reissingeri* (Harris) Morbey 1975, *Leptolepidites argenteaeformis* (Bolkhovitina) Morbey 1975, *Limbosporites lundbladii* Nilsson 1958, *Lunatisporites rhaeticus* (Schulz) Warrington 1974, *Microreticulatisporites fuscus* (Nilsson) Morbey 1975, *Ovalipollis pseudoalatus* (Thiergart) Schuurman 1976, *Protohaploxypinus* cf. *microcorpus* (Schaarschmidt) Clarke 1965, *Rhaetipollis germanicus* Schulz 1967, *Ricciisporites tuberculatus* Lundblad 1954, *Tsugaepollenites? pseudomassulae* (Mädler) Morbey 1975, *Vesicaspora fuscus* (Pautsch) Morbey 1975

- Acritarchs – *Cymatiosphaera* cf. *polypartita* Morbey 1975, *?Dictyotidium* sp., *Micrhystridium* sp., *Veryhachium* sp.
- Dinoflagellate cysts – *Dapcodinium priscum* Evitt 1961, *Rhaetogonyaulax rhaetica* (Sarjeant) Loeblich & Loeblich emend. Harland, Morbey & Sarjeant 1975

The miospores *O. pseudoalatus*, *R. germanicus* and *R. tuberculatus*, an association indicating a Rhaetian age, and the dinoflagellate cyst *R. rhaetica*, indicating the Rr Biozone, of Rhaetian age (Woollam & Riding 1983), were recorded throughout the section examined; other taxa occurred less regularly.

Borehole 75/44, Loch Indaal Basin

Fauna (35.55–50.30 m)
- Foraminifers
- Brachiopods – *Discinisca* sp., *Piarorhynchia* sp.
- Scaphopod – *Dentalina* sp.
- Gastropods
- Bivalves – *Antiquilima* sp., *Camptonectes* sp., *Cardinia* sp., *Chlamys* sp., *Dacryomya?*, *Grammatodon* sp., *Gryphaea arcuata* Lamarck, *?Laevitrigonia troedssoni* Melville, *Modiolus hillanus* J. Sowerby, *Oxytoma* sp., *Palaeoneilo galatea* (d'Orbigny), *Palaeonucula* sp., *Plagiostoma* sp., *Protocardia* sp., *Pseudolimea* sp., *Pseudopecten* sp., *Rollieria bronni* (Andler), *Ryderia* sp., *Tutcheria* sp.
- Ammonites – *Asteroceras* sp., *Cymbites laevigatus* (J. de C. Sowerby), *Eparietites?*, *Promicroceras* sp., *Xiphoceras* sp.
- Belemnite
- Ostracods

Palynomorphs (35.70–50.30 m)
- Miospores – *Acanthotriletes ovalis* Nilsson 1958, *Alisporites thomasii* (Couper) Nilsson 1958, *Chasmatosporites* cf. *apertus* (Rogalska) Nilsson 1958, *C. magnolioides* (Erdtman) Nilsson 1958, *Classopollis torosus* (Reissinger) Balme 1957, *Deltoidospora neddeni* (Potonié) Orbell 1973, *Kraeuselisporites reissingeri* (Harris) Morbey 1975, *Lycopodiacidites rugulatus* (Couper) Schulz 1967, *?Neoraistrickia* sp., *Osmundacidites wellmanii* Couper 1953, *Quadraeculina anellaeformis* Maljavkina 1949, *Tsugaepollenites mesozoicus* Couper 1958, *Vesicaspora fuscus* (Pautsch) Morbey 1975
- Acritarchs – *Micrhystridium lymense* var. *?gliscum* Wall 1965, *M. lymense* var. *rigidum* Wall 1965
- Tasmanitid alga – *Tasmanites suevicus* (Eisenack) Wall 1965

The Mesozoic and Tertiary history of the Irish Sea

JOHN C. W. COPE

Department of Earth Sciences, University of Wales Cardiff, PO Box 914,
Cardiff CF1 3YE, UK

Abstract: Uplift centred on the Irish Sea in latest Cretaceous times accompanied by a locally raised geothermal gradient is supported by apatite fission track analyses from the area, the outcrop patterns of the Mesozoic rocks of England and the positive gravity anomalies over the whole of the Irish Sea region. This conclusion allows more precise modelling of the history of the region and the lost Mesozoic cover can be reconstructed, with varying degrees of certainty, by combining the results of extrapolation from existing land and submarine outcrops and subcrops, with the amount of cover lost. Stratigraphical considerations suggest the occurrence of both Late Cimmerian and latest Cretaceous inversions of the Irish Sea area. Following rapid erosion of the central part of the domed area in the Palaeogene, alluvial plain sedimentation was established over the Irish Sea region, developing around valleys cut through the Mesozoic sediments of the area. Deflation of the dome and accompanying faulting allowed the sea to enter the area, initially primarily through St George's Channel in the early Oligocene and into the whole of the basin in Neogene times.

The geology and basin development of the Irish Sea have long proved to be an enigma and it is only over the past thirty years, with the advent of submarine sampling techniques and the search for hydrocarbons on the continental shelves of Northwest Europe that much of the geology has become understood. The first suggestions that there were Mesozoic sediments under the Irish Sea arose from OT Jones' suggestion (1956) that the curved shoreline of Cardigan Bay was very similar to the curved eastern margin of the Permo-Triassic Cheshire Basin. Jones argued that Cardigan Bay could well be a Triassic Basin with a fault-defined eastern margin. Geophysical surveys supported the existence of a thick sequence of flat-lying sediments and showed that the Cardigan Bay Basin extended onto the Welsh landmass at one point, to the west of the Cambrian Harlech Dome, with which it was in fault contact; thick Pleistocene cover, however, precluded any knowledge of what lay beneath. A borehole at this point proved, instead of the expected thick Triassic, a sequence of fluviatile and lacustrine Oligocene sediments overlying the thickest and most complete succession of Lower Jurassic rocks known in the British area (Woodland 1971). This discovery heralded a major period of investigation into the geology of the Cardigan Bay area, during which it was discovered that although there were considerable thicknesses of Jurassic rocks in several areas, Cretaceous rocks appeared to be absent to the north of St George's Channel. Further to the north, exploration work proceeded in the East Irish Sea Basin, spurred on by knowledge of oil seeps on the Lancashire coast

(Cope 1939; Wray & Cope 1948) and commercial quantities of gas, together with some oil, were found. Exploration continues to the present day. For this area of the Irish Sea, the youngest rocks are of early Jurassic age (Warrington 1997), whilst most of the sea bed is floored by Permian and Triassic rocks. Higher Jurassic and Cretaceous rock are unknown, although they are known further to the south in Cardigan Bay (Jurassic) and the St George's Channel Basin (Jurassic and Cretaceous).

The nearest onshore Cretaceous rocks are in Northern Ireland. There, the Chalk is well known as it is extensively preserved beneath the Palaeogene basalts of the Antrim Plateau, but there remains controversy as to whether it was ever deposited over the Irish Sea area. For instance Tucker & Arter (1987) decided that the Chalk had been deposited over Cardigan Bay, whilst Smith (1992) was of the opinion that the absence of Chalk over the whole Irish Sea area was due to non-deposition, rather than to erosion. Palaeogeographical reconstructions of this area for the Cretaceous period are somewhat equivocal, and several models are based on the idea that the Irish Sea would have been land during the Lower Cretaceous and would only have slowly been submerged during the rising sea-levels of the Late Cretaceous. For example, Hancock (*in* Cope *et al.* 1992) concluded that the whole Irish Sea area was land during the Cenomanian, but by the Campanian, was fully submerged. These models are, however, based on two principal tenets: that there is no Lower Cretaceous in Northern Ireland (and even the Cenomanian is of restricted occurrence), and that there are no Cretaceous rocks

From Meadows, N. S., Trueblood, S. P., Hardman, M. & Cowan, G. (eds), 1997, *Petroleum Geology of the Irish Sea and Adjacent Areas,* Geological Society Special Publication No. 124, pp. 47–59.

                                    J. C. W. COPE

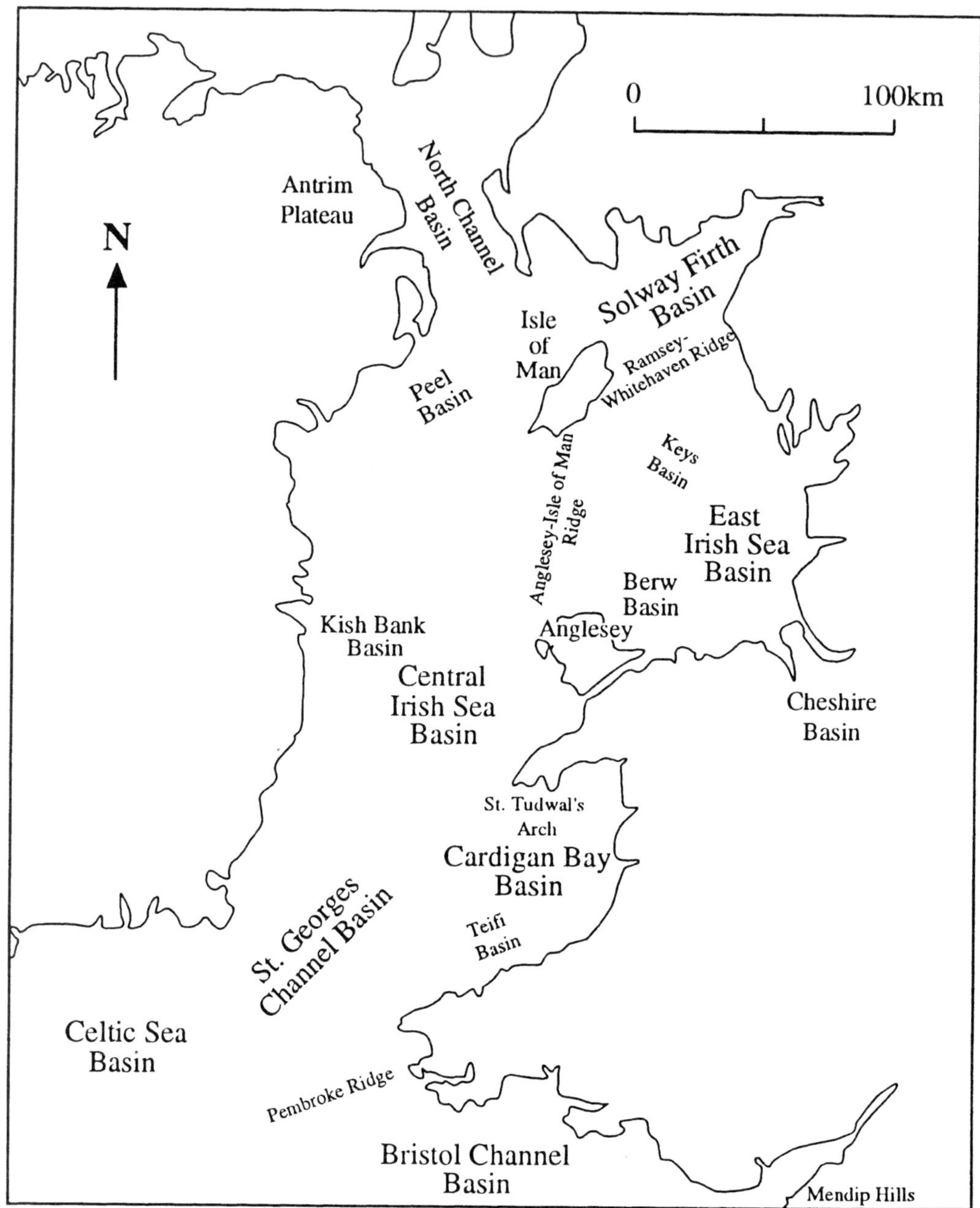

Fig. 1. Map of the Irish Sea and environs showing localities referred to in the text.

known in the Irish Sea, north of St George's Channel. The latter, however, may not be an original feature, and the alternative hypothesis, that there was formerly a significant Cretaceous succession in the area, should be considered.

Doming of the Irish Sea area

Cope (1994) has recently suggested that the Irish Sea area was the focus of a major thermally induced doming during the closing phases of the

Cretaceous Period and the model he proposed suggested that the Irish Sea had a considerable cover of Jurassic and late Cretaceous rocks, which were removed rapidly during the Palaeogene. If this hypothesis can be substantiated, it has considerable impact on the models produced by the petroleum industry on burial (and thermal) histories of the source and reservoir rocks in the area. It therefore seems appropriate to examine in some depth the nature of the evidence on which the doming model was based.

Apatite Fission Track (AFT) analysis is a tool used widely in the exploration industry. It provides a method of reconstructing the temperatures to which rocks have been subjected, by measuring the degree of annealing of fission tracks caused by fission of uranium present in apatite crystals. Annealing starts at $50°$ and is complete at $110°C$, so that this method covers a temperature range which is very useful to the petroleum geologist. There is clearly a relationship between such temperatures and the depth of burial to which a rock has been subjected, and it is from this basis that a number of attempts have been made to reconstruct the likely cover of rocks which have been removed in the geologically recent past from parts of western and northern Britain, which are now largely devoid of a Mesozoic cover.

Green (1986) and Lewis et al. (1992) had suggested, on the basis of AFT analyses, that a cover of up to 3.3 km had been lost from northern parts of Britain. These results were so at variance with the sort of cover which might reasonably have been predicted, that they were not accepted by most stratigraphers. Holliday (1993) suggested that both Green and Lewis et al. had underestimated both the thermal conductivity of the eroded cover and the ambient surface temperature during the Palaeogene. He showed that the amount of cover removed from areas of northern England, based on his recalculation of the earlier data, was in the range of 1.4–1.7 km. Holliday (1993) also demonstrated that this range of figures could be accommodated satisfactorily by the likely missing cover, based on extrapolation from known thicknesses of Mesozoic sediments in the surrounding areas. On this basis he suggested a minimum cover of Chalk over the Pennine area of northern England of 300 m.

The differences between the figure of Lewis et al. (1992) and Holliday (1993) demonstrate the major problem of interpretation of AFT analyses – that it is difficult to distinguish the proportions of the temperatures recorded contributed by depth of burial and by the local geothermal gradient. This problem could be particularly acute in parts of the Irish Sea, where recent studies have shown that Anglesey and the adjacent part of mainland North Wales and their offshore environs have a large number of Palaeogene minor intrusions (Bevins et al. 1996); similar problems affect parts of the Irish Sea in the vicinity of the Fleetwood Dyke (Arter & Fagin 1993). Clearly in the vicinity of such intrusions AFT analyses would be considerably affected, but outside of these areas, it has been assumed that heat flow during the Palaeogene was not significantly raised.

Extrapolating from these areas where there appears to be agreement that a considerable amount of cover has been lost, Cope (1994) argued that there were clearly areas in southern Britain which had lost little significant cover through the Tertiary, and that if a geometrical reconstruction were made from areas where known cover loss could be closely calculated (either from AFT analyses or by simple extrapolation from the nearest outcrops) to those areas without loss of cover, a regional dip could be computed. In this way a picture emerged suggesting that northwestern parts of Britain had suffered the greatest loss of cover, and that this amount decreased both southwards and eastwards (Cope 1994, fig.1).

AFT analysis figures from the East Irish Sea Basin (EISB), support this conclusion, and suggested to Hardman et al. (1993) that parts of the basin have lost up to a maximum of 10 000 feet (3 km) of cover. These figures were, however, calculated on the same basis as those of Lewis et al. (1992); using the figure of Holliday (1993) a figure of 2 km of maximum lost cover appears more likely. In either event, the figures suggest that the cover lost from the EISB is greater than that from adjacent parts of northern England. This is in accord with Cope's (1994) conclusion that the centre of the uplift could be located in the Irish Sea, at a point to the NW of Anglesey and to the SW of the Isle of Man.

One of the reasons for this suggested location was that that the gravity map of England and Wales (BGS 1986) showed a northwesterly-increasing positive Bouguer anomaly into the Irish Sea. There was a discrete area of anomaly greater than 40 mgals NW of Anglesey and SW of the Isle of Man, which could (Cope suggested) represent the area of maximum crustal thinning after erosion, or could be the site of a buried basic intrusion. On the other hand, the gravity map of Ireland (Murphy 1981) shows the reversal of the picture over England and Wales. Although the Irish picture is locally substantially modified by thick basinal sequences of Old Red Sandstone and Carboniferous rocks, there is a positive gravity anomaly on the eastern Irish coast, decreasing westwards to neutral values, before rising again towards the western coast (reflecting the Tertiary uplift there).

Another part of the reasoning to place the central part of the dome in the Irish Sea was the curvature of the outcrop patterns of the Mesozoic rocks over the southern and eastern English Midlands. Cope (1994) showed that if the Cheshire Basin had lost 2 km of cover and if one assumed that the Chiltern Hills had lost no cover of significance, then uplift of up to 2 km to the NW would impart a south-easterly regional dip to England and Wales amounting to 2°, which is in fact the observed dip. Thus the uplift in the Irish Sea was sufficient to give southern Britain its southeasterly tilt and impart the regional easterly dip to the Permian and Triassic outcrops to the east of the Pennines. Extrapolation from the radius of curvature of the Jurassic outcrops in the south and east Midlands suggested that the centre of uplift lay in the Irish Sea. Furthermore, the curvature was closely parallel to that observed in the change from neutral to positive Bouguer gravity anomalies (Lee *et al.* 1990). Thus it is to be expected that the amount of cover lost in the EISB reported by Hardman *et al.* (1993) will be exceeded to the west of the basin where, it is assumed, uplift would have been greater; however, no AFT analysis figures appear to be available for this area.

The establishment of the uplift in the area concatenates the AFT analysis figures for the EISB with those of northern England; both are in accord with Cope's (1994) model. What is important from the petroleum point of view is the timing of the doming. Burial history figures produced by Hardman *et al.* (1993, p.816) suggest that the rocks in the EISB reached their maximum burial in Cretaceous times and that cooling from elevated temperatures began in late Cretaceous–early Tertiary times. Cope (1994) showed that there is a restricted time interval into which the thermal updoming can be placed and that Northern Ireland is the place where temporal constraints are tightest. The youngest rocks upon which the Palaeocene lavas rest provide the downward constraints, whilst the timing is constrained upwards by the date of the oldest lavas themselves. In the Antrim plateau region of Northern Ireland the lavas rest in places on Lower Maastrichtian rocks. The date of the overlying lavas is, however, not well constrained and there is need for a thorough appraisal of the age of the basal flows. Published dates suggest that the lavas were erupted between 63 and 58 Ma but, as noted by Cope (1994), although the date of 65.5 Ma for the Tardree Rhyolite (Fitch & Hurford 1977) has been superseded by more recent dates (Mussett *et al.* 1988), it may not be too far from correct. Cope (1994) argued that because erosion would quickly strip off the Chalk as soon as it was uplifted, it could not have proceeded for long, since it had eroded no further than the Lower Maastrichtian; he thus suggested that initial uplift occurred in the latter part of the Late Maastrichtian

at around 67 Ma. If the first lavas followed soon after, at around 66–65 Ma, then the Tardree Rhyolite (which is later than the Lower Lavas – Lyle 1985) could well be as old as 64 Ma. It is interesting to note that dykes in North Wales were dated as having a minimum age of 63.5 Ma (Evans *et al.* 1973) whilst the recently discovered Fleetwood Dyke in the EISB has given dates of 65.5 ± 1.0 Ma and 61.5 ± 0.8 Ma (Arter & Fagin 1993). Clearly, accurate dates for the basal lavas could constrain the possible time interval. Lewis *et al.* (1992) suggested that the initial uplift was in the region of 300–400 m and this, coupled with a sea-level fall in the Late Maastrichtian (Hancock & Kauffman 1979) would have been sufficient to lift the Chalk of the Irish Sea above sea-level and begin the subaerial erosion which produced the Clay-with-Flints which lies between the Antrim Chalk and the overlying basalts in places (Wilson 1972).

Further support for Cope's (1994) model of uplift in the Irish Sea is provided by the occurrence of substantial pockets of sour gas in the EISB. Hardman *et al.* (1993, p.820) have shown that the most convincing way to explain the fact that sweet and sour gases occur in close proximity is to envisage a hot flush basin model involving thermal sulphate reduction of hydrocarbon accumulations that originally included moderate to high sulphur levels; this latter process requires temperatures in the range of 100–140°C and is higher than the temperatures over which biogenic (bacterial) sulphate reduction occurs (<92°) (Machel 1987). A raised thermal gradient associated with uplift in the Irish Sea would very simply provide the necessary temperatures for the hot flush model and thus the thermal sulphate reduction; these temperatures are supported by fluid inclusion analysis (Hardman *et al.* 1993, pp. 814–816).

Thomson (1995) argued that the principal flaw in Cope's (1994) model was that if there was thermal doming affecting a large part of the western British Isles, as the thermal anomaly decayed, so the post-thermal subsidence of the area should have matched the amount of erosion, and most of the region should have been below sea-level at present. Thomson suggested that another mechanism must be involved, such as igneous underplating as suggested by Brodie & White (1994). However, one major factor which has hitherto been overlooked in this argument, and which I overlooked in my reply to Thomson (Cope 1995*a*) is that the uplift occurred at the very time in the latest Cretaceous when sea-levels in Britain were at their maximum. Hancock & Kauffman (1979) suggested that the Maastrichtian sea-level may have been 600 m higher than at present, although later (Hancock 1987) this figure was pronounced as too high; a figure of 350 m (Pitman 1978) to 400 m may be more reasonable. Such a figure would result in a large proportion (*c.* 90%) of the area suggested to be affected by doming and erosion (Cope 1994, fig. 1) being below sea-level, should today's sea-level be as high as that of the latest Cretaceous. Haq *et al.* (1987) suggested that the highest sea-levels in the Cretaceous were in the Cenomanian and that these were some 250 m above present levels. However, it is clear that the northwest European continental margins must have continued to subside through much of the late Cretaceous, as sea-levels continued to rise over the area, despite the eustatic fall in sea-level recorded by Haq *et al.* (1987). This is why, for example, Campanian Chalk rests in places directly on Dalradian in Northern Ireland, and why Maastrichtian Chalk rests directly on the Precambrian basement in Cotentin. In addition to comparisons with the latest Cretaceous sea-levels, there is one further point to consider, and that is the possible present isostatic anomalies of the area, which could have a major impact on any assessment of the relative sea-level heights of the latest Cretaceous and the present day.

The composition of the lost cover

The thickness of rocks lost by erosion in the EISB as indicated by AFT analyses ranges from 300 m up to 3 km (Hardman *et al.* 1993); this figure was produced by using the figures for ambient surface temperature and thermal conductivity of the lost cover used by Lewis *et al.* (1992). Holliday (1993) suggested that the figures assumed for both of these factors were incorrect and that if recalculations were made, based on more likely figures, the results accorded more feasibly with stratigraphical predictions of lost cover. Holliday's (1993) figure is supported by the Irish Sea doming model of Cope (1994) which suggests that the maximum figure lost in the EISB is somewhat over 2 km.

Studies of regional stratigraphy and possible palaeogeographical reconstructions allow a model of what has been eroded in the area to be constructed. Two other factors need to be taken into account in production of such a model. Firstly, there is the variation in amount of cover lost indicated by AFT studies. At first sight, it seems difficult to reconcile the wide variations reported by Hardman *et al.* (1993) with the 'blanket' uplift and erosion suggested by Cope, which suggests that all areas have lost similar amounts of cover. Secondly there is the possible occurrence of a discrete Late Cimmerian event to be taken into account. In fact, as discussed below, these two events can be considered together. Before discussing this problem, however, it is necessary to consider the nature of the lost cover.

The Triassic

Triassic rocks are known widely throughout the Irish Sea. In most areas the basal Triassic Sherwood Sandstone Group overlaps the margins of the Late Permian sediments to rest directly on pre-Permian rocks. In turn, the margins of the Mercia Mudstone Group extend beyond those of the Sherwood Sandstone Group and in places rest directly on the Palaeozoic basement. Permian and Triassic sediments accumulated during a period of crustal extension and sedimentation patterns became established around major depocentres separated by areas of slower sediment accumulation. In some parts of the basins very great thicknesses accumulated, such as developments of the Mercia Mudstone Group of 2138 m in Block 110 in the EISB or 3200 m in the Keys Basin (Jackson *et al.* 1987) or total Triassic thickness of 4375 m in the EISB (Jackson & Mulholland 1993). Such rapid accumulations require deposition to be controlled by growth faults. It is also clear that growth faults played a major part in the accumulation of some very thick sequences recorded in the overlying Jurassic rocks (see below). Between these basinal successions, however, thicknesses of Permo-Triassic are more modest and the succession in places consists of only Mercia Mudstone Group rocks resting upon a pre-Permian surface, reflecting the progressive covering of the eroded Variscan (or earlier) relief. In some of the areas of the Irish Sea now floored by Carboniferous or older rocks the Permian and Triassic cover may well have been less than 500 m in total thickness and its feather-edge is preserved around the margins of some of the Palaeozoic massifs such as on the Isle of Man, where the north coast lies at the southwestern end of the Solway Firth Basin.

The Jurassic

Jurassic rocks are well known on land around the margins of the Irish Sea in the Cheshire Basin, around Carlisle (northwest England), and in the Antrim Plateau region of Northern Ireland; there are more extensive outcrops around the Inner Hebrides. Offshore, Jurassic is well known from St George's Channel northwards into the Cardigan Bay Basin, in the Kish Bank Basin, within the EISB and in the Solway Firth Basin. These basinal occurrences coincide with areas of thick Mercia Mudstone Group sedimentation and it is apparent that Lower Jurassic sedimentation represented a continuation of the same extensional regime as the Late Triassic. Present erosion has separated these basinal thicknesses of Jurassic rocks so that they now occur as discrete outliers separated by areas in which there are older rocks at outcrop. During the

Jurassic, however, the outcrops were much more extensive; between the basins there were areas of attenuated successions, swells, such as are well known onshore. There is no reason to doubt that the basin and swell picture of Jurassic sedimentation continued over the whole Irish Sea area as it did over the whole of the northwest European continental shelf.

Lower Jurassic. The Lower Jurassic consists entirely of the Lias Group, of which the basal few metres are of latest Triassic age and is well developed over the whole area. In Cardigan Bay its minimum thickness is found towards the southwestern flank of the St George's Channel Basin, where Borehole 103/2-1 penetrated 393 m of Lias Group in a position close to the northwestern margin of the Pembroke Ridge. The Mochras Borehole proved the thickest Lias known in Britain of 1305 m (Woodland 1971). The Kish Bank Basin offshore of Dublin (Naylor *et al.* 1993) has been known to contain Lias Group rocks for some time (Dobson & Whittington 1979), but more recently Broughan *et al.* (1989) have suggested that the thickness of Lias, close to the southwestern margin Bray Fault, may be as much as 2700 m thick – a truly remarkable thickness of Lower Jurassic rocks and one which suggests the possibility that some of the mudrocks which dominate the succession could be of later Jurassic age. It is clear that the thicknesses at Mochras and in the Kish Bank Basin developed in proximity to growth faults. Within the EISB there are poorly known areas of Lias Group rocks in the Keys Basin and possibly in the Berw Basin (Jackson & Mulholland 1993). More extensive outcrops occur in the Solway Firth Basin. With thick Lias to the SW of the EISB in the Prees Outlier on the southern edge of the Cheshire Basin (where the Hettangian to Pliensbachian is preserved and totals 600 m), thick Lias at Mochras and in the Kish Bank Basin, and with Lias outcrops in Northern Ireland and as outliers under the Irish Sea it seems inevitable that the whole of the Irish Sea area would have had a cover of Lias Group rocks. Dobson & Whittington (1987, fig. 11) suggested that a subsiding trough ran across North Wales from Mochras towards Prees. However, Cope (1984, 1994) argued that there were frequently swells between Jurassic basins and that North Wales had all the characteristics associated with a swell; in that case, it had probably accumulated an attenuated and somewhat condensed Lower Jurassic sequence. In similar manner, it is to be expected that the Lower Jurassic sequences of the Irish Sea would be variable, with the major Triassic depocentres continuing as basins into the Jurassic Period, whilst other areas, such as the Ramsay–Whitehaven Ridge or the ridge extending southwards from the Isle of Man to Anglesey, could well have acted as swells, separating basinal areas and may well have only accumulated a modest thickness of Lower Jurassic rocks. Thus within the whole area of the Irish Sea, there are likely to have been considerable thickness variations in the Lias, but it seems a virtual certainty that Lias Group sediments once mantled the whole of the area.

Middle Jurassic. With Middle Jurassic rocks, there is much more room for controversy, and their development to the north of the Cardigan Bay Basin must be largely speculative, but there are certain pointers which may narrow the realms of the speculation. In the St George's Channel and Cardigan Bay basins, the Middle Jurassic (Aalenian to Callovian stages) is as much as 1000 m thick and it seems possible that the succession is without breaks. Towards the basin margins, however, the succession is more incomplete and there is evidence of erosion. Thus, for example, around the St Tudwal's Arch the Toarcian and Pliensbachian have been eroded, so that possible Bajocian or Bathonian rocks rest directly on Sinemurian and Pliensbachian (Penn & Evans 1976). There are no Middle Jurassic rocks known in the remainder of the Irish Sea; they are not known on land either in northern England or Northern Ireland, and it is not until the Hebrides Basin is reached that Middle Jurassic rocks are seen again at outcrop. It might be argued that there is thus no evidence to suggest that the Middle Jurassic rocks were ever deposited over, for instance, the EISB, but there are some important arguments in favour of its former presence.

The Middle Jurassic Inferior Oolite of Dorset and Somerset consists of a succession of condensed limestones (often totalling 3 m or less in thickness) which vary considerably in thickness and in completeness from one locality to the next. The limestones are well known for a rich fauna of beautifully preserved fossils, including abundant ammonites. In contrast to these successions, in the Cotswold Hills, the Inferior Ooolite is much thicker (up to 100 m), but is frequently devoid of ammonites. Much further to the north, the Middle Jurassic rocks reappear in the Inner Hebrides, where the Bearreraig Sandstone Formation, which is up to 550 m in thickness, carries an abundant ammonite fauna (Parsons *in* Cope *et al.* 1980). It is clear that the ammonites, which include southern European forms, were able to migrate freely from southern Britain up to the Hebridean area throughout the Aalenian and Bajocian, the presence of facies clearly inimical to ammonites would have precluded their migration across the English Midlands. The Irish Sea area would provide the easiest route for such migration, otherwise a longer route around Ireland, most of which was probably land (Bradshaw & Cripps *in* Cope *et al.* 1992) would be required. Bradshaw & Cripps (*in* Cope *et al.* 1992, fig. J4b) suggested that the Irish Sea was likely to have been an alluvial plain during the late Aalenian, but Underhill & Partington (1993) have suggested that the effects of the Toarcian–Aalenian thermally-induced doming in the North Sea did not affect sedimentation as far west as the Irish Sea area, and Bradshaw & Cripps' maps were produced before the causes and limits of the late Aalenian regression in the North Sea were fully appreciated. On this basis it seems quite probable that there was an open marine connection across the Irish Sea at this time.

The late Bajocian transgression (the Vesulian transgression) is well documented across southern Britain; it can be dated as being initiated in the Acris Subzone of the Garantiana Zone. Around the Mendip Hills, Upper Inferior Oolite rests with marked angular discordance on Dinantian limestones and a similar overstepping relationship is envisaged around the margins of Palaeozoic South Wales. In the Cotswold area, the Upper Trigonia Grit of Acris Subzone age rests disconformably on Lower Bajocian or Aalenian portions of the Inferior Oolite (Parsons *in* Cope *et al.* 1980, fig. 3b). Further afield, in Normandy, Late Bajocian limestones rest directly on the Precambrian at the western extremity of the Paris Basin. Tappin *et al.* (1994, p.40) record that throughout the Cardigan Bay area there is a widespread break in late Toarcian or early Mid-Jurassic time and that Bajocian–Bathonian sediments rest unconformably on the Lower Jurassic. In a more widespread context, I believe this trangression is the same late Bajocian one and on this basis it appears most likely that the St Tudwal's Arch succession recorded by Penn & Evans (1976) is of Late Bajocian age at its base. I thus conclude that the basinal regions of the Irish Sea would have had a reasonably complete succession of Bajocian rocks (possibly greater than 50 m thick in some areas), and around its margins the Upper Bajocian would have overlapped the older Bajocian and Aalenian to rest on Lias in eastern Ireland, North Wales and northern England. Holliday (1993) suggested that northern England probably had a maximum cover thickness of 100 m of Middle Jurassic–Lower Cretaceous rocks, but that figure could well be at least doubled by reducing the postulated thickness of Chalk in the area. Thus Holliday (1993) suggested that there was some 300 m maximum Chalk cover over the Pennines; reducing this figure to 150 or 200 m would allow the thickness of Middle Jurassic–Lower Cretaceous to be increased to 200–250 m in the area. These figures appear to be more likely to reflect the original Middle and Upper Jurassic successions in the area. Apatite

Fission Track data have at least suggested that the Pennines should now be regarded as a Jurassic swell (which may well have been largely submerged beneath the sea) rather than the traditional Pennine High of palaeogeographical maps. There is likely to have been at least a marginal Late Bajocian cover. In the Inner Hebrides, the Late Bajocian trangression started a little earlier and the Garantiana Clay (base of Sequence E of Morton (1992)), probably encompassing the whole Garantiana Zone, rests upon youngest Lower Bajocian Humphriesianum Zone (Parsons *in* Cope *et al.* 1980).

The margins of Bathonian sedimentation were in general somewhat wider than the Bajocian, despite a eustatic sea-level fall; this has been attributed to more rapid shelf subsidence, producing an effective rise of sea-level (Vail & Todd 1981). In St George's Channel the Bathonian is up to 800 m thick (Penn & Evans 1976), considerably thicker than on land. Increased silty horizons and an impoverished fauna, together with lagoonal deposits and carbonaceous horizons suggest that the upper part of the Cardigan Bay Bathonian was deposited away from marine influence. Palaeogeographically, this implies that the more northerly parts of the Irish Sea may well have been a low-lying alluvial plain at this time; the extent of any sedimentation in the area is unknown. To the north, lagoonal sedimentation proceeded in the Hebrides and it is quite possible that similar conditions obtained over at least the basinal regions of the Irish Sea with accumulation of a modest thickness of sediments.

By Callovian times transgression would have restored marine conditions across the Irish Sea area. The Kellaways Beds and Lower and Middle Oxford Clay are recorded from St George's Channel and the thickness totals *c.* 50 m (Tappin *et al.* 1994, fig. 34). This is thicker than the Hebridean successions of c. 30 m, but comparable with that of the English Midlands (Cope *et al.* 1980). It can be concluded that the Irish Sea area would have received at least 30 m of Callovian sediments and possibly 50 m or more in the basinal areas.

Upper Jurassic. The same pattern of sedimentation would have continued into the Oxfordian, with the Upper Oxford Clay and overlying West Walton and Ampthill Clay formations (or possibly Corallian facies sediments) over the Irish Sea area. In the Cardigan Bay Basin the Oxfordian rocks are over 300 m thick. The upper part contains sandstones interpreted as coastal plain sediments and was interpreted as lying at the eastern end of a large area of alluvial plain sediments extending across the whole of the Celtic Sea and Bristol Channel basins (Cope *et al.* 1992). Wales may have acted as a landmass supplying these sediments into the Bristol

Channel (Evans & Thompson 1979; Cope 1984) and it is likely that material could have been shed northwards too, into the northern basins of the Irish Sea. Oxfordian sediments total over 100 m in the Hebrides (Cope *et al.* 1980) and it seems probable that the northern part of the Irish Sea area received a similar thickness.

A renewed transgression in the Kimmeridgian restored marine sedimentation across much of Britain. Thicknesses of sediment vary greatly. In the Wessex Basin some 500 m of Kimmeridge Clay accumulated in east Dorset, whilst in East Anglia the total thickness, though the top has been removed by erosion (Cope 1974) is *c.* 100 m. In western Scotland some 24 m of the top of the Staffin Shale can be assigned to the Kimmeridgian, although it seems clear that much more was originally deposited as only the two basal zones are fully represented. In the St George's Channel Basin some 958 m were encountered in well 106/24-1 with the top two zones missing; the succession includes some sandstones, together with glauconite and lignite (Tappin *et al.* 1994). The presence of freshwater microfossils suggests a lagoonal setting, but normal Kimmeridge Clay organic shale facies in the Bristol Channel (Evans & Thompson 1979) indicates that Wales was unlikely to have been a landmass at this time. On the other hand, the fact that in the Irish Sector of the North Celtic Sea Basin there is a considerable development of anhydrite in the Kimmeridgian (Tappin *et al.* 1994) indicates uplift of the southeastern portion of the Irish Massif and this may well have influenced sedimentation in the St George's Channel Basin. Figures for the Irish Sea are purely speculative, but in rapidly subsiding basins such as those of the EISB, a figure of 300 m for the Kimmeridge Clay appears to be a conservative estimate; at least 500 m may have accumulated.

Portlandian sediments are recorded in the Bristol Channel Basin and in St George's Channel Basin. However, they are known no further to the north and it can be fairly safely assumed that none were deposited over the Irish Sea area. Palaeogeographically, there is a major change in the latest part of the Kimmeridgian (now more correctly referred to the Bolonian Stage (Cope 1993, 1995*b*)). This change can be seen even in the thickest basinal successions, as in east Dorset, where silt is introduced into the clays of the Rotunda Zone, reflecting the uplift of areas to the north. The overlying Fittoni Zone has yet to be recorded outside of south Dorset (Cope 1978) and in the succeeding basal Portlandian Albani Zone sedimentation was confined to a restricted area comprising chiefly southern and eastern Britain (see Cope *et al.* 1992, maps J10 and J11a). The whole Irish Sea area is interpreted as being land at this time.

The Cretaceous

Exactly when sedimentation recommenced in the Cretaceous Period in the Irish Sea is difficult to ascertain, but it seems at least safe to assume that there was no marine sedimentation in the area throughout the Neocomian and Aptian. It is, however, just possible that there was localized non-marine sedimentation in the Neocomian. The earliest Cretaceous sediments in Northern Ireland and western Scotland are of Cenomanian age; these are not widely distributed. However, there seems no reason to suppose that the Albian trangression did not reach at least the southern part of the EISB. In southern Britain, thanks to the erosion of the Cretaceous rocks, we have a prejudiced view of the Cretaceous transgressions. Hancock & Kauffman (1979) have shown that sea-levels in the Cretaceous Period rose to their maximum in the Campanian and then commenced to fall in the later part of the Maastrichtian. At the time of publication of this paper, they believed that this picture could be applied globally, but it has susequently become clear that the maximum sea-levels were attained in the Cenomanian, with subsequent falls (Haq *et al.* 1987). The continued late Cretaceous rise in sea-levels recorded on the northwest European continental margins must therefore be the consequence of regional subsidence of the area exceeding the eustatic sea-level fall. Thus, where erosion has not removed marginal Cretaceous sediments, successions show progressive overlap, with the Campanian frequently resting on much older rocks. In Northern Ireland, for example, the earliest Cretaceous rocks recorded are of Cenomanian age, whilst Campanian rests in places directly on the Dalradian schists, reflecting the nature of the transgression; in southern England, however, the Cenomanian Chalk always rests on Albian rocks and Campanian always on Santonian. It seems clear that there must have been a Late Cretaceous cover over the Irish Sea, and this is likely to have started with rocks as early as the Cenomanian, if no earlier. A probable thickness of Chalk cover over the area is in the range of 200–300 m, though locally thicknesses may have been greater. Holliday (1993) suggested that a Chalk cover of 300 m was indicated by AFT analysis figures for the north of England and it should not be forgotten that some Jurassic swells were inverted in the Late Cretaceous, so that the Jurassic–Early Cretaceous Market Weighton Swell became a Late Cretaceous basin and accumulated some 500 m of Chalk. Cope (1994) suggested it possible that North Wales could have behaved similarly. The thicknesses of Chalk preserved in Northern Ireland is up to 120 m (Wilson 1972) and taking this figure as a guide, some workers have suggested that it is the maximum thickness of Chalk which could have been deposited over the Irish Sea area. Indeed, as we have seen, some authors still believe that the Irish Sea never received any Chalk sediment. I here take the opposing view, developed from that which I first proposed some years ago (Cope 1984) when I suggested that most, if not all of Wales was inundated by the latest Cretaceous. I now take the view that most of the Irish Sea (with the possible exception of horst blocks such as the Isle of Man) would have accumulated a complete Chalk succession, with a thickness of at least 200–300 m possibly overlying (at least in the EISB) a thin Albian succession. The development of the Chalk in Northern Ireland is, I believe, a one which has caused us to draw the wrong conclusions for a long time; it is merely an accident that the very place where the basalts have so effectively protected the Chalk is where it was originally the most incomplete. According to the model proposed by Cope (1994) Chalk would have been deposited across Ireland and some authors (e.g. Naylor 1992) are now willing to concede the possibility that most of Ireland may have had a thick Chalk cover.

The Tertiary

Following from doming in the latest Cretaceous (Cope 1994), rapid subaerial erosion would have started, producing a radial drainage centred on the Irish Sea. Initially, much of the material eroded would have been Chalk, but as the Chalk cover was rapidly stripped, rivers started to cut into the older rocks beneath, transporting clastic debris to the margins of the British Isles. The most rapid erosion would have been through the thick Mesozoic successions of the Irish Sea and rivers would have most quickly become mature through this region, so that rapid accumulations of Palaeogene sediments, in areas peripheral to the domed region, probably started in the late Palaeocene and continued through into the mid-Eocene; beyond this, erosion continued at a progressively reducing rate from the late Eocene through to the late Oligocene or earliest Miocene. Deflation of the Cretaceous doming led to influx of the sea into this lowland area; the earliest recorded incursion is in the Teifi Basin, where sediments dated as late Eocene to mid-Oligocene contain glauconite (Warrington & Owens 1977). Within the Cardigan Bay Basin a meandering river crossed a swampy floodplain (Dobson & Whittington 1987). The principal marine flooding of the Irish Sea seems to have been a post-Oligocene event, since post-Oligocene movement can be demonstrated along some of the faults around the Irish Sea. By inference, similar movement took place contemporaneously along other faults in the region. Neogene sediments may occur in the Central Irish Sea Basin.

Igneous activity in the Irish Sea area includes the basalts and other igneous rocks of Northern Ireland, and the intrusion of dykes such as the Fleetwood dyke (Arter & Fagin 1993) and a series of dykes in North Wales and the adjacent area of the Irish Sea (Evans *et al.* 1973; Bevins *et al.* 1996).

Tectonic history

Burial histories for rocks from the EISB have revealed that rocks reached their maximum burial depth in the Cretaceous and that there was a major inversion in the latest Cretaceous or earliest Tertiary. What is less certain is whether there was a Late Cimmerian inversion too. Bushell (1986) suggested that there had been a Cimmerian inversion episode during which something approaching 1000 m of inversion had taken place. On the other hand Hardman *et al.* (1993) were unable to detect a Late Cimmerian inversion and suggested that if it took place, it was less than 1000 feet (300 m). At first reading, these two figures are totally irreconcilable. On purely regional stratigraphical grounds, I believe a Late Cimmerian inversion over the Irish Sea to have been inevitable. As argued above, it is probable that no rocks were deposited over the Irish Sea region, between the Rotunda Zone of the Bolonian and the base of the Middle Albian – a time interval of *c.* 40 Ma. Over that interval of time, considerable erosion would have taken place. However, let us consider the very different Jurassic successions which would have been developed over the Irish Sea. In the basins a thick succession would have included probably 1400–1800 m of Jurassic rocks, whilst the interbasinal swells might have had as little as 500 m. This latter figure takes into account the fact that erosional episodes during the Jurassic significantly reduced the total thickness accumulated up to that point. For example, we have seen that the Late Bajocian transgression cut down to at least the Sinemurian in places, whilst in the basins a complete succession was preserved. A loss of 300 m from a basinal succession might not even completely remove the Kimmeridgian, whilst a loss of 300 m from a reduced interbasinal swell succession could easily remove everything down to the Lower Lias (as in Northern Ireland). I believe this allows a reconciliation of the figure produced by Bushell (1986) with that of Hardman *et al.* (1993). These differences also provide a means of production of the varying AFT analysis figures mentioned by Hardman *et al.* (1993). The basinal figures would show up to 1300 m (4350 feet) more cover than the swell areas.

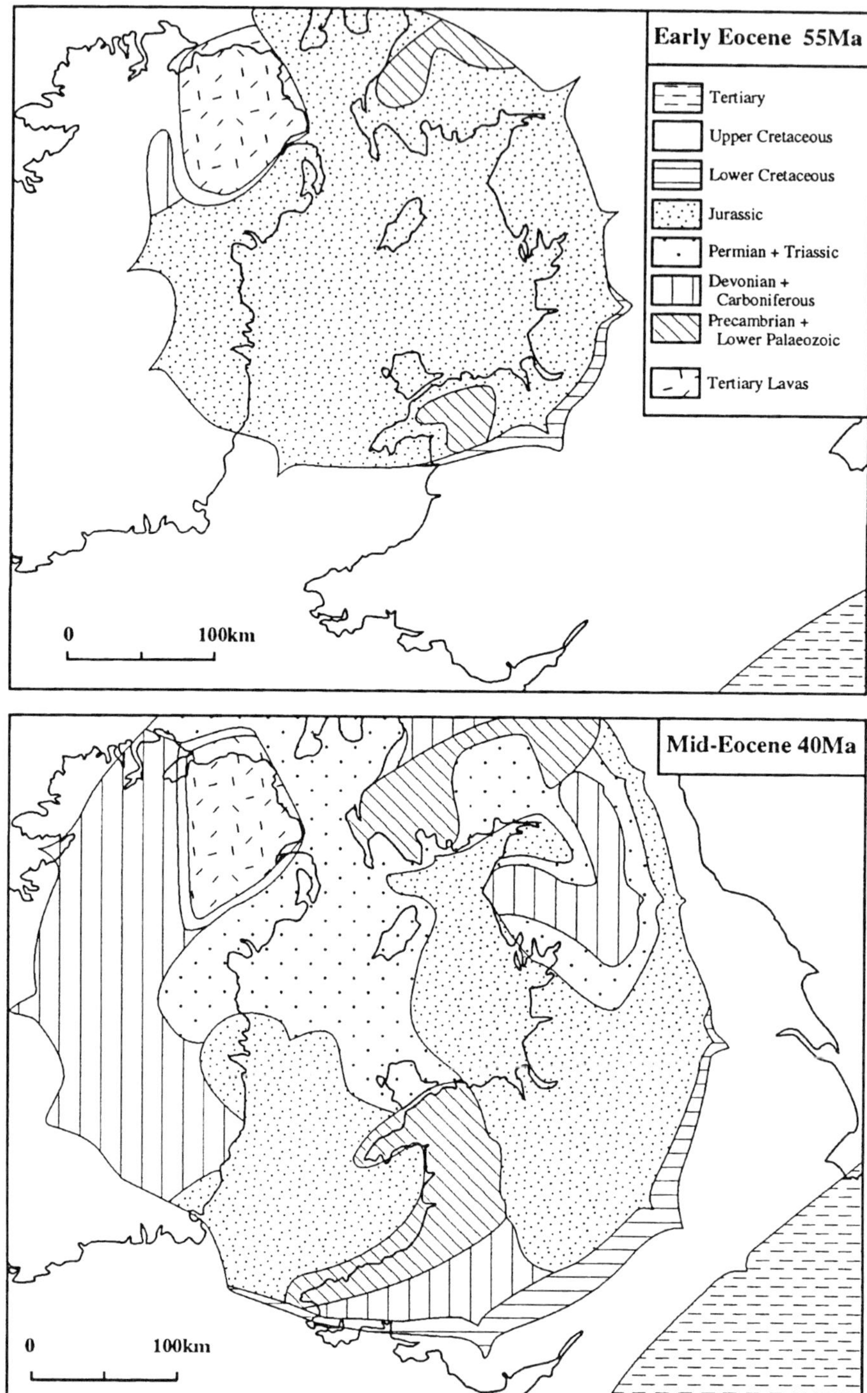

Fig. 2. Postulated palaeogeological maps to illustrate the unroofing of the Irish Sea area during the Palaeogene. **Early Eocene**: map shows initial stage of erosion of Irish Sea dome, exposing largely Jurassic rocks, except in the Southern Uplands and North Wales where late Cimmerian erosion is likely to have completely removed attenuated Jurassic successions. Note possible Lower Cretaceous (Albian) over the southeastern margin of the eroded area. **Mid-Eocene**: the erosional area was considerably extended and has exposed areas of Upper Palaeozoic rocks in Ireland,

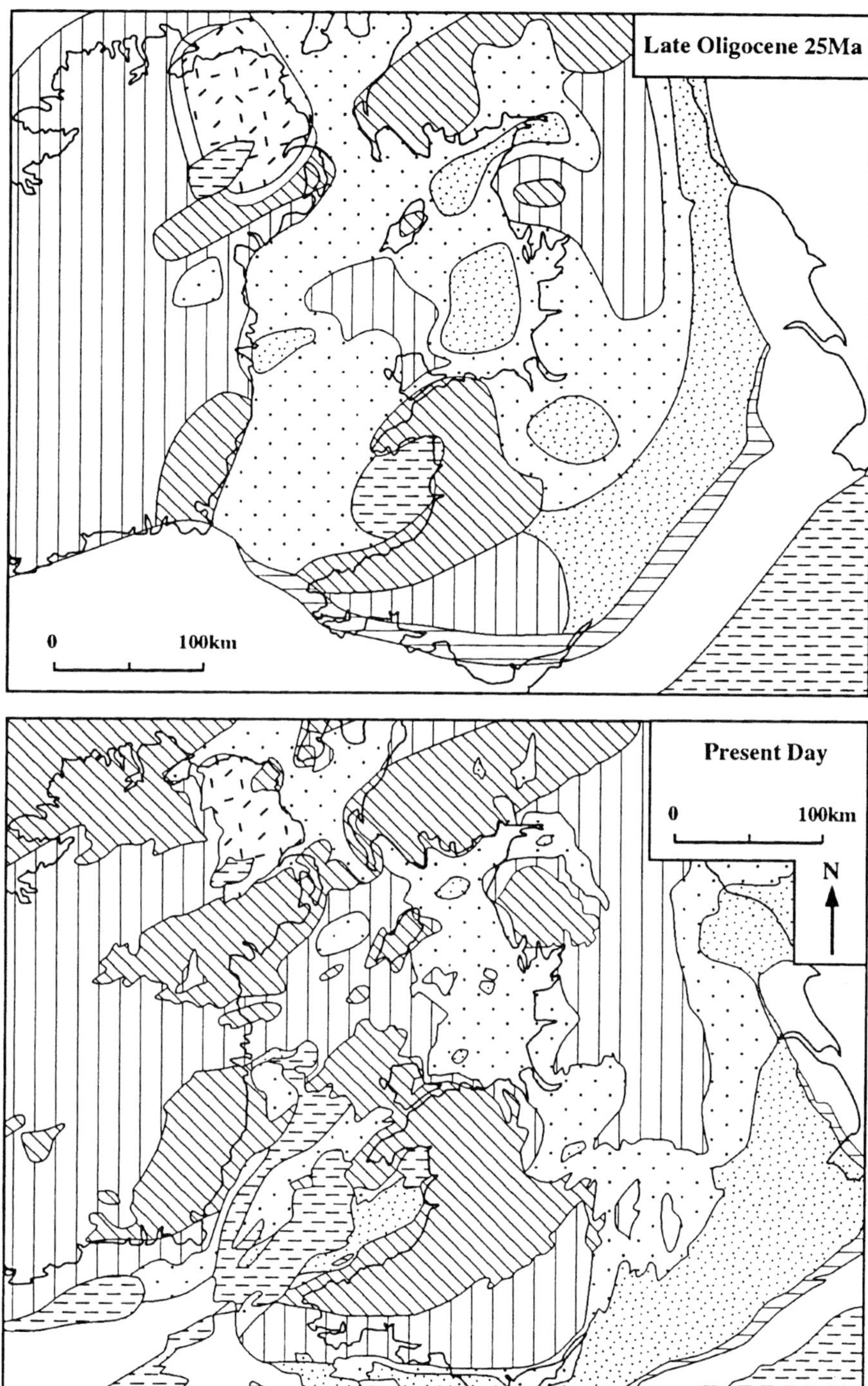

the Northern Pennines and South Wales. To the southeast the Mid-Eocene shoreline has transgressed in a
northwesterly direction. **Late Oligocene**: by this time the Permo-Triassic cover has been breached in some areas of
the Irish Sea and significant Jurassic outliers persist in the Irish Sea. In Ireland the Chalk has been stripped
everywhere apart from those areas protected by the Palaeogene basalt cover. **Present day**: for comparison. Note
diminutive Jurassic outliers in the Cheshire Basin (at Prees), in the Solway Basin and in the East Irish Sea Basin.

Following the deposition of the Upper Cretaceous cover over the area, the whole region was updomed in the very latest Cretaceous, initiating a period of very rapid erosion in the Palaeogene, probably by a series of pulsed uplifts (Cope 1994). This is when the Chalk was stripped off the Irish Sea area, followed by stripping of the Jurassic cover from most of the area. Where the Jurassic cover was thinner, the Permo-Triassic sequence was exposed to erosion and then the underlying pre-Permian basement subsequently eroded.

Post-Oligocene downfaulting around the margins of the Irish Sea, as the uplifted dome deflated, caused incursion of the sea and production of the modern Irish Sea by early Neogene times.

The assistance of Mrs D. G. Evans in drafting of the figures is gratefully acknowledged.

References

ARTER, G. & FAGIN, S. W. 1993. The Fleetwood Dyke and the Tynwald fault zone, Block 113/27, East Irish Sea Basin. *In*: PARKER, J. R. (ed.) *Petroleum Geology of Northwest Europe: Proceedings of the 4th Conference*. Geological Society, London, 835–843.

BEVINS, R. E., HORAK, J. M. , EVANS, A. D. & MORGAN, R. 1996. A Palaeogene dyke swarm in NW Wales: evidence for Cenozoic sinistral fault movement. *Journal of the Geological Society, London*, **153**, 177–180.

BRITISH GEOLOGICAL SURVEY. 1986. *1:625 000 Bouguer anomaly map of the British Isles*. British Geological Survey, Keyworth.

BRODIE, J. & WHITE, N. 1994. Sedimentary basin inversion caused by igneous underplating. *Geology*, **22**, 147–150.

BROUGHAN, F. M., NAYLOR, D. & ANSTEY, N. A. 1989. Jurassic rocks in the Kish Bank Basin. *Irish Journal of Earth Sciences*, **10**, 99–106.

BUSHELL, T. P. 1986. Reservoir Geology of the Morecambe Field. *In*: BROOKS, J., GOFF, J. C. & VAN HOORN, B. (eds) *Habitat of Palaeozoic Gas in N.W.Europe*. Geological Society, London, Special Publications, **23**, 189–208.

COPE, F. W. 1939. Oil occurrences in south-west Lancashire. *Bulletin of the Geological Survey of Great Britain*, **2**, 18.

COPE, J. C. W. 1974. Upper Kimmeridgian ammonite faunas of the Wash area and a subzonal scheme for the lower part of the Upper Kimmeridgian. *Bulletin of the Geological Survey of Great Britain*, **47**, 29–37.

——, 1978. The ammonite faunas and stratigraphy of the upper part of the Upper Kimmeridge Clay of Dorset. *Palaeontology*, **21**, 469–544.

——, 1984. The Mesozoic history of Wales. *Proceedings of the Geologists' Association*, **95**, 373–385.

——, 1993. The Bolonian Stage: an old answer to an old problem. *Newsletters on Stratigraphy*, **28**, 151–156.

——, 1994. A latest Cretaceous hotspot and the southeasterly tilt of Britain. *Journal of the Geological Society, London*, **151**, 905–908.

——, 1995a. Reply to K. Thomson: Discussion on a latest Cretaceous hotspot and the southeasterly tilt of Britain. *Journal of the Geological Society, London*, **152**, 729–731.

——, 1995b. Towards a unified Kimmeridgian Stage. *Petroleum Geoscience*, **1**, 351–354.

——, DUFF, K. L., PARSONS, C. F., TORRENS, H. S., WIMBLEDON, W. A. & WRIGHT, J. K. 1980. *A correlation of Jurassic rocks in the British Isles. Part Two: Middle and Upper Jurassic*. Geological Society, London, Special Report, **15**.

——, INGHAM, J. K. & RAWSON, P. F. (eds) 1992. *Atlas of Palaeogeography and Lithofacies*. Geological Society, London, Memoir, **13**.

DOBSON, M. R. & WHITTINGTON, R. J. 1979. The geology of the Kish Bank Basin. *Journal of the Geological Society, London*, **136**, 243–249.

—— & ——, 1987. The geology of Cardigan Bay. *Proceedings of the Geologists' Association*, **98**, 331–353.

EVANS, A. L. FITCH, F. J. & MILLER, J. A. 1973. Potassium–argon age determinations on some British Tertiary igneous rocks. *Journal of the Geological Society, London*, **129**, 419–443.

EVANS, D. J. & THOMPSON, M. S. 1979. The geology of the central Bristol Channel and the Lundy area, South Western Approaches, British Isles. *Proceedings of the Geologists' Association*, **90**, 1–14.

FITCH, F. J. & HURFORD, A. J. 1977. Fission track dating of the Tardree rhyolite, Co. Antrim. *Proceedings of the Geologists' Association*, **88**, 267–274.

GREEN, P. F. 1986. On the thermo-tectonic evolution of Northern England: evidence from fission track analysis. *Geological Magazine*, **123**, 493–506.

HANCOCK, J. M. 1987. Sea-level changes in the British region during the Late Cretaceous. *Proceedings of the Geologists' Association, London*, **100**, 565–594.

—— & KAUFFMAN, E. G. 1979. The great transgressions of the late Cretaceous. *Journal of the Geological Society, London*, **136**, 175–186.

HAQ, B. U., HARDENBOL, J. & VAIL, P. R. 1987. Chronology of fluctuating sea-levels since the Triassic. *Science*, **235**, 1156–1167.

HARDMAN, M., BUCHANAN. J., HERRINGTON, P. & CARR, A. 1993. Geochemical modelling of the East Irish Sea Basin: its influence on predicting hydrocarbon type and quality. *In*: PARKER, J. R. (ed.) *Petroleum Geology of Northwest Europe: Proceedings of the 4th Conference*. Geological Society, London, 809–821

HOLLIDAY, D. W. 1993. Mesozoic cover over northern England: interpretation of fission track data. *Journal of the Geological Society, London*, **150**, 657–660.

JACKSON, D. I. & MULHOLLAND, P. 1993. Tectonic and stratigraphic aspects of the East Irish Sea Basin and adjacent areas: contrasts in their post-Carboniferous structural styles. *In*: PARKER, J. R. (ed.) *Petroleum Geology of Northwest Europe: Proceedings of the 4th Conference*. Geological Society, London, 791–808.

——, ——, JONES, S. M. & WARRINGTON, G. 1987. The geological framework of the East Irish Sea Basin. *In*: BROOKS, J. & GLENNIE, K. W. (eds) *Petroleum Geology of North West Europe*. Graham & Trotman, London, 191–203.

JONES, O. T. 1956. The geological evolution of Wales and adjacent regions. *Quarterly Journal of the Geological Society of London*, **111**, 323–352.

LEE, M. J., PHARAOH, T. C. & SOPER, N. J. 1990. Structural trends in central Britain from images of gravity and aeromagnetic fields. *Journal of the Geological Society, London*, **147**, 241–255.

LEWIS, C. L. E., GREEN, P. F., CARTER, A. & HURFORD, A. J. 1992. Elevated K/T palaeotemperatures throughout Northwest England: three kilometres of Tertiary erosion? *Earth and Planetary Science Letters*, **112**, 131–145.

LYLE, P. 1985. The petrogenesis of the Tertiary basaltic and intermediate lavas of northeast Ireland. *Scottish Journal of Geology*, **21**, 71–84.

MACHEL, H. G. 1987. Some aspects of diagenetic sulphate-hydrocarbon redox reactions. *In*: MARSHALL, J. D. (ed.) *Diagenesis of Sedimentary Sequences*. Geological Society, London, Special Publications, **36**, 15–28.

MORTON, N. 1992. Dynamic stratigraphy of the Triassic and Jurassic of the Hebrides Basin, NW Scotland. *In*: PARNELL, J. (ed.) *Basins on the Atlantic Seaboard: Petroleum Geology, Sedimentology and Basin Evolution*. Geological Society, London, Special Publications, **62**, 97–110.

MURPHY, T. 1981. Geophysical evidence. *In*: HOLLAND, C. H. (ed.) *A geology of Ireland*. Scottish Academic Press, Edinburgh, 225–229.

MUSSETT, A. E., DAGLEY, P. & SKELHORN, R. R. 1988. Time and duration of activity in the British Tertiary Igneous Province. *In*: MORTON, A. C. & PARSON, L. M. (eds) *Early Tertiary Volcanism and the opening of the NE Atlantic*. Geological Society, London, Special Publications, **39**, 337–348.

NAYLOR, D. 1992. The post-Variscan history of Ireland. *In*: PARNELL, J. (ed.) *Basins on theAtlantic Seaboard: Petroleum Geology, Sedimentology and Basin Evolution*. Geological Society, London, Special Publications, **62**, 255–275.

——, HAUGHEY, G., CLAYTON, G. & GRAHAM, J. R. 1993. The Kish Bank Basin, offshore Ireland. *In*: PARKER, J. R. (ed.) *Petroleum Geology of Northwest Europe*: *Proceedings of the 4th Conference*. Geological Society, London, 845–855.

PENN, I. E. & EVANS, C. D. R. 1976. *The Middle Jurassic (mainly Bathonian) of Cardigan Bay and its palaeogeographical significance*. Report of the Institute of Geological Sciences, **76/6**.

PITMAN, W. C. 1978. Relationship between eustasy and stratigraphic sequences of passive margins. *Bulletin of the Geological Society of America*, **89**, 1389–1403.

SMITH, A. J. 1992. Offshore Geology. *In*: DUFF, P. McL. D. & SMITH, A. J. (eds) *Geology of England and Wales*. Geological Society, London, 445–487.

TAPPIN, D. R., CHADWICK, R. A., JACKSON, A. A., WINGFIELD, R. T. R. & SMITH, N. J. P. 1994. *The geology of Cardigan Bay and the Bristol Channel*. United Kingdom Offshore Regional Report, British Geological Survey, HMSO, London.

THOMSON, K. 1995. Discussion on a latest Cretaceous hotspot and the southeasterly tilt of Britain. *Journal of the Geological Society, London*, **152**, 729.

TUCKER, R. M. & ARTER, G. 1987. The tectonic evolution of the North Celtic Sea and Cardigan Bay basins with special reference to tectonic inversion. *Tectonophysics*, **137**,291–307.

UNDERHILL, J. R. & PARTINGTON, M. A. 1993. Jurassic thermal doming and deflation in the North Sea: implications of the sequence stratigraphic evidence. *In*: PARKER, J. R. (ed.) *Petroleum Geology of Northwest Europe: Proceedings of the 4th Conference*. Geological Society, London, 337–345.

VAIL, P. R. & TODD, R. G. 1981. Northern North Sea Jurassic unconformities, chronostratigraphy and sea-level changes from seismic stratigraphy. *In*: ILLING, L. V. & HOBSON, G. D. (eds) *The geology of the continental shelf of North-West Europe*, Heyden & Son, London, 216–235.

WARRINGTON, G. 1997. The Penarth Group–Lias Group succession (Late Triassic–Early Jurassic) in the East Irish Sea Basin and neighbouring areas: a stratigraphical review. *This volume*.

—— & OWENS, B. 1977. *Micropalaeontological biostratigraphy of offshore samples from southwest Britain*. Report of the Institute of Geological Sciences, **77/7**.

WILSON, H. E. 1972. *Regional Geology of Northern Ireland*. HMSO, London.

WOODLAND, A (ed.) 1971. *The Llanbedr (Mochras Farm) Borehole*. Institute of Geological Sciences, Report, **71/18**.

WRAY, D. A. & COPE, F. W. 1948. *Geology of Southport and Formby*. Memoir of the Geological Survey of Great Britain.

Lower crustal structure of the East Irish Sea from deep seismic reflection data

R. W. ENGLAND[1] & N. J. SOPER[2]

[1] BIRPS, Bullard Laboratories, Madingley Road, Cambridge CB3 0EZ, UK
[2] Department of Earth Sciences, Dainton Building, Brookhill, Sheffield S3 7HF, UK

Abstract: Deep near-normal incidence reflection profiles across the East Irish Sea show that the middle and lower crust are dominated by dipping reflectors with a NE–SW strike. The northward dipping Iapetus suture can be traced and contoured beneath the Solway Firth where it projects upward beneath the Peel and Solway Firth basins. South of the suture there are both northward and southward dipping reflectors. These are considered to be related to Early Devonian (Acadian) deformation. Although the Carboniferous was deposited in basins with a broadly NE–SW trend there is no clear evidence of reactivation of earlier structures during either Carboniferous sedimentation or Variscan inversion. The faults that controlled Permian and Mesozoic basin development in the East Irish Sea are at a high angle to Caledonian and Acadian trends and are considered to be juvenile structures which developed in the late Westphalian.

The lower crustal structure of the East Irish Sea has received little attention in the literature while the structure and petroleum geology of the late Palaeozoic and Mesozoic basins is becoming increasingly well established (Jackson & Mulholland 1993). Some of the earliest BIRPS deep near-normal incidence reflection profiles, recorded to 15 s two-way travel time (twtt), were shot through the Irish Sea as part of the WINCH program (Brewer *et al.* 1983). These profiles cross to the west of the Isle of Man (Fig. 1), passing through the Peel, Kish Bank, Central Irish Sea, Caernarvon Bay and Cardigan Bay basins (Brewer *et al.* 1983; Klemperer & Hobbs 1992). On WINCH 2 a reflector that dips northward from the middle to the lower crust was interpreted as the Iapetus Suture, representing the boundary between Laurentian crust to the north and Avalonian crust to the south. This reflector projects to the surface a few kilometres north of the Isle of Man beneath the Peel Basin. On the basis of correlation between this and similar patterns of reflectivity seen on BIRPS' NEC and MOBIL profiles off the northeast coast of England the suture has been traced beneath the north of England (Beamish & Smythe 1986; Freeman *et al.* 1988; Soper *et al.* 1992*a*). Soper *et al.* (1992*a*) drew attention to northward dipping lower crustal reflectors south of the Iapetus Suture which were attributed to Acadian (Early Devonian) deformation that produced the British slate belts (Soper *et al.* 1987). It has been suggested that the structures developed in the basement as a result of this deformation determined the development and structural grain of the Carboniferous and subsequent basins developed in northern England

(Chadwick & Holliday 1991) and in the Southern North Sea Gas Basin. This contribution describes the reflectivity south of the Iapetus Suture on the WINCH 2, 3, 3A and 4 profiles and data recorded to 10 s twtt supplied to BIRPS by JEBCO Seismic Limited, and discusses the age and origin of this reflectivity and its relationship to the development of the late Palaeozoic and Mesozoic basins of the East Irish Sea.

Migration and depth conversion of BIRPS' WINCH data

The BIRPS' WINCH data presented here have a revised post-stack processing sequence which is different to that employed in previously published versions. The original raw stacks produced for BIRPS by GECO-PRAKLA are contaminated by side-swipe reflections from the coast or from the uneven sea-floor topography and water-borne noise, which are resistant to pre-stack processing. During migration this noise becomes smeared across the data obscuring reflections at long two-way travel times. In order to remove it, the data were first constant-velocity f–k migrated (Stolt 1978) at 1480 m s^{-1} to collapse diffractive energy with water velocity but leave the common mid-point stack virtually unmigrated. The water velocity was carefully optimized to completely collapse water-borne diffractive energy to high amplitude peaks. These peaks were muted prior to final migration by applying an automatic gain correction (agc) with a short (250 ms) operator length. This method was preferred to surgical

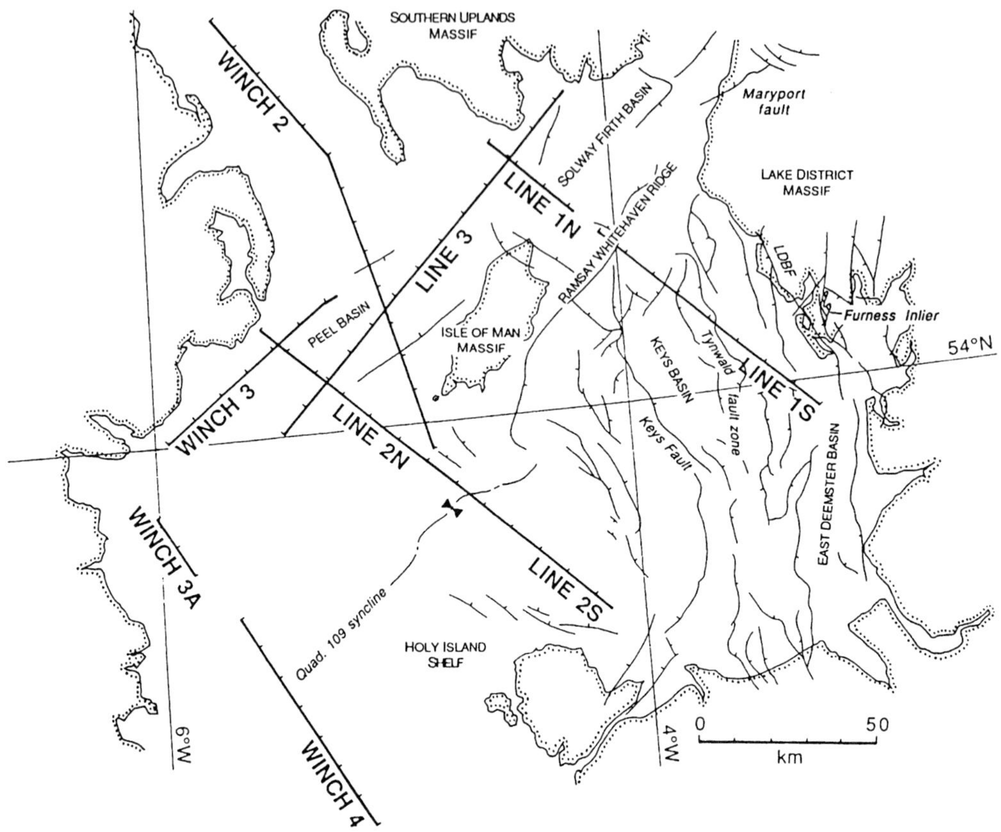

Fig. 1. Map showing the location of the seismic reflection profiles presented here and their relationship to late Palaeozoic and Mesozoic basin structure. Ticks on faults indicate downthrow side. Ticks on line tracks are at 10 km intervals. LBDF – Lake District Boundary Fault.

muting of the high amplitude peaks because in places the side-swipe reflections were so numerous that muting would have left large gaps in the data. The data were then *f–k* migrated using the velocity function determined from the CSSP wide-angle refraction experiment across northern England (Lewis 1986). The previously applied agc was then removed and the data depth converted using the same velocity function as that used for migration. This processing produces an evenly grey looking section which can be interpreted with careful reference to the unmigrated data (Fig. 2). Migration of these data improved lateral coherency of lower crustal reflectivity and correctly located the numerous cross-cutting north and south dipping events visible on the data (Fig. 2). This proved useful in interpreting the unmigrated data supplied by JEBCO which also contained similar reflectivity, since it is possible by comparison between the un-migrated and migrated WINCH data and the

JEBCO data to discriminate between reflectors in the lower crust and artefacts.

The WINCH data presented here are reproduced as depth converted stacks with superimposed line drawings. The data supplied by JEBCO, hereafter referred to as Lines 1, 2 and 3 are presented as line drawings of the lower crustal reflectivity taken directly from paper sections. The contours in twtt on Fig. 3 are constructed from BIRPS and JEBCO data migrated by hand.

Deep seismic reflection profiles in the East Irish Sea

The data supplied by JEBCO Seismic Limited fill an important gap in knowledge of the deep structure beneath the East Irish Sea (Fig. 1). When combined with the BIRPS data it is possible to provide good constraints on the geometry of the

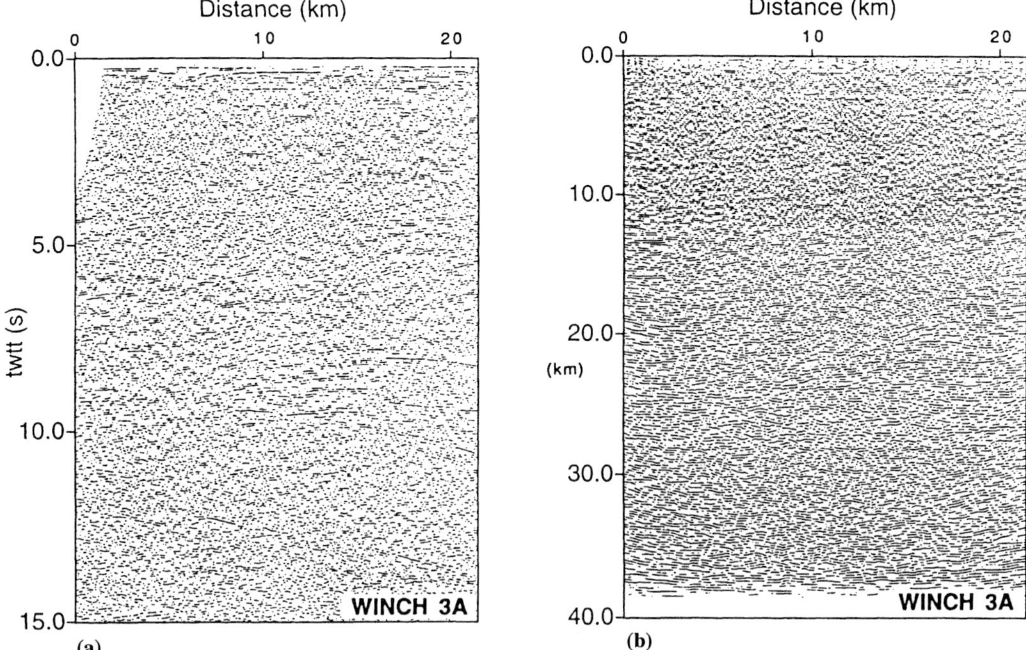

Fig. 2. (a) Unmigrated and (b) migrated and depth converted versions of WINCH 3A for comparison of the effects of processing on the diffractive reflectors in the middle and lower crust.

lower crustal reflectors in the areas where the lines intersect in the region of the Isle of Man massif. There is still a lack of deep reflection data in the southern part of the East Irish Sea and as a consequence it is not possible to accurately constrain the geometry of reflectors between the southern ends of Lines 1 and 2. Interpretations in this area have been made on the basis of similar patterns of reflectivity and the strike of structure established where a number of profiles intersect.

Patterns of middle and lower crustal reflectivity beneath the East Irish Sea

A northward inclined boundary that separates unreflective middle crust to the north from reflective middle crust to the south is visible on WINCH 2 (A–A' on Fig. 4) and has been interpreted as the Iapetus Suture since it occurs at approximately the location predicted by surface geology (Brewer *et al.* 1983; Soper *et al.* 1992a). This boundary can be traced to up to 3 s twtt on WINCH 2 (Fig. 4). No such event can be seen on Line 2N (Fig. 5). However, on Line 1N a northward dipping reflector is present between 2 and 3.5 s twtt (B–B' on Fig. 6). Line 3 shows a generally reflective middle and lower crust, particularly at the

NE end of the profile. The pattern of reflectivity on Line 3 is similar to that seen on WINCH 3 (not reproduced here) which was interpreted by Soper *et al.* (1992a) as a characteristic of the Avalonian crust immediately south of the suture. Similar reflectivity is seen on Line 1N beneath the northward dipping reflector B–B'. On the basis of these observations contours were drawn on the top of this surface and through A–A' and B–B' (Fig. 3) and we regard this as the best control on the location of the Iapetus Suture determined in this area.

South of the northward dipping boundary interpreted as the Iapetus Suture, the middle and lower crust show a varying amount of reflectivity. These variations can be directly correlated with changes in the thickness of the sediments in the late Palaeozoic and Mesozoic basins. Where the sediments are thickest, attenuation has resulted in poor penetration of seismic energy into the lower crust and consequently little reflectivity is seen. This effect is particularly noticeable on Line 3 where the profile crosses from the Peel Basin to the Southern Uplands massif at its NE end (Fig. 7) and on Line 2S (Fig. 8) which shows good lower crustal reflectivity only beneath the Holy Island shelf; it is largely due to the use of a smaller source for shooting the JEBCO data.

                     R. W. ENGLAND & N. J. SOPER

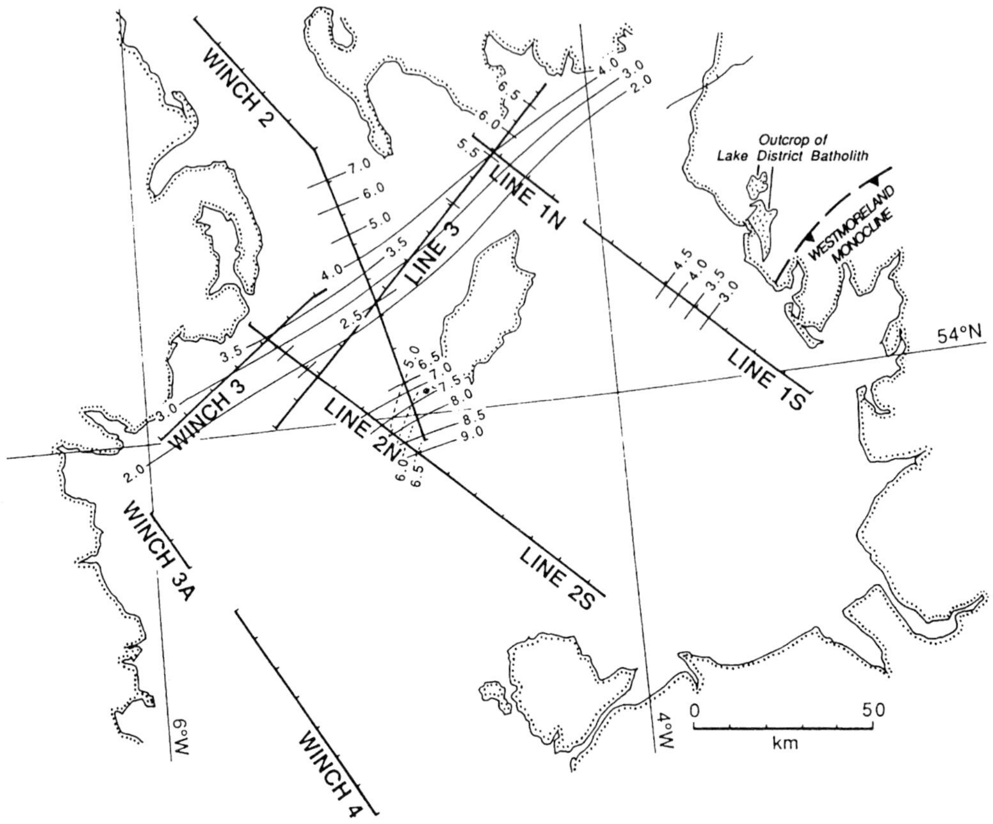

Fig. 3. Map showing contours (in s twtt) drawn on the upper surface of the reflective middle crust and the reflector B–B′ on Line 1N. Additional contours are drawn on reflector C–C′ on Line 1S and on packages of reflections which can be correlated between WINCH 2 and Line 2N. Ticks on line tracks are at 10 km intervals.

The results of the migration of the BIRPS data suggest that much of the diffractive dipping energy seen in the middle and lower crust in both datasets is relatively stable and does not move significantly on migration (Fig. 2). This reflectivity consists of both north and south dipping packages of reflectors. At the southern ends of WINCH 2 and Line 2N (Figs 4 & 5) two southward dipping packages of reflectors can be correlated between the two profiles and have been contoured (Fig. 3). The southern end of Line 2S is dominated by sub-horizontal reflectors beneath the Holy Island shelf (Fig. 8). A similar pattern of reflectivity is seen on WINCH 4 60–70 km south of the north end of the profile (Fig. 9) although a direct correlation is not realistic due to the distance separating the two profiles. A north dipping event beneath the Caernarvon Bay Basin is visible on WINCH 4 (E–E′ on Fig. 9) and was identified by Brewer *et al.* (1983) as the South Irish Sea Lineament; it proves to be a reflection from a fault plane.

Line 1S shows packages of north and south dipping reflectors in the middle and lower crust and two prominent northward dipping reflectors which continue into the upper crust (C–C′ and D–D′ on Fig. 10). Reflector D–D′ is of lower amplitude than C–C′ on the reflection section and can be shown from analysis of two-way travel times to known shallow faults to be a reflection from a fault plane within the Selker Rocks anticline. Reflector C–C′ cannot be related to known shallow faults and is interpreted as a mid-crustal reflector.

No NE–SW profile is available across the Keys Basin so it is not possible to trace reflectors between Lines 1 and 2. Neither is it possible to determine the strike of reflectors in this region independently. However, on the basis of the comparability of the patterns of reflectivity seen on Lines 1 and 2 and WINCH 4 and the general trend of regional gravity and magnetic anomalies (Lee *et al.* 1990), it is thought that the the reflectors imaged on the data are likely to have a general NE–SW

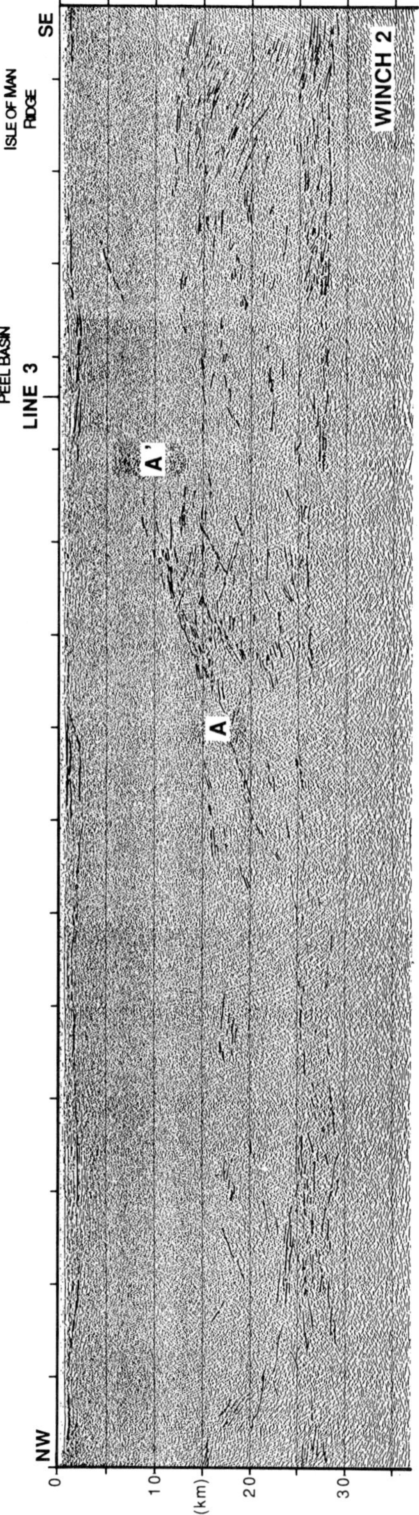

Fig. 4. WINCH 2 line drawing superimposed on migrated and depth converted seismic data originally recorded to 15 s twtt. Ticks on top edge of profile are at 10 km intervals.

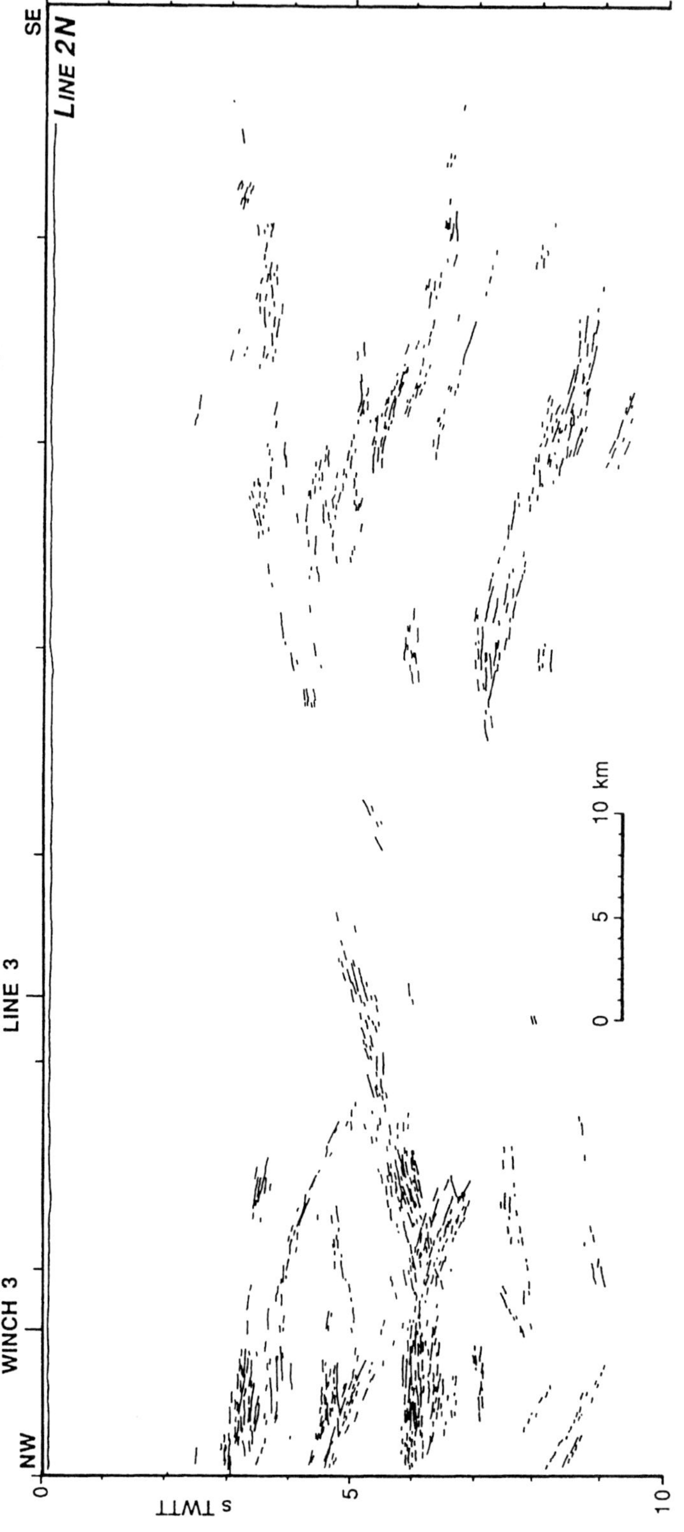

Fig. 5. Line 2N line drawing of unmigrated seismic data recorded to 10 s twtt. Data courtesy of JEBCO Seismic Limited. Ticks on top edge of profile are at 10 km intervals.

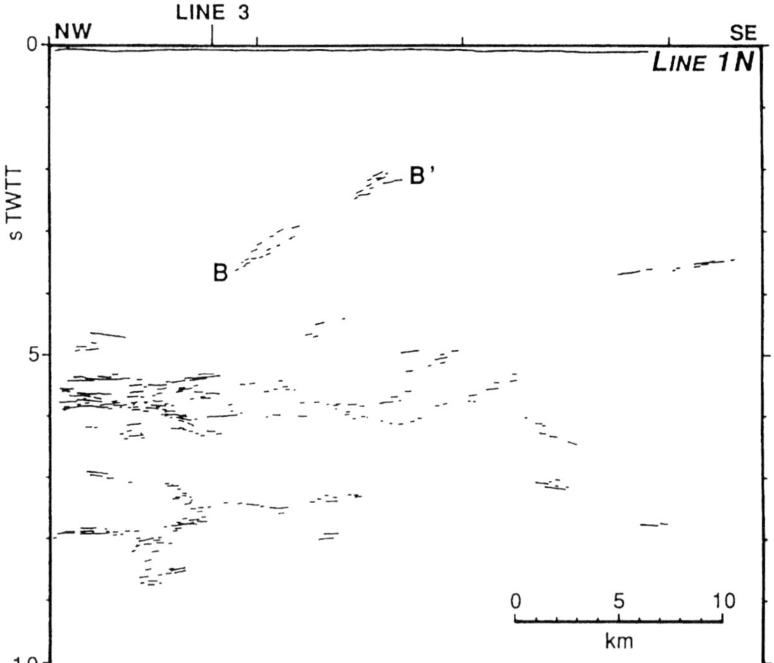

Fig. 6. Line 1N line drawing of unmigrated seismic data recorded to 10 s twtt. Data courtesy of JEBCO Seismic Limited. Ticks on top edge of profile are at 10 km intervals.

strike, similar to that seen beneath the Isle of Man massif and the Solway Firth basin in the north of the region.

Reflections from late Palaeozoic and Mesozoic sediments in the Peel (WINCH 2, Fig. 4) and the Kish Bank, Central Irish Sea, Caernavon Bay and Cardigan Bay Basins (WINCH 4, Fig. 9) are all truncated at the sea floor. In the Central Irish Sea Basin a thin layer of Tertiary sediments overlies the Mesozoic rocks. The Kish Bank Basin is described as having a similar stratigraphy by Jenner (1981) and Naylor *et al.* (1993).

Interpretation

The Iapetus Suture

The contours drawn on the top of the lower and mid-crustal reflectivity and through reflectors A–A′ and B–B′ on Lines 1, 2 and WINCH 2 (Fig. 3) are parallel to the strike of, and broadly coincident with, the Iapetus Suture as defined on faunal grounds, and from previous attempts to map the sub-surface geometry of the suture using geophysical data (Beamish & Smythe 1986; Soper *et al.* 1992*a*), lending support to the existing interpretations. An upward continuation of the

contours projects to the floor of the Peel Basin to the north of the Isle of Man. There is no evidence that the suture coincides with the fault that defines the edge of the basin seen on WINCH 2 (Fig. 4). Contours on the top of the reflective crust on WINCH 3 project to meet the surface at the Slane Fault. This fault lies at the southern edge of the Grangegeeth inlier which is the most southerly terrane containing Laurentian fauna (Owen *et al.* 1992).

Age of reflectivity south of the suture

Soper *et al.* (1992*a*) attributed northward dipping reflectors seen to the south of the Iapetus Suture on the BIRPS' NEC profile and the reflective mid-crust south of the suture on WINCH 2 to Acadian crustal imbrication. The reflectivity seen on Lines 1N, 2N, 3 could be also attributed to Acadian deformation in that it appears to have a NE–SW strike parallel to that of the Acadian structures mapped in the Lake District Massif which are also parallel to trends in onshore and offshore gravity and magnetic data (Lee *et al.* 1990). Alternatively, the reflections may originate from closure of the Iapetus ocean in the Silurian (Soper *et al.* 1992*b*) or

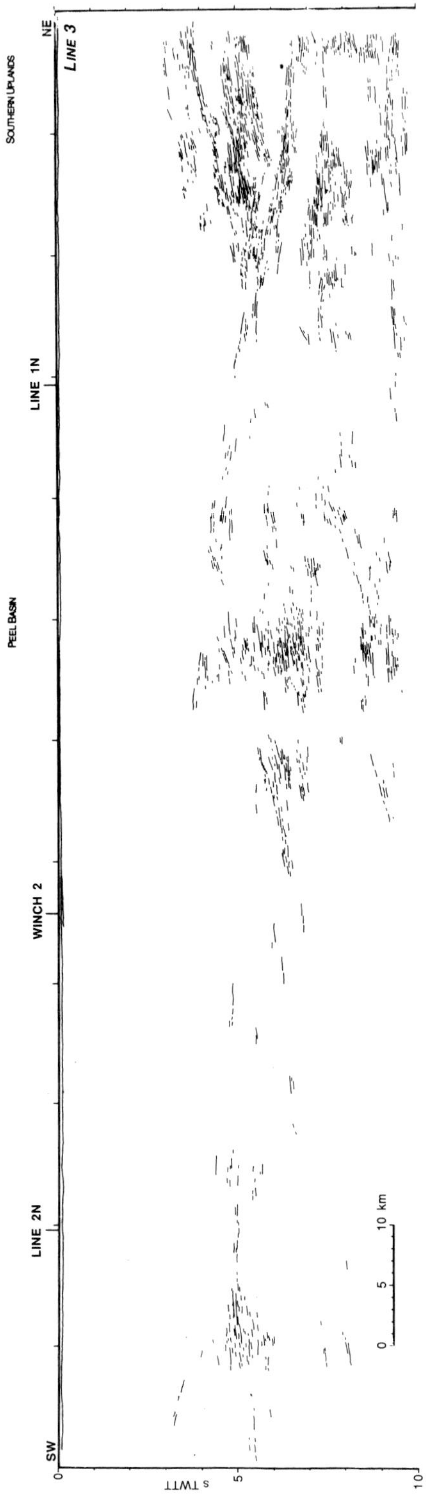

Fig. 7. Line 3 line drawing of unmigrated seismic data recorded to 10 s twtt. Data courtesy of JEBCO Seismic Limited. Ticks on top edge of profile are at 10 km intervals.

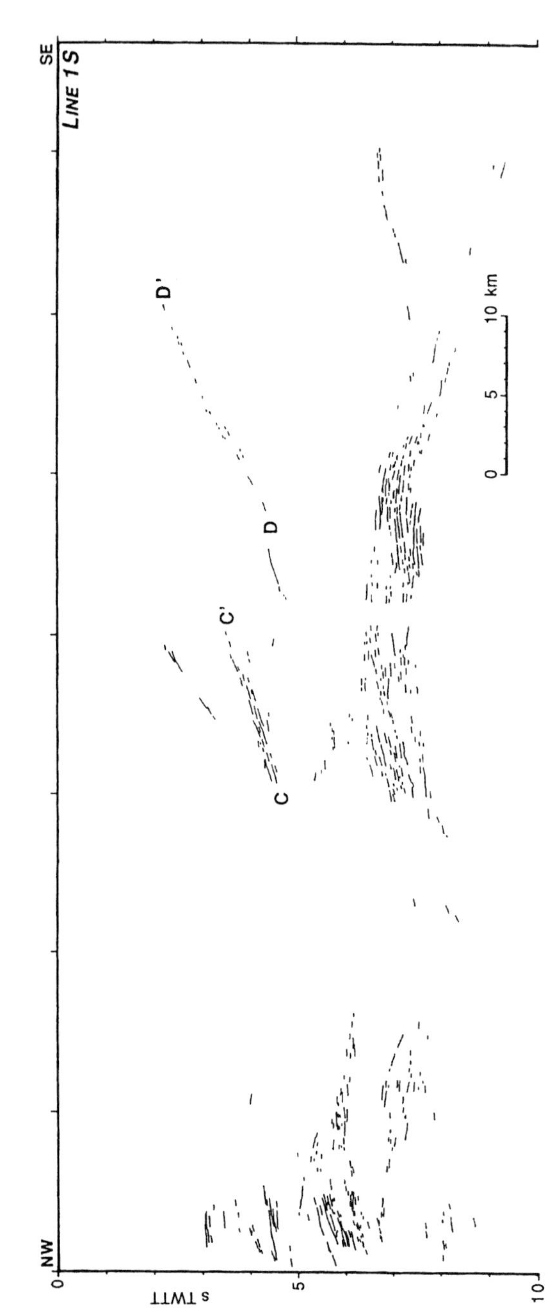

Fig. 8. Line 2S line drawing of unmigrated seismic data recorded to 10 s twtt. Data courtesy of JEBCO Seismic Limited. Ticks on top edge of profile are at 10 km intervals.

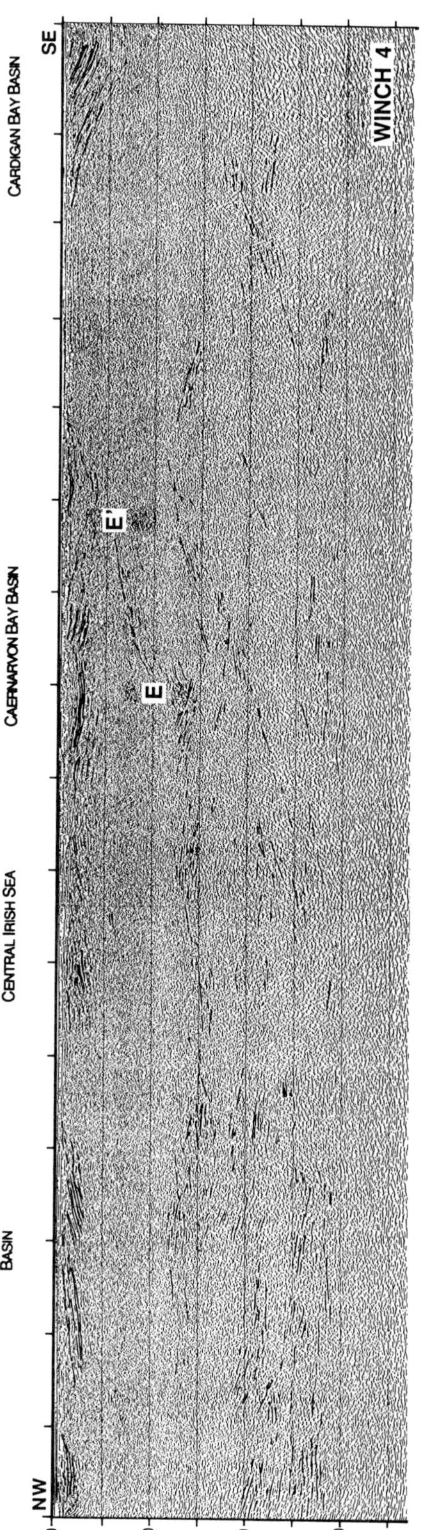

Fig. 9. WINCH 4 line drawing superimposed on migrated and depth converted seismic data originally recorded to 15 s twtt. Ticks on top edge of profile are at 10 km intervals.

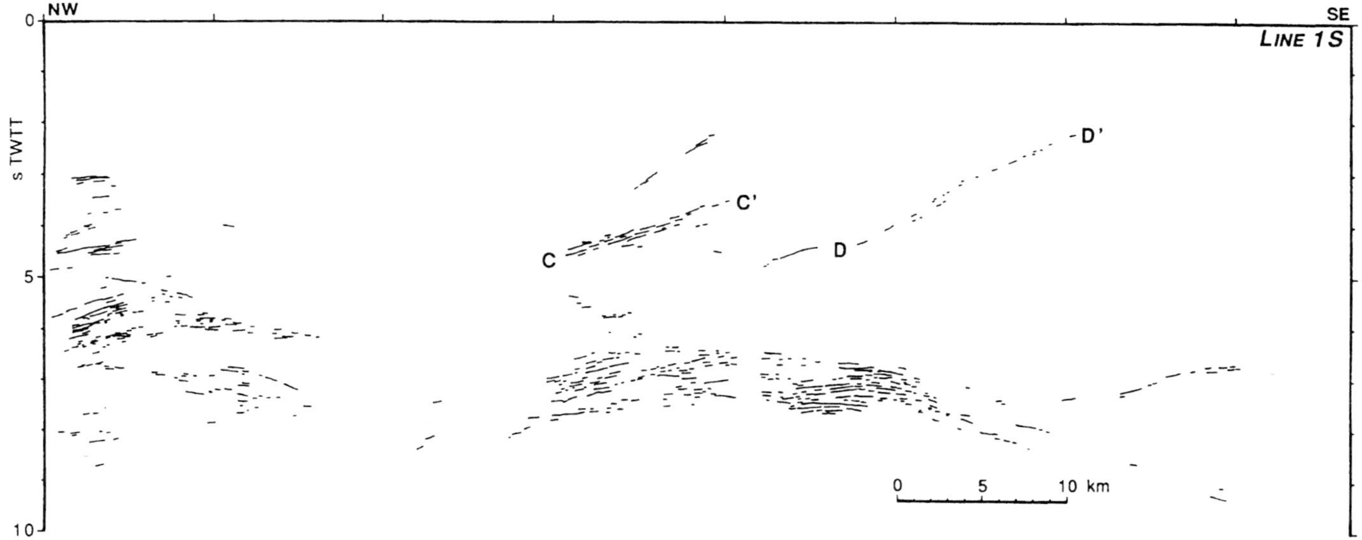

Fig. 10. Line 1S line drawing of unmigrated seismic data recorded to 10 s twtt. Data courtesy of JEBCO Seismic Limited. Ticks on top edge of profile are at 10 km intervals.

they could be even earlier, perhaps Pan-African which would explain the truncation of mid-crustal reflectivity at the suture. However, a younger (Acadian) age is preferred on the basis that later deformation would overprint earlier structures and would produce the more laterally coherent reflections observed on the profiles.

The extensive lower Tertiary unconformity, which is seen as a distinct truncation of the Mesozoic sediments at the sea floor across the Kish Bank, Central Irish Sea and Cardigan Bay Basins on WINCH 4 (Fig. 9) has been attributed by Brodie & White (1994) and Cope (1994) to erosion of Mesozoic sediments across northwestern Britain during the early Tertiary as a result of uplift following underplating of the crust with basaltic magma produced by melting within a mantle plume. However, lower crustal reflectivity patterns on WINCH 4 do not resemble that of the 'layered lower crust' commonly associated with the intrusion or underplating of basaltic magma close to the Moho (Warner 1990; Singh & McKenzie 1993). Where present, lower crustal reflectors are of too low an amplitude and are not flat and continuous over distances of 5 to 10 km. Also, there is no reason why such underplating should terminate at the Iapetus Suture if it is responsible for a regional unconformity. Chadwick *et al.* (1993) described erosion of Permo-Triassic and younger rocks as a result of Tertiary contractional reactivation of the Maryport fault caused by Alpine compression. This weak compressional event, and earlier Variscan inversion, may result in reinforcement of reflectivity patterns from reactivated structures but it is unlikely to have formed substantial new structures. Consequently NE–SW striking middle and lower crustal reflectivity seen on the profiles presented here is considered to largely originate from structures formed during the Acadian orogeny and not from later tectonic or magmatic events.

Acadian structure beneath the East Irish Sea

The opposing northward and southward dip of the packages of reflectivity beneath the East Irish Sea suggest that the crust last deformed in a NW–SE direction by shortening or extension on conjugate structures. If the structures developed as a result of the Acadian compression, the reflectors must represent thrust faults. Lee (1989), on the basis of potential field data, proposed that the Lake District batholith had been displaced southwards over magnetic basement on a north dipping crustal-scale thrust. Kneller & Bell (1993) associated this putative thrust with the south vergent Westmoreland monocline, which is responsible for a major change of structural level in the southern

Lake District. The dipping reflector C–C′ on Line 1S (Figs 3 & 8) projects upwards and along-strike to the northeast towards the core of the monocline, and may thus represent the offshore continuation of the controlling thrust fault.

The relationship between lower and middle crustal structure and basin development in the East Irish Sea

Gibbs (1987) suggested that the Solway and Peel basins may have developed in the hangingwall of the Iapetus Suture following extensional reactivation. However, as previously noted by Beamish & Smythe (1986), there is no evidence that the suture coincides with any of the faults that bound these basins. On WINCH 2 the suture appears to project to the north of the fault at the southern edge of the Peel Basin. Similarly, upward continuation of the contours drawn on the Iapetus Suture beneath the Solway Firth Basin project to the north of the Maryport fault (Chadwick *et al.* 1993) (Fig. 3), so it cannot root into the suture. However, the South Irish Sea Lineament (E–E′ on Fig. 9), which appears to control deposition in the Caernarvon Bay Basin, may be a reactivated Acadian structure, rather than one of Silurian age, related to the suture, as proposed by Brewer *et al.* (1983).

Rather than a direct control by fault reactivation the geometry of Carboniferous depocentres appears to have been broadly related to the distribution of buoyant blocks cored by Caledonian granites. The Ramsay–Whitehaven Ridge–Isle of Man Massif remained a resistant barrier between the Solway Firth and East Irish Sea Basins throughout late Palaeozoic and Mesozoic deposition, with each basin retaining a distinct depositional facies (Jackson & Mulholland 1993). In the East Irish Sea Basin, Carboniferous deposition occurred along the NE–SW axis of the Quad 109 synform (Fig. 1). This pattern only broke down during the late Carboniferous (Westphalian) when updoming along the NE–SW axis of the Quad 109 antiform (as a result of Variscan inversion?) was followed by development of the N–S striking Keys fault and adjacent basin (Jackson & Mulholland 1993; BGS 1994). Jackson & Mulholland (1993) argued that the Quad 109 antiform forms an important barrier between the northern and southern parts of the East Irish Sea Basin, separating fault systems of different trends, NNW–SSE to the north and N–S to the south, and dividing the region into two sub-basins (Knipe *et al.* 1993). On this basis it is thought that the Quad 109 antiform continues beneath the East Irish Sea Basin and across the Lake District Boundary fault where it crops out as the Furness inlier, south of the Westmoreland monocline (Fig. 1) (Jackson & Mulholland 1993). However, there is no evidence on Lines 1 and 2 or WINCH 4 for any major structures rooting in the core of the antiform which could have been related to Variscan inversion.

Permian and Mesozoic extension occurred in a NE–SW to ENE–WSW direction orthogonal to the Carboniferous basins. Only the Ramsay–Whitehaven and Isle of Man Ridge remained a positive Caledonian feature. The major NNW–SSE and N–S trending faults developed during rifting (e.g. the Keys Fault) bear no relationship to the NE–SW trends of the Caledonian and Acadian structures imaged on the deep seismic profiles. An E–W deep profile across the south of the basin, joining Lines 1 and 2 would address the issue of whether these structures rooted into deeper features of the basement on the edge of Avalonia. Onshore, field evidence in the southern Lake District demonstrates that Acadian deformation is partitioned by NNE–SSW and N–S trending faults which were reactivated during the Permian extensional phase (Soper 1993) (Fig. 1), but the NNW–SSE structures must be juvenile.

Conclusions

The deep structure of the East Irish Sea Basin, as interpreted from deep seismic reflection data, is dominated in the north by the Iapetus Suture (Silurian) and south of the suture by Acadian contractional structures (Early Devonian).

The absence in the seismic data of obvious structures in the middle crust beneath the East Irish Sea Basin does not support arguments in favour of the control of Carboniferous deposition south of the Ramsay–Whitehaven Ridge by Caledonian or Avalonian structures, or Variscan inversion of these basins. The South Irish Sea Lineament bounding the southern edge of the Caernavon Bay Basin is interpreted as an Acadian contractional structure reactivated as an extensional fault during the Carboniferous. The faults controlling sedimentation in Permian and Mesozoic basins are of two sets. Those with N–S trends may be reactivated Acadian or older, while the NNW–SSE set that are orthogonal to the Caledonian structures are likely to be juvenile, and must have initiated during the late Westphalian.

The lower crust in this region is not as strongly reflective as those areas which are postulated to have been underplated with basaltic magma. Consequently, models attempting to explain the early Tertiary regional uplift and erosion of basins as a result of igneous underplating within the lower crust may not be supported by observation.

RWE gratefully acknowledges permission to use and publish the data provided by JEBCO Seismic Limited. Richard Hobbs provided invaluable assistance in processing of the WINCH data. The authors are grateful for the comments of Neil Meadows, Paul Ryan and an anonymous reviewer, which resulted in improvements in the manuscript. BIRPS is funded by the Natural Environment Research Council and the BIRPS Industrial Associates (Amerada Hess Ltd, BP Exploration Operating Co. Ltd, British Gas Exploration and Production Ltd, Chevron (UK) Ltd, Conoco (UK) Ltd, Lasmo North Sea plc, Mobil North Sea Ltd, Shell (UK) Ltd, Statoil (UK) Ltd). Cambridge Earth Science Contribution number 4541.

References

BEAMISH, D. & SMYTHE, D. B. 1986. Geophysical images of the deep crust: the Iapetus suture. *Journal of the Geological Society, London,* **143**, 489–497.

BREWER, J. A., MATTHEWS, D. H., WARNER, M. R., HALL. J., SMYTHE, D. K. & WHITTINGTON, R. J. 1983. BIRPS deep seismic reflection studies of the British Caledonides – the WINCH profile. *Nature,* **305**, 206–210.

BRITISH GEOLOGICAL SURVEY. 1994. *East Irish Sea.* 1:250 000 Special Edition Sheet, HMSO.

BRODIE, J. & WHITE, N. 1994. Sedimentary basin inversion caused by igneous underplating: Northwest European continental Shelf. *Geology,* **22**, 147–150.

CHADWICK, R. A & HOLLIDAY, D. W. 1991. Deep crustal structure and Carboniferous basin development within the Iapetus convergence zone, northern England. *Journal of the Geological Society, London,* **148**, 41–53.

——, EVANS, D. J. & HOLLIDAY, D. W. 1993. The Maryport fault: the post-Caledonian tectonic history of southern Britain in microcosm. *Journal of the Geological Society, London,* **150**, 247–250.

COPE, J. C. W. 1994. A latest Cretaceous hotspot and the southeasterly tilt of Britain. *Journal of the Geological Society, London,* **151**, 905–908.

FREEMAN, B., KLEMPERER, S. L. & HOBBS, R. W. 1988. The deep structure of northern England and the Iapetus Suture zone from BIRPS deep seismic reflection profiles. *Journal of the Geological Society, London,* **145**, 727–740.

GIBBS, A. 1987. Development of extension and mixed mode sedimentary basins. *In:* COWARD, M. P., DEWEY, J. F. & HANCOCK, P. L. (eds) *Continental Extensional Tectonics.* Geological Society, London, Special Publication, **28**, 19–33.

JACKSON, D. I. & MULHOLLAND, P. 1993. Tectonic and stratigraphic aspects of the East Irish Sea Basin and adjacent areas: contrasts in their post-Carboniferous structural styles. *In:* PARKER, J. R. (ed.) *Petroleum*

Geology of Northwest Europe: Proceedings of the 4th conference.. Geological Society, London, 791–808.

JENNER, J. K. 1981. The structure and stratigraphy of the Kish Bank Basin. *In*: ILLING, L. V. & HOBSON, G. D. (eds) *Petroleum Geology of the Continental Shelf of North West Europe.* Heyden, London, 426–431.

KLEMPERER, S. L. & HOBBS, R. W. 1992. *The BIRPS Atlas: Deep seismic reflection profiles around the British Isles.* Cambridge University Press.

KNELLER, B. C. & BELL, A. M. 1993. An Acadian mountain front in the English Lake District: the Westmoreland monocline. *Geological Magazine,* **130,** 203–213.

KNIPE, R., COWAN, G. & BALENDRAN, V. S. 1993. The tectonic history of the East Irish Sea Basin with reference to the Morecambe fields. *In*: PARKER, J. R. (ed.) *Petroleum Geology of Northwest Europe: Proceedings of the 4th Conference.* Geological Society, London, 857–866.

LEE, M. K. 1989. *Upper crustal structure of the Lake District from modelling and image processing of potential field data,* BGS Technical Report, **WK/89/1.**

——, PHARAOH, T. C. & SOPER, N. J. 1990. Structural trends in central Britain from images of gravity and aeromagnetic fields. *Journal of the Geological Society, London,* **147,** 241–258.

LEWIS, A. H. J. 1986. *The deep seismic structure of northern England and adjacent marine areas from the Caledonian Suture Seismic Project.* PhD thesis, University of Durham.

NAYLOR, D., HAUGHEY, N., CLAYTON, G. & GRAHAM, J. R. 1993. The Kish Bank Basin, offshore Ireland. *In*:

PARKER, J. R. (ed.) *Petroleum Geology of Northwest Europe: Proceedings of the 4th Conference.* Geological Society, London, 845–855.

OWEN, A. W., HARPER, D. A. T. & ROMANO, M. 1992. The Ordovician biogeography of the Grangegeeth terrace and the Iapetus suture zone in eastern Ireland. *Journal of the Geological Society, London,* **149,** 3–6.

SINGH, S. C. & MCKENZIE, D. 1993. Layering in the lower crust. *Geophysical Journal International,* **113,** 622–628.

SOPER, N. J. 1993. *The Silurian rocks of 1:25,000 sheet SD 38, Low Furness, Cumbria.* British Geological Survey Technical Report, Onshore Geology Series, **WA/93/16.**

——, ENGLAND, R. W., SNYDER, D. B. & RYAN, P. D. 1992*a.* The Iapetus suture zone in England, Scotland and eastern Ireland: a reconciliation of geological and deep seismic data. *Journal of the Geological Society, London,* **149,** 697–700.

——, STRACHAN, R. A., HOLDSWORTH, R. E., GAYER, R. A., & GREILING, R. O. 1992*b.* Sinistral transpression and the Silurian closure of Iapetus: *Journal of the Geological Society, London,* **149,** 871–880.

——, WEBB, B. C. & WOODCOCK, N. H. 1987. Late Caledonian (Acadian) transpression in north-west England: timing, geometry and geotectonic significance. *Proceedings of the Yorkshire Geological Society,* **46,** 175–192.

STOLT, R. H. 1978. Migration by Fourier transform. *Geophysics,* **43,** 23–48.

WARNER, M. R. 1990. Basalts, water or shear zones in the lower continental crust? *Tectonophysics,* **173,** 163–174.

Variation in thermal history styles around the Irish Sea and adjacent areas: implications for hydrocarbon occurrence and tectonic evolution

PAUL F. GREEN[1], IAN R. DUDDY[1] & RICHARD J. BRAY[2]

[1] Geotrack International Pty Ltd, 37 Melville Road, Brunswick West, Victoria 3055, Australia

[2] Geotrack International UK Office, Unit 14, Crewkerne Business Park, South Street, Crewkerne, Somerset TA18 7HJ, UK

Abstract: Thermal history is a major factor in determining hydrocarbon prospectivity in the Irish Sea region. Early application of Apatite Fission Track Analysis (AFTA®) to exploration wells revealed early Tertiary palaeotemperatures around 110°C or more. More recent studies in the Irish Sea and adjacent areas, using a combination of AFTA and vitrinite reflectance (VR), have revealed at least three additional palaeo-thermal episodes: pre-Permian (>290 Ma), late Permian to mid-Triassic (260 to 220 Ma) and Early Cretaceous (140 to 110 Ma). Other evidence of early paleo-thermal effects, for which timing is only constrained to the interval 300 to 150 Ma, may reflect these or additional episodes.

In some areas, early effects involved extremely high palaeogeothermal gradients. Early Cretaceous gradients are unconstrained due to a lack of wells in which palaeotemperatures reflect this episode. Early Tertiary palaeotemperatures derived from AFTA and VR in wells from the southern part of the East Irish Sea Basin define a consistent linear depth profile with a gradient of only 10°C km^{-1}. Wells from the north of the region also define a broadly consistent profile but with a much higher gradient. The low gradients are thought to reflect flow of hot fluids while the high gradients to the north may be associated with igneous activity in that region.

Our preferred interpretation of the early Tertiary palaeo-thermal effects involves a combination of burial by around 1.5 to 2 km of post-Triassic sediment and the thermal effects of hot fluid flow, probably from north to south and associated with igneous activity. This interpretation is consistent with geological evidence and is supported by structural data, sonic velocity data and geochemical observations.

Different combinations of these episodes lead to a wide variety of thermal history styles across the region. The southern part of the East Irish Sea Basin, containing major accumulations of oil and gas, is characterized by the most suitable thermal history conditions for generation and preservation of hydrocarbons to the present day. In adjacent regions, thermal histories are more complex, and those identified to date are generally less suitable particularly for preservation of hydrocarbons to the present day. Future exploration in these areas will require a thorough understanding of the thermal history of Carboniferous source rocks in order to define the most prospective regions.

The East Irish Sea Basin is the most prolific offshore hydrocarbon province in the UK outside the North Sea. Huge gas reserves exist in the Morecambe Field and recently several commercial oil discoveries have also been made in the basin. Thermal history is a major factor in determining hydrocarbon prospectivity in the Irish Sea region. Early application of Apatite Fission Track Analysis (AFTA®) to Triassic and Permian samples from exploration wells in the Irish Sea revealed a major episode of Tertiary cooling, from early Tertiary palaeotemperatures around 110°C or higher in samples from shallow depths. These results suggested that Carboniferous source rocks at deeper levels were within the gas generation window immediately prior to the onset of cooling

in the early Tertiary. They also implied kilometre-scale Tertiary erosion in the Irish Sea region, and confirmed the regional aspect of this episode, previously recognized onshore in northern England.

More recent studies in the Irish Sea and adjacent areas, using a combination of AFTA from the Permo-Triassic and Carboniferous sections with vitrinite reflectance (VR) data from Carboniferous samples, have revealed a number of additional palaeo-thermal episodes, and a complex variation in thermal history styles for Carboniferous source rocks across the region which is crucial to the occurrence of hydrocarbons. In this paper we review evidence for different styles of thermal history undergone by potential source rocks within

From Meadows, N. S., Trueblood, S. P., Hardman, M. & Cowan, G. (eds), 1997, *Petroleum Geology of the Irish Sea and Adjacent Areas,* Geological Society Special Publication No. 124, pp. 73–93.

the Carboniferous section. We then discuss constraints on palaeogeothermal gradients during individual heating episodes, and consider the origin of the palaeo-thermal episodes where possible. Finally we provide a general guide to regional thermal history trends and their implications for hydrocarbon exploration.

Hydrocarbon generation in the Irish Sea: the importance of thermal history reconstruction

A number of papers (Bushell 1986; Woodward & Curtis 1987; Macchi *et al.* 1990; Stuart & Cowan, 1991) have contributed to a model for hydro-carbon generation in the Morecambe Field, based primarily on the study of diagenetic patterns within the reservoir section. This model suggests that hydrocarbon generation from Carboniferous source rocks probably began in the early Jurassic, with hydrocarbons migrating into a low-relief structure formed by early Cimmerian movements (late Triassic to early Jurassic). Hydrocarbon generation continued through the Jurassic until late Cimmerian (late Jurassic to early Cretaceous) uplift and erosion at which time the structure was breached and the hydrocarbons were lost. Renewed burial in the Cretaceous then led to further generation of gas, which accumulated in structures resealed by salt movement in the overlying Mercia Mudstone Group. Gas generation was terminated by uplift and

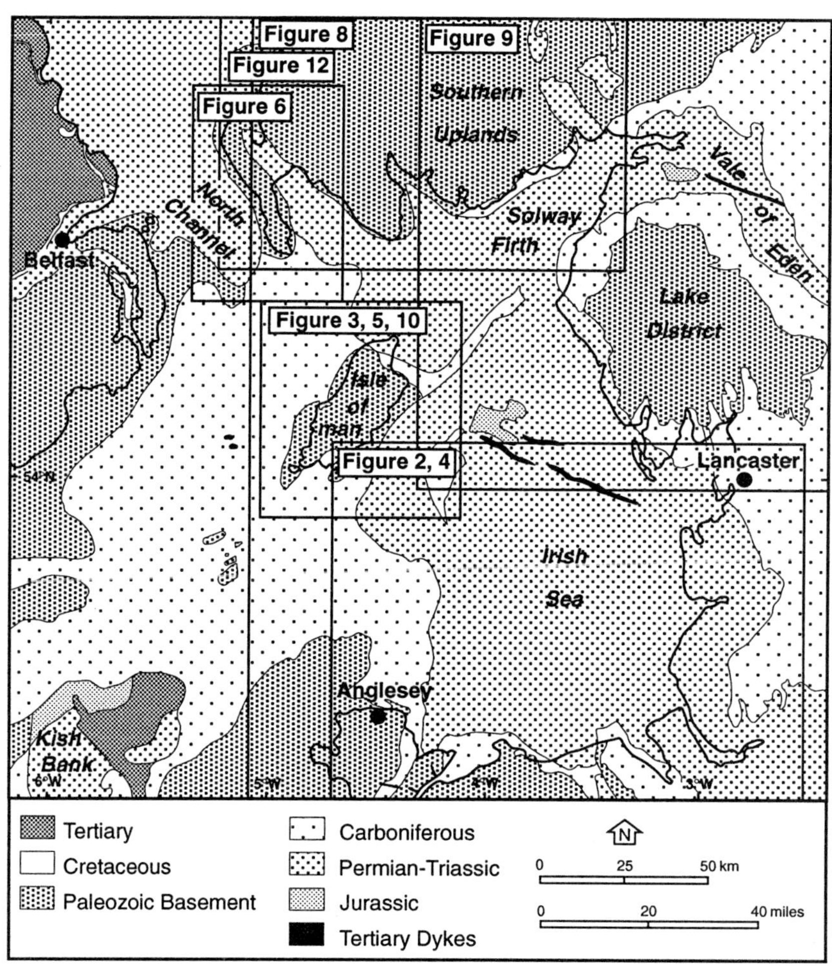

Fig. 1. Solid geology (simplified) of the East Irish Sea Basin and adjacent areas, as discussed in this paper, showing areas covered by later figures.

erosion in the early Tertiary, which led to the present configuration and caused some expansion of the accumulated gas due to the pressure reduction caused by uplift.

Carboniferous strata with good source potential, similar to those thought to have sourced the gas in Morecambe Bay, together with Permo-Triassic sands with reservoir potential, are all anticipated in areas surrounding the East Irish Sea Basin (Fraser *et al.* 1990; Hardman *et al.* 1993). In understanding and predicting regional trends in hydrocarbon generation and preservation, and in testing and extending the above model over the surrounding region, two critical aspects are the degree of maturation of the Carboniferous source rocks and the timing of peak maturity development in relation to formation of potential trapping structures. The sedimentary basins of the Irish Sea and adjacent areas (Jackson *et al.* 1987; Jackson & Mulholland 1993) are characterized by a conspicuous lack of sediments younger than Early Jurassic over most of the region (Fig. 1) and, following the model outlined above, all important events occurred subsequent to deposition of the youngest preserved sediment. Therefore structural and stratigraphic evidence from the preserved section allows only partial insight into the evolution of the region.

In addition, over much of the Irish Sea region the presence of thick Permian and Triassic red beds precludes assessment of thermal history based on organic petrology or geochemistry, which are usually applicable only in the Carboniferous section, only a small thickness of which is usually penetrated in hydrocarbon exploration wells within the Irish Sea. In the past, this has caused particular difficulty in obtaining constraints on palaeo-geothermal gradients, and in developing regional thermal history models.

In such situations, thermal history reconstruction using integrated AFTA and VR (e.g. Bray *et al.* 1992; Green *et al.* 1995) allows more complete reconstruction of the thermal and tectonic history of the region. Sands within the red bed sequences contain abundant detrital apatite grains and AFTA data from these allow palaeotemperature assessment throughout the section. Integration of AFTA and VR data can establish unique thermal history constraints, including definition of major episodes of heating and cooling, resolution of multiple heating episodes, estimation of palaeotemperatures during each episode, independent estimation of the timing of cooling, estimation of palaeogeothermal gradients, assessment of the nature of each episode and estimation of amounts of section removed by erosion, where appropriate. This approach is particularly useful in the Irish Sea region, as explained in the following.

Previous thermal history studies of the Irish Sea and adjacent areas

Using AFTA, Green (1986) showed that samples of outcropping Caledonian basement from the Lake District block reached palaeotemperatures between *c.* 70° and *c.* 110°C (or higher in places) prior to cooling which began at around 60 Ma. As this study involved only outcrop samples, no constraints were possible on Early Tertiary palaeo-geothermal gradients. Green (1986) concluded that although the cause of the early Tertiary palaeo-thermal effects identified from AFTA was not clear, any likely explanation must require kilometre-scale Tertiary uplift and erosion.

Green (1989) subsequently applied AFTA to outcrop samples from the onshore East Midlands Shelf (EMS) and the Pennines, and samples of Jurassic to Carboniferous age from five wells from the onshore EMS. This study also showed consistent evidence of a period of elevated palaeo-temperatures from which cooling began in the early Tertiary. Bray *et al.* (1992) reported that VR data from the five wells indicated palaeotemperatures consistent with those derived from AFTA. Analysis of the palaeotemperature profiles derived from AFTA and VR suggests palaeogeothermal gradients between 28 and 41°C km^{-1} in the five wells (close to present gradients between 26 and 41°C), implying removal of around 1 to 2 km of section by Tertiary uplift and erosion (Bray *et al.* 1992).

Lewis *et al.* (1992) reported that samples from throughout northwest England and also from exploration wells in the Irish Sea showed the effects of elevated palaeotemperatures prior to early Tertiary cooling, although none of the samples analysed provided any constraints on palaeogeothermal gradients. Lewis *et al.* (1992) suggested that early Tertiary palaeogeothermal gradients were probably close to present values and that cooling to present temperatures involved removal of up to 3 km of section. This conclusion was based on the regional nature of the palaeo-thermal effects, the observation of early Tertiary palaeogeothermal gradients close to present values in the onshore EMS wells, VR data from Irish Sea wells suggesting that palaeogeothermal gradients were close to present values (Roberts 1989), studies of shale sonic velocity studies (Colter 1978) which support estimates of 2 to 3 km of eroded cover from the East Irish Sea Basin, and the lack of any positive evidence to suggest that early Tertiary gradients were significantly elevated

Whereas these previous studies have discussed only early Tertiary cooling, more recent studies have revealed a variety of heating and cooling episodes ranging in time from end-Carboniferous to early Tertiary. Evidence for these palaeo-thermal

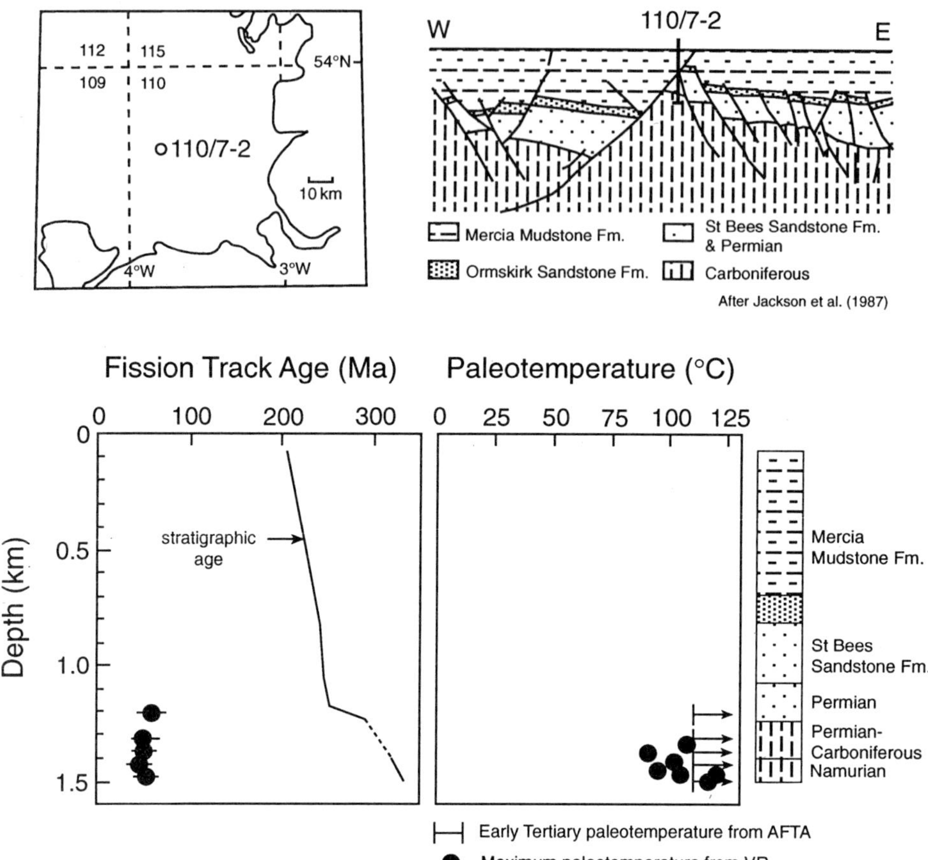

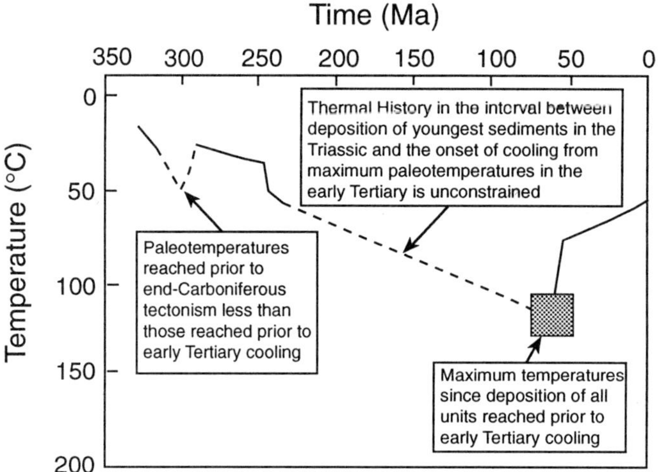

Fig. 2. Thermal history reconstruction in Irish Sea well 110/7-2 based on AFTA and VR data suggests a thermal history in which samples cooled from palaeotemperatures around 110°C in the early Tertiary (*c.* 60 Ma). The cross-section at top right is taken from Jackson *et al.* (1987).

episodes is reviewed in the following section. All palaeotemperature estimates refer to heating rates around 1°C Ma^{-1}.

Styles of thermal history in the Irish sea and adjacent areas

Style A: Maximum palaeotemperatures immediately prior to early Tertiary cooling

AFTA data across much of the Lake District block and northwest England are dominated by early Tertiary cooling from palaeotemperatures generally around 100°C or higher (Green 1986; Lewis *et al.* 1992; Green *et al.* 1993*a,b*). Samples of palaeozoic granites from Carrock Fell and the Skiddaw region all give results which are characteristic of samples which cooled rapidly below *c.* 110°C between 60 and 70 Ma. Other samples of Palaeozoic basement from the Lake District, as well as Carboniferous sediments from the Pennine Block and Permian and Triassic sediments from the Vale of Eden and the Fylde region, cooled from slightly lower palaeo-temperatures (mostly between 80 and 100°C) at this time.

Early Tertiary cooling is also revealed by AFTA data from exploration wells in the East Irish Sea Basin. Data from well 110/7-2 (Fig. 2) are typical of the region. Fission track ages are generally between 40 and 50 Ma in samples of Permian to Carboniferous stratigraphic age, and rigorous interpretation of these data shows that the samples cooled from palaeotemperatures around 110°C or higher at *c.* 60 Ma (early Tertiary). Track length distributions in these samples further suggest that, by 50 Ma, the samples had cooled to temperatures less than 80 to 100°C since which time slow cooling has brought the samples to their present temperatures. This two-stage cooling history is typical of data from the region, as also discussed by Lewis *et al.* (1992).

VR data from the Carboniferous section in this well suggest maximum palaeotemperatures around 90 to 120°C as shown in Fig. 2. The consistency between these palaeotemperatures and those estimated from AFTA shows that these were the maximum palaeotemperatures reached by the Carboniferous section at any time since deposition. The resulting thermal history reconstruction for the Carboniferous section in this well is illustrated in Fig. 2. Note that the history prior to the onset of cooling in the early Tertiary is not constrained, except that palaeotemperatures were less than the maximum values reached immediately prior to the onset of cooling.

Style B: Maximum palaeotemperatures immediately prior to early Cretaceous cooling

AFTA data in a Carboniferous sample from Shallow Borehole A on the Isle of Man (Fig. 3) suggest a maximum palaeotemperature between 105 and 115°C prior to cooling which began between 140 and 100 Ma (early Cretaceous). Two VR determinations from the shales in this borehole indicate maximum palaeotemperatures close to 105°C. The consistency between palaeotemper-atures derived from AFTA and VR show that both systems reflect the same event, and reveal an episode of early Cretaceous cooling. AFTA data from various units across the Isle of Man and in nearby offshore boreholes show that this episode affected a wide area.

The resulting thermal history reconstruction for the Carboniferous section in this borehole is illustrated in Fig. 3. The history prior to the onset of cooling in the early Cretaceous is not constrained, except that palaeotemperatures were less than the maximum values reached immediately prior to the onset of cooling. The detail of the track length distribution in this sample suggests that the palaeotemperature at *c.* 60 Ma in this sample was between 50 and 80°C. This may reflect a discrete episode of early Tertiary heating and cooling, or may alternatively reflect an intermediate stage on a more or less continuous cooling path.

Style C: Maximum palaeotemperatures in an early event and a lower peak prior to early Tertiary cooling

AFTA data in a Namurian sandstone from Onshore Well X, drilled on the Carboniferous outcrop on the eastern flank of the East Irish Sea Basin (Fig. 4), are dominated by the effects of early Tertiary cooling from a palaeotemperature of *c.* 100°C, consistent with published evidence for regional early Tertiary cooling in northwest England (Green, 1986; Lewis *et al.* 1992; Green *et al.* 1993*a,b*). VR data from similar depths in this well indicate maximum palaeotemperatures around 30°C greater than those derived from AFTA, showing that these palaeotemperatures must have been reached in an earlier event. AFTA data in this sample allow only a broad constraint on the time of cooling in this earlier event, to between 300 and 150 Ma. Possible origins for this earlier event are discussed later. Deeper AFTA samples were totally annealed (palaeotemperatures $\geq$ 110°C) prior to early Tertiary cooling and preserve no evidence of the earlier episode.

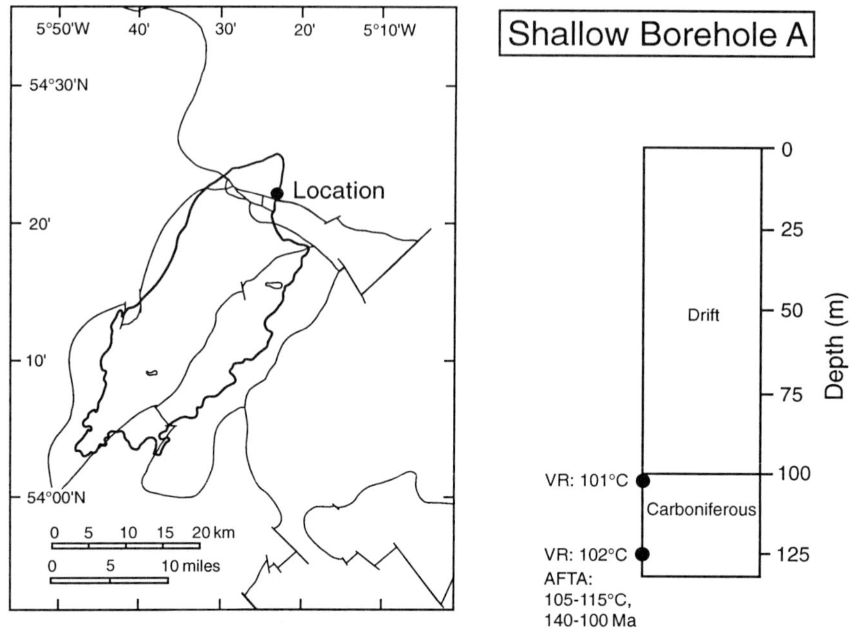

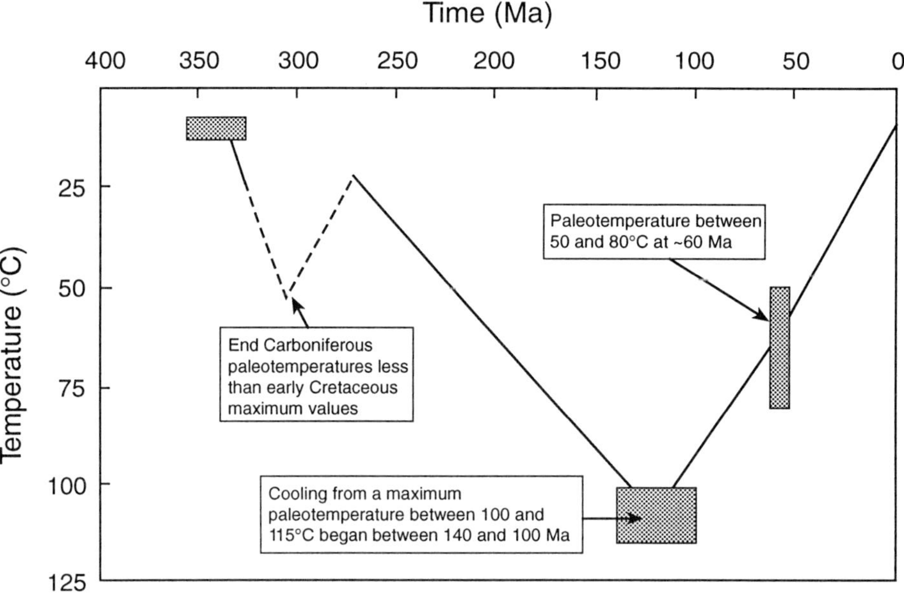

Fig. 3. AFTA and VR data from Shallow Borehole A, Isle of Man, show clear evidence of cooling from maximum palaeotemperatures of 100 to 115°C at some time between 140 and 100 Ma (early Cretaceous). AFTA data also suggest a palaeotemperature between 50 and 80°C at *c.* 60 Ma. Whether this reflects a subsequent episode of heating and cooling, or simply an intermediate step in a protracted cooling history (as shown), is not clear.

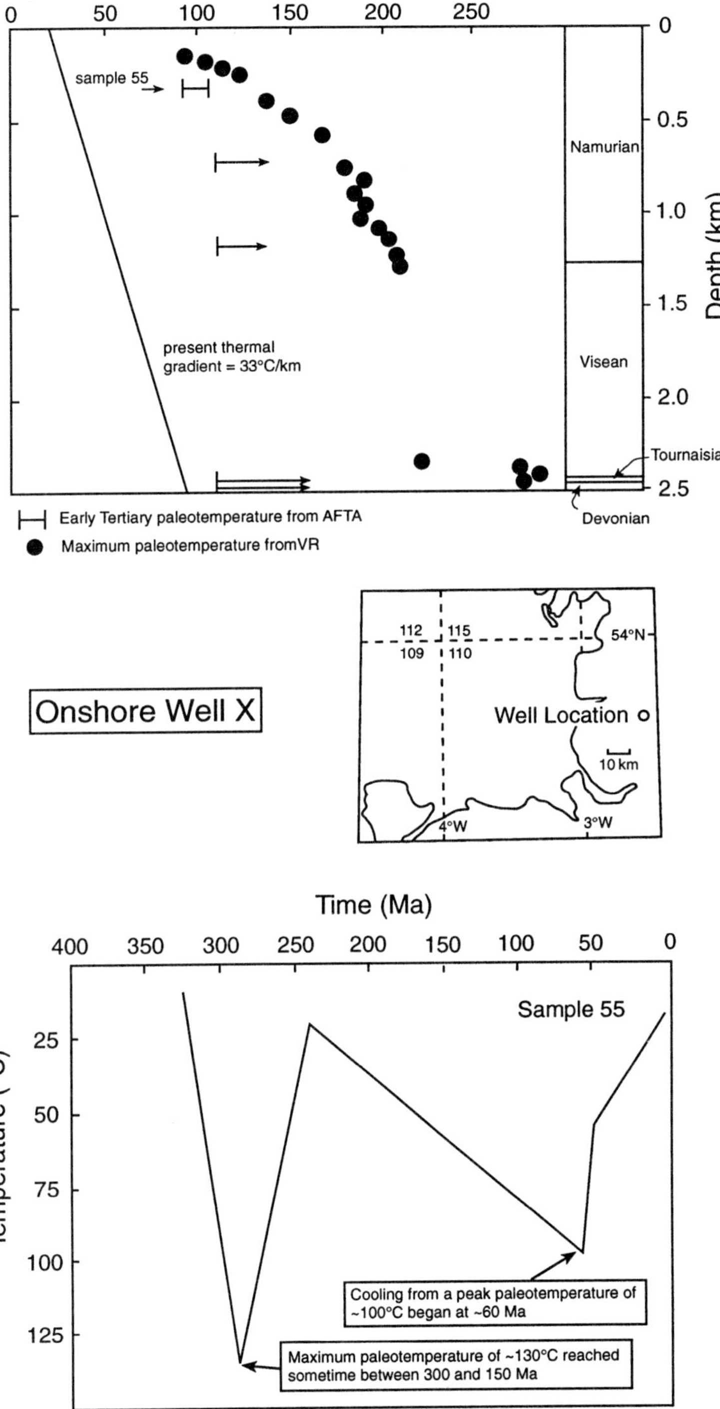

Fig. 4. Thermal history reconstruction in the Onshore Well X well based on AFTA and VR data suggests two episodes of heating and cooling, with maximum palaeotemperatures reached during an earlier episode (between 300 and 150 Ma) and lower palaeotemperatures reached during a later peak (at *c.* 60 Ma). VR data from this well define extremely high palaeogeothermal gradients during the earlier episode, the origin of which is presently unknown.

Style D: Maximum palaeotemperatures in an early event and a lower peak prior to early Cretaceous cooling

AFTA data in a Permian sample from Isle of Man Shallow Borehole B indicate a maximum palaeotemperature of *c.* 105°C prior to the onset of cooling between 150 and 100 Ma (Fig. 5). VR values from the underlying Carboniferous section indicate maximum palaeotemperatures between 124 and 127°C, for any reasonable timing for the onset of cooling. The difference of *c.* 20°C between the palaeotemperature indicated by the AFTA data and VR data, is greater than can be explained by the resolution of kinetic models etc., and must be regarded as a real difference. Assuming that an unreasonably high geothermal gradient of *c.* 300°C km⁻¹ can be discounted, the VR data must record maximum palaeotemperatures reached during an earlier event than that expressed in the AFTA data. Other data from locations nearby confirm this pattern. Therefore cooling from the maximum palaeotemperatures in the Carboniferous section revealed by the VR data must have occurred prior to deposition of the overlying Permian and Triassic section. The preferred thermal history reconstruction is shown in Fig. 5. The detail of the track length distribution in this sample suggests that at *c.* 60 Ma the temperature of this sample was between 70 and 80°C. As before, whether this denotes a discrete episode of early Tertiary heating and cooling, or merely an intermediate step on the cooling history, is unclear at present.

Style E: Peak palaeotemperatures prior to late Permian to mid-Triassic cooling

AFTA data from the western end of the Solway Firth coast of southern Scotland (Fig. 6) show a rather different style of thermal history to those discussed so far. Samples of Ordovician grits, Devonian granite and Early Permian sands all show clear evidence of an episode of cooling which began between 260 and 220 Ma. The earlier history of the Palaeozoic samples is not constrained because the data are dominated by the history since the onset of cooling in the late Permian to mid-Triassic. Local geological evidence suggests that the Palaeozoic samples would have been at or near the surface in Carboniferous and early Permian times, and therefore the episode of heating prior to the cooling revealed by the AFTA was presumably a discrete event, rather than simply an intermediate stage in the cooling history of the Palaeozoic samples. Details of the track length distributions in all the samples shown in Fig. 6 are compatible with a more or less continuous cooling history, and

suggest an acceleration in the rate of cooling at *c.* 60 Ma from palaeotemperatures of *c.* 60 or 70°C. Regional data discussed in more detail in a later section suggests that the apparently continuous cooling history actually represents the superposition of a number of discrete episodes of cooling, possibly interspersed with prior heating.

Evidence for samples cooling below *c.* 110°C during a similar time interval is seen in AFTA data from North Wales, e.g. from the Lleyn Peninsula, which again suggest more or less continuous cooling but in this case with much lower early Tertiary palaeotemperatures (≤ 50°C) than along the southern Scottish coast. Data in a number of samples from North Wales also suggest palaeotemperatures around 90°C between 200 to 150 Ma (Jurassic), which may suggest another discrete episode of cooling, or may simply reflect an intermediate stage of cooling.

Palaeogeothermal gradients during pre-Tertiary episodes and origins of heating and cooling

Earlier cooling episodes

The above discussion has referred to earlier (pre-Cretaceous) episodes of cooling in Onshore Well X and in the Isle of Man, and also to a late Permian to mid-Triassic onset of cooling in southern Scotland. With the exception of the latter example, precise constraints on the timing of the onset of cooling in earlier episodes are difficult to obtain, because AFTA data from the region are normally dominated by early Cretaceous or early Tertiary effects. However it is clear that in some locations, early heating and cooling is associated with end-Carboniferous events (e.g. in Shallow Borehole B, Fig. 5), while in others (e.g. Onshore Well X, Fig. 4) cooling is only constrained to the interval 300 to 150 Ma).

VR data from Shallow Borehole C from the north of the Isle of Man (Fig. 7) show erratic variation over a small vertical distance. One possible explanation of this is that the VR data reflect local heating effects associated with either fluid movement or igneous activity. Comparison of these VR data with AFTA data from nearby locations shows that the VR data reflect maximum palaeotemperatures during an early episode, thought to be end-Carboniferous in age on the basis of data from Shallow Borehole B (Fig. 5). No other constraints are available on palaeotemperature profiles definitely dated as end-Carboniferous and it is presently unclear whether all palaeo-thermal effects at this time are due to local effects or whether a number of different causes were operative.

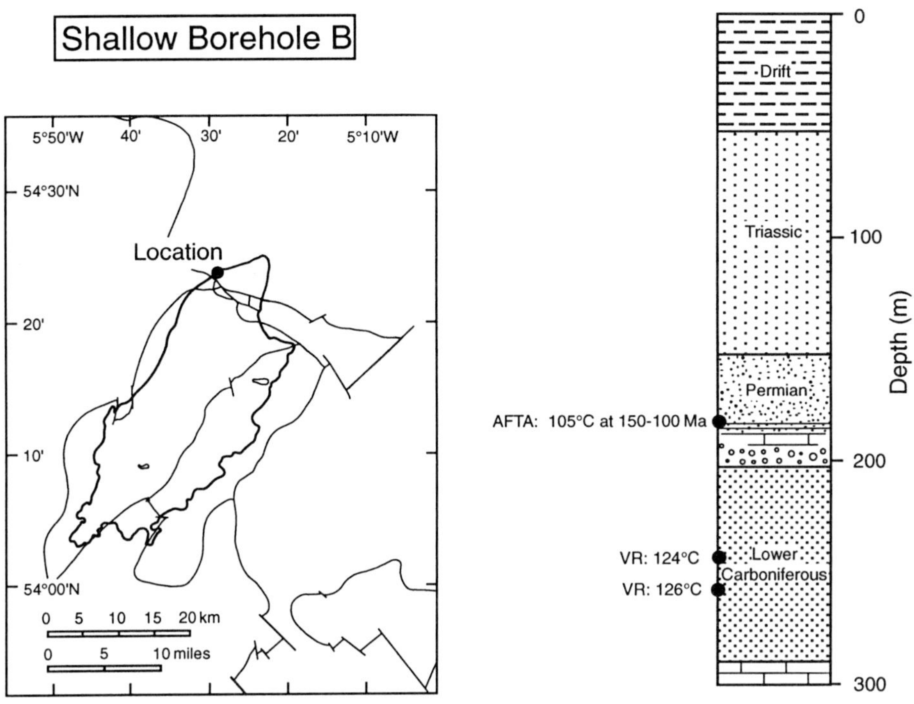

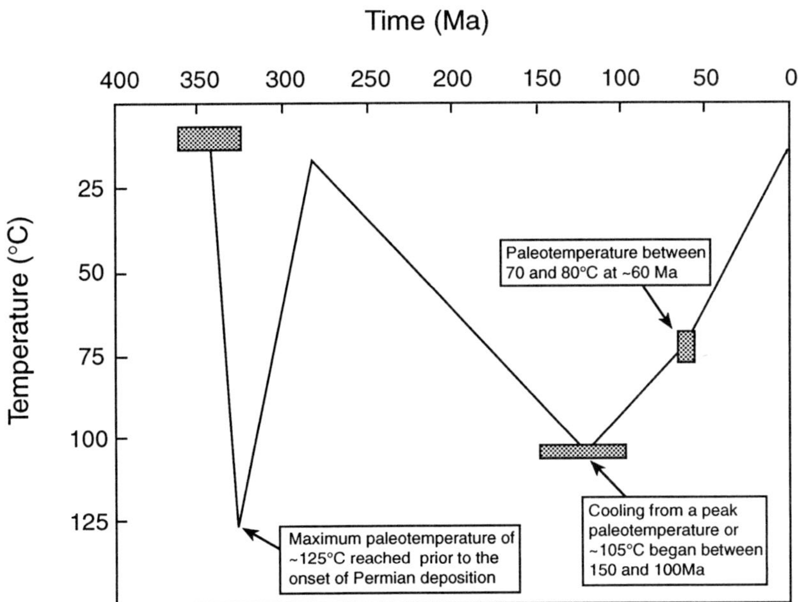

Fig. 5. AFTA and VR data from Isle of Man Shallow Borehole B suggest cooling from maximum palaeotemperatures of 125°C prior to the onset of deposition of overlying Permian sediments, and a subsequent peak palaeotemperature of *c.* 105°C from which cooling began between 150 and 100 Ma. AFTA data also suggest a palaeotemperature between 70 and 80°C at *c.* 60 Ma. Whether this reflects a subsequent episode of heating and cooling, or simply an intermediate step in a protracted cooling history (as shown), is not clear.

However, the widespread occurrence of possible end-Carboniferous effects suggests a regional mechanism.

In an earlier section we discussed how AFTA and VR data from Onshore Well X combined to reveal an earlier episode of heating and cooling dated to the interval 300 to 150 Ma. It is not at present clear if this reflects end-Carboniferous effects or a rather later episode perhaps similar to the late Permian to mid-Triassic event identified to the northwest, or perhaps a Cimmerian event. Whatever the timing of this episode, maximum palaeotemperatures derived from VR data throughout this well (Fig. 4) show that the palaeogeothermal gradient immediately prior to cooling was extreme. Data over the shallower 800 m of the

section define a palaeogeothermal gradient of *c.* 150°C km^{-1}, with a gradient of *c.* 60°C km^{-1} through the deeper section.

AFTA and VR data from other wells in the vicinity of Onshore Well X also show evidence of an earlier episode in which the palaeogeothermal gradient was *c.* 70°C km^{-1}. Again the AFTA data do not allow tighter constraint on the timing than 300 to 150 Ma, because of the severity of early Tertiary effects. However, it seems clear that in this region of Lancashire, an early episode of heating and cooling occurred, involving extremely high palaeogeothermal gradients. The origin of this episode is not clear, but it appears to explain a notable 'hot-spot' in the coal rank map of the UK (National Coal Board 1960).

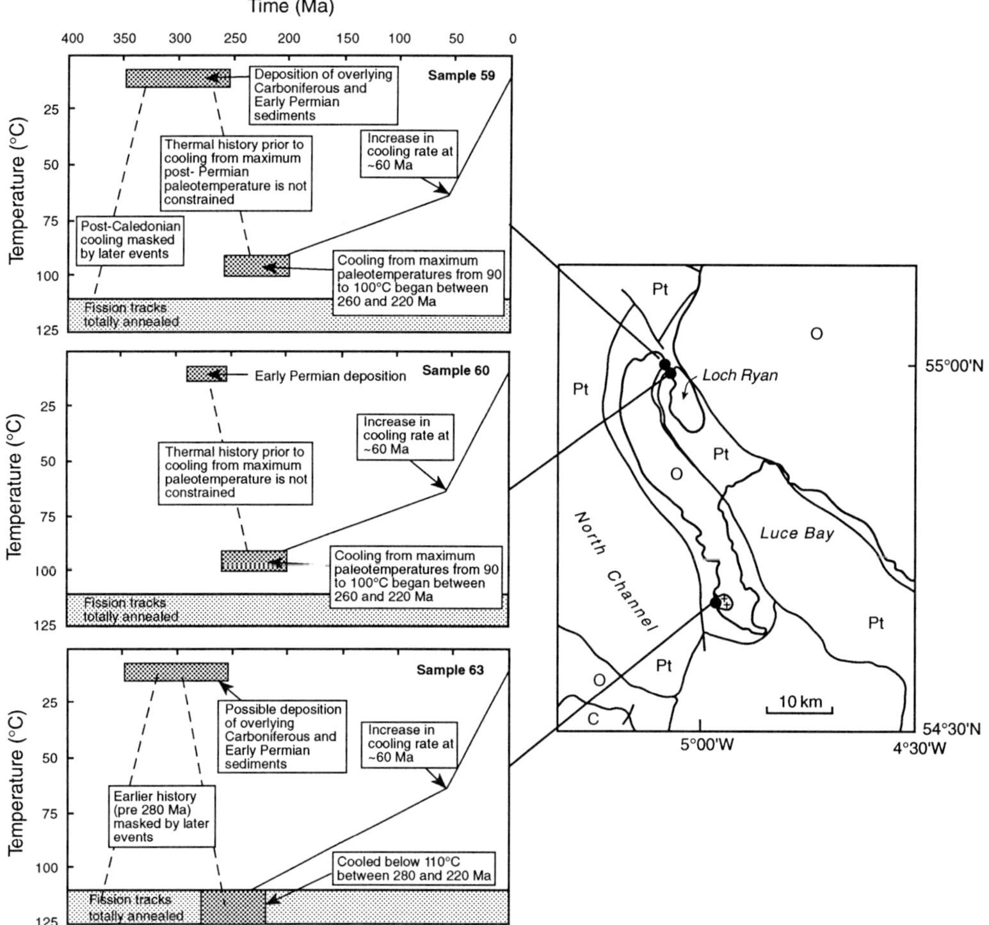

Fig. 6. AFTA data in samples of Permian to Ordovician age from southwest Scotland show consistent evidence of a cooling episode between 260 and 220 Ma (combining evidence from all three samples). All samples require an increase in the cooling rate around the early Tertiary, as shown.

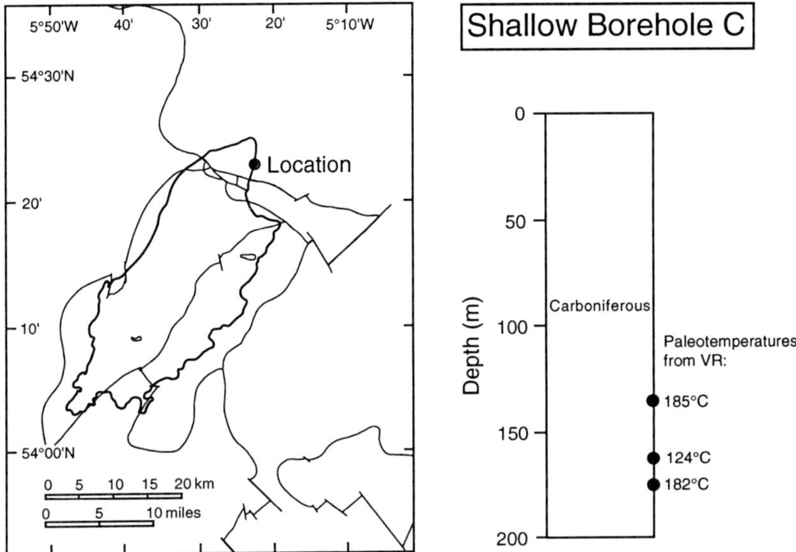

Fig. 7. VR data from Shallow Borehole C, Isle of Man, show pronounced variation over a narrow range of depths, suggesting localized heating effects possibly due to igneous activity or passage of hot fluids.

No constraints are available on palaeogeothermal gradients immediately prior to late Permian to mid-Triassic cooling seen in southwest Scotland, due to a lack of wells in this region which might provide palaeotemperature coverage over a range of depths. In this area, the lack of obvious igneous activity at the appropriate time again appears to require that heating was largely due to burial, with subsequent cooling due to uplift and erosion. Similar comments apply to Mesozoic cooling in North Wales. Until samples are available in these areas over a range of depths, further insight into the origin of these cooling episodes must rely on inference rather than direct measurement.

Early Cretaceous

A discrete episode of Early Cretaceous cooling from palaeotemperatures generally around 100°C or above has been identified throughout the Isle of Man and adjacent regions, and may also have affected the region south of the Isle of Man into Anglesey. Due to the lack of deep wells in the region, no constraints are available on palaeo-geothermal gradients immediately prior to this cooling episode, and the origin and nature of this episode and any heating which preceded it are largely speculative at present. Early Cretaceous cooling was clearly regional in its effects, the full extent of which is yet to be defined.

The lack of recognized Early Cretaceous igneous activity anywhere in the region leads to the conclusion that cooling must have been due largely to uplift and erosion, and presumably heating prior to the onset of cooling was due largely to additional depth of burial.

Early Tertiary palaeogeothermal gradients

Although AFTA data from northwest England and the Irish Sea have revealed extensive evidence of early Tertiary cooling from palaeotemperatures around 100°C or above, constraints on palaeogeo-thermal gradients prior to the onset of cooling have been difficult to establish in earlier studies due to the lack of discrete palaeotemperature estimates over a range of depth. The main reason for this is that deep exploration wells from which such constraints could be obtained are restricted largely to the East Irish Sea Basin and onshore margins, and in these wells, AFTA samples were either totally, or almost totally, annealed prior to cooling and allow only minimum estimates of the palaeo-temperature prior to cooling, while VR data are usually available over only a very restricted depth range.

More recently, analysis of samples from a number of wells which intersected thick Carbon-iferous sections, combined with compilation of early Tertiary palaeotemperature estimates from both AFTA and VR in a number of wells, has provided evidence of major regional differences in palaeogeothermal gradient which could have major implications for causal mechanisms for heating and cooling.

Figure 8a shows a compilation of palaeotemperature estimates from AFTA and VR in six wells from the southern part of the Irish Sea (south of the Morecambe gas fields) and adjacent onshore area to the east, plotted against sample depth. Only AFTA data corresponding to samples which were not totally annealed are shown in this plot, as these allow discrete palaeotemperature estimates whereas samples which were totally annealed provide only a minimum estimate (usually > 110°C). It should be borne in mind that palaeotemperature estimates from non-totally annealed AFTA samples are normally quoted as a range (e.g. 100 to 110°C), but in Fig. 8a only the mid-point value is plotted. VR data interpreted as reflecting an earlier heating episode (e.g. those from Onshore Well X, shown in

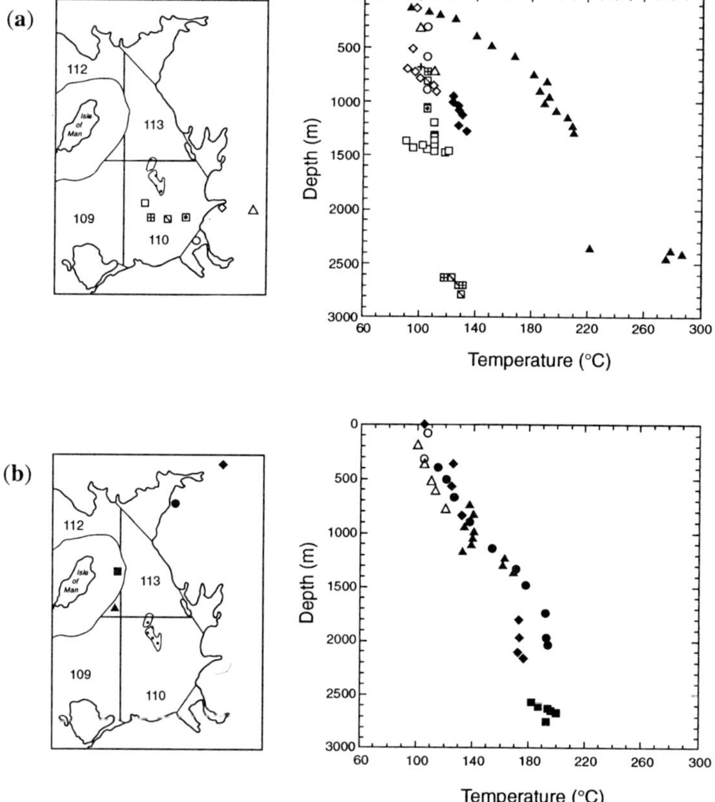

Fig. 8. (a) Palaeotemperatures derived from AFTA and VR data in samples from six wells from the southern part of the East Irish Sea Basin and onshore flanks, plotted against sample depth. AFTA data which give only a minimum estimate of maximum palaeotemperature are not shown, but are consistent with other data. Values derived from VR data considered to reflect an earlier heating episode (e.g. data from Onshore Well X, from Fig. 4) are shown as solid symbols. Early Tertiary palaeotemperatures from AFTA and VR data are represented as open symbols, and define a consistent, linear profile with a gradient of only *c.* 10°C km⁻¹, which is taken to represent the early Tertiary palaeogeothermal gradient throughout this region. (b) Palaeotemperatures derived from AFTA and VR data in samples from four wells from the northern part of the East Irish Sea Basin and the eastern Solway Basin, plotted against sample depth. AFTA data which give only a minimum estimate of maximum palaeotemperature are not shown, but are consistent with other data. Early Tertiary palaeotemperatures derived from AFTA are shown as open symbols, while values derived from VR are shown as solid. Although differences certainly exist between different wells, these data all define a broadly consistent profile, with a much higher gradient than defined by the southern wells. This may reflect a major regional difference in early Tertiary palaeogeothermal gradients, or alternatively it is possible that most of the VR data in these wells may reflect an earlier episode similar to that seen in two of the southern wells.

Fig. 4) are represented as solid symbols, while early Tertiary palaeotemperatures from both VR and AFTA are represented using open symbols. Excluding the solid data, early Tertiary palaeotemperature estimates from the six wells define a consistent, linear profile from palaeotemperatures around 100°C near sea bed to *c.* 130°C at depths between 2500 and 3000 m, suggesting a palaeogeothermal gradient of only *c.* 10°C km^{-1}.

Figure 8b shows a similar plot of palaeotemperature estimates from AFTA and VR in four wells from the northern Irish Sea and eastern Solway Basin. Palaeotemperatures derived from AFTA, reflecting values reached immediately prior to Late Cretaceous or early Tertiary cooling, are shown as open symbols, while values derived from VR are shown solid (note this is different to the symbolism in Fig. 8a). Although differences certainly exist between different wells, data from the four wells define a broadly consistent profile, very different to that from the southern wells shown in Fig. 8a.

One interpretation of these data is illustrated in Fig. 9, which shows in more detail palaeotemperature estimates from AFTA and VR data in Onshore Well Y (one of the four wells shown in Fig. 8b), located on the northern fringe of the Lake District block between the Solway Firth and Carlisle basins. AFTA data in five samples from this well all show clear evidence of cooling at *c.* 60 Ma. A shallow Triassic sample reached a maximum palaeotemperature between 105 and 110°C while samples from the underlying Dinantian section all reached palaeotemperatures of *c.* 110°C or above immediately prior to the onset of cooling. VR data from the Dinantian section provide estimates of maximum palaeotemperature between 115 and 194°C, which are consistent with the AFTA data and which form a co-linear profile with the estimate from AFTA in the Triassic section. Track length data in the five AFTA samples suggest a similar style of cooling history as seen elsewhere in samples dominated by early Tertiary cooling (e.g. well 110/7-2, Fig. 2).

The simplest interpretation of these observations is that both the AFTA and VR data throughout the well reflect maximum palaeotemperatures immediately prior to early Tertiary cooling. Using methods outlined by Bray *et al.* (1992), the combined palaeotemperature data define a palaeogeothermal gradient of *c.* 50°C km^{-1}, and extrapolating this gradient to a palaeo-surface temperature of 10°C suggests that 1.75 km of overlying section has been removed by Tertiary uplift and erosion. For an early Tertiary palaeo-surface temperature of 20°C, the estimated amount of section removed is reduced by *c.* 200 m. The resulting reconstructed thermal history for this well is shown in Fig. 9. In this reconstruction, palaeotemperatures in the Carboniferous section during any earlier episodes of heating must have been less than those reached immediately prior to early Tertiary cooling, but are otherwise unconstrained.

This interpretation provides a consistent, viable interpretation of both AFTA and VR data from the four wells, and in the absence of any other considerations might be accepted without comment. However one crucial aspect of the data from the northern wells is that in none of the wells do palaeotemperature constraints from VR and non-totally annealed AFTA samples overlap in depth. In other words, wherever AFTA and VR constraints overlap, the AFTA data only provide minimum estimates of the maximum palaeotemperature (typically > 110°C), while VR-derived palaeotemperatures are always greater than 110°C. With this in mind, the evidence of a low early Tertiary palaeogeothermal gradient from the southern Irish Sea allows an alternative interpretation of data from the northern region, in which the early Tertiary palaeogeothermal gradient there was also low, similar to that identified to the south, and the higher gradient defined by the VR data reflects an earlier event similar to that seen in Onshore Well X. At present, no data are available to discriminate between these rival scenarios, which have major implications for hydrocarbon occurrence.

Origins of early Tertiary palaeo-thermal effects

Although more data are required before a definitive description of the regional variation in early Tertiary palaeogeothermal gradients can be established, the consistency of the separate palaeotemperature trends shown in Fig. 8 is suggestive of a model involving low gradients in the southern part of the East Irish Sea Basin and higher gradients to the north (while evidence from areas further away from the Irish Sea suggests gradients close to present-day values; Bray *et al.* 1992).

The presence of a Tertiary intrusive centre within the northern part of the Irish Sea (Kirton & Donato 1985; Arter & Fagin 1993) suggests a possible explanation of the higher gradients to the north, in terms of increased basal heat flow associated with igneous activity. The lower gradients to the south cannot be explained solely in terms of conductive heat flow through a Permo-Triassic section dominated by salt and sandstones with high thermal conductivities, as the section is essentially similar over the entire region (Jackson & Mulholland 1993), and the low gradient also spans the Carboniferous section which is less homogeneous. The only viable explanation of such low gradients at

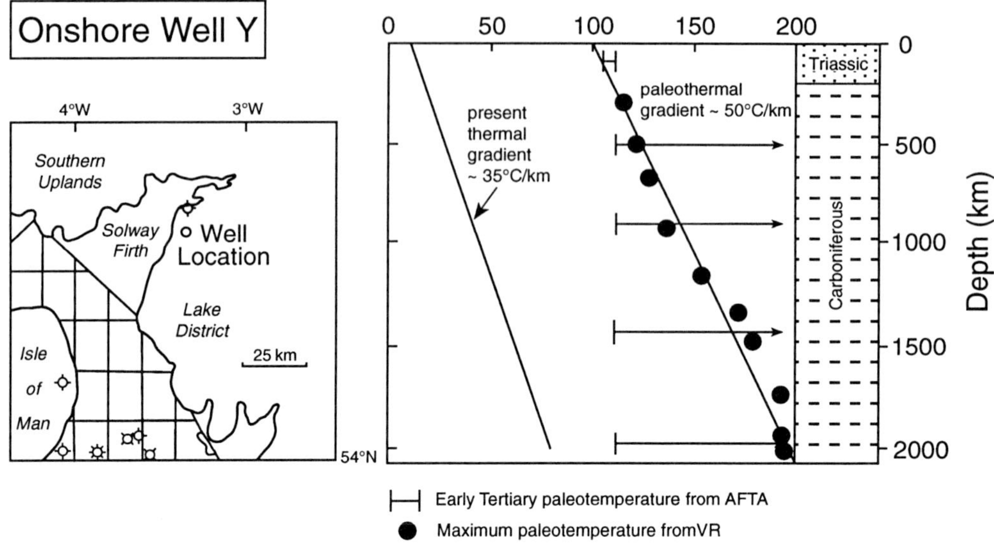

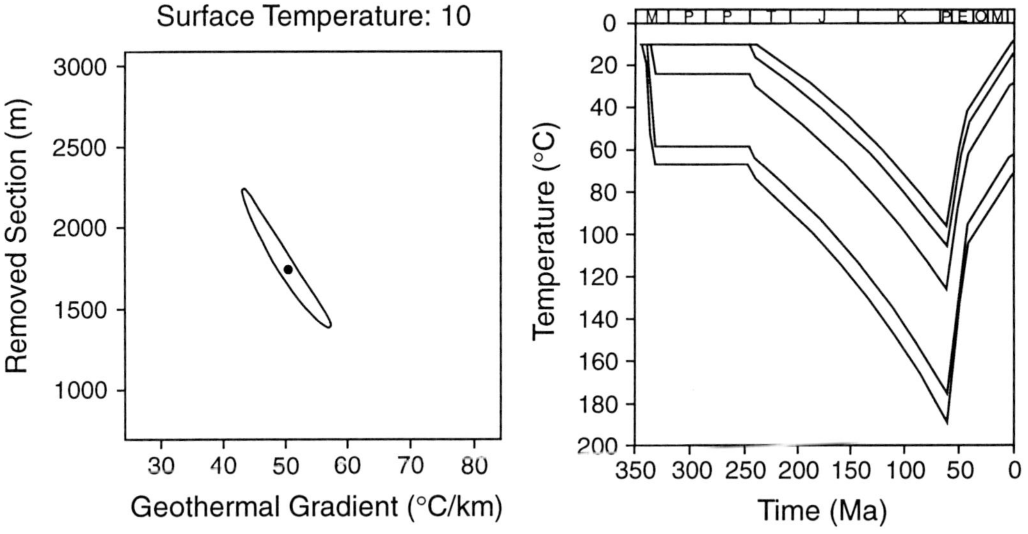

Fig. 9. Thermal history reconstruction in Onshore Well Y (eastern Solway Basin). AFTA and VR data can be interpreted in terms of a single episode of heating and cooling, with maximum palaeotemperatures reached immediately prior to early Tertiary cooling (beginning at *c.* 60 Ma) as shown. Maximum palaeotemperatures determined from AFTA and VR define a palaeogeothermal gradient of *c.* 50°C km^{-1}, and extrapolating this to a palaeo-surface temperature of 10°C suggests *c.* 1.75 km of section removed by Tertiary uplift and erosion. However, an alternative interpretation of these data is also possible, with a low early Tertiary palaeogeothermal gradient and VR data from this well reflecting an earlier heating episode as seen in a number of other wells.

present appears to be flow of hot fluids shallow in the section (Duddy *et al.* 1994). The pattern of gradients suggests transport of hot fluids from north to south within the basin, which would be consistent with an origin for the fluids associated with igneous activity. Rowe *et al.* (1993) have also reported evidence for movement of hot fluids in the Irish Sea region based on geochemical observations.

In an alternative scenario, early Tertiary gradients could have been low across the entire region, and the VR data from the northern region which dominate the palaeotemperatures in Fig. 8b could reflect the earlier episode also identified in some of the southern wells. In this scenario, the low gradients must again be due to passage of heated fluids.

The geographic variation in the magnitude of early Tertiary palaeotemperatures in presently outcropping rocks (Fig. 10) shows that the highest values occur in the Skiddaw and Carrock Fell region of the Lake District and at the eastern end of the Solway Firth. Palaeotemperatures decrease only slowly to the east and south into the Pennines and the Cheshire Basin, but decrease rapidly to the south and northeast into the Palaeozoic massifs of the Southern Uplands and North Wales (not shown in Fig. 10). This may reflect focusing of fluid movements within Carboniferous to Triassic basinal regions (Green *et al.* 1993*b*). In particular, the Welsh Massif could have provided a southerly buttress, constraining the heated fluids to the basinal region and causing a downward recirculation resulting in the observed low palaeogeothermal gradients. Heterogeneity of

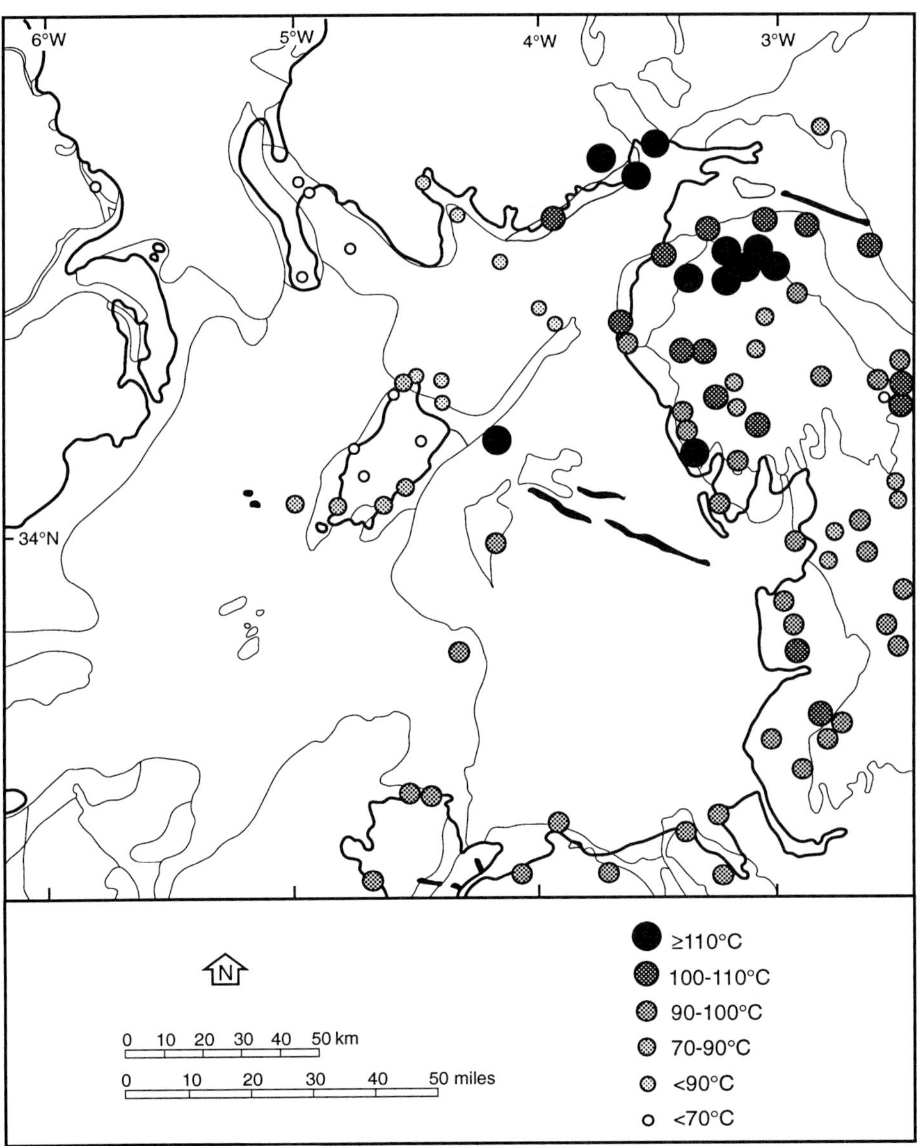

Fig. 10. Maximum palaeotemperatures prior to early Tertiary cooling, estimated from AFTA data from outcrop and shallow borehole samples in the Irish Sea Basin and adjacent areas.

palaeotemperature estimates across the Lake District block may also reflect local fluid movements.

Although the Tertiary Igneous Province has previously been considered to represent a likely focus of early Tertiary heating, early Tertiary palaeotemperatures at outcrop decrease towards Northern Ireland, where pronounced early Tertiary palaeo-thermal phenomena are restricted to local contact effects related to recognized igneous bodies. In this region, no evidence has yet been found for significant early Tertiary uplift and erosion, which is consistent with local preservation of thin Jurassic and Cretaceous units below the early Tertiary basalts.

If we accept the first thermal gradient scenario outlined above, estimates of eroded section in all wells in the northern part of the Irish Sea are generally between 1.5 and 3 km (for a palaeo-surface temperature of 10°C), with most values around 2 km. The low palaeogeothermal gradients from the southern region clearly cannot be extrapolated to a palaeo-surface temperature to estimate the amount of section removed by erosion. However, estimates around 2 km have also been obtained from wells to the southeast in Staffordshire and the East Midlands, and bearing in mind the uniformity of palaeotemperature estimates in outcrop samples across this region (Fig. 10) it seems likely that a similar amount of cover may have been removed from most of the region. This interpretation is consistent with geological evidence (Chadwick *et al.* 1993; Holliday 1993; Jackson & Mulholland 1993) and is supported by structural data (Knipe *et al.* 1993) and sonic velocity data (Colter 1978; Hardman *et al.* 1993).

At present, our preferred interpretation of the early Tertiary palaeotemperatures detected across northwest England and in the Irish Sea therefore involves a combination of burial by around 1.5 to 2 km of post-Triassic sediment and the thermal effects of hot fluid flow, probably from north to south and associated with igneous activity. We stress that although this model is consistent with the observed pattern of early Tertiary palaeotemperatures, it requires further testing (involving assessment of palaeotemperature profiles over as wide a region as possible) before it can be widely accepted. In particular, the possibility that VR data from northern wells reflects an earlier episode must be examined in detail.

Regional variation in thermal history styles

Having identified at least four episodes of cooling in the Irish Sea and adjacent areas, the possible combinations of these episodes lead to a wide variety in thermal history styles across the region. Figure 11 illustrates some of the variation in thermal history styles for Carboniferous source rocks around the region. One particular feature to note is the variation along the southern coast of Scotland from west (dominated by more or less continuous cooling since the late Permian to mid-Triassic) to east (where maximum post-Palaeozoic palaeotemperatures were reached in the early Tertiary), through an intermediate region in which maximum post-Palaeozoic palaeotemperatures were reached in the Cretaceous. This variation, over a relatively small distance, suggests that the whole region was affected by at least three episodes of heating and cooling, with the magnitude of palaeotemperatures in each episode varying in each region, as illustrated in Fig. 12. Similar comments are probably applicable to the whole region shown in Fig. 11, and actual thermal histories possibly involved a larger number of cooling episodes, only a small number of which can be resolved.

Implications for hydrocarbon occurrence

One of the most important aspects of regional hydrocarbon prospecivity is the timing of hydrocarbon generation in relation to the formation of potential trapping structures and periods of tectonic activity. Only those structures which exist when hydrocarbons are generated can be filled, and later tectonic movements may modify pre-existing structures, leading to remigration and potential loss of trapped hydrocarbons. Uplift of trapped hydrocarbons may lead to expansion, phase changes and possible seal rupture. Clearly the ideal situation occurs when generation takes place towards the end of the history, after trap formation and after most, if not all, of the tectonic pulses which might lead to loss of accumulated hydrocarbons.

Considering the variation in thermal history styles outlined in this paper, together with the structural history of the East Irish Sea Basin and adjacent areas, shows that the most favourable conditions in respect of the relative timing of hydrocarbon generation and trap formation occur within the East Irish Sea Basin, coinciding with most of the recognized hydrocarbon reserves within the region. Throughout the southern part of the East Irish Sea Basin the thermal history of Carboniferous source rocks was such that oil and gas were generated in the last major episode of heating that affected the region. Some gas expansion occurred in the Morecambe gas field during Tertiary uplift, as noted earlier, but because of the regional nature of the uplift, little tilting would have occurred which might have led to loss of accumulated hydrocarbons.

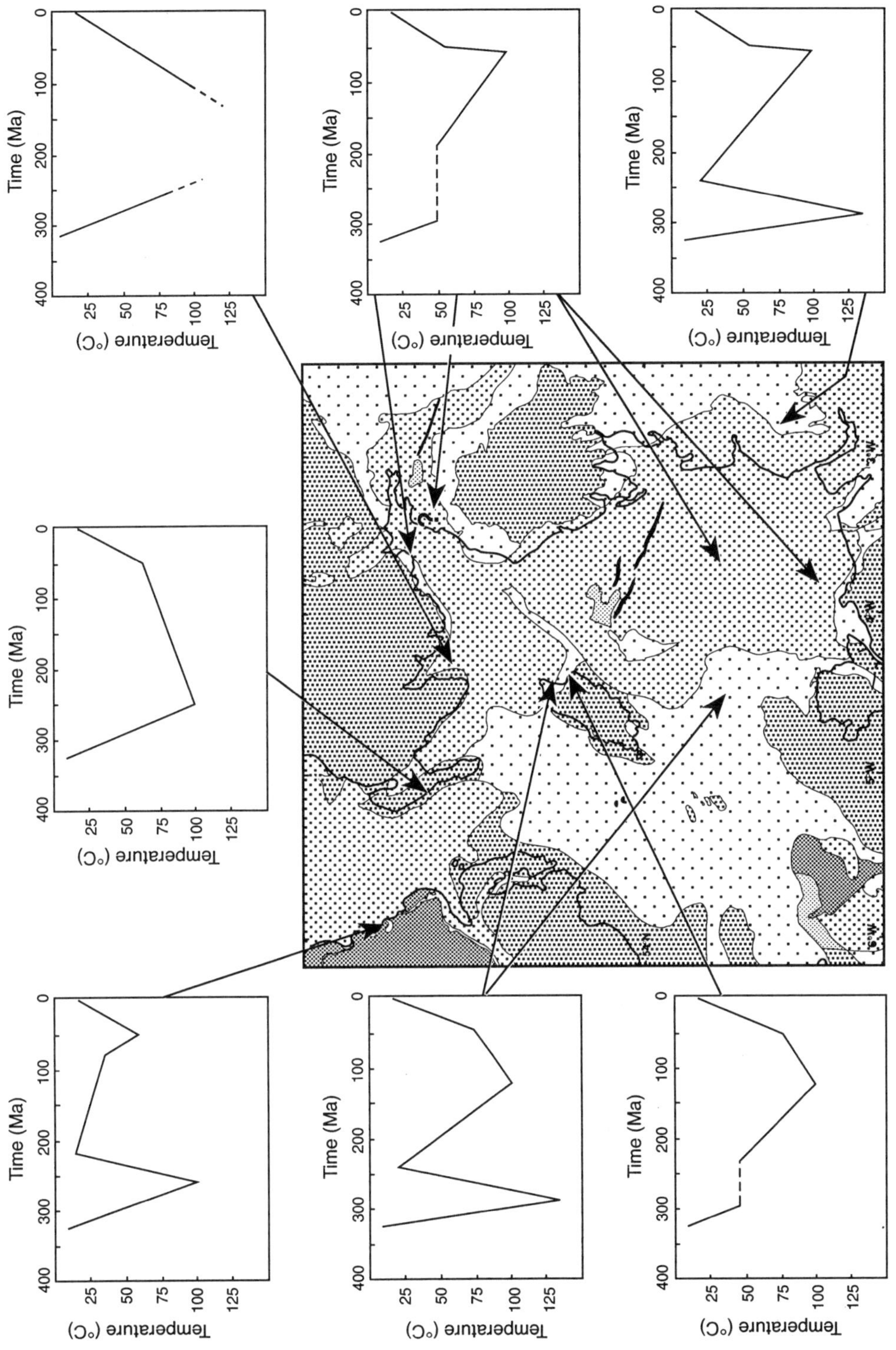

Fig. 11. Variation in thermal history styles around the East Irish Sea Basin and adjacent areas. The histories illustrated are the simplest which satisfy the AFTA and VR data from each locality. In reality, the actual thermal histories are likely to have been more complicated, involving a series of heating and cooling episodes, but approximating to the histories shown.

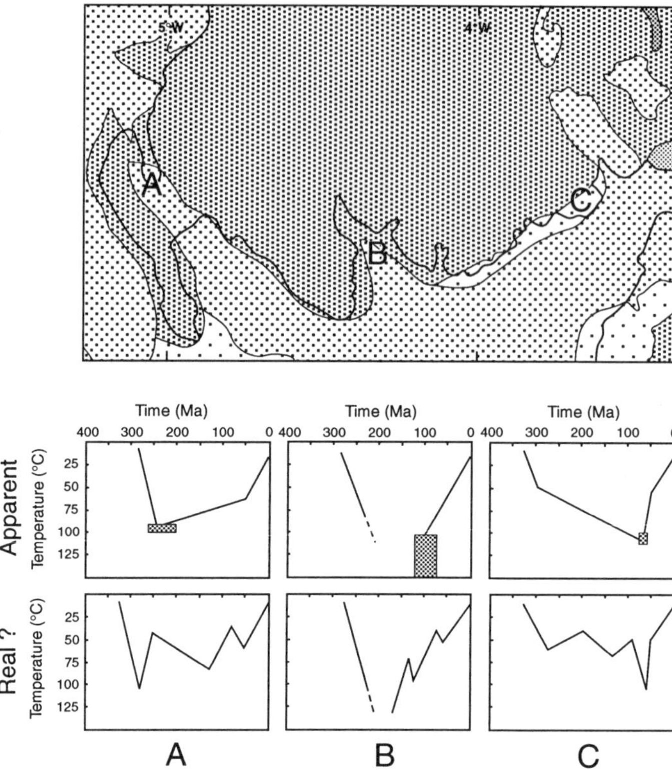

Fig. 12. Local variation in thermal history styles inferred from AFTA data in outcrop and shallow borehole samples along the southern Scottish coast is suggestive of a series of discrete cooling episodes as shown. Samples from the three localities require different thermal history styles, involving cooling from maximum palaeotemperatures at different times. The real histories at the three localities probably involve all three episodes, but with the magnitude of peak palaeotemperatures in each episode varying cross the region.

Elsewhere, maximum palaeotemperatures, and therefore peak hydrocarbon generation occurred earlier in the history, in many areas earlier than the main (Cimmerian) episode of trap formation. Locally, peak generation would have occurred during early Cretaceous heating, after formation of Cimmerian structures, but the likelihood of a discrete Late Cretaceous episode of burial followed by early Tertiary uplift and erosion in this region means that any hydrocarbons trapped in this region would probably have been lost, or at least re-migrated. Since early Cretaceous cooling, presumably reflecting uplift and erosion, appears to have been concentrated to the west of the East Irish Sea Basin, along a roughly north–south axis through the Isle of Man region, Cretaceous uplift could have produced tilting in the East Irish Sea Basin which might have led to the loss of the initial charge of the Morecambe Field as described in the model outlined earlier (but perhaps slightly later than described there).

Figure 13 shows an attempt at predicting regional trends in timing of hydrocarbon generation, based on all available data. This is probably over-simplistic, and considerable uncertainty still exists, particularly in respect of the possible influence of earlier events. In addition, considerable local variation exists in the maturity of the Carboniferous section in places, particularly to the west and northeast of the Isle of Man, and sparse data coverage precludes any definitive conclusions which must await analyses from deep wells in this region.

In summary, the southern part of the East Irish Sea Basin, containing major accumulations of oil and gas, is characterized by the most suitable thermal history conditions for generation and preservation of hydrocarbons to the present day. In adjacent regions, thermal histories are complex, and those identified to date are generally less suitable particularly for preservation of hydrocarbons to the present day. Exploration in these

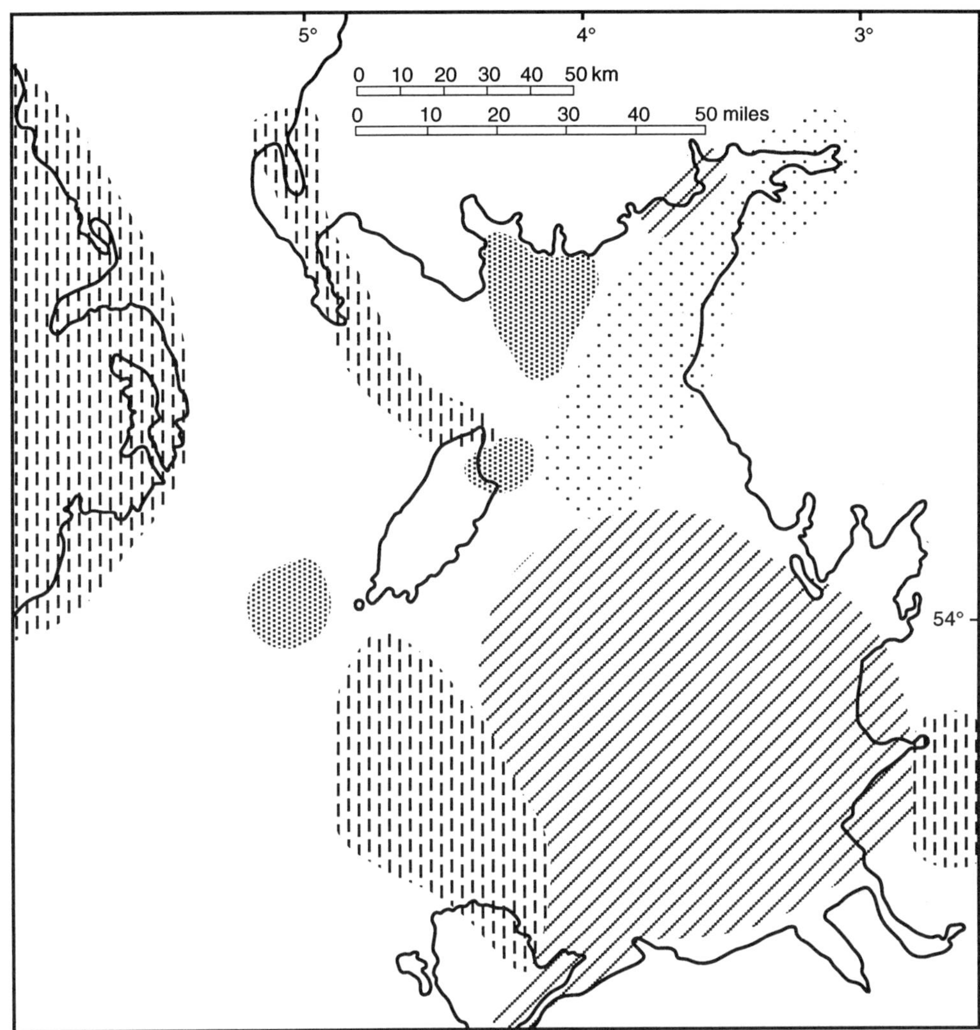

Thermal history reconstruction shows that peak hydrocarbon generation occurred in the late Mesozoic, immediately prior to the early Tertiary cooling episode. Of all the thermal history styles recognised, this presents the most favourable conditions for preservation of hydrocarbons to the present day.

Peak generation occurred immediately prior to early Cretaceous cooling. Hydrocarbons generated and trapped during this time would need to survive subsequent Cretaceous and Tertiary tectonism.

Peak generation possibly occurred immediately prior to early Tertiary cooling, but the possibility exists that an early episode affected this region.

Thermal history reconstruction shows that peak hydrocarbon generation occurred during earlier (pre-Cretaceous) heating episodes. Hydrocarbons generated during these episodes are less likely to have survived subsequent tectonic and thermal events to the present day.

Fig. 13. Regional trends in hydrocarbon prospectivity based on variation in thermal history styles identified in this paper. This map shows trends in prospectivity based solely on thermal history considerations, and takes no account of other significant factors such as source, reservoirs, seals and traps.

areas will require a thorough understanding of the thermal history of Carboniferous source rocks in order to define the most prospective regions.

We would like to thank Enterprise Oil PLC, Amoco (UK) Exploration Company, The Manx Museum and National Trust, the Department of Trade and Industry and the British Geological Survey for providing samples from exploration wells and shallow boreholes described in this paper. We also thank BG Exploration and Production Ltd, BHP Petroleum and the Geochem Group for permission to release data. Thanks are also due to Michael Smith, of Geotrack's UK office, for helpful comments on the manuscript. AFTA® is the registered trademark of Geotrack International Pty Ltd.

References

ARTER, G. & FAGIN, S. W. 1993. The Fleetwood dyke and the Tynwald fault zone, Block 113/27-1, East Irish Sea Basin. *In*: PARKER, J. R. (ed.) *Petroleum Geology of Northwest Europe: Proceedings of the 4th Conference*, Geological Society, London, 835–843.

BRAY, R. J., GREEN, P. F. & DUDDY, I. R. 1992. Thermal History Reconstruction using apatite fission track analysis and vitrinite reflectance: a case study from the UK East Midlands and the Southern North Sea. *In*: HARDMAN, R. F. P. (ed.) *Exploration Britain: Into the next decade*. Geological Society, London, Special Publication, **67**, 3–25.

BUSHELL, T. P. 1986. Reservoir geology of the Morecambe field. *In*: BROOKS, J. & VAN HOORN, B. (eds) *Habitat of paleozoic Gas in N.W. Europe*. Geological Society, London, Special Publication **23**, 189–208.

CHADWICK, R. A., EVANS, D. J. & HOLLIDAY, D. W. 1993. The Maryport Fault: the post-Caledonian tectonic history of southern Britain in microcosm. *Journal of the Geological Society, London*, **150**, 247–250.

COLTER, V. S. 1978. Exploration for gas in the Irish Sea. *Geologie en Mijnbouw*, **57**, 503–516.

DUDDY, I. R., GREEN, P. F., BRAY, R. J. & HEGARTY, K. A. 1994. Recognition of the thermal effects of fluid flow in sedimentary basins. *In*: PARNELL, J. (ed.) *Geofluids: origin, migration and evolution of fluids in sedimentary basins*. Geological Society, London, Special Publication, **78**, 325–345.

FRASER, A. J., NASH, D. F., STEELE, R. P. & EBDON, C. C. 1990. A regional assessment of the intra-Carboniferous play of Northern England. *In*: BROOKS, J. (ed.) *Classic Petroleum Provinces*. Geological Society, London, Special Publication, **50**, 417–440.

GREEN P. F. 1986. On the thermo-tectonic evolution of Northern England: evidence from fission track analysis. *Geological Magazine*, **123**, 493–506.

—— 1989. Thermal and Tectonic history of the East Midlands shelf (onshore U.K.) and surrounding regions assessed by apatite fission track analysis. *Journal of the Geological Society, London*, **146**, 755–773.

——, DUDDY, I. R., BRAY, R. J. & LEWIS, C. L. E. 1993*a*. Elevated palaeotemperatures prior to early Tertiary cooling throughout the UK region: implications for hydrocarbon generation. *In*: PARKER, J. R. (ed.) *Petroleum Geology of Northwest Europe: Proceedings of the 4th Conference*. Geological Society, London, 1067–1074.

——, —— & —— 1993*b*. Early Tertiary heating in Northwest England: fluids or burial (or both?) (extended abstract). *In*: PARNELL, J., RUFFELL, A. H. & MOLES, N. R. (eds) *Geofluids '93: Contributions to an international Conference on fluid evolution, migration and interaction in rocks*, 119–123.

——, —— & —— 1995. Applications of Thermal History Reconstruction in inverted basins. *In*: BUCHANAN, J. G. & BUCHANAN, P. G. (eds) *Basin Inversion*. Geological Society, London, Special Publication, **88**, 149–165.

HARDMAN, H., BUCHANAN, J., HERRINGTON, P. & CARR, A. 1993. Geochemical Modelling of the East Irish Sea: its influence on predicting hydrocarbon type and quality. *In*: PARKER, J. R. (ed.) *Petroleum Geology of Northwest Europe: Proceedings of the 4th Conference*. Geological Society, London, 809–821.

HOLLIDAY, D. W. 1993. Mesozoic cover over northern England: interpretation of apatite fission track data. *Journal of the Geological Society, London*, **150**, 657–660.

JACKSON, D. I. & MULHOLLAND, P. 1993. Tectonic and stratigraphic aspects of the East Irish Sea Basin and adjacent areas: contrast in their post-Carboiniferous structural styles. *In*: PARKER, J. R. (ed.) *Petroleum Geology of Northwest Europe: Proceedings of the 4th Conference*. Geological Society, London, 791–808.

——, ——, JONES, S. M. & WARRINGTON, G. 1987. The geological framework of the East Irish Sea Basin. *In*: BROOKS, J. & GLENNIE, K. (eds) *Petroleum Geology of North West Europe*, 191–203.

KIRTON, S. R. & DONATO, J. A. 1985. Some buried Tertiary dykes of Britain and surrounding waters deduced by magnetic modelling and seismic reflection methods. Journal of the Geological Society, London, **142**, 1047–1057.

KNIPE, R. J., COWAN, G. & BALENDRAN, V. S. 1993. The tectonic history of the east Irish Sea Basin with reference to the Morecambe Fields. *In*: PARKER, J. R. (ed.) *Petroleum Geology of Northwest Europe: Proceedings of the 4th Conference*. Geological Society, London, 857–866.

LEWIS, C. L. E., GREEN, P. F., CARTER, A. & HURFORD, A. J. 1992. Elevated late Cretaceous to Early Tertiary paleotemperatures throughout North-west England: three kilometres of Tertiary erosion? *Earth and Planetary Science Letters*, **112**, 131–145.

MACCHI, L., CURTIS, C. D., LEVISON, A., WOODWARD, K. & HUGHES, C. R. 1990. Chemistry, morphology and distribution of illites from Morecambe Gas Field, Irish Sea, Offshore United Kingdom. *AAPG Bulletin*, **74**, 296–308.

NATIONAL COAL BOARD 1960. *The Coalfields of Great*

Britain. Variation in coal rank. NCB Scientific Department, Coal Survey.

ROBERTS, D. G. 1989. Basin inversion in and around the British Isles. *In*: COOPER, M. A. & WILLIAMS, G. D. (eds) *Inversion Tectonics.* Geological Society, London, Special Publication, **44**, 131–150.

ROWE, J., BURLEY, S., GAWTHORPE, R., COWAN, G. & HARDMAN, M. 1993. paleo-fluid flow in the East Irish Sea Basin and its margins (extended abstract). *In*: PARNELL, J., RUFFEL, A. H. & MOLES, N. R. (eds) *Geofluids '93: Contributions to an international Conference on fluid evolution, migration and interaction in rocks,* 358–362.

STUART, I. A. & COWAN, G. 1991. The south Morecambe Field, blocks 110/2a, 110/3a, 110/8a, UK East Irish Sea. *In*: ABBOTS, I. L. (ed) *United Kingdom Oil and Gas Fields, 25 Years Commemorative Volume.* Geological Society, London, Memoir, **14**, 527–541.

WOODWARD, K. & CURTIS, C. D. 1987. Predictive modelling for the distribution of production-constraining illites – Morecambe Gas Field, Irish Sea, Offshore UK. *In*: BROOKS, J. & GLENNIE, K. (eds) *Petroleum Geology of North West Europe,* 205–215.

The geology and hydrocarbon prospectivity of the North Channel Basin

STEVEN J. MADDOX, RICHARD A. BLOW & SEAN R. O'BRIEN

British Gas Exploration and Production Ltd, 100 Thames Valley Park Drive, Reading RG6 1PT, UK

Abstract: The geology and hydrocarbon prospectivity of the offshore North Channel has been evaluated, integrating data from a number of onshore wells with seismic, gravity, magnetic and onshore outcrop studies. Interpretation of these data suggest that the North Channel can be divided into two sub-basins, separated by the offshore extension of the Southern Uplands Fault Zone (SUFZ). The narrow, southern, Portpatrick sub-basin has a NNW–SSE trend and consists of a series of simple tilted fault-blocks which dip east, towards the basin-bounding Portpatrick Fault. The northern sub-basin widens considerably into the Larne–Firth of Clyde basin complex, which has a predominantly Caledonide (NE–SW) orientation and is an extension of the Scottish Midland Valley. This sub-basin is characterised by westerly dipping fault blocks, the reversal in basin polarity apparently accommodated along the SUFZ.

The primary reservoir interval is expected to occur in the Lower Triassic Sherwood Sandstone Group, although secondary targets may exist in the Permian and Carboniferous. Shallow coring and seismic facies analysis suggests that the Mercia Mudstone Group subcrops Quaternary–Recent deposits at sea bed and overlies the Sherwood Sandstone. Halites and shales within the Mercia Mudstone Group should provide an effective regional top-seal, except in areas where the Sherwood Sandstone is close to the surface. Large tilted-fault block and rollover anticline traps have been identified on seismic data, although the Carboniferous source rock potential and burial history are poorly understood and exploration risk is high. Regional outcrop patterns and the compaction state of the Mesozoic strata, together with vitrinite measurements suggest the basin was uplifted in the early Tertiary. In common with the East Irish Sea, maturity was probably reached prior to the end of the Cretaceous. This implies that source rocks within the basin will not currently be generating hydrocarbons, having effectively been 'switched' off by Tertiary uplift. Charge mechanics are uncertain but must involve early pre-uplift migration or remigration.

The North Channel Basin is an unexplored (apart-from the recent British Gas 112/15-1 well) Late Palaeozoic to Mesozoic basin complex which is believed to be stratigraphically analogous to the East Irish Sea Basin (EISB) hydrocarbon province to the south east (Fig. 1). This paper presents an evaluation of the geology and hydrocarbon prospectivity of the area, based on onshore exploration activity and offshore seismic and potential field data.

Exploration history

No deep wells or boreholes have been drilled in the offshore North Channel, although a number of deep boreholes have been drilled onshore Northern Ireland. In the periods 1960–61 and 1978–9 a series of boreholes were commissioned by the Northern Ireland Government. The first hydrocarbon exploration well, Newmill-1, was drilled by Marathon in 1971 under a farm-in agreement with Shell. This was plugged and abandoned as a dry hole, with no significant hydrocarbon shows. Subsequent wells have been drilled by Fynegold Petroleum in 1990 (Ballytober-1) and Nuevo Energy in 1993 (Annaghmore-1) and 1994 (Ballynamullan-1). These wells are currently unreleased and the results confidential. It is not known whether these wells tested valid traps.

At present only three onshore licences exist. These are operated by Nuevo Energy Company north of Lough Neagh and Mustang Oil and Antrim Resources in east Antrim. Onshore exploration to date has been hampered by the difficulty of acquiring seismic data beneath the Tertiary basalt cover which covers approximately $4000\,\mathrm{km^{-2}}$ of potentially prospective acreage (Griffith 1983).

Two offshore North Channel licences were awarded to British Gas Exploration & Production Limited (BG E&P) as part of the UK 14th Round of Licensing in 1993. The northern licence (P869) includes Blocks 111/03, 111/04 and 111/08, whilst the southern licence (P870) comprises Blocks 111/10 and 111/15 (Fig. 2). All of these blocks are previously unallocated.

From Meadows, N. S., Trueblood, S. P., Hardman, M. & Cowan, G. (eds), 1997, *Petroleum Geology of the Irish Sea and Adjacent Areas*, Geological Society Special Publication No. 124, pp. 95–111.

 S. J. MADDOX *ET AL.*

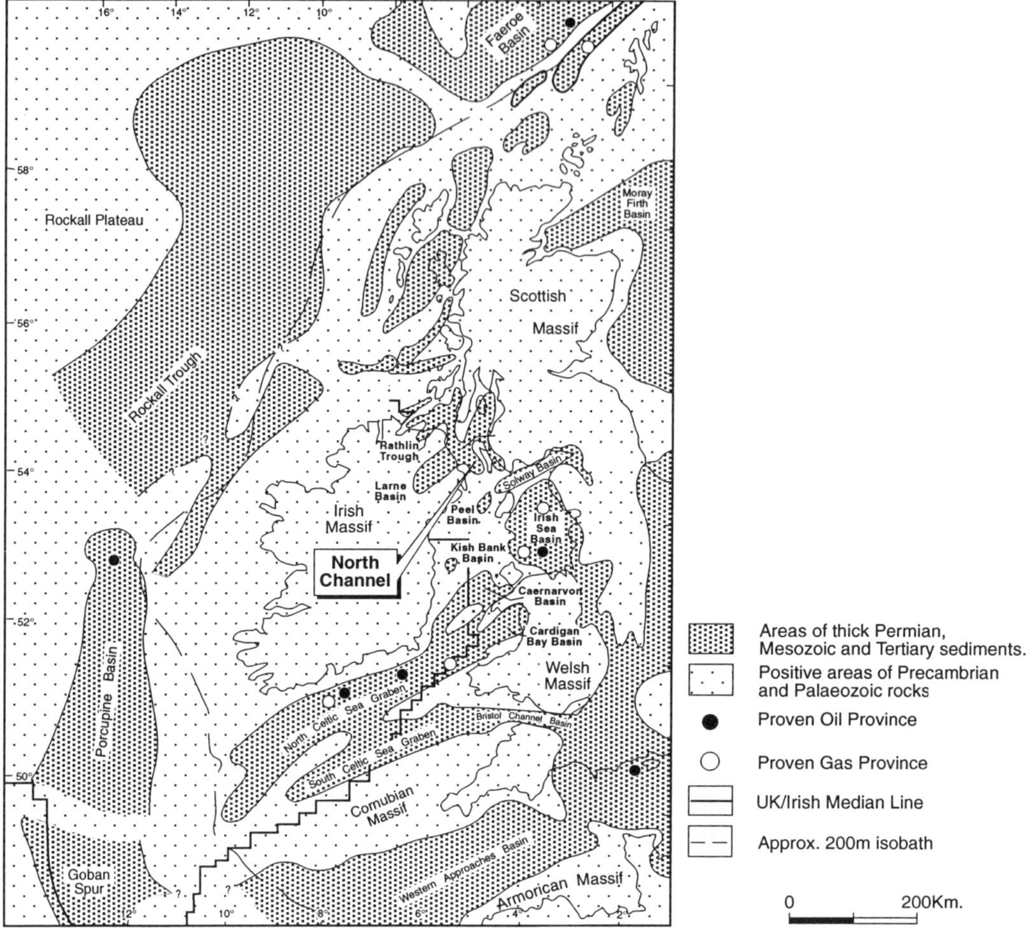

Fig. 1. Map showing locations of sedimentary basins west of Britain with the North Channel highlighted. Adapted from Naylor & Shannon (1982).

Well and geophysical database

BG E&P has access to a number of wells drilled onshore Northern Ireland that are considered representative of the offshore North Channel stratigraphy. The most relevant are a series of wells drilled along the Antrim coast, adjacent to the North Channel seaway. The most important of these is the British Geological Survey (BGS) Larne-2 well, which encountered a thick Permo-Triassic section and was drilled in 1981 as a geothermal test. Other nearby wells include Larne-1 drilled by the BGS in 1962/3, Marathon's Newmill-1 and Fynegold's Ballytober-1. Offshore shallow bore-hole data and onshore outcrops and boreholes in Northern Ireland and south west Scotland have also been evaluated and incorporated within this study.

BG E&P's regional offshore seismic database consists of approximately 3600 km of 2D data, half of which is proprietary data acquired in September 1994. Selected onshore seismic data have been purchased to tie the onshore Larne-2 and Newmill-1 wells. These comprise vibroseis lines acquired in 1981 along the Antrim coast by the former Department of Commerce of Northern Ireland. Proprietary gravity and magnetic studies have also been undertaken, utilizing all BGS marine and land gravity and aeromagnetic data, with additional marine gravity data from Jebco's 1992 and the BG E&P 1994 seismic surveys. Jebco/Ark/Air Atlantic high resolution aeromagnetic data acquired in 1993/1994 over the North Channel and surrounding areas were also incorporated.

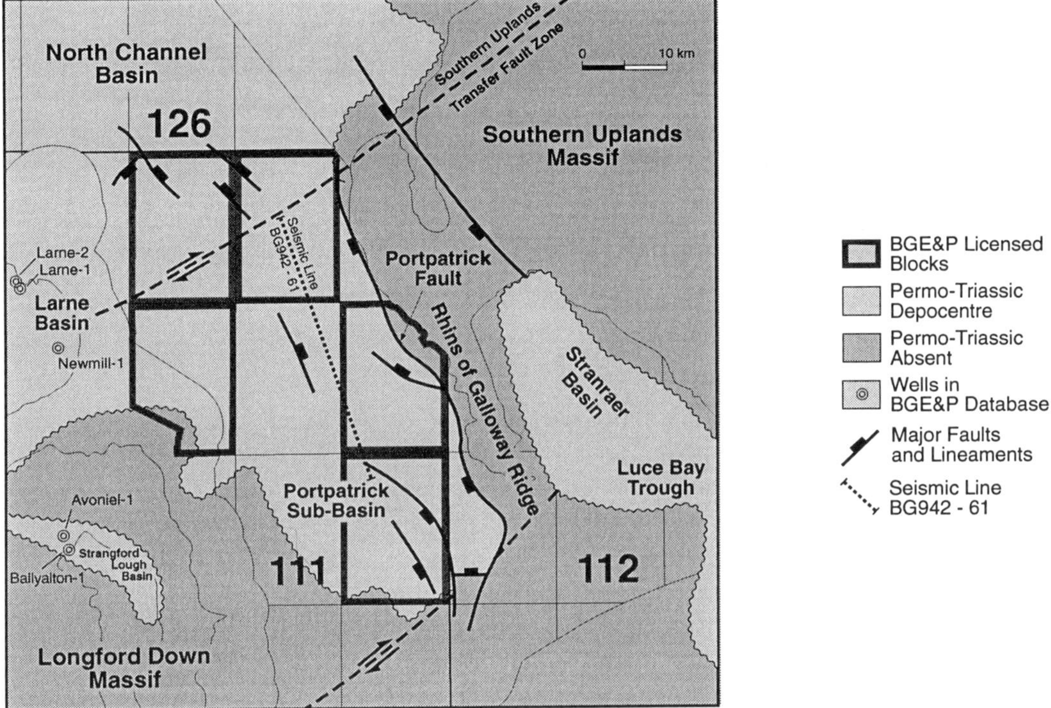

Fig. 2. Major tectonic elements of the North Channel and location of British Gas E&P licences.

Regional and tectonic setting

The North Channel lies between the Antrim area of Northern Ireland and the Rhins of Galloway peninsula of southwest Scotland (Fig. 2). The seaway is characterized by a major bathymetric depression, the Beaufort's Dyke, which is a NW–SE linear glacial channel feature locally increasing water depths to in excess of 300 m. The underlying North Channel Basin is part of a related set of basins, including the EISB, which are remnants of a more extensive late Palaeozoic–mid-Mesozoic basin complex evolving under the influence of the opening of the North Atlantic and pre-existing Caledonian and Variscan trends (McLean 1978; Jackson & Mulholland 1993; Shelton 1995; Maddox *et al.* 1995).

The North Channel Basin can be subdivided into southern and northern sub-basins, the depocentres of which are characterized by well-defined Bouguer gravity anomaly lows. These sub-basins are separated by the offshore extension of the Southern Uplands Fault Zone (SUFZ) which is clearly imaged by the gravity, aeromagnetic and seismic data (Fig. 3). The narrow southern sub-basin (Portpatrick Basin) trends NNW–SSE and is

a simple half-graben, dipping east towards a depocentre in the hangingwall of the basin-bounding Portpatrick Fault. The northern sub-basin widens considerable into the Larne–Firth of Clyde basin complex, which has a predominantly Caledonide (SW–NE) orientation. Unlike the southern basin, the northern sub-basin (Larne Basin) consists of a westerly dipping half-graben, the reversal in polarity apparently accommodated along the SUFZ. A similar change in basin polarity is observed to the south in the EISB, where SW–NE orientated Caledonian lineaments appear to have acted as transfer zones (cf. Morley *et al.* 1990) during predominantly ESE–WNW directed Permo-Triassic extension (Knipe *et al.* 1993).

The aeromagnetic data indicate the presence of intrusive igneous bodies within both the Larne and Portpatrick sub-basins. These are inferred from their predominant NW–SE strike to be of Tertiary age. On the 1994 proprietary seismic, igneous intrusives are clearly evident as high amplitude events, which are both cross-cutting and also bedding-parallel (Fig. 4). Sidescan sonar data (Caston 1975) suggest that igneous dykes in the offshore North Channel have a well-defined northwesterly orientation. In areas of the North

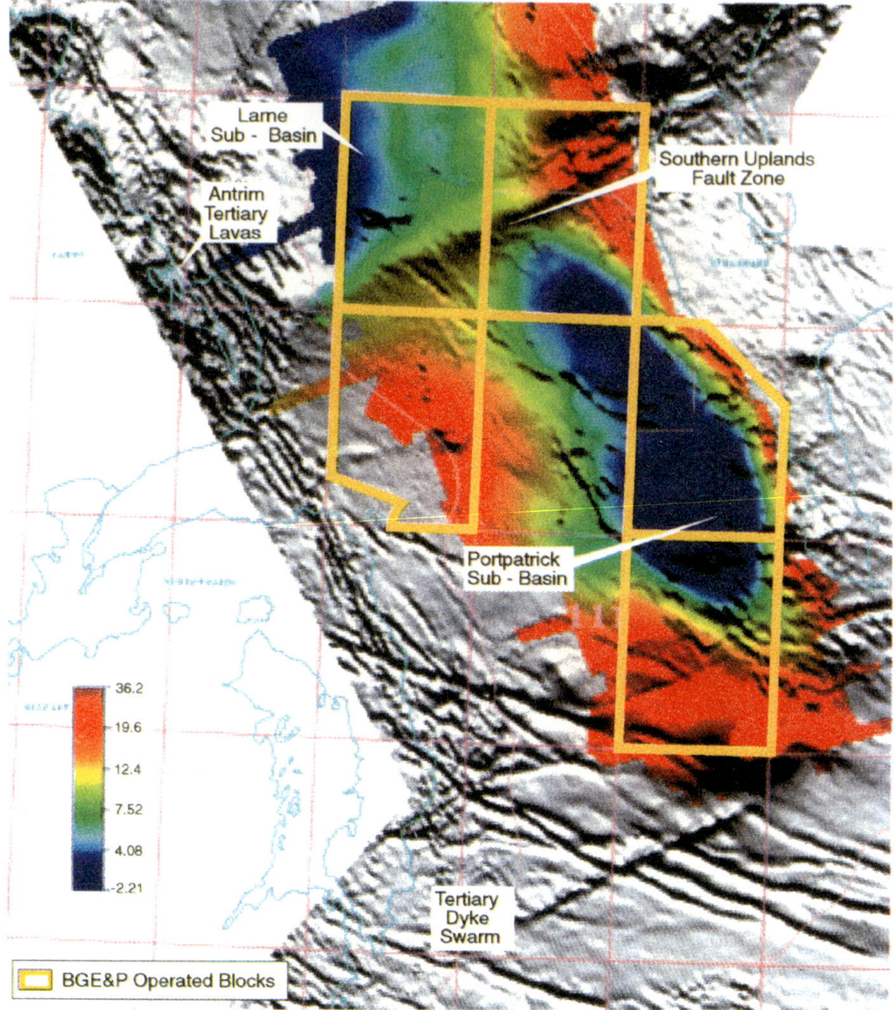

Fig. 3. High resolution aeromagnetic map with Bouguer gravity overlay showing location of the Portpatrick and Larne sub-basins and their relationship to the Southern Uplands Fault. Aeromagnetic data courtesy of Jebco/Ark/Air Atlantic.

Channel where there is significant rock outcrop at sea bed, dykes have been observed from submarines (C.Stoneham pers. comm. 1993) and sidescan sonar data (Caston 1975, 1976) to form 'walls' up to 30 m above the base of tidally scoured hollows in the Mercia Mudstone host rock.

Three major fault trends exist in the North Channel area (NNW–SSE, NW–SE and SW–NE). These are observed on gravity, magnetic and seismic data and correlate with onshore trends and those recorded in other Permo-Triassic basins in NW Britain, suggesting a strong underlying basement control (Anderson *et al.* 1995). During Mesozoic rifting, extension appears to have occurred either along a Caledonoid NE–SW trend or near normal to it (NNW–SSE).

Tectono-stratigraphic evolution

Pre-Carboniferous

The basement rocks of the North Channel area comprise Lower Palaeozoic (Ordovician and Silurian) rocks of the Southern Uplands/Longford Down Massifs. These are probably overlain by Carboniferous and younger rocks over much of the

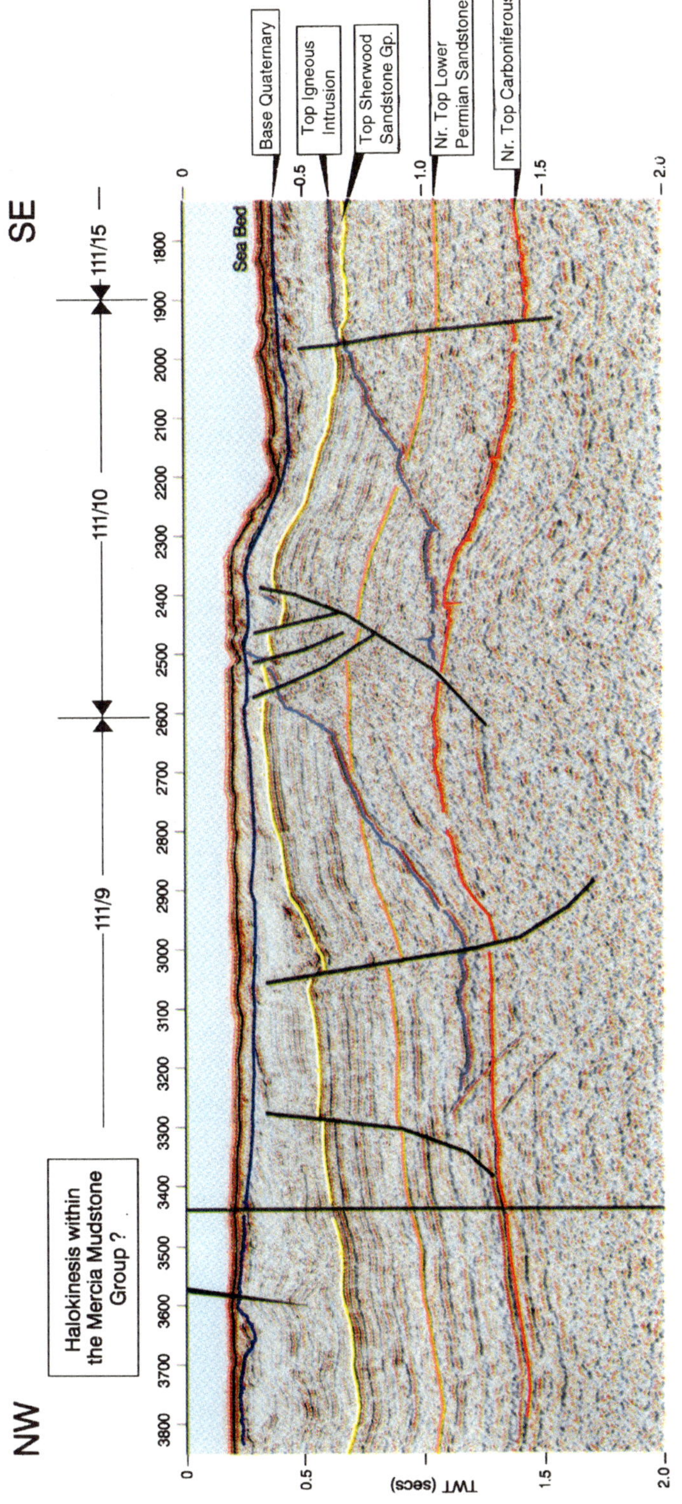

Fig. 4. Regional NW–SE seismic line (BG942-61) through Portpatrick sub-basin showing identified mapped events, Tertiary igneous dykes/sills and halokinesis within the Mercia Mudstone.

offshore area, although Lower Old Red Sandstone rocks are known to the north at outcrop on the Antrim coastline. The Ballyalton borehole drilled by the Geological Survey of Northern Ireland (GSNI) at Newtownards, to the southeast of Belfast, proved a Lower Palaeozoic basement section of greywackes and mudstones, overlain by Permian breccias (Bazley 1975).

Carboniferous

From late Devonian to mid-Namurian, several phases of moderate crustal extension probably affected the North Channel–Northern Ireland area, associated with syndepositional faulting, half-graben development and sporadic volcanism. Dinantian rocks in the area record fluvio-deltaic and marine interaction until the late Namurian–early Westphalian, when extension waned and an overall regressive clastic regime was established under the influence of regional post-rift thermal subsidence. In the last major period of movement of the Variscan Orogeny during the late Westphalian, regional inversion resulted in the erosion and reddening of significant thicknesses of Carboniferous; and between 300–1000 m of Variscan uplift and erosion is predicted for the North Channel area (Mitchell 1992; Wang 1992).

Very few wells and boreholes in the onshore area flanking the North Channel have penetrated the Carboniferous, and most have terminated in the Permo-Triassic. The Ballyalton and Langford Lodge boreholes on the east side of Lough Neagh (Manning *et al.* 1970) failed to prove Carboniferous, encountering Permian rocks unconformably overlying Lower Palaeozoic basement (Fig. 5). Areas of Carboniferous outcrop and subcrop are known from the Strangford and Stranraer–Luce Bay half-graben of Northern Ireland and southwest Scotland (Fig. 2). These basins trend NW–SE and parallel the offshore Portpatrick Basin. All have major controlling faults along the eastern margin and a sediment-fill dipping and thickening to the east. In the Strangford Basin and along the southern shore of Belfast Lough at Cultra, Lower Carboniferous limestones with basal conglomerates and sandstones are known (Clayton 1986), and the Belfast Harbour Borehole cored 200 m (unbottomed) of ?Chadian carbonates, mudstones and siltstones, subcropping the Variscan unconformity (Smith 1986). In the Stranraer–Luce Bay Basin, approximately 30 m of Lower Carboniferous and ?Namurian rocks crop out. These consist of mudstones, sandstones and basalt lava flows (Fuller 1958).

To the west of Lough Neagh in the Fermanagh area, the Lower Carboniferous is up to 3500 m thick, consisting of a mixed clastic–carbonate succession. Volcanic-prone Dinantian outcrops are also present at Ballycastle in north Antrim, north of the Highland Border Ridge. Conformable Namurian sections occur in Fermanagh, Tyrone (where the succession is up to 650m thick) and Ballycastle (Parnell 1991, 1992*a*). Westphalian rocks occur only at Coalisland (Tyrone) 8 km west of Lough Neagh, where there is a Westphalian A section about 250 m thick (Eagar 1975). It is possible, however, that a complete Westphalian A to D succession was formerly present across the north of Ireland (Sevastopulo 1981), which may have been up to 800 m thick (Parnell 1991, 1992*a*). Carboniferous rocks are thought to be present beneath much of eastern Antrim, based upon onshore seismic evidence (Illing & Griffith 1986).

Permian–Jurassic

Permian to Jurassic extension resulted in a complex series of half-graben, with over 3 km of Permo-Triassic preserved in the Larne sub-basin. A similar thickness probably occurs in the Portpatrick sub-basin depocentre. The Permo-Triassic was a period of largely continental sedimentation characterized by the development of two reservoir–seal couplets (Fig. 6).

The lower of the two potential reservoir–seal couplets consists of Lower Permian sandstones, breccias and conglomerates up to 450 m thick, overlain by Upper Permian shales and evaporites. The Lower Permian reservoir unit displays important lateral facies and thickness changes and is locally volcanic-prone, with over 620 m of lavas and tuffs proven by the Larne-2 well (Penn *et al.* 1983). These volcanics have yet to be proven elsewhere in the area, although the well database is limited. A model for Lower Permian deposition based on the East Irish Sea (Jackson *et al.* 1987) involves basin margin alluvial fanglomerates passing basinwards into fluvial and aeolian sandstones. These are inferred to grade further into basin centre shaley and evaporitic playa lake deposits (Jackson *et al.* 1987).

The Upper Permian seal unit is over 200 m thick in Larne-2 (Penn 1981) and consists of the Belfast Harbour Evaporite Formation overlain by the Connswater Marl Formation. In the nearby Newmill-1 well, the Upper Permian is only 10 m thick and the major halite horizons of the Larne-2 well are absent. This may be due to faulting (Illing 1982), although other offset wells outside the Larne Basin depocentre also lack major halite development. This suggests that Upper Permian salt deposition may have been relatively localized towards the centre of the Larne Basin (Jackson & Mulholland 1993), although the undrilled Portpatrick Basin depocentre may have also been

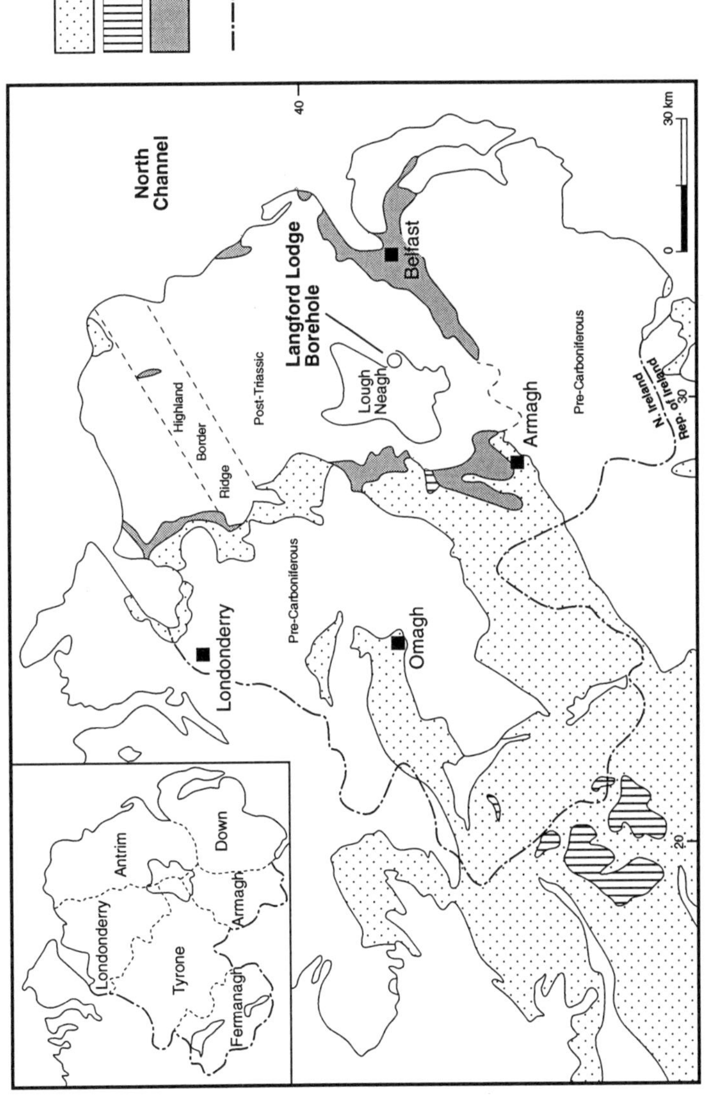

Fig. 5. Map of Northern Ireland showing surface geology and location of Langford Lodge Borehole.

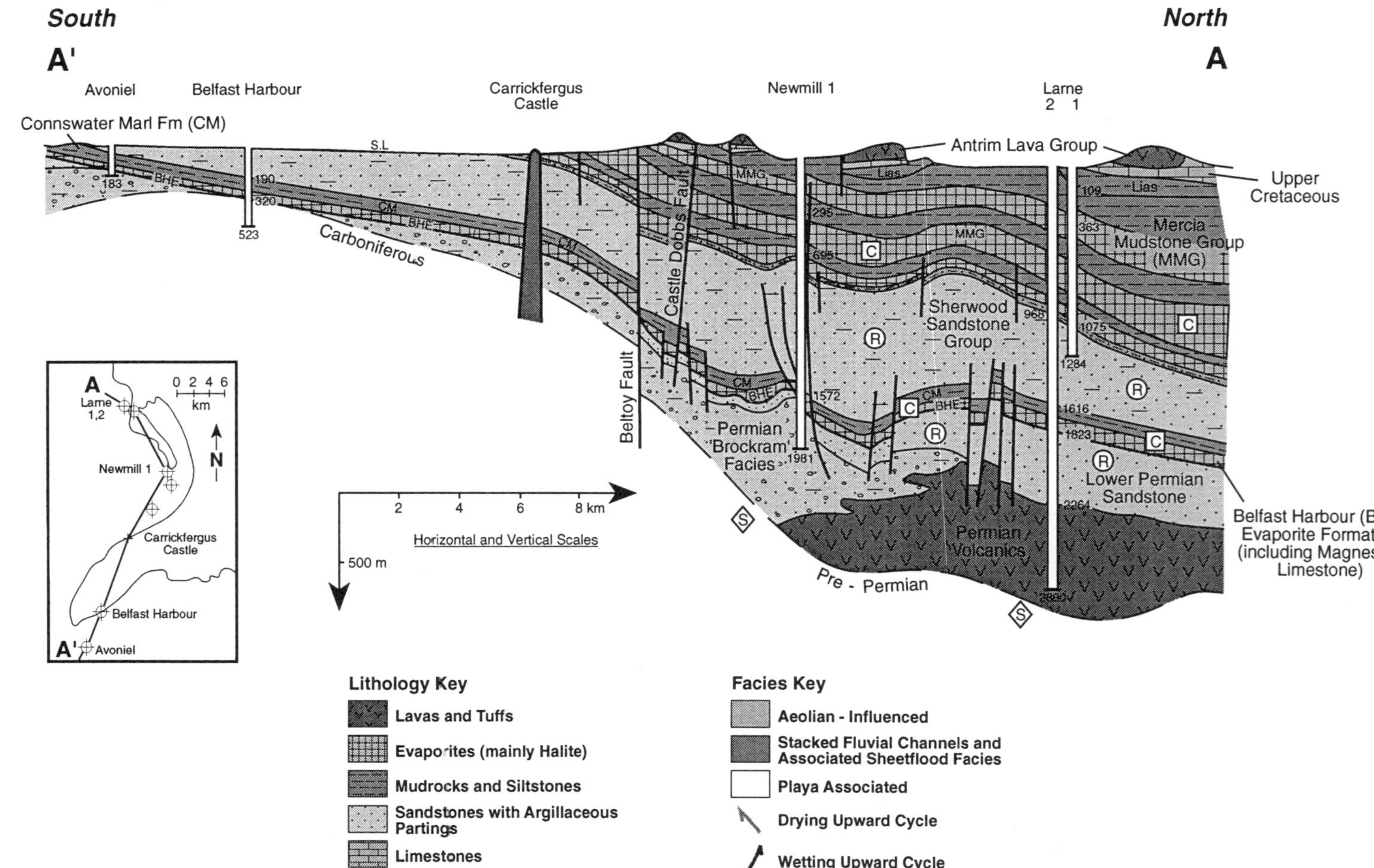

Fig. 6. South–north geological cross-section of the east Antrim coast incorporating deep well, shallow borehole and outcrop data (adapted from McCann 1990 and McCaffrey & McCann 1992). ®, Potential reservoirs; ©, caprocks; Ⓢ, source horizons.

the site of a similar salt basin. This is supported by gravity and seismic data.

The upper reservoir–seal couplet comprises Lower Triassic sandstones overlain by Upper Triassic halites, siltstones and shales. The reservoir unit, the Sherwood Sandstone Group, comprises fluvial and aeolian sediments deposited basinwide with generally good reservoir quality. It displays pronounced thickness variations, from over 850 m in Newmill-1 to less than 300 m at outcrop and outside the Larne Basin depocentre (Parnell 1992b). The sealing Mercia Mudstone Group consists of shales, siltstones and thin sandstones interbedded with three major halite units. In the Larne Basin depocentre it is 951 m thick in the Larne-2 well. Outside the Larne Basin depocentre, the Mercia Mudstone Group is predominantly shale-prone and major halite intervals are absent. Offshore, seismic character suggests that halites are probably present in the Portpatrick Basin depocentre, and this is supported by the work of Jackson & Mulholland (1993).

In the North Channel, interpretation of seismic data suggests that post-Triassic rocks are absent, with the exception of Tertiary igneous intrusives, and the Mercia Mudstone Group is unconformably overlain by Quaternary–Recent deposits. Onshore, Lower Jurassic shallow marine mudstones with minor limestones overlie the Triassic and are generally 25–100 m thick (McCaffrey & McCann 1992). In the Port More Borehole of the Rathlin Trough, however, almost 250 m of Lower Jurassic mudstones were encountered (Wilson 1981). These have been dated as Hettangian to Pliensbachian using macrofossil zonation (Wilson 1972).

Post-Jurassic

The extent of former post-Pliensbachian sedimentation, prior to Cretaceous deposition is unknown. Following Cimmerian uplift and erosion, Cenomanian to Maastrichtian greensands and chalks, 40–100 m in thickness, were deposited on the tilted and faulted pre-Cretaceous surface (Wilson 1972; Shelton 1995). Clasts of middle and upper Lias age occur in the basal Cretaceous conglomerate in north Antrim (Wilson 1972), and outliers of Liassic age occur onshore northern England and in the East Irish Sea Basin (Jackson & Mulholland 1993). These suggest a formerly thicker Liassic cover.

Extremely rapid pre-Palaeocene uplift and erosion is suggested by the karstification of the Late Cretaceous chalk surface, onto which Palaeocene basalt lavas, up to 800 m thick were extruded. These occur in fault-controlled basins (Fyfe *et al.* 1993) and related dyke and sill intrusions were also emplaced in the offshore section. The basaltic lavas of Antrim form the largest onshore area of extrusives in the British Tertiary Volcanic Province (Fyfe *et al.* 1993). The Tertiary extrusive volcanics correlate with an area of high frequency aeromagnetic anomalies which does not extend far offshore. This suggests that extrusive lavas are largely absent offshore and is corroborated by seismic data, shallow sea-bed coring and geological studies which suggest that they were probably not originally deposited (Fyfe *et al.* 1993).

Following the volcanism, the late Eocene and early Oligocene were characterized by uplift and erosion. Localized subsidence during the Oligocene led to the development of the non-marine Lough Neagh Basin onshore Northern Ireland. Similar lacustrine basins developed in many regions of western Britain and examples include the onshore Bovey Tracey Basin in Devon and the offshore Cardigan Bay Basin, west of Wales (Dobson & Whittington 1987). In the Lough Neagh Basin the succession consists of up to 380 m of clays and silts, with lignite beds up to 80 m thick and pebble beds developed in the west (Wilkinson *et al.* 1980; Geological Survey of Northern Ireland 1984).

Reconstructed cross-sections, flattened on the base of the lacustrine Oligocene deposits of the Lough Neagh Basin, show post-basalt, Eocene deformation, possibly controlled by inversion of Late Jurassic (Cimmerian) extensional faults (George 1967; Roberts 1989). No Eocene sediments are known from Northern Ireland, Scotland or the inner Scottish shelf, although great thicknesses of sediment were deposited on the outer Hebridean shelf and in the Rockall Trough, which had begun to subside. (Wood & Van Horn 1987; Fyfe *et al.* 1993). Local warping of the Oligocene freshwater clays and lignites suggests possible Miocene inversion (Roberts 1989), leading to deformation, uplift and erosion, prior to Quaternary–Recent deposition and erosion. No Miocene or Pliocene deposits are recorded in onshore Northern Ireland.

In the northern part of the North Channel and along the margins bordering the coasts there is significant rock outcrop at sea bed (Caston 1976; Fyfe *et al.* 1993). Here, glacial deposits have been removed and all superficial deposits are mobile. This is probably due to erosion associated with strong tidal currents in the range of 1.5 m s^{-1} to 2.3 m s^{-1}, from surface to bottom. Elsewhere in the North Channel, Recent sea-bed sediments (sands and gravels) are underlain by a Quaternary succession of variable thickness, with possible kettle holes and erosive channel features (Caston 1976). The Quaternary consists of firm to stiff gravelly clays underlain by till, although in places till outcrops at sea bed, forming local bathymetric highs. The gravelly clays predominantly infill steep-sided channel features and Caston (1976) has suggested that the North Channel must have been subjected to particularly intense glaciation. Bathymetric deeps such as the 'Beaufort Dyke' are therefore believed to be the result of glacial erosion of softer Mesozoic sediments (Fyfe *et al.* 1993). Although BGS data suggest that Quaternary thicknesses in the North Channel area do not exceed 50–80 m, proprietary British Gas studies utilizing high resolution shallow seismic indicate that thicknesses of around 150 m occur.

Play concepts

Three potential plays in the Triassic, Permian and Carboniferous have been identified in the North Channel, where the potential for prospectivity has been compared to the East Irish Sea (Parnell 1992*a,b*). The primary and lowest risk play is the Triassic, with secondary potential in the Permian. Further secondary potential is recognized in the Carboniferous, although this play is considered an exceptionally high risk. The Triassic play fairway is defined as the area where the Sherwood Sandstone Group is sealed by shales and evaporites of the Mercia Mudstone Group and charged by Carboniferous source rocks. The secondary Permian play fairway is defined as the area where Lower Permian reservoir facies are sealed by Upper Permian shales and evaporites and charged by Carboniferous source rocks. The Carboniferous play fairway is considered to occur throughout the area and relies on either intraformational mudstones, evaporites (Ward, this volume) or overlying Lower Permian caprock facies, to seal possible clastic and carbonate reservoirs. Migration of hydrocarbons from Carboniferous source horizons into adjacent reservoirs is not regarded as problematical.

Reservoirs

The Lower Triassic Sherwood Sandstone Group is believed to extend across the offshore North Channel into the Larne and Portpatrick sub-basins. Onshore Northern Ireland, in the landward extension of the Larne sub-basin, several wells have encountered reservoir quality sands (Parnell 1992*b*). The key Larne-2 well penetrated 648 m of Sherwood Sandstone Group, the upper 286 m of which correlates with the Ormskirk Sandstone Formation (cf. Meadows & Beach 1993) of the East Irish Sea. Log analysis of this interval suggests porosities in excess of 20% and a net:gross ratio of 65%. Core data towards the base of the unit indicate porosities of 23% and permeabilities of 13–63 mD (Fig. 7). In the nearby Newmill-1 well, drilled by Marathon in 1971, a core close to the top of the

Sherwood Sandstone Group displayed permeabilities of up to 1 Darcy (averaging 306 mD), with porosities of 21% and a net:gross of 90%. Although the presence of an effective reservoir is considered low risk, the Sherwood Sandstone Group may be downgraded in quality by burial-diagenetic effects and the occurrence of local igneous intrusions (Preston 1962; Parnell 1992*b*).

Lower Permian sandstones are also believed to extend into the offshore sub-basins of the North Channel, although the lack of well control leads to uncertainty in predicting facies variations. Onshore Northern Ireland, Larne-2 penetrated 441 m of Lower Permian sandstones underlain by 671 m of lavas, tuffs and basal 'brockram facies' breccio-conglomerates (Penn *et al.* 1983). Core data from Larne-2 Permian sandstones indicate average porosities of 7%, with permeabilities ranging from 0.2 mD to 1 mD. These contrast with data from the nearby Newmill-1 well, where average porosities and permeabilities were 16% and 212 mD respectively, with permeabilities as high as 1 Darcy (Fig.8).

Source

Onshore Northern Ireland and southwest Scotland, potential source rocks have been identified in the Lower, Middle and Upper Carboniferous (Fig. 9). The prospectivity of the offshore depends on the presence of these facies beneath the thick Permo-Triassic succession. Westphalian B coals and mudstones from Ayrshire and Coalisland (N. Ireland) have good quality, gas-prone and oil-prone source potential, whilst Brigantian shales and coals from Antrim and Kintyre have good quality oil-prone and very good quality gas-prone source potential (Parnell 1991; 1992*a*).

Around the onshore margins of the offshore North Channel, Carboniferous samples are immature to peak oil-window mature, exhibiting vitrinite reflectance values of 0.35–0.90% (averaging 0.6%). Offshore, buried beneath the thick Permo-Triassic succession, extrapolation from onshore trends suggests that source beds will be mature for oil and some gas generation (Parnell 1992*a*).

Burial history modelling of the North Channel Basin is problematical due to the absence of post-Triassic section and significant well data. Unpublished apatite fission track data from around the onshore basin margins and adjacent areas, however, suggest several thermal 'events' which may have led to hydrocarbon generation. The most significant of these appears to have been 'early' and may be related to Variscan deformation and/or Permo-Triassic rifting and volcanism. This may have predated major trap development associated

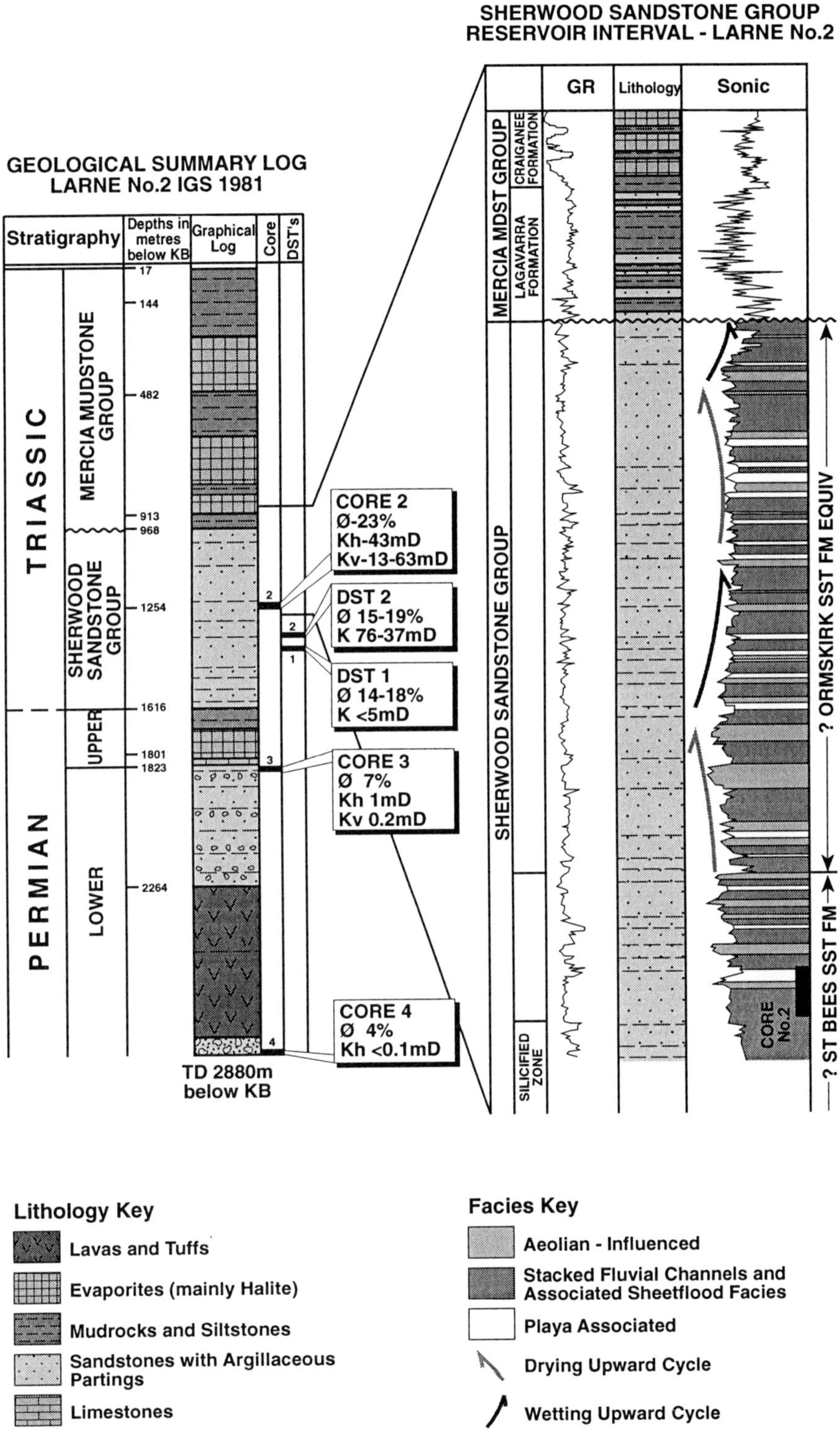

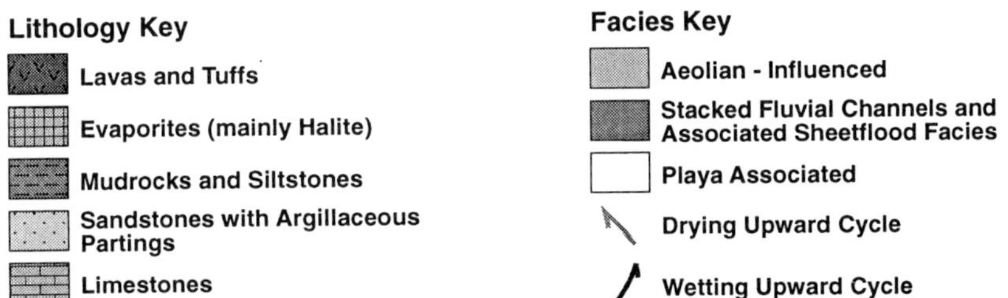

Fig. 7. Geological summary log of Larne-2 (adapted from Penn 1981) and facies interpretation of the upper part of the Sherwood Sandstone.

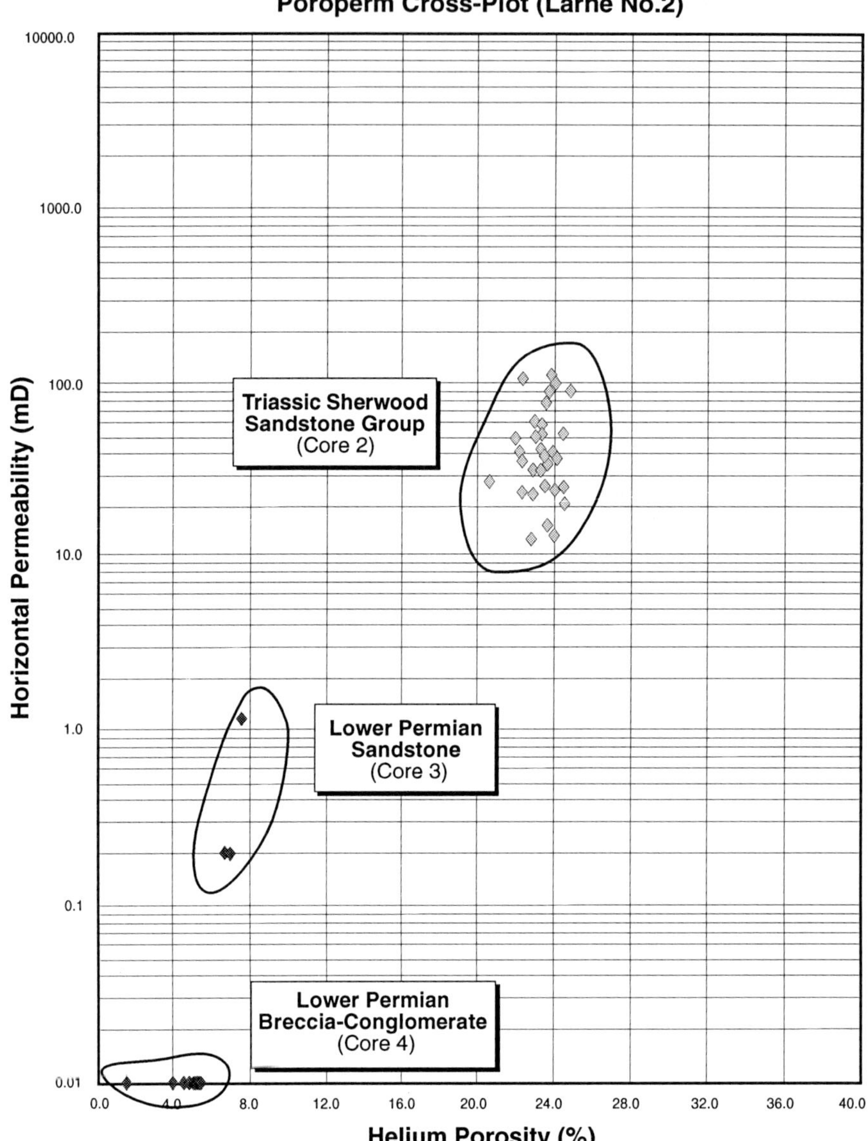

Fig. 8. Poroperm cross-plot of horizontal permeability versus helium porosity for Larne-2 Permo-Triassic core data.

with Permian to Jurassic extensional faulting and is a critical exploration risk.

Subsequent thermal and burial events in the Jurassic and Tertiary may, however, have resulted in further hydrocarbon generation and migration, leading to the charging of trapping structures. This is believed to have been the case in the analogous East Irish Sea Basin, to the southeast, where several phases of hydrocarbon generation, punctuated by

uplift and breaching are inferred (Hardman *et al.* 1993; Stuart & Cowan 1991).

Seals

Offshore, seismic data suggest that the Middle–Upper Triassic Mercia Mudstone Group overlies the Sherwood Sandstone Group over much of the Larne and Portpatrick sub-basins. Onshore

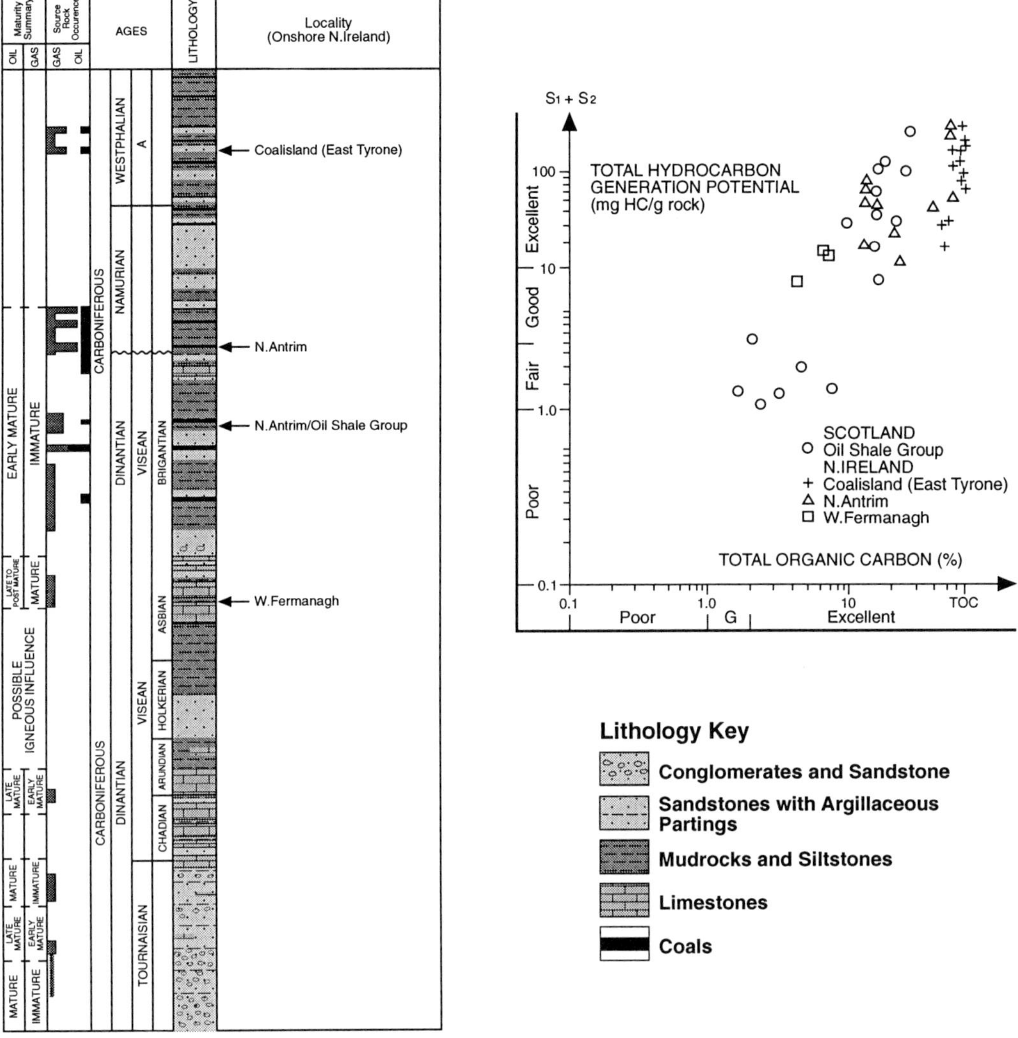

Fig. 9. Composite Carboniferous section from Northern Ireland showing potential source rock horizons and TOC versus pyrolysis data for Northern Ireland samples and Scottish oil shales (adapted from Parnell 1991 and 1992*a*).

Northern Ireland, in Larne-2, the Mercia Mudstone Group is 951 m thick and contains three major halite units interbedded with mudstones, abundant siltstones, anhydrites and thin sandstones (Fig. 6). These coarser clastics may locally downgrade the side- and top-sealing capacity of the Mercia Mudstone Group. Major halites are also present in Newmill-1, but are absent away from the Larne Basin depocentre, where the Mercia Mudstone Group consists predominantly of claystones. Seal effectiveness is downgraded where Tertiary uplift has brought the Sherwood Sandstone Group close to surface, resulting in a limited thickness of the Mercia Mudstone Group. In the analogous East Irish Sea Basin to the southeast, numerous shallow hydrocarbon accumulations have been discovered, and in-house British Gas studies suggest that significant hydrocarbon accumulations (e.g. South Morecambe Field) may be reservoired at sub-sea depths as shallow as 600–800 m (Maddox *et al.* 1995). Another potential risk relates to the presence of a 'waste zone' between the top of the Sherwood Sandstone Group and the base of the lowest effective halite seal (cf. Maddox *et al.* 1995). This is characterized by thin sandstones which may act as 'thief' zones, causing leakage of hydrocarbons.

A waste zone may also degrade structural closures, reducing net pay, by acting as an uppermost, low net-to-gross 'reservoir' horizon. In Larne-2 the waste zone corresponds to the Lagavarra Formation which is 55 m thick.

Onshore Northern Ireland, in Larne-2, potential Lower Permian reservoirs are overlain by a 207 m thick Upper Permian section with significant sealing potential. This consists of a basal, 22 m thick Magnesian Limestone, with a dolomitic cap, overlain by the Belfast Harbour Formation, comprising 113 m of halite and 6 m of anhydritic siltstone. The upper part of the section (Connswater Marl Formation) consists of 66 m of red-brown mudstones, siltstones and very-fine sand stringers, broadly equivalent to the St Bees Shales of the Solway Firth Basin. In the offshore North Channel, Upper Permian facies similar to those in Larne-2 should provide an effective seal.

Geophysical evaluation and trapping structures

The quality and resolution of the BG E&P 1994 proprietary 2D seismic is very high with intensive processing successfully removing the sea-bed multiples, which were a major problem in some of the earlier surveys acquired in the North Channel area. It is thus now possible to identify horizons down to the base Permian unconformity, and also within the underlying Carboniferous. Most significantly, the proprietary data, integrated with high resolution aeromagnetics, have allowed Tertiary dykes and sills to be recognized and mapped more accurately, and a better understanding has emerged of their effect on data quality. In the vicinity of these igneous bodies, and particularly underlying bedding-parallel sills, the quality of data deteriorates, with significant attenuation of higher frequency events. Away from the Tertiary volcanics, however, data quality is excellent.

The nearest well control is provided by the onshore coastal Larne-2 and Newmill-1 wells in Northern Ireland. These can be tied to the onshore vibroseis seismic and a 'jump correlation' can be made to two offshore seismic 'tie lines' that were shot within 5 km of the well locations. Horizon identification is further aided by seismic character, but a great deal of uncertainty exists away from well control, with event correlation across major faults and in the vicinity of Tertiary volcanics being particularly difficult. Utilizing these well data and seismic character correlations, the following events have been identified and mapped: Base Quaternary, Top Sherwood Sandstone Group, Top Lower Permian Sandstone/Magnesian Limestone and Base Permian Unconformity (Fig. 4). The Top Sherwood Sandstone event is mapped at the base of a halokinetically deformed Mercia Mudstone package, particularly well defined on the BG E&P 1994 dataset.

Significant extensional faulting is recognized both at Top Sherwood Sandstone and Top Lower Permian Sandstone levels (Shelton 1995). The two structural styles which predominate are rollover anticlines and tilted fault blocks, both of which represent exploration targets (Fig. 10). Rollover anticlines are developed in the hangingwalls of major NNW–SSE orientated listric faults with ramp-flat geometries. These are interpreted as early features, related to Permian to Jurassic extension, although some later, compressional 'tightening' may also have occurred. Tilted fault block features are also related to Permian to Jurassic extensional faulting and are mainly developed in the footwalls of NNW–SSE trending faults.

Potential hydrocarbon traps recognized in the offshore North Channel are relatively high risk due to uncertainty of the timing of trap formation versus hydrocarbon generation and the shallow depths to Top Sherwood Sandstone. Apatite fission track data suggest several thermal events which may have led to hydrocarbon generation. The most significant of these probably predated trap formation, although later phases of generation probably occurred after the development of trapping structures during Permian to Jurassic extensional episodes. A further and perhaps more fundamental risk relates to Carboniferous source rock presence and quality. Although potential source rocks have been identified onshore Northern Ireland and in the Midland Valley of Scotland, these may not be present beneath the North Channel, although the basin is believed to have originated during the Devonian.

Conclusions

The geology and hydrocarbon prospectivity of the North Channel has been investigated using all available data, which includes 1994 seismic data proprietary to British Gas and high resolution aeromagnetic data acquired by Jebco/Ark/Air Atlantic. The main conclusions are as follows.

- The North Channel is a narrow seaway underlain by two sub-basins, separated by the offshore extension of the Southern Uplands Fault Zone. A reversal in dip polarity between the sub-basins occurs across the SUFZ, suggesting that it has acted as a transfer fault.
- Regional geological evaluation suggest that the North Channel originated as a Devono-Carboniferous basin, which underwent significant extension and subsidence during ENE–WSW directed rifting in the Permo-Triassic. The post-

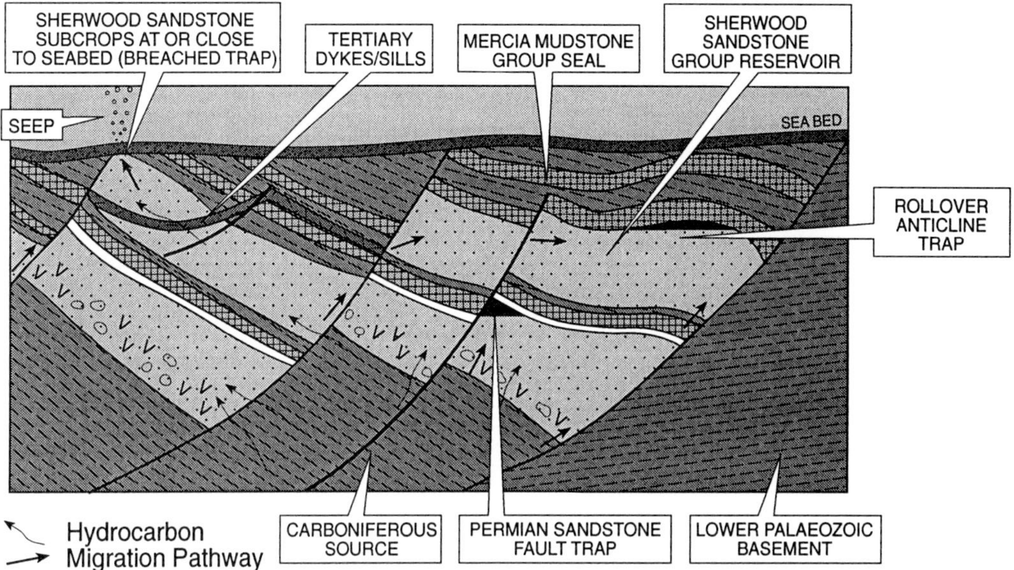

Fig. 10. Hydrocarbon play summary for the North Channel showing structural styles, play risks, possible hydrocarbon migration pathways and play elements.

Triassic history of the area is poorly understood, although it is likely that several uplift and erosion events in the late Mesozoic and Tertiary have resulted in the removal of significant thicknesses of post-Triassic section. This former cover is in part recorded onshore Northern Ireland, where Jurassic, Cretaceous and Tertiary deposits are preserved.

- The North Channel is presently undrilled (apart from the recent British Gas 112/15-1 well), but offset well data onshore Northern Ireland suggests that the basin fill is analogous to the East Irish Sea. The primary play therefore comprises Sherwood Sandstone reservoirs, sealed by Mercia Mudstone halites and shales and sourced by underlying Carboniferous shales and coals. Secondary reservoir and sealing potential is also recognized in the Permian.

- Significant tilted fault block and rollover anticline structures are observed at Triassic and Permian levels on seismic data, and these potential trapping structures represent targets for future exploration drilling.

- Exploration risks relate to the uncertainty regarding the presence of Carboniferous source rocks and the present limited understanding of the timing of hydrocarbon generation versus trap formation. Other risks include the shallow depths of the Triassic reservoir targets and the possible detrimental effects of Tertiary igneous bodies intruded into reservoirs.

We would like to thank the directors of British Gas Exploration and Production for permission to publish this paper. Colleagues at British Gas involved in the North Channel exploration programme are also thanked, particularly Alex Lewicka, James Buchanan, Martin Hardman, John Field, Joanna Bradney, Greig Cowan and Vanessa Lloyd. The manuscript also benefited from the constructive comments of referees Glen Cayley of Esso Exploration and Production UK Limited and John Parnell of the Queen's University of Belfast.

We are grateful to the cartography department at BG E&P for the drafting of diagrams and to Jebco Ltd and Ark/Air Atlantic for allowing us to use the high resolution aeromagnetic data.

References

ANDERSON, T. B., PARNELL, J. P. & RUFFELL, A. H. 1995. Influence of basement on the geometry of Permo-Triassic basins in the northwest British Isles. *In*: BOLDY, S. A. R. (ed.) *Permian and Triassic Rifting in Northwest Europe.* Geological Society, London, Special Publication, **91**, 103–122.

BAZLEY, R. A. B. 1975. The Tertiary igneous and Permo-Triassic rocks of the Ballyalton Borehole, County Down. *Bulletin of the Geological Survey of G.B.* **50**, 71–101.

CASTON, G. F. 1975. Igneous dykes and associated scour

hollows of the North Channel, Irish Sea. *Marine Geology*, **18**, M77–M85.

—— 1976. The floor of the North Channel, Irish Sea: a side-scan sonar survey. *Report of the Institute of Geological Sciences*, No.76/7.

CLAYTON, G. 1986. Late Tournaisian miospores from the Ballycultra Formation at Cultra, County Down, Northern Ireland. *Irish Journal of Earth Sciences*, **8**, 73–79.

DOBSON, M. R. & WHITTINGTON, R. J. 1987. The geology of Cardigan Bay. *Proceedings of the Geologists' Association*, **98**, 331–353.

EAGAR, R. M. C. 1975. Neuere Arbeiten uber das Westfal in Irland. *Zentralblatt fur Geologieund Palaontologie*, **1**, 291–308.

FULLER, J. G. C. M. 1958. The petrology of the Carboniferous rocks near Stranraer, Wigtownshire. *Proceedings of the Geological Association*, **69**, 16–174.

FYFE, J. A., LONG, D. & EVANS, D. 1993. *United Kingdom offshore regional report: the geology of the Malin–Hebrides sea area.* HMSO for the British Geological Survey, London.

GEOLOGICAL SURVEY OF NORTHERN IRELAND. 1984. *Mineral Exploration Programme.* Vols 1–9, GSNI, Belfast.

GEORGE, T. N. 1967. Landform and structure in Ulster. *Scottish Journal of Geology*, **3**, 413–448.

GRIFFITH, A. E. 1983. The Search for Petroleum in Northern Ireland. *In*: BROOKS, J. (ed.) *Petroleum Geochemistry and Exploration of Europe.* Geological Society, london, Special Publication, **12**, 213–222.

HARDMAN, M., BUCHANAN, J., HERRINGTON, P. & CARR, A. 1993. Geochemical modelling of the East Irish Sea Basin: its influence on predicting hydrocarbon type and quality. *In*: PARKER, J. R. (ed.) *Petroleum Geology of Northwest Europe: Proceedings of the 4th Conference.* Geological Society, London, 809–821.

ILLING, L. V. 1982. An Assessment of Petroleum Prospects near Larne Northern Ireland (on behalf of the Department of Economic Development Northern Ireland). Unpublished Report, V. C. Illing and Partners, Surrey, September 1982.

—— & GRIFFITH, A. E. 1986. Gas prospects in the 'Midland Valley' of Northern Ireland. *In*: BROOKS, J., GOFF, J. C. & VAN HOORN, B. (eds) *Habitat of Palaeozoic Gas in NW Europe.* Geological Society, London, Special Publication, **23**, 73–84.

JACKSON, D. I. & MULHOLLAND, P. 1993. Tectonic and stratigraphic aspects of the East Irish Sea Basin and adjacent areas: contrasts in their post-Carboniferous structural styles. *In*: PARKER, J. R. (ed.) *Petroleum Geology of Northwest Europe: Proceedings of the 4th Conference.* Geological Society, London, 791–808.

——, ——, JONES, S. & WARRINGTON, G. 1987. The Geological framework of the East Irish Sea Basin. *In*: BROOKS, J. & GLENNIE, K. (eds) *Petroleum Geology of North-West Europe.* Graham and Trotman, London, 191–204.

KNIPE, R. J., COWAN, G. & BALENDRAN, V. S. 1993. The tectonic history of the East Irish Sea Basin with reference to the Morecambe Fields. *In*: PARKER, J. R. (ed.) *Petroleum Geology of Northwest Europe: Proceedings of the 4th Conference.* Geological Society, London, 857–866.

MADDOX, S. J., BLOW, R. A. & HARDMAN, M. 1995. Hydrocarbon prospectivity of the Central Irish Sea Basin with reference to Block 42/12, offshore Ireland. *In*: CROKER, P. F. & SHANNON, P. M. (eds) *The Petroleum Geology of Ireland's Offshore Basins*, Geological Society, london, Special Publication, **93**, 59–77.

MANNING, P. I., ROBBIE, J. A. & WILSON, H. E. 1970. *Geology of Belfast and the Lagan Valley.* Memoir of the Geological Survey of Northern Ireland.

MEADOWS, N. S. & BEACH, A. 1993. Controls on reservoir quality in the Triassic Sherwood Sandstone of the Irish Sea. *In*: PARKER, J. R. (ed.) *Petroleum Geology of Northwest Europe: Proceedings of the 4th Conference.* Geological Society, London, 823–833.

McCAFFREY, R. J. & McCANN, N. 1992. Post-Permian basin history of northeast Ireland. *In*: PARNELL, J. (ed.) *Basins on the Atlantic Seaboard: Petroleum Geology, Sedimentology and Basin Evolution.* Geological Society, London, Special Publication, **62**, 277–290.

McCANN, N. 1990. The Subsurface Geology between Belfast and Larne, Northern Ireland. *Irish Journal of Earth Sciences*, **9**, 71–78.

McLEAN, A. C. 1978. Evolution of fault controlled ensialic basins in north western Britain. *In*: BOWES, D. R. & LEAKE, B. E. (eds) Crustal Evolution in northwestern Britain and adjacent regions. *Geological Journal Special Issue*, **10**, 325–346.

MITCHELL, W. I. 1992. The origin of Upper Palaeozoic sedimentary basins in Northern Ireland and relationships with the Canadian Maritime Provinces. *In*: PARNELL, J. (ed.) *Basins on the Atlantic Seaboard: Petroleum Geology, Sedimentology and Basin Evolution.* Geological Society, London, Special Publication, **62**, 191–202.

MORLEY, C. K., NELSON, R. A., PATTON, T. L. & MUNN, S. G. 1990. Transfer Zones in the East African Rift System and Their Relevance to Hydrocarbon Exploration in Rifts. *American Association of Petroleum Geologists Bulletin*, **74**, 1234–1253.

NAYLOR, D. & SHANNON, P. M. 1982. *The Geology of Offshore Ireland and West Britain.* Graham & Trotman, London.

PARNELL, J. 1991. Hydrocarbon potential of Northern Ireland: Part I. Burial histories and source rock potential. *Journal of Petroleum Geology*, **14**, 65–78.

—— 1992a. Burial histories and hydrocarbon source rocks on the North West Seaboard. *In*: PARNELL, J. (ed.) *Basins on the Atlantic Seaboard: Petroleum Geology, Sedimentology and Basin Evolution.* Geological Society, London, Special Publication, **62**, 3–16.

—— 1992b. Hydrocarbon Potential of Northern Ireland: Part III. Reservoir potential of the Permo-Triassic. *Journal of Petroleum Geology*, **15**, 51–70.

PENN, I. E. 1981. Larne No.2 Geological well completion report. *Report of the Institute of Geological Sciences*, London, 81/6.

——, HOLLIDAY, D. W., KIRBY, G. A., KUBALA, M., SOPER, R. A. *ET AL.* 1983. The Larne borehole No.2: discovery of a new Permian volcanic centre. *Scottish Journal of Geology*, **19**, 333–346.

PRESTON, J. 1962. Explosive volcanic activity in the Triassic sandstone of Scrabo Hill, Co. Down. *Irish Naturalist's Journal*, **14**, 45–51.

ROBERTS, D. G. 1989. Basin inversion in and around the British Isles. *In*: COOPER, M. A. & WILLIAMS, G. D. (eds) *Inversion Tectonics*. Geological Society, London, Special Publication, **44**, 131–150.

SEVASTOPULO, G. D. 1981. Upper Carboniferous. *In*: HOLLAND, C. H. (ed.) *A geology of Ireland*. Scottish Academic Press, Edinburgh, 173–187.

SHELTON, R. 1995. Mesozoic basin evolution of the North Channel: preliminary results. *In*: CROKER, P. F. & SHANNON, P. M. (eds) *The Petroleum Geology of Ireland's Offshore Basins*. Geological Society, London, Special Publication, **93**, 17–20.

SMITH, R. A. 1986. Permo-Triassic and Dinantian rocks of the Belfast Harbour Borehole. *Report of the British Geological Survey*, **18**.

STUART, I. A. & COWAN, G. 1991. The South Morecambe Field, Blocks 110/2a, 110/3a, 110/8a, UK East Irish Sea. *In*: ABBOTTS, I. L. (ed.) *United Kingdom Oil and Gas Fields, 25 Years Commemorative Volume*. Geological Society, London, Memoir, **14**, 527–541.

WANG, W. H. 1992. Origin of reddening and secondary porosity in Carboniferous sandstones, Northern Ireland. *In*: PARNELL, J. (ed.) *Basins on the Atlantic Seaboard: Petroleum Geology, Sedimentology and Basin Evolution*. Geological Society, London, Special Publication, **62**, 243–254.

WARD, J. 1997. Early Dinantian evaporites of the Easton-1 well, Solway Basin, onshore Cumbria. *This volume*.

WILKINSON, G. C., BAZLEY, R. A. B. & BOULTER, M. C. 1980. The geology and palynology of the Oligocene Lough Neagh Clays, Northern Ireland. *Journal of the Geological Society, London*, **137**, 65–75.

WILSON, H. E. 1972. *British Regional Geology: Northern Ireland*. HMSO for Geological Survey of Northern Ireland, Belfast.

—— 1981. Permian and Mesozoic. *In*: HOLLAND, C. H. (ed.) *A geology of Ireland*. Scottish Academic Press, Edinburgh, 201–212 .

WOOD, M. V. & VAN HOORN, B. 1987. Post-Mesozoic differential subsidence in the northeast Rockall Trough related to volcanicity and sedimentation. *In*: BROOKS, J & GLENNIE, K. W. (eds) *Petroleum Geology of North West Europe*. Graham and Trotman, London, 677–685.

Tectonic evolution of the Larne Basin

ROB SHELTON

Department of Geology, School of Geosciences, The Queen's University of Belfast, Belfast BT7 1NN, Northern Ireland
Present address: Shell Exploration and Production Namibia BV, Postbus 663, 2501 CR Den Haag, The Netherlands

Abstract: The Larne Basin lies in the Midland Valley terrane between Northern Ireland and SW Scotland. A proportion of the basin lies onshore Northern Ireland, with an offshore continuation into the North Channel seaway. The basin is orientated NE–SW, and the sedimentary basin fill thickens to the west. Extensional faults in the south of the basin dip generally to the east, those in the north to the northwest. Preserved sediments range in age from Carboniferous to Tertiary, but are predominantly Permo-Triassic. Integration of seismic interpretations, well data and outcrop fault and fracture measurements has revealed a number of Mesozoic and Cenozoic tectonic events. Late Carboniferous N–S compression was followed by a major phase of regional early Permian ENE–WSW/NE–SW extension. ENE–WSW extension followed in the late Triassic and early Jurassic with post-early Jurassic to late Cretaceous minor subsidence and uplift resulting in a major erosional unconformity between the Lower Lias and Upper Cretaceous Greensand and Chalk. Early Tertiary (Palaeocene) uplift and NE–SW extension led to extensive basalt extrusion, and later NW–SE to NNE–SSW extension and faulting resulted in the deposition of the Upper Oligocene Lough Neagh Clays in the neighbouring Lough Neagh Basin. The late Tertiary (Miocene) saw a phase of compression, leading to uplift and erosion. AFTA® analyses of Permian and Triassic samples indicate *c.* 450 m of Tertiary uplift and erosion. Sonic velocity analysis of the Mercia Mudstone Group and Sherwood Sandstone Group suggests uplift of 0.5–1.1 km, possibly reflecting maximum burial at an earlier ?Jurassic time.

The Larne Basin lies at the intersection of the Midland Valley terrane and the North Channel, both of which have significant thicknesses of Devonian to Recent sediments preserved in them. A proportion of the basin is onshore Northern Ireland, with an offshore continuation into the North Channel seaway. The Larne Basin forms part of a tripartite system of basins in the North Channel, the others being the Portpatrick and Clyde group of basins. The Clyde Basin consists of the NE and SW Arran troughs, the Plateau and the Carrick Basin (Fig. 1). The Larne Basin has been considered part of the Larne–Lough Neagh Basin complex, which experienced the formation of a Tertiary sedimentary depocentre in the Lough Neagh area. The Larne Basin was not part of this Tertiary sedimentation, and gravity modelling by Carruthers *et al.* (1987) shows an area of Permo-Triassic sedimentary thinning between the two basins, possibly controlled by a basement high. Therefore the Larne Basin is considered to be separate in this paper. The available seismic data and detailed structural information in this study are therefore restricted to the Larne Basin.

Nearly all the previous research on the Larne Basin has been focused onshore (e.g. McCann 1990, 1991; McCaffrey & McCann 1992), owing to a lack of offshore data. A number of seismic datasets has been acquired in the offshore, most in the early 1980s, e.g. Horizon Exploration 1981 Hex survey, BIRPS 1982 WINCH Deep Seismic survey (Hall *et al.* 1984), and the Geco 1985 survey. However seismic data of much improved quality and line density were acquired in 1992 by JEBCO Seismic Ltd as part of the 14th Licensing Round. In addition, a good quality Vibroseis onshore survey was acquired by Fynegold Petroleum Ltd in 1988. Access has been granted to both the JEBCO and Fynegold data, totalling 4000 km line length. In addition 4 deep boreholes have been drilled in the onshore Larne Basin. It is the ability to link the tectonic features and stratigraphic development of the onshore and offshore which forms the basis of this paper.

The Larne Basin has been the focus of hydrocarbon exploration on a number of occasions. In 1968 Marathon obtained an exploration licence in the SE Antrim area, and in 1971 the Newmill No.1 well was drilled under a farm-in agreement with Shell. The well was unsuccessful and the Marathon licence was subsequently relinquished (Griffith 1983). In 1986 Fynegold Petroleum Ltd and partners were granted UK Licence Concession PL6/86 in east Antrim, which resulted in the Ballytober No.1 well. This well was also dry. In south Antrim a group headed by Mustang obtained

From Meadows, N. S., Trueblood, S. P., Hardman, M. & Cowan, G. (eds), 1997, *Petroleum Geology of the Irish Sea and Adjacent Areas,* Geological Society Special Publication No. 124, pp. 113–133.

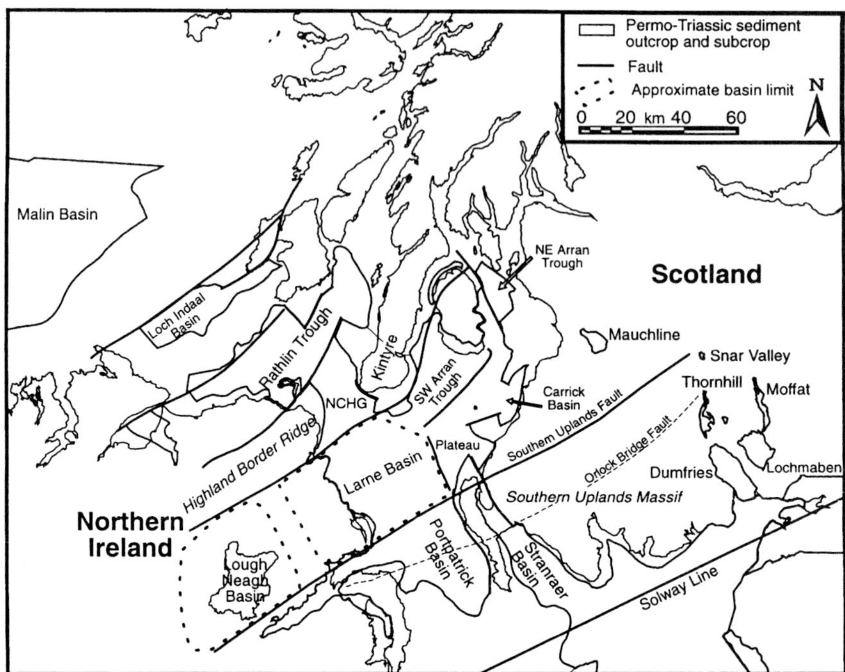

Fig. 1. Location map showing North Channel Region, the Larne Basin and major Caledonian faults. Note the long axes of the basins switch from from NE–SW Solway Basin to NW–SE Portpatrick Basin, NE–SW Larne Basin and finally the NW–SE trending NCHG (North Channel Half Graben) on the Highland Border Ridge, abutting the Devonian basement of Kintyre and down throwing to the SW.

Licence PL8/88 in 1988. This did not result in any drilling activity but the licence has been renewed recently. In the 14th UK Licensing Round conducted in 1993, British Gas Exploration and Production Ltd was awarded two 100% licences in the North Channel. Blocks 111/3, 4 and 8 cover part of the offshore extension of the Larne Basin. Finally Antrim Resources were granted a licence in 1995 roughly covering the area previously held by Fynegold. In the Lough Neagh Basin a group led by Lough Neagh Energy have drilled two exploration wells in an area to the north of Lough Neagh in 1993 and 1994, but details are presently confidential.

Regional structural setting

A simplified interpreted deep seismic WINCH profile (Hall *et al.* 1984) is presented as Fig. 4, and shows the Larne Basin in its regional context. Thickening of Carboniferous and Permo-Triassic can be seen south of the Highland Border Ridge into the Larne Basin. There is a zone of Carboniferous and Permo-Triassic thinning towards the southerly limit of the Larne Basin due

to (?Tertiary) uplift and erosion, before thickening again into the Portpatrick Basin.

The Larne Basin is bounded to the south by the Southern Uplands Massif (SUM), and to the north by the Highland Border Ridge (Fig. 1). Both are Caledonian Massifs. The trace of the Highland Boundary Fault (HBF) is taken to follow that of Hutton (1987). The Southern Uplands Fault (SUF) is taken after Illing & Griffith (1986), which is supported by the aeromagnetic work of Lee *et al.* (1990). However, the SUF is best considered as a zone rather than a discrete lineament, as it is not clearly imaged on seismic and appears as three *en echelon* lineaments on magnetic images (Lee *et al.* 1990). The trace of the Orlock Bridge Fault (OBF) is regarded to follow that of Anderson & Oliver (1986) and McKerrow & Elders (1989). Finally the southerly margin of the SUM is taken as the Solway Line after Cope *et al.* (1992)

Along the SUF there is a 'transfer zone', separating the Larne and Portpatrick basins (Fig. 1). To the south of the transfer zone the Portpatrick Basin thickens to the east, with major extensional faults dipping to the west. To the north the Larne Basin thickens to the west in the offshore with major synthetic faults dipping to the east. A similar

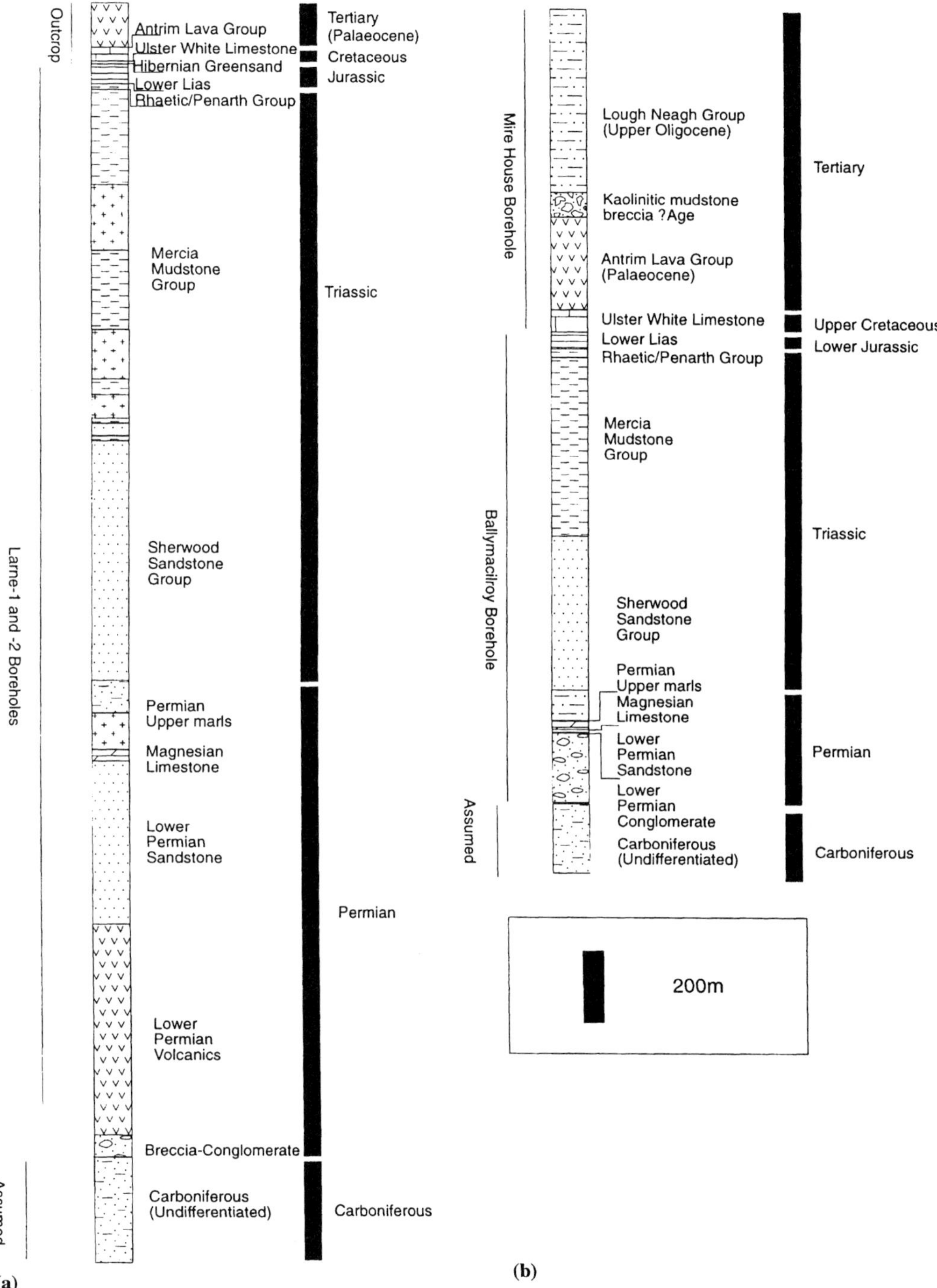

Fig. 2. Composite stratigraphic column for (**a**) the Larne Basin, based on outcrop in the Larne area and the Larne No.1 and No. 2 boreholes (after Manning & Wilson 1975; Penn *et al.* 1983); (**b**) the Lough Neagh Basin, based on the Mire House (Fowler & Robbie 1961) and Ballymacilroy (Thompson 1979) boreholes, together with information from Mitchell (1984).

feature exists in the East Irish Sea Basin (Maddox *et al.* 1997).

There does appear to be a regular pattern of alternating offshore basin orientations in the North Channel region (Fig. 1). The long axes of the basins switch from NE–SW (Solway Basin), NW–SE (Portpatrick Basin), NE–SW (Larne Basin) and finally NW–SE (small NE-dipping half-graben on the Highland Border Ridge, abutting the Devonian basement of Kintyre). Crustal inhomogeneity has been proposed for the formation of the Clyde Belt, which extends to the NW as far as the HBF and to the SE as far as the Vale of Clwyd and Cheshire, and is up to 150 km wide (McLean 1978; Jackson & Mulholland 1993). It has been associated with the opening of the North Sea Basin in Westphalian or Namurian times (McLean 1978), the product of a NE–SW tension similar in orientation and age to the stress which affected other areas of NW Europe.

It seems likely, therefore, that such heterogeneity is, at least in part, responsible for the switching, particularly given the degree of basement control on Permo-Triassic basin geometry (Anderson *et al.* 1995).

Stratigraphy

Larne No.2 was drilled in 1981 as a geothermal test well just SE of Larne town (Penn 1981), and is considered the type well for the offshore part of the basin. A composite stratigraphic column for the whole basin is given in Fig. 2a, based on outcrop and information from Larne No. 1 (Manning & Wilson 1975), and Larne No. 2 (Penn 1981). A combination of data sources has been chosen in order to fully represent the stratigraphy of the onshore and offshore basin areas. Preserved sediments are dominantly Permo-Triassic, with the

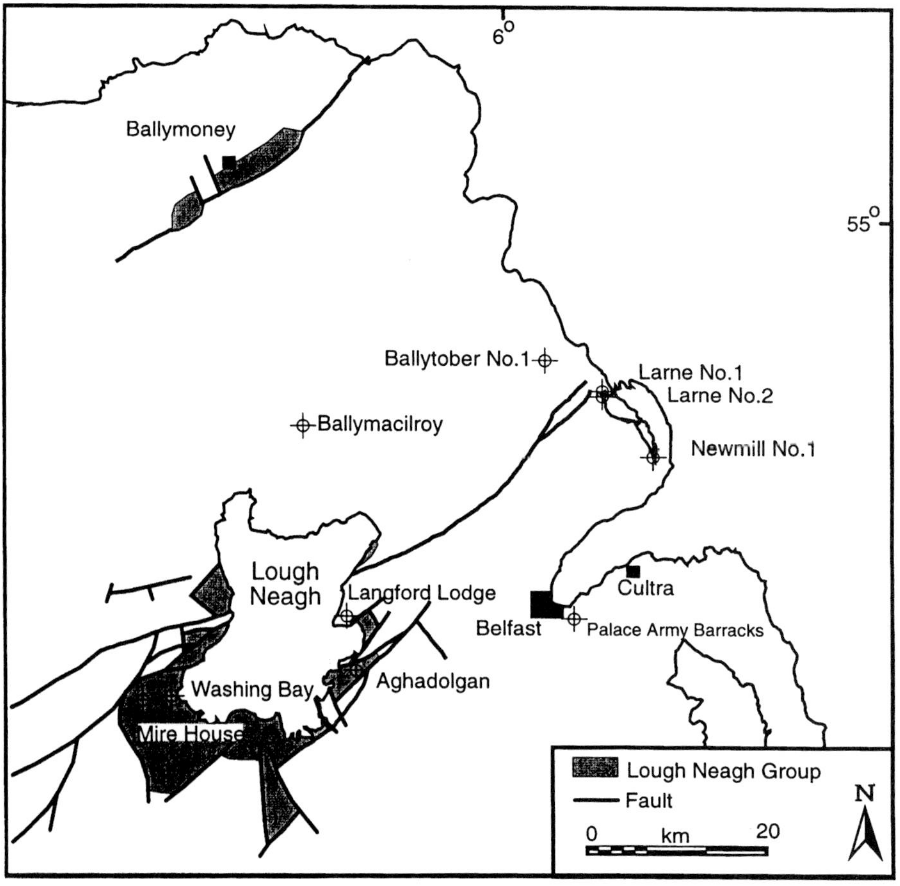

Fig. 3. Map of locations discussed in the text. Lough Neagh Group at Ballymoney redrawn in part after Parnell *et al.* (1989).

major difference between the onshore and offshore stratigraphy being the absence of sediments more recent than late Triassic offshore. This is due to a combination of onshore preservation by the basalts, and possibly greater uplift and erosion offshore. No strata thought to be older than Permian have been proved by drilling anywhere in the Larne Basin. However there are a number of lines of evidence to suggest the presence of Carboniferous strata in the Larne Basin:

(1) faulted Lower Carboniferous outcrop at Cultra on the south shore of Belfast Lough (see Fig. 4, Grid Ref. 422 816);

(2) discovery of reworked Upper Carboniferous (Namurian–Westphalian) miospores in Triassic mudstones from the Palace Army Barracks borehole east of Belfast (Griffith & Wilson 1982). See Fig. 3;

(3) the imaging of reflection terminations against an unconformity at base Lower Permian level on onshore seismic data tied to the Larne No.2 well (Illing & Griffith 1986). A similar major unconformity is seen on offshore data, which is taken to represent the basal Permian unconformity.

The Larne No. 2 well encountered a basal section of ?Permian lavas and tuffs, some 600 m thick (Penn *et al.* 1983), which suggests that they are more than local in their extent. No other wells drilled in the basin have penetrated such volcanics *in situ*.

Figure 2b shows a composite stratigraphic column for the Lough Neagh Basin based on the Ballymacilroy (Thompson 1979) and Mire House (Fowler & Robbie 1961) boreholes, with information from Mitchell (1984). Ballymacilroy proved the Permian–Jurassic section and Mire House penetrated the Cretaceous–Recent section. The reason for the combination of these two stratigraphic penetrations is to compensate for the fact that the Lough Neagh depocentre has not been drilled. In order that time-subsidence plots for the two basins are comparable, the greatest proven thicknesses from the Lough Neagh Basin have been used as described above. The 62 m section of kaolinitic mudstone breccia reported by Mitchell (1984) is also the thickest drilled in the Lough

Neagh Basin. Note the similarity in stratigraphy between the Lough Neagh and Larne basins, until the Oligocene.

Methodology

Interpretation of onshore and offshore seismic data defines the basin structure and major tectonic events. Outcrop fracture and fault measurements provide information on past stress regimes. Seismic and deep boreholes provide information on stratigraphy, with the latter also providing a means of uplift quantification.

Tectonic evolution – evidence

Onshore data

(I) Outcrop. A number of detailed fracture studies have been conducted in the Larne Basin and adjacent areas. The most important of these are Kerr (1987) from the North Antrim area, and Millar (1990) who, although primarily concerned with the NW Carboniferous Basin of Ireland, studied a certain amount of the Antrim and Down region. Figure 5 shows the stress histories of the two authors, together with information from onshore and offshore seismic (see later section).

(II) Shales-with-beef. 'Shales-with-beef' *sensu* Stoneley (1983) i.e. fibrous calcite veins within a mudstone host rock have been discovered outcropping west of Belfast, and are restricted to the Lower Lias. There have been a number of reports of shales-with-beef elsewhere in the UK (e.g. Marshall 1982, Stoneley 1983), but never previously in Northern Ireland.

The depth of formation of the 'beef' veins is relatively unconstrained (Rukin 1990), however, by analogy there is reason to believe it is still of use to the study. Rukin cites Coniglio 1985 (*in* Hesse 1986) who believes that displacive fibrous calcites form within the top tens of metres of burial, although no evidence for this theory is presented. El-Shahat & West (1983) report that 'beef in Britain is present only in rocks that have been buried beneath several hundred metres of strata'. They imply, therefore, that significant burial is a

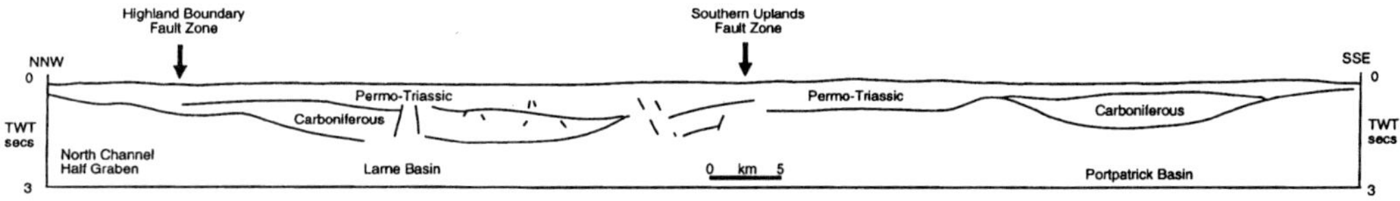

Fig. 4. An interpreted WINCH profile for the Firth of Clyde and North Channel, into which the Larne Basin extends offshore (after Hall *et al.* 1984).

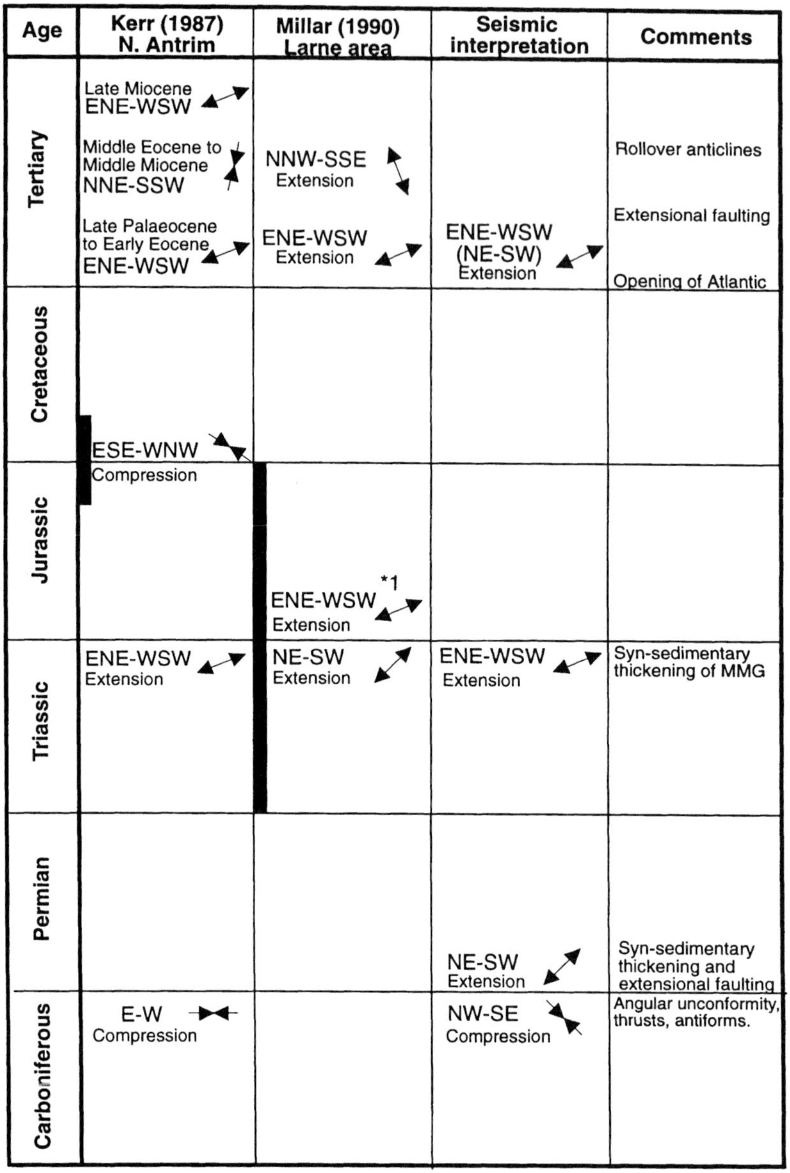

Age	Kerr (1987) N. Antrim	Millar (1990) Larne area	Seismic interpretation	Comments
Tertiary	Late Miocene ENE-WSW Middle Eocene to Middle Miocene NNE-SSW Late Palaeocene to Early Eocene ENE-WSW	NNW-SSE Extension ENE-WSW Extension	ENE-WSW (NE-SW) Extension	Rollover anticlines Extensional faulting Opening of Atlantic
Cretaceous	ESE-WNW Compression			
Jurassic		ENE-WSW *1 Extension		
Triassic	ENE-WSW Extension	NE-SW Extension	ENE-WSW Extension	Syn-sedimentary thickening of MMG
Permian			NE-SW Extension	Syn-sedimentary thickening and extensional faulting
Carboniferous	E-W Compression		NW-SE Compression	Angular unconformity, thrusts, antiforms.

*1 The Lias in the Larne area has been comprehensively overprinted with N-S, E-W trends associated with the Sixmilewater Fault (Millar 1990).

Fig. 5. Diagram showing stress regimes reported for outcrop in Northern Ireland by Kerr (1987), Millar (1990). In addition observations from onshore and offshore seismic interpretation have been included.

contributory factor in the generation of these calcitic beds.

The total Jurassic burial of the material studied by Rukin is 789 m. 'Beef' from the Vectis Formation on the Isle of Wight has similarly been buried to approximately 700 m, (Ruffell, pers. comm. 1994). Indeed, all of the occurrences of

'beef' in the south of England lie on a Cretaceous–Tertiary inversion axis (the Purbeck–Wight monocline). There is some relationship of the occurrence of 'beef' to sites or axes of inversion (e.g. Davison 1995).

With regard to the Northern Ireland material, it has been under a pre-Quaternary minimum burial of 150–200 m of Lower Greensand, Ulster White Limestone and Tertiary Basalt. There is a major unconformity between the Lias and Greensand, thus for the 'beef' to form at greater than tens of metres depth, one of two things may have happened as follows.

(1) Lower Jurassic deposition of the order of 300 m or so, followed by an extended period of non-deposition (Naylor 1992). This is possible given the preservation of 270 m of argillaceous Lias in the Port More borehole in the Rathlin Trough (McCann 1987), thus showing that significant Lias sedimentation occurred in the north of Ireland. Subsequently erosion of 2–300 m or so of Lias occurred in the late Jurassic/early Cretaceous estimated by comparison with the Port More (c. 270 m) and Larne-1 (c. 50 m) sections. However, soft shales have a low preservation potential when exposed, and it seems remarkable that only 2–300 m of erosion took place in 60 Ma.

or

(2) Deposition of an extensive section of Lower and Middle or Lower and Upper Jurassic. The latter is suggested in fig. 10 of Morton (1992). Either could have subsequently been uplifted and eroded during the 'Late Cimmerian' phase of tectonics (*sensu* Rawson & Riley 1982), possibly simultaneously with the formation of the 'beef'. This would create a situation analogous to the southern Britain 'beef-bearing' section, where the beef formed during the Jurassic (Lower Lias, West Dorset) or Cretaceous (Purbeck and Vectis formations). Timing of formation of the West Dorset material was constrained by using cross-cutting fault relationships (Rukin 1990). However a number of problems exist with this interpretation:
  (a) the only fossil evidence for Upper Lias deposition in Northern Ireland is the presence of Upper Lias fossils in the basal Cretaceous conglomerate on the Highland Border Ridge (Wilson 1972);
  (b) where has the eroded Jurassic material been transported to?

The Northern Ireland material lies on strike of the previously mentioned Southern Uplands Fault transfer zone, which in itself demonstrates Permo-Triassic reactivation along the SUF. Therefore, although many believe the Southern Uplands Fault was active in post-Silurian times (e.g. Hutton 1987; McKerrow & Elders 1989), the occurence of 'beef' in the Lias adjacent to this fault suggests that there was even later reactivation, either during the Jurassic or Tertiary. Preliminary isotope work is equivocal due to the possibility of a meteoric water influence on the oxygen isotope ratios, associated with uplift. Further work is required to constrain this. Movement of the OBF has been prevented due to the emplacement of several Devonian granitic plutons in SW Scotland, which has had the effect of binding the two terranes together and thus lessens the likelihood of reactivation. Conversely the SUF has no such binding plutons (T.B. Anderson, pers. comm. 1995).

In conclusion, hypothesis (1) is favoured as there is no sedimentary evidence for deposition of an extensive Middle and/or Upper Jurassic sequence, although the possibility does remain. Finally there is no reason to suspect that 'beef' cannot form at relatively modest burial depths.

(III) Seismic. Figure 6 shows two onshore geoseismic sections from the Larne Basin. Figure 6(a) orientated WNW–ESE, shows a major fault dipping to the east, parallel with the dominant trend offshore. To the west of the section another major fault is imaged, but downthrowing to the west. Figure 6(b) shows a NW–SE orientated geoseismic section, showing the tie to the Larne No. 2 well. Probable Carboniferous reflectors are imaged, truncated against the base Permian unconformity. The Sixmilewater Fault is clearly imaged, and although it has a Caledonoid trend has obviously been active in the late Triassic, with a downthrow of around 500 m at near Top Sherwood Sandstone level, and possible thickening of the Mercia Mudstone Group into the fault. That the Sixmilewater Fault has also been active in post Palaeocene times is demonstrated by the faulted juxtaposition of Upper and Lower Basalts (*Geological Map of Northern Ireland,* 1:250 000, Solid Edition).

(IV) Volcanic rocks. Permian volcanicity is of importance in the area: the discovery of 617 m of volcanics in the Larne No.2 borehole (Penn *et al.* 1983), suggests an important phase of volcanicity, possibly associated at this time with rift initiation. These olivine–pyroxene basaltic lavas and tuffs have been isotopically dated at 245 ± 13 Ma (Penn *et al.* 1983) i.e. approximately at the Permian: Triassic boundary. This is around 50 Ma younger than stratigraphically equivalent lavas in southwest Scotland (Forster & Warrington 1985) and Islay (Upton *et al.* 1986), possibly suggesting more than one rift event in the region (see Fig. 7). The stratigraphic evidence, however, tends to suggest that the Permian:Triassic date is in error, as they underlie over 600 m of Permian clastic and evaporitic sediments and include tuffs which are a result of extrusive rather than intrusive volcanic activity. Therefore it seems more likely that they

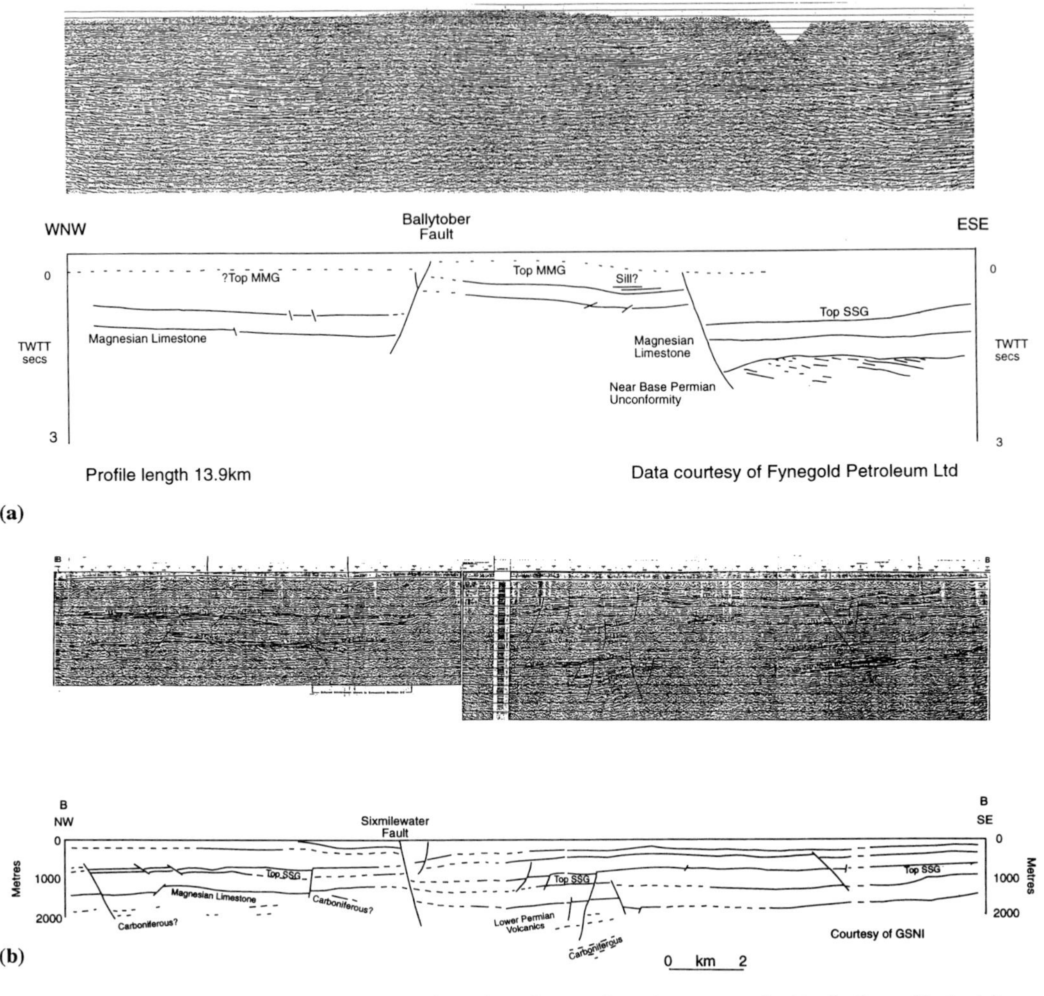

Fig. 6. Migrated seismic and simplified geoseismic sections from onshore datasets acquired in the Larne Basin. (a) Roughly ENE–WSW section showing the major westerly-dipping NNE–SSW orientated Ballytober Fault. Note the thickening of the MMG into the fault, and the thinning on the horst. Data courtesy of Fynegold Petroleum Ltd. (b) NW–SE orientated migrated seismic and interpreted geoseismic section, showing the tie with Larne No.2. Note in particular the truncated reflectors at base Permian level, labelled as Carboniferous. In addition the Sixmilewater Fault is clearly imaged, with 500 m of southerly downthrow at Top Sherwood level. After Illing & Griffith (1986).

are Lower Permian in age. The volcanics have been linked to a tripartite early Permian rift system in the area (Francis 1978, 1991, see Fig. 7).

Northern Ireland forms part of the British Tertiary Igneous Province (BTIP), the product of an extensive phase of extrusive volcanic activity. The thick basalts result primarily from the presence of the Icelandic hotspot in the area of continental break-up prior to rifting in the northern North Atlantic, as most of the igneous rocks predate the rifting between Iceland and Europe (White 1992). Later dyke emplacement evidences early Tertiary

(Palaeocene) extension resulting from this rifting (Macintyre *et al.* 1975). The basalts have been Ar–Ar dated at 58 to 61 Ma (Thompson 1985), and consist of the Lower Basalt Formation, Interbasaltic Formation and the Upper Basalt Formation. The Lower Basalt Formation has a maximum preserved thickness in the Lough Neagh area (531 m were recorded in the Langford Lodge borehole, Manning *et al.* 1970, see figure 3), although it is likely that they erupted from a number of extrusive sites. The Interbasaltic Formation represents a gap in igneous activity

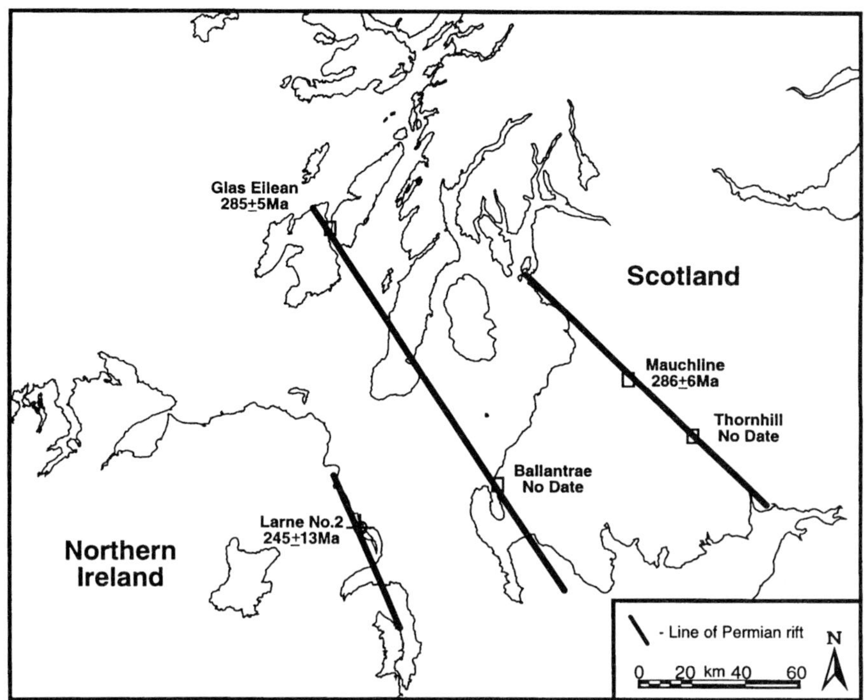

Fig. 7. Permo-Carboniferous rift axes in the region, their location and orientation in relation to the Larne Basin (after Francis 1978, 1991).

before a second cycle of eruptions formed the Upper Basalt Formation. An unknown amount of basalt has been removed by a variety of erosional processes (McCaffrey & McCann 1992).

(V) Isopach maps. A number of onshore isopach maps in the Larne–Lough Neagh area have been published by McCaffrey & McCann (1992). They have been redrawn and are presented as Figs 8(a) to (d). However, they are based on data from sparse drilling, and all are possibly affected by erosive tops to isopachs which reflect preservational rather than depositional thicknesses. Caution should therefore be exercised in their interpretation. In addition a Tertiary basalt isopach map for the north of Ireland has been redrawn based on Tomkeiff (1964) *in* Preston (1981). This is presented as Fig. 8(e).

Sherwood Sandstone Group (SSG). The SSG isopach map (Fig. 8a) suggests that there are two discrete depocentres of the Larne and Lough Neagh basins. The Larne Basin has a 400 m thicker maximum section than the Lough Neagh Basin, possibly suggesting a more rapid rate of subsidence. However there is the *caveat* that the depocentre of the Lough Neagh Basin has not been

drilled, its location indicated by gravity modelling (Carrruthers *et al.* 1987) is under the present day Lough Neagh. Consequently a proper comparison cannot be made.

Mercia Mudstone Group (MMG). The shallowest parts of the onshore seismic data are of poor quality, and thus correlating the top and base MMG is particularly difficult. Offshore, there has been extensive erosion of the MMG as a result of the absence of the Tertiary basalt cover. It is apparent, therefore, that an integrated onshore and offshore MMG isopach map is of little use.

The MMG isopach map (Fig. 8b) shows similar thickening trends to the Sherwood Sandstone i.e. two discrete depocentres of the Larne and Lough Neagh basins, suggesting continuity of depositional trends i.e. continued subsidence in both areas. Again the MMG is of the order of 400 m thicker in the Larne Basin, similar to the SSG. However the *caveat* for the SSG also applies to the MMG. In addition there is evidence for differential erosion of the MMG on fault blocks, resulting in localised thinning. Such thinning is imaged on Fig. 6a.

Lower Lias. The Jurassic (Lower Lias) isopach map (Fig. 8c), shows a localized thickening in the

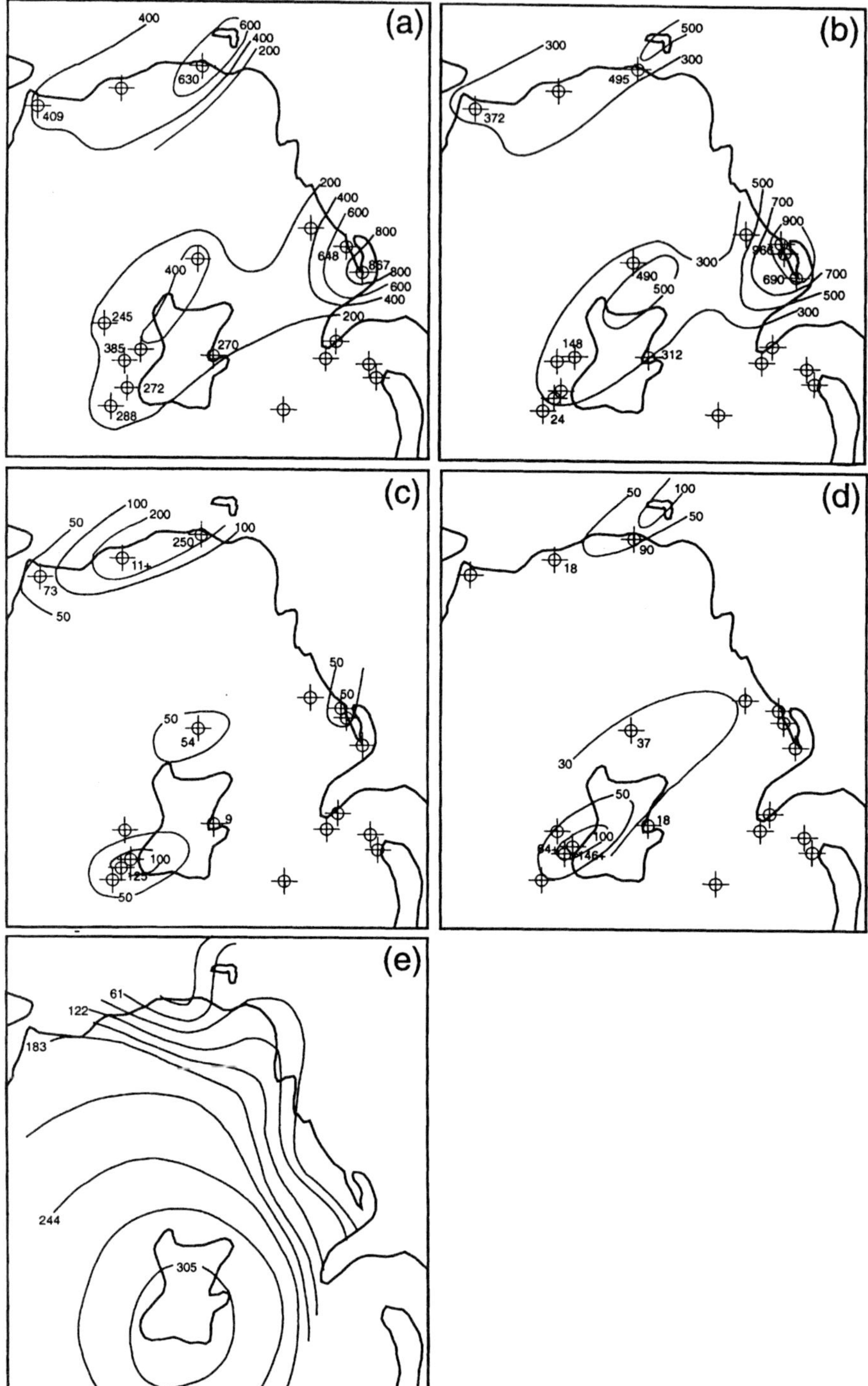

Fig. 8. Onshore isopach maps for the Larne Basin based on borehole data. (**a**) Sherwood Sandstone Group; (**b**) Mercia Mudstone Group; (**c**) Lower Lias (Jurassic); (**d**) Cretaceous Ulster White Limestone. (Figs 8a–d are redrawn after McCaffrey & McCann 1992). (**e**) Isopach map for the Tertiary Basalt (redrawn after Tomkeieff 1964 *in* Preston 1981).

Larne area as it is the site of the depocentre. The greater preserved thickness apparent to the west of Lough Neagh may be due to a greater amount of early Cretaceous uplift and erosion in the Larne area. This is in agreement with the *South of Scotland Regional Geology Guide* (Greig 1971), where it is suggested that Rhaetic and Lower Lias sedimentation occurred over the whole of the south of Scotland, prior to deposition of the Upper Cretaceous Greensand and Chalk. Preservation of these strata in Northern Ireland is attributed to the Tertiary Antrim Basalts overlying and protecting them from erosion (Wilson 1972). Should this have been the case it would seem probable that similar deposition occurred in the offshore part of the Larne Basin, although there is no evidence for their occurrence offshore based on seismic mapping or shallow boreholes. Subsequent erosion of the Lower Lias in the offshore may have been in early Cretaceous or Tertiary times along with the Greensand and Chalk (see Cope, 1997). However, the Liassic section is similar to the MMG in that there has been differential erosion on fault blocks, resulting in localized areas of thinning.

Cretaceous Chalk. The Cretaceous Ulster White Limestone isopach map (Fig. 8d) shows a greater preserved Chalk section in the Lough Neagh district. Preservation is a function of Lower Basalt distribution which postdated Chalk erosion in the Larne area. That the Chalk was exposed for a considerable time is demonstrated by the report of up to 15 m of topography at the top of the Chalk (Preston 1981). As with the Lower Lias, it is probable that early Tertiary uplift and erosion was of greater magnitude in the Larne area than around Lough Neagh. Erosion of the Chalk in the offshore may have occurred during early Tertiary uplift, or associated with the later post-Basalt Tertiary uplift event.

Tertiary Basalts. Figure 8e is a total Basalt Isopach map (after Tomkeieff 1964 *in* Preston 1981) and shows the thickest section centred on Lough Neagh. The thickening is due to a combination of preservation and deposition. Information from sparse drilling shows the estimated thicknesses used in the isopach map to be in error, for example 770 m of basalts were reported for the Ballymacilroy borehole (Thompson 1979). However, it is considered that the gross thickening trends indicated are realistic. Note that separation of Upper and Lower Basalts at outcrop is very problematic, thus only a total isopach map has been produced to date (J. Wallace, pers. comm.).

(VI) Estimation of uplift using sonic velocities. It has been conventional to use sonic velocities from the Lower Triassic Brockelschiefer shale in uplift estimation in the Southern North Sea. The minimum uplift suffered by a previously deeply buried basin can be calculated by comparing the maximum palaeo-depth derived from sonic velocities and the present depth of burial of the Triassic Bunter Shale (Marie 1975). However, more recent work has used the Sherwood Sandstone, Mercia Mudstone Groups, and the Lias (Bulat & Stoker 1987). Hillis (1991, 1993) also employed the Cretaceous Chalk in other areas.

Published burial curves from the Southern North Sea (SNS) have been used in this study, as only four wells in the study area have sonic velocity logs. There are no published curves for the East Irish Sea Basin (EISB), and those for the Cheshire Basin are for specific Sherwood Sandstone members which cannot be correlated reliably to the Larne Basin. This is clearly an inadequate number of wells on which to define normal compaction curves. In contrast, the SNS database comprises over 400 hundred wells and has been the focus of a large volume of work which has meant the velocity/depth plots are far better constrained. It is appreciated that there are a number of problems in using curves from a remote area.

(1) A major facies difference between the *Lower Triassic* Brockelschiefer or Bunter shale (for which the Marie (1975) and Bulat & Stoker (1987) curve was created) of the *Southern North Sea,* and the *Upper Triassic* Mercia Mudstone Group in the *North Channel area.* The MMG is very heterogeneous in the Larne No. 2 borehole in particular, with a number of halites which show high sonic velocities (typically $5 \, \text{km s}^{-1}$) and not always reliably identified on the composite log. However, the Ballytober No. 1 and Ballymacilroy wells were found not to contain halite.

(2) The SSG may, in part, be reflecting local porosity characteristics due to a number of diagenetic processes (Chadwick *et al.* 1994).

However, curves from the SNS have already been used with apparent success in the EISB, e.g. Colter (1978).

Velocity data for the stratigraphic intervals were derived from well logs in the standard manner used in the oil and gas industry, i.e. extraction of depth and sonic slowness values for the top and base of each interval from composite and velocity logs and application of the equation: velocity = interval thickness (feet)/ one-way interval transit time (seconds). Velocity data have been used in the manner documented by Marie (1975), Bulat & Stoker (1987) and Hillis (1993) for the MMG and SSG. The results are presented as Tables 1 and 2 respectively.

Table 1. *Uplift values for the Mercia Mudstone Group (equivalent to the Haisborough Group, Southern North Sea) using the sonic logs from a number of boreholes in Northern Ireland (see Fig. 3 for locations)*

Basin	Well	Bulat & Stoker (1987)	Marie (1975)	Marie (1975) See *1 below
Larne	Larne No.2	1931 m[*2]	2267 m[*2]	1154 m
Larne	Ballytober No.1	No data[*3]	No data[*3]	871 m
Larne	Newmill No.1			762 m
Lough Neagh	Ballymacilroy	1745 m	1098 m	1067 m

[*1] In addition to obtaining sonic slowness values from the formation tops and bottoms, several average values were taken from portions of the section with good quality sandstones. These were plotted at the depth from which they were obtained on the Marie curve. Note, however, that they were not the same unit correlated in each well.

[*2] Note that the sonic log was not run in the top 198 m of the Mercia Mudstone Group.

[*3] The sonic log was run only in the bottom 65 m of the MMG.

There is broad agreement between the MMG and SSG uplift values derived from the Marie (1975) and Hillis (1993) curves, with values averaging around a kilometre. The Bulat & Stoker (1987) SSG curve gives values approximately 500 m less than that of Hillis (1993), due to differences in the selection of wells used to derive normal compaction curves. A number of other points should be taken into account in interpreting the data.

(1) Contact baking of sands adjacent to igneous intrusions, but not penetrated by the wells may have occurred and would have affected the results by decreasing the sonic slowness. However, the interval transit times are very consistent throughout the section, and so thermal effects are considered to be local and thus insignificant.

(2) The complete MMG sequence may not be present in every borehole, as discussed in the isopach section. This is definitely the case in Ballytober No.1, as thinning of the MMG is imaged on seismic over a horst on trend from the well (Fig. 6a).

(3) The sonic log was not run in the upper 198 m of the MMG section of the Larne No.2 well. Therefore, the average sonic velocity plotted at the depth of the formation mid-point has had the effect of showing a faster velocity than would have been the case in a complete log run, i.e. an apparently greater degree of compaction at a shallower depth. This has led to an over-estimation of the uplift.

(4) Significant problems have arisen in correlating the few widely spaced boreholes. Particularly as the Ballytober No.1 and Ballymacilroy wells contain no halite marker beds. Consequently, thin, discrete intervals could not be reliably identified in all wells, thus the values using discrete mudstones do not come from the same unit across the study area.

Finally the uplift estimates for wells in the Larne Basin are similar to the well drilled in the Lough Neagh Basin.

(VII) Apatite Fission Track Analysis (AFTA®). AFTA® data from Permian and Triassic sandstones in Larne No.2 have been provided by courtesy of Geotrack International. The data reveal cooling from high palaeotemperatures at some time in the range 100 Ma to 50 Ma. This range includes the age of the Tertiary igneous rocks in and around the well and it seems most likely that the AFTA® data reflect cooling in the early Tertiary around 60 Ma. The palaeotemperature data derived from the AFTA® define an early Tertiary palaeogeothermal

Table 2. *Uplift values for the Sherwood Sandstone Group, using sonic velocities from a number of boreholes drilled in Northern Ireland (see Fig. 3 for locations)*

Basin	Well	Hillis (1993)	Bulat and Stoker (1987)
Larne	Larne No.2	1146 m	590 m
Larne	Ballytober No.1	1293 m	770 m
Lough Neagh	Ballymacilroy	1122 m	655 m

gradient of around 15°C km^{-1} which is considerably lower than the present day value of 30°C km^{-1}. High palaeotemperatures with low geothermal gradients have been recognized in a number of settings and are interpreted to be caused by lateral heat transfer due to fluid movement (e.g. Green *et al.* 1993; Duddy *et al.* 1994). In the case of Larne No.2 the numerous early Tertiary igneous intrusions would have provided a ready source of heat shallow in the section.

Because of the low palaeogeothermal gradient defined by AFTA® in the Larne No.2 well, it is not possible to use the normal method of extrapolating the palaeogeothermal gradient to a palaeo-surface temperature to estimate the amount of uplift and erosion. However, the data show that the deepest sample analysed (*c.* 2880 m) was not more than about 15°C hotter than its present temperature. If it is assumed that the 'background' geothermal gradient in the absence of the heating effects due to the igneous activity was 30°C km^{-1}, the 15°C equates to about 450 m of uplift and erosion. However, it is emphasized that this assumption may not be valid and rigorous estimates of uplift cannot be made from palaeotemperature data where geothermal gradients are known to be perturbed (R. Bray, pers. comm. 1995).

The palaeotemperatures derived from AFTA®are considered to have an error bar of ± 5°C. This introduces an uncertainty of about 300 m in estimates of uplift.

In the western Solway Firth area, there is AFTA® evidence for a much earlier heating in probable late Permian to late Triassic times (Green *et al.* 1997). However, the AFTA® data in the Larne No.2 borehole is dominated by the Tertiary heating event.

The figure of *c.* 450 m is therefore for *Tertiary* uplift and erosion. The value of *c.* 1 km obtained from sonic velocity analysis may be reflecting maximum burial at an earlier ?Jurassic time i.e. post-Lias/pre-late-Cretaceous uplift and erosion. Bulat & Stoker (1987) cite an error bar of ± 2000 ft (610 m) when comparing the burial histories of different areas within their SNS study area. This is approximately the same as the disparity between the sonic velocity and AFTA® values. Finally Bray *et al.* (1992) have reviewed the sonic velocity uplift estimate technique, as applied in the Southern North Sea (e.g. Marie 1975, Hillis 1993). They recognized a lack of appreciation by the earlier workers that their baseline reference wells which were assumed to be at maximum depth of burial had in fact also been uplifted. This was also suggested by Hillis (1993). The Bray *et al.* (1992) study found closest agreement with the Hillis curves, and deemed the baseline wells to be less affected by uplift. Consequently the Hillis (1993)

curve should be considered a more reliable indicator of normal compaction. With this in mind the Bulat & Stoker (1987) curves would be underestimating the true amount of uplift. However, they are closest to the 700 m estimate from the shales-with-beef, implied by analogy with other sections in southern Britain, although it must be stressed that the latter is a much less well constrained estimate, and is not favoured by the author.

Offshore

(1) Seismic. Figure 9 shows four raw seismic and simplified geoseismic cross-sections of the North Channel, showing the offshore structure of the basin. A section running WSW–ENE from near the depocentre of the Larne Basin towards the Rhins of Galloway (Fig. 9a), shows a general westerly thickening of the basin fill, with major extensional faults dipping to the east. In addition the Lower Permian volcanics are imaged in the Larne area. In the section crossing from the Larne Basin to the Plateau (Fig. 9b), rollover occurs on the easterly bounding fault (a continuation of the Loch Ryan Fault). To the west there are several listric faults dipping to the east. The most easterly fault (the here named Dounan Bay Fault) shows a synsedimentary package, and represents the easterly limit of the Larne Basin. Section three (Fig. 9c) shows thickening of preserved sediment into the Larne Basin. Further south of a poorly imaged major normal fault, the section crosses a tilted fault block of the Portpatrick Basin. The northerly continuation of Fig. 9c shows the switch in fault orientation across the apparent line of the Plateau Fault, thinning of the MMG before crossing the trace of the HBFZ (Fig. 9d).

From integration of these data and published geological maps (*Geological Map of Northern Ireland,* 1:250 000, Solid Edition, *Isle of Man.* Sheet 54°N–06°W. 1:250 000 Series, Solid Geology, *Clyde.* Sheet 55°N–06°W. 1:250 000 Series, Solid Geology), Fig. 10 has been constructed, showing a fault map for the entire Larne Basin. Two fault trends predominate. The first is of NE–SW (Caledonoid) orientation, the second is orientated NW–SE (Variscan or Tertiary trend). Offshore we can see that the basin can be roughly split into two fault domains. The northerly area is characterized by NE–SW orientated faults, the southerly part by NW–SE faults. The switch in fault orientations is apparently coincident with the southwesterly continuation of the Plateau Fault (see also Fig. 9c). Brian Bluck (pers. comm. 1996) has suggested this may be due to an anticlinal feature which runs along the axis of the Scottish Midland Valley from Ochil to Heads of Ayr (see also Bluck 1984). However, the basement of the Midland

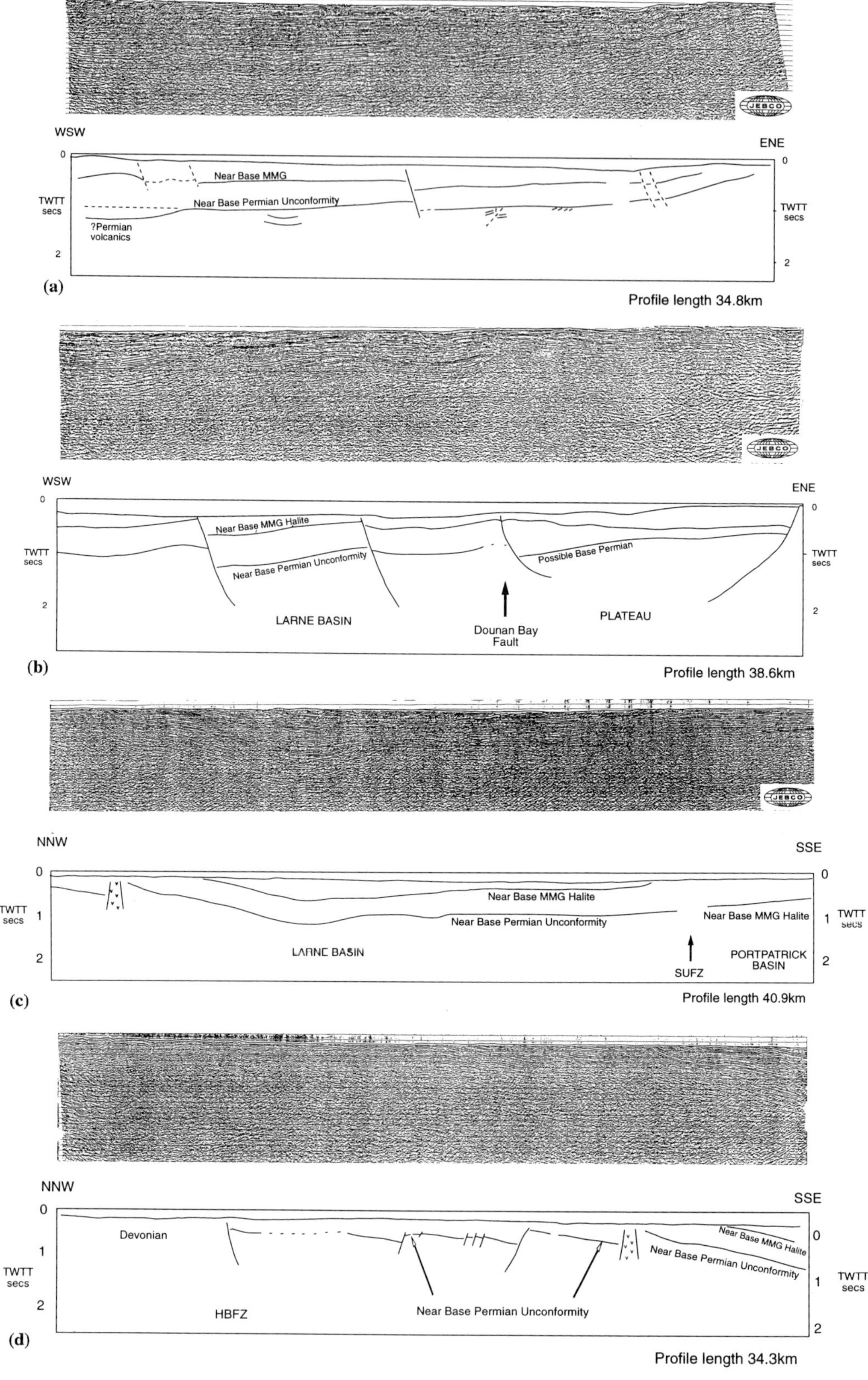

WSW
ENE
Near Base MMG
Near Base Permian Unconformity
?Permian volcanics
TWTT secs
TWTT secs
(a)
Profile length 34.8km

WSW
ENE
Near Base MMG Halite
Near Base Permian Unconformity
Possible Base Permian
LARNE BASIN
Dounan Bay Fault
PLATEAU
TWTT secs
TWTT secs
(b)
Profile length 38.6km

NNW
SSE
Near Base MMG Halite
Near Base Permian Unconformity
Near Base MMG Halite
LARNE BASIN
SUFZ
PORTPATRICK BASIN
TWTT secs
TWTT secs
(c)
Profile length 40.9km

NNW
SSE
Devonian
Near Base MMG Halite
Near Base Permian Unconformity
Near Base Permian Unconformity
HBFZ
TWTT secs
TWTT secs
(d)
Profile length 34.3km

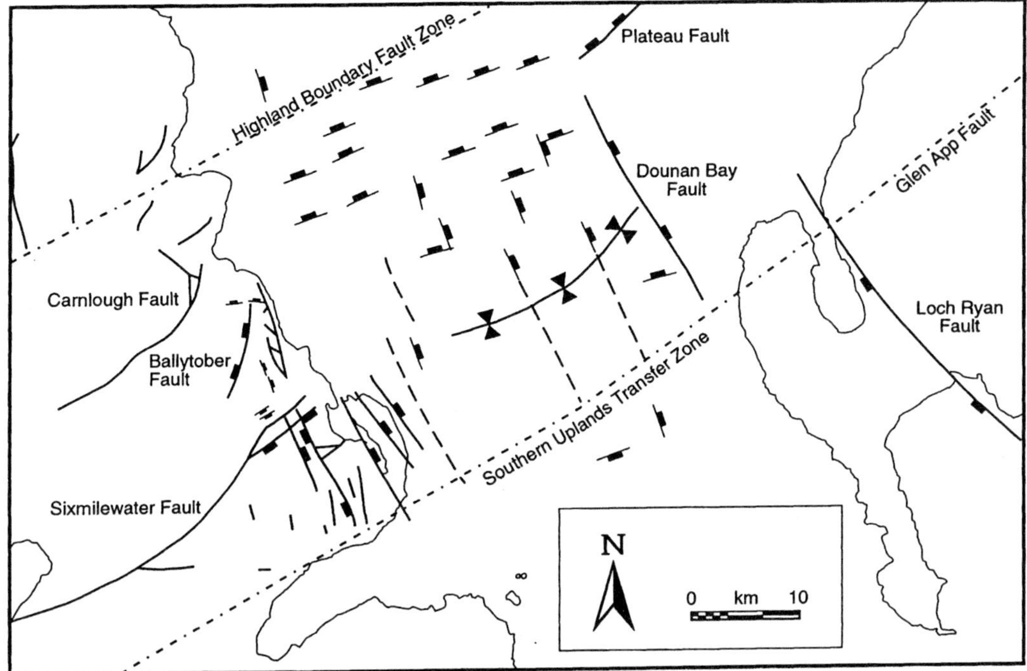

Fig. 10. Fault map for the Larne Basin, based on seismic data and published geological maps. Note that Caledonoid and Tertiary trending faults predominate, with the exceptions of the Carnlough and Ballytober faults.

Valley is not well understood, particularly in the continuation into Northern Ireland, where there is no exposure. Orientations mapped onshore match those offshore with the exception of the Ballytober Fault. The Ballytober Fault is anomalous as it is orientated NNE–SSW and dips to the west (the offshore trend is for major synthetic faults to dip to the east). The Ballytober Fault has a downthrow of over 0.5 s TWTT (*c.* 750 m) at Magnesian Limestone level (see Fig. 6a), and apparently curves round to become sub-parallel to the Carnlough Fault. The MMG thickens into the fault indicating movement in the late Triassic. The Caledonoid fault trends suggest that there has been Mesozoic reactivation of Caledonian basement tectonics. A good example of this is the Sixmilewater Fault (see Fig. 6b), which also shows synsedimentary thickening of the MMG into the hangingwall. There is also evidence from basalt outcrop of post-Palaeocene movement (*Geological Map of Northern Ireland,* 1:250 000, Solid Edition).

Integration – a tectonic history of the Larne Basin

During the Caledonian Orogeny the Grampian Terrane was displaced and eventually sutured to the Midland Valley Terrane along the Highland Boundary Fault (Hutton 1987). The Southern Uplands Massif was subject to sinistral strike-slip movements along the Southern Uplands Fault until the Devonian, when it too was sutured to the Midland Valley Terrane (Hutton 1987). The latest movement on the Orlock Bridge Fault was sinistral in late Silurian or early Devonian times (Anderson & Oliver 1986). Therefore, the primary structural framework of the North Channel region has been in place since the mid-Devonian. Thereafter, the bounding faults of the major structural elements (see Fig. 1), were relatively inactive (McCaffrey & McCann 1992). However, as previously discussed, the presence of a transfer zone apparently along the

Fig. 9. Simplified geoseismic sections from the offshore. (**a**) WSW–ENE from near the depocentre of the Larne Basin towards the Rhins of Galloway. (**b**) WSW–ENE section crossing from the Larne Basin to the Plateau. (**c**) NNW–SSE section from the northerly limit of the Portpatrick Basin north into the Larne Basin. (**d**) NNW–SSE orientated continuation of (**c**) showing the northerly limit of the Larne Basin, the HBFZ. Data courtesy of JEBCO Seismic Ltd.

the SUF, and the occurrence of shales-with-beef, suggests later Mesozoic reactivation.

According to Francis (1991) there are three components of tectonism which affected sedimentation in the Midland Valley throughout the Carboniferous. The first is the NE–SW to NNE–SSW orientation of basins from the Dinantian onwards. The second component are NE–SW orientated faults with strike-slip elements. The HBF and SUF are thought to be in this category. Thirdly a synsedimentary system of E–W fractures was active throughout the Carboniferous. There is little new data on the pre-Permian tectonic history of the Larne Basin in this paper. However, as previously discussed, it is assumed that there are Carboniferous strata beneath an angular unconformity imaged on onshore and offshore seismic data (Figs 6 and 9a).

Late Carboniferous

Kerr (1987) documents a late Carboniferous E–W compressive event in North Antrim. A number of compressive features are imaged in Carboniferous strata on offshore seismic data, e.g. antiforms, small thrusts (see Fig. 9a). Mitchell (1992) favours sinistral transpression associated with N–S compression, resulting in thrusting and strike-slip along Caledonoid (NE–SW) trending faults for Upper Palaeozoic basins in Northern Ireland. This is similar to the NW–SE to NNW–SSE direction of maximum shortening found by Corfield *et al.* (1996) for similarly orientated (NE–SW) basins to the south. N–S compression would account for the features observed in the WSW–ENE orientated Fig. 9a, and would appear to be related to the Variscan front to the south (Coward 1990).

Early Permian

Permian NW–SE trending half-grabens are developed in areas marginal to the Larne Basin. They are the Portpatrick Basin, the Stranraer Basin and the North Channel Half Graben on the Highland Border Ridge (Fig. 1). On seismic data from the Portpatrick Basin syn-tectonic packages are imaged resting on the base Permian unconformity. Although neither of the above are found in the Larne Basin itself, there is evidence to support a Permian rift here:

(1) there are the thick Permian extrusive volcanics in the Larne No.2 well (Penn *et al.* 1983);
(2) volcanics are interpreted on the onshore seismic data (Fig. 6b), which has been tied to the Larne No.2 well (Illing & Griffith 1986);
(3) there is thickening of a thick volcanic package,

in addition to Permian halites in Fig. 9a, (see fig. 5 in Jackson & Mulholland 1993);
(4) thickening of Lower Permian strata into a easterly dipping fault imaged on the seismic section shown in Fig. 6a;
(5) over 60 m of basal Permian breccia-conglomerate and sandstone was drilled beneath the volcanics in the Larne No.2 borehole, the base of which was not penetrated;
(6) the large number of closely spaced NNW–SSE Tertiary faults in the Larne district, not found elsewhere in the basin suggests control by, or exploitation of, a previous similarly orientated structural grain of presumed Permian age or older.

These points in combination strongly suggest a major ENE–WSW to NE–SW extensional event in Permian times.

Other Permian half-graben in the region are bounded to the east by a Lower Palaeozoic basement high i.e. the Portpatrick Basin is bounded by the Rhins of Galloway and the North Channel half-graben by the Mull of Kintyre (Fig. 1). The absence of such a significant high to the west of the Larne Basin explains why there has been a rift along a NNW–SSE axis (Fig. 7) with little or no extensional sedimentary geometry observed, as there is a lack of a single major controlling fault. The area of thinning discussed earlier suggests a more gradual downthrow by a series of smaller faults. It is possible that the Ballytober horst shown on Fig. 6a represents some kind of palaeohigh, since neither Permian nor Triassic halites were proven in the Ballytober-1 well spudded on the horst. It is also more N–S in orientation than the Rhins of Galloway, and a much smaller feature. It is unlikely therefore, that it represents a basin bounding feature.

Late Permian

The deposition of a significant section of Permian strata with no apparent disconformities suggests continued subsidence throughout the Permian in this area. The late Permian transgression of the 'Bakevellia Sea' is evidenced by the deposition of a marine dolomitic limestone, a thick halite (although probably restricted to the Larne Basin depocentre), and saliferous marls (Wilson 1972).

Triassic

The basin depocentre initiated in the Permian did not change location during deposition of the Sherwood Sandstone and Mercia Mudstone groups indicating little or no major change in subsidence or sedimentation patterns during this time. In addition,

Permian and SSG thicknesses along the basin axis are relatively uniform (see Fig. 9b). Changes perpendicular to this axis are more likely to be due to later erosion than deposition. This suggests that the Triassic was a time of continued subsidence.

However, it is hard to rule out any depositional factor in the thickness changes in the MMG, i.e. it may always have been thicker in the basin centre, and thinning at the basin margins has been enhanced by later, ?Tertiary erosion. The burial: subsidence plot for the Larne Basin is presented as Fig. 11a. The chronostratigraphy of the MMG has been taken as that from the Larne-1 borehole (Warrington & Harland 1975). Given the relative rates of accumulation of argillaceous versus arenaceous sediments, the plot shows relatively rapid subsidence in the MMG possibly suggesting a syn-MMG extensional event. The MMG is known to become arenaceous next to synsedimentary faults in the Rathlin Trough (Magilligan Borehole, J.

Parnell pers. comm.) and conglomeratic in the Wessex Basin (Ruffell 1991). The MMG in the Larne No.2 borehole is arenaceous in part at the base (Lagavarra Formation), and sporadically throughout. Figure 6a and b both show thickening of the MMG into faults. Kerr (1987) and Millar (1990) document ENE–WSW extension which has resulted in fractures restricted to Triassic rocks (see Fig. 5). There is no evidence onshore for significant synsedimentary thickening of the SSG. Finally there is the late Permian to late Triassic heating event in the western Solway region indicated by AFTA® analysis. On balance it seems most likely that the late Triassic was a major phase of ENE–WSW extension, rather than post-rift thermal subsidence following the Permian extension event.

Post-early Jurassic to late Cretaceous

Only a very thin section of Lower Jurassic is preserved beneath the Upper Cretaceous unconformity, with little evidence to suggest deposition of any other Middle to Upper Jurassic. In addition Kerr (1987) reports an ESE–WNW compressional event in Triassic and Jurassic strata from North Antrim. This suggests an extensive period of minor subsidence in the Jurassic and uplift in the early Cretaceous.

Early Tertiary (Palaeocene)

Up to 15 m of karst topography on the top Chalk surface indicates a prolonged period of uplift related exposure. NW–SE (WNW–ESE) extensional faulting is imaged on seismic from the onshore and offshore Larne Basin. Tertiary dykes are orientated NW–SE onshore Northern Ireland (*Geological Map of Northern Ireland*, 1:250 000, Solid Edition), NNW–SSE to N–S offshore (Caston 1976). Kerr (1987) and Millar (1990) document ENE–WSW extension, the former dating the extension as late Palaeocene to early Eocene. This evidence suggests there was NE–SW (ENE–WSW) early Tertiary (Palaeocene) uplift followed by extension.

Upper Oligocene

There is evidence for a change in tectonic history between the Larne and Lough Neagh basins after Palaeocene volcanism waned. A 300 m interbedded sequence of basalt weathering products, pyroclastics and lacustrine sediments resting on basalt was penetrated at Ballymoney (Fig. 3) to the north of Lough Neagh (Griffith *et al.* 1987; Parnell *et al.* 1989). A borehole drilled at Aghadolgan, just east of Lough Neagh (Fig. 3) penetrated 62 m of kaolinitic mudstone breccia, with a bauxitic

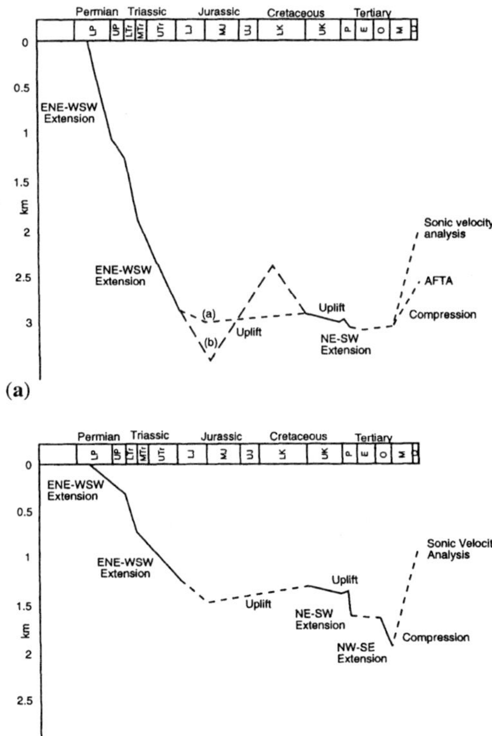

Fig. 11. (a) Burial history plot for the Larne Basin, based on the composite stratigraphy presented in Fig. 2a. Important tectonic events are labelled. (b) Burial history plot for the Lough Neagh Basin, using the composite stratigraphy shown in Fig. 2b.

composition (Mitchell 1984). Although the breccia has initially been ascribed to the Interbasaltic Formation, the thickness of the breccia is greater than the maximum recorded Interbasaltic thickness in outcrop in Northern Ireland. Overlying these are siliciclastics and lignite-rich sediments of the Lough Neagh Group, deposited in a lacustrine depositional environment. They are 300 m thick in the Mire House Borehole, and interpreted as Upper Oligocene in age (Wilkinson *et al.* 1980). They reach a maximum proved thickness of 350 m in the Washing Bay Borehole (Mitchell 1981, see Fig. 3). It should be pointed out, however, that deposition of the Lough Neagh Group and the underlying basalt weathering products seems to be fault controlled for two reasons. Firstly the basalt weathering material has only been found adjacent to faults, with boreholes drilled further away from the faults penetrating the Lough Neagh Group followed by fresh basalt. This represents a possible time gap of over 20 Ma between the Palaeocene basalts and Upper Oligocene Lough Neagh Group. Secondly the Lough Neagh Group is arenaceous at the base, and is largely fault bounded at the present day. It is possible that the basalt weathering product is base Upper Oligocene in age, and so should be considered as part of the Lough Neagh Group.

Creation of the accommodation space for the Lough Neagh Group to accumulate was probably due to faulting of now cooled basalts. The Tertiary basalts are thickest in the Lough Neagh area, and there is a regional dip of the basalts towards Lough Neagh, either due to a greater amount of thermal subsidence or increased compaction of the underlying Mesozoic sediments. Both suggest Lough Neagh to be the location of an important extrusion site. This, however, appears to be unimportant for two reasons. Collapse of a magma chamber under Lough Neagh 20 Ma after extrusion of the basalts is unrealistic. Secondly, England (1992) has shown that the thickest Tertiary basalts do not underlie the Tertiary sedimentary depocentre in Skye. Therefore it seems that the Upper Oligocene sediments accumulated relatively independently of the underlying basalts. Consequently the Larne Basin became relatively elevated with respect to the Lough Neagh Basin and may have become a sediment source area. Therefore the Upper Oligocene was a time of extensional faulting in the Lough Neagh area. The direction of extension should be regarded as approximate at best. England (1988) documents NW–SE extension normal to the dominant Caledonoid structural grain of the region. However, Kerr (1987) favours NNW–SSE in north Antrim and Millar (1990) NNE–SSW in the Larne area. Therefore a broadly northerly direction seems most appropriate.

Late Tertiary (Miocene)

Deducing tectonic events in the Miocene is made difficult by the lack of any preserved sediments younger than Upper Oligocene. Triassic sediments subcrop Quaternary and recent deposits in the North Channel (*Isle of Man.* Sheet 54°N–06°W. 1:250 000 Series. Solid Geology, *Clyde.* Sheet 55°N–06°W. 1:250 000 Series, Solid Geology), and an unknown amount of Tertiary basalt has also been eroded (McCaffrey & McCann 1992). A phase of NNE–SSW compression has been dated as Middle Eocene to Middle Miocene by Kerr (1987). Roberts (1989) cites George (1967) for local warping of the Upper Oligocene clays in Lough Neagh as evidence for Miocene inversion. A hangingwall anticline with fold axis trending NW–SE in the Portpatrick Basin to the south formed at least in part as a result of reverse reactivation of a normal fault (fig. 4(a) in Shelton 1995). However, this is an unreliable stress indicator due to the combined influence of the previously discussed basement high, and rollover associated with synsedimentary fault movement. *In situ* stress analysis of two boreholes onshore Northern Ireland (method after Evans & Brereton 1990) reveals a N–S compressive stress at present, although it is not clear whether this reflects Miocene or earlier compressive stress. Finally Boldreel & Andersen (1993) report NW–SE compressional tectonics in the Faroe–Rockall area in the Miocene. However, the crust of the two regions is likely to be significantly different. In addition Boldreel & Anderson (1993) document a N–S compressive phase in the Palaeocene, which possibly relates to the *in situ* stress in Northern Ireland. In conclusion, it is probable that the compression had at least some northerly component. The variation in orientation could reflect combined effects of Alpine and Atlantic tectonics.

Summary – burial history plots

Figure 11 shows burial history plots for the Larne and the Lough Neagh basins, based on Figs 2a and b respectively and incorporating the information given above.

In the Larne Basin (Fig. 11a), there was rapid burial during the Permian associated with cooling and subsidence after the early Permian extension. Following early Triassic subsidence, a major late Triassic extensional event occurred, resulting in the rapid deposition of a large volume of argillaceous sediment. During the Jurassic there are two possibilities, as discussed above: (a) deposition of Lias only, followed by non-deposition and a small amount of uplift and erosion in the late Jurassic to early Cretaceous; or (b) deposition of Middle and/ or Upper Jurassic strata, subsequently completely eroded during uplift in the early Cretaceous. The early Tertiary saw a period of uplift, followed by Lower and Upper Basalt deposition on what is now onshore Northern Ireland. There is no evidence to deduce tectonic events in the Eocene and early Oligocene. In the late Oligocene the Larne Basin was relatively uplifted in relation to the subsiding Lough Neagh and probably became a sediment source area. The Larne Basin experienced its last inversion episode in the Miocene, of the order of 0.5–1.1 km.

In contrast, the Lough Neagh Basin (Fig. 11b), became a depocentre in the late Oligocene, due to extensional faulting. In the absence of proof that younger pre-Quaternary sediments are preserved in the Lough Neagh Basin, and the local warping of the Upper Oligocene Lough Neagh Group (Roberts 1989), it seems most likely that the Lough Neagh Basin was also inverted in the Miocene. To date this uplift has only been quantified in the Ballymacilroy Borehole as around 0.6–1.1 km using sonic velocities (Tables 1 and 2). As already discussed the borehole occupies a marginal location and therefore may not be indicative of absolute uplift in the basin centre.

The author gratefully acknowledges support from BG Exploration and Production Ltd. Thanks are due to JEBCO Seismic Limited, Fynegold Petroleum Limited and partners for permission to publish their seismic data. Geotrack International are thanked for allowing access and publication of their AFTA® analyses. The manuscript benefited greatly from discussion with Alastair Ruffell and Derek Reay and the comments of the two reviewers, Steve Knott and Brian Haig.

References

ANDERSON, T. B. & OLIVER, G. J. H. 1986. The Orlock Bridge Fault: a major late Caledonian sinistral fault in the Southern Uplands Terrane, British Isles. *Transactions of the Royal Society of Edinburgh: Earth Sciences*, **77**, 203–222.

——, PARNELL, J. & RUFFELL, A. 1995. Influence of basement on the geometry of Permo-Triassic basins in the North-West British Isles. *In:* BOLDY, S. (ed.) *Permian and Triassic Rifting in Northwest Europe.* Geological Society, London, Special Publication, **91**, 103–122.

BLUCK, B. J. 1984. Pre-Carboniferous history of the Midland Valley of Scotland. *Transactions of the Royal Society of Edinburgh: Earth Sciences*, **75**, 275–295.

BOLDREEL, L. O. & ANDERSEN, M. S. 1993. Late Paleocene to Miocene compression in the Faeroe–Rockall area. *In:* PARKER, J. R. (ed.) *Petroleum Geology of North West Europe. Proceedings of the 4th Conference.* Geological Society, London, 1025–1034.

BRAY, R. J., GREEN, P. F. & DUDDY, I. R. 1992. Thermal history reconstruction using apatite fission track analysis and vitrinite reflectance: a case study from the UK East Midlands and Southern North Sea. *In:* HARDMAN, R. P. F. (ed.) *Exploration Britain: Geological Insights for the Next Decade.* Geological Society, London, Special Publication, **67**, 3–25.

BULAT, J. & STOKER, S. J. 1987. Uplift determination from interval velocity studies, UK southern North Sea. *In:* BROOKS, J. & GLENNIE, K. (eds) *Petroleum Geology of North West Europe.* Graham and Trotman, London, 293–305.

CARRUTHERS, R. M., CORNWELL, J. D., TURNBULL, G., WALKER, A. S. D. & BENNETT, J. R. P. 1987. Interpretation of the Bouguer gravity anomaly data for Northern Ireland. *Regional Geophysics Research Group British Geological Survey.* No. RG 87/5.

CASTON, G. F. 1976. The floor of the North Channel, Irish Sea: a sidescan sonar survey. *Report of the Institute of Geological Sciences,* **76/7.**

CHADWICK, R. A., KIRBY, G. A. & BAILY, H. E. 1994. The post-Triassic structural evolution of north-west England and adjacent parts of the East Irish Sea. *Proceedings of the Yorkshire Geological Society,* **50**, 91–102.

COLTER, V. S. 1978. Exploration for gas in the Irish Sea. *Geologie en Mijnbouw,* **57**, 503–516.

COPE, J. C. W. 1997. The Mesozoic and Tertiary history of the Irish Sea. *This volume.*

——, INGHAM, J. K. & RAWSON, P. F. 1992. *Atlas of Palaeogeography and Lithofacies.* Geological Society, London, Memoir **13**.

CORNFIELD, S. M., GAWTHORPE, R. L., GAGE, M., FRASER, A. J. & BESLEY, B. M. 1996. Inversion tectonics of the Variscan foreland of the British Isles. *Journal of the Geological Society, London,* **153**, 17–32.

COWARD, M. P. 1990. The Precambrian, Caledonian and Variscan framework to NW Europe. *In:* HARDMAN, R. F. P. & BROOKS, J. (eds) *Tectonic Events Responsible for Britain's Oil and Gas Reserves.* Geological Society, London, Special Publication, **55**, 1–34.

DAVISON, I. 1995. Fault slip evolution determined from crack-seal veins in pull-aparts and their implications for general slip models. *Journal of Structural Geology* **17**(7), 1025–1034.

DUDDY, I. R., GREEN, P. F., BRAY, R. J. & HEGARTY, K. A. 1994. Recognition of the thermal effects of fluid flow in sedimentary basins. *In:* PARNELL, J. (ed.) *Geofluids: Origin, Migration and Evolution of Fluids in Sedimentary Basins.* Geological Society, London, Special Publication, **78**, 325–345.

EL-SHAHAT, A. & WEST, I. 1983. Early and late lithification of aragonitic bivalve beds the Purbeck formation (Upper Jurassic–Lower Cretaceous) of southern England. *Sedimentary Geology,* **35**, 15–41.

ENGLAND, R. W. 1988. The early Tertiary stress regime in NW Britain: evidence from the patterns of volcanic activity. *In:* MORTON, A. C. & PARSONS, L. M. (eds) *Early Tertiary Volcanism and the Opening of the North Atlantic.* Geological Society, London, Special Publication, **39**, 381–389.

—— 1994. The structure of the Skye lava field. *Scottish Journal of Structural Geology*, **30** (1) 33–37.

EVANS, C. J. & BRERETON, N. R. 1990. In situ crustal stress in the United Kingdom from borehole breakouts. *In:* HURST, A., LOVELL, M. A. & MORTON, A. C. (eds) *Geological Applications of Wireline Logs.* Geological Society, London, Special Publication, **48**, 327–338.

FORSTER, S. C. & WARRINGTON, G. 1985. Geochronology of the Carboniferous, Permian and Triassic. *In:* SNELLING, N. J. (ed.) *The Chronology of the Geological Record.* Geological Society, London, Memoir, **10**, 99–113.

FOWLER, J. & ROBBIE, J. A. 1961. *Geology of the Country around Dungannon.* Memoir of the Geological Survey of Northern Ireland, HMSO, Belfast.

FRANCIS, E. H. 1978. The Midland Valley as a rift, seen in connection with the late Palaeozoic European rift system. *In:* RAMBERG, I. B. & NEUMANN, E. R. (eds) *Tectonics and Geophysics of Continental Rifts.* Reidel, Dordrecht, Holland, 133–147.

—— 1991. Carboniferous–Permian igneous rocks. *In:* CRAIG, G. Y. (ed.) *A Geology of Scotland.* Scottish Academic Press, Edinburgh, 393–420.

GEORGE, T. N. 1967. Landform and structure in Ulster. *Scottish Journal of Geology*, **3** (3), 413–448.

GREEN, P. F., DUDDY, I. R. & BRAY, R. J. 1993. Early Tertiary heating in Northwest England: fluids or burial (or both?). *Geofluids '93 Conference Extended Abstract.*

——, —— & —— 1997. Variation in thermal history styles around the Irish Sea and adjacent areas: implications for hydrocarbon occurrence and tectonic evolution. *This volume.*

GREIG, D. C. 1971. *The South of Scotland.* British Regional Geology Guide. I.G.S. HMSO, Edinburgh.

GRIFFITH, A. E. 1983. The search for petroleum in Northern Ireland. *In:* BROOKS, J. (ed.) *Petroleum Geochemistry and Exploration of Europe.* Geological Society, London, Special Publication, **12**, 213–222.

—— & WILSON H. E. 1982. *Geology of the country around Carrickfergus and Bangor.* Memoir of the Geological Survey of Northern Ireland, HMSO, Belfast.

——, LEGG, I. C. & MITCHELL, W. I. 1987. Mineral Resources. *In:* BUCHANAN, R. H. & WALKER, B. M. (eds) *Province, City and People: Belfast and its Region.* Greystone Books, Belfast, 43–58.

HALL, J., BREWER, J. A., MATTHEWS, D. H. & WARNER, M. R. 1984. Crustal structure across the Caledonides from the 'WINCH' seismic reflection profile: influences on the evolution of the Midland Valley of Scotland. *Transactions of the Royal Society of Edinburgh: Earth Sciences*, **75**, 97–109.

HESSE, R. 1986. Diagenesis 11. Early diagenetic pore water/sediment interaction: Modern offshore basins. *Geoscience Canada*, **13**, 165–196.

HILLIS, R. R. 1991. Chalk porosity and Tertiary uplift, Western Approaches Trough, SW UK and NW French continental shelves. *Journal of the Geological Society, London*, **148**, 669–679.

—— 1993. Tertiary erosion magnitudes in the East Midlands Shelf, onshore UK. *Journal of the Geological Society, London*, **150**, 1047–1050.

HUTTON, D. H. W. 1987. Strike slip terranes and a model for the evolution of the British and Irish Caledonides. *Geological Magazine*, **124**, 405–425.

ILLING, L. V. & GRIFFITH A. E. 1986. Gas prospects in the 'Midland Valley' of Northern Ireland. *In:* BROOKS, J., GOFF, J. C. & VAN HOORN, B. (eds) *Habitat of Palaeozoic Gas in N.W. Europe.* Geological Society, London, Special Publication, **23**, 73–84.

JACKSON, D. I. & MULHOLLAND, P. 1993. Tectonic and stratigraphic aspects of the East Irish Sea Basin and adjacent areas: contrasts in their post–Carboniferous structural styles. *In:* PARKER, J. R. (ed.) *Petroleum Geology of North West Europe. Proceedings of the 4th Conference.* Geological Society, London, 791–808.

KERR, I. D. V. 1987. *Basement/Cover Structural Relationships in the North Antrim Area, Ireland.* Unpublished PhD thesis, The Queen's University of Belfast.

LEE, M. K., PHARAOH, T. C. & SOPER, N. J. 1990. Structural trends in central Britain from images of gravity and aeromagnetic fields. *Journal of the Geological Society, London*, **147**, 241–258.

MACINTYRE, R. M., McMENAMIN, T. & PRESTON, J. 1975. K–Ar results from Western Ireland and their bearing on the timing and siting of Thulean Magmatism. *Scottish Journal of Geology*, **11**, 227–249.

MADDOX, S. J., BLOW, R. A. & O'BRIEN, S. R. 1997. The geology and hydrocarbon prospectivity of the North Channel Basin. *This volume.*

MANNING, P. I. & WILSON, H. E. 1975. The stratigraphy of the Larne Borehole, Co. Antrim. *Bulletin of the Geological Survey of Great Britain*, **50**, 1–50.

——, ROBBIE, J. A. & WILSON, H. E. 1970. *Geology of Belfast and the Lagan Valley.* Memoir of the Geological Survey of Northern Ireland.

MARIE, J. P. P. 1975. Rotliegendes stratigraphy and diagenesis. *In:* WOODLAND, A. W. (ed.) *Petroleum and the Continental Shelf of North-west Europe. Vol. 1, Geology.* Applied Science Publishers, London, 205–210.

MARSHALL, J. 1982. Isotopic composition of displacive fibrous calcite veins: reversals in pore water composition trends during burial diagenesis. *Journal of Sedimentary Petrology*, **52**, 615–630.

McCAFFREY, R. J. & McCANN, N. 1992. Post-Permian basin history of northeast Ireland. *In:* PARNELL, J. (ed.) *Basins on the Atlantic Seaboard: Petroleum Geology, Sedimentology and Basin Evolution.* Geological Society, London, Special Publication, **62**, 277–290.

McCANN, N. 1988. An assessment of the subsurface geology between Magilligan Point and Fair Head, Northern Ireland. *Irish Journal of Earth Sciences*, **9**, 71–78.

—— 1990. The subsurface geology between Belfast and Larne, Northern Ireland. *Irish Journal of Earth Sciences*, **10**, 157–173.

—— 1991. Subsurface geology of the Lough Neagh–Larne Basin, Northern Ireland. *Irish Journal of Earth Sciences*, **11**, 53–64.

McLEAN, A. C. 1978. Evolution of fault-controlled ensialic basins in northwestern Britain. *In:* BOWES, D. R. & LEAKE, B. E. (eds) *Crustal Evolution in Northwestern Britain and Adjacent Regions.* Seel House Press, Liverpool, 325–346.

McKERROW, W. S. & ELDERS, C. F. 1989. Short Paper: Movements on the Southern Upland Fault. *Journal of the Geological Society. London,* **146**, 393–395.

MILLAR, G. 1990. *Fracturing in the NW Carboniferous Basin, Ireland.* Unpublished PhD thesis, The Queen's University of Belfast.

MITCHELL, G. H. 1981. Other Tertiary events. *In:* HOLLAND, C. H. (ed.) *A Geology of Ireland.* Scottish Academic Press, Edinburgh, 231–234.

MITCHELL, W. I. 1984. Mineral exploration in the Lough Neagh Group – Results of the Phase I Drilling Programme – 1983/84, in the Lough Beg–Tolans Point area of Lough Neagh, Co. Antrim. *Geological Survey of Northern Ireland Open File Report,* **70.**

—— 1992. The origin of Upper Palaeozoic sedimentary basins in Northern Ireland and relationships with the Canadian Maritime Provinces. *In:* PARNELL, J. (ed.) *Basins on the Atlantic Seaboard: Petroleum Geology, Sedimentology and Basin Evolution.* Geological Society, London, Special Publication, **62**, 191–202.

MORTON, N. 1993. Potential reservoir and source rocks in relation to Upper Triassic to Middle Jurassic sequence stratigraphy, Atlantic margin basins of the British Isles. *In:* PARKER, J. R. (ed.) *Petroleum Geology of North West Europe. Proceedings of the 4th Conference.* Geological Society, London, 285–297.

NAYLOR, D. 1992. The post-Variscan history of Ireland. *In:* PARNELL, J. (ed.) *Basins on the Atlantic Seaboard: Petroleum Geology, Sedimentology and Basin Evolution.* Geological Society, London, Special Publication, **62**, 255–276.

PARNELL, J., SHUKLA, B. & MEIGHAN, I. G. 1989. The lignite and associated sediments of the Lough Neagh Basin. *Irish Journal of Earth Sciences,* **10**, 76–88.

PENN, I. E. 1981. Larne No.2 geological well completion report. *British Geological Survey Deep Geology Unit Report 81/6.*

——, HOLLIDAY, D. W., KIRBY, G. A., KUBALA, M., SOBEY, R. A. *ET AL.* 1983. The Larne No.2 borehole:

Discovery of a New Permian Volcanic Centre. *Scottish Journal of Geology,* **19**, 333–346.

PRESTON, J. 1981. Tertiary Igneous Activity. *In:* HOLLAND, C. H. (ed.) *A Geology of Ireland.* Scottish Academic Press, Edinburgh, 213–223.

RAWSON, P. F. & RILEY, L. A. 1982. Latest Jurassic–early Cretaceous events and the 'Late-Cimmerian Unconformity' in the North Sea area. *American Association of Petroleum Geologists Bulletin,* **66** (12), 2628–2648.

ROBERTS, D. G. 1989. Basin inversion in and around the British Isles. *In:* COOPER, M. A. & WILLIAMS, G. D. (eds) *Inversion Tectonics.* Geological Society, London, Special Publication, **49**, 131–152.

RUFFELL, A. 1991. Palaeoenvironmental analysis of the late Triassic succession in the Wessex Basin and correlation with surrounding areas. *Proceedings of the Ussher Society,* **7**, 402–407.

RUKIN, N. 1990. *The diagenesis of the Shales-with-beef of the Lower Lias, West Dorset.* Unpublished PhD thesis, University of Liverpool.

STONELEY, R. 1983. Fibrous calcite veins, overpressures and primary oil migration. *American Association of Petroleum Geologists Bulletin,* 1427–1428.

THOMPSON, P. 1985. *Dating the British Tertiary Igneous Province in Ireland by the ^{40}Ar–^{39}Ar stepwise degassing method.* Unpublished PhD thesis, University of Liverpool.

THOMPSON, S. J. 1979. Preliminary Report on the Ballymacilroy No.1 Borehole, Ahoghill, Co. Antrim. *Geological Survey of Northern Ireland, Open File Report* **63**.

TOMKEIEFF, S. I. 1964. Petrochemistry and Petrogenesis of the British Tertiary Igneous Province. *In Advancing Frontiers of Geology and Geophysics*, p. 327.

UPTON, G. J., FITTON, J. G. & MACINTYRE, R. M. 1986. The Glas Eilean lavas: evidence of a Lower Permian volcano-tectonic basin between Islay and Jura, Inner Hebrides. *Transactions of the Royal Society of Edinburgh: Earth Sciences,* **77**, 289–293.

WHITE, R. S. 1992. Crustal structure and magmatism of North Atlantic continental margins. *Journal of the Geological Society, London,* **149**, 841–854.

WILKINSON, G. C., BAZLEY, R. A. B. & BOULTER, M. C. 1980. The geology and palynology of the Oligocene Lough Neagh Clays, Northern Ireland. *Journal of the Geological Society, London,* **137**, 65–75.

WILSON, H. E. 1972. *Regional Geology of Northern Ireland.* HMSO, Belfast.

Structural evolution of the Isle of Man and central part of the Irish Sea

D. G. QUIRK[1] & G. S. KIMBELL[2]

[1]*School of Construction and Earth Sciences, Oxford Brookes University, Gipsy Lane, Oxford OX3 0BP, UK*

[2]*British Geological Survey, Nicker Hill, Keyworth, Nottinghamshire NG12 5GG, UK*

Abstract: Gravity, aeromagnetic, seismic, borehole and outcrop data from the Isle of Man area have been used to study the structural framework and tectonic evolution of the central part of the Irish Sea since the Ordovician. The area has been affected by two major compressional events (Caledonian and Variscan), three rifting events (early Carboniferous, early Permian and ?late Jurassic) and a period of thermal uplift (early Tertiary). A large part of the Isle of Man consists of early Ordovician turbidites ('basement') interpreted to have been deposited in a back arc or fore arc basin that was inverted during the Caledonian orogeny. Major NE–SW and subsidiary NW–SE strike-slip faults were initiated at this time, some of which were later reactivated to form important basin controlling faults in the Irish Sea. Post-Caledonian basin subsidence was initiated in the early Carboniferous with a period of NW–SE directed extension followed by thermal sag which ceased at the end of the Carboniferous as a result of Variscan compression. E–W extension in the early Permian led to the development of new extensional faults which have influenced the thickness of the Collyhurst Sandstone in the offshore. Post-rift thermal sag continued throughout the late Permian and the Triassic. A late Jurassic period of extension has been proposed in order to explain post-Triassic offset across major normal faults east of the Isle of Man although direct evidence of syn-rift or post-rift strata has been removed by erosion during uplift in the early Tertiary. This uplift was due to thermal doming of the crust, an event which was associated with the intrusion of a large number of dolerite dykes. The structural evolution of the Isle of Man has also been influenced by a major NW–SE trending basement fracture known as the 'Central Valley Lineament' which extends into the offshore to the SE and NW of the island. The lineament appears to be important in defining areas of different hydrocarbon prospectivity in the Irish Sea.

The Isle of Man occupies a central position in the Irish Sea. It is bounded to the east by the Eubonia and Lagman basins, to the west by the Peel Basin and by the Solway Basin at the NE end of the island (Fig. 1). The island has a population of approximately 70 000 and covers an area of 572 km². The Isle of Man is a self-governing dependency of the British Crown and is not part of the UK nor the European Union. The Manx Government extended its Territorial Sea from 3 to 12 miles in September 1992 by purchasing it from the UK Government for £800 000. A first hydrocarbon licensing round was announced on 6 December 1994 and closed on 28 March 1995. To date 11 blocks and part blocks covering the entire NE sector of Manx Territorial Sea have been licensed, including several blocks awarded in two out-of-round awards announced on 12 February 1996. The area is important for fishing and recreation and contains diverse marine life including several species of sea mammals, basking sharks and shellfish. Consequently certain environmental restrictions will apply to seismic and drilling activities.

Very little geological work has been published about the Isle of Man. To date, the best source of information is a field guide by Ford (1993). This guide serves as a useful summary of the previous (limited) observations and ideas on the geology exposed onshore. However, the current paper represents the first published attempt to describe the structural evolution of the Isle of Man and aims to provide some new ideas on the hydrocarbon prospectivity in the central part of the Irish Sea.

The paper is based on a study between longitudes 3°50′W and 5°20′W and latitudes 53°52′N and 54°32′N where one of the authors (Kimbell) generated and manipulated Bouguer gravity anomaly images and the other author (Quirk) interpreted these data in conjunction with 2D seismic, high resolution aeromagnetic data, onshore and offshore well information and field observations (Table 1). The Bouguer gravity data are shown in Figs 2a and b and a summary of the integrated interpretation is presented in the form of a structural lineaments map in Fig. 3 and a tectono-stratigraphic column in Fig. 4. The main part of the text and Fig. 6 aim to

From Meadows, N. S., Trueblood, S. P., Hardman, M. & Cowan, G. (eds), 1997, *Petroleum Geology of the Irish Sea and Adjacent Areas,* Geological Society Special Publication No. 124, pp. 135–159.

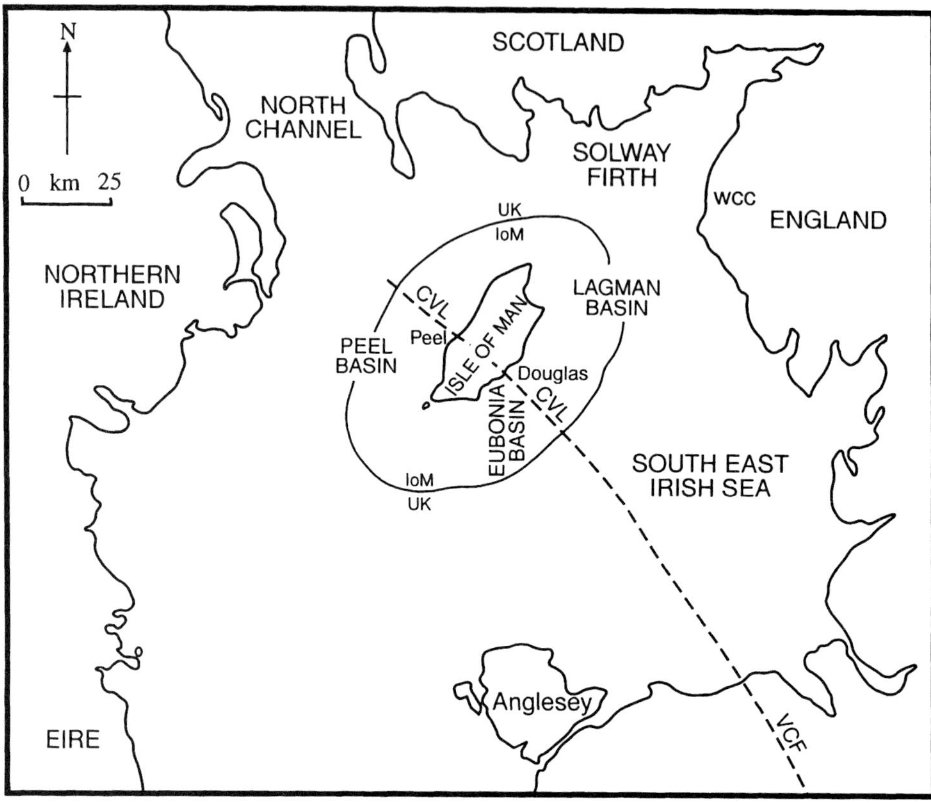

Fig. 1. Location map of the central part of the Irish Sea showing the boundary between Isle of Man and UK territorial sea (CVL = Central Valley Lineament; VCF = Vale of Clywd Fault; WCC = West Cumbrian Coalfield).

Table 1. *Main data used in study*

Data type	Details
Bouguer gravity anomaly and derivatives	Onshore data – 360 stations; offshore data – typical line spacing 3–6 km (data owned by Western Geophysical & BGS)
High resolution aeromagnetics and derivatives	400 m traverse spacing; 125 m image pixel size (data owned by World Geoscience)
2D migrated seismic lines	300 km of data mostly along SE coast of Isle of Man (data owned by Western Geophysical) + 200 km of oil company proprietary data + various regional lines
Onshore borehole cores	103 m of core from three wells in north of Isle of Man (Ballavarkish, Shellag Point & Cranstal) drilled by Riofinex in 1985–1987 (now stored by Manx National Heritage)
Released offshore well formation	Composite logs, TWT information and palynological data from wells 112/25-1 and 112/30-1
Field observations	15 weeks fieldwork over 5 years undertaken by one author (Quirk)

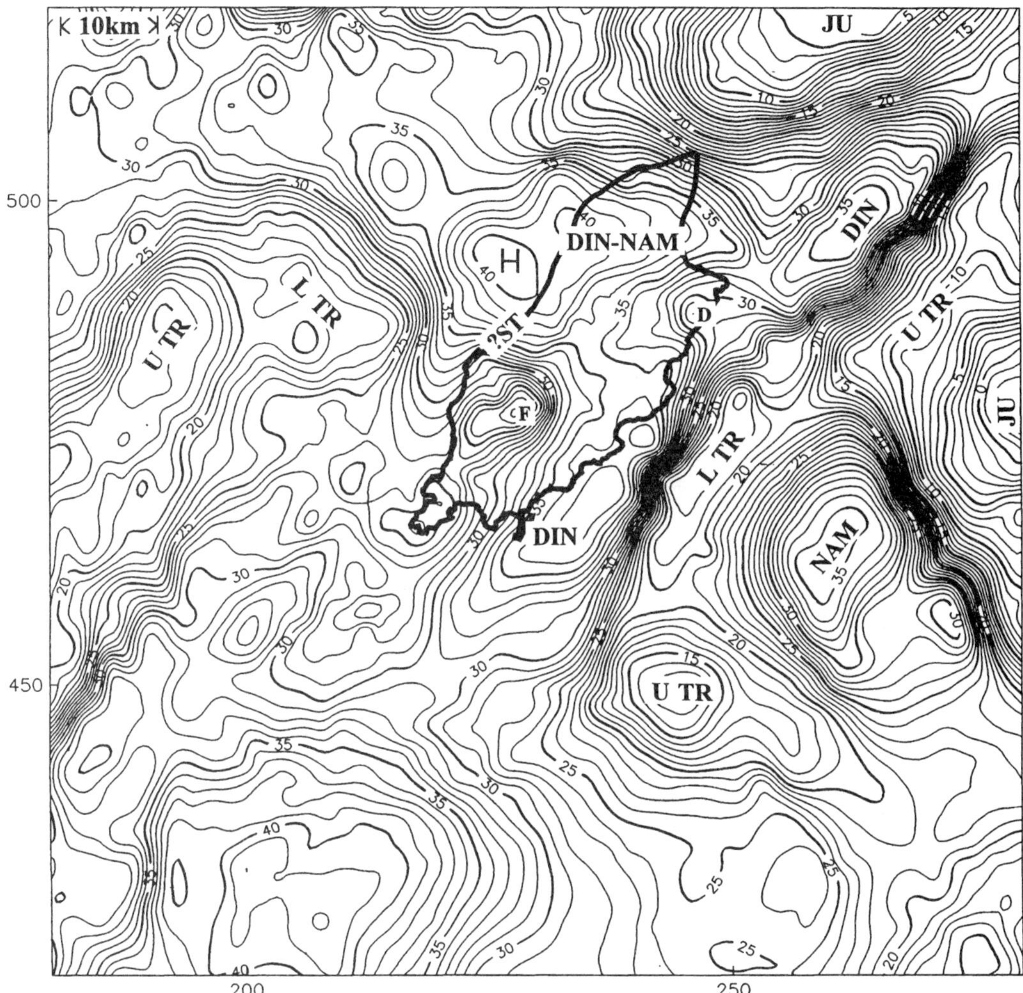

Fig. 2. (a) Bouguer gravity anomaly map of the Isle of Man and central part of the Irish Sea, courtesy of BGS and Western Geophysical; contour interval = 1 Mg (1 milligal); Bouguer reduction density = 2.20 Mg m^{-3} (offshore), 2.70 Mg m^{-3} (onshore). Note the correlation between gravity anomaly values and the exposed solid geology where DIN = Dinantian, NAM = Namurian, L TR = Lower Triassic, U TR = Upper Triassic and JU = Lower Jurassic, ? ST = Upper Carboniferous or Devonian. F = Foxdale Granite; D = Dhoon Granite; H = gravity high north of Peel.

show when the main lineaments were active since the early Palaeozoic and are based mostly on seismic and field evidence.

Lower Palaeozoic (Caledonian) basement

Eighty per cent of the exposed rock on the Isle of Man consists of low grade Lower Palaeozoic turbidites, related meta-sediments and early intermediate intrusions known collectively as the Manx Group. The sediments are commonly bioturbated, indicating that oxic conditions prevailed during deposition. Based on a limited collection of acritarchs and graptolites, these rocks are dated as early Ordovician (Downie & Ford 1966; Molyneux 1979; Rushton 1993) and the acritarchs are Avalonian in affinity (Molyneux 1979). Recent fieldwork on the Manx Group by Quirk & Ford (1994) and a reappraisal of the acritarch data (Molyneux 1979; Cooper *et al.* 1995) have shown that many of the structural and stratigraphic interpretations of the early Palaeozoic of the Isle of Man made by Lamplugh (1903), Gillott (1956) and Simpson (1963, 1968) are difficult to justify.

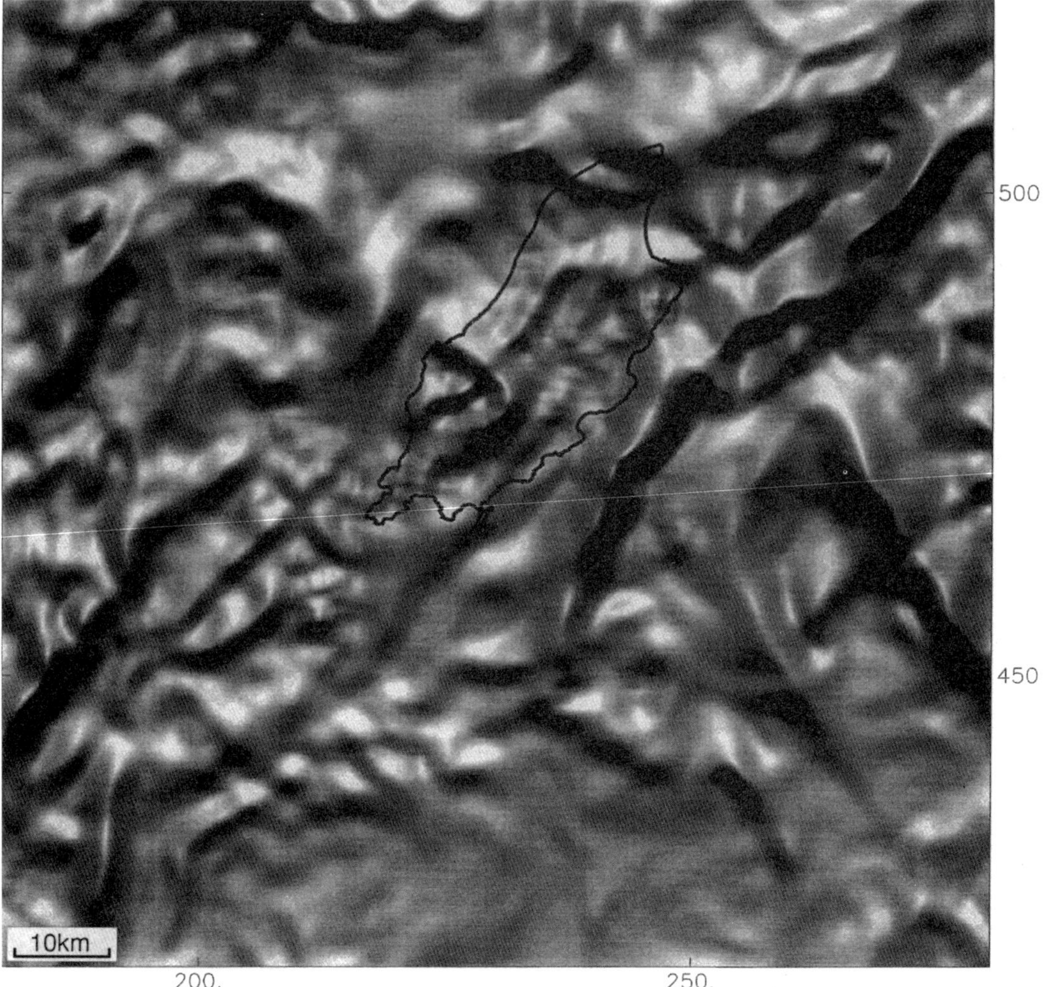

Fig. 2. (b) Scalar horizontal gradient of Bouguer gravity anomaly displayed as pseudo-relief illuminated from the north.

However, Simpson's (1963) observation still holds true that a remarkable thickness of Manx Group sediment (?5000 m) was deposited over a relatively short period of time (?10 Ma) based on minimum estimates by Molyneux (1979).

The NW coast of the Isle of Man lies within a few kilometres of the SE edge of the Caledonian Iapetus suture (McKerrow & Soper 1989; Todd *et al.* 1991; Quirk & Ford 1994) which fully closed in the late Silurian (Soper *et al.* 1992a). Caledonian compressional structures in the form of cleavage, folds and thrust faults are orientated ENE–WSW on the Isle of Man. NW–SE and NE–SW oriented vertical shear zones are also exposed on the NW coast and are attributed to Caledonian strike slip. In addition, syn-tectonic quartz veining is common and syn- and post-tectonic granitic bodies are present.

Back arc basin

Current Caledonian plate reconstructions would place the Isle of Man at the NW edge of Eastern Avalonia, the most northerly of the landmasses on the southern side of Iapetus (e.g. Soper & Woodcock 1990; Todd *et al.* 1991; Soper *et al.* 1992b). The SE coast of the Isle of Man consists

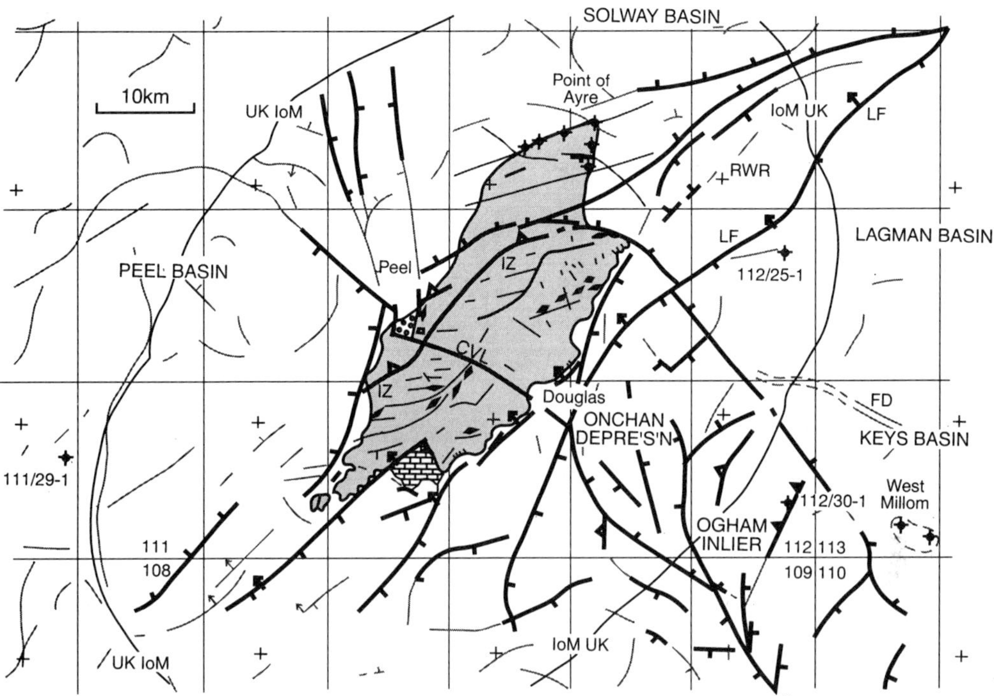

Fig. 3. Interpreted structural lineaments in the Isle of Man and central part of the Irish Sea (brick ornament = outcrop of Castletown Group; dotted ornament = outcrop of Peel Group; CVL = Central Valley Lineament; RWR = Ramsey–Whitehaven Ridge; IZ = imbricate zone; FD = Fleetwood Dyke; LF = Lagman Fault; thick black lines = major faults & fractures; single grey lines = inferred faults; ticks on faults = normal down-throw direction; triangles on faults = reverse up-throw direction; arrows on faults = steeply dipping strata on up-thrown side; double grey lines = dolerite dykes; grey pod shapes on the Isle of Man = felsite dykes & granite intrusions); the positions of seismic lines shown in Figs 10 & 19 are also indicated as double black lines.

mostly of sand-rich meta-sediments including pure quartzites with NW-directed palaeocurrents which support these plate reconstructions. The centre of the island is dominated by fine grained meta-sediments which are interpreted as deep water, distal deposits. However, along the NW coast of the Isle of Man, medium to coarse grained greywackes are common which contain rare flute casts indicating SE-directed palaeoflow from an unknown land source to the NW (Quirk & Ford 1994). Mud-supported breccias with diverse sedimentary clasts occur, particularly to the NE of Peel (Fig. 3), which are interpreted to have been deposited as fault-induced debris flows from areas of shallower water. Juvenile fragments of intermediate-mafic rocks are found as lithic clasts within the greywackes and conglomerates along the NW coast and rare tuffaceous horizons are also present. Intermediate sills and dykes are widespread throughout the Isle of Man and many of these appear to have been intruded into soft sediment.

It therefore appears that, rather than representing a passive margin on the NW side of Eastern Avalonia, the SE and NW coasts of the Isle of Man represent two sides of a tectonically active basin of early Ordovician age. The source of sediment from the NW is thought to be either an island arc or an accretionary prism implying that the Manx Group was deposited within a back arc or fore arc basin. The basin was inverted during Caledonian orogenesis at the end of the Silurian but has continued to uplift and subside several times since then. It is therefore worth bearing in mind that, by virtue of the widespread occurrence of coarse-grained meta-sediments, quartz veins and granitic rocks within the Manx Group, these rocks represent a potential source of sand for younger intervals.

Sinistral transpression

High resolution aeromagnetic data over the Isle of Man reveals a left-stepping *en echelon* structure or

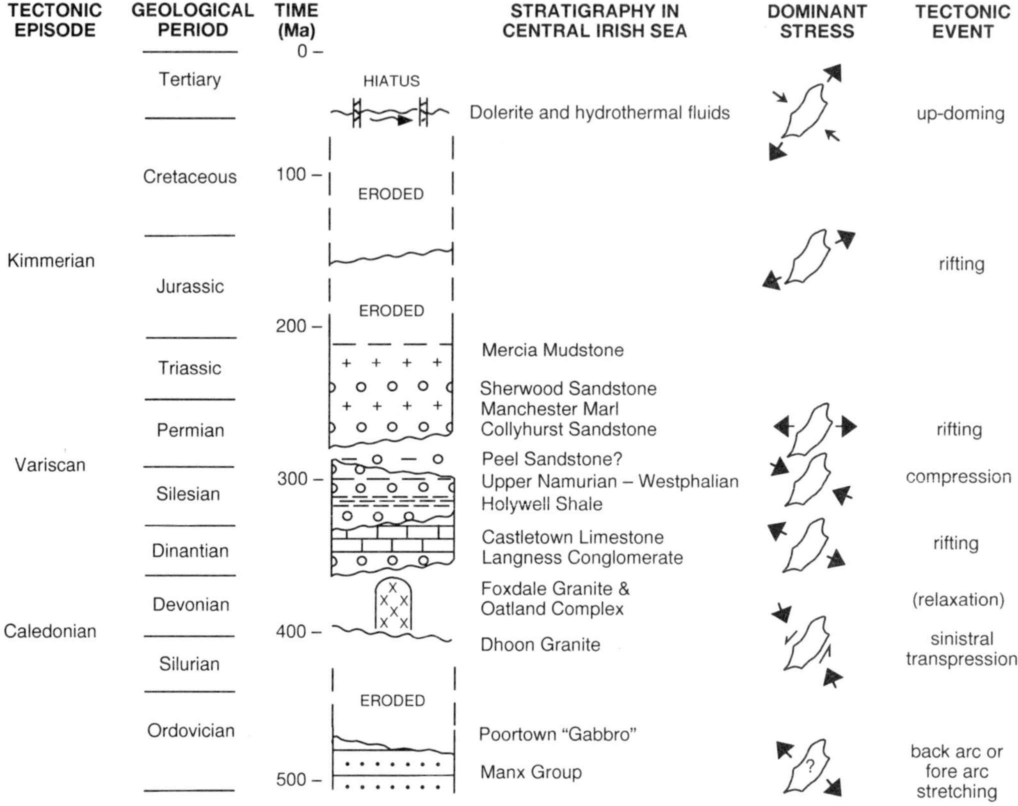

Fig. 4. Summary stratigraphy of the central part of the Irish Sea in relation to main tectonic events.

imbricate zone about 5 km wide and running for at least 40 km along the length of the island (Fig. 5). Total field magnetic anomalies have a pronounced negative component indicating the presence of strong reversed magnetization in the Manx Group within this zone. The margins of the imbricate zone are orientated NE–SW and the imbricate pattern itself is produced by ENE–WSW to E–W trending lineaments, 1–5 km long, which are spaced at intervals of approximately 2 km (Fig. 3). The structure is also visible on a horizontal gradient display of the Bouguer gravity anomaly (Fig. 2b) but is less clearly delineated than on the higher resolution aeromagnetic data (Fig. 5).

At the northern end of the imbricate zone, a prominent escarpment marks the boundary between uplands to the south, comprising Manx Group metasediments, and lowlands to the north, consisting of recent glacial sediments which are underlain progressively away from the escarpment by Carboniferous, Permian and Triassic strata. On the NW coast, an extension of this escarpment crops out as an obvious fault affecting both Manx

Group strata and overlying Pleistocene sediments. There are even recent reports of minor earthquakes close to the escarpment at Ramsey (Fig. 6a). It is therefore thought to mark the position of a major fault which crosses the southern end of the Ramsey–Whitehaven Ridge (Fig. 3).

Outcrop is fairly limited within the imbricate zone except in the SW where it intersects the western coast of the Isle of Man between Port Erin and Niarbyl (Fig. 6b). Even here access is difficult because of the presence of precipitous cliffs except close to the NW margin of the imbricate zone south of Niarbyl. Here a set of north-dipping, NE–SW to E–W shear zones and faults are exposed which coincide with the orientation of the main lineaments on the aeromagnetic data (Figs 3, 5, & 7). Faults orientated E–W to ENE–WSW consist mostly of shallow- to moderately dipping thrusts whereas faults trending closer to NE–SW are generally steeper and kinematic indicators suggest that they have moved in a sinistral-reverse sense of oblique slip. These field observations coupled with the left-stepping nature of the imbrication (Fig. 3)

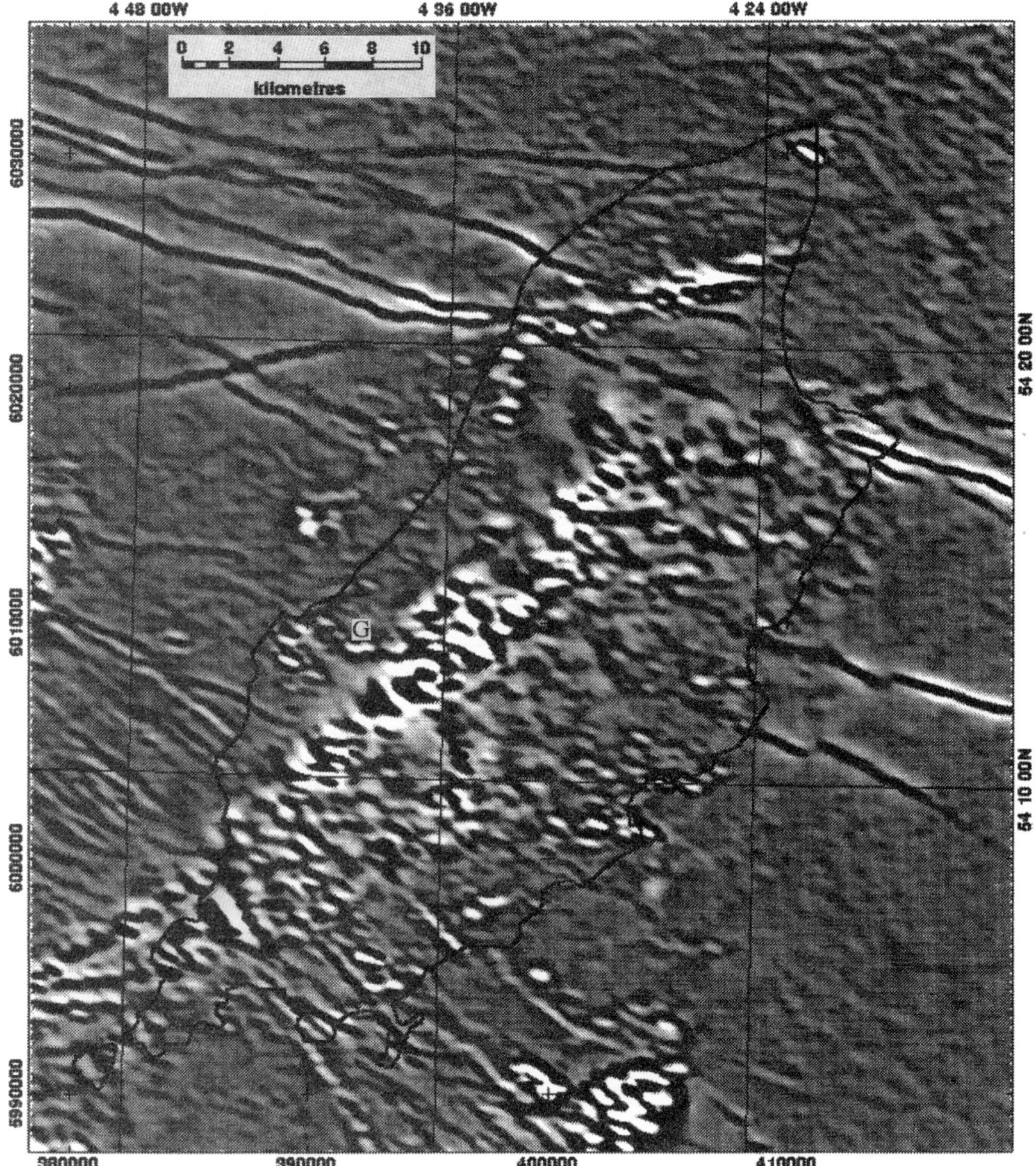

Fig. 5. Grey scale image of first vertical derivative of reduced-to-pole magnetic field from high resolution aeromagnetic survey, courtesy of World Geoscience. The data have been filtered to accentuate short wavelength anomalies from near-surface sources (black = low; white = high). The most obvious features are NW–SE trending lineaments which correspond with Tertiary dolerite dykes but note also the NE–SW oriented imbricate structure running through the centre of the Isle of Man. G = Poortown Gabbro.

suggests that the zone represents a fault duplex formed by sinistral transpression (Fig. 6b). The widespread occurrence of ENE–WSW trending Caledonian compressional structures on the Isle of Man indicates that the duplex developed during final NNW–SSE directed closure of Iapetus. Evidence for sinistral transpression on late Caledonian structures in other areas along the NW margin of Eastern Avalonia (e.g. Soper *et al.* 1992a) supports this view.

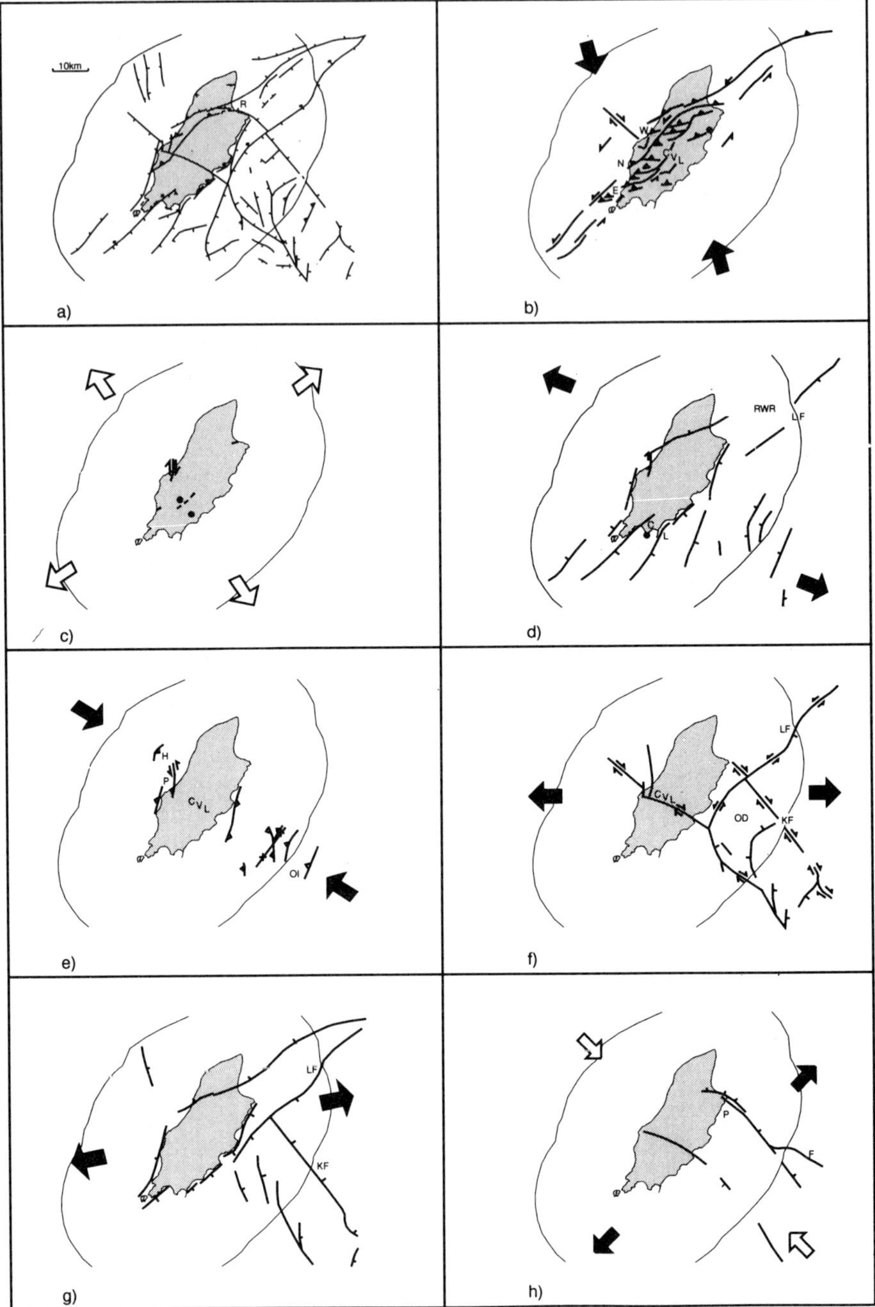

Fig. 6. Interpreted structural history of the central part of the Irish Sea: (**a**) present day fault pattern (R = Ramsey); (**b**) late Silurian (Caledonian) tectonic activity (N = Niarbyl, E = Port Erin, W = Will's Strand, CVL = Central Valley Lineament); (**c**) Devonian tectonic activity (solid circles and dotted lines = granitic & felsitic intrusions); (**d**) Dinantian–earliest Namurian tectonic activity (C = Castletown, L = Langness, solid circle = volcanic pile, RWR = Ramsey–Whitehaven Ridge, LF = Lagman Fault); (**e**) end Carboniferous (Variscan) tectonic activity (P = Peel, OI = Ogham Inlier, H = Bouguer gravity high, CVL = Central Valley Lineament); (**f**) early Permian tectonic activity (OD = Onchan Depression, LF = Lagman Fault, KF = Keys Fault, CVL = Central Valley Lineament); (**g**) late Jurassic (Kimmerian) tectonic activity (LF = Lagman Fault, KF = Keys Fault); and (**h**) early Tertiary tectonic activity but excluding most of the Tertiary dolerite dykes visible in Fig. 5 (P = Port Mooar, F = Fleetwood Dyke).

Fig. 7. Part of a major fault exposed at Niarbyl (see Fig. 6b). Note NNW-dipping thrust plane adjacent to hammer which separates deformed mudstones in the footwall from fairly undeformed greywackes in the hanging wall.

It is interesting to speculate why the imbricate zone is associated with anomalous reversed remanent magnetization. Pre-deformation intermediate dykes are orientated mostly ENE–WSW, sub-parallel to the imbricate zone. However, these are fairly evenly distributed throughout the Manx Group, are rarely vertical and are generally less than 1 m thick. They are therefore unlikely to be observed on aeromagnetic data. A highly sheared mafic rock (the 'Poortown Gabbro') is quarried 5 km east of Peel and a similar body is exposed on the coast nearby at Will's Strand (Fig. 6b). These can be resolved as weak aeromagnetic features but are located outside the imbricate zone (Fig. 5). Many granitic bodies and felsite dykes occur within and on the SE side of the imbricate zone (Fig. 3) but most of these post-date shearing and they are unlikely to have an anomalous magnetic response. In contrast, most of the early Tertiary dolerite dykes which run through this part of the Irish Sea show a marked reversed magnetic response similar to the anomalies present within the imbricate zone. However, these trend NW–SE, perpendicular to the imbricate zone, and do not appear to be concentrated within it. Instead one possible explanation for the anomalous magnetism can be found at the NE end of the imbricate zone on the Isle of Man where a significant amount of iron mineralisation is present, some of which has been mined in the past. The iron mostly occurs as haematite and siderite although pyrrhotite has also been reported (Lamplugh 1903) and magnetite may develop at depth. These minerals occupy fissure veins and zones of metasomatic replacement. A 5 m wide zone of mineralization is exposed at Port Mooar (Fig. 6h) in the NE wall of a major Tertiary dyke which can be traced to the Fleetwood Dyke in the offshore (Fig. 3). This consists of metasomatic haematite and siderite, and dolomite in fractures between brecciated material. Furthermore, where rocks are exposed elsewhere along the imbricate zone, they consist predominantly of pelites with pyrite porphyroblasts, siderite concretions and related metalliferous minerals. It is therefore possible that the reversed magnetic anomalies present within the imbricate zone are due to high metal concentrations within the Manx Group, some of which may have been redistributed by hydrothermal fluids associated with dyke intrusion in the early Tertiary.

Central Valley Lineament

The imbricate zone is transected and offset slightly by a NW–SE trending lineament (Fig. 5) which coincides with the position of the Central Valley at surface on the Isle of Man. This valley is an important geomorphological feature; it separates the northern uplands from the southern uplands and links the two main bays on the NW and SE coasts (Peel and Douglas, respectively) (Fig. 3). Hence, the lineament is called the 'Central Valley Lineament' in this paper. The lineament is clearly imaged by Bouguer gravity data, where it can be traced for some distance offshore both to the NW and SE of the island (Fig. 2a), and, on seismic data, a linked set of NE- and SW-dipping faults which offset Carboniferous and Permo-Triassic strata occur along the trend (Fig. 3). It crosses the East Irish Sea and apparently links up with the Vale of Clwyd Fault in North Wales (Fig. 1). To date, no hydrocarbon discoveries have been made on the SW side of the Central Valley Lineament, suggesting that it has had a fundamental influence on basin prospectivity.

Although the Manx Group is not exposed anywhere along the Central Valley Lineament, some smaller NW–SE Caledonian shear zones are exposed at other places along the coast. Also, a marked change in the lithofacies of the Manx Group occurs to the north and south of Peel, approximately where the Central Valley Lineament is projected to lie, suggesting that it may have influenced sedimentation during the early Ordovician on the NW side of the Isle of Man. However, at least some of this change is probably due to post-Caledonian sinistral offset and so any early Palaeozoic movement on the Central Valley Lineament remains speculative.

Later reactivation

Offshore, in Upper Palaeozoic and Mesozoic rocks of the central part of the Irish Sea, large NE–SW and NW–SE faults are fairly common (Fig. 3). These are thought to represent Caledonian shear zones similar to those described on the Isle of Man which have been reactivated during later basin evolution. In contrast, an ENE–WSW to E–W trend, which represents the general strike of Caledonian thrusts on the Isle of Man, is comparatively rare in the offshore. This may reflect the fact that the thrusts are shorter and less steeply dipping than the Caledonian strike-slip faults (Fig. 6b) and were consequently more difficult to reactivate.

Post-inversion relaxation

Caledonian compression ceased in the early Devonian (Soper *et al.* 1987) at approximately the time when a suite of granitic and dioritic rocks were intruded on the Isle of Man (Lamplugh 1903; Brown *et al.* 1968; Cornwell 1972) mostly to the south of the Central Valley Lineament (Fig. 6c).

These rocks are uncleaved and show evidence of extension during the later stages of intrusion as, for example, in NW–SE orientated pegmatite veins and NE–SW orientated granitic dykes. It appears, therefore, that post-orogenic relaxation was associated with some crustal melting and possible upper mantle fractionation during a period when maximum principal stress was vertical and minimum principal stress was orientated either NE–SW or NW–SE (Figs 4 & 6c).

A package of continental red beds known as the Peel Group crops out on the northern side of Peel (Ford 1993). On the basis of their similarity with Old Red Sandstone facies these rocks have until recently been assigned a Devonian age (Allen & Crowley 1983; Crowley 1985). However, mostly on structural considerations, the present authors believe that there is a possibility that the Peel Group is latest Carboniferous in age and is hence discussed later.

Dinantian–early Namurian rifting

Lower Carboniferous onshore

The earliest firmly dated post-Caledonian sediments on the Isle of Man are a succession of Lower Carboniferous (Dinantian) limestones and associated rocks that crop out in the south of the Island around Castletown (Fig. 6d). These range in age from Arundian to Brigantian and are known as the Castletown Group (Dickson *et al.* 1987). The succession is more than 350 m thick and consists predominantly of interbedded pale-coloured and dark-coloured limestones with minor dolomites and shales. These occur within an outlier approximately 5 km wide that is fault-bounded to the NW and is dip controlled to the SE where the oldest rocks are exposed. Evidence is seen in the strata for syn-depositional tectonism (e.g. Quirk *et al.* 1990) and the entire outlier is interpreted as representing an exhumed half graben that was first active during the early Carboniferous (Figs 6d & 8).

The base of the Castletown Group is marked by a conglomerate of Tournaisian age at Langness (Dickson *et al.* 1987, Fig. 6d). Langness is an elongate, NE–SW trending peninsula which is joined to the mainland by a narrow natural causeway. The SE side of the peninsula consists entirely of rocks belonging to the Manx Group which the conglomerate directly overlies on the NW side (Fig. 8). A major SE-dipping normal fault is inferred to exist to the SE immediately offshore (Fig. 3).

The conglomerate at Langness is sedimentologically immature and contains angular clasts of locally derived Manx Group metasediments. Within the section, which is approximately 25 m

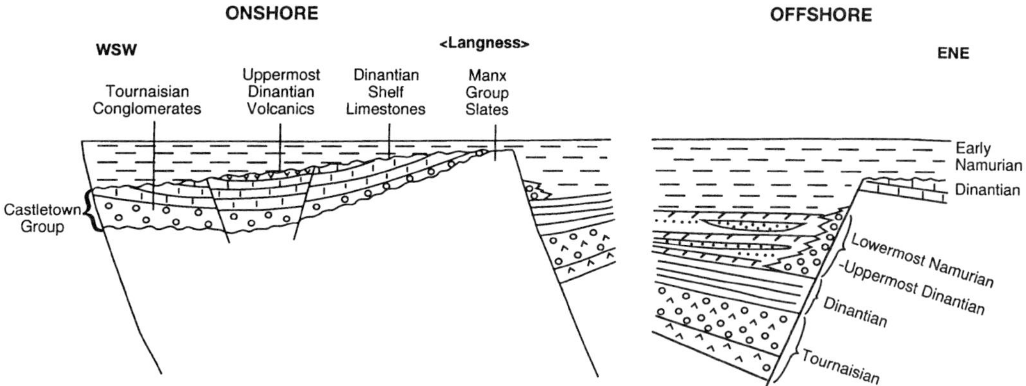

Fig. 8. Schematic geological cross-section through Lower Carboniferous strata onshore and offshore Isle of Man.

thick, syn-depositional thickening can be seen to occur on the down-thrown side of various minor faults. The faults trend N–S, E–W and NE–SW and are thought to have developed due to the effects of gravity in the up-thrown foot-wall of the major fault. On the down-thrown side of this and other major NE–SW faults a thicker, hanging-wall sequence is likely to occur (Fig. 8).

At the very base of the conglomerate, a thin stratiform veinlet consisting of chalcopyrite, quartz, calcite and minor galena is developed in the most easterly part of the outcrop on Langness. Larger, vertical veins with a similar mineralogy occur in the underlying Manx Group and it is possible therefore that they have an early Carboniferous or younger age.

Most of the younger strata within the Castletown Group consist of rather monotonous alternations between pale coloured calcilutites-bioclastic calcarenites and dark coloured calcilutites-mudstones. Small algal bioherms are developed in places. Many of the mudstones are organic rich and terrestrial plant remains, brackish water bivalves and marine goniatites are preserved at certain levels. In some of the dark limestones all porosity is filled with bitumen and in places veins of calcite within the pale limestones contain hard droplets of bitumen.

The youngest part of the Castletown Group consists of a pile of late Brigantian basaltic volcanic rocks, the base of which is erosive and below which the underlying limestone is baked. These rocks are difficult to resolve on aeromagnetic data, suggesting that the pile is not particularly thick. They consist mostly of thick sheets of volcanic breccia with obvious flow textures. A large proportion of the breccia clasts consist of limestone and occasional limestones are interbedded with the breccia. In addition, pillow lavas,

slump structures and volcanic sand are present in the lower part of the volcanic pile indicating that at least some of it was deposited sub-aqueously. At the SE end of the outcrop, the volcanic pile is faulted against Asbian limestones where various volcanic vents and fissures are exposed. Faults, fissures and dykes within the volcanic rocks form a concentric pattern, sub-parallel with the margins of the volcanic pile, suggesting that it represents part of a collapsed caldera.

The Castletown Group is interpreted as having been deposited during a period of rifting associated with NW–SE extension (Fig. 6d). Coarse-grained bioclastic limestones and dolomitized limestones of Dinantian age have been encountered in boreholes below the glacial cover in the north of the Island (Fig. 9) indicating that subsidence also occurred to the north of the Ramsey–Whitehaven Ridge during the early Carboniferous (Fig. 6d).

Lower Carboniferous–basal Namurian offshore

By correlation with Well 112/25-1 (Fig. 11), the top of the Dinantian on seismic data consists of a high amplitude increase in acoustic impedance expressed either as a single reflection or the uppermost of a set of high amplitude reflections (e.g. Fig. 10). A significant part of the seismic data that was interpreted in detail during the study was from the Ogham Inlier and an adjacent offshore area east of Douglas named the 'Onchan Depression' (Fig. 3). Within this region, a package of discontinuous reflections is seen overlying the top of the Dinantian which varies in thickness from below seismic resolution up to 350 m in thickness, using an estimated seismic interval velocity of 4000 m s^{-1}. This package clearly onlaps the top of

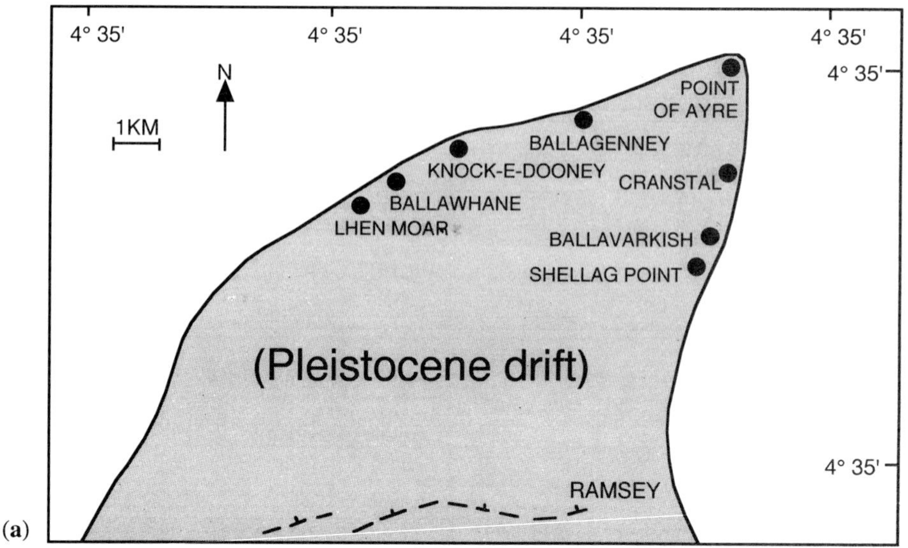

Fig. 9. (a) Location of shallow boreholes in the north of the Isle of Man shown in (b) and Fig. 18. (b) Sedimentary logs (made by N. S. Jones and D. G. Quirk) of cores stored at the Manx Museum from: (i) Cranstal borehole; (ii) Ballavarkish borehole; (iii) Shellag Point borehole.

the Dinantian (Fig. 10) and is thickest in the hanging wall of NE–SW to N–S trending normal faults. It correlates with a thin sequence or an unconformity in the footwall. The package is absent in Well 112/25-1 due to erosion and lies below TD in Well 112/30-1 but these wells help constrain its age as earliest (basal) Namurian. The reflection configurations suggest that it represents a late syn-rift sequence which corresponds approximately with the volcanic rocks exposed onshore in the Isle of Man (Fig. 8).

In some places where the basal Namurian is thick on the hanging wall side of syn-depositional normal faults it is anomalously reflective. By analogy with work by Bott & Johnson (1967), Leeder (1987), Leeder & Strudwick (1987) and Collier (1991), it is suggested that this fault-controlled seismic facies may correspond with 'Yoredale'-type sediments (Fig. 12) which comprise cyclothems of shallow water limestones, pro-delta mudstones, fluvio-deltaic sandstones and delta top coals of Dinantian–earliest Namurian age. A possible source for fluvio-deltaic sand in the late Dinantian has been identified in Anglesey where fluvial channel sandstones cut through Brigantian shelf limestones (see Walkden & Davies 1983). Palaeocurrents are to the north and NE in Anglesey; i.e. towards the central part of the Irish Sea. A second more obvious potential source of sand is the Isle of Man itself. It is difficult to know whether the Manx Group was exposed at this time. However, the general NE–SW trend of the present day coastline, parallel to Lower

Carboniferous faults (Fig. 6d), and the presence of Manx Group clasts within the basal conglomerate at Langness suggests that the Isle of Man may have been up-faulted during the Dinantian. Seismic data indicates that to the NE of the Isle of Man, the Ramsey–Whitehaven Ridge also formed an active fault block during the early Carboniferous bounded by the Lagman Fault to the SE (Fig. 6d). It is overlain by a thin Dinantian succession displaying a divergent stratal geometry similar to that interpreted for the Castletown Group onshore (Fig. 8).

Away from the immediate hanging wall of the Lagman Fault (Fig. 3), Well 112/25-1 encountered Upper Dinantian limestones with occasional thin sandstones and mudstones becoming common towards the top of the interval (Fig. 11). These may represent the distal correlatives of 'Yoredale' cyclothems that consequently should become more sand-rich towards the Lagman Fault.

The possible existence of sand-prone 'Yoredale' cyclothems in the immediate hanging wall of early Carboniferous normal faults represents a potential new hydrocarbon play in the central part of the Irish Sea. Fluvio-deltaic sandstones and fractured limestones may form reservoirs sourced from interbedded mudstones and coals and sealed by the mudstones or overlying Holywell Shale (see below). End Carboniferous inversion structures around the Ogham Inlier form particularly attractive targets (e.g. Fig. 12). Oil shows were encountered in Well 112/30-1 within Upper Carboniferous sandstones (Fig. 13), in an off-structural

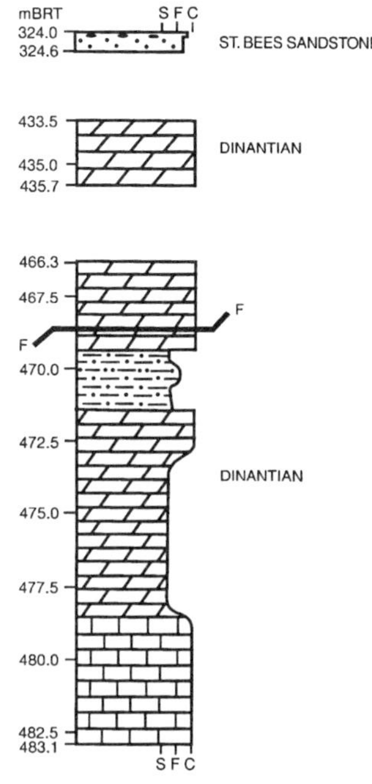

(i)   CRANSTAL (NX 466 027)
mBRT   S F C
324.0
324.6   ST. BEES SANDSTONE
433.5
435.0   DINANTIAN
435.7
466.3
467.5
F   F
470.0
472.5   DINANTIAN
475.0
477.5
480.0
482.5
483.1
S F C

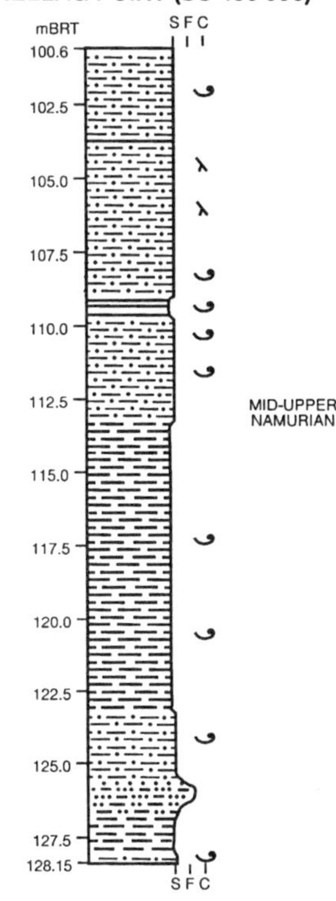

(iii)   SHELLAG POINT (SC 456 996)
mBRT   S F C
100.6
102.5
105.0
107.5
110.0
112.5   MID-UPPER
            NAMURIAN
115.0
117.5
120.0
122.5
125.0
127.5
128.15
S F C
(b)

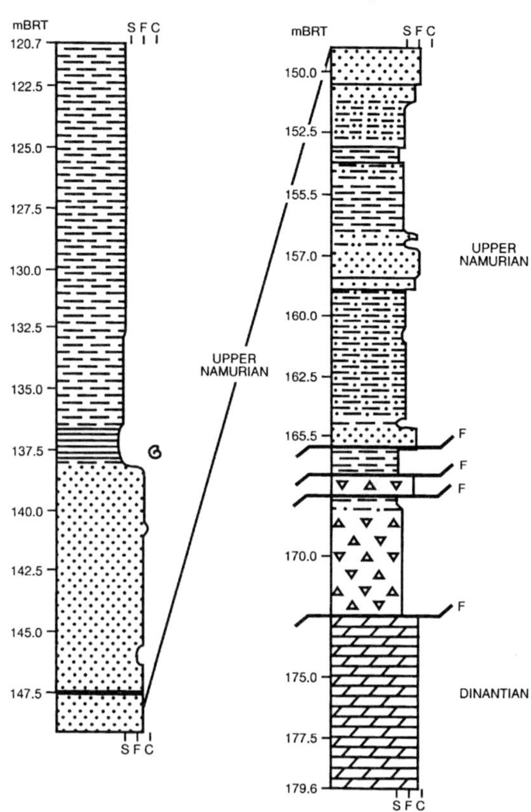

(ii)   BALLAVARKISH (SC 462 007)
mBRT   S F C
120.7
122.5
125.0
127.5
130.0
132.5   UPPER
            NAMURIAN
135.0
137.5   G
140.0
142.5
145.0
147.5
S F C

mBRT   S F C
150.0
152.5
155.5
157.0   UPPER
            NAMURIAN
160.0
162.5
165.5   F
            F
            F
170.0
            F
175.0   DINANTIAN
177.5
179.6
S F C

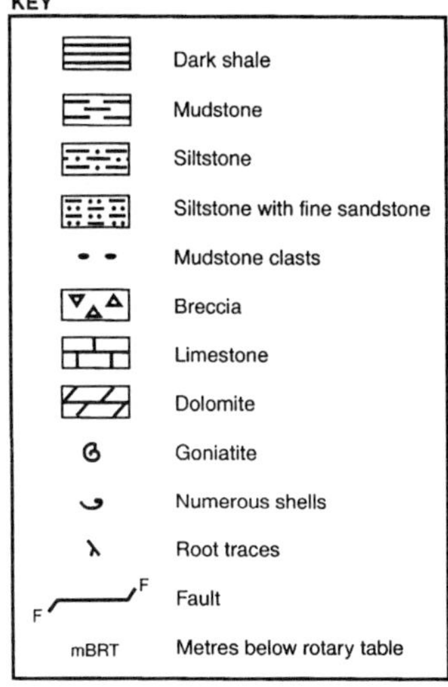

KEY
Dark shale
Mudstone
Siltstone
Siltstone with fine sandstone
Mudstone clasts
Breccia
Limestone
Dolomite
G   Goniatite
        Numerous shells
        Root traces
F   Fault
mBRT   Metres below rotary table

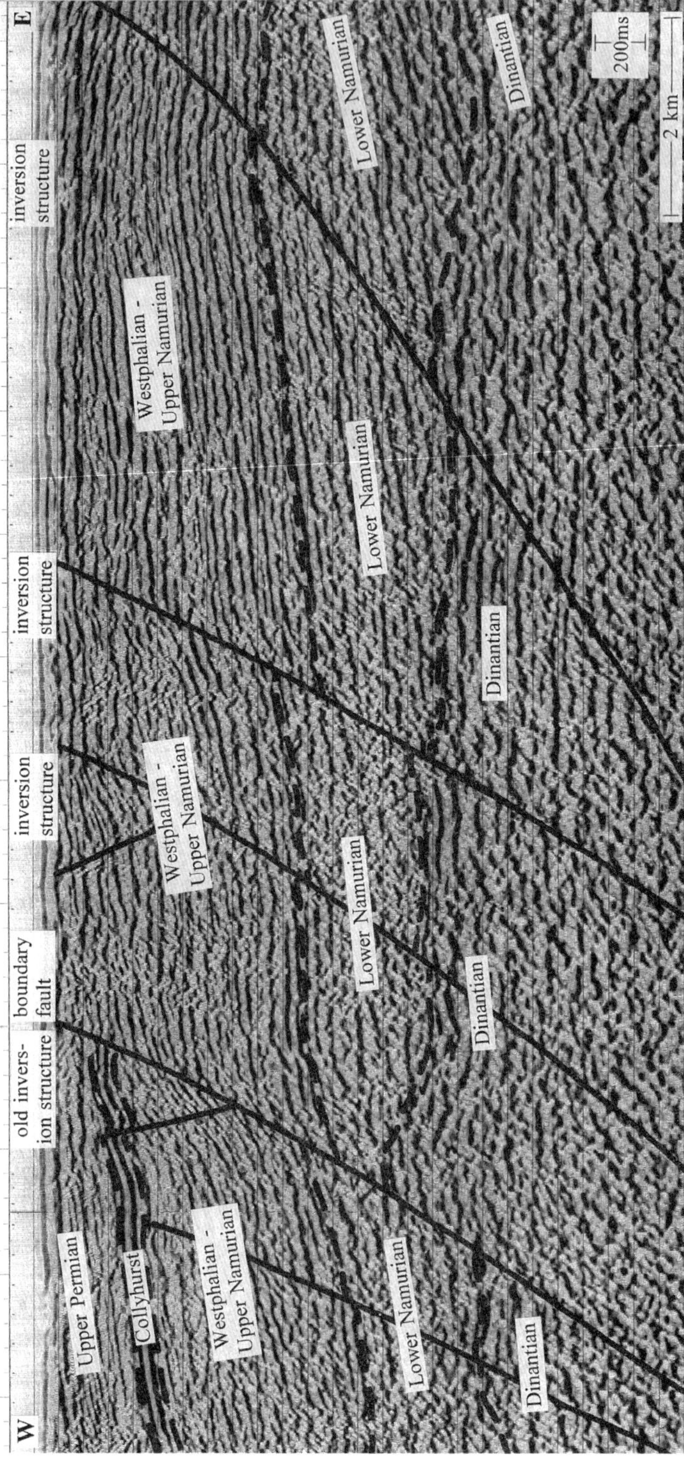

Fig. 10. Migrated 2D seismic line, courtesy of Western Geophysical, across the boundary fault separating the Onchan Depression (W) from the Ogham Inlier (E) in order to show the major faults affecting Carboniferous and Permian strata. Note normal and reverse fault movement at different levels. The Lower Namurian consists mostly of the Holywell Shale; the Collyhurst Sandstone is abbreviated here to Collyhurst.

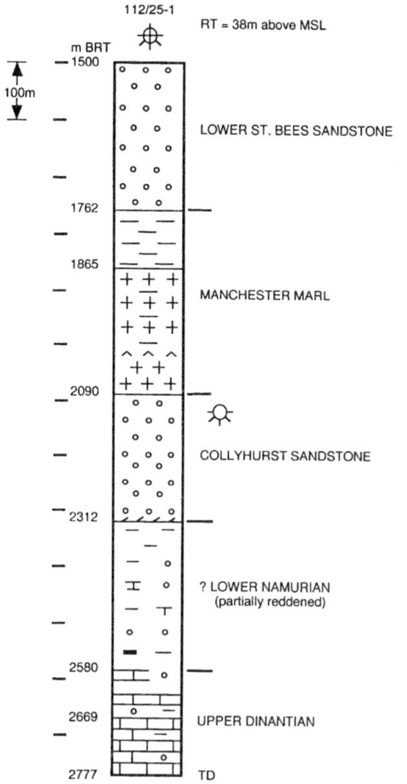

Fig. 11. Simplified stratigraphic log of the lower part of Well 112/25-1.

position, suggesting that hydrocarbon charge has occurred in the area.

Silesian sag

The basal Namurian interval is overlain by the Lower Namurian Holywell Shale, the upper part of which was encountered in Well 112/30-1 (Fig. 13). It has a characteristic appearance on seismic sections comprising discontinuous reflections that become more coherent towards the top where subtle clinoforms are sometimes visible. The Holywell Shale is a thick, organic rich mudstone which is the main hydrocarbon source rock in the SE Irish Sea area (e.g. Armstrong *et al.* this volume). The lower part was deposited during transgression associated with the onset of rapid thermal subsidence after rifting ceased. The end of transgression is marked by a maximum flooding surface interpreted as Chokierian (H1) in age based on data in Ramsbottom *et al.* (1978). This maximum flooding surface is developed as an organic-rich shale which has high natural gamma radiation and is oil-prone. The middle and upper parts of the Holywell Shale contain increasing amounts of land-derived material and are gas-prone (Armstrong *et al.* this volume), thought to reflect the onset of deltaic progradation. Around the Ogham Inlier and Onchan Depression, where it has been mapped using seismic (e.g. Fig. 14), the Holywell Shale varies in total thickness from approximately 200 to 1000 m based on an interval velocity of 4000 m s^{-1}

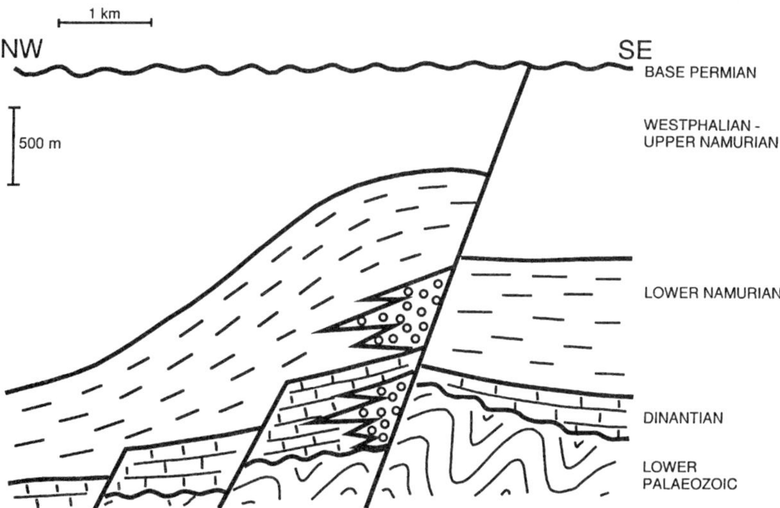

Fig. 12. Idealized geological cross-section through a major Variscan inversion structure present in the Ogham Inlier based on confidential seismic data. Circles indicate possible Yoredale facies, brick ornament indicates carbonates and dashes indicate mudstone.

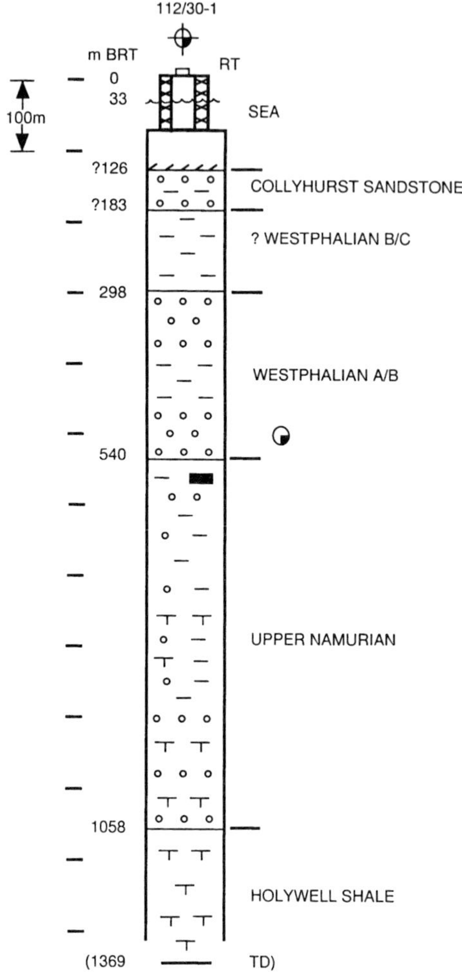

Fig. 13. Simplified stratigraphic log of upper and middle part of Well 112/30-1.

derived from well data. Reflection characteristics indicate that this variation is mostly due to the infilling of earlier rift-related topography (see Fig. 10).

On the Isle of Man, Silesian rocks have only been encountered in boreholes below the glacial cover in the north of the Island (Fig. 9). The oldest of these are Kinderscoutian in age and are mud-dominated (Fig. 9b(iii)). These mudstones contain both plant and shelly material and are thought to represent the uppermost part of the Holywell Shale. TOCs of more than 4% have been measured by one author (Quirk) with vitrinite reflectance values in the order of 0.6%.

Above the Holywell Shale and below the base Permian unconformity, most of the Silesian (Upper Namurian–Westphalian strata) has a tramline-like appearance on seismic sections; i.e. local thickness variations are not observed (e.g. Fig. 10). This indicates that by the late Namurian the area was unaffected by normal faults and was subsiding passively due to the effects of thermal subsidence. Close to the top of the Upper Namurian, apparent truncation of underlying reflections is sometimes observed. A subtle unconformity at approximately the same level has also been observed in the Southern North Sea (Quirk 1993) and onshore UK (J. F. Aitken pers. comm. 1995).

Well 112/30-1 encountered an overall regressive sequence from the Upper Namurian to the Lower Westphalian (Fig. 13). The sediments in the upper part of this interval are fairly typical of the Westphalian in that they consist of lacustrine-deltaic facies and include coarse-grained fluvial sandstones and occasional thin coals. In total, 350 m of Westphalian strata were penetrated in Well 112/30-1 below the base Permian uncon-formity. In the West Cumbrian Coalfield (Fig. 1), a thickness of approximately 500 m of Westphalian A to Westphalian C/D is present. A 30 m thick, coarse-grained fluvial sandstone known as the Whitehaven Sandstone occurs near the top of the preserved section (Jones 1992) and this or equiv-alent sand bodies may have reservoir potential in the Irish Sea as they do in the Southern North Sea (e.g. Quirk 1993) and onshore UK (e.g. Rothwell & Quinn 1987; Fraser & Gawthorpe 1990; Storey & Nash 1993).

Seismic data across the Onchan Depression indicate that, below the base Permian uncon-formity, more than 1000 m of Westphalian section is present above the interval penetrated in Well 112/30-1 (Fig. 15), based on an interval velocity of 4000 m s^{-1} derived from well data. By analogy with the Southern North Sea, this section is thought to be Westphalian C–early Stephanian in age (see Quirk 1993). The area where these young Silesian rocks are present is confined to a pre-Permian syncline, the axis of which trends NE–SW (Fig. 15).

End Carboniferous (Variscan) compression

Where Silesian rocks occur in the offshore, they are gently folded below the base Permian unconformity (e.g. Fig. 15). NE–SW to N–S trending reverse faults are also present which seem to represent reactivated Dinantian normal faults (e.g. Figs 10 & 12). These compressional structures are interpreted as having formed during WNW–ESE-directed Variscan (end Carboniferous) shortening (Fig. 6e). The Ogham Inlier is a Carboniferous inversion

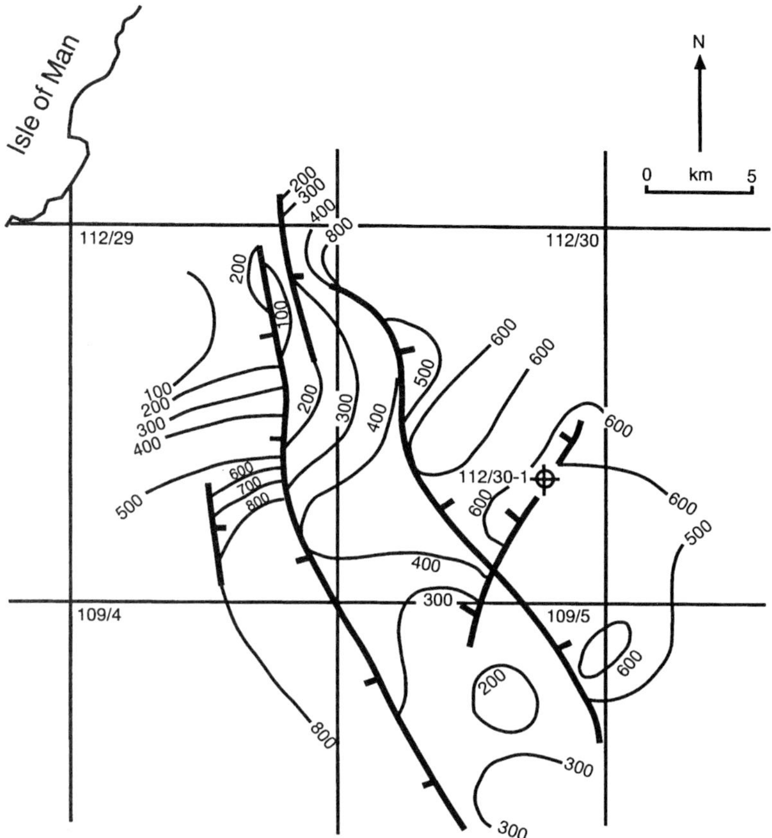

Fig. 14. Isotime thickness map of the Lower Namurian (top Dinantian–top Holywell Shale) around the Onchan Depression and Ogham Inlier. Isochore contour values are in milliseconds two way time.

structure which was initiated at this time (e.g. Fig. 12), as is the pre-Permian syncline mentioned above.

Ogham Inlier

The Ogham Inlier is clearly imaged as a relative Bouguer gravity anomaly high (Fig. 2a). The highest part of the inversion structure occurs to the SW of Well 112/30-1 (Fig. 15) where most Westphalian strata were removed by erosion at the end of the Carboniferous. It is possible that some of the eroded material may have been redeposited as molasse within the adjacent pre-Permian syncline (Fig. 15). To support this idea, a comparison is made here with the Peel Group on the opposite side of the Isle of Man.

Peel Group

The Peel Group on the NW side of the Isle of Man occurs in a similar structural position relative to the Central Valley Lineament as the pre-Permian syncline in the Onchan Depression on the SE side (Figs 3 & 6e). About 1000 m of continental strata ('red beds') are exposed (Ford 1993) along the coast NE of Peel (Fig. 3) including fluvial sandstones, interfluvial mudstones and calcretes. Most of the sandstones contain a large proportion of lithic fragments. However, in the lower part of the section, medium grained sandstones with some reservoir potential are present. These consist of well rounded quartz grains with calcite cement.

The Peel Group is juxtaposed at the northeastern end of the outcrop against Manx Group metasediments by a boundary fault that is oriented N–S and inclined steeply to the west. The overall structure of the Peel Group is that of a northwards plunging, eastwards verging monocline in that close to the boundary fault the beds dip gently to the north and further away from the fault they dip steeply to the NW. The monocline probably developed as an inversion structure as a result of reverse movement along the boundary fault.

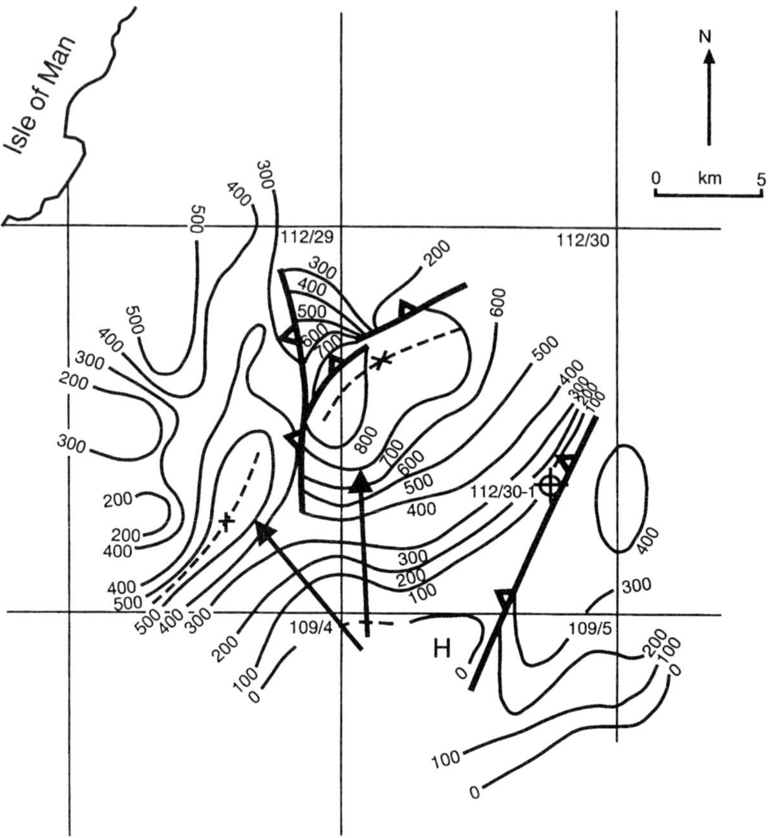

Fig. 15. Isotime thickness map of the Westphalian C/D to early Stephanian (top Westphalian B–base Permian) around the Onchan Depression and Ogham Inlier. Isochore contour values are in milliseconds two way time. Dotted line with cross indicates axis of pre-Permian syncline; large arrows indicate possible transport direction of molasse-type sediment from the highest part of the Variscan inversion structure in the Ogham Inlier (H).

Similar structures are observed on seismic sections within Uppei Carboniferous strata at the margins of the pre-Permian syncline in the Onchan Depression (Figs 10 & 12), where they are assigned a Variscan age (see above).

The Peel Group is completely devoid of palynomorphs although some calcified root structures and one large sandstone burrow have been identified. These sediments have been interpreted by Allen & Crowley (1983) to be early Devonian in age but, unlike the basal conglomerate within the Castletown Group, they contain no clasts of Manx Group material. It is instead tentatively suggested, by analogy with the offshore, that the Peel Group is late Carboniferous in age and therefore equivalent to red beds of Westphalian C/D to early Stephanian age onshore UK (e.g. Besly 1988).

A set of asymmetric folds, extensional faults, glide planes and slump structures occur within the Peel Group at the northern end of the outcrop (e.g. Ford 1972). Most of these indicate movement to the SE and are thought to have developed as a result of gravitational processes. The type of deformation displayed within these sediments (e.g. Fig. 16) suggests that tilting occurred due to tectonism during or soon after sedimentation. One possibility is that the boundary fault was active during deposition. However, adjacent to the boundary fault, the Peel Group is brecciated and, judging from the angular nature of the clasts, fault movement appears, at least in part, to post-date lithification of the sediment. Furthermore, no thickening of the strata is observed towards the fault and the grain size of the sediments seems to be unrelated to the fault. There is therefore no evidence for syn-depositional movement on the boundary fault nor for the existence of a growth fault further to the SE. Instead, a Bouguer gravity

Fig. 16. SE-directed slump-like fold in the Peel Group near Will's Strand. Note hammer for scale.

anomaly high has been identified to the NW of Peel which is similar to the gravity signature of the Variscan inversion structure forming the Ogham Inlier (Figs 2a & 6e). It is suggested, therefore, that the gravity-related structures observed in the Peel Group developed due to syn-depositional over-thrusting in the area to the NW (Fig. 6e) and the sediments themselves represent molasse-type deposits. The boundary fault was probably formed during the later stages of compression, associated with folding and over-steepening of the molasse deposits. This model does not exclude the possibility that the sediments are Devonian in age and were affected by both late Acadian and Variscan compressional events.

Early Permian rifting

In the offshore many normal faults can be seen on seismic reflection data that post-date the Westphalian but pre-date the base Permian unconformity (e.g. Fig. 10). Where these normal faults have been mapped around the Ogham Inlier and Onchan Depression, they fall into two main sets orientated NE–SW and WNW–ESE to NNW–SSE (Fig. 6f). These faults offset earlier Variscan compressional structures and seem to record a phase of rifting and associated strike-slip during the early Permian. The direction of extension during rifting is estimated as E–W (Fig. 6f) which, on a more regional scale, corresponds with the orientation of a chain of N–S trending half graben basins of Permo-Triassic age which extend to the NW and SE of the Isle of Man (see McClean 1978; Jackson & Mulholland 1993). In the vicinity of the Isle of Man, the Central Valley Lineament, the Lagman Fault and the Keys Fault seem to have been active at this time (Fig. 6f). Early Permian rifting is also recorded in the Southern North Sea as the Saalian event where the overall direction of extension was N–S (Quirk 1993; Quirk & Aitken 1997)

Bouguer gravity anomaly lows around the Isle of Man highlight where thick, low density Permo-Triassic strata are present (Fig. 2a). The lower part of the Permo-Triassic section is represented by a red continental sandstone known as the Collyhurst Sandstone (Jackson & Mulholland 1993) which is assigned an early Permian age. Approximately 220 m of Collyhurst Sandstone was encountered in Well 112/25-1 (Fig. 11) on the down-thrown side of the Lagman Fault (Fig. 6f). In contrast, in Well 112/30-1 (Fig. 13) and shallow boreholes onshore Isle of Man (Fig. 18), the Collyhurst Sandstone is

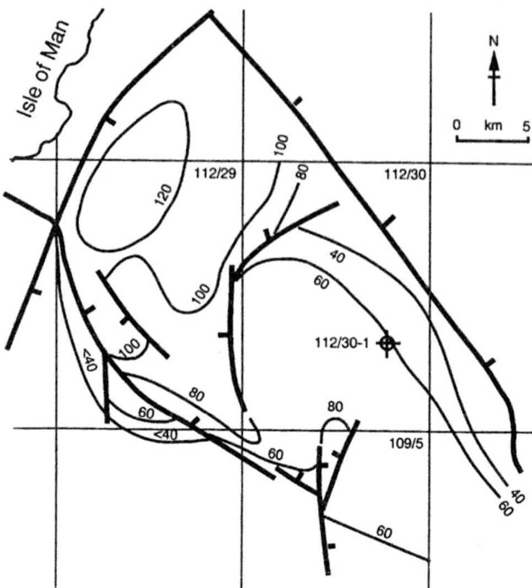

Fig. 17. Map showing normal faults active during the early Permian around the Onchan Depression and Ogham Inlier with isotime thickness of the Collyhurst Sandstone superimposed (isochore contour values in milliseconds two way time).

either thin or absent. On seismic sections the Collyhurst Sandstone can be seen as a strong positive reflection (decrease in acoustic impedance) at the top and as a discordancy between underlying (Carboniferous) reflections and overlying (Permian) reflections at its base (Fig. 10). The Collyhurst Sandstone is thickest on the hanging wall side of early Permian normal faults (Fig. 17) where it can reach thicknesses of up to 300 m based on an interval velocity of 4500 m s^{-1} derived from well data. However, reflections within the Collyhurst Sandstone appear to be fairly parallel and the variations in thickness that are observed are due to onlap rather than divergence of strata; i.e. no evidence of syn-depositional growth faulting is recorded. The Collyhurst Sandstone is therefore interpreted as having infilled remanent graben and half graben after rifting ceased. Similar to the Southern North Sea (see Quirk & Aitken 1997), early Permian syn-rift sediments have not been identified on seismic sections, suggesting that the overall effects of thermal uplift during rifting were greater than those of tectonic subsidence caused by extension. However, early Permian volcanic rocks have been discovered in the Irish Sea in Well 110/2-10 (Hardman 1992).

Well 112/25-1 encountered gas bearing Collyhurst Sandstone which failed to produce on test due to very low permeabilities. However, elsewhere around the Isle of Man, particularly in areas close to normal faults active during the early Permian (Fig. 6f), coarser grained, reservoir-prone fluvial and/or aeolian sandstones may have accumulated. Sand may have been derived from the Isle of Man or further to the NW or SE along the Central Valley Lineament (Fig. 1).

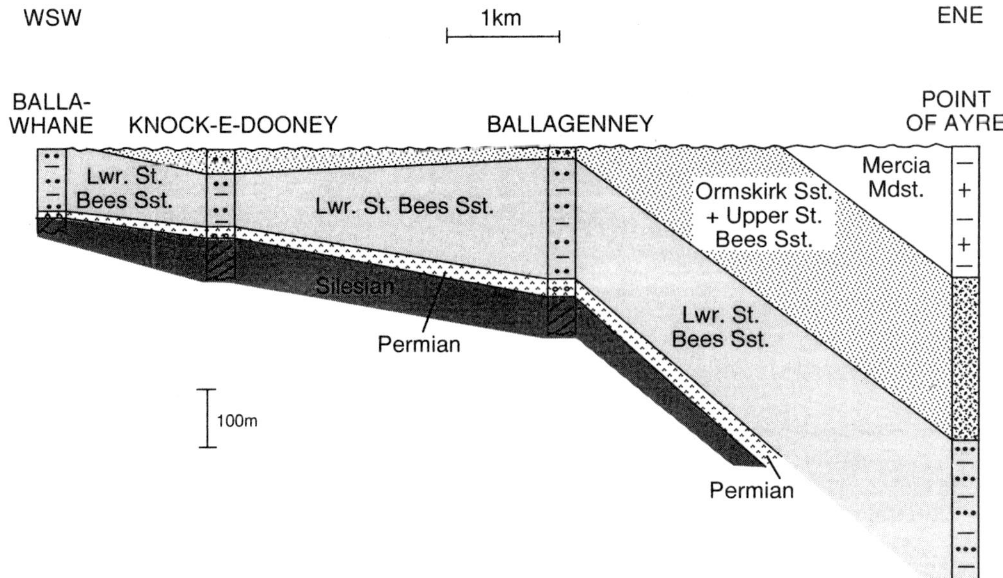

Fig. 18. Well correlation of boreholes drilled at the northernmost edge of the Isle of Man (after Gregory 1920).

Late Permian–early Jurassic sag

After deposition of the Collyhurst Sandstone, thermal sag continued on through the late Permian, the Triassic and early Jurassic when the Manchester Marl, Sherwood Sandstone, Mercia Mudstone and Lias groups were deposited (Fig. 4). Upper Permian–Lower Jurassic strata, which represent the youngest preserved solid succession in the central part of the Irish Sea, have a general layer cake expression on seismic data, apart from the presence of some halokinetic structures (e.g. Fig. 19). Hence, the Bouguer gravity anomaly contours in the central part of the Irish Sea give a general indication of the solid geology that crops out at the sea bed (Fig. 2a). No obvious divergence of reflections nor any fault-related lapping geometries are observed, although Permo-Triassic sequences are often thicker in the hanging wall of some of the major faults than those preserved in the foot wall. These thickness variations are thought to reflect isostatic readjustment on faults formed during the early Permian rather than representing evidence for renewed rifting in the Triassic.

Permo-Triassic sediments are present where the Solway Basin extends onshore beneath glacial sediments at the northern end of the Isle of Man (Fig. 18). Highly porous Ormskirk Sandstone was encountered in the northernmost borehole at the Point of Ayre (Gregory 1920).

?Late Jurassic–Cretaceous rift-sag

More than 1000 m of post-Triassic offset is recorded on some of the large NW–SE, N–S and NE–SW faults in the central part of the Irish Sea such as the Keys Fault (e.g. Fig. 19). It is tentatively suggested that these faults were active during the late Jurassic and represent a phase of Kimmerian rifting similar to that recorded over large parts of NW Europe (e.g. Ziegler 1990). The direction of extension during rifting is estimated as approximately ENE–WSW (Fig. 6g) based on the amount of post-Triassic throw present on the main faults. Approximately the same direction of extension is recorded in the North Sea at this time (Bartholomew *et al.* 1993).

A thickness of at least 500 m of sediment is estimated to have been deposited in the area during the Cretaceous as a result of renewed thermal sag following the end of Kimmerian rifting (cf. Cope, this volume and Warrington, this volume). Subsidence was terminated during the early Tertiary when the basin began to invert (Green *et al.* this volume). The Permo-Triassic rocks now exposed at the sea bed (e.g. Fig. 2a) are thought, therefore, to represent exhumed Kimmerian structures, including a major pop-up horst which underpins the present day Isle of Man (Fig. 6g).

Early Tertiary event

It is now generally accepted that the Irish Sea experienced a large amount of uplift (1.5–2.0 km) during the early Tertiary, beginning about 60 Ma ago (e.g. Holliday 1993; Green *et al.* this volume). The uplift was associated with intrusion of NW–SE orientated dolerite dykes as well as a postulated phase of hydrothermal fluid flow (Green *et al.* this volume). Hydrothermal fluids may have affected reservoir quality and are likely to have influenced source rock maturity as well as caused the removal of salt by dissolution.

Aeromagnetic data (Fig. 5) reveal a large number of linear reversed polarity anomalies crossing the Irish Sea which are due to Tertiary dolerite dykes. Several tens of such dykes have been identified on the Isle of Man. The largest of these, the Fleetwood Dyke, comes onshore on the Isle of Man at Port Mooar (Fig. 6h). The dyke here is approximately 12 m wide and trends 125°/90°. The Manx Group forms the country rock and on the NE side of the dyke this has been metasomatically replaced by haematite and quartz. It is difficult to prove whether this metasomatism occurred before, during or after intrusion but it is possible that it records early Tertiary hydrothermal fluid flow related to the formation of the dyke.

A few small NE–SW trending reverse faults are seen to cross-cut some of the Tertiary dykes exposed on the Isle of Man, for example, in an area north of Port Mooar (Fig. 6h). Evidence of small amounts of compression during the Tertiary is also recorded on some of the major NE–SW to N–S oriented faults on the eastern side of the Irish Sea (e.g. Chadwick *et al.* 1993; Haig *et al.* this volume).

Therefore, during the early Tertiary, the Irish Sea was affected by uplift, igneous and hydrothermal activity, NE–SW extension and a limited amount of NW–SE compression. These are probably related to one major tectonic event resulting from upwelling of the asthenosphere at the end of the Cretaceous which caused crustal doming, high heat flow, tectonic stretching and gravitational collapse (Figs 4 & 6h). Strangely, the area has not been affected by thermal subsidence since doming ceased in the late Tertiary suggesting that the crust was underplated by mantle-derived igneous material.

During the late Cretaceous and Tertiary, many other parts of NW Europe were affected by uplift and igneous activity (see Ziegler 1990). The early Tertiary event recorded in the Irish Sea probably only forms part of a wholesale reorganization of

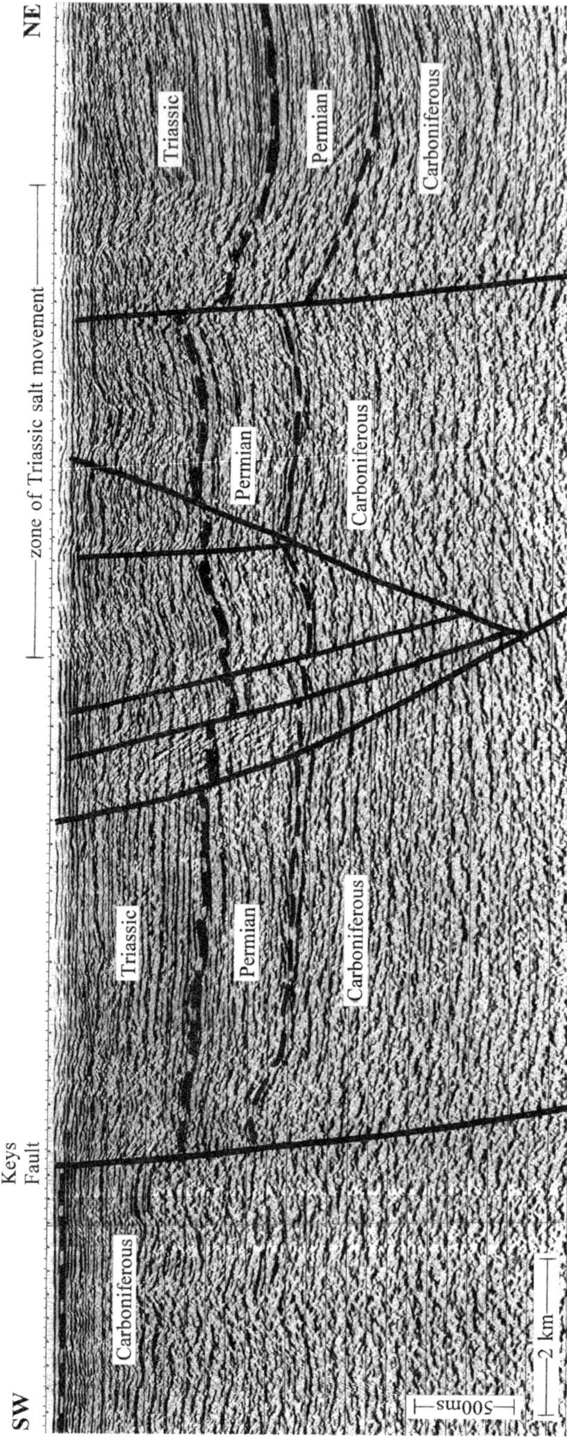

Fig. 19. Migrated 2D seismic line, courtesy of Western Geophysical, across the Keys Fault which separates the Ogham Inlier (SW) from the Keys Basin (NE). Note the layer cake appearance of Triassic strata and evidence of halokinesis.

mantle circulation patterns associated with the formation of the North Atlantic.

Conclusions

The central part of the Irish Sea reveals a structural history similar to other areas in NW Europe, namely (i) Caledonian orogenesis during the late Silurian; (ii) rifting in the early Carboniferous; (iii) thermal sag during the late Carboniferous; (iv) Variscan inversion at the end of the Carboniferous; (v) rifting during the early Permian; (vi) thermal sag from the late Permian to the early Jurassic; (vii) rifting during the late Jurassic; (viii) thermal sag during the Cretaceous; and (ix) doming in the early Tertiary. In an overall sense, each successive tectonic event has seen an apparent anticlockwise rotation in the direction of minimum principal stress such that from the early Carboniferous to the early Tertiary, the direction of extension has changed in intermittent steps from NW–SE to NE–SW. A newly identified basement structure known as the Central Valley Lineament has been active during many of the tectonic episodes and is thought to influence the hydrocarbon prospectivity over a large part of the Irish Sea. Close to the Central Valley Lineament round the Ogham Inlier and Onchan Depression it is proposed that potential hydrocarbon targets may exist at Lower Permian and Carboniferous levels.

The authors wish to thank World Geoscience, Western Geophysical, Manx National Heritage and an anonymous oil company who provided data or allowed access to material for this study. Help and advice was gratefully received from Richard Morgan, Michael Keaveny, Martin Millwood Hargrave, Paul Green, Simon Deadman, Janice Weston, Neil Jones, Ian Longley, Richard Collier, Stewart Molyneux, Trevor Ford, David Kelly, Jeanette Williams, Andrew Johnson, Wendy Horn, Roger Sims, Fred Radcliffe and Greig Cowan. Most of the diagrams were expertly drafted by Lisa Hill. Fieldwork by DGQ was partly funded by Oxford Brookes University. This paper is published with the permission of the Director, British Geological Survey (NERC).

References

ALLEN, J. R. L. & CROWLEY, S. F. 1983. Lower Old Red Sandstone fluvial dispersal systems in the British Isles. *Transactions of the Royal Society of Edinburgh, Earth Science*, **74**, 61–68.

ARMSTRONG, J. P., SMITH, J., D'ELIA, V. A. A. & TRUEBLOOD, S. P. 1997. The occurrence and correlation of oils and Namurian source rocks in the Liverpool Bay–North Wales area. *This volume*.

BARTHOLOMEW, I. D., PETERS, J. M. & POWELL, C. M. 1993. Regional structural evolution of the North Sea: oblique slip and the reactivation of basement lineaments. *In*: PARKER, J. R. (ed.) *Petroleum Geology of Northwest Europe: Proceedings of the 4th Conference*. Geological Society, London, 1109–1122.

BESLY, B. M. 1988. Palaeogeographic implications of late Westphalian to early Permian red beds, central England. *In*: BESLY, B. M. & KELLING, G. (eds) *Sedimentation in a Synorogenic Basin Complex: the Upper Carboniferous of Northwest Europe*. Blackie, Glasgow, 200–221.

BOTT, M. H. P. & JOHNSON, G. A. L. 1967. The controlling mechanism of Carboniferous cyclic sedimentation. *Quarterly Journal of the Geological Society, London*, **122**, 421–442.

BROWN, P. E., MILLER, J. A. & GRASTY, R. L. 1968. Isotopic ages of late Caledonian granitic intrusions in the British Isles. *Proceedings of the Yorkshire Geological Society*, **36**, 251–276.

CHADWICK, R. A., EVANS, D. J. & HOLLIDAY, D. W. 1993. The Maryport fault: the post-Caledonian tectonic history of southern Britain in microcosm. *Journal of the Geological Society, London*, **150**, 247–250.

COLLIER, R. E. Ll. 1991. The Lower Carboniferous Stainmore Basin, N. England: extensional basin tectonics and sedimentation. *Journal of the Geological Society, London*, **148**, 379–390.

COOPER, A. H., RUSHTON, A. W. A., MOLYNEUX, S. G., HUGHES, R. A., MOORE, R. M., & WEBB, B. C. 1995. The stratigraphy, correlation, provenance and palaeogeography of the Skiddaw Group (Ordovician) in the English Lake District. *Geological Magazine*, **132**, 185–211.

COPE, J. C. W. 1997. The Mesozoic and Tertiary history of the Irish Sea. *This volume*.

CORNWELL, J. D. 1972. A gravity survey of the Isle of Man. *Proceedings of the Yorkshire Geological Society*, **39**, 93–106.

CROWLEY, S. F. 1985. Lithostratigraphy of the Peel Sandstones, Isle of Man. *Mercian Geologist*, **10**, 73–76.

DICKSON, J. A. D., FORD, T. D. & SWIFT, A. 1987. The stratigraphy of the Carboniferous rocks around Castletown, Isle of Man. *Proceedings of the Yorkshire Geological Society, London*, **46**, 203–229.

DOWNIE, C. & FORD, T. D. 1966. Microfossils from the Manx Slate Series. *Proceedings of the Yorkshire Geological Society*, **35**, 307–322.

FORD, T. D. 1972. Slump structures in the Peel Sandstones Series, Isle of Man. *Isle of Man Natural History and Antiquarian Journal*, **7**, 440–448.

—— 1993. *The Isle of Man*. Geological Association Guide, **46**.

FRASER, A. J. & GAWTHORPE, R. L. 1990. Tectono-stratigraphic development and hydrocarbon habitat of the Carboniferous in northern England. *In*: HARDMAN, R. F. P. & BROOKS, J. (eds) *Tectonic Events Responsible for Britain's Oil and Gas Reserves*. Geological Society, London, Special Publications, **55**, 49–86.

GILLOTT, J. E. 1956. Structural geology of the Manx slates. *Geological Magazine*, **93**, 301–313.

GREEN, P. F., DUDDY, I. R. & BRAY, R. J. 1997. Variation

in thermal history styles around the Irish Sea and adjacent areas: implications for hydrocarbon occurrence and tectonic evolution. *This volume.*

GREGORY, J. M. 1920. The red rocks of a deep bore at the north end of the Isle of Man. *Transactions of the Institute of Mining Engineers of Scotland,* **59**, 156–168.

HAIG, D. B., PICKERING, S. C. & PROBERT, R. 1997. The Lennox oil and gas field. *This volume.*

HARDMAN, M. 1992. The East Irish sea Basin. Direct evidence for Permo-Carboniferous rifting. *In: Abstracts with Programme Conference on Permo-Triassic rifting.* Geological Society, London, October 1992.

HOLLIDAY, D. W. 1993. Mesozoic cover over northern England: interpretation of apatite fission track data. *Journal of the Geological Society, London,* **150**, 657–660.

JACKSON, D. I. & MULHOLLAND, P. 1993. Tectonic and stratigraphic aspects of the East Irish Sea Basin and adjacent areas: contrasts in their post-Carboniferous structural styles. In: PARKER, J. R. (ed.) *Petroleum Geology of Northwest Europe: Proceedings of the 4th Conference.* Geological Society, London, 791–808.

JONES, N. S. 1992. *Sedimentology of Westphalian Coal-bearing Strata and Applications to Opencast Coal Mining, West Cumbrian Coalfield.* PhD Thesis, Oxford Brookes University.

LAMPLUGH, G. W. 1903. *The Geology of the Isle of Man.* Memoir of the Geological Survey of England & Wales.

LEEDER, M. R. 1987. Tectonic and palaeogeographic models for Lower Carboniferous Europe. *In:* MILLER, J., AADAMS, A. E. & WRIGHT, V. P. (eds) *European Dinantian Environments.* Wiley, Oxford, 1–20.

—— & STRUDWICK, A. E. 1987. Delta–marine interactions: a discussion of sedimentary models for Yoredale-type cyclicity in the Dinantian of northern England. *In:* MILLER, J., ADAMS, A. E. & WRIGHT, V. P. (eds) *European Dinantian Environments.* Wiley, Oxford, 115–130.

McCLEAN, A. C. 1978. Evolution of fault controlled ensialic basins in north western Britain. *In:* BOWES, D. R. & LEAKE, B. E. (eds) *Crustal Evolution in Northwestern Britain and Adjacent Regions.* Geological Journal, Special Issue, **10**, 325–346.

McKERROW, W. S. & SOPER, N. J. 1989. The Iapetus suture in the British Isles. *Geological Magazine,* **126**, 1–8.

MOLYNEUX, S. 1979. New evidence for the age of the Manx Group, Isle of Man. *In:* HARRIS, A. L., HOLLAND, C. H. & LEAKE, B. E. (eds) *The Caledonides of the British Isles – Reviewed.* Geological Society, London, 415–421.

QUIRK, D. G. 1993. Interpreting the Upper Carboniferous of the Dutch Cleaver Bank High. *In:* PARKER, J. R. (ed.) *Petroleum Geology of Northwest Europe: Proceedings of the 4th Conference.* Geological Society, London, 697–706.

—— & AITKEN, J. F. 1997. Structure of the Westphalian in the northern part of the Southern North Sea. *In:* ZIEGLER, K., TURNER, P. & DAINES, S. R. (eds) *Petroleum Geology of the Southern North Sea: Future Potential.* Geological Society, London, Special Publication, **123**, 143–152.

—— & FORD, T. D. 1994. The early Ordovician Manx Group and its relationship to the Iapetus Suture. In: *Caledonian Terrane Relationships in Britain.* Programme with abstracts, 27–28 September 1994, British Geological Survey, Keyworth, 20.

——, ——, KING, J. A., ROBERTS, I. L., POSTANCE, R. B., & ODELL, I. 1990. Enigmatic boulders and syn-sedimentary faulting in the Carboniferous Limestone of the Isle of Man. *Proceedings of the Yorkshire Geological Society,* **48**, 99–113.

RAMSBOTTOM, W. H. C., CALVER, M. A., EAGER, R. M. C., HODSON, F., HOLLIDAY, D. W., STUBBLEFIELD, C. J. & WILSON, R. B. 1978. *A Correlation of Silesian Rocks in the British Isles.* Geological Society of London Special Report, **10**.

ROTHWELL, N. R. & QUINN, P. 1987. The Welton Oilfield. *In:* BROOKS, J. & GLENNIE, K. (eds) *Petroleum Geology of North West Europe.* Graham & Trotman, London, 181–189.

RUSHTON, A. W. A. 1993. Graptolites from the Manx Group. *Proceedings of the Yorkshire Geological Society,* **49**, 259–262.

SIMPSON, A. 1963. The stratigraphy and tectonics of the Manx Slate Series, Isle of Man. *Quarterly Journal of the Geological Society, London,* **119**, 367–400.

—— 1968. The Caledonian history of the north-eastern Irish Sea region and its relation to surrounding areas. *Scottish Journal of Geology,* **4**, 135–163, with discussion on pp. 375–385.

SOPER, N. J. & WOODCOCK, N. H. 1990. Silurian collision and sediment dispersal patterns in southern Britain. *Geological Magazine,* **127**, 527–542.

——, ENGLAND, R. W., SNYDER, D. B. & RYAN, P. D. 1992b. The Iapetus suture zone in England, Scotland and eastern England: a reconciliation of geological and deep seismic data. *Journal of the Geological Society, London,* **149**, 697–700.

——, STRACHAN, R. A., HOLDSWORTH, R. E., GAYER, R. A. & GREILING, R. O. 1992a. Sinistral transpression and the Silurian closure of Iapetus. *Journal of the Geological Society, London,* **149**, 871–880.

——, WEBB, B. C. & WOODCOCK, N. H. 1987. Late Caledonian (Acadian) transpression in north west England: timing, geometry and geotectonic significance. *Proceedings of the Yorkshire Geological Society,* **46**, 175–192.

STOREY, M. W. & NASH, D. F. 1993. The Eakring Dukeswood oil field: an unconventional technique to describe a field's geology. *In:* PARKER, J. R. (ed.) *Petroleum Geology of Northwest Europe: Proceedings of the 4th Conference.* Geological Society, London, 1527–1537.

TODD, S. P., MURPHY, F. C. & KENNAN, P. S. 1991. On the trace of the Iapetus suture in Ireland and Britain. *Journal of the Geological Society, London,* **148**, 869–880.

WALKDEN, G. M. & DAVIES, J. R. 1983. Polyphase erosion of subaerial omission surfaces in the late Dinantian of Anglesey, North Wales. *Sedimentology,* **30**, 861–878.

WARRINGTON, G. 1997. The Penarth Group–Lias Group

succession (Late Triassic–Early Jurassic) in the East Irish Sea Basin and neighbouring areas: a stratigraphical review. *This volume.*

ZIEGLER, P. A. 1990. *Geological Atlas of Western and Central Europe* (2nd edition). Shell Internationale Petroleum Maatschappij, The Hague.

The hydrocarbon potential of the Cheshire Basin

P. W. MIKKELSEN & J. B. FLOODPAGE

Brabant Petroleum Ltd, Suffolk House, 154 High Street, Sevenoaks, Kent TN13 1XE, UK

Abstract: Permo-Triassic basins, both onshore and offshore UK, typically have good reservoir rocks and frequently good seals, but their continental depositional environment requires that source beds of a different age must be invoked if they are to be prospective. In the Dorset–English Channel Basin, the overlying Liassic shales provide an oil prone source, while in the Southern Gas Basin of the North Sea and the East Irish Sea, the underlying Carboniferous has generated large volumes of gas and some oil. Other Permo-Triassic basins, onshore, are generally devoid of hydrocarbons and this can in large part be explained by an absence of an adequate source rock.

The Cheshire Basin lies adjacent to the East Irish Sea Basin and contains a similar, thick Permo-Triassic sequence. The presence of outcropping Coal Measures on three sides of the Basin gives encouragement as to sourcing potential. However, the six exploration wells drilled in the Basin to date (four during the last ten years) have proved dry.

Review of recent seismic data, particularly mid-late 1980s vibroseis surveys, suggests that the Basin developed as a Permo-Triassic half-graben over a Hercynian inversion. Therefore, as one moves from the western edge towards the basin centre progressive erosion of Westphalian strata below the base Permian event may be observed, culminating in a likely Lower Palaeozoic subcrop in the southeastern part of the Basin. A large, faulted high in the centre of the Basin, in many ways similar to the Morecambe Field structure, was proven dry by the Burford-1 well, apparently due to an absence of an underlying source. Nevertheless, seismic data in the northern part of the Basin, supported by the Knutsford-1 penetration, suggest the presence of underlying Coal Measures. However, regional projections suggest that the thick, Dinantian–Namurian bituminous shales which appear to have sourced the East Irish Sea fields, are absent from much of the Basin.

A further difference between the Cheshire and East Irish Sea Basins is the presence, in the former, of the Tarporley 'Siltstone' Formation above the main reservoir objective of the Helsby Sandstone Formation. The former may be regarded as a likely non-sealing waste zone over much of the central and eastern parts of the Basin, but in the SW where the coarser Malpas Sandstone facies is developed, and in the north, it becomes a viable reservoir target in its own right, sealed by the Bollin Mudstone Formation and, ultimately, by the Northwich Halite.

It is therefore proposed that the successful exploration for hydrocarbons in the Basin is likely to depend on finding a combination of sub-cropping Coal Measures and reservoir-grade Tarporley Siltstone, sufficiently deeply buried to retain an effective Mercia Mudstone and preferably halite seal. These conditions appear to prevail in the NW part of the Basin, where the search for suitably large, closed structures is currently in progress.

One of the main limiting factors in the exploration for hydrocarbons onshore UK has been the frequent poor quality of the prospective reservoir horizons. However, the Triassic sandstones, generally of Scythian–Anisian age (Warrington *et al.* 1980; Benton *et al.* 1994), are an exception to this rule, having a widespread distribution with generally good reservoir characteristics. Nevertheless, onshore success for Triassic targets has been limited to the Wytch Farm oilfield, although offshore more numerous discoveries have been recorded. In each case, the key to success has been the presence of mature source rock, in either the Lower Jurassic section, or the deeper, pre-Hercynian Palaeozoic rocks. No source potential has yet been identified in the Permo-Triassic continental red bed sequences.

Figure 1 shows the main basinal areas which contain thick Triassic sandstones at depth, in and adjacent to onshore England. In Southern England (to the south of the post-Hercynian platforms of St George's Land and the London–Brabant Massif) thick, organic-rich mudstones are present in the Lower Lias, but only in the Dorset–English Channel Basin do they appear to have achieved maturity. The underlying pre-Hercynian strata there are either low-grade Variscan metamorphic rocks or generally organically lean rocks from the Devonian or Lower Carboniferous.

Further north, the Worcester Graben forms a N–S rift in the post-Hercynian platform and preserves no Jurassic strata, but has an extremely thick (*c.* 3300 m) sequence of Permo-Trias. However, despite the presence of outcropping Coal Measures to the northeast and south of the basin, all the

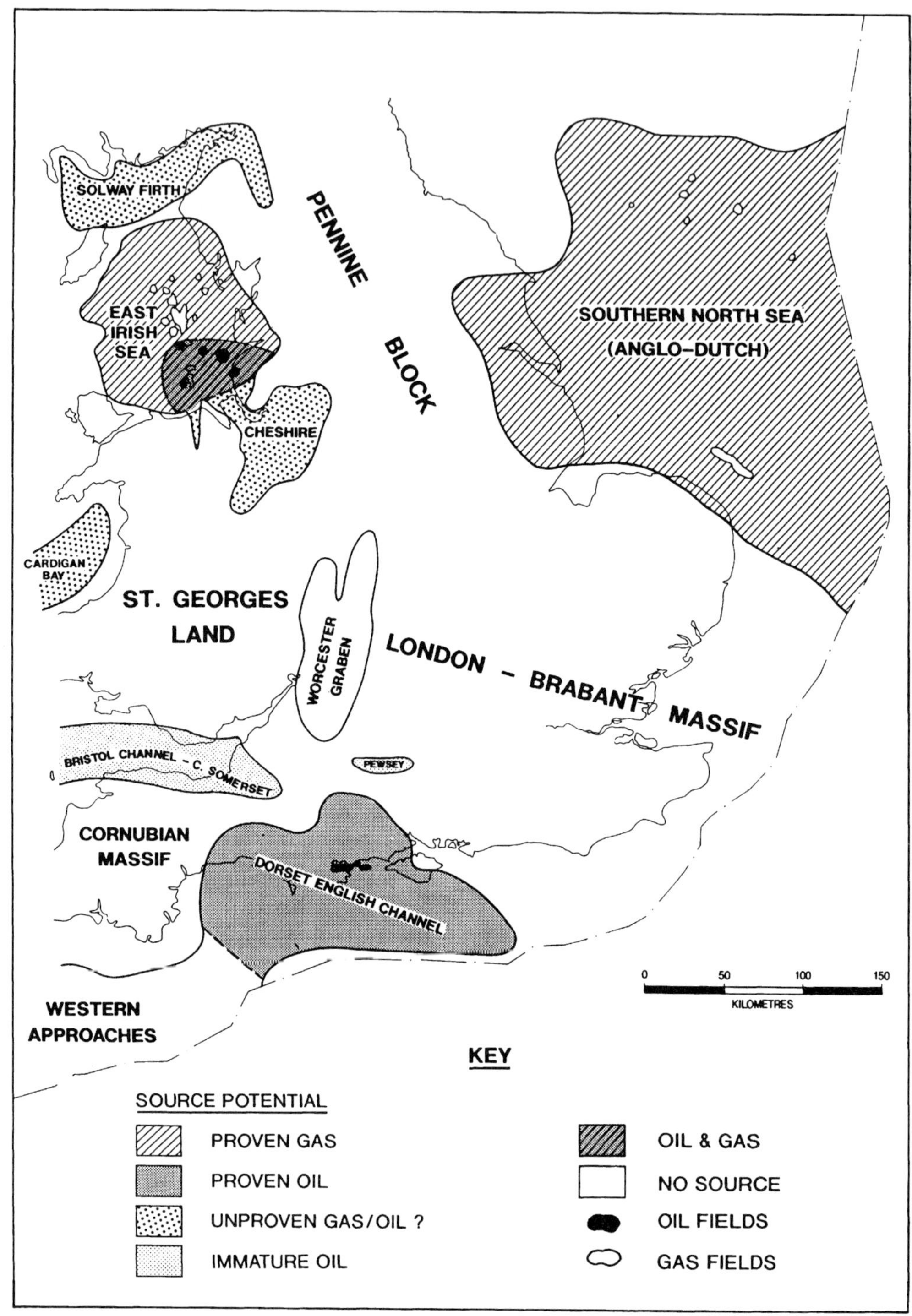

Fig. 1. Major Permo-Triassic basins of England and Wales.

current evidence points to a barren Lower Palaeozoic subcrop more akin to that exposed in the Malvern Hills on the western margins of the graben.

By contrast the large, mostly offshore, basins north of the platforms are underlain by thick Carboniferous strata, which have demonstrably generated major volumes of hydrocarbons. In the Southern North Sea a wide subcrop of Coal Measures has charged numerous basal Permian (Leman Sandstone) traps with gas. The efficacy of the overlying Zechstein Salt as a seal prevents further vertical migration unless the latter is breached, mainly by withdrawal associated with major fault zones. Where this has occurred, towards the northern and southern margins of the basin, gas accumulations have been discovered in the 'Bunter' sandstones.

In the East Irish Sea Basin (hereafter EISB), erosion following the Hercynian orogeny appears to have largely truncated the Westphalian section, but very thick, organic-rich shales have been preserved in the Namurian, and are thought to be the main hydrocarbon source. Potentially, such shales may also exist in the Dinantian; indeed the latter would seem the most prospective source rock in the largely undrilled Solway Basin. In the absence of a substantial thickness of Upper Permian evaporites, hydrocarbons migrated vertically through a thick coarse clastics section of the Permo-Trias to be trapped in the Anisian Helsby Sandstone Formation (Benton *et al.* 1994). A regional topseal is then provided by shales and evaporites of the Mercia Mudstone Group (dating from the Anisian onwards; Jackson *et al.* 1987; Jackson and Mulholland 1993).

The Cheshire Basin appears to share some of the characteristics of the offshore basins north of the St George's Land / London–Brabant Massif platform and also of the intervening Worcester Graben. The stratigraphy of the Permo-Triassic is similar to that of the EISB, to which it is connected by a narrow neck, straddling the Wirral. No equivalent of the thick, lower- to mid-Carboniferous shale sequences found to the north and east seems to exist, although a minor sub-basin may exist north of the Bala–Llanelidan fault zone (Fig. 14), and the main potential sources lie in the Westphalian. Seismic data demonstrate that the Coal Measures, which outcrop north of the Basin, are progressively truncated southwards (Fig. 16) and, in the southern part of the Basin the pre-Hercynian subcrop seems more comparable to that beneath the Worcester Graben.

History of exploration

Following the successful period of major discoveries in the Southern Gas Basin of the North Sea in the mid–late 1960s, exploration in the Cheshire Basin commenced in the early 1970s. Deep wells, located by a loose, low-fold dynamite seismic grid were drilled at Prees (Trend 1973) and Knutsford (British Gas 1974) and these remain the sole penetrations of the pre-Hercynian section (see Figs 2 & 19 for well locations). Much of the Basin is a difficult seismic area and the early data were generally of very poor quality; neither well is now thought to have been located on a closed structure.

The late 1970s and early 1980s saw a concerted attempt, principally by Shell, to establish a seismic grid of reasonable quality in the central and northern parts of the Basin. Some of the first onshore UK vibroseis data was acquired here and the region was a test-bed for establishing nationally acceptable environmental parameters for vibroseis. Data quality continued to be variable, however, although increasing fold of stack in later surveys provided clear benefit.

The final spell of drilling lasted from 1987 to 1992 with three shallow wells testing promising Triassic sandstone reservoired traps, at Burford (Shell 1987), Elworth (Mobil 1988) and Blakenhall (Hamilton 1992), in the centre of the Basin. The failure of these saw the withdrawal of major companies, so that fewer areas within the basin are licensed today and coal-bed methane specialists predominate around the fringes. Only 40 km of seismic data have been acquired since 1990 (by Brabant), but reprocessing of selected late 1970s data has indicated that substantial improvements can be made, as the comparison between the 1979 and 1994 versions of seismic line 2 demonstrates (Fig. 3). Following the recent discoveries of oil in the southern area of the EISB, the exploration of the Cheshire Basin looks due for a modest upturn, with both seismic and drilling activity forecast over the next year or two. The first result of which, Boots Green-1 (Brabant), was unfortunately P&A as a dry hole in Spring 1996.

Structural evolution

With only two penetrations of pre-Hercynian strata beneath the Basin, hard evidence of its early history is very limited and must largely be gleaned from surrounding outcrops and various indirect sources including magnetic and gravity data. The presence of cleaved Ordovician–Silurian mudstone in Prees-1 suggests similarity to the North Wales succession, with Devonian strata probably absent due to the presence of a post-orogenic Caledonoid high, which did not begin to subside until the Lower Carboniferous, resulting in limestone directly onlapping onto Lower Palaeozoic rocks as in North Wales (the stratigraphy of the Basin is summarized in Figs 2 & 4).

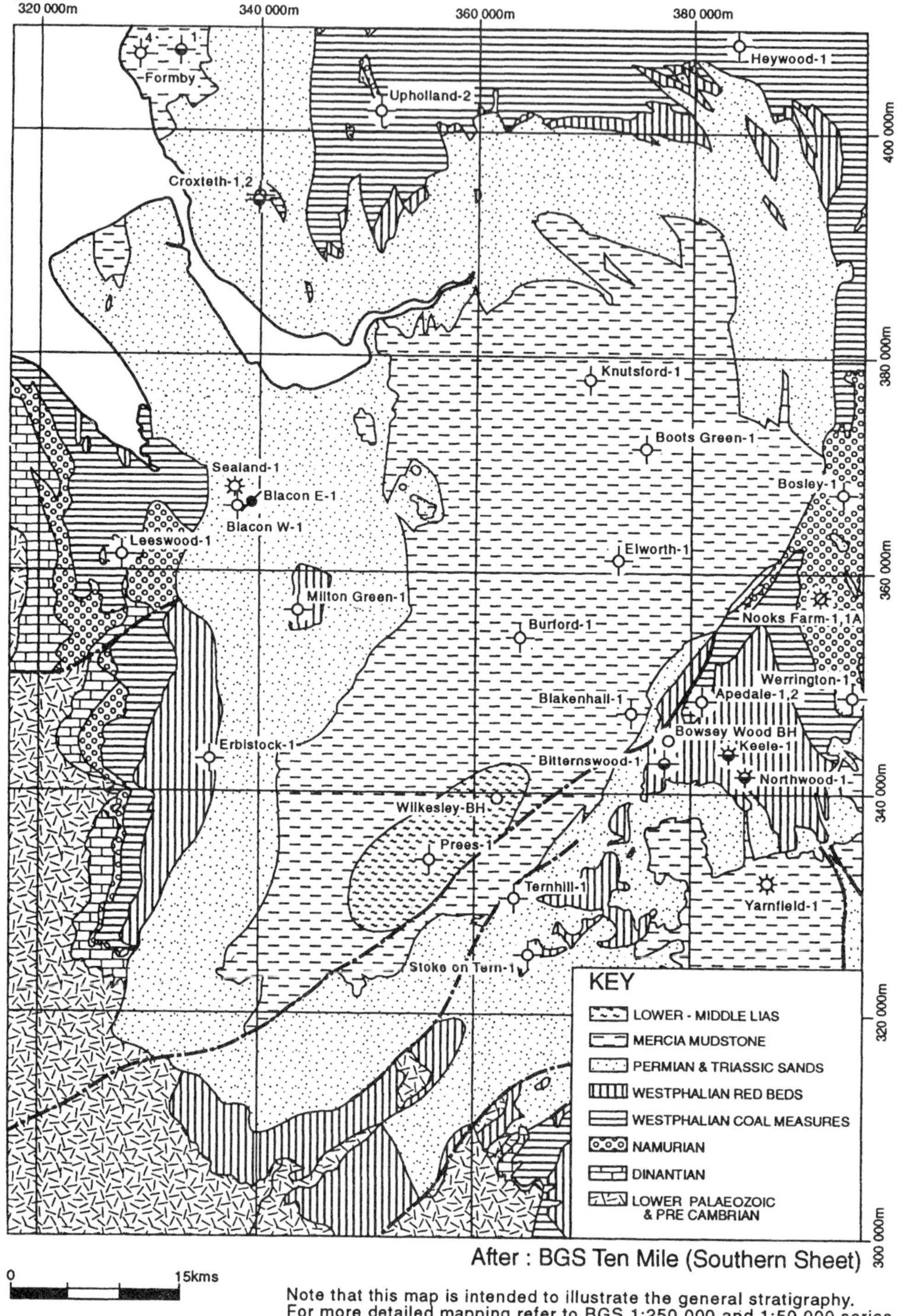

Fig. 2. Generalized surface geology map of the Cheshire Basin area, annotated with exploration wells and selected boreholes (after BGS Ten Mile Map South Sheet 1:625 000 scale 1979).

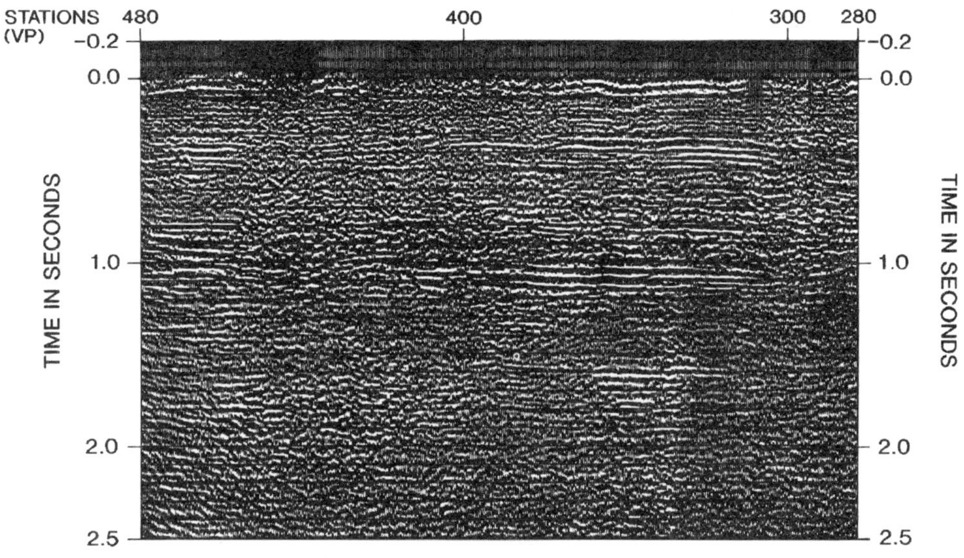

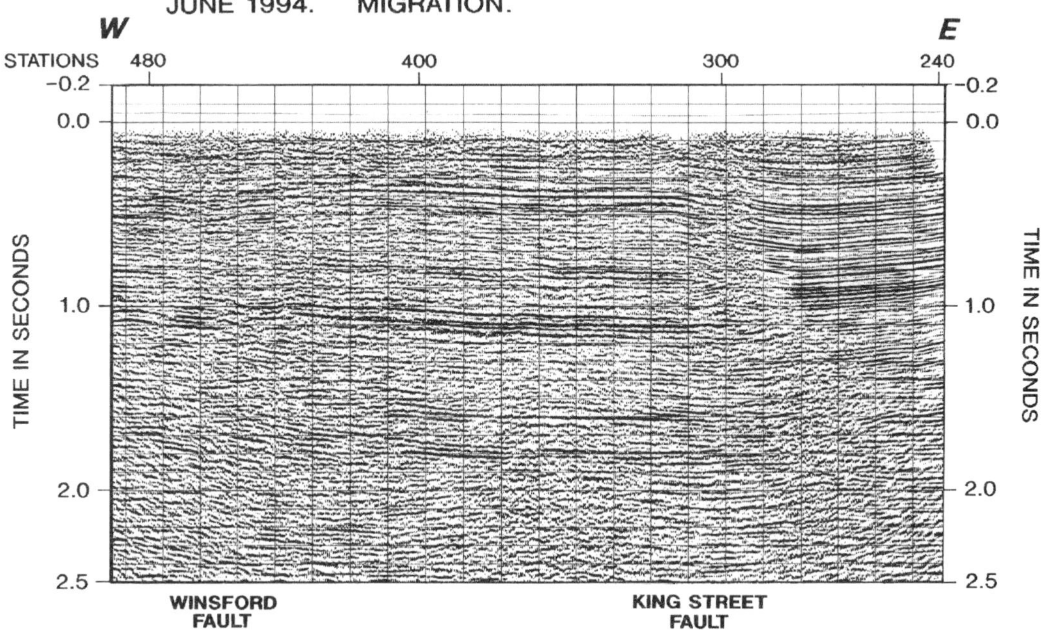

Fig. 3. Comparison of original seismic processing from the 1970s with modern reprocessed data. Considerable improvement in reflector resolution and continuity is observed, particularly at shallow levels (<0.5 s). (See Fig. 19 for location.)

P. W. MIKKELSEN & J. B. FLOODPAGE

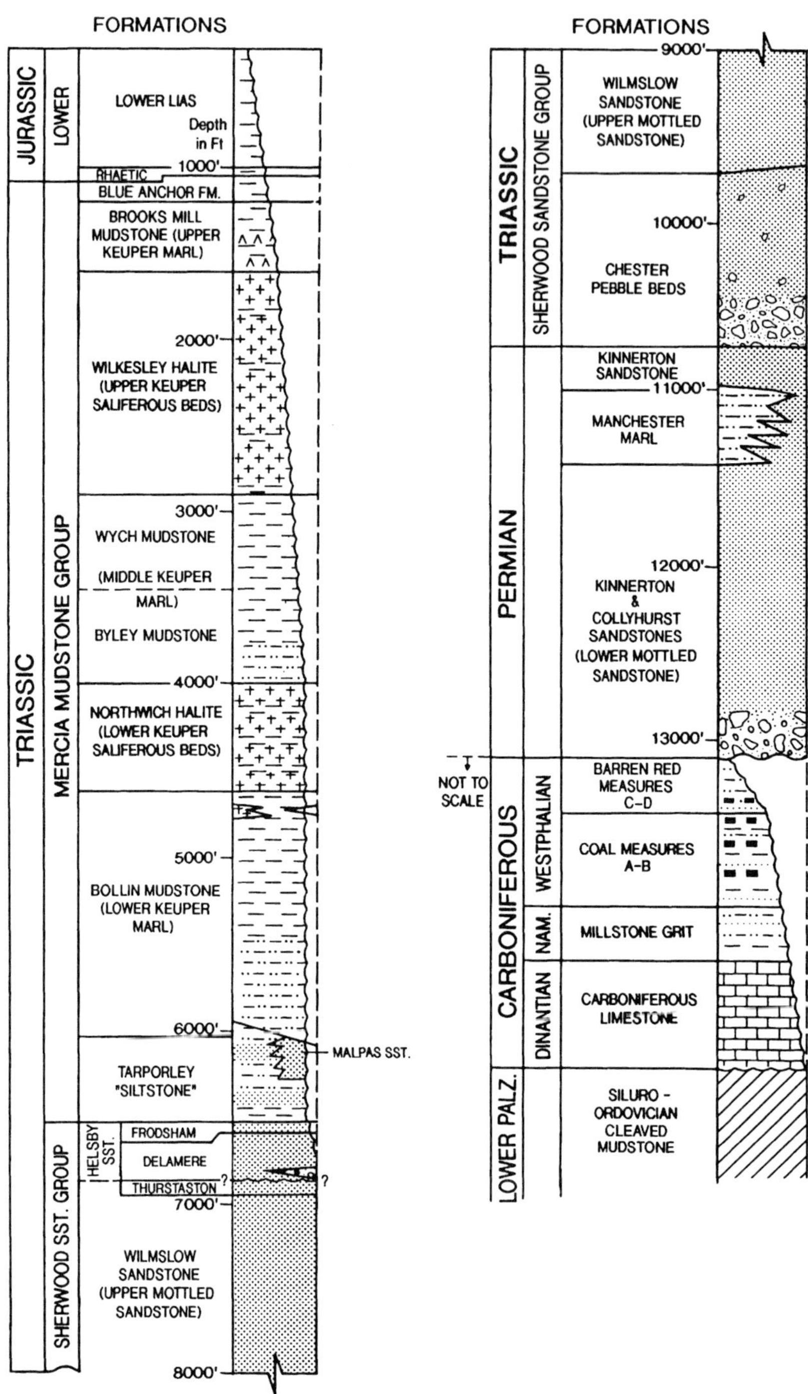

Fig. 4. Summary of Cheshire Basin stratigraphy, compiled from borehole data.

Platform-type conditions appear to have continued into the Namurian over much of the CB, in contrast to the deeper water environments characteristic of the rapidly subsiding Pennine sub-basins seen to the north and east. Outcropping Coal Measures from Westphalian A to C are found immediately west, east and north of the basin and have been proved at Knutsford-1. Localized Upper Carboniferous red measures are also probably preserved beneath the Basin, for example the, possibly faulted, section encountered in Prees-1. Seismic evidence suggests that the Hercynian orogeny uplifted the Palaeozoic rocks beneath the present basin, such that erosion increased towards the southeast (Fig. 5). The foregoing is at variance with much of the published literature (Gale *et al.*. 1984; Fraser & Gawthorpe 1991; Chadwick this volume) which has tended to invoke minimal Hercynian uplift of the Basin (<500 m; Fraser & Gawthorpe 1990, fig. 15).

Collapse of the Variscan orogen and related initiation of extension (regionally NW–SE, but locally NE–SW in the EISB–CB, Coward 1995; Chadwick & Evans 1995) led to the development of the Basin, in early Permian times, part of a suite of *en echelon* rifts roughly parallel to the putative North Atlantic. The Basin formed as an asymmetric half-graben, oriented primarily north–south, with the deepest axis running near the eastern side,

downthrown from the basin-bounding Red Rock–Wem Fault system. Syndepositional faulting along other major north-south trends (e.g. King Street, Mobberley, Winsford and Burford) can be identified (see Figs 7, 10 & 11) and isochron mapping, performed by Brabant but not presented here, demonstrates movement during Permian and early Triassic times. Evans *et al.* (1993) have presented evidence of an intra-Triassic uplift, inducing up to 900 m of erosion, possibly equivalent to the Hardegsen Unconformity in Germany. Significant footwall uplift, of this age, appears to have occurred along the basin flanks, but seismic evidence for an angular unconformity within the basin is weak.

An apparently continuous stratigraphic sequence is observed in the SE part of the Basin culminating in Middle Liassic strata, preserved as a small outlier around Prees, near Wem. Comparison with other Triassic basins, particularly in the East Midlands, suggests that modest thicknesses of a generally conformable Jurassic rocks were deposited until late-Kimmeridgian times. This was followed by uplift and development of a Cimmerian unconformity, caused an unknown, but probably moderate, amount of erosion.

From regional considerations it is likely that Upper Cretaceous transgressions covered the Cheshire Basin, with several hundred feet of Chalk probably deposited (Hardman *et al.* 1993). Tertiary

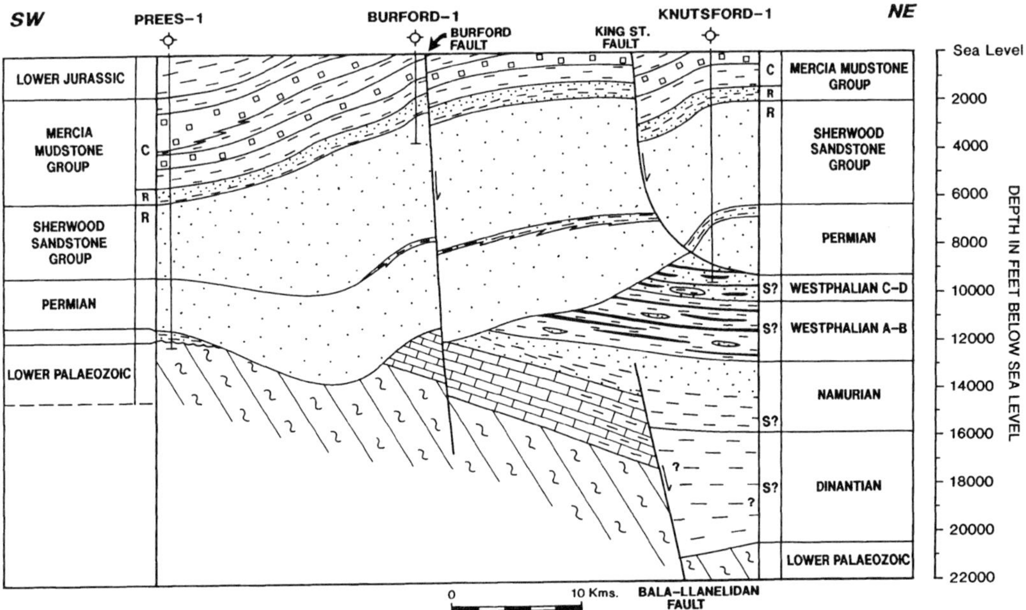

Fig. 5. Diagramatic cross-section (SW–NE) through the Cheshire Basin, based on well and regional seismic data, illustrating the relationship between source (S) reservoir (R), cap-rock seal (C) and the dry holes at Prees-1 and Burford-1 (Knutsford-1 is off-structure).

dyke activity of Palaeocene and Eocene age, documented to the north (Fleetwood; Arter and Fagin 1993) and west (Anglesey/Snowdonia) suggests that regional uplift, inversion and fault reactivation had commenced by then. Continuing uplift (sea regression off Lancashire) and current earthquake activity implies that these movements are still active. Shale velocity trends from the Bollin Mudstone and Lias (confidential), indicate that, locally, the Cheshire Basin may have undergone a maximum uplift of over 2000 m since the late Cretaceous. Considerable debate is currently taking place about the magnitude and timing of Tertiary inversion in NW England (Green *et al.* 1993; Hillis 1993).

Trap characterization

Four main fault trends can be observed in the Basin (see Figs 6 & 19):

- In the south a Caledonoid SW–NE trend is prominent, with the Wem–Red Rock fault system being fundamental to the half-graben geometry and forming the SE boundary.
- In the central part a N–S fault orientation predominates and many of these lineaments demonstrate active growth during the Permo-Trias (see Figs 7–11). This would appear to be the main extensional axis of the Basin.
- In the northern part a NW–SE trend is also evident and some of these faults appear to have been reactivated in the Tertiary, and involved in strike-slip faulting.
- Finally a more subtle series of E–W faults, of generally restricted extent, can be observed in the central part of the Basin. These have relatively small throws, no evidence of reactivation and it is speculated that movement was of late Triassic to Jurassic age.

The Burford-1 well (1987) lies near the intersection of the second and fourth trends, located on the footwall of the N–S oriented Winsford Fault, with northward closure formed by a subsidiary E–W fault. Seismic line 5 (Fig. 7; for line locations see Fig. 19) shows that the Winsford Fault breaks into several anastamosing faults at this point, where the total throw at base Helsby level exceeds 1200′ (200 ms). No obvious Lower Triassic syndepositional growth can be seen across the fault zone, with the top Manchester Marl to top Helsby isochron stable at around 700 ms (>4500′). However, the shape of the hanging-wall syncline of Mercia Mudstone sediments suggests subsequent growth, although seismic resolution on the footwall is inadequate to confirm this. Seismic line 4 (Fig. 8) similarly implies that the E–W fault is of relatively late age, with only regional northward thinning (top

Helsby Sandstone to top Manchester Marl) being evident across the line.

The location and large size of the Burford structure near the basin centre, south of the major horst trend formed by the King Street and Winsford Faults, is rather analogous to the position of the Morecambe Field within the EISB. The absence of any hydrocarbon shows in Burford-1, despite good Malpas and Helsby sandstone reservoir and thick topseal of Northwich Halite, strongly suggests an absence of an effective source rock and/or migration paths.

The third category of faulting is demonstrated by seismic line 1 (Fig. 9) towards the NE part of the basin. The N–S Mobberley Fault (Figs 6 & 19) is displaced 2 km westwards, at Top Manchester Marl level (mapping not presented here), at its junction with the NW–SE Boots Green Fault. At Top Helsby level this displacement is made more complex by the associated subsidiary faulting, which has all the appearance of a strike-slip-induced flower structure. The cause of the strike-slip movement is considered to be (roughly) N–S Tertiary compression. The offset of the Mobberley Fault would tend to indicate dextral slip, contrary to previously reported sinistral slip on the Keys Fault in the EISB during the Early Tertiary (Knipe *et al.* 1993).

Seismic line 2 (Fig. 10) shows the King Street Fault at the eastern end of the section, with line 3 (Fig. 11) providing an approximate easterly continuation, commencing 4 km to the south. This N–S trending fault is by far the largest feature within the Basin, with a maximum displacement down to the east of 300 ms (about 1800′) at Helsby level, increasing to 550 ms (over 3600′) on the Manchester Marl. Syndepositional movement is obvious, with considerable thickening evident on the eastern side for every interval between base Northwich Halite and top Manchester Marl (except the Tarporley Siltstone Formation) totalling over 1250′ overall between these horizons. The large displacements of the youngest Mercia Mudstone Group reflectors suggest that the fault may have also been active during the Jurassic, but its Tertiary history is unclear, with no signs of inversion or wrenching.

The second type of faulting is typically associated with the major lineaments in the basin, such as the King Street Fault and one such was tested by the Elworth-1 borehole in 1988. Seismic line 3 (Fig. 11) shows an antithetic fault to the east of the King Street Fault, with maximum displacement at Helsby/Wilmslow levels quickly diminishing in the higher section and probably being absorbed within the Wilkesley Halite, without reaching the surface. About 150′ (25 ms) of thickening can be observed between the top Northwich Halite and the Helsby, from the structural crest to

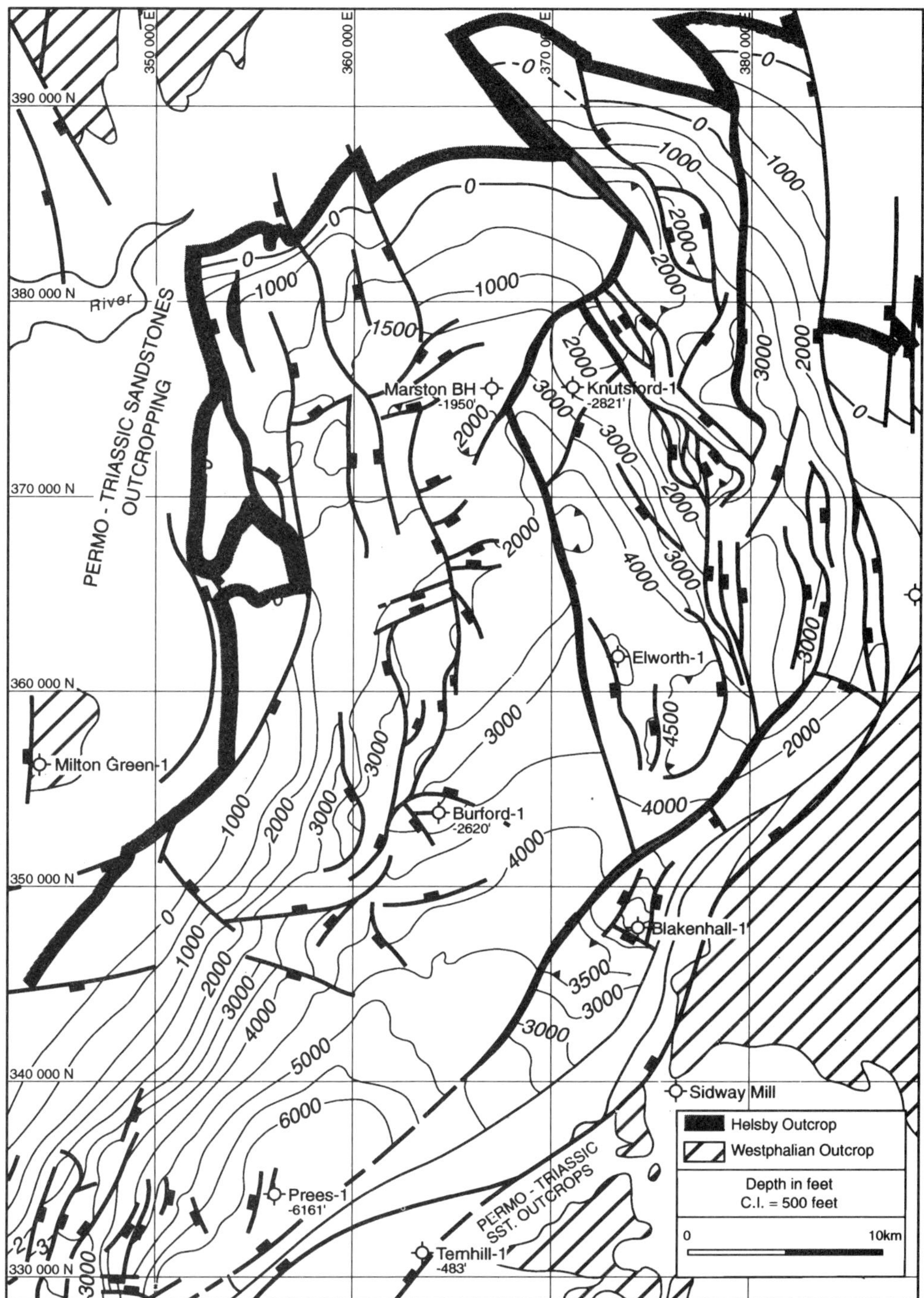

Fig. 6. Regional Top Helsby Sandstone depth structure map compiled from seismic and surface geological mapping.

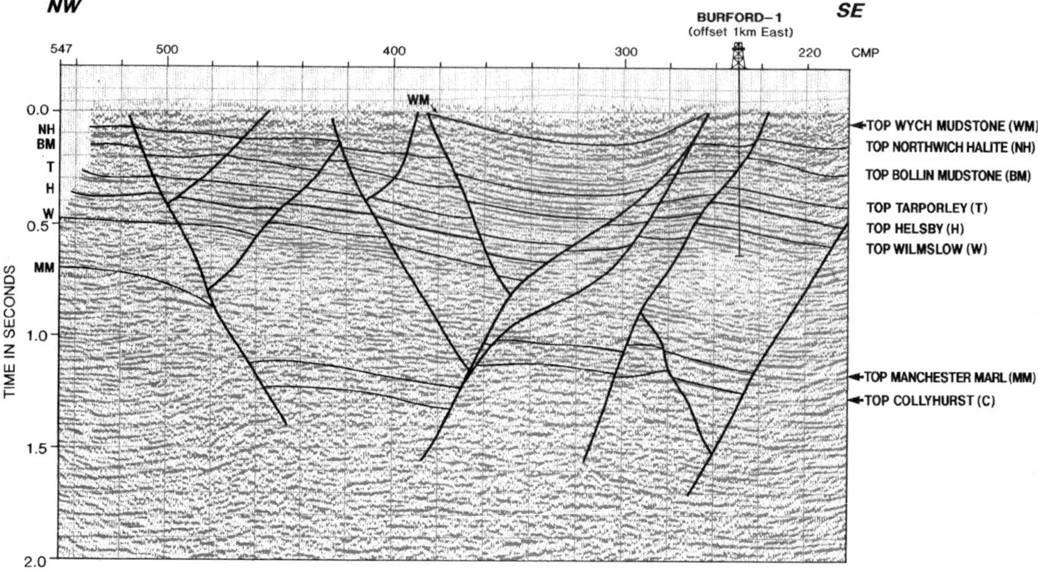

Fig. 7. Seismic line 5 runs NW–SE through the Burford structure (see Fig. 19), which is bounded to the east by the Winsford Fault.

the synclinal axis some 2 km to the east. That Elworth-1 was drilled on an old and well sealed structure is not in doubt and yet the well had no gas shows in either the (admittedly poor) Tarporley, or (better) Helsby reservoirs. As with Burford-1, the most convincing explanation is an absence of underlying Westphalian source rocks.

The central regional horst, bounded by the Winsford and King Street Faults, runs for about 15 km in a N–S direction, with a width of around

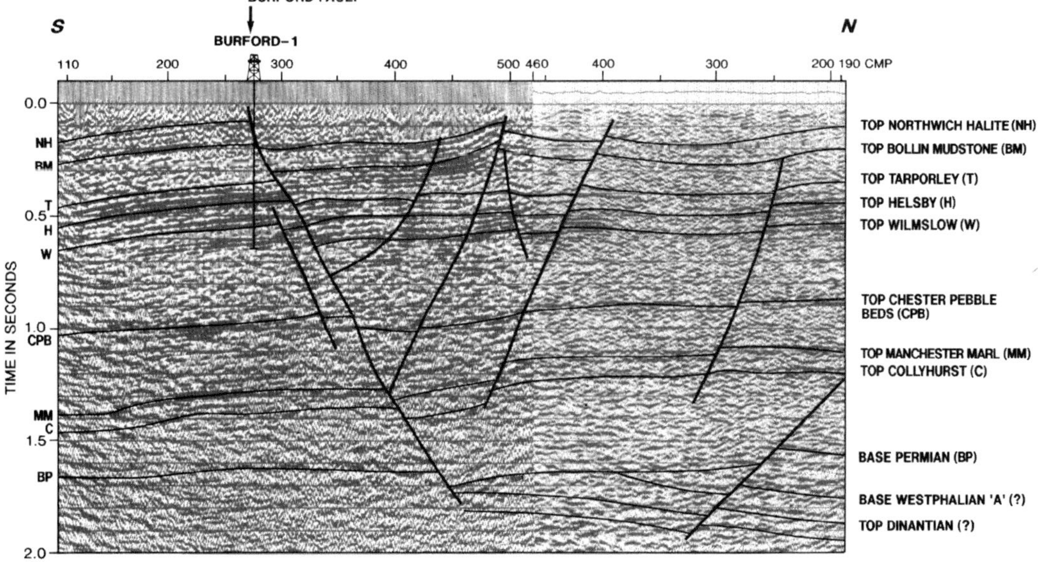

Fig. 8. Seismic line 4 (N–S) passes through the E–W Burford Fault on the north side of the Burford structure (see Fig. 19). There is no evidence of growth, indicating a post-Triassic age for the fault.

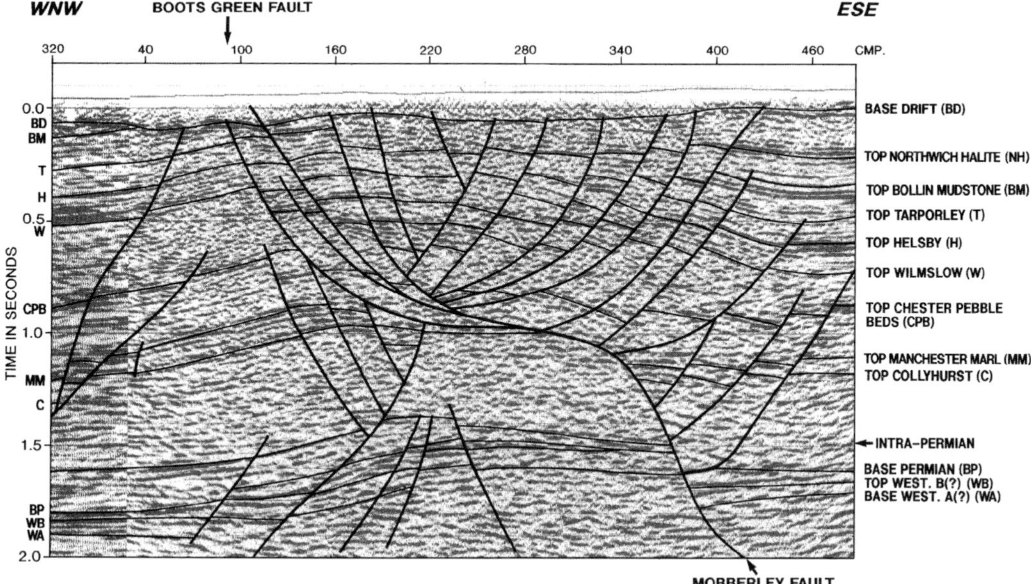

Fig. 9. Seismic line 1 (WNW–ESE) illustrates a complex negative flower structure at Triassic level overlying a prominent basement high, coincident with a 2 km offset of the N–S Mobberley Fault (see Figs 6 & 19). The origin of the flower structure is probably Tertiary strike slip along the NW–SE Boots Green fault.

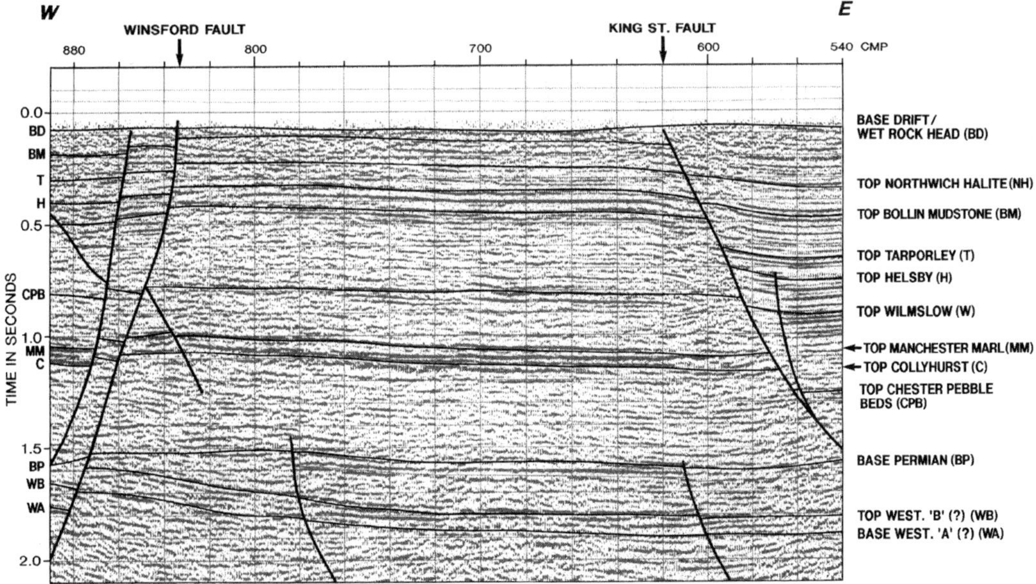

Fig. 10. Seismic line 2 (W–E) crosses the central horst of the basin, bounded to the west by the Winsford Fault and to the east by the King Street Fault (see Fig. 19). Considerable syn-sedimentary growth is evident across the latter.

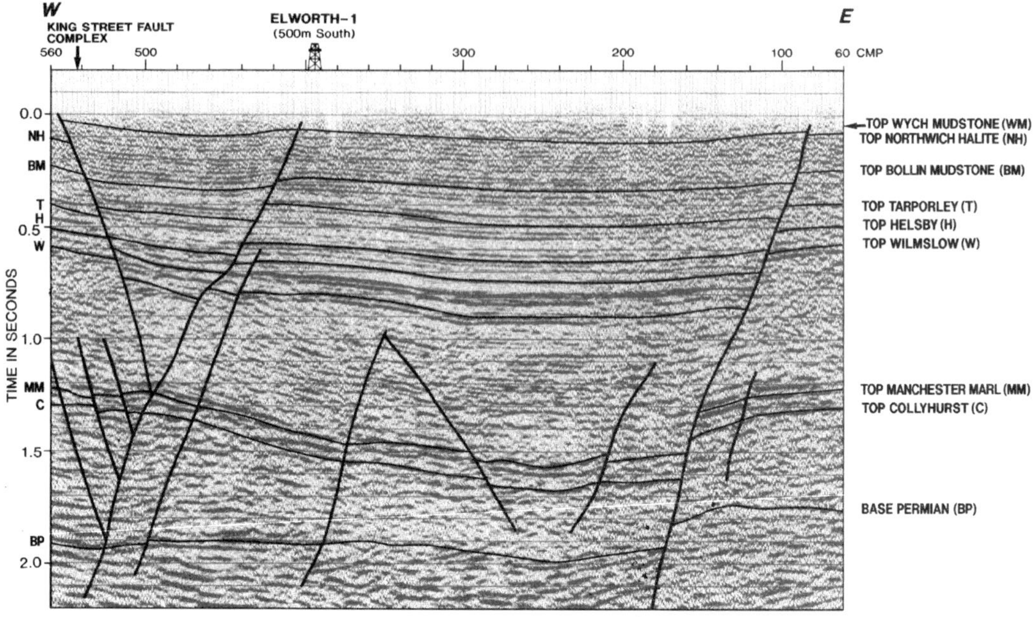

Fig. 11. Seismic line 3 (W–E) which traverses the King Street Fault approximately 4 km south of line 2, demonstrates antiquity of the feature drilled by Elworth-1. Twenty-five milliseconds of growth is evident on the eastern side of the structure between the Top Helsby and Top Northwich Halite reflectors, whilst the western bounding fault appears to die out into the Wilkesley Halite.

6 km. On seismic line 2 (Fig. 10) this feature appears as a broad, stable high, but lines perpendicular to line 2 (confidential, not shown) show subtle E–W faulting, which generally dies out into the younger evaporitic section, similar to that observable on the northern part of line 4 (Fig. 8). These block-faulted structures should be adequately sealed and have favourable timing for late Mesozoic hydrocarbon migration. Since most of the regional horst is thought to be underlain by Westphalian strata, these traps should provide some of the lowest risk prospects in the basin, albeit of rather modest size (Fig. 6).

Reservoir potential

Potential hydrocarbon reservoirs exist within the Cheshire Basin at Carboniferous, Permian and Triassic levels. However, porosity measurements from outcrop and boreholes around the periphery of the basin indicate that Carboniferous sandstones are generally tight and are therefore unlikely to retain porosity at burial depths exceeding 10 000′ in the basin centre. Permian sandstones have excellent reservoir quality in the shallower areas of the Cheshire Basin, but the overlying Manchester Marl is not thought to provide a viable seal. Conse-

quently, Brabant has focused its attention upon the Triassic play and upon the variation in Triassic reservoir quality.

The Triassic of the Basin is divided into two groups; the lower, sand-rich Sherwood Sandstone and the upper, mudstone-evaporite-rich Mercia Mudstone (Fig. 4). The Sherwood Sandstone is sub-divided into three formations: the basal, Chester Pebble Beds (fluvial), the Wilmslow Sandstone (fluvial at the base, more aeolian at the top) and the fluvial/aeolian Helsby Sandstone (Thompson 1970; Warrington *et al.* 1980). Sandstones equivalent to those of the Helsby Formation reservoir the large gas accumulations of the Morecambe and Hamilton fields in the East Irish Sea, where the basal evaporites of the Mercia Mudstone Group act as top seal.

An important difference in the Cheshire Basin is that the basal Mercia Mudstone comprises a thick sequence of interbedded sands and silty mudstones termed the Tarporley Siltstone Formation (absent from the EISB; Jackson *et al.* 1987). Between 600 and 750′ are present within the basin proper, thinning towards the margins; only 30′ are present south of the Wem–Red Rock Fault at Grinshill (Thompson 1985) whilst between 150′ and 390′ are present at outcrop around the northern periphery

(Ireland *et al.* 1978). These sediments may be too sandy to form a reliable topseal to the Helsby Sandstone Formation, but locally have much poorer reservoir quality than the Helsby, potentially constituting a waste zone in the central areas of the Basin.

The reservoir quality of the Tarporley Siltstone Formation varies considerably across the Basin and two distinct sandstone units are evident on well log correlations, separated by argillaceous sediments (Figs 12 & 13): the lower unit appears to be restricted to the northern part of the Cheshire Basin, with net pay (10% porosity cut-off) diminishing rapidly from 78′ (160′ gross) in Knutsford-1 to 5′ (118′ gross) in Elworth-1 (Fig. 13). The lower unit comprises thinly interbedded, micaceous sandstones, siltstones and mudstones which give a striped appearance to the rocks. The mudstones commonly have pseudomorphs after halite and the siltstones often display wave ripple marks on their upper surfaces, with desiccation cracks being common throughout. These sediments were deposited in marginal marine, probably estuarine environments varying from low-intertidal sandflats to high-intertidal mud flats (Earp & Taylor 1986). Further south, where the lower sand unit has largely shaled-out, the time-equivalent basal Tarporley mudstones of Burford-1 and Prees-1 were deposited, possibly in a wholly terrestrial, alluvial plain environment.

The upper sandstone unit is thickest in the extreme south of the basin, where it is largely aeolian and locally termed the Malpas Sandstone, achieving 366′ net sand (449′ gross) in Prees-1. The Malpas Sandstone appears to be restricted to the axis of the basin, shaling out eastwards into the Red Rock Fault Zone and is missing from Blakenhall-1. Net sand diminishes northwards from 235′ (350′ gross) in Burford-1 to 67′ (384′ gross) in Elworth-1 and 93′ (246′) in Knutsford-1. Deterioration in reservoir quality probably reflects a change in depositional facies north of Burford-1, from aeolian to predominantly fluvial–intertidal sediments similar to those of the lower sand unit.

Another trend, of increasing shaliness westwards, in also evident in the north of the basin. In the extreme NE, the upper and lower sand units appear to coalesce; in Ashley BH, the Tarporley Siltstone comprises an uninterrupted 895′ sequence of banded sandstones and siltstones, with only modest intervening mudstones (Evans *et al.* 1968). Westward from Ashley BH, the tripartite subdivision of the Tarporley becomes clearer and, between the Marston and Crabtree Green boreholes, the upper sandstone shales-out, becoming indistinguishable from the Bollin Mudstone in the latter (Earp & Taylor 1986).

The underlying Helsby Sandstones in Cheshire have similar reservoir qualities to the equivalent section within the EISB (where it is termed the Ormskirk Sandstone). However, there is a strong risk that the argillaceous component of the overlying Tarporley Siltstone Formation would be insufficiently tight to provide a viable top-seal, particularly in the northern parts of the basin where the lower sand unit is present. Further south, where the lower sand unit is reduced or missing, the Helsby Sandstone could be a valid reservoir target, particularly within structures bounded by faults of modest throw. Fault throws of greater than *c.* 250′ (the thickness of the Tarporley argillaceous unit in Burford-1) risk juxtaposition of the Helsby against the Tarporley upper sand unit.

Source potential

Potential source horizons are present within the Carboniferous rocks of the Cheshire Basin. The overlying Permian–Triassic sequence is of predominantly continental origin, whilst the underlying Ordovician–Silurian (Earp and Taylor 1986) is generally lean and has been tectonized by Caledonian and Hercynian orogenic episodes.

The basin is fringed by outcropping Carboniferous rocks to the west, north, east and SE. In the extreme SW the Carboniferous is missing and Triassic strata lie upon Lower Palaeozoic shales (Wills 1956; fig. 4). Potential source horizons at outcrop are present within the Westphalian, Namurian and Dinantian. Generalized depositional trends during the Dinantian and Namurian are presented in Figs 14 and 15, with approximate depositional thicknesses of strata, based on well data (for more detail of the evolution of Carboniferous palaeoenvironments refer to Fraser *et al.* 1990; Fig. 10). Dinantian and Namurian basinal bituminous mudstones crop out in the Goyt Trough to the east of the Cheshire Basin. However, boreholes on the western (Milton Green-1, Erbistock-1) and eastern margins (Bowsey Wood-1, Earp 1961) indicate that a carbonate platform environment extended across much of the central–southern parts of the Basin during the Dinantian (Fig. 14). There is a marked facies change and three-fold thickening of the Dinantian succession northward from Milton Green-1 to Blacon East-1. Fraser and Gawthorpe (1990; fig. 2) have proposed that the Bala–Llanelidan Fault may extend ENE (towards Knutsford-1), acting as the hingeline to a Dinantian–Namurian trough traversing the NW Cheshire Basin (Figs 5, 14 & 15). Although the Dinantian–lower Namurian in Blacon East-1 comprise organically-lean, gas-prone siltstones, a more basinal facies (similar to the Holywell Shale which sources the EISB play) might be anticipated to the NE, more distant from the probable N. Wales palaeoshore (Fraser *et al.* 1990; fig. 10). Recently

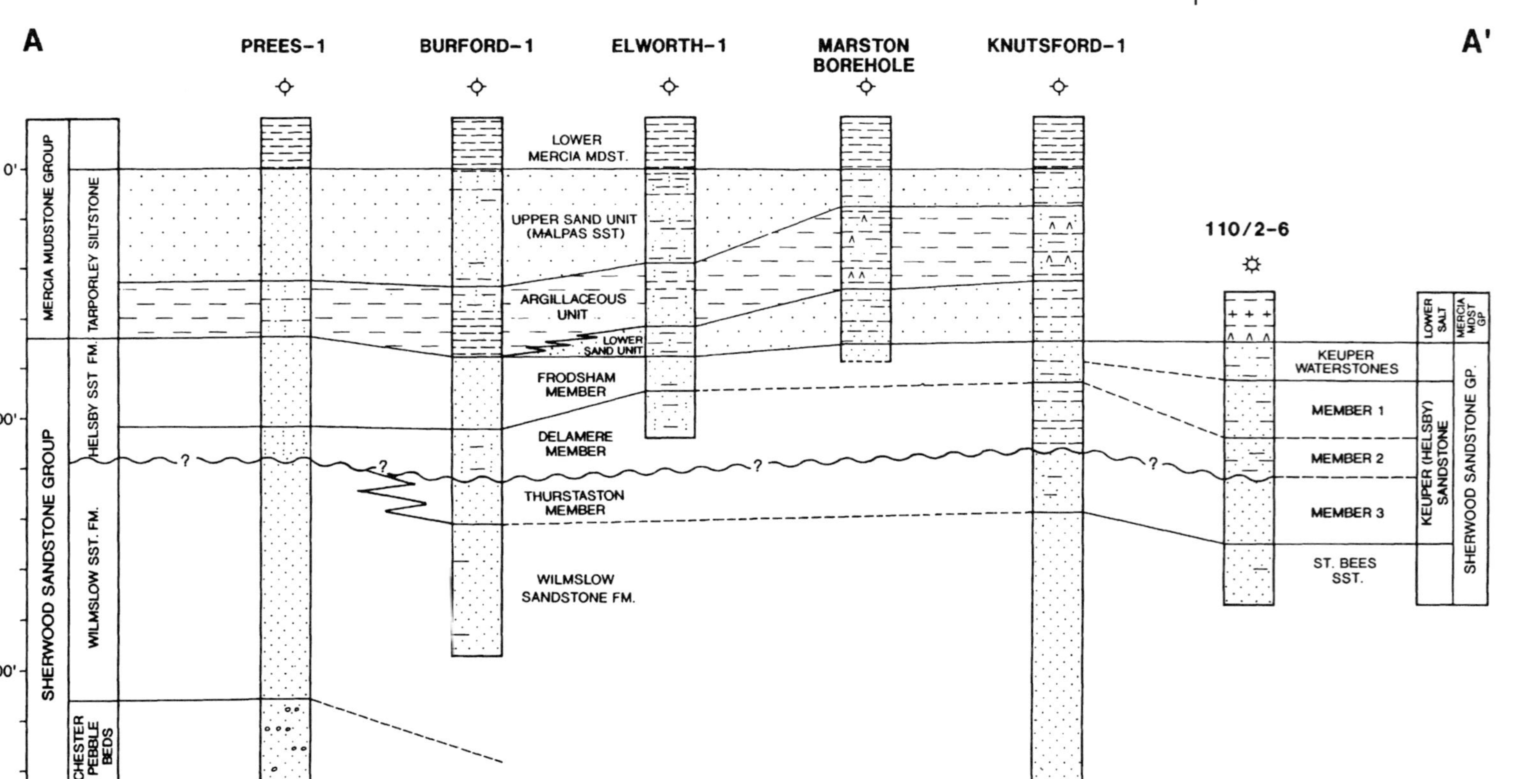

Fig. 12. Lithological correlation of the upper part of the Triassic sandstone sequence, based (largely) upon wireline log interpretation. An unconformity may exist at base Helsby Sandstone Delamere Member level (Evans *et al.* 1993). The Helsby Sandstone Thurstaston Member (*sensu* Thompson 1970) may pass laterally into Wilmslow Sandstone (as indicated) or may be eroded off at the unconformity between Burford-1 and Prees-1.

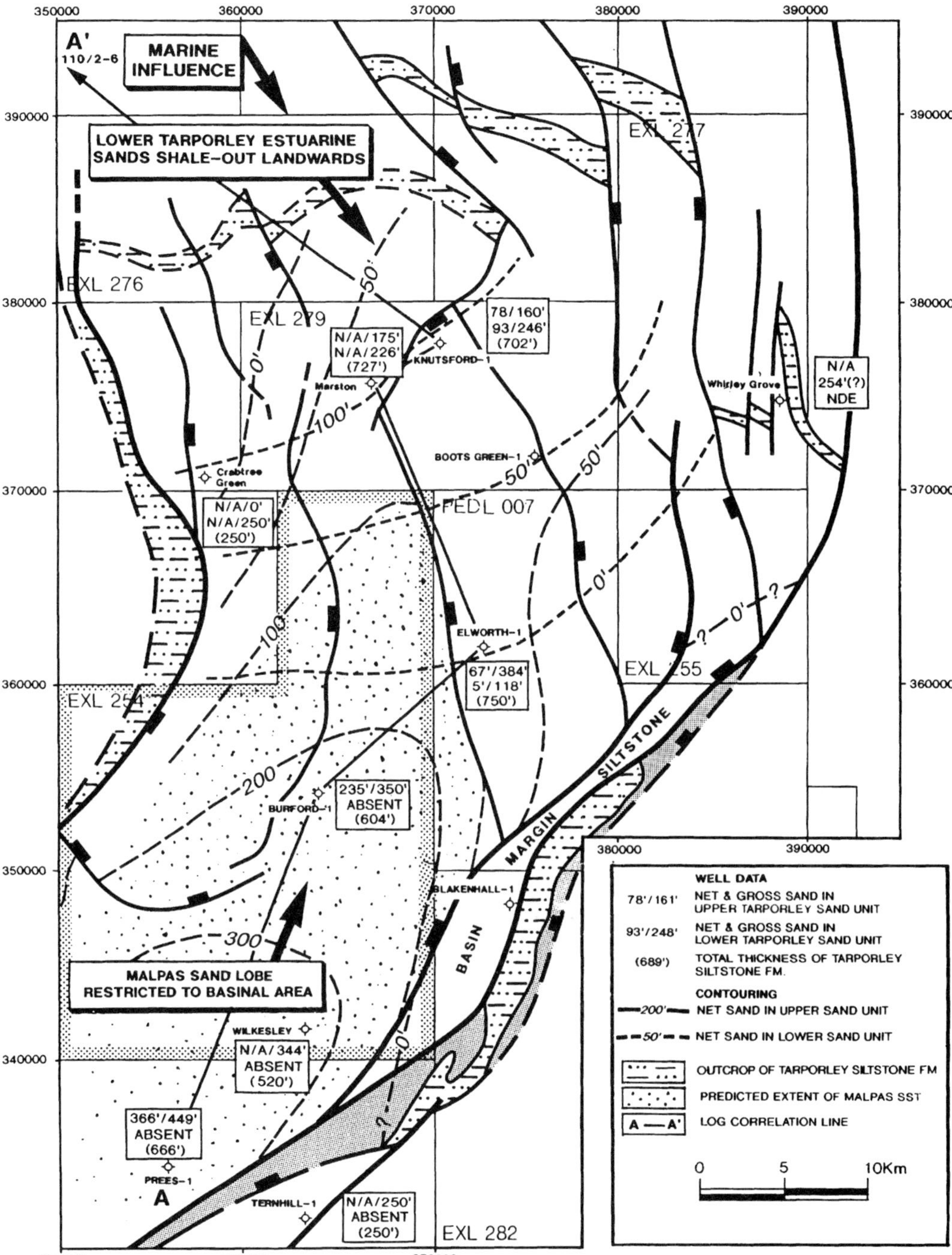

Fig. 13. Depositional trends and net pay map of the sand units within the Tarporley Siltstone Formation, based primarily upon well log data.

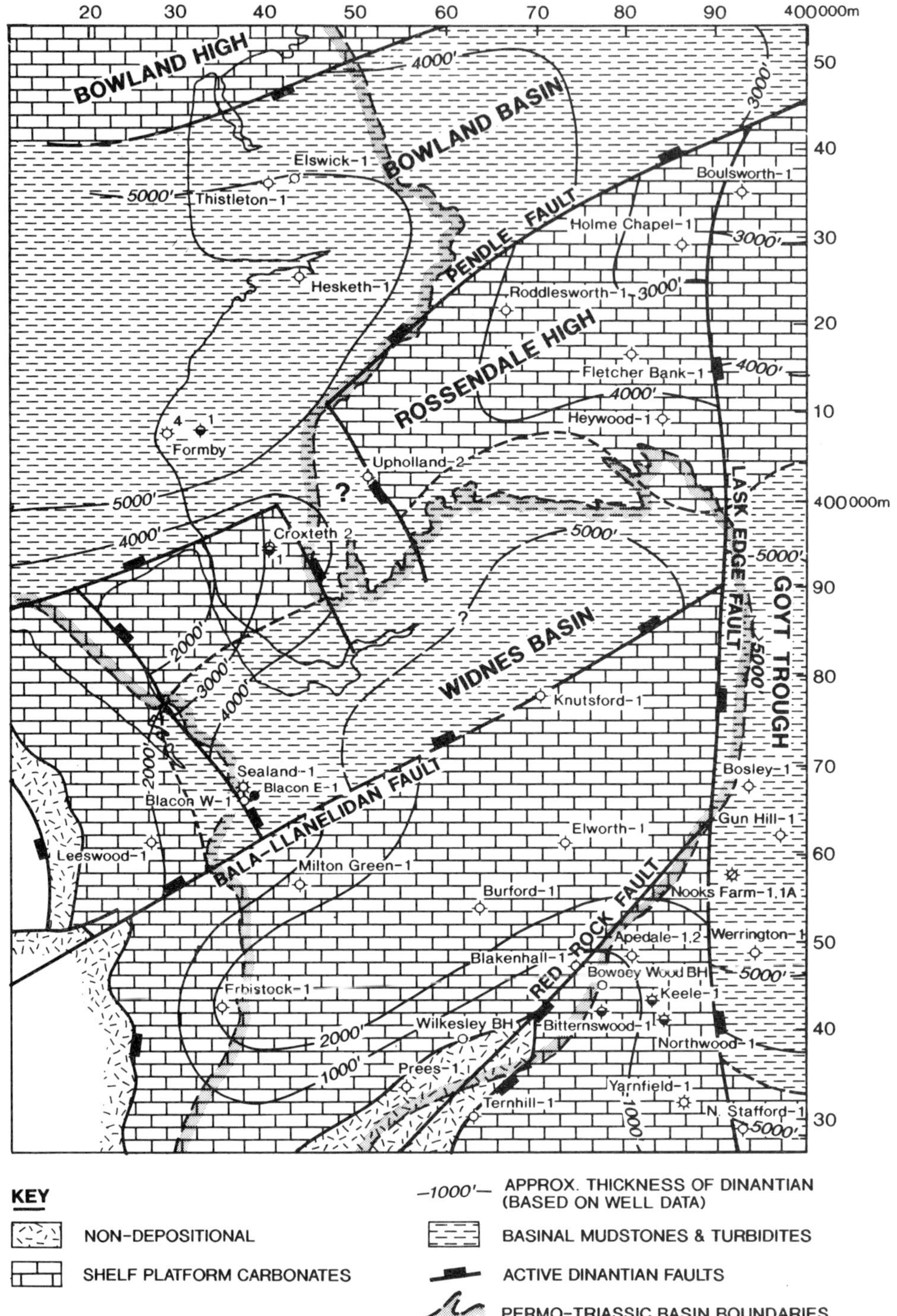

Fig. 14. Generalized Dinantian depositional trends and regional isopach of NW England compiled from confidential well data and published sources. Platform limestone deposition predominated within the Cheshire Basin area, but Holywell Shales source rock facies may be present within the inferred Widnes Basin.

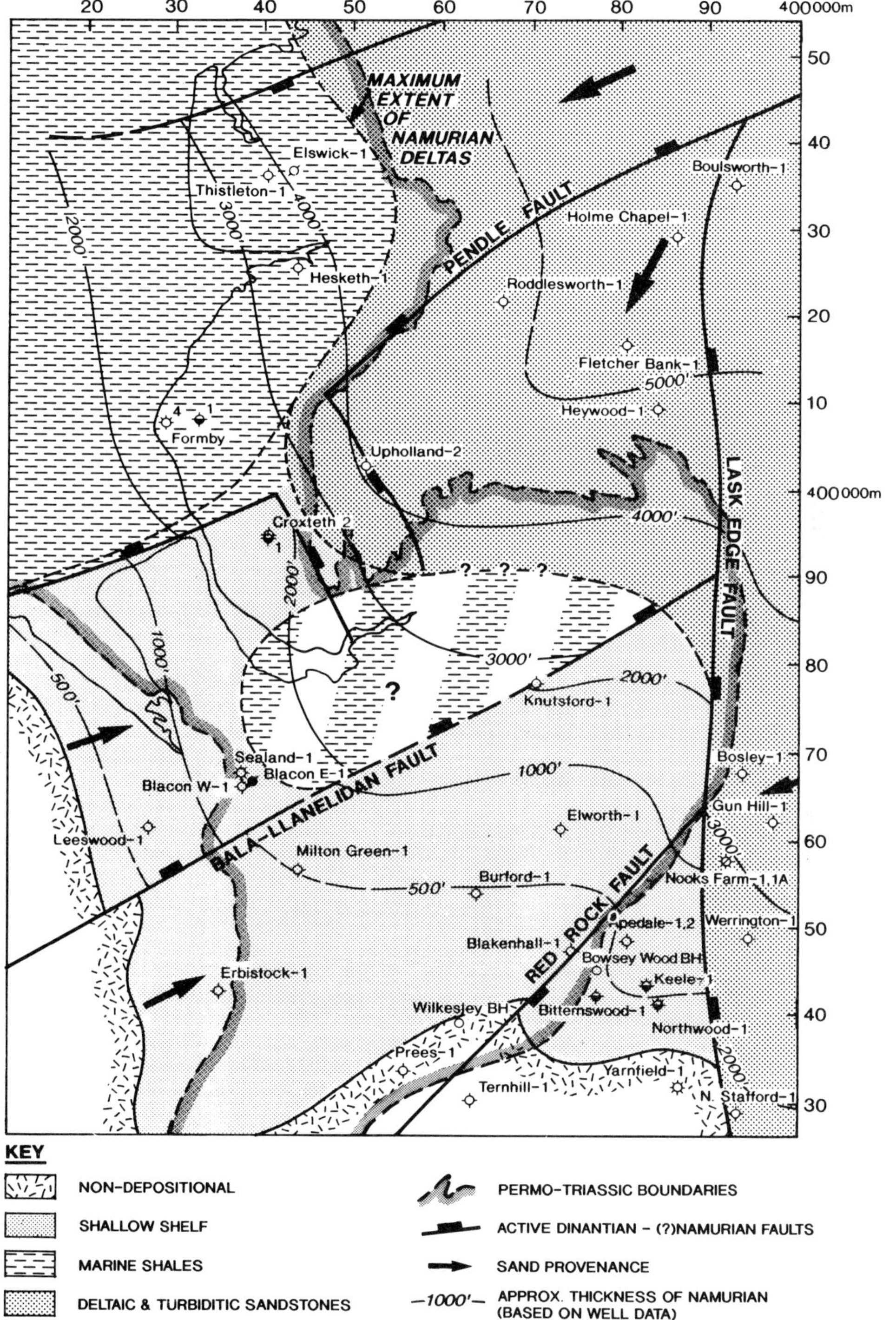

Fig. 15. Generalized Namurian depositional trends and regional isopach of NW England compiled from confidential well data and published sources.

acquired proprietary seismic (confidential) strongly supports the existence of a fault-bounded Dinantian sub-basin beneath the north–central CB.

Westphalian outcrops surround most of the Basin and geochemical analyses have demonstrated that the coals and shales of stages A to C ('Coal Measures') are variable-quality, predominantly gas-prone, source rocks and that oil seepages from intervening sandstones (Fig. 16) are probably sourced by locally rich bituminous shales within the A sequence.

The subcrop of the Coal Measures has been proved in the west and north of the Basin by the Erbistock-1, Blacon East-1 and Knutsford-1 bore-holes. Coal Measures were absent from Prees-1, the only deep well on the eastern side of the Basin, which discovered 700′ of Westphalian D–Stephanian redbeds overlying Silurian basement. A line marking the truncation of the Coal Measures has been interpreted from seismic data to lie approximately midway between Milton Green-1 and Prees-1, trending roughly NE across the Basin (Fig. 16). The angular unconformity between the Carboniferous and the Permo-Triassic is particularly marked on seismic line 6 (Fig. 17), where the Coal Measures reflectors display characteristically high amplitude. Westphalian strata, originally deposited to the south and east of this

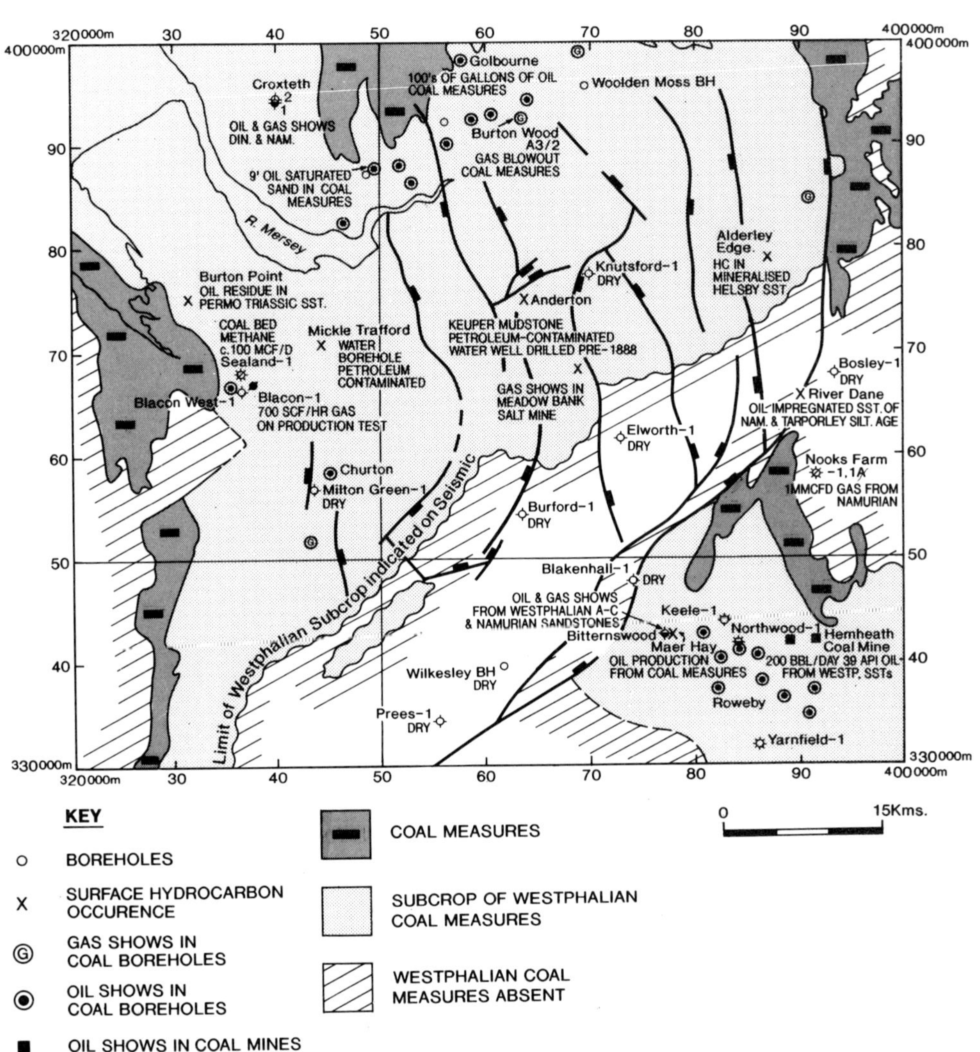

Fig. 16. Hydrocarbon occurrences and Westphalian outcrop within the Cheshire Basin area.

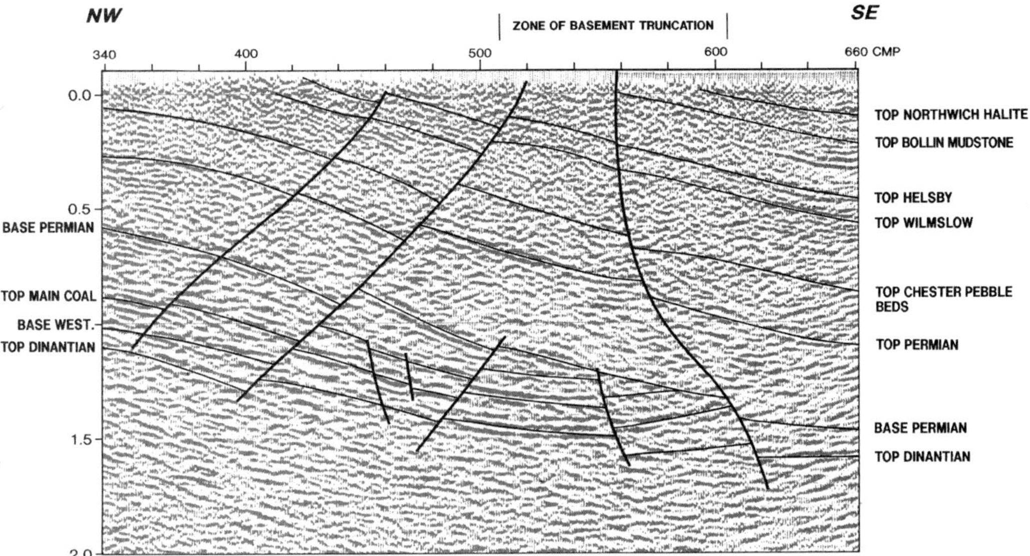

Fig. 17. Seismic line 6 (NW–SE) illustrates the eastward truncation of the Westphalian beneath the base Permian unconformity (see Fig. 19). The seismic has been directly tied to the velocity log of the Milton Green-1 well 8 km to

truncation line, were eroded as a result of Hercynian uplift. The redbeds encountered by Prees-1 are interpreted to be post-orogenic, deposited in basins formed above the eroded core of the Hercynian uplift (Besly 1988).

Maturation and timing of maturation

There have been numerous oil seeps reported from the Carboniferous within coal mines and boreholes around the periphery of the Basin, while uneconomic quantities of gas have been produced from Carboniferous sandstones in Blacon East-1 and Nooks Farm-1A. Vitrinite reflectance measurements, taken from Westphalian strata outcropping around the Basin, typically average 0.7–0.8%, indicating peak oil maturity (Figure 16).

Isolated examples of seepage from Permo-Triassic sandstones have been reported in the literature from Burton Point (Macchi & Meadows 1987), Mickle Trafford and Alderley Edge (Thompson pers. com.). However, within the area of the Basin overlain by Mercia Mudstone, there have only been two reported occurrences of hydrocarbons:

- A petroleum-contaminated 19th century water well into the 'Keuper Mudstone' (Bollin Mudstone) at Anderton near Northwich referred to by Selley (1992).
- Release of methane from small cavities at the base of the Northwich Halite has been recorded

in the Meadowbank Salt Mine by the operators, Salt Union.

At Knutsford-1, vitrinite reflectance measurements of 1.18–1.32% (9771′–9991′) from coals of Westphalian C age indicate late-oil to early-gas maturity. A burial history plot for this well (Fig. 18) has been constructed, using a kinetic model (Lawrence Livermore National Laboratory – LLNL). The model indicates that the base of the Westphalian may have reached peak oil maturity (R_o>0.7%) during the mid-Triassic, commenced gas generation during the early Jurassic, and achieved peak gas maturity (R_o >1.3%) during middle Jurassic times, generating predominantly dry gas thereafter until terminated by early Tertiary uplift.

Knutsford-1 reached TD in upper Westphalian and the thicknesses of unpenetrated Westphalian, Namurian and Dinantian strata have been reconstructed from regional mapping of well data (Figs 14 & 15). The missing Upper Triassic and Lower Lias section has been assumed to be of similar thickness (4400′) to that of the Prees-1 well. The Middle Jurassic to Upper Cretaceous section was estimated to have totalled about 2000′, on the basis outlined earlier, with the constraint that Mercia Mudstone shale velocities, when compared to a normal burial curve from the Southern Gas Basin (Marie 1975), indicate about 6200′ of additional burial at Knutsford-1.

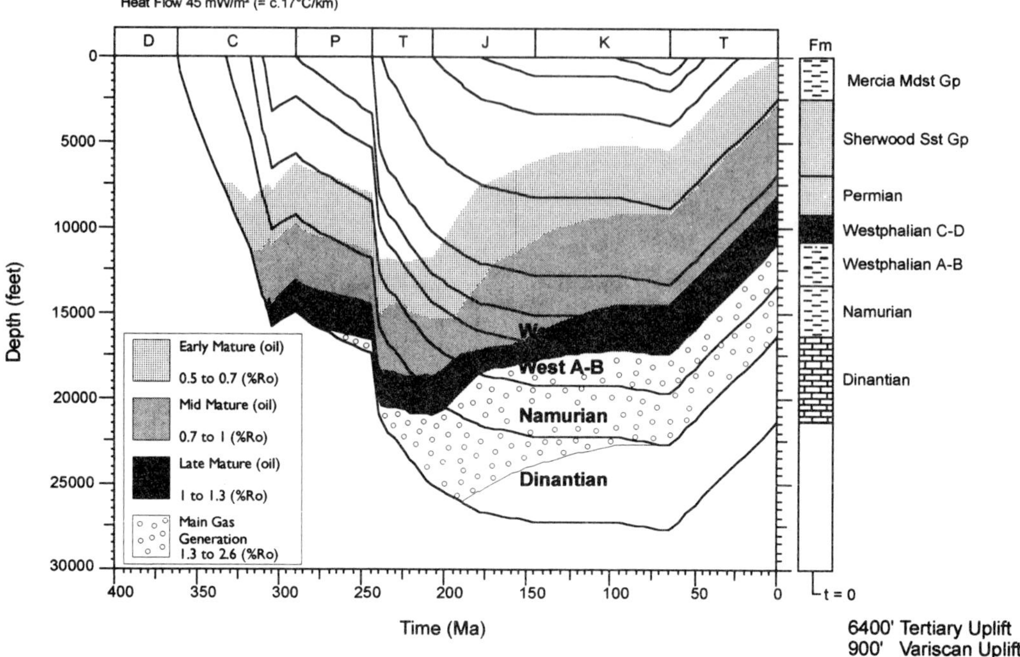

Fig. 18. Burial history model of the Knutsford-1 well, computed using a Platte River BasinMod™ kinetic model (LLNL), calibrated to vitrinite reflectance data from the Westphalian. Thickness of Westphalian A–B , Namurian and Dinantian estimated from regional trends.

Heat flow has been estimated, at about 45 mW m^{-2} from the BHTs of Knutsford-1; close to the present-day average heat flow of 50 mW m^{-2} for the Cheshire Basin (Gale *et al.* 1984). Although higher heat transmission may have occurred in the past, during the Carboniferous, early Triassic rifting and early Tertiary dyke intrusion, vitrinite reflectance data from Knutsford-1 is consistent with relatively low average heat flux.

Seal potential

The Tarporley Siltstone Formation is top sealed by the Mercia Mudstone Group. This is a thick cyclic sequence of mudstones and evaporites which were deposited in a shallow epicontinental sea (Warrington *et al.* 1980), although some authors consider that parts of the succession were deposited sub-aerially (Arthurton 1980; Earp & Taylor 1986). The Group comprises six formations: Bollin Mudstone, Northwich Halite, Byley Mudstone, Wych Mudstone, Wilkesley Halite and Brooks Mill Mudstone. In the EISB success in the Triassic play has generally coincided with the presence of overlying halite sequences. Intense faulting during Tertiary uplift could have caused seal failure in areas with thin mudstone seal, whereas faults passing through salt sequences would have been rapidly annealed, preventing large-scale trap failure.

The Northwich Halite Formation underlies much of the central and southern Basin (Fig. 19) between the Bickerton–Bulkley Hill–East Delamere and Red Rock Faults, but has been eroded from the northern parts of the basin and from the footwalls of major normal faults (e.g. Boots Green and Mobberley faults). The previous extent of the halite is inferred from the wide distribution of a characteristic collapse breccia, developed by salt dissolution, beneath Quaternary drift.

A deeper, potential Permian play is dependent upon the presence of a Manchester Marl seal to the Collyhurst Sandstone (Fig. 4). However, the 486' thick Manchester Marl penetrated by Knutsford-1 is probably too sandy to provide an effective seal, though it may extend as far south as Whitchurch (Gale *et al.* 1984). By comparison with the EISB, a more evaporitic Manchester Marl may be anticipated towards the basin axis, although any such deposits would only be capable of sealing those fault-bounded structures with relatively small throw.

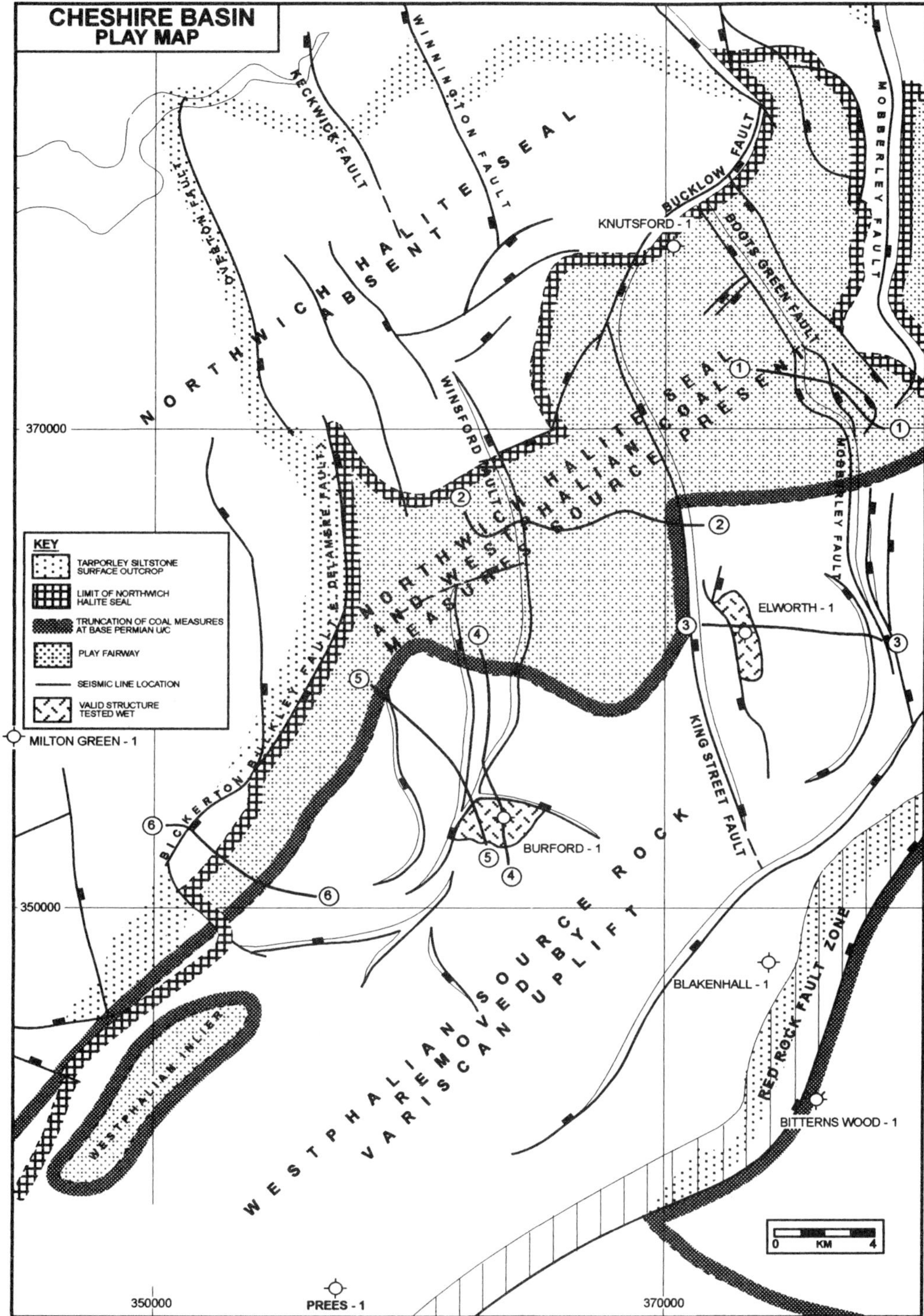

Fig. 19. Play summary map for the Cheshire Basin illustrates the main elements controlling prospectivity in the Cheshire Basin; (**a**) erosion of the Westphalian source rock from the SE of the Basin following Hercynian uplift and (**b**) Tertiary erosion of the Northwich Halite seal from the NW. The play fairway lies between these two lines of truncation, explaining the dry holes on apparently closed structures at Elworth and Burford.

Conclusions

Given the proximity and similar geology of recent discoveries of oil and gas in the southern part of the East Irish Sea, the Cheshire Basin cannot be written off on the basis of only two structurally valid Triassic tests over an area of some 1800 km². The key ingredients for potential success appear to be the identification of preserved Westphalian subcrop, in order to provide an adequate source and the retention of an effective reservoir topseal in an area which has undergone large-scale Tertiary fault reactivation and uplift (Fig. 19).

These factors appear to exist in a band across the NW to central parts of the basin, between the proven Westphalian source rocks in Knutsford-1 and the seismically delineated Westphalian truncation line to the south. To the north, the increasingly shallow Helsby (and more particularly Tarporley) reservoir becomes increasingly difficult to seal and those areas retaining overlying Northwich Halite should offer the least risk of breaching. It would seem to be no coincidence that the only documented hydrocarbon shows within the Cheshire Basin proper lie within this fairway.

Thanks are expressed to Edinburgh Oil & Gas (acting on behalf of Enterprise Oil plc), Mobil North Sea Ltd and Shell UK Expro Ltd for their permission to present the seismic lines in this paper and to Brabant's partners in the Cheshire Basin – Canyon Resources Company and Hamilton Oil Company Ltd – for their agreement to publish and their helpful comments. The authors would also like to acknowledge David Thompson for his thorough refereeing of the paper and constructive criticism.

References

ARTER, G. & FAGIN, S. W. 1993. The Fleetwood Dyke and the Tynwald fault zone, Block 113/27, East Irish Sea Basin. *In*: PARKER J. R. (ed.) *Petroleum Geology of Northwest Europe: Proceedings of the of the 4th Conference.* Geological Society, London, 835–843.

ARMSTRONG, J. P., D'ELIA, A. A. & LOBERG, R. 1995. Holywell Shale: a potential source of hydrocarbons in the East Irish Sea. *In*: CROKER, P. F. & SHANNON, P. M. (eds) *The Petroleum Geology of Ireland's Offshore Basins.* Geological Society, London, Special Publications, **93**, 37–38.

ARTHURTON, R. S. 1980. Rhythmic sedimentary sequences in the Triassic Keuper Marl (Mercia Mudstone Group) of Cheshire, northwest England. *Geological Journal*, **15**, 43–58.

BENTON, M. J., WARRINGTON, G., NEWALL, A. J. & SPENCER, P. S. 1994. A review of the British mid-Triassic tetrapod faunas. *In*: FRASER, N. C. & SUES, H. D. (eds) *In the Shadow of the Dinosaurs: Early Mesozoic Tetrapods.* Cambridge University Press, Cambridge, 131–160.

BESLY, B. 1988. Palaeogeographic implications of late Westphalian to early Permian red beds, Central England. *In*: BESLY, B. M. & KELLING, G. (eds) *Sedimentation in a Synorogenic Basin Complex: the Carboniferous of North West Europe.* Blackie, Glasgow, 200–221.

BUSHELL, T. P 1986. Reservoir geology of the Morecambe Field. *In*: BROOKS, J., GOFF, J. & VAN HOORNE, B. (eds) *Habitat of Palaeozoic Gas in N.W. Europe.* Geological Society, London, Special Publications, **23**, 189–207.

CHADWICK, R. A. 1997. Fault analysis of the Cheshire Basin, NW England. *This volume.*

—— & EVANS, D. J. 1995. The timing and direction of Permo-Triassic extension in southern Britain. *In*: BOLDY, S. A. R. (ed.) *Permian and Triassic Rifting in Northwest Europe.* Geological Society, London, Special Publications, **91**, 161–192.

COWARD, M. P. 1995. Structural and tectonic setting of the Permo-Triassic basins of northwest Europe. *In*: BOLDY, S. A. R. (ed.) *Permian and Triassic Rifting in Northwest Europe.* Geological Society, London, Special Publications, **91**, 7–39.

EARP, J. R. & TAYLOR, B. J. 1986. Geology of the country around Chester and Winsford. British Geological Survey, Memoir **109**.

EVANS, D. J., REES, J. G. & HOLLOWAY, S. 1993. The Permian to Jurassic stratigraphy and structural evolution of the central Cheshire Basin. *Journal of the Geological Society, London,* **150**, 857–870.

EVANS, W. B., WILSON, A. A., TAYLOR, B. J. & PRICE, D. 1968. Geology of the Country around Macclesfield, Congleton, Crewe and Middlewich, British Geological Survey, Memoir **110**.

FRASER, A. J. & GAWTHORPE, R. L. 1990. Tectono-stratigraphic development and hydrocarbon habitat of the Carboniferous in Northern England. *In*: HARDMAN, R. P. F. & BROOKS, J. (eds) *Tectonic Events Responsible for Britain's Oil and Gas Reserves.* Geological Society, London, Special Publications, **55**, 49–86.

——, NASH, D. F., STEELE, R. P. & EBDON, C. C. 1990. A regional assessment of the intra-Carboniferous play of Northern England. *In*: BROOKS, J. (ed.) *Classic Petroleum Provinces.* Geological Society, London, Special Publications, **50**, 417–440.

GALE, I. N., EVANS, C. J., EVANS, R. B., SMITH, I. F., HOUGHTON, M. T. & BURGESS, W. G. 1984. The Permo-Triassic aquifers of the Cheshire and West Lancashire Basins. *In*: *Investigation of the Geothermal Potential of the UK.* British Geological Survey.

GREEN, P. F. 1989. Thermal and tectonic history of the East Midlands shelf (onshore UK) and surrounding regions assessed by fission track analysis. *Journal of the Geological Society, London,* **146**, 755–773.

——, DUDDY, I. R., BRAY, R. J. & LEWIS, C. L. E. 1993. Elevated palaeotemperatures prior to Early Tertiary cooling throughout the UK region: implications for hydrocarbon generation. *In*: PARKER, J. R. (ed.)

Petroleum Geology of Northwest Europe: Proceedings of the of the 4th Conference. Geological Society, London, 1067–1074

HARDMAN, M., BUCHANAN, J., HERRINGTON, P. & CARR, A. 1993. Geochemical modelling of the East Irish Sea Basin: its influence on predicting hydrocarbon type and quality. *In*: PARKER, J. R. (ed.) *Petroleum Geology of Northwest Europe: Proceedings of the of the 4th Conference.* Geological Society, London, 809–821.

HILLIS, R. 1993. Tertiary erosion magnitudes in the East Midlands shelf, onshore UK. *Journal of the Geological Society, London,* **150**, 1047–1050.

IRELAND, I. J., POLLARD, J. E., STEEL, R. J. & THOMPSON, D. B. 1978. Intertidal sediments and trace fossils from the Waterstones (Scythian–Anisian?) at Daresbury, Cheshire. *Proceedings of the Yorkshire Geological Society,* **41**, 399–436.

JACKSON, D. J. & MULHOLLAND, P. 1993. Tectonic and stratigraphic aspects of the East Irish Sea Basin and adjacent areas: contrasts in their post-Carboniferous structural styles. *In*: PARKER, J. R. (ed.) *Petroleum Geology of Northwest Europe: Proceedings of the 4th Conference.* Geological Society, London, 791–808.

——, MULHOLLAND, P., JONES, S. M. & WARRINGTON, G. 1987. The geological framework of the East Irish Sea Basin. *In*: BROOKS, J. & GLENNIE, K. W. (eds) *Petroleum Geology of North West Europe.* Graham & Trotman, London, 191–203.

KNIPE, R., COWAN, G. & BALENDRAN, V. S. 1993. The tectonic history of the East Irish Sea Basin with reference to the Morecambe Fields. *In*: PARKER, J. R. (ed.) *Petroleum Geology of Northwest Europe: Proceedings of the of the 4th Conference.* Geological Society, London, 857–866.

LEE, A. G. 1988. Carboniferous basin configuration of central and northern England modelled using gravity data. *In*: BESLY, B. M. & KELLING, G. (eds) *Sedimentation in a Synorogenic Basin Complex: The Carboniferous of North West Europe.* Blackie, Glasgow, 69–84.

MARIE, J. P. P. 1975. Rotliegendes stratigraphy and diagenesis. *In*: WOODLAND, A. W. (ed.) *Petroleum and the Continental Shelf of North-west Europe, Vol. 1, Geology,* Applied Science Publishers (for the Institute of Petroleum), 213–223.

MAACHI, L. & MEADOWS, N. S. 1987. *Field Excursion Guide to the Permo-Triassic of Cumbria and Cheshire.* Excursion Guide 12, Poroperm-Geochem Limited, Chester, UK.

POOLE, E. G. & WHITEMAN, A. J. 1966. *Geology of the Country around Nantwich and Whitchurch.* Memoirs of the Geological Survey, Great Britain: HMSO, London.

SELLEY, R. C. 1992, Petroleum seepages and impregnations in Great Britain. *Marine and Petroleum Geology,* **9**, 226–244

THOMPSON, D. B. 1970. The stratigraphy of the so-called Keuper Sandstone Formation (Scythian–?Anisian) in the Permo-Triassic Cheshire Basin. *Quarterly Journal of the Geological Society, London,* **126**, 151–181.

—— 1985. *Field Excursion to the Permo-Triassic of the Cheshire–East Irish Sea–Needwood and Stafford Basins.* Excursion Guide **4**, Poroperm-Geochem Limited, Chester, UK.

TUCKER, R. M. & TUCKER, M. E. 1981. Evidence of syn-sedimentary tectonic movements in the Triassic halite of Cheshire. *Nature,* **290**, 495–496.

WARREN, P. T., PRICE, D., NUTT, M. J. C. & SMITH, E. G. 1984. *Geology of the Country around Rhyl and Denbigh.* NERC, HMSO, London.

WARRINGTON, G., AUDLEY-CHARLES, M. G., ELLIOTT, R. E. *ET AL.* 1980. *A Correlation of Triassic Rocks of the British Isles.* The Geological Society, London, Special Reports, **13**.

WILLIAMS, G. D. & EATON, G. P. 1993. Stratigraphic and structural analysis of the Late Palaeozoic–Mesozoic of NE Wales and Liverpool Bay: implications for hydrocarbon prospectivity. *Journal of the Geological Society, London,* **150**, 489–499.

WILLS, L. J. 1956. *Concealed Coalfields.* Blackie, Glasgow.

—— 1970. The Triassic succession of the Central Midlands in its regional setting. *Quarterly Journal of the Geological Society, London,* **126**, 225–283.

WILSON, A. A. 1982. Structure. *In*: JAMES, J. W. C. (ed.) *The Sand and Gravel Resources of the Country around Prees, Shropshire.* Institute of Geological Sciences, Mineral Assessment Report, **134.**

ZIEGLER, P. A. 1990. *Geological Atlas of Western and Central Europe 1990.* Drukkerij Verweij B.V., Mijdrecht.

Real and relict direct hydrocarbon indicators in the East Irish Sea Basin

A. FRANCIS[1], M. MILLWOOD HARGRAVE, P. MULHOLLAND[2] & D. WILLIAMS

IKODA Limited, 5 Old Lodge Place, St Margaret's, Twickenham, Middlesex TW1 1RQ, UK

[1] *Present address: Lasmo plc, 101 Bishopsgate, London EC2M 3XH, UK*

[2] *Present address: Agip (UK) Ltd, Wellington Circle, Redmoss, Aberdeen AB12 3JG, UK*

Abstract: Seismic amplitude anomalies associated with the presence of hydrocarbon gas now play an important role in the exploration of the East Irish Sea Basin. Tilted flat spots were first observed on data acquired by JEBCO Seismic Ltd. in 1986 within the productive Triassic reservoir of the Sherwood Sandstone Group located in Quadrant 110 (offshore North Wales). With hindsight Direct Hydrocarbon Indicators (DHIs) can now be identified on seismic data acquired in earlier phases of the exploration of the East Irish Sea Basin, although many anomalies were not always identified as such at that time.

This study determines the geophysical parameters by which amplitude anomalies associated with diagenetic variations could be separated from anomalies associated with the presence of gas in the reservoir. The predicted model results are compared with the seismic response of proven fields, discoveries and relict or palaeo-DHIs around the basin. Practical implications for effective exploration, development and production methods in the Sherwood Sandstone are suggested.

Anomalous seismic reflections and amplitude variations associated with the presence of hydrocarbon fluids have been observed over the last 20 years in many localities world-wide and described by various authors (Backus & Chen 1975; White 1977; Brown 1986). The presence of DHIs on seismic data acquired in the East Irish Sea was first reported by geoscientists working the basin in the mid and late 1970s (V. Colter, pers. comm. 1991).

The existence of relict or palaeo gas–water contacts associated with the distribution of diagenetic platy illite in the sub-surface, had been determined previously by mapping borehole data for the Morecambe Gas field (Bushell 1986). The hypothesis that some of the observed seismic amplitude anomalies in this basin are associated with diagenetic boundaries in the Triassic sandstones (palaeo-DHIs or direct porosity indicators) rather than the presence of present day gas was postulated.

The seismic reflection character of the main lithostratigraphic units is well established in the East Irish Sea Basin (EISB) (Ross & Flack 1992). Seismic markers for the major formation and member boundaries of the top Ormskirk, Calder, Rottington and Manchester Marls and also the Hercynian unconformity have been recognised and used across the basin by various authors (Arter & Fagin 1993; Knipe *et al.* 1993).

The Permo-Triassic structure and stratigraphy of the EISB is well known and has been described by various authors (Colter 1978; Jackson *et al.* 1987; Jackson & Mulholland 1993; Jackson *et al.* this

volume). The basin is linked to the contiguous onshore West Lancashire basin where similar stratigraphic sequences have been described (Warrington *et al.* 1980; Smith *et al.* 1974) and similar structural styles exist.

The Formby Point platform is a major structural element of the study area (Fig. 1). The feature is associated with a positive Bouguer gravity anomaly which defines the platform's extent along the Lancashire coast (Deegan 1977). It is bounded on the east and west sides by gravity lows which are associated with Permo-Triassic basins that contain Mercia Mudstone Group sediments at outcrop (Chesher & Jackson 1994), as proved by wells Formby No.1 onshore (Kent 1945) and 110/9-1 offshore (Jackson *et al.* 1987).

Prior to the removal of the Mercia Mudstone Group seal during the Tertiary, it is probable that the Formby Point platform, like the nearby Deemster platform in Block 110/8, was a major hydrocarbon trap (Fig. 1). Well composite logs for borehole 110/8-1 on the Deemster platform record the existence of live and dead oil staining in the Sherwood Sandstone Group from 450 ft down to 1900 ft sub-sea level (137–580 m), implying either a hydrocarbon fluid migration route or the remnants of a previously filled trap.

Observed amplitude anomalies

Figure 2 shows an east–west orientated 2D seismic line located offshore in Quadrant 110. (For reasons

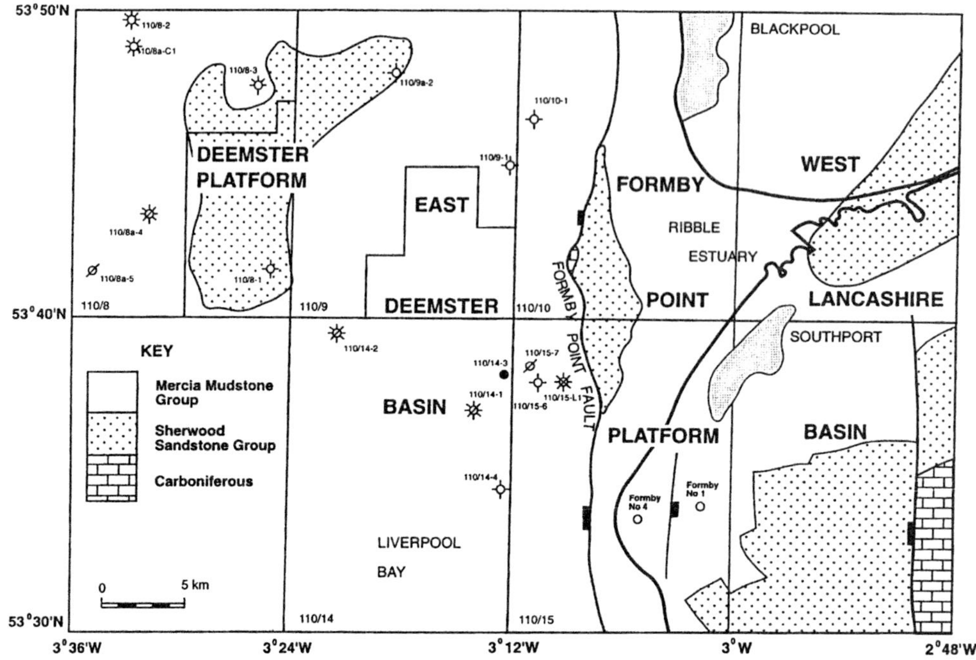

Fig. 1. Location map of the study area.

of commercial confidence the exact location of this line cannot be shown.) The line illustrates four of the significant structural elements of this area; the West Lancashire basin, the tilted fault block of the Formby Point platform; the Formby Point fault and the East Deemster basin (part of the main ESIB).

Two significant seismic amplitude anomalies are apparent on Fig. 2. To the east, on the Formby Point platform, a strong anomaly at 280 ms (estimated depth 1640 ft or 500 m) cross-cuts the dipping reflectors of the Triassic Sherwood Sandstone Group. To the west at 580 ms a bright spot is apparent in the Ormskirk Sandstone Formation on the crest of the roll-over anticline, that contains the Lennox field, next to the Formby Point fault (Haig *et al.* this volume).

The possibility that the amplitude anomaly located on the Formby Point platform is due to a present day gas–water contact within the Triassic Sherwood Sandstone Group can be discounted for the following reasons:

(i) The platform was tested onshore by the Formby No. 4 well located near Formby Point (Fig. 1). This well was spudded in the Mercia Mudstone Group and penetrated the water-wet Triassic Sandstone, above the level of the offshore seismic anomaly, at a depth of 720 ft (220 m) below Ordnance Datum (Falcon & Kent 1960).

(ii) Offshore mapping to the north of Formby Point demonstrates that the Sherwood Sandstone Group sub-crops the recent marine sediments in the mouth of the Ribble Estuary (Chesher & Jackson 1994) thereby breaching any potential platform-wide trap (Fig. 1).

(iii) Seismic interpretation of Fig. 2 confirms that the Sherwood Sandstone Group strata sub-crop the Quaternary offshore and so lack a competent enough seal above the anomaly (cf. well 110/8-1 on the Deemster platform).

It is proposed here that the most likely explanation for the anomalous seismic event on the Formby Point platform seen on Fig. 2 is that it arises from an acoustic impedance contrast caused by a diagenetic boundary within the Triassic sandstones. Above this boundary, the formations has had a different diagenetic history to those parts of the same formation lying below the event. A different diagenetic history can best be explained by the existence of a former hydrocarbon fill in the sandstones of at least 500 m (1640 ft) present thickness, and estimated to be originally 1100 m (3600 ft) pre-Tertiary erosion.

Model construction

To determine the likely rock property contrast necessary to produce the observed seismic anomaly

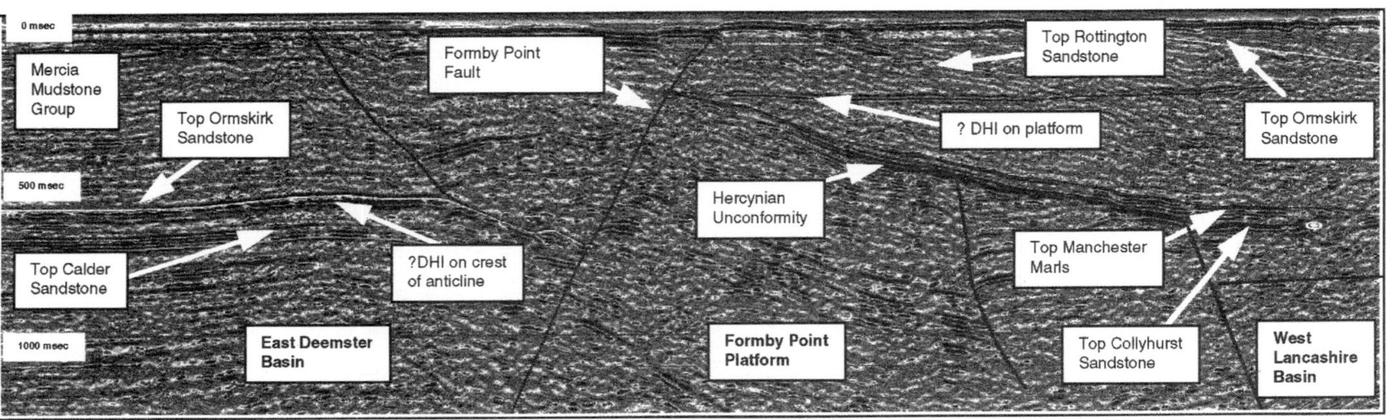

Fig. 2. East–west seismic section across the Formby Point platform in Quadrant 110.

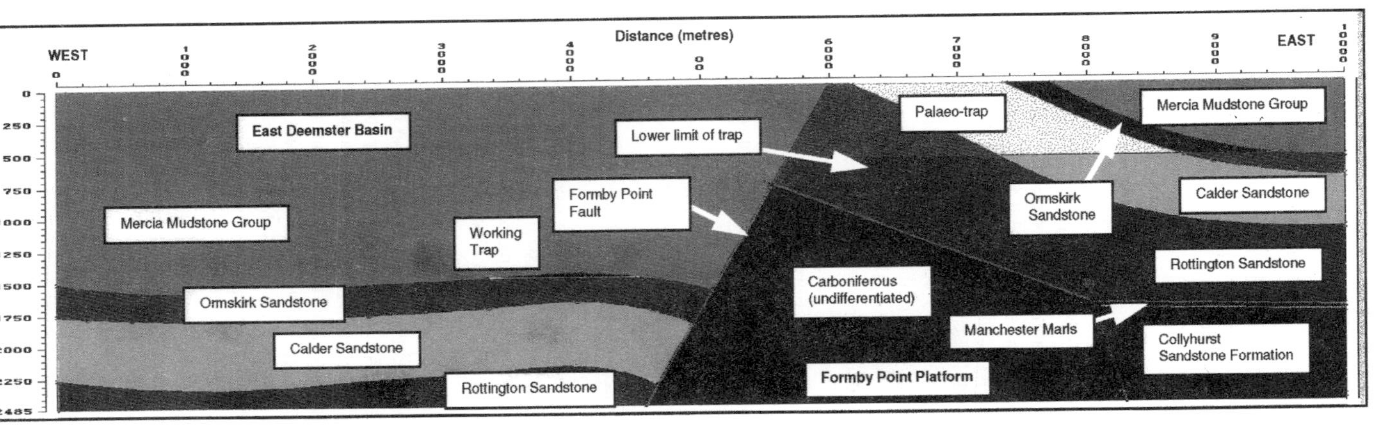

Fig. 3. Model depth cross-section (metres).

a simple synthetic structural model was built. This was populated with petrophysical parameters derived from data for the Permo-Triassic formations within the basin.

Figure 3 shows the geometric template used in the seismic modelling analysis and is based on the seismic interpretation of Fig. 2. The depth parameters for the Permo-Triassic intervals in the model were derived from well 110/9-1 for the East Deemster basin, Formby No. 4 for the Formby Point platform and Formby No. 1 for the West Lancashire basin. The Formby Point fault was modelled as a planar surface dipping 60° from horizontal (Jackson & Mulholland 1993).

Table 1 lists the petrophysical parameters of interval velocity and formation density used for this study. These data were derived from onshore and offshore wells which prove the Carboniferous, Permian and Triassic stratigraphy of the basin (Jackson *et al.* 1987) combined with the detailed rock property values of Poisson's Ratio (Ross & Flack 1992). The calculated acoustic impedance values were used to populate the model's geometric template (Fig. 3) and produce an acoustic impedance section. Formation porosities for the Sherwood Sandstone Group have been estimated using a pure quartz matrix with a density of 2.65 g cm^{-3} (Table 1).

Using GX Technology's forward modelling package GXII the synthetic seismic section (Fig. 4) was generated by convolving a zero-phase Ricker wavelet (peak frequency = 30 Hz) with the modelled reflectivity series. In Fig. 4 the synthetic traces are shown using a wiggle trace variable area normal polarity display to illustrate the stratigraphic boundaries. An acoustic impedance increase down section is represented by a black peak (after Ross & Flack 1992).

The synthetic section is displayed with the fault plane reflections turned off to optimise the amplitude scaling for the horizon reflectors. This procedure is necessary because the largest reflection coefficient (and hence the greatest seismic amplitude) occurs between the water wet Ormskirk Sandstone Formation and the undifferentiated Carboniferous strata across the dipping Formby Point fault (Table 2).

The acoustic impedance parameters for the model were varied to determine their sensitivity to both the timing, shape and amplitude strength of the reflections produced on the synthetic section. The parameters listed in Table 1 produce a model section that has a good visual match to the original seismic data (Fig. 2), both in terms of two-way time (TWT), which relates to the interval velocity component and the reflection amplitude character, which also relates to the acoustic impedance contrasts in the model.

In order to study the effects of gas charge on the seismic properties of the Ormskirk Sandstone Formation reservoir a set of synthetic gathers were generated using GXII for the anticline trap at the foot of the Formby Point fault (Fig. 2). Using the Hampson Russell seismic amplitude versus offset modelling package AVO, the trap was modelled to contain either a gas accumulation or a water filled high porosity zone, set against a background water-wet porosity of 17% for the rest of the reservoir. The modelled synthetic gathers (Fig. 5) were analysed to identify changes with offset of amplitude strength and phase to establish if AVO is a valid technique for differentiating gas filled structures from empty palaeo-traps in this case.

Results

Palaeo-DHI

Figure 4 successfully models the suspected palaeo-DHI located over the Formby Point platform. On the original seismic section (Fig. 2) the anomaly has a 'concave up' time profile. The simple model has replicated this TWT geometry as well as the varying amplitude strength seen along the profile as the boundary passes from Rottington Sandstone Member stratigraphically upwards, into the Calder Sandstone Member and then into the Ormskirk Sandstone Formation, to the east.

The concave shape of this seismic profile reflects the varying travel time within the model as the steeply dipping Ormskirk Sandstone, with its lower velocity (3400 m s^{-1}), is progressively buried to the east by higher velocity Mercia Mudstone Group (4360 m s^{-1}) located in the near surface. The eastward decrease in strength of the palaeo-DHI on the Formby Point platform (Fig. 2) can be explained by changes in reflection coefficient of the anomalous boundary (from 0.0866 to 0.0542 to 0.0328) as it passes up sequence through the separate sandstone units of the Sherwood Sandstone Group (Table 2).

Genuine DHI

Figure 4 shows a simple model that effectively explains the amplitude brightening caused by seismic tuning seen in the Lennox field anticline at the foot of the Formby Point fault (Haig *et al.* this volume). This feature was modelled assuming a gas-filled trap with a horizontal gas–water contact within the Ormskirk Sandstone. The model parameters for a water-filled palaeo-trap also produce a similar, though less pronounced, seismic anomaly which suggests that this anomaly relates to a working trap with a gas fill.

Table 1. *Representative petrophysical parameters input into the model*

Layer	ΔT (μs ft^{-1})	V_p (m s^{-1})	V_p (ft s^{-1})	Density (g cm^{-3})	Poisson's ratio	Acoustic impedance (m/s*g/cm^3)	Porosity (%) (estimated)
Mercia Mudstone Group	70 0	4360	14 300	2.40	0.28	10 464	
Ormskirk Sandstone Formation (gas-charged)	96.0	3180	10 400	2.33	0.12	7409	20
Ormskirk Sandstone Formation (palaeo-trap)	90.0	3400	11 150	2.34	0.30	7956	19
Ormskirk Sandstone Formation (water-wet)	85.0	3600	11 800	2.36	0.30	8496	17
Calder Sandstone Member (palaeo-trap)	80.2	3800	12 500	2.38	0.30	9044	16
Calder Sandstone Member (water-wet)	65.0	4200	13 800	2.40	0.30	10 080	15
Rottington Sandstone Member (palaeo-trap)	74.0	4100	13 450	2.50	0.30	10 250	9
Rottington Sandstone Member (water-wet)	65.0	4690	15 400	2.60	0.30	12 194	3
Manchester Marls Formation	55.4	5500	18 000	2.64	0.28	14 520	
Collyhurst Sandstone Formation	55.4	5500	18 000	2.60	0.30	14 300	3
Carboniferous (undifferentiated)	50.8	6000	19 700	2.60	0.30	15 600	

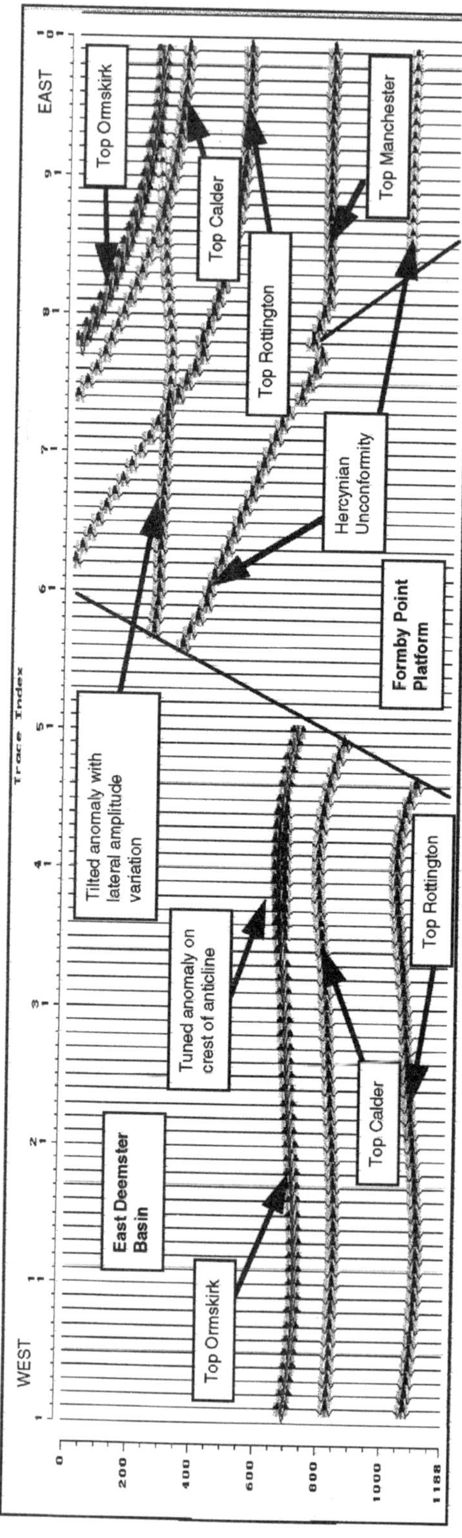

Fig. 4. Model P-wave seismic reflections.

Table 2. *Reflection coefficients for the main model boundaries (both stratigraphic and fault plane)*

Layer Above : Layer below	Mercia Mudstone Group	Ormskirk Sandstone Formation (gas-charged)	Ormskirk Sandstone Formation (palaeo-trap)	Ormskirk Sandstone Formation (water-wet)	Calder Sandstone Member (palaeo-trap)	Calder Sandstone Member (water-wet)	Rottington Sandstone Member (palaeo-trap)	Rottington Sandstone Member (water-wet)	Manchester Marls Formation	Collyhurst Sandstone Formation
Ormskirk Sandstone Formation (gas-charged)	−0.1709									
Ormskirk Sandstone Formation (palaeo-trap)	−0.1362									
Ormskirk Sandstone Formation (water-wet)	−0.1038	0.0683	**0.0328**							
Calder Sandstone Member (palaeo-trap)			0.0640							
Calder Sandstone Member (water-wet)				0.0853	**0.0542**					
Rottington Sandstone Member (palaeo-trap)	−0.0103				0.0625					
Rottington Sandstone Member (water-wet)	0.0764					0.0949	**0.0866**			
Manchester Marls Formation	0.1623							0.0871		
Collyhurst Sandstone Formation					−				−0.0076	
Carboniferous (undifferentiated)	0.1971			0.2948		0.1833		0.1139	0.0359	0.0435

AVO modelling

AVO modelling of the bright spot in the rollover anticline (Fig. 2) produced the following results:

(i) Synthetic gathers for the brine-filled palaeo-trap in the Ormskirk Sandstone show no significant lateral change in top reservoir trough amplitude and no phase rotation of the wavelet with increasing offset angle for the Top Ormskirk reflector (Fig. 5a).

(ii) In contrast, the synthetic gathers for the gas-charged Ormskirk Sandstone demonstrate a lateral change in trough amplitude strength for the Top Ormskirk and a significant associated phase rotation of the wavelet with increasing offset angle (Fig. 5b).

These observations suggest that gas-charged traps can, in principle, be distinguished from palaeo (water-wet) traps for the Ormskirk Sandstone Formation of the EISB. The significant Poisson's Ratio variation observed between a gas-charged reservoir in a present day working trap and enhanced porosity preserved in a breached palaeo-trap allows AVO analysis to be a successful discriminator for the Ormskirk Sandstone reservoir.

Fault modelling

In the seismic model, the Formby Point fault was drawn as a planar boundary dipping at 60° to the horizontal (Fig. 3). Major faults seen in the EISB produce a boundary that may be imaged on a reflection seismic survey. The reflection coefficient data (Table 2) show that the fault plane can produce significant seismic reflections depending on the lithological contrast across the fault. Within the model, the strongest observed fault plane boundary lies between the water-wet Ormskirk Sandstone formation and the Carboniferous (reflection coefficient 0.2948). The weakest boundary modelled (reflection coefficient −0.0103) lies between the Mercia Mudstone Group and the palaeo-gas charged portion of the Rottington Sandstone member (Table 2). The model explains the major variations in fault plane reflection character seen on real seismic data within the EISB.

Conclusions

The seismic anomaly of the Formby Point platform which cuts across the tilted Sherwood Sandstone stratigraphy is a palaeo-DHI. One possible mechanism for this feature is that preserved reservoir porosity exists within the former trap. However, the preserved porosity in the Rottington Sandstone would need to be over 6% higher than in the un-charged part of the former trap (Table 2). This explanation requires a significant and physically unrealistic contrast in matrix density at the boundary of the trap as the seismic amplitude of the palaeo-DHI is very pronounced (Fig. 2). In addition, there is little supporting evidence for such a hypothesis, in that diagenesis is often related in the EISB to permeability loss due to the presence of platy illite (e.g. South Morecambe field) but not to a porosity change in the range suggested.

Other diagenetic explanations for the observed seismic response are possible; e.g. a thin trap-wide tar mat or an extensive layer of preferentially cemented sandstone at the base of the trap. Our preferred model postulates the presence of an iron sulphide precipitate which could provide the necessary density contrast to produce the observed seismic amplitude strength and also peak over trough signature typical of a thin bed (Fig. 2).

The mobility of deep formation waters resulting in the emplacement of diagenetic haematite ore in the Carboniferous limestones of the Furness peninsular is well documented as a feature of the fluid history of the EISB (Rose & Dunham 1977). This movement of iron-rich formation waters could also account for the creation of a layer of iron cemented sandstone at the base of the Formby Point platform trap, prior to the loss of the hydrocarbon fill. Reaction between iron-rich formation waters and trapped hydrocarbons containing a postulated high sulphur content (e.g. up to 400 ppm H_2S for Lennox gas cap) could result in precipitation at a palaeo-OWC of high density minerals, such as pyrite.

The horizontal nature of the depth surface used to model the Formby Point platform palaeo-DHI demonstrates that no significant tilting of the platform has occurred after the trap was breached. Local uplift of the platform along the Formby Point fault in the Tertiary might have resulted in block rotation, so tilting the relic boundary. Absence of tilt suggests that during the regional uplift the relationship between the platform and its associated basins was maintained.

Distinguishing between palaeo and genuine DHI events carries a significant exploration risk in the EISB. The AVO modelling results suggest that this technique can be used to identify gas-charged traps in Triassic sandstone reservoirs. The technique will be limited to locations where the Ormskirk Sandstone can be clearly imaged and the geometry of the overlying Mercia Mudstone group seal allows recovery of sufficient offset information to permit successful modelling of the ray paths.

The authors would like to thank JEBCO Seismic Ltd for permission to use the seismic line displayed in this paper.

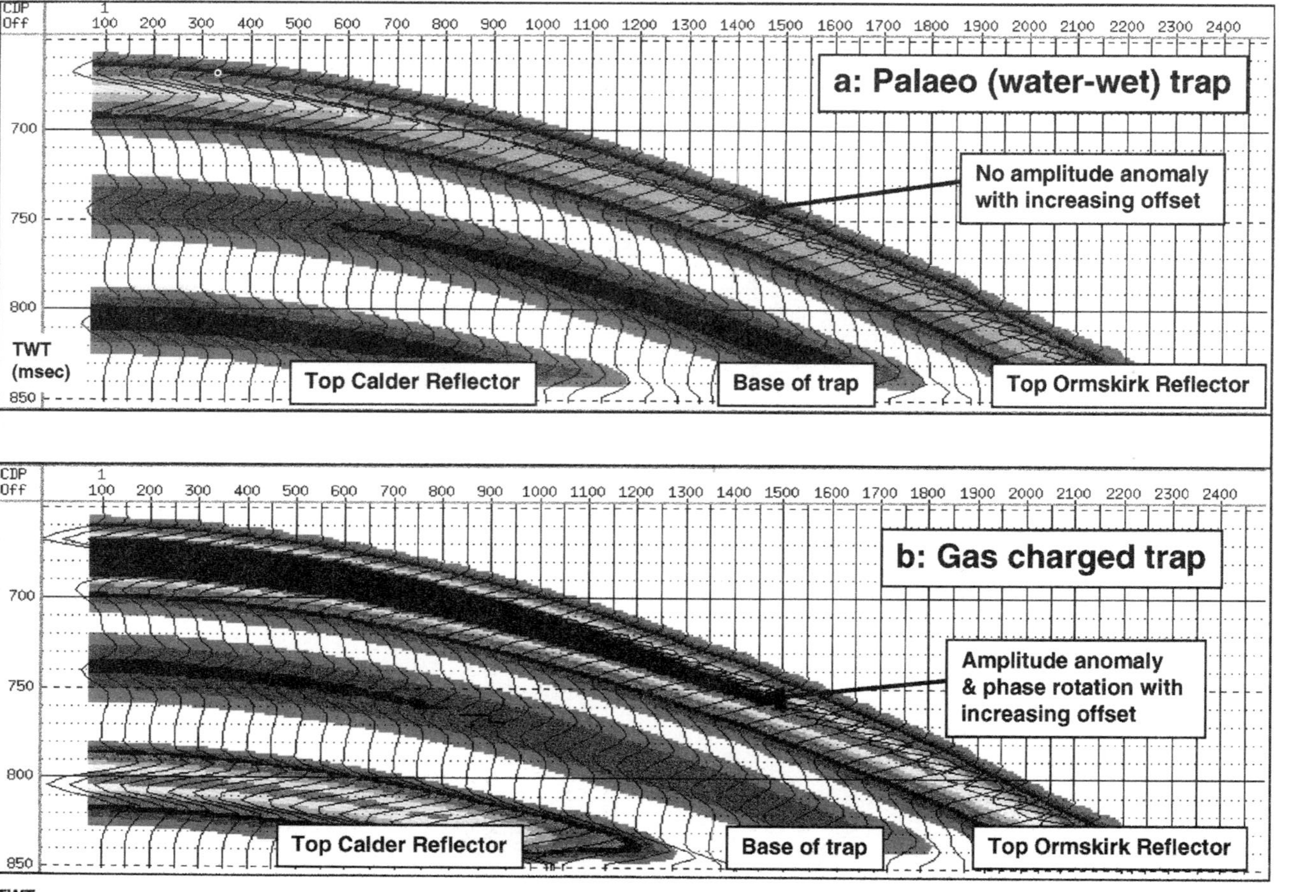

Fig. 5. Synthetic AVO gathers for the Ormskirk Sandstone Formation.

References

ARTER, G. & FAGIN, S. W. 1993. The Fleetwood Dyke and the Tynwald fault zone, Block 113/27, East Irish Sea Basin. *In*: PARKER, J. R. (ed.) *Petroleum Geology of Northwest Europe: Proceedings of the 4th Conference*, Vol. 2. The Geological Society, London, 835–843.

BACKUS, M. M. & CHEN, R. L. 1975. Flat spot exploration. *Geophysical Prospecting*, **23**, 533–577.

BROWN, A. R. 1986. Bright spot validation. *In*: *Interpretation of Three-dimensional Seismic Data*. AAPG Memoir, **42**, 105–127.

BUSHELL, T. P. 1986. Reservoir geology of the Morecambe Field. *In*: BROOKS, J., GOFF, J. C. & VAN HOORN, B. (eds) *Habitat of Palaeozoic Gas in N.W. Europe*, Geological Society, London, Special Publication, **23**, 189–208.

CHESHER, J. A. & JACKSON, D. I. (Compilers) *1994: 1:250,000 East Irish Sea Solid Geology (Special Sheet Edition)*. The British Geological Survey.

COLTER, V. S. 1978. Exploration for gas in the Irish Sea. *In*: VAN LOON, A. J. (ed.) *Key Notes of the MEGS-II (Amsterdam 1978)*. Geologie en Mijnbouw, **57**, 503–516.

DEEGAN, S. E. 1977. *Liverpool Bay Bouguer anomaly map, 1:250,000 series*. Institute of Geological Sciences, Natural Environment Research Council.

FALCON, N. L. & KENT, P. E. 1960. Geological results of petroleum exploration in Britain, 1945–57. *Geological Society, London, Memoir*, **2**, 56 pp.

HAIG, D. B., PICKERING, S. C. & PROBERT, R. 1997. The Lennox oil and gas field. *This volume*.

JACKSON, D. I. & MULHOLLAND, P. 1993. Tectonic and stratigraphic aspects of the East Irish Sea Basin and adjacent areas: contrasts in their post-Carboniferous structural styles. *In*: PARKER, J. R. (ed.) *Petroleum Geology of Northwest Europe: Proceedings of the 4th Conference*, Vol. 2. Geological Society, London, 791–808.

——, JOHNSON, H. & SMITH, N. J. P. 1997. Stratigraphical relationships and a revised lithostratigraphical nomenclature for the Carboniferous, Permian and Triassic rocks of the offshore East Irish Sea Basin. *This volume*.

——, MULHOLLAND, P., JONES, S. M. & WARRINGTON G. 1987. The geological framework of the East Irish Sea Basin. *In*: BROOKS, J. & GLENNIE, K. (eds) *Petroleum Geology of North West Europe*, Vol. 1. Graham and Trotman, 191–203.

KENT, P. E. 1948. A deep borehole at Formby, Lancashire. *Geological Magazine*, **85**, 253–264.

KNIPE, R. J., COWAN, G. & BALENDRAN, V. S. 1993. The tectonic history of the East Irish Sea Basin with reference to The Morecambe Fields. *In*: PARKER, J. R. (ed.) *Petroleum Geology of Northwest Europe: Proceedings of the 4th Conference*, Vol. 2. Geological Society, London, 857–866.

ROSE, W. H. C. & DUNHAM, K. C. 1977. *Geology and Hematite Deposits of South Cumbria*. Memoirs of the Geological Survey of Great Britain, 179 pp.

ROSS, C. P. & FLACK, S. D. 1992. Predictive seismic reservoir modelling in the UK Manx Basin. *Energy Exploration and Exploitation*, **10**, 321–334.

SMITH, D. B., BRUNSTROM, R. G. W., MANNING, P. I., SIMPSON, S. & SHOTTON, F. W. 1974. *A Correlation of Permian Rocks in the British Isles*. Geological Society, London, Special Report, **5**.

WARRINGTON, G., AUDLEY-CHARLES, M. G., ELLIOTT, R. E. *ET AL.* 1980. *A Correlation of Triassic Rocks in the British Isles*. Geological Society, London, Special Report, **13**.

WHITE, R. S. 1977. Seismic Bright Spots in the Gulf of Oman. *Earth and Planetary Science Letters*, **37**, 29–37.

The occurrence and correlation of oils and Namurian source rocks in the Liverpool Bay–North Wales area

JAMES P. ARMSTRONG[1], JANET SMITH[1], VICTOR A. A. D'ELIA[2] &
STEPHEN P. TRUEBLOOD[3]

[1] *James Armstrong and Associates, 30 Linden Walk, Prestatyn,*
Denbighshire LL19 9EB, UK
[2] *Present address: Littleover Community School, Pastures Hill, Littleover,*
Derby DE23 7BD, UK
[3] *BHP Petroleum, Devonshire House, Mayfair Place, Piccadilly,*
London SW1, UK

Abstract: Along the northeast coast of Wales, the Holywell Shale is a significant lithological unit of about 150 m thickness. In the field, the Holywell Shale appears to be lithologically fairly uniform. Geochemical analysis of field samples has demonstrated highly variable features within the Holywell Shale. These pertain to organic richness and hydrocarbon source potential. Some samples exhibit good to rich potential for oil and gas, with extractable hydrocarbons akin to those of a light or medium gravity crude oil. Richer examples also exhibit higher hydrogen indices. These data point to oil-prone sources which have realised some of their potential. By contrast, the other samples taken from the section, whilst not poor in organic carbon, have low S_1 and S_2 and HI signatures. It is anticipated, from the data collected that cyclicity is present within the Holywell Shale sediments.

The isotope geochemical data also highlights another, more gradual variation. Sources towards the base of the Holywell Shale formation have generated hydrocarbons which are isotopically lighter than those from the middle and upper sections of the formation. This is attributed to middle/upper sections being more proximal to land and consequently, having higher proportions of terrestrially derived organic matter.

Two oils (110/13-10 and 110/15-6) from the Douglas and Lennox fields discovered within the Liverpool Bay area have been analysed. Although the two oils are isotopically similar, chromatography analyses reveal that one oil comprises two hydrocarbon phases whereas the other oil has three envelopes. Extracts taken from the Lower Holywell Shale sources and Holywell bitumen are isotopically similar to both oils. Organically rich sources from higher up in the Holywell Shale formation yield more condensate-like hydrocarbons upon extraction and are isotopically heavier than the oils.

Namurian sediments are preserved along the eastern margins of the former County of Clwyd in northeast Wales. The areal occurrence of the sediments is highlighted in Fig. 1. Figure 2 graphically depicts the nature of Namurian sediments in this area. To the south, sequences are condensed and generally arenaceous (Ramsbottom 1974). To the north the Namurian is generally thicker and more argillaceous. This sedimentary facies variation can be attributed to the proximity to St. George's Land (to the south and west of the outcrop area). The northern sequences represent more basinal, distal sedimentary facies.

The generalized stratigraphy of the region with the most commonly used formation names is noted in Fig. 2. The 'lower' Namurian sandstone in Clwyd is commonly referred to as the Cefn-y-fedw Sandstone. This is only developed in the southern parts of Clwyd, sourced from the denudation of St George's land. These sandstones are locally, coarse, gritty and almost universally, quartzitic (Ramsbottom 1969). The transition from the underlying limestones into the arenaceous sequences of the Cefn-y-Fedw Sandstone is best exposed along the Panorama Walk to the north of Llangollen (Somerville 1979). To the north, in the Holywell area the arenaceous facies of the Cefn-y-Fedw are replaced by argillaceous sediments of the Lower Holywell Shale, which rest directly on the Gronant Chert at Pentre Quarry, Gronant (Figs 1 and 2). Continuity between the Holywell Shales and the Gronant Chert cannot be proved and the lowest part

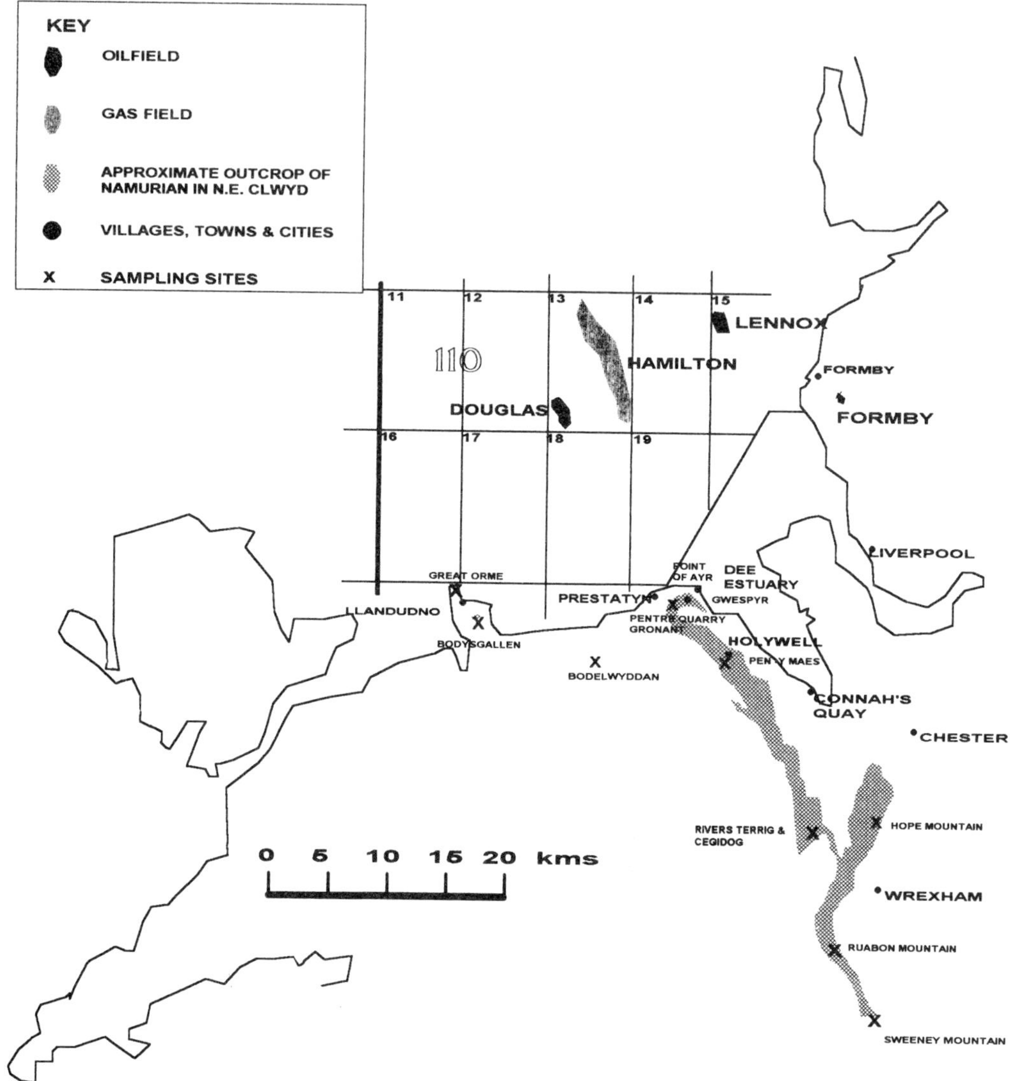

Fig. 1. Map showing the general location of Namurian sediments, oil and gas fields in the study area.

of the Namurian in this area may be condensed (Ramsbottom 1974).

The Holywell Shale is developed between the top of the Pendleian (E_1) marked by the *Cravenoceras malhamense* (E1c) marine band and the lower Yeadonian (G_1) represented by the *Gastrioceras cancellatum* marker (Ramsbottom 1969). Consequently, it represents much of Namurian stratigraphy in areas where it is best developed. It also represents a distal, marine facies (Fig. 3) where the fossil record is dominated by non-benthonic goniatites and bivalves. There are

occasional developments of thin argillaceous limestones which have preserved the remains of benthonic organisms in parts. These appear to be best developed within the *G. cancellatum* zone (i.e. Upper Holywell Shale). This feature is also recorded along the northern shore of St George's Land in Lincolnshire (Ramsbottom 1969).

The Gwespyr Sandstone overlies the Holywell Shale and comprises a thick, fine grained felspathic sequence of sandstones of uppermost Namurian/ Westphalian age (Ramsbottom 1969, 1974). This is well developed along the southern flanks of the Dee

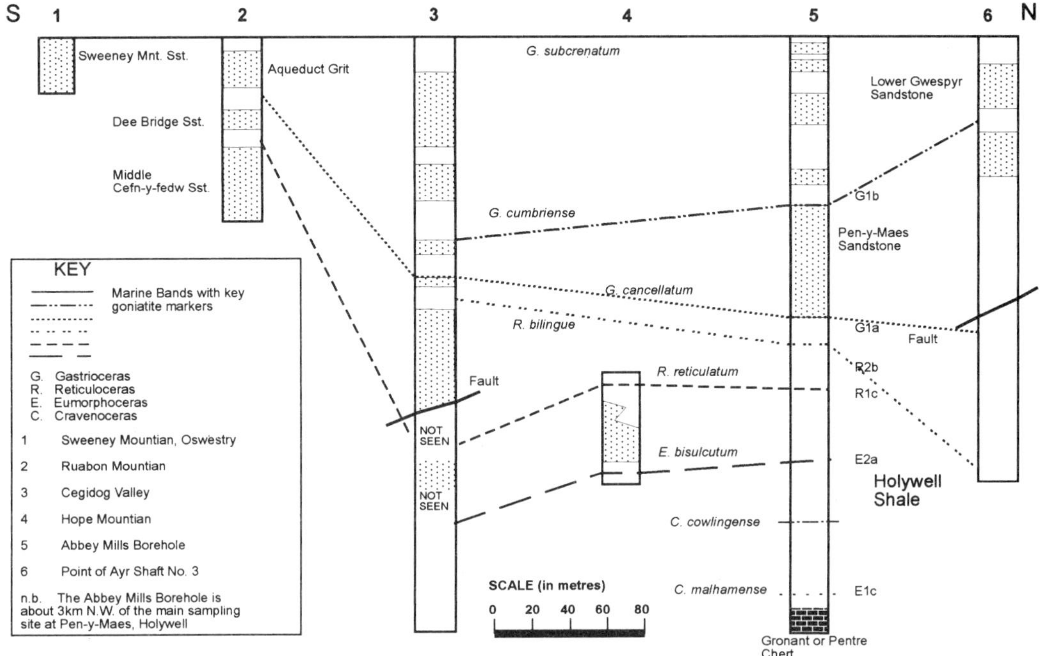

Fig. 2. Comparative sections and goniatite markers for the Namurian of notheast Clwyd (after Ramsbottom).

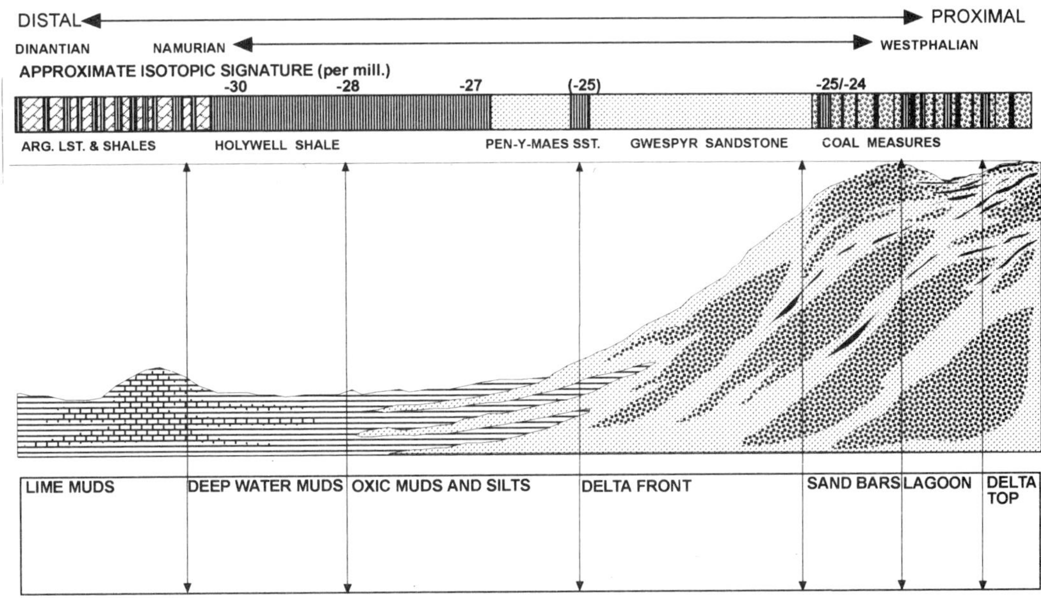

Fig. 3. Time–sedimentation diagram for the Carboniferous in the Holywell area of Clwyd.

Estuary (Fig. 1). To the south, lateral equivalents are referred to as the Aqueduct Grit (Fig. 2) which in turn is considered to be an equivalent of the Rough Rock of the Pennine region (N. Meadows per. comm.). These arenaceous sediments are developed above the *Gastrioceras cumbriense* marine band. In the Holywell area, another thick arenaceous sequence, unnamed in the literature, but referred to here as the Pen-y-Maes Sandstone occurs between the *G. cancellatum* and the *G. cumbriense* markers (Fig. 2). This sandstone is very fine grained, indeed silty in parts and is considered to represent delta front sedimentation (Fig. 4). Its position in the succession would make it a lateral

equivalent of the Lower Haslingden Flags in the Rossendale area of Lancashire.

Further to the west, the occurrence of Namurian age sediments is not conclusive. At Bodelwyddan, some 9 km west of Gronant (Fig. 1), Coal Measures rest directly on Visean carbonates (Warren *et al.* 1984) and the Namurian would appear to be missing. Gritty sandstones which are exposed to the north of Bodysgallen Hall, Llandudno (Fig. 1), are possibly of Namurian age and may represent lateral equivalents of the Cefn-y-Fedw Sandstone (Stratham 1885). Others place these sediments within the Permo-Trias (Williams & Eaton 1993) or even the Dinantian (Warren *et al.* 1984).

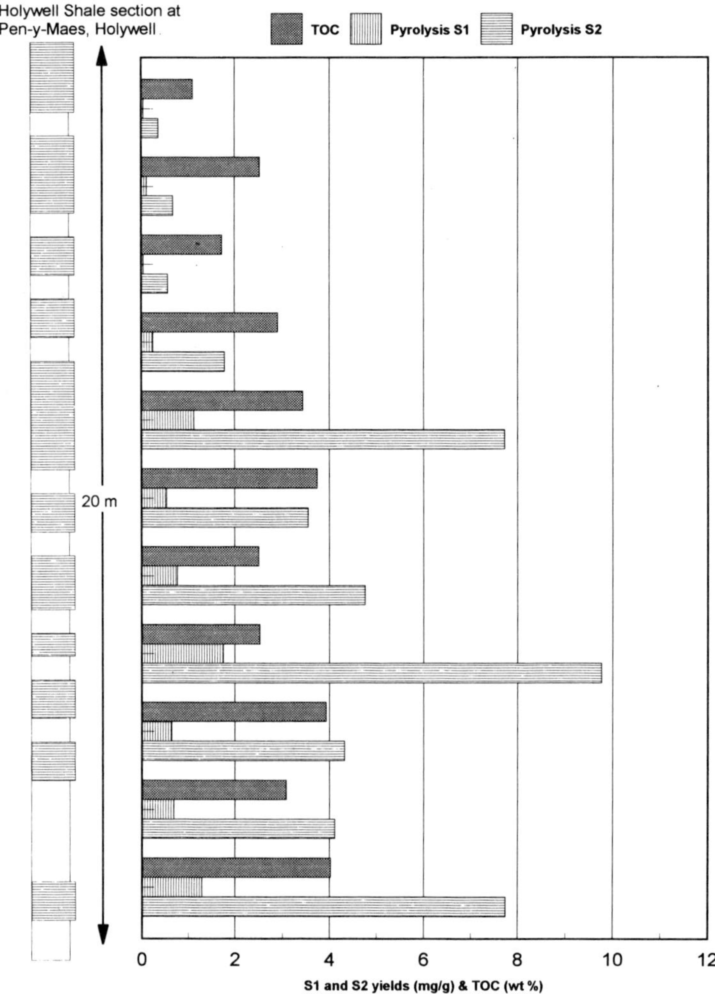

Fig. 4. Total organic carbon (TOC) and pyrolysis yields for the Holywell Shale section at Coed Pen-y-Maes, Holywell, Clwyd.

Source quality and richness

For some time the Holywell Shale has been considered to be a potential source for hydrocarbons (Hardman *et al.* 1993; Armstrong *et al.* 1995). Verification of this assessment has been made by sampling 12 outcrops of Holywell Shale in and around the town of Holywell. These outcrops have allowed analyses to be performed on examples from the lowermost Holywell Shale (i.e. that directly overlying the Gronant Chert) to samples taken from within the *G. cancellatum* zone (Fig. 2). Superficially, the Holywell Shale appears to be an homogeneous dark grey claystone sequence. Closer examination reveals a number of differing lithologies comprising thinly laminated claystones, fissile claystones, massive claystones, argillaceous limestones and argillaceous dolomites. Lithologies near the base of the Holywell Shale tend to be well laminated and occasionally, fissile. These have been sampled at a number of localities (Fig. 1). Sampling of the middle and upper portions of the Holywell Shale was undertaken around the Pen-y-Maes area of Holywell. Here significant outcrops of Holywell Shale and overlying Pen-y-Maes Sandstone may be found in the banks of a

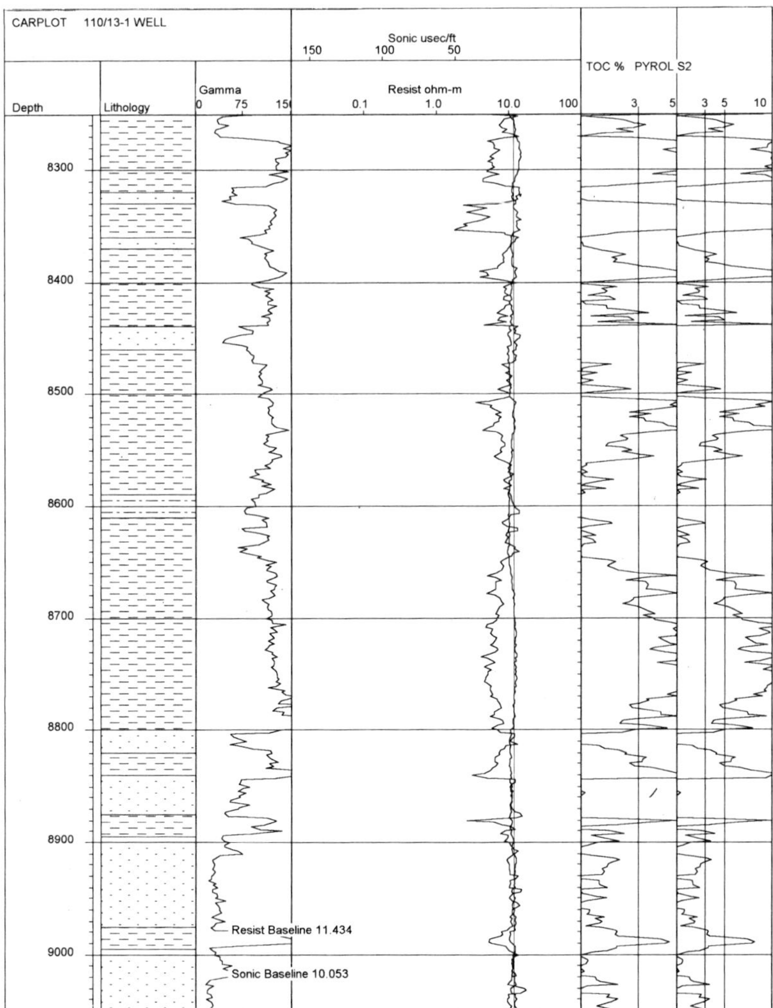

Fig. 5. Analysis of electric logs: indications of variation in source richness of the Namurian section in 110/13-1 well.

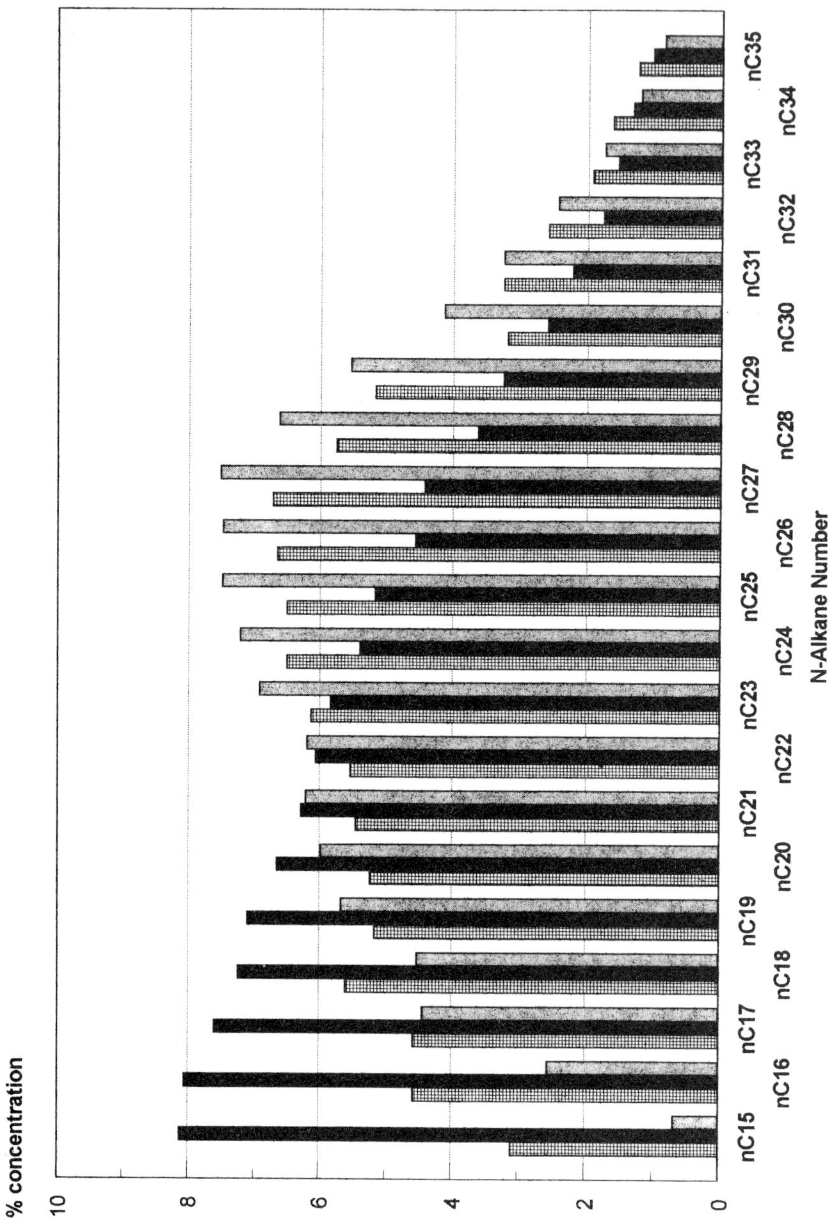

Fig. 6a. Normalized gas chromatograms: oils, bitumen and lower Holywell Shale samples. HS = Holywell Shale.

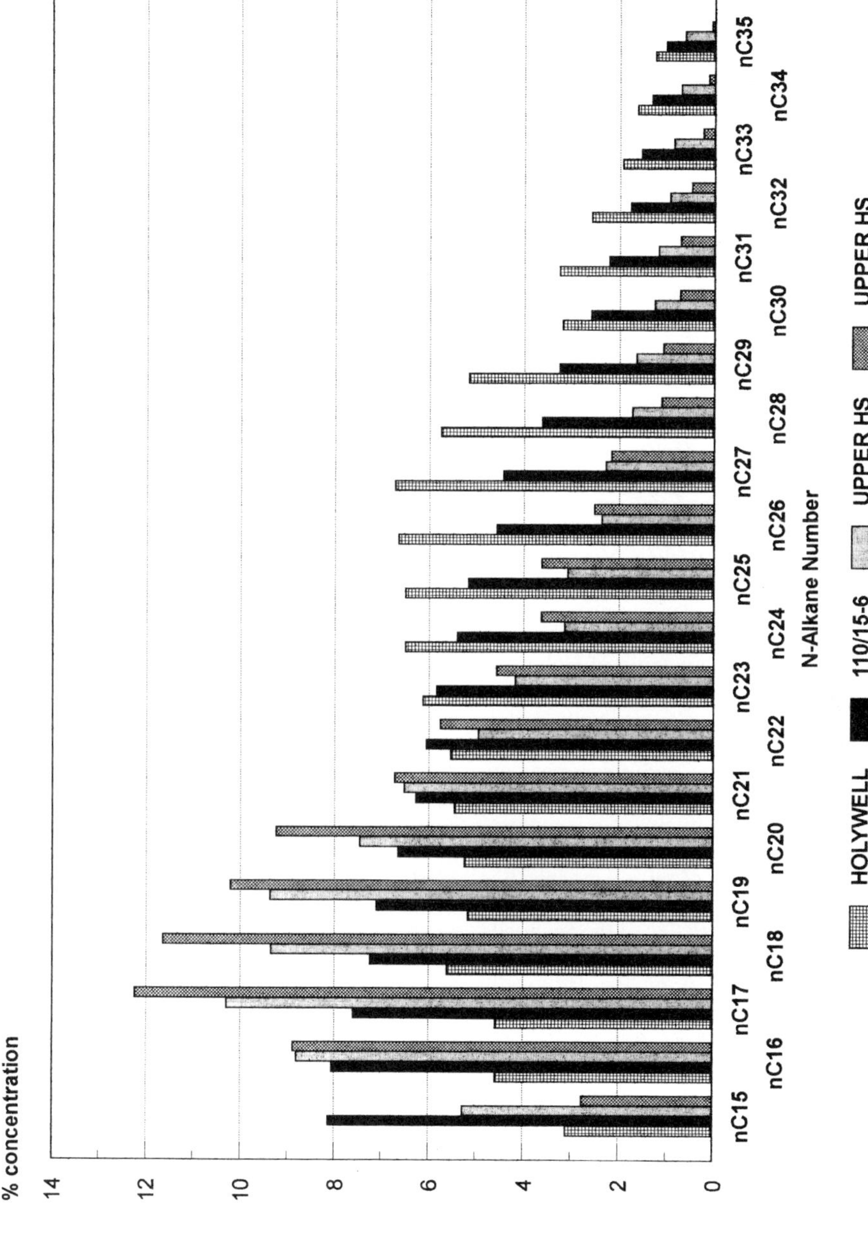

Fig. 6b. Normalized gas chromatograms: oils, bitumen and upper Holywell Shale samples. HS = Holywell Shale.

deeply incised culvert running through Coed Pen-y-Maes. Other outcrops of Holywell Shale also occur immediately to the east of this location. In the Pen-y-Maes area, sampling has concentrated on an exposure of 20 m thickness to the east of Holywell High School. In this area, the Holywell Shale is about 120 m thick, although most outcrops reveal only one or two metres of the sequence.

Total organic carbon and pyrolysis analysis of the Holywell Shale samples reveal variations in richness. Organic carbon contents for the 65 outcrop samples analysed range from less than 0.7% TOC to over 5.0% TOC. The average value for the samples analysed is around 2.1% TOC. Similarly, pyrolysis analyses shows a wide range of values for the S_2 (pyrolysate) parameter. These range between 0.9 kg ton^{-1} and 13.1 kg ton^{-1}, conclusively proving that the Holywell Shale is not organically homogeneous. Lithologies which show the richest concentrations of organic carbon (3% TOC) and better pyrolysate yields (7.0 kg ton^{-1}) are the well laminated more fissile claystones within the Lower Holywell Shale and massive, calcareous rich lenses from the middle and upper Holywell Shale. The latter are frequently associated with marine bands containing high proportions of non-benthonic fauna. Variations in organic carbon contents and pyrolysate yields occur over even a relatively small depth range. This is illustrated by the ranges in such data over the 20 metre interval of the main outcrop of Holywell Shale in Coed Pen-y-Maes as recorded in Fig. 4. This rich-poor variation is considered to be repeated several times over the entirety of the Holywell Shale. Investigation and assessment of the electric log data taken from the Namurian section of the 110/13-1 well by use of a computer program based on the methodology of Passey *et al.* (1990), shows the possibility of several organically rich lenses of up to 30 m thickness and similar variation in organic richness to that of the analysed onshore equivalents (Fig. 5).

The organic heterogeneity of the Holywell Shales is further emphasized by detailed geochemical analyses of the richer sources. Lower Holywell Shale samples exhibit an extensive development of *n*-alkanes for C$_{15}$+ gas chromatograms[1] (Fig. 6a). These range from nC$_{15}$ to nC$_{35}$+, displaying well developed heavier (nC$_{25}$+) *n*-alkanes. Indeed, some gas chromatograms show a waxy distribution of alkanes. By contrast, gas chromatograms of sources from the middle/upper parts of the Holywell Shale show a more limited development of *n*-alkanes, specifically, heavier (nC$_{25}$+) species (Fig. 6b). When compared, the lower Holywell Shale samples also show poor development of isoprenoids pristane and phytane relative to the middle/upper examples. This implies a more restricted supply of oxygen to the environment in which the lower Holywell Shale sediments accumulated.

Perhaps the most variable data obtained from the analyses of Holywell Shale outcrop samples is that of carbon isotope determinations. These range (for saturated hydrocarbon fractions) from −30.0 per mill. in the lower Holywell Shale to −27.0 per mill. for samples taken from within the *G. cancellatum* zone (middle/upper Holywell Shale) at Pen-y-Maes (Fig. 7). These changes are considered to reflect shifts within the environmental settings for sediment accumulation as depicted in Fig. 3. More distal settings result in isotopically light (more negative) hydrocarbons. Middle/upper Holywell Shale sediments, specifically, those within the *G. cancellatum* zone, are more proximal. This is recognized by the development of accumulations of woody debris and less negative, heavier isotopic signatures. This progression is also evident from the analyses of 'Holywell Shale' from the Hope Mountain area (Figs 1 & 2) undertaken by Hamilton Oil (now BHP Petroleum). Here, shales which are equivalent to the lower Holywell Shale of the Holywell area (*Eumorphoceras bisulcatum* zone) have isotopic signatures of circa −25.5 per mill. for their saturated fractions. These shales were deposited in more proximal settings to their lateral equivalents in the Holywell areas. Hence, their considerably heavier isotopic values.

In addition, analysis of coals from the Westphalian which have been undertaken in previous proprietary studies, exhibit isotopically heavy extracts of −25.0 to −24.0 per mill. These sources have yet to be correlated with known accumulations of hydrocarbons.

With the exception of the extremely rich and oil-prone nature of Westphalian cannel coals, the best oil-prone facies in the study area are represented by organically richer elements of the lower Holywell Shale, probably from below the *E. bisulcatum* marker. Some elements of the middle/upper Holywell Shale are also rich and oil-prone. These will, at optimum maturity produce hydrocarbons of a different nature and differing isotopic signature to those of the lower Holywell Shale. The Holywell Shale from eastern Clwyd is heterogeneous in terms of richness, potential product and isotopic signature. To that extent it is probably similar to Yeadonian claystones of South Wales (Bloxham & Thomas, 1969).

---

[1] Normalized gas chromatograms are used in these illustrations as these data have been collected over a period of time using several laboratories and slightly differing equipment. Analytical conditions remain the same, but the nature of the output chromatograms differ. Hence, these have been normalized.

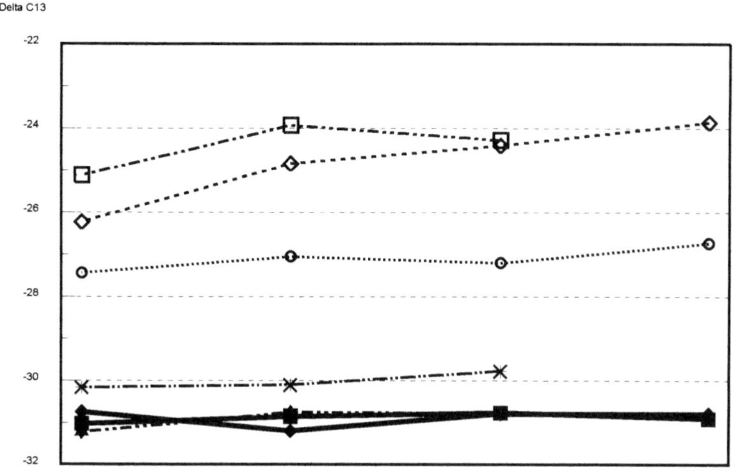

Fraction		Saturates	Aromatics	NSOs	Asphaltenes
110/13-10 Oil	■	-31.02	-30.85	-30.76	-30.90
110/15-6 Oil	◆	-30.74	-31.19	-30.78	-30.79
Holywell Bitumen	✱	-31.22	-30.75	-30.77	-30.85
Gt Orme Bitumen (biodegraded)	☐	-25.12	-23.93	-24.28	
Cannel Coal Point of Ayr	◇	-26.23	-24.84	-24.41	-23.86
Lower HS sample 1	✕	-30.16	-30.10	-29.77	
Upper HS sample 1	○	-27.44	-27.05	-27.20	-26.73

Fig. 7. Carbon isotope determinations: oils, bitumen and potential source rock horizons in northeast Wales.

Occurrence and nature of hydrocarbons

Several wells in blocks 110/13 and 110/15 have proved the presence of oil and gas in commercial quantities. This study concentrates on the oils recovered from 110/13 (Douglas) and 110/15 (Lennox). In addition, bitumen samples recovered from the area around St. Winefride's well, Holywell and the Great Orme, Llandudno are also included in this discussion. The former is remarkable in its lack of biodegradation (Fig. 6) and preservation. The Great Orme (Fig. 1) bitumen by contrast, is heavily altered and no normalisation of its gas chromatogram is possible.

Whole oil gas chromatograms (Fig. 8) of the two oil samples exhibit the influence of more than one expulsion pulse. In the case of the 110/13-10 oil, high proportions of waxy n-alkanes are noted in association with a 'trimodal' paraffin envelope. This oil possesses a dominant light end (nC_4 to nC_7) envelope, with a depletion in the nC_8 to nC_{11} range. The next main preference range covers nC_{12} to nC_{16} alkanes, followed by a depletion at carbon numbers 17 and 18. The final envelope comprises a waxy distribution in the nC_{20} to nC_{31} range (see also Fig. 6).

The 110/15-6 oil also possesses a similar front end hydrocarbon preference to the 110/13-10 crude. It does not display the waxy nC_{20} to nC_{31} envelope (Fig. 8), instead a smoother, concave paraffin envelope is present. The distribution of n-alkanes in this sample suggests two phases of generation.

The possible phases of generation linked to these oils are summarized below:

- Waxy, heavy hydrocarbons phase (nC25 to nC35).
- Medium molecular weight generative phase (nC15 to nC25).
- Mature condensate phase giving rise to abundant gasoline range (nC_4 to nC_7) hydrocarbons.

As already stated, the 110/15-6 crude has a differing distribution of n-alkanes compared with the 110/13-10 oil. Initially, this was interpreted to be due to either differing sources for the crudes or maturation differences between the two oils. Further investigation (see below) has dispensed with both of these theories. The occurrence of waxy hydrocarbons now appears to be a localized feature associated with the lower Holywell Shale. Such hydrocarbons also appear to be generated at similar

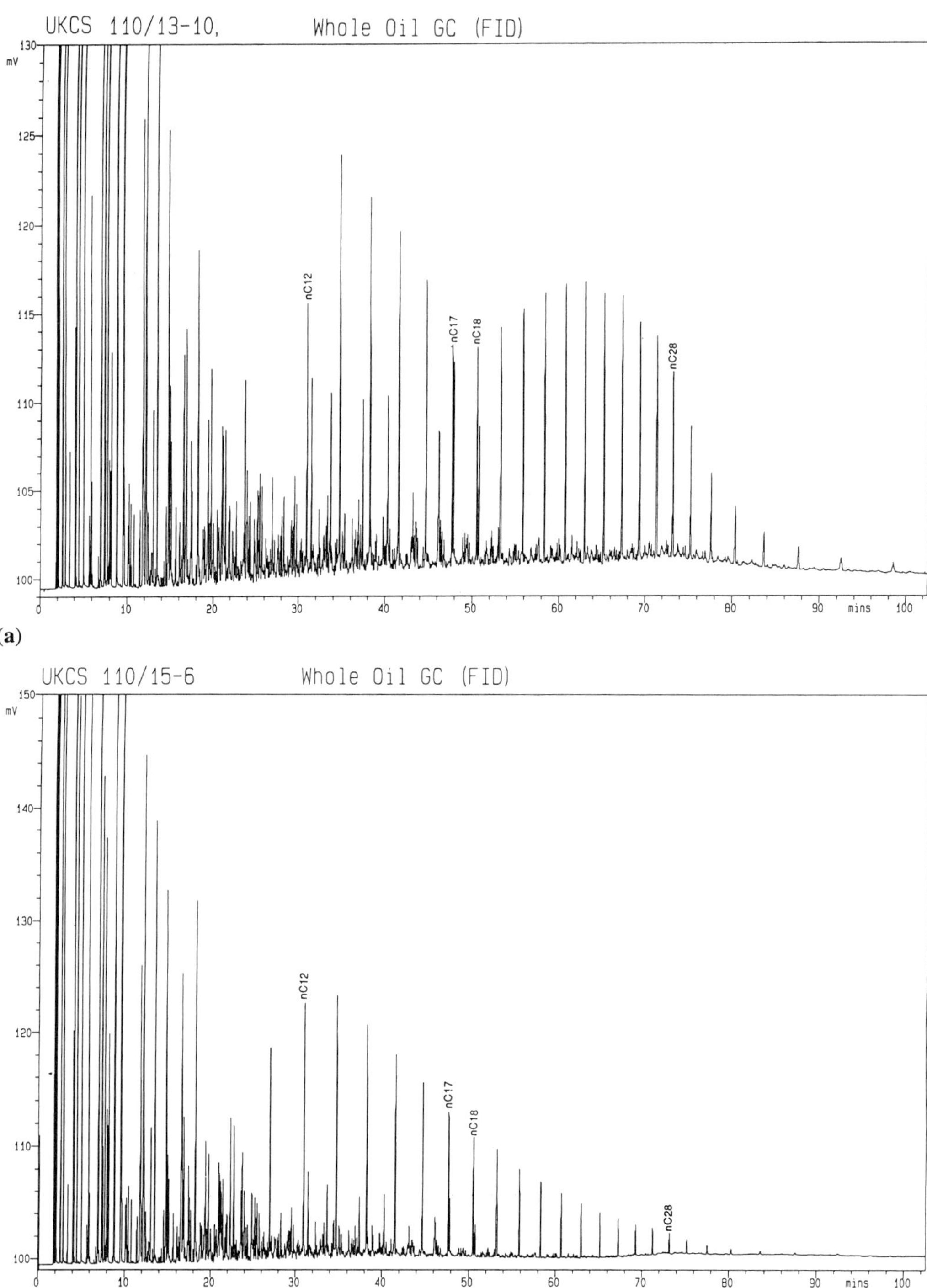

(a)

(b)

Fig. 8. Whole oil gas chromatograms. **(a)** 110/13-10 oil, **(b)** 110/15-6 oil.

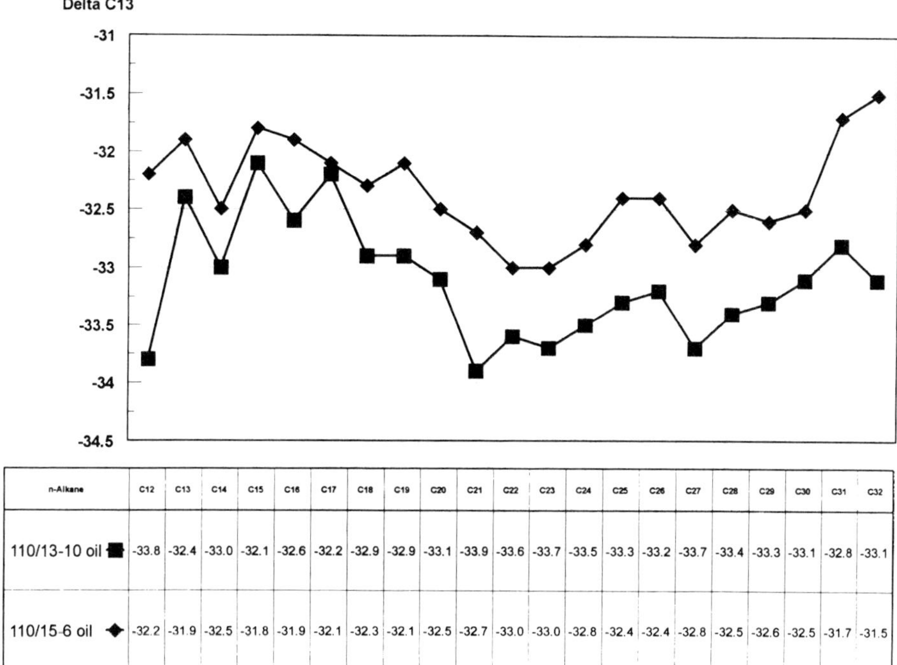

n-Alkane	C12	C13	C14	C15	C16	C17	C18	C19	C20	C21	C22	C23	C24	C25	C26	C27	C28	C29	C30	C31	C32
110/13-10 oil ■	-33.8	-32.4	-33.0	-32.1	-32.6	-32.2	-32.9	-32.9	-33.1	-33.9	-33.6	-33.7	-33.5	-33.3	-33.2	-33.7	-33.4	-33.3	-33.1	-32.8	-33.1
110/15-6 oil ◆	-32.2	-31.9	-32.5	-31.8	-31.9	-32.1	-32.3	-32.1	-32.5	-32.7	-33.0	-33.0	-32.8	-32.4	-32.4	-32.8	-32.5	-32.6	-32.5	-31.7	-31.5

(a)

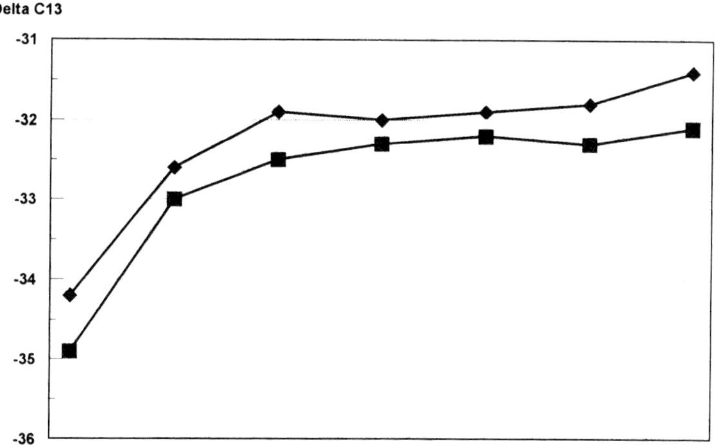

iso-alkane	iC13	iC14	iC15	iC16	iC18	Pristane	Phytane
110/13-10 oil ■	-34.9	-33.0	-32.5	-32.3	-32.2	-32.3	-32.1
110/15-6 oil ◆	-34.2	-32.6	-31.9	-32.0	-31.9	-31.8	-31.4

(b)

Fig. 9. Isotopic profile from GC-IRMS analysis. (**a**) 110/13-10 and 110/15-6 oils (*n*-alkanes), (**b**) 110/13-10 and 110/15-6 oils (isoprenoids).

(a)

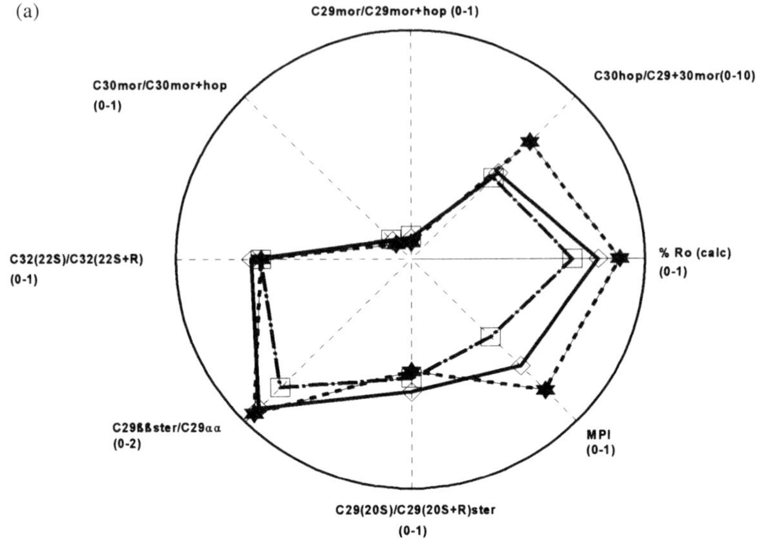

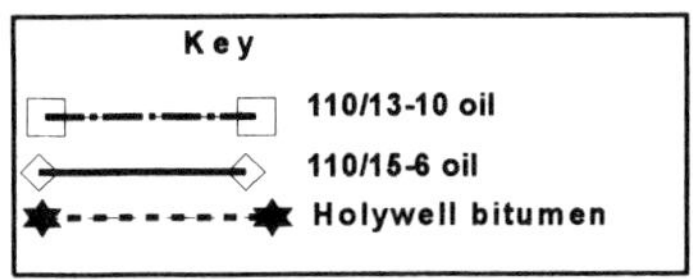

Ratio	*Meaning*
C29mor/C29mor+hop (0–1)	Ratio of C_{29} moretane to C_{29} moretane and C_{29} hopane (scale of 0 to 1), from *m/z* 191 mass fragmentogram.
C30hop/C29mor+C30mor (0–10)	Ratio of C_{29} hopane to C_{29} moretane and C_{30} moretane (scale of 0 to 10), from *m/z* 191 mass fragmentogram.
% Ro (calc) (0–1)	Equivalent % Ro vitrinite reflectance calculated from various biomarker ratios (scale 0 to 1.0% Ro).
MPI (0–1)	Methylphenathrene Index calculated from appropriate biomarker ratios (scale 0 to 1).
C29(20S)/C29(20S+R)ster (0–1)	Ratio of C_{29} (20S epimer) sterane to C_{29} (20S and 20R epimer) steranes (scale 0 to 1) from *m/z* 217 mass fragmentogram.
C29ββster/C29αα (0–2)	Ratio of $C_{29}\beta\beta$ steranes to $C_{29}\alpha\alpha$ steranes (scale 0 to 2) from *m/z* 217 mass fragmentogram.
C32(22S)/C32(22S+R) (0–1)	Ratio of C_{32} (22S epimer) hopane to C_{32} (20S and 20R epimer) hopanes (scale 0 to 1) from *m/z* 191 mass fragmentogram.
C30mor/C30mor+hop (0–1)	Ratio of C_{30} moretane to C_{30} moretane and C_{30} hopane (scale of 0 to 1), from *m/z* 191 mass fragmentogram.

Fig. 10(a)

maturation levels to medium molecular weight products.

In addition to the two whole oil gas chromatograms (Fig. 8), two saturated gas chromatograms (Fig. 6) are presented to facilitate direct comparison with bitumen and source rock samples which have been solvent extracted prior to analysis, hence only nC_{15}+ are represented.

The Extractable Organic Matter (EOM) gas chromatogram of the Holywell bitumen is biased towards waxy end members, the dominant *n*-alkane envelope being in the nC_{24} to nC_{34} range. There is only minor alteration of this crude and its overall *n*-alkane distribution is akin to that of the waxy phase of the 110/15-10 oil. By contrast, the sample taken from a bitumen stain on the Great Orme is severely biodegraded. As both are present in association with near surface occurrences of mineralization, the biodegradation at Great Orme is not unexpected. The level of preservation of *n*-alkanes within the Holywell bitumen is most remarkable. We have no explanation of this, save

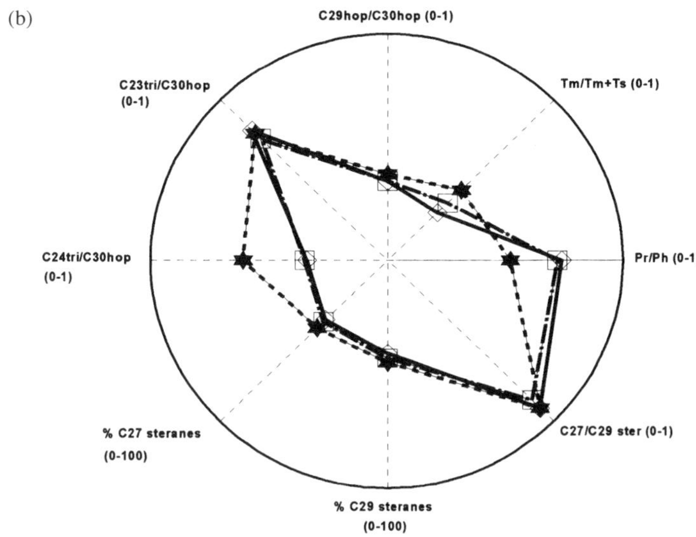

Ratio	Meaning
C29hop/C30hop (0–1)	Ratio of C_{29} hopane to C_{30} hopane (scale of 0 to 1), from m/z 191 mass fragmentogram.
Tm/Tm+Ts (0–1)	Ratio of $17\alpha(H)$ trisnorhopane (Tm) to $17\alpha(H)$ trisnorhopane and $18\alpha(H)$ trisnorneohopane (scale of 0 to 1), from m/z 191 mass fragmentogram.
Pr/Ph (0–1)	Pristane to Phytane ratio (scale 0 to 1) from gas chromatography data.
C27/C29 ster (0–1)	Ratio of C_{27} to C_{29} steranes (scale 0 to 1), from m/z 217 mass fragmentogram.
% C29 steranes (0–100)	Percentage of C_{29} steranes to Total sterane (scale 0 to 100), from m/z 217 mass fragmentogram.
% C27 steranes (0–100)	Percentage of C_{27} steranes to Total sterane (scale 0 to 100), from m/z 217 mass fragmentogram.
C24tri/C30hop (0–1)	Ratio of C_{24} tricyclic terpane to C_{30} hopane (scale 0 to 1), from m/z 191 mass fragmentogram.
C23tri/C30hop (0–1)	Ratio of C_{23} tricyclic terpane to C_{30} hopane (scale 0 to 1), from m/z 191 mass fragmentogram.

Fig. 10. Spider diagram showing comparison of (**a**) maturity-related and (**b**) source-related biomarker ratios between 110/13-10, 110/15-6 oils, Holywell bitumen.

for the obvious proximity of the sample to St Winefride's Well, the 'Lourdes of Wales', a sight of fabled safe keeping and preservation.

Carbon isotope data (Fig. 7) show the close association between the two crudes and the Holywell bitumen. For example, for saturated hydrocarbon fractions these have a narrow range of −30.74 to −31.22 per mill. The Great Orme bitumen is isotopically heavy at *c.* −25.0 per mill for the various fractions analysed. This may result from a differing source facies, for example a localized claystone source within the Carboniferous limestone sequence. Alternatively, this may reflect the effects of biodegradation resulting in a very heavy isotopic signature.

Analysis of the two oils (110/13-10 and 110/15-6) by GC–IRMS reveal generally similar profiles in the nC_{11}–nC_{32} range (Fig. 9). The 110/15-6 oil appears to be slightly heavier (i.e. less negative). This most likely is a function of maturity.

GC–MS biomarkers (Fig. 10) also show close correlation between the three samples. The m/z191, mass fragmentograms are characterized by the presence of strong tricyclic terpanes with a C_{23} bias, indicating a marine influence (Aquino Neto *et al.* 1983). Appreciable quantities of C_{21} tricyclic terpane are also present. Evidence for the input of terrestrially derived organic matter is provided by the strong C_{31} and C_{32} homohopanes (Quirk *et al.* 1984) and the presence of C_{24}, C_{25}, and C_{26} tetracyclic terpanes (Philp & Gilbert, 1986).

Sterane distributions favour C_{27} species although there are significant amounts of C_{29} compounds present. Overall, these data point to a marine source with a terrestrial overprint. In addition biomarker data give an indication of the maturity of the source(s) at the time of expulsion. Various maturity controlled ratios are displayed in a spider diagram (Fig. 10a). These combine to illustrate that the maturity levels for the two oils and the Holywell bitumen are only slightly different. The waxy products characterized by the heavy end envelope of the 110/13-10 oil and the Holywell bitumen show similar maturation characteristics to the 110/15-6 oil. The latter is probably a slightly more mature product. It appears that hydrocarbons of the heavy waxy or medium molecular ranges are all produced at maturation levels of approximately 0.75–0.85% Ro. This level of maturation, therefore defines the expulsion window for the first phase of hydrocarbon generation in the Liverpool Bay area.

The condensate phase identified in the whole oil gas chromatograms is considered to be the product of a second phase of expulsion at a higher maturation level.

Oil source correlation

An oil to source correlation appraisal has been undertaken principally through comparison of gas chromatography, carbon isotope and GC–MS biomarker data.

As noted above, gas chromatograms pertaining to richer source horizons within the lower Holywell Shale (Fig. 6a) have strong distributions of higher molecular weight *n*-alkanes. Indeed, localized examples are associated with the *in situ* generation of rather waxy products. These have accumulated locally as the Holywell bitumen (Fig. 6a). Lower Holywell Shale samples showing waxy hydrocarbon generation and the site of the Holywell bitumen are about 1 km apart. Moreover, when compared, the generated products of the middle and upper Holywell Shales are deficient in heavier hydrocarbons (i.e. $nC_{25}+$ species) which are characteristic of the oils and bitumen. Indeed, the products of the middle/upper Holywell Shale are best described as light oils/condensates. Less rich sections within the middle/upper Holywell Shale have generated only condensate-like hydrocarbons (Fig. 6b).

Analysed Holywell Shale samples have maturities of between 0.6% Ro and 0.7% Ro vitrinite reflectance. Consequently, some generation has taken place *in situ*. Expulsion in the Holywell area manifests itself as the Holywell bitumen. As noted above, this has a strong *n*-alkane 'finger print' similar to examples of lower Holywell Shale.

Carbon isotope data amplify the correlation between examples of Lower Holywell Shale, Douglas and Lennox oils and the Holywell bitumen (Fig. 7). These claystones are in some cases about

1 per mill heavier than the oils and bitumen (–29.8 to –30.2 per mill. compared with –30.7 to –31.2 per mill.). Other, lower Holywell Shale sources are isotopically heavier at –28.5 to –29.0 per mill. These are probably too heavy to be considered as sources of the reservoired hydrocarbons. The middle and upper Holywell Shale examples are even heavier at –26.5 to –27.5 per mill. (Figs 3 & 7). These too, are improbable sources of the oils and bitumen. Heavier still is the highly oil-prone Westphalian 'A' cannel coal from Point of Ayr Colliery (Fig. 7). This has a similar, heavy isotopic signature to the Great Orme bitumen. The latter is however, a highly altered product. Biodegradation may have substantially altered the isotopic signature of this sample.

Spider diagrams of specific parameters (Fig. 10) from the GC–MS biomarker data sets for the oils and bitumens have already been used to illustrate the very close links between these three samples in terms of source character and maturity. Figure 11a illustrates the high level of correlation between the oil data and similar data pertaining to lower Holywell Shale samples. These source facies controlled data do show a good degree of correlation which is not evident from other source rock analysed (Fig. 11b). Middle and upper Holywell Shale samples exhibit clear differences in concentrations of isoprenoid pristane and C_{27}, C_{29} sterane abundances (Fig. 11b). These claystones are considered to have formed in more oxic settings. In the middle/upper Holywell Shale, better quality sources are associated with lime-rich sediments.

The cannel coal, although highly oil-prone, shows no apparent correlation in terms of source with reservoired hydrocarbons (Fig. 11b).

Conclusions

Analysis of the lower portions of the Holywell Shale indicate good correlation in terms of hydrocarbon product, isotopic signature and source facies with the reservoired hydrocarbons of the Douglas and Lennox fields. These commercial accumulations of hydrocarbons show multiple phases of generation and it is the isotopically heavier, oily phases which show this strong correlation with elements of the Lower Holywell Shale. There is also very good correlation between the Douglas oil and the Holywell bitumen which is present in mineral veins at localities close to the outcrops of analysed Holywell Shale.

The middle and upper sections of the Holywell Shale do not have strong potential for the generation of heavier products. These sources have good to rich potential for light oil and condensate. Moreover, middle and upper Holywell Shales sediments have a differing isotopic signature to the

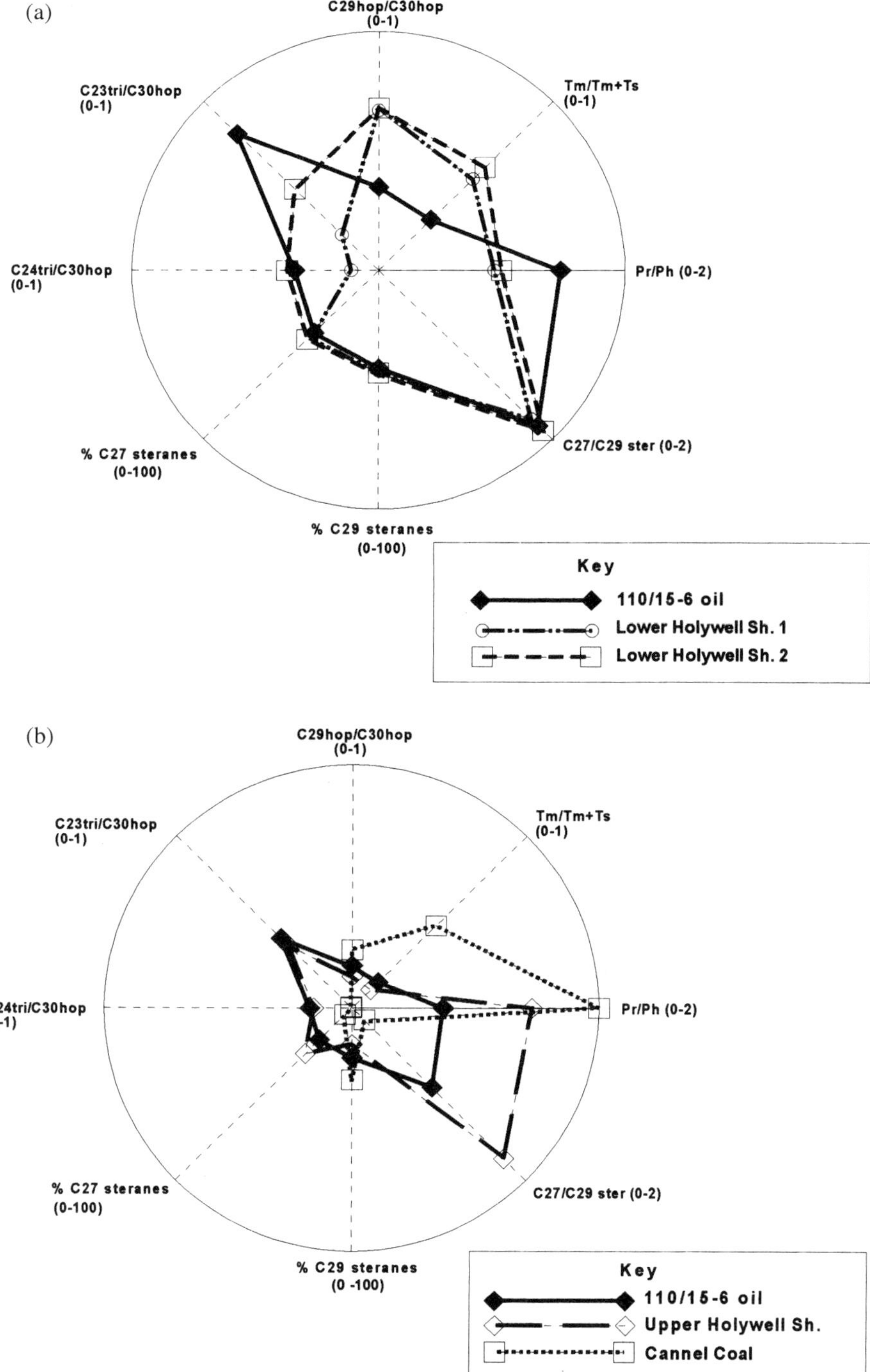

Fig. 11. Spider diagram showing comparison of source related biomarker ratios between oils, bitumen and (**a**) lower and (**b**) upper Holywell Shale. See Fig. 10 for abbreviation of ratios.

Lower Holywell Shale and consequently, the heavier phases reservoired hydrocarbons.

It is possible that all of the Holywell Shale was responsible for the second major generative pulse of condensate/light oil which is present in both the Douglas and Lennox accumulations. This is considered to represent a higher maturity product. The oil phase was expelled at a maturation level of about 0.8% Ro, while the condensate phase was at levels in excess of 1.0% Ro.

The very rich and oil-prone Westphalian 'A' cannel coals which are common in northeast Clwyd are associated with hydrocarbon products which differ markedly from the Douglas and Lennox oils. There is no correlation between cannel coal and the oils.

The bitumen found on the Great Ormes Head, Llandudno is isotopically heavy but highly biodegraded. Its geochemical characteristics have probably been greatly modified by its alteration. Consequently, it is not possible to attempt a correlation between this hydrocarbon occurrence and other samples examined during the course of this study.

This paper could not have been produced without the benevolent assistance of BHP Petroleum and the laboratory personnel at Geolab UK and Geolab Nor in performing analysis of rock and oil samples. We are also indebted to Ian Carr, Nick Miles and Brian Pim who assisted in the collection of field samples. The assessment of electric logs, utilizing the CARPLOT Programme was made courtesy of Mike Sauer. The manuscript was typed by Vivienne Armstrong who also assisted in the creation of many of the figures. Finally, the authors wish to thank Kevin Dale (Marathon), Neil Meadows (Geochem) and an anonymous referee who thoroughly reviewed the manuscript, thus significantly improving its quality.

Appendix – analytical techniques

Total organic carbon (TOC) and total carbon analysis

This analysis is performed using a LECO CS125 Carbon Analyser.

Hand-picked lithologies from cuttings samples are crushed with a mortar and pestle and approximately 200 mg (50 mg for coals) are accurately weighed into LECO crucibles. The samples are then treated four times with hot 10% hydrochloric acid too remove oxidized (carbonate) carbon, and washed four times with distilled water. The samples are dried in an oven at 60°C before analysis of total organic carbon. Total carbon is also analysed on the same instrument using approximately 200 mg of untreated crushed whole rock. Oxidized (carbonate) carbon is calculated by weight difference.

Rock-Eval pyrolysis

This analysis is performed by using a Rock-Eval II Pyrolyser. Approximately 100 mg crushed whole rock is analysed. The sample is first heated at 300°C for three min in an atmosphere of helium to release the free hydrocarbons present (S_1 peak) and then pyrolysed by increasing the temperature from 300°C to 550°C (temp. Gradient 25°C min^{-1}) (S_2 peak). Both the S_1 and S_2 yields are measured using a flame ionization detector (FID). In the temperature interval between 300 and 390°C, the released gases are split and a proportion passed through a carbon dioxide trap, which is connected to a thermal conductivity detector (TCD). The value obtained from the TCD corresponds to the amount of oxygen contained in the kerogen of the sample and is reported as the S_3 peak.

Solvent extraction of organic matter (EOM)

The samples are extracted using a Tecator Soxtec HT-System. Carefully weighed samples are taken in a pre-extracted thimble. Some activated copper is added to the extraction cup and dichloromethane is used as an extraction solvent. The samples are boiled for one hour and then rinsed for two hours. If the samples contain more than 10% TOC, then the whole procedure is repeated once. The resulting solution is filtered and the solvent removed by rotary evaporation (200 mb, 30°C). The amount of EOM is gravimetrically established.

Removal of asphaltenes

Asphaltenes are removed from the EOM by precipitation in *n*-pentane. *n*-pentane is added to the EOM and the solution is then stored in the dark and at ambient temperature for at least eight hours. The solution is then filtered (Baker 10-spe system) and the precipitated asphaltenes dissolved in dichloromethane are returned to the original flask. The solvent is removed by rotary evaporation (200 mb and 30°C).

Chromatographic separation of deashpaltened EOM

Chromatographic separation is performed using a MKW-16 MPLC system. The EOM (minus asphaltenes) is injected into the MPLC and separated using hexane as an eluent. The saturated and aromatic hydrocarbon fractions are collected and the solvent removed using a rotary evaporator at 30°C. The fractions are then transferred to small pre-weighed vials and evaporated to dryness in a stream of nitrogen. The vials are re-weighed to obtain the weights of both the saturated and the aromatic fractions. The weight of the NSO fraction which is retained on the column, is obtained by weight difference.

Gas chromatographic analyses

Gas chromatographic analyses were performed on a number of differing instruments and by different laboratories over a three year period. The main analyses were performed under the following conditions:

Saturated hydrocarbon fractions. The instrument used for this analysis is a Hewlett Packard 5890 series II Gas Chromatograph equipped with an FID detector and an OV1 column. The carrier gas is helium and the temperature program runs from 80 to 300°C at a rate of

4°C min^{-1}. Final hold time is 20 min. The saturated hydrocarbon fraction is diluted by 1:30 and a 1 µl aliquot of this is injected into the instrument.

Whole oil/whole extract. Whole oil chromatograms are determined on a Hewlett Packard 5890A Gas Chromatograph fitted with a split injector, 40 m SE 543 capillary column and effluent splitter connected to FID and FPD detectors allowing simultaneous determination of hydrocarbons and sulphur compounds. Approximately 0.1 µl of whole oil are injected and the temperature program on the chromatograph runs from −10 to 300°C at 4°C min^{-1}.

Combined gas chromatography–mass spectrometry (GC–MS)

The GC–MS analyses are performed on a VG TS250 system interfaced to a Hewlett Packard 5890 gas chromatograph. The GC is fitted with a fused silica SE54 capillary column (40 m × 0.22 mm i.d.) directly into the ion source. Helium (12 psi) is used as carrier gas and the injections are performed in splitless mode. The GC oven is programmed from 45 to 150°C at 35°C min^{-1}, at which point the programme rate is 2°C min^{-1} up to 310°C where the column is held isothermally for 15 min. For the aromatic hydrocarbons, the GC oven is programmed from 50 to 310°C at 5°C min^{-1} and held isothermally at 310°C for 15 min. The mass spectrometer is operated in electron impact (EI) mode at 70 eV electron energy, a trap current of 500 µA and a source temperature of 220°C. The instrument resolution used is 1500 (10% value).

The data system used is a VG PDP11/73 for acquiring data, and a VAX station 3100 for peak processing the data. The samples are analysed in multiple ion detection mode (MID) at a scan cycle tine of approximately 1.1 s.

Calculation of peak ratios is performed from peak heights in the appropriate mass fragmentograms.

Saturated fractions

Terpanes. The most commonly used fragment ions for detection of terpanes are m/z 163 for detection of 25, 28, 30 trisnormoretane or 25, 28, 30 trisnorhopane, m/z 177 for detection of demethylated hopanes or moretanes, m/z 191 for detection of tricyclic, tetracyclic and pentacyclic terpanes and m/z 205 for methylated hopanes or more-tanes., The molecular ions m/z 370 and 384 are also recorded for identification of C_{27} and C_{28} triterpanes, respectively.

Steranes. The most commonly used fragment ions for detection of steranes are m/z 149 to distinguish between 5α and 5β steranes, m/z 189 and 259 for detection of rearranged steranes, m/z 217 for detection of re-arranged and normal steranes and m/z 218 for detection of 14(H) 17(H) steranes.

The m/z 231 fragment ion is used to detect possible aromatic contamination of the saturated fraction, It is also used for detection of methyl steranes.

References

AQUINO NETO, F. R., TRENDEL, J. M., RESTLE, A., CONNAN, J. & ALBRECHT, P. A. 1983. Occurrence and formation of tricyclic and tetracyclic terpanes in sediments and petroleum. *In*: BJOROY, M. (ed.) *Advances in Organic Geochemistry 1981.* Wiley, New York, 659–676.

ARMSTRONG, J. P. D'ELIA, V. A. A. & LOBERG, R. 1995. Holywell Shale: a potential source of hydrocarbons in the East Irish Sea. In: Croker, P. F. & Shannon, P. M. (eds) *The Petroleum Geology of Ireland's Offshore Basins.* Geological Society, London, Special Publications, **93**, 37–38.

BLOXHAM, T. W. & THOMAS, R. L. 1969. Palaeontological and geochemical facies in the *Gastrioceras subcrenatum* marine band and associated rocks from the North Crop of the South Wales coalfield. *Quarterly Journal of the Geological Society, London,* **124**, 239–281.

HARDMAN, M., BUCHANAN, J., HERRINGTON, P. & CARR, A. 1993. Geochemical modelling of the East Irish Sea Basin: its influence on predicting hydrocarbon type and quality. *In*: PARKER, J. R. (ed.) *Petroleum Geology of Northwest Europe: Proceedings of the 4th Conference.* Geological Society, London, 809–821.

PASSEY, Q. R., CREANY, S., KULLA, J. B., MORETTI, F. J. & STROUD, J. D. 1990. A practical model for organic richness from porosity and resistivity logs. *AAPG Bulletin,* **74**, 1777–1794.

PHILP, R. P. & GILBERT, T. D. 1986. Biomaker distribution in oils predominantly derived from terrigenous source material. *In*: LEYTHAEUSER D. & RULLKOLLER J. (eds) *Advances in Organic Geochemistry 1985*, Pergamon, Oxford, 73–84.

QUIRK, M. M., WARDROPER, A. M. K., WHEATLEY, E. & MAXWELL, J. R. 1984. Extended hopanoids in peat environments. *Chemical Geology,* **42**, 25–43.

RAMSBOTTOM, W. H. C. 1969. The Namurian of Britain. *C. R. Congress on Stratigraphic Geology Carboniferous*, Sheffield 1967, **1**, 219–232.

—— 1974. The Namurian of North Wales. *In*: OWEN, T. R. (ed.) *The Upper Palaeozoic and post-Palaeozoic rocks of Wales.* University of Wales Press, Cardiff, 161–167.

STRATHAM, A. 1885. Geology of the coasts adjoining Rhyl, Abergele and Colwyn. *Memoirs of the Geological Survey of Great Britain.*

SOMERVILLE, I. D. 1979. Minor cyclicity in the late Absian (Upper D1) limestones in the Llangollen district of North Wales. *Proceedings of the Yorkshire Geological Society,* **42**, 317–341.

WARREN, P. T. PRICE, D., NUTT, M. J. C. & SMITH, E. G. 1984. Geology of the country around Rhyl and Denbigh. *Memoirs of the Geological Survey of Great Britain.* 217 pp.

WILLIAMS, G. D. & EATON, G. P. 1993. Stratigraphic and structural analysis of the Late Palaeozoic–Mesozoic of NE Wales and Liverpool Bay: implications for hydrocarbon prospectivity. *Journal of the Geological Society,* **150**, 489–499.

Fluid migration history in the north Irish Sea–North Channel region

JOHN PARNELL

School of Geosciences, Queen's University, Belfast BT7 1NN, UK

Abstract: Basins in the north Irish Sea region have experienced numerous stages of fluid migration. Carboniferous and Permo-Triassic sandstones both experienced rapid cementation due to high sedimentation/burial rates. Both successions also experienced syn-sedimentary faulting which focused cement precipitation and created permeability barriers. Further fracture-based diagenesis occurred post-lithification, and accompanying hydrothermal activity associated with Tertiary intrusions. The extensional activity which helped to define Permo-Triassic basins in the Southern Uplands terrane also channelled mineralizing fluids in the same terrane. These fluids included hydrocarbons in at least two districts. Along the North Solway Fault Zone, basinal fluids from the Carboniferous Solway Basin interacted with metalliferous groundwaters to precipitate metalliferous bitumens. The Solway Carboniferous outcrop also contains intergranular oil residues.

The history of fluid flow through rocks is important to an understanding of the petroleum geology of a region. Fluid flow influences diverse aspects of geology, including the diagenetic evolution of reservoir rocks, migration pathways for hydrocarbons, heat transport and the behaviour of fracture systems.

Information about geological fluids can be inferred from authigenic minerals in the rocks through which the fluids passed, including diagenetic phases in reservoir sandstones and infillings in faults and other fractures. This account reviews the information about fluid flow history available from outcrops in the north Irish Sea–Solway–North Channel region. It forms part of a broader study of basin evolution in North West Britain undertaken at Queens University (Parnell 1992*a,b*, 1995*a,b*; McCaffrey & McCann 1992; Anderson *et al.* 1995; Fitzsimons & Parnell 1995, Shelton this volume).

Geological setting

The Caledonian basement is divisible into several NE–SW terranes (Fig. 1), passing southwards from the Grampian terrane through the Midland Valley and Southern Uplands to the Lake District. The observations reported here are mostly from the Midland Valley and Southern Uplands terranes.

On the basement, Devonian (Old Red Sandstone) rocks have a very limited distribution, related to the northern part of the region in Co. Antrim–Kintyre–Clyde, i.e. in the Midland Valley terrane. The Carboniferous is more widely distributed, particularly on the north Solway shore, scattered outcrops in the Southern Uplands terrane, and in north Antrim. Permian rocks occur in several fault-bound basins in the Southern Uplands terrane (often directly on the Caledonian basement) and more widely to the north and south. The Triassic is very widespread offshore, and also forms a thick basin-fill in the Lough Neagh–Larne Basin (Ulster Basin). The Jurassic–Cretaceous sequences are thin where preserved in Northern Ireland and in places missing altogether due to pre-Tertiary erosion. Tertiary rocks are dominated by the thick basaltic lavas in Northern Ireland and a dyke swarm across the entire region.

There is evidence for contemporaneous fault control during each of the main episodes of basinal sediment accumulation, in the Devonian (Simon 1984), Carboniferous (Ord *et al.* 1988) and Permo-Triassic (Anderson *et al.* 1995). In each case, faulting was by reactivation of Caledonian basement structures. Basin-bounding and other prominent faults may also have been preferential sites for reactivation after sedimentation, and therefore preferential sites for fluid flow. It is possible to examine rocks at the faulted margins of Carboniferous and Permo-Triassic basins in southern Scotland.

Carboniferous basins

Carboniferous sedimentation may have been widespread in the North Channel region of the Midland Valley terrane, judging by the extensive deposits in the Midland Valley of Scotland, but subsequent burial beneath the Permo-Triassic

From Meadows, N. S., Trueblood, S. P., Hardman, M. & Cowan, G. (eds), 1997, *Petroleum Geology of the Irish Sea and Adjacent Areas*, Geological Society Special Publication No. 124, pp. 213–228.

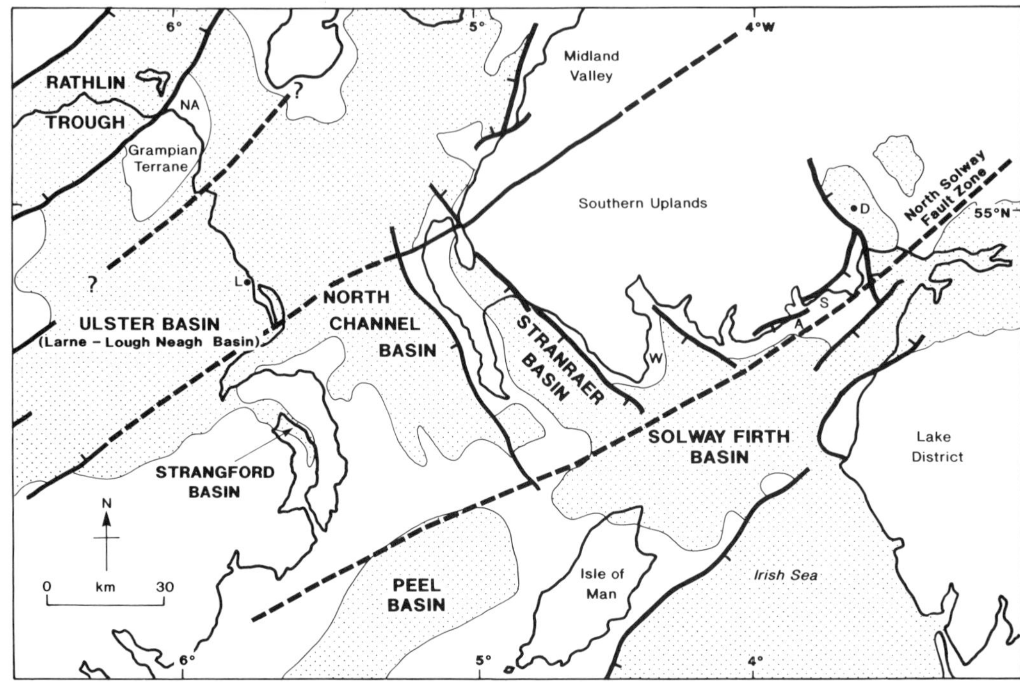

Fig. 1. Map of north Irish Sea–North Channel region, showing main Permo-Triassic basins (stippled) and faults (simplified from Jackson & Mulholland 1993). Dashed lines represent approximate boundaries of Caledonian basement terranes. Carboniferous onshore sections: NA, North Antrim; S, Solway shore. Other localities: A, Auchencairn Bay; D, Dumfries; L, Larne; W, Wigtown Bay.

makes this difficult to confirm. Seismic evidence in the Larne region suggests that Carboniferous *does* occur at depth (Illing & Griffith 1986). A thin fringe of Carboniferous is preserved on the south shore of Belfast Lough; the restricted thickness probably reflects transgression across the northern margin of the Southern Uplands terrane. Isolated Carboniferous outcrops occur in Southern Uplands fault-bounded basins (see below), but no marginal facies have been identified which would indicate syn-depositional faulting, and sedimentation may have been widespread across the terrane before later (pre-Permian) uplift and erosion.

At the southern margin of the Southern Uplands, the ENE–WSW to NE–SW (Caledonoid)-trending North Solway Fault acted as a northern bounding fault to the Solway Basin during Dinantian sedimentation (Deegan 1973; Ord *et al.* 1988). Facies distributions and soft-sediment deformation features clearly indicate that the fault was active during the Dinantian. Coarse alluvial wedges thicken into the fault, and contemporaneous parallel faults within the Carboniferous sandstones show similar wedging of adjacent beds and fault-drape structures.

Permo-Triassic basins

The Permian basins of the Southern Uplands are bounded by normal faults which have a predominantly NNW–SSE to NW–SE trend (Fig. 1). The Carboniferous Sanquhar Basin is also bounded by NW–SE faulting. Some extensional structures on this orientation may date back to the Devono-Carboniferous when they developed orthogonally to the main strike-slip faults which traverse northern Britain along a Caledonoid trend (Gibbs 1989).

The particular distribution of NNW–SSE basins in the Southern Uplands is related to a pronounced structural fabric in the Lower Palaeozoic basement (Anderson *et al.* 1995). Bedding and cleavage in the basement are tectonically aligned to produce a strong ENE–WSW high-angle anisotropy, cross-cut by strike-slip faults including a prominent dextral set trending NNW–SSE. Crustal stretching would have occurred normal or parallel to the anisotropy. Permo-Triassic basins formed due to ENE–WSW extension, as a series of half-graben accommodated by dip-slip reactivation of the Caledonian strike-slip faults. The faults are at a consistent spacing across the Southern Uplands, although not all have onshore basin-fills preserved. NW–SE faults which have clear geomorphological expressions as coastal inlets and inland valleys occur in Wigtown Bay and Auchencairn Bay. These two faults bound Permo-Triassic offshore in Wigtown Bay (Jackson & Mulholland 1993) and displace the NE–SW North Solway Fault (Ord *et al.* 1988), respectively.

In the Midland Valley terrane, the Lough Neagh–Larne Basin (Ulster Basin) includes a thick (>2 km) sequence of Permo-Triassic rocks (Illing & Griffith 1986). The basin has a weakly defined NE–SW trend, constrained by the adjacent Southern Uplands and Grampian terranes. Early (syn-sedimentary?) faults in Co. Antrim have an approximately E–W trend, and the outcrops are dissected by Tertiary faults with a NNW–SSE trend. The degree to which Tertiary NNW–SSE faulting influenced the Southern Uplands terrane, which had already experienced Permian NNW–SSE faulting, is difficult to assess. However the distribution of Tertiary intrusions within the Southern Uplands is much more limited than in the Midland Valley terrane: the Southern Uplands appears to have resisted widespread intrusion and associated faulting.

Diagenetic histories

Carboniferous

Diagenesis of the Carboniferous sandstones at outcrop along the north Solway coast outcrop is dominated by early dolomite cementation in some sections, which is to be expected given the occurrence of limestone beds within those sections (Deegan 1973). The dolomite is so pervasive in some beds that grain replacement has altered sandstone to an impure dolomitic limestone. Dolomite cementation was followed by kaolinite and quartz overgrowth precipitation (Fig. 2). Some samples show a subsequent calcite cement, the precipitation of which overlapped with illite/illite–smectite precipitation. The sandstones exhibit a patchy secondary porosity following dissolution of the carbonate cements. The secondary pores have irregular shape and distribution. This porosity was available to oil migration (see below), after which diagenesis progressed no further except for recent weathering effects and mineralization of the bitumen by pyrite and calcite.

Carboniferous (Dinantian–Namurian) sandstones in North Antrim are also dominated by dolomite cementation, along with quartz and kaolinite. Most cementation and diagenesis was largely completed during the Carboniferous due to a high sedimentation rate which led to rapid burial, before late Carboniferous/Permian uplift which caused weathering, dedolomitization, reddening and generation of secondary porosity (Wang 1992).

Permo-Triassic

The diagenesis of potential reservoir sandstones of Permo-Triassic age has been reviewed in each of

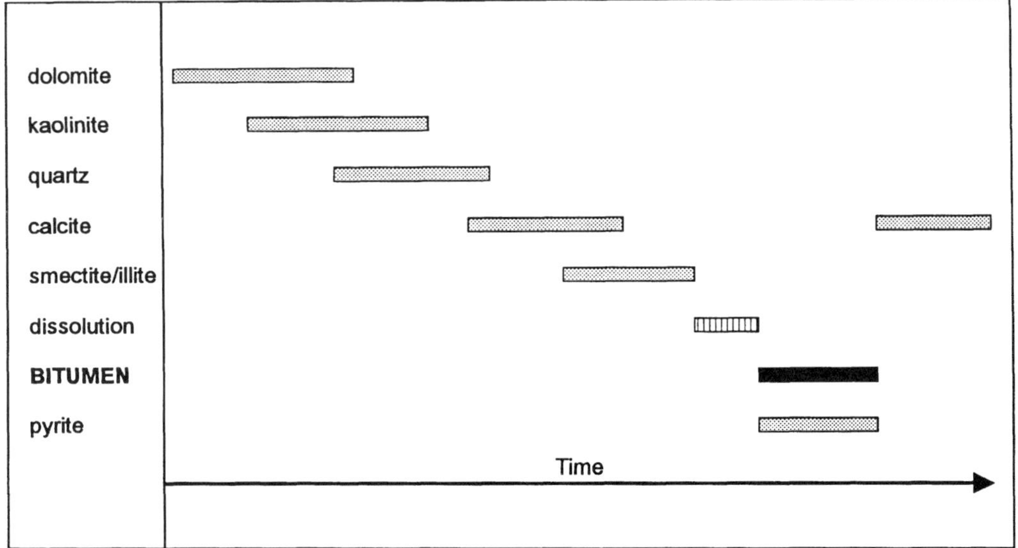

Fig. 2. Summary diagenetic sequence for Carboniferous sandstones of the north Solway shore outcrop. Oil residues (bitumen) were emplaced after a major dissolution event.

the terranes in the region; the Rathlin Trough in the Grampian terrane (Fitzsimons & Parnell 1995), the Lough Neagh-Larne Basin in the Midland Valley (Parnell 1992*b*) and the basins within the Southern Uplands terrane (McKeever 1992). The diagenetic sequences are similar throughout the region, and also to the eastern Irish Sea and elsewhere in the British Permo-Triassic (Burley 1984). They each exhibit the phases typical of red bed diagenesis in a semi-arid climate (Walker 1976). An important aspect of this diagenesis is that many of the processes and mineral precipitates occur at a relatively early stage, i.e. during very shallow burial. The most complete diagenetic sequences are those for Triassic sandstones in Northern Ireland, which include extensive evaporitic cements (Fig. 3).

Fracture planes

Diagenetic sequences give a general idea of the evolution of formation waters, whose chemistry controls the composition of widely distributed cements. Just as important are phases deposited in cross-cutting faults and other fractures, although they are of very restricted distribution. These planar structures compartmentalize the rocks and restrict intraformational fluid flow.

Fracturing and associated fluid flow can be divided into three stages:

(i) Faults associated with active basin extension, particularly at or near basin margins. Syn-sedimentary faults are generally healed during early diagenesis.

(ii) Faults which occur after sediment lithification (Fig. 4). These faults may be reactivations of (i) but are characterized by more brittle deformation, and are often unhealed.

(iii) In northwest Britain, fracturing occurred in association with Tertiary igneous activity, particularly the NNW–SSE to NW–SE trending dyke swarm which crosses the whole region.

Examples from each stage show how the fractures focused mineral precipitation and created permeability barriers:

(i) Syn-sedimentary faults exhibit a concentration of clay minerals (smectite–illite) in a millimetre-scale zone about the fault plane. In Permo-Triassic sandstones, which experienced mineral precipitation at an early (shallow) stage of diagenesis, the clays post-date grain-coating smectite and feldspars which characterize the silicate cement phase of Fig. 3.

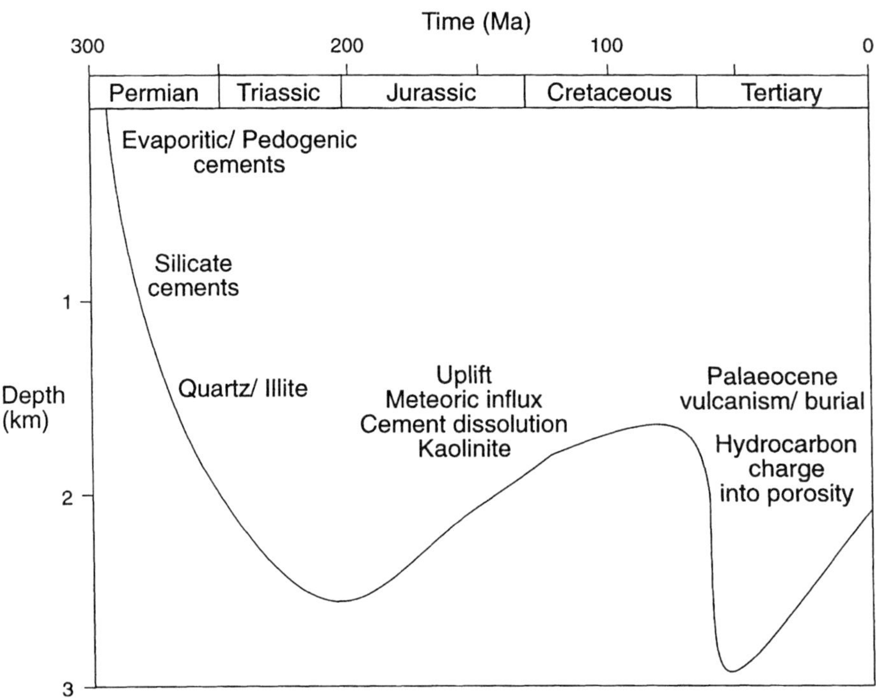

Fig. 3. Summary diagenetic evolution of Permo-Triassic sandstones related to burial pathway within Lough Neagh–Larne Basin.

Fig. 4. Fault plane cutting Permian sandstones, Locharbriggs Quarry, Dumfries. Slickensides parallel to hammer indicate dip-slip motion.

(ii) Later faults exhibit similar concentrations of clays to (i) along the fault plane, but the clays may be more pervasive. Figure 5 compares a clay-impregnated plane in a fault zone with a higher porosity 2 mm beyond the plane, in Permian sandstones near Dumfries.

(iii) The dykes of the Tertiary swarm are likely to be barriers to lateral permeability. Major faults along the same trend in East Antrim (Fig. 6) must also effectively compartmentalize the Mesozoic rocks. Exposures in the vicinity of the Cretaceous Chalk (Ulster White Limestone) show that the faults are sealed by calcite. On a smaller scale, fractures on the same trend show extensive cementation, e.g. the Carboniferous sandstones of North Antrim are cut by dolomite veins parallel to Tertiary dykes.

The Solway Carboniferous sediments show a multi-stage history of fracturing, including active faulting at the margin with Caledonian basement, liquefaction features including sandstone dykes/volcanoes and contorted bedding increasing with proximity to the North Solway Fault, syn-sedimentary faults in semi-lithified sediment which were later healed, and post-lithification brittle faults.

Ore distribution

The basement rocks of the Southern Uplands are host to numerous ore mineral deposits, including base metal sulphide veins. The sulphide veins are hosted particularly in NNW–SSE to NW–SE trending structures which exhibit fault displacement, shearing and brecciation. The age of the veins has been a matter of debate; K–Ar dating has yielded Permo-Carboniferous ages (Ineson & Mitchell 1974), and it seems likely that the dextral faults were both conduits for mineralizing fluids and controlled basin subsidence during Upper Palaeozoic (and possibly Triassic) extension.

A closer examination of ore distribution emphasises this relationship with basin subsidence. Three of the most prominent group of mineral veins in the Southern Uplands are the lead–zinc veins at Leadhills–Wanlockhead, lead–zinc/copper veins of the Blackcraig group, and a set of uranium/copper-mineralized veins in the Southwick district on the Solway coast (Fig. 7). The Leadhills veins are associated with a group of NW–SE to N–S strike-slip faults (Temple 1956), the trace of which extends directly southward into a set of faults of similar trend defining the northern end of the Carboniferous–Permian Thornhill Basin (Fig. 8). The Blackcraig group of veins are almost all within NW–SE structures (Wilson 1921), and as a group form a NW–SE zone of deposits a few kilometres to the north of the NW–SE fault which delimits Permo-Triassic sedimentation in Wigtown Bay (Fig. 9). The veins west of Southwick trend NW–SE to NNW–SSE (Fig. 10a) on the coast either side of Auchencairn Bay, in which more prominent NNW–SSE faults displace the North Solway Fault (Ord *et al.* 1988). The mineral veins also displace shear planes associated with the North Solway Fault and are therefore younger than the Lower Carboniferous rocks which the fault bounds. Some of the veins extend across the North Solway Fault into the Carboniferous rocks, but most are in hornfelsed Lower Palaeozoic which is a rheologically more suitable host. The veins exhibit lateral (strike-slip) displacements and are structurally comparable with those of the Blackcraig group (Miller & Taylor 1966).

In Co. Down, the largest ore deposit in the Lower Palaeozoic basement is a lead–zinc vein at Conlig (Fig. 7), which occupies a brecciated NNW–SSE fault (Smith *et al.* 1991) a few kilometres north of the Newtownards Fault which bounds the Strangford Basin.

                    J. PARNELL

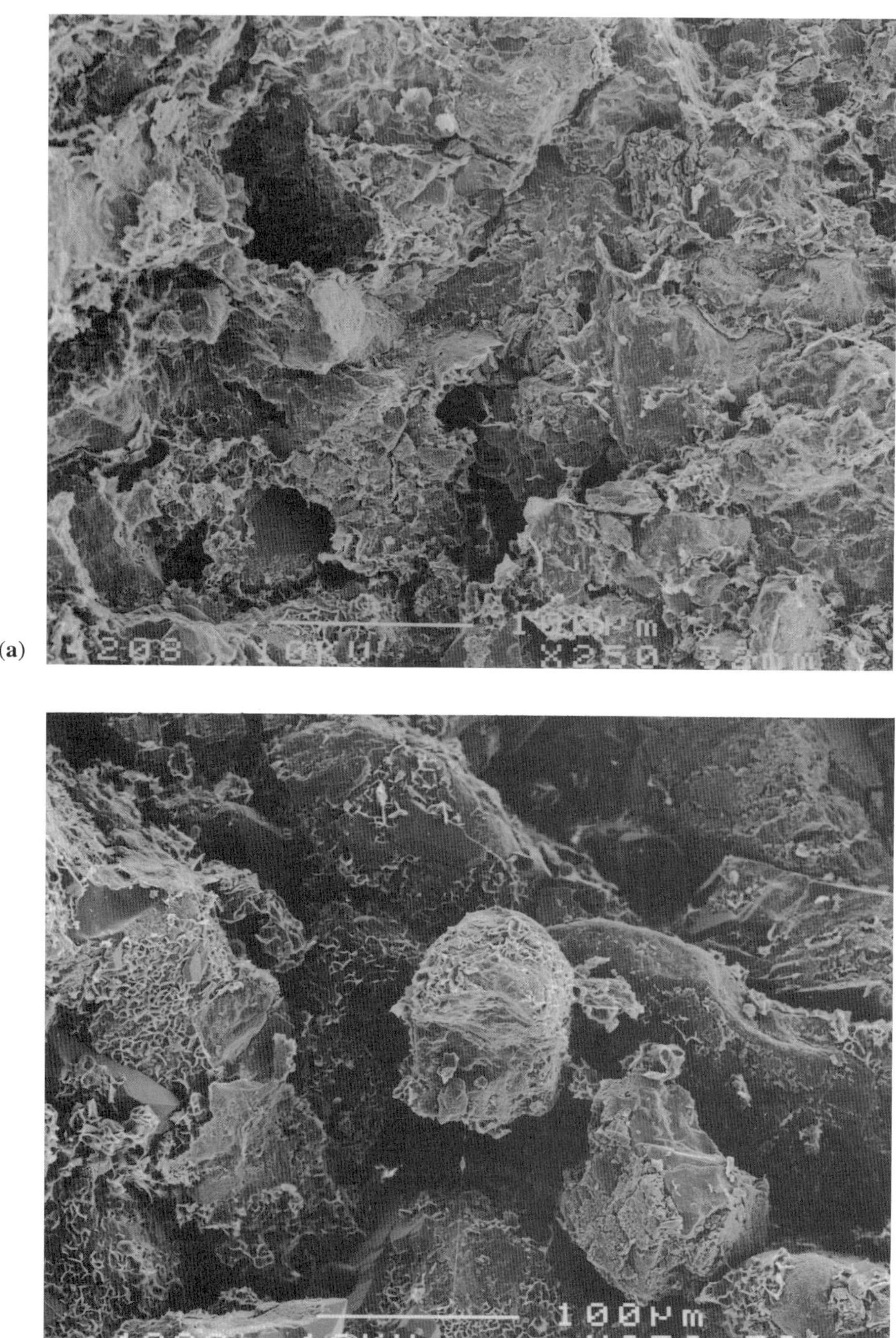

Fig. 5. Electron micrographs of sandstone in Fig. 4. (**a**) From fault plane, impregnated with clay; (**b**) beyond fault plane, with higher porosity. Field widths 375 µm.

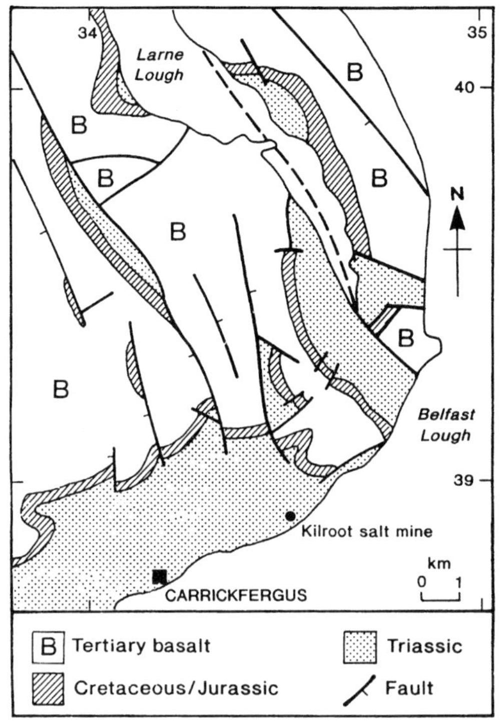

Fig. 6. Geology of East Antrim coastal district, highlighting NNW–SSE trending faults which cut Tertiary basalts and are therefore of Tertiary age. Many Tertiary dykes (omitted) follow similar trend across district. Thick halite mined from upper Triassic at Kilroot.

Ore fluids

The ore fluids are of interest because the models for ore formation involve the circulation of saline fluids derived from adjacent sedimentary basins. Fluid inclusion data are available for both the Leadhills veins and the Blackcraig group (Samson & Banks 1988). Both groups yield ice melting temperatures which are equivalent to highly saline fluids, up to 30 equiv. wt.% NaCl. The homogenization temperatures are low (<150°C, Leadhills peak 60–80°C), which suggests an origin from basin brines. A likely source for saline fluids is in the Lower Carboniferous, which contains evidence for evaporites (Ward this volume). There is no outcrop evidence for evaporites in the Scottish Permian, but boreholes in Northern Ireland show that the Permian can contain thick evaporite beds and sulphate cements, as does the Triassic (Parnell

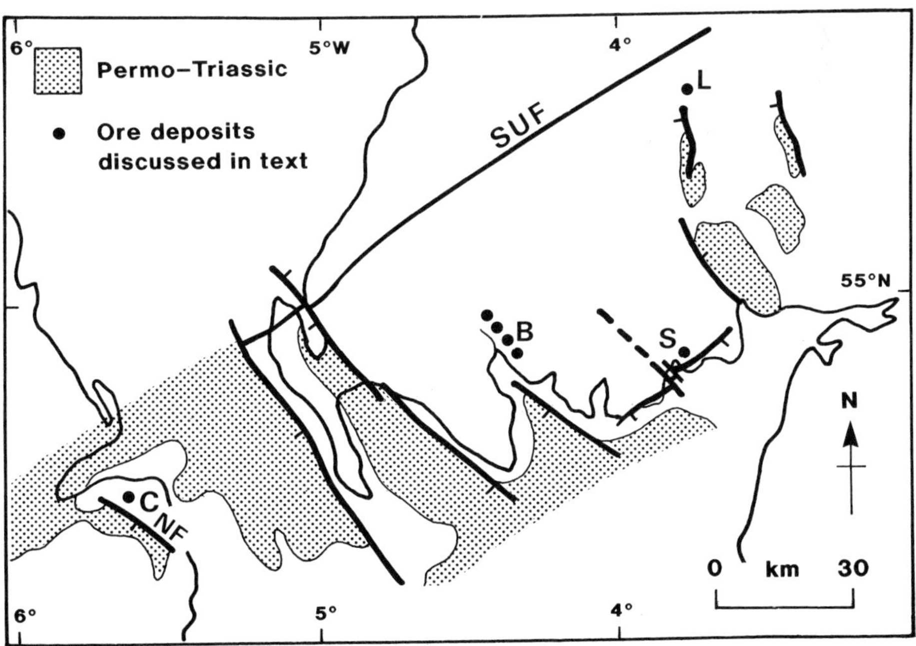

Fig. 7. Map of Southern Uplands terrane showing fault-controlled Permo-Triassic basins, and location of ore deposits discussed in text which exhibit similar structural control. B, Blackcraig group, C, Conlig; L, Leadhills; NF, Newtownards Fault; S, Southwick; SUF, Southern Uplands Fault.

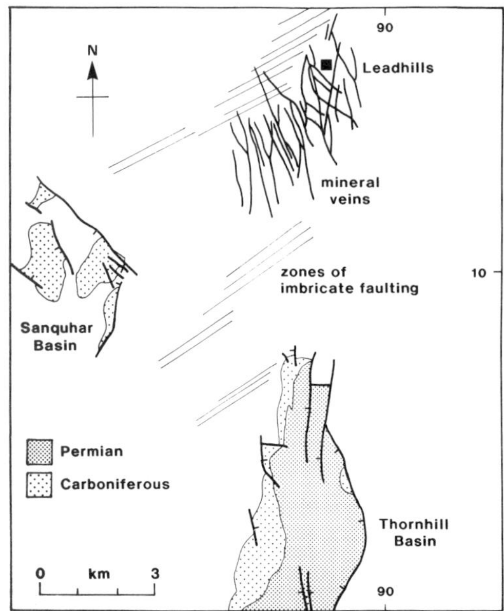

Fig. 8. Map of Leadhills region showing lead–zinc veins; trend of veins traces southwards to faults at north end of Thornhill Basin. Structures are near-normal to trend of imbricate fault zones.

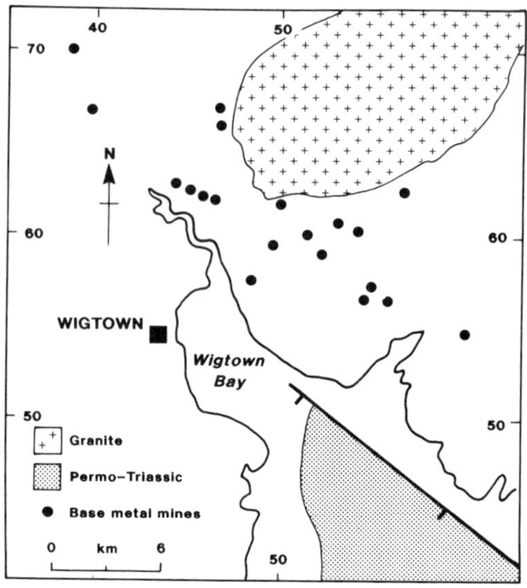

Fig. 9. Map of Wigtown Bay district showing group of base metal mines within a NW–SE zone to the north of a NW–SE fault delimiting Permo-Triassic basin.

1992*b*; Fig. 6). Hydrogen/oxygen isotope data reported by Samson & Banks (1988) indicate a strong meteoric signature, which would be consistent with fluid evolution during the latest Carboniferous/Permian when deep oxidative weathering occurred in the region (Wang 1992).

Measurements from primary fluid inclusions in quartz from a vein in the Southwick region show homogenization temperatures mostly in the range 100–120°C (Fig. 11), which is similarly low, consistent with an origin from basin brines in a

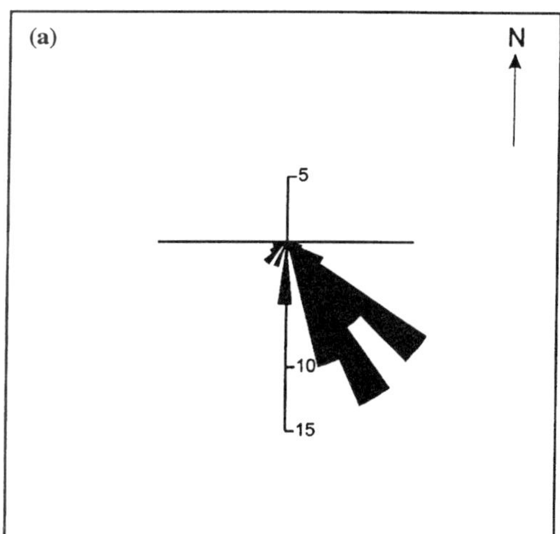

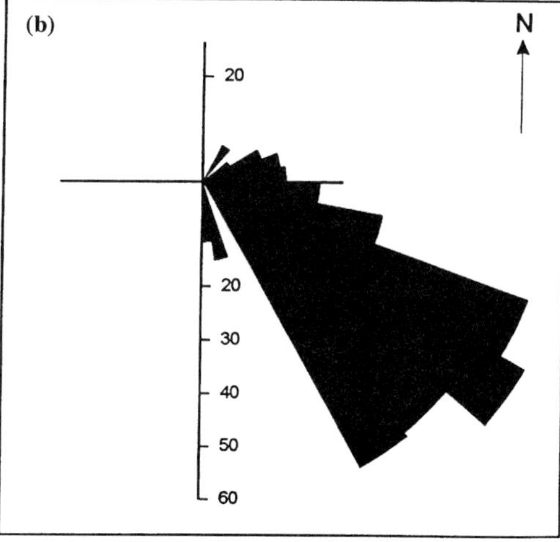

Fig. 10. Orientations of (a) uranium-mineralized veins in the Southwick district, north Solway shore (after Miller & Taylor 1966); (b) dextral Caledonian faults in the Southern Uplands Lower Palaeozoic (after Anderson *et al.* 1995). The marked NNW–SSE to NW–SE trend of the Caledonian faults was reactivated to define Permian basins and control mineralizing fluids.

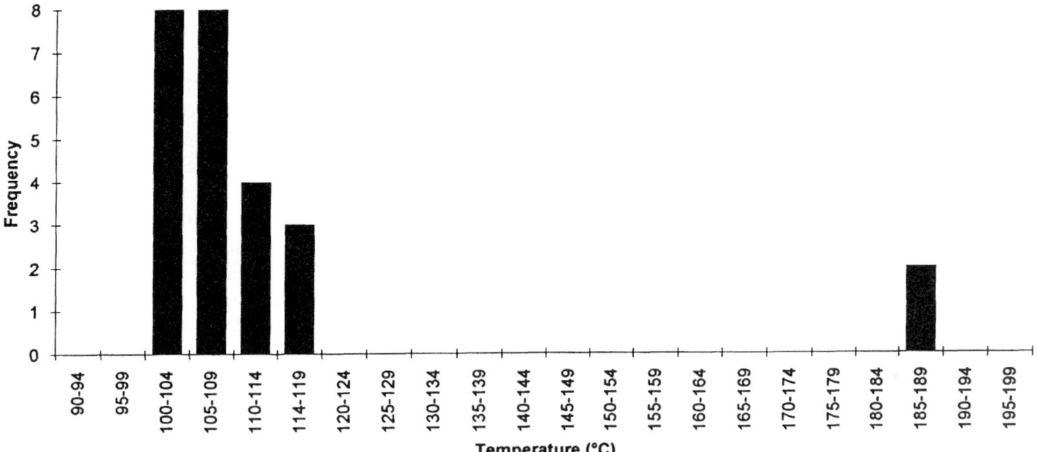

Fig. 11. Homogenization temperatures from primary two-phase aqueous fluid inclusions in quartz from vein in Southwick district.

hydrothermal system. Sulphur isotope compositions for barite in the region are consistent with derivation from Carboniferous seawater (Pattrick & Russell 1989).

Ore mineralization at Leadhills and Southwick did not occur during the initial vein mineralization stage. At both localities an early quartz vein-fill was followed by later ore precipitation; this may also have been the case elsewhere but insufficient data are available.

At Conlig, fluid inclusion data from quartz also indicates a two-phase history. Homogenization temperatures are bimodal, in two distinct populations of aqueous inclusions (Fig. 12). A set of crystallographically controlled primary inclusions is cross-cut by trails of secondary inclusions. The primary inclusions yield uncorrected homogenization temperatures of 225–250°C, while the secondary inclusions yield lower homogenization temperatures in the range 145–200°C. The higher temperatures represent quartz veining during Caledonian deformation, while the secondary

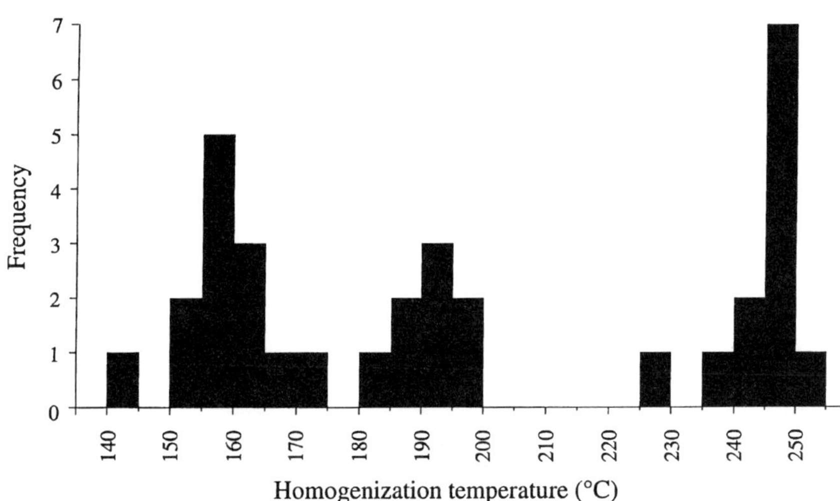

Fig. 12. Homogenization temperatures from two-phase aqueous fluid inclusions in quartz from Conlig ore deposit.

inclusion temperatures represent mineralizing fluids during later ore deposition. This is consistent with the observed vein paragenesis which suggests that sulphide deposition post-dated quartz precipitation (Griffith & Wilson 1982).

Hydrocarbons

The Southwick data are particularly interesting because the mineral veins contain solid hydrocarbons (bitumens), which are coeval with the quartz (Fig. 13). Bitumens occur in several mineralized structures along the length of the North Solway Fault (Parnell 1988a), although it should be emphasized that the structures cross-cut the fault and are probably localized along it because the boundary between Carboniferous and Lower Palaeozoic rocks conferred changes in fluid chemistry (e.g. pH) which triggered ore precipitation. The bitumens occur as millimetre-scale masses; they are uranium-rich and contain inclusions (Fig. 14a) of uraninite and the uranium arsenates zeunerite and trogerite (Parnell 1988b; Braithwaite & Knight 1990). The association with uranium suggests that hydrocarbons derived from the adjacent Solway Basin were precipitated by the effects of irradiation from the uranium. Irradiation induces polymerization/condensation reactions in fluid hydrocarbons and the consequent precipitation of solids (process discussed in Parnell et al. 1990).

The Carboniferous sandstones along the north Solway outcrop contain other evidence for hydrocarbon migration (Fig. 15). Samples from several exposures exhibit traces of intergranular bitumen (Parnell 1995a), although nowhere has this been found to be abundant. Many of the sandstones in which bitumen occurs are of limited porosity (<5%). A second type of radioelement-rich bitumen in the sandstones is in the form of thoriferous bitumen nodules at several localities (Veale & Parnell 1996). The nodules occur on a sub-millimetre scale and represent a phenomenon which is widespread in the British Carboniferous

and elsewhere, particularly in mineralogically immature sandstones (Parnell et al. 1990). The bitumen contains inclusions of thorite, which are in some cases authigenic and in other cases appear to be a detrital heavy mineral residue. As in the case of the uraniferous bitumens, the thoriferous bitumens are interpreted as a product of the interaction between migrating hydrocarbons from the Solway Basin and metal-rich groundwaters/detritus derived from nearby Caledonian granites.

The timing of hydrocarbon migration through the Solway margin rocks has been assessed elsewhere (Parnell 1995a). Burial history modelling suggests that Carboniferous source rocks in the Solway Basin may have entered the oil window during burial beneath a thick Permo-Triassic cover. This constrains the maximum possible age of migration. The creation of the secondary porosity in which intergranular bitumens occur could relate to the next phase of uplift, which may have been during the Jurassic (Barrett 1988). Chemical age dating of uraninite inclusions in the bitumen, precipitation of which occurred shortly after uranium–hydrocarbon interaction, also yields Jurassic ages (Parnell 1995a). A Jurassic timing of hydrocarbon migration would be consistent with hydrocarbon occurrence in the NNW–SSE fault structures cutting the North Solway Fault, which are probably of Permian age (see above).

No hydrocarbon-bearing inclusions have been identified in the quartz of the Southwick region. However, in the thicker Carboniferous succession to the east in the Borders region, some carbonate veins contain hydrocarbon inclusions (Fig. 14b). Figure 16 shows an example from a calcite/dolomite/quartz-filled fracture trending approximately N–S across Kershope Burn in the Middle Border Group (locality recorded in Parnell 1995a).

Uraniferous and thoriferous bitumens also occur elsewhere in the northern Irish Sea region (Fig. 17). On the opposite side of the Solway Firth, thoriferous bitumen nodules occur in feldspathic sandstones of the Namurian Hensingham Group near Maryport. They have not so far been detected

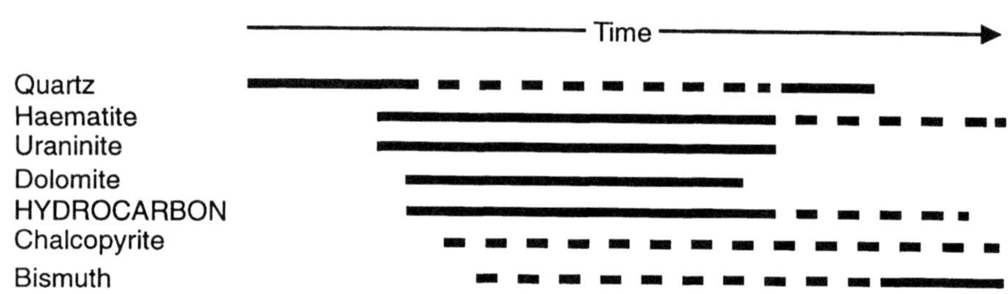

Fig. 13. Paragenesis of uranium/bitumen-bearing veins in NNW–SSE faults in Southwick district.

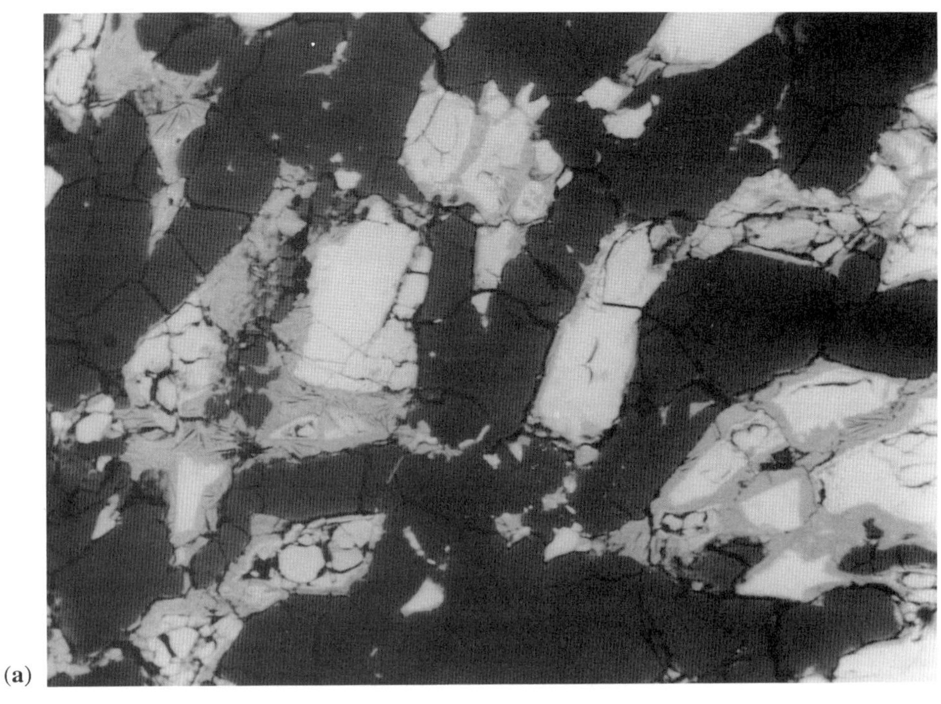

(a)

(b)

Fig. 14. Electron micrographs of fracture-bound bitumens. (a) Bitumen (black) containing crystals of uraninite/uranium arsenates (bright) from vein, Southwick district. (b) Bitumen (black) within calcite vein (grey) cutting mudrock, Kershope Burn, Borders. Field widths (a) 360 µm; (b) 1.7 mm.

J. PARNELL

Fig. 15. Electron micrographs of bitumens in Carboniferous sandstones, Solway shore outcrop. (**a**) bitumen (black) in dissolution pore in replacive calcite/clay (C) cements between detrital grains. (**b**) bitumen nodule (black) containing thorite crystals (bright) in sandstone. Field widths (a) 470 μm; (b) 1.1 mm.

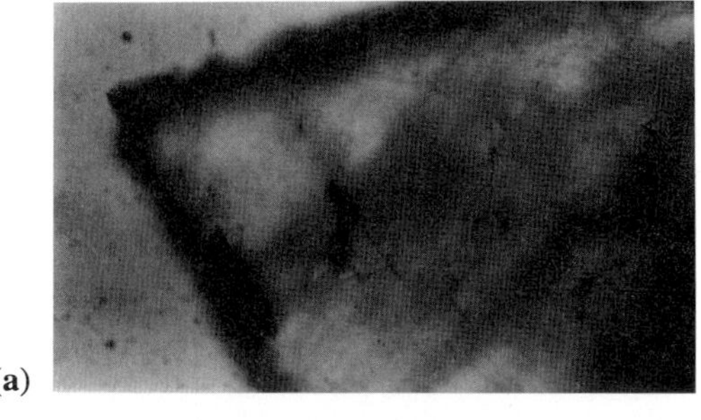

(a)

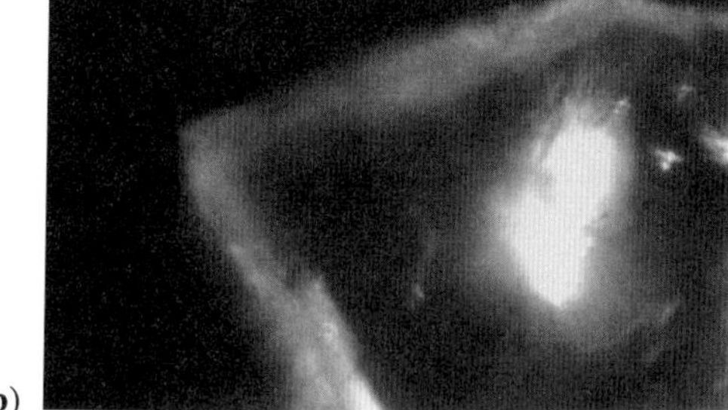

(b)

in Dinantian or Westphalian sandstones in north-west Cumbria, which are mineralogically more mature and less susceptible to nodule precipitation. In the eastern Lake District, traces of bitumen have been recorded in the porous wallrock of a mineralized quartz vein within the Grainsgill greisen. The bitumen is directly associated with uraninite (Dawson & Harrison 1966). In the Laxey lead–zinc mines, Isle of Man, bitumen was encountered within a copper lode (Quiggins 1852; Davidson & Bowie 1951). The bitumen contains abundant inclusions of uraninite, chalcopyrite and diverse other sulphides/arsenides (Parnell 1988*b*). In each case the radioelement-rich bitumens occur in settings updip from Carboniferous rocks (the assumed hydrocarbon source) and down-dip from

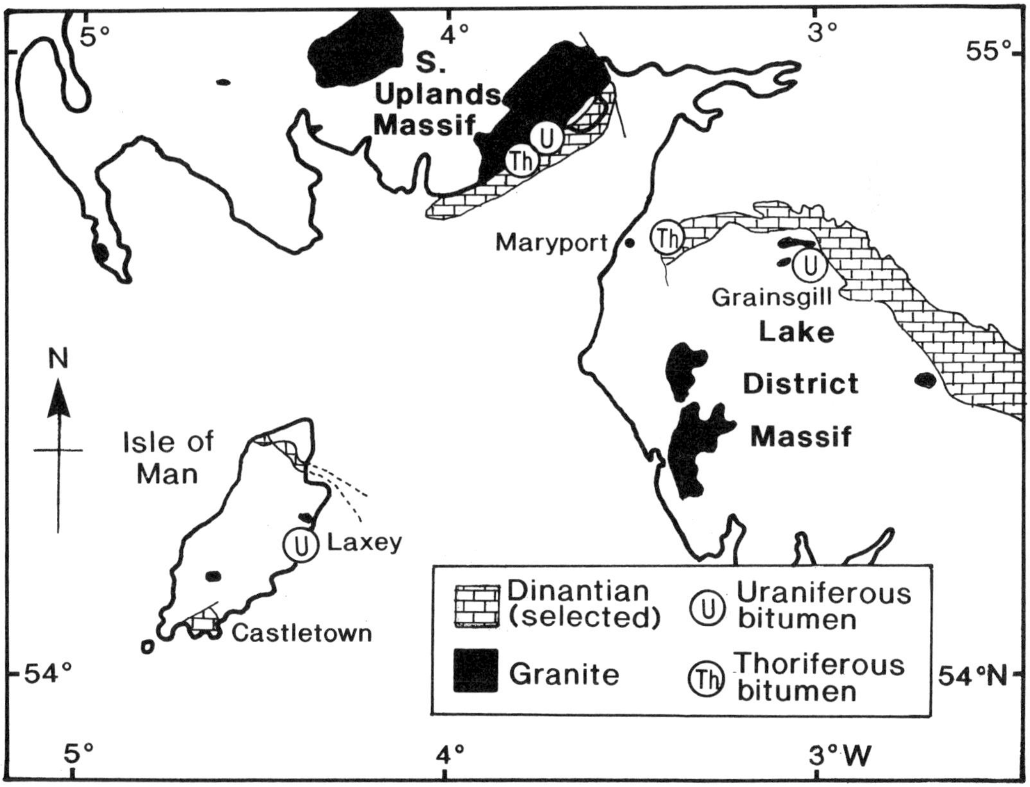

Fig. 16. Fluid inclusion in calcite vein, Kershope Burn, Borders. (**a**) Plane-polarized light; (**b**) ultraviolet light, highlighting hydrocarbon inclusion which fluoresces. Field widths 500 µm.

Fig. 17. Occurrences of uraniferous and thoriferous bitumens relative to Caledonian granites (U/Th source) and selected Dinantian outcrops.

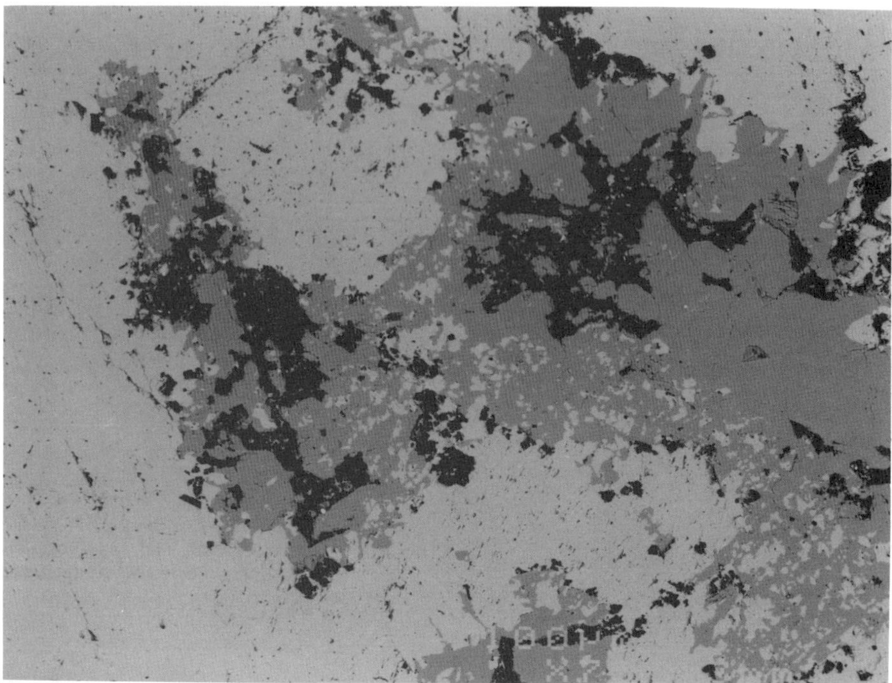

Fig. 18. Backscattered electron micrograph of calcite vein (light) containing coeval bitumen (black) and quartz (dark grey), Poyllvaaish Bay, Isle of Man. Field width 940 μm.

Caledonian granite (the assumed radioelement source), i.e. at sites of fluid interaction.

Hydrocarbons have been recorded in a sample of quartz vein rock from the mineralized structure at Conlig, Co. Down (Parnell 1993). They could be derived from a Carboniferous source in the Lough Neagh–Larne Basin a few kilometres to the north, but the origin of the hydrocarbon show is not yet clear.

The most convincing evidence for hydrocarbon migration related to Tertiary activity is in bitumens within NW–SE trending fractures parallel to Tertiary dykes cutting the Carboniferous section at Castletown, Isle of Man (section described in Dickson *et al.* 1987). The bitumens are coeval (intergrown) with authigenic quartz crystals within calcite veins (Fig. 18). Fluid inclusions in the quartz yield temperatures in the range 200–220°C (P. Carey pers. comm.). Such high temperatures strongly suggest hydrothermal fluids directly associated with igneous activity.

Hydrothermal activity related to Tertiary igneous intrusion is particularly evident at the north end of the Strangford Basin where intruded Triassic sandstones possess a pervasive actinolite cement (Parnell 1992*b*) and partial bleaching of iron oxide cement. Bleaching of Permo-Triassic sandstones elsewhere in Northern Ireland is rare: some oolitic Sherwood Sandstone Group in the Lough Neagh district is bleached and porous, but at present it is not possible to state whether this is related to hydrocarbon migration updip from the Lough Neagh depocentre.

The timing of hydrocarbon migration over the region as a whole was of course highly variable due to differences in burial history. In general, areas of present Carboniferous outcrop experienced maximum burial, and hydrocarbon generation, before late Carboniferous basin inversion, whereas the Permo-Triassic basins experienced maximum burial during the Mesozoic (or Palaeocene beneath basalt lavas in Co. Antrim).

Conclusions

The region experienced many stages of fluid evolution affecting sedimentary basins. The following lists the most important stages, omitting the detail of diagenetic sequences, minor uplift events, and accepting some uncertainty/overlap between stages:

1. Syn-sedimentary faulting and dewatering during Dinantian sedimentation (Solway).

2. Diagenetic modification of Carboniferous sandstones through the Carboniferous.
3. Permian deep oxidative weathering of Carboniferous (and other) rocks.
4. Mineralization of NNW–SSE extensional (?Permian rifting) fractures (Southern Uplands).
5. Diagenetic evolution of Permo-Triassic sandstones.
6. Clay mineralization of fault planes in Permo-Triassic sandstones.
7. Hydrocarbon migration through Solway Carboniferous and fractures in adjacent basement.

8. Regional basin inversion (late Mesozoic); some meteoric weathering.
9. Hydrothermal activity associated with Tertiary intrusion.

G. Alexander, P. Carey and the staff of the QUB Electron Microscope Unit are thanked for skilled technical support. Samples from Southwick and Conlig were kindly supplied by the British Geological Survey and Ulster Museum, respectively. Fluid inclusion data were made available by P. Carey and A. Rodgers. The manuscript benefited from comments by P. C. Barnard and K. Maguire.

References

ANDERSON, T. B., PARNELL, J. & RUFFELL, A. H. 1995. Influence of basement on the geometry of Permo-Triassic basins in the North-West British Isles. *In*: BOLDY, S. (ed.) *Permo-Triassic Rifting*. Geological Society, London, Special Publications, **91**, 103–122.

BARRETT, P. A. 1988. Early Carboniferous of the Solway Basin: a tectonostratigraphic model and its bearing on hydrocarbon potential. *Marine and Petroleum Geology*, **5**, 271–281.

BRAITHWAITE, R. S. W. & KNIGHT, J. R. 1990. Rare minerals, including several new to Britain, in supergene alteration of U–Cu–As–Bi–Co mineralization near Dalbeattie. *Mineralogical Magazine*, **54**, 129–131.

BURLEY, S. D. 1984. Patterns of diagenesis in the Sherwood Sandstone Group (Triassic), United Kingdom. *Clay Minerals*, **19**, 403–440.

DAVIDSON, C. F. & BOWIE, S. H. U. 1951. On thucholite and related hydrocarbon–uraninite complexes. *Bulletin of the Geological Survey of Great Britain*, **3**, 1–19.

DAWSON, J & HARRISON, R. K. 1966. Uraninite in the Grainsgill greisen, Cumberland. *Bulletin of the Geological Survey of Great Britain*, **25**, 91.

DEEGAN, C. E. 1973. Tectonic control of sedimentation at the margin of a Carboniferous depositional basin in Kirkcudbrightshire. *Scottish Journal of Geology*, **9**, 1–28.

DICKSON, J. A. D., FORD, T. D. & SWIFT, A. 1987. The stratigraphy of the Carboniferous rocks around Castletown, Isle of Man. *Proceedings of the Yorkshire Geological Society*, **46**, 203–229.

FITZSIMONS, S. & PARNELL, J. 1995. Diagenetic history and reservoir potential of Permo-Triassic sandstones in the Rathlin Basin. *In*: CROKER, P. F. & SHANNON, P. M. (eds) *The Petroleum Geology of Ireland's Offshore Basins*. Geological Society, London, Special Publications, **93**, 21–35.

GIBBS, A. D. 1989. Structural styles in basin formation. *In*: TANKARD, A. J. & BALKWILL, H. (eds) *Extensional Tectonics and Stratigraphy of the North Atlantic Margins*. American Association of Petroleum Geologists Memoir, **46**, 81–93.

GRIFFITH, A. E. & WILSON, H. E. 1982. *Geology of the Country around Carrickfergus and Bangor*. Memoir of the Geological Survey of Northern Ireland. HMSO, Belfast.

IILLING, L. V. & GRIFFITH, A. E. 1986. Gas prospects in the 'Midland Valley' of Northern Ireland. *In*: BROOKS, J., GOFF, J. C. & VAN HOORN, B. (eds) *Habitat of Palaeozoic Gas in NW Europe*. Geological Society, London, Special Publications, **23**, 73–84.

INESON, P. R. & MITCHELL, J. G. 1974. K–Ar isotopic age determinations from some Scottish mineral localities. *Transactions of the Institution of Mining and Metallurgy*, **83**, B13–B18.

JACKSON, D. I. & MULHOLLAND, P. 1993. Tectonic and stratigraphic aspects of the East Irish Sea Basin and adjacent areas: contrasts in their post- Carboniferous structural styles. *In*: PARKER, J. R. (ed.) *Petroleum Geology of Northwest Europe: Proceedings of the 4th Conference*. Geological Society, London, 791–808.

MCCAFFREY, R. J. & MCCANN, N. 1992. Post-Permian basin history of northeast Ireland. *In*: PARNELL, J. (ed.) *Basins on the Atlantic Seaboard: Petroleum Geology, Sedimentology and Basin Evolution*. Geological Society, London, Special Publications, **62**, 277–290.

MCKEEVER, P. J. 1992. Petrography and diagenesis of the Permo-Triassic of Scotland. *In*: PARNELL, J. (ed.) *Basins on the Atlantic Seaboard: Petroleum Geology, Sedimentology and Basin Evolution*. Geological Society, London, Special Publications, **62**, 71–96.

MILLER, J. M. & TAYLOR, K. 1966. Uranium mineralization near Dalbeattie, Kirkcudbrightshire. *Bulletin of the Geological Survey of Great Britain*, **25**, 1–18.

ORD, D. M., CLEMMEY, H. & LEEDER, M. R. 1988. Interaction between faulting and sedimentation during Dinantian extension of the Solway basin, SW Scotland. *Journal of the Geological Society, London*, **145**, 249–259.

PARNELL, J. 1988a. Migration of biogenic hydrocarbons into granites: a review of hydrocarbons in British plutons. *Marine and Petroleum Geology*, **5**, 385–396.

——1988b. Mineralogy of uraniferous hydrocarbons in Carboniferous- hosted mineral deposits, Great Britain. *Uranium*, **4**, 197–218.

——1992*a*. (ed.) *Basins on the Atlantic Seaboard: Petroleum Geology, Sedimentology and Basin Evolution*. Geological Society, London, Special Publications, **62**.

——1992*b*. Hydrocarbon potential of Northern Ireland: III. Reservoir potential of the Permo-Triassic. *Journal of Petroleum Geology*, **15**, 51–70.

——1993. Metal enrichments in Bitumens from the Carboniferous of Ireland: potential in exploration for ore deposits. *In*: PARNELL, J., KUCHA, H. & LANDAIS, P. (eds) *Bitumens in Ore Deposits*. Springer, New York, 475–489.

——1995*a*. Hydrocarbon migration in the Solway Basin. *Geological Journal*, **30**, 25–38.

——1995*b*. Sulphide vein mineralization in Northern Ireland: constraints on fluid sources. *In*: PASAVA, J., KRIBEK, B. & ZAK, K. (eds) *Mineral Deposits: From Their Origin to Their Environmental Impacts*. Balkema, Rotterdam, 373–376.

——, MONSON, B. & TOSSWILL, R. J. 1990. Petrography of thoriferous hydrocarbon nodules in sandstones, and their significance for petroleum exploration. *Journal of the Geological Society, London*, **147**, 837–842.

PATTRICK, R. A. D. & RUSSELL, M. J. 1989. Sulphur isotope investigation of Lower Carboniferous vein deposits of the British Isles. *Mineralium Deposita*, **24**, 148–153.

QUIGGINS, M. A. 1852. *Illustrated Guide and Visitors Companion through the Isle of Man*. Simpkin & Marshall, London.

SAMSON, I. M. & BANKS, D. A. 1988. Epithermal base-metal vein mineralization in the Southern Uplands of Scotland: nature and origin of the fluids. *Mineralium Deposita*, **23**, 1–8.

SHELTON, R. 1996. Tectonic evolution of the Larne Basin. *This volume*.

SIMON, J. B. 1984. Sedimentation of a small complex alluvial fan of possible Upper Old Red Sandstone age, northeast County Antrim. *Irish Journal of Earth Sciences*, **6**, 109–119.

SMITH, R. A., JOHNSTON, T. P. & LEGG, I. C. 1991. *Geology of the Country around Newtownards*. Memoir of the Geological Survey of Northern Ireland. HMSO, Belfast.

TEMPLE, A. K. 1956. The Leadhills–Wanlockhead lead and zinc deposits. *Proceedings of the Royal Society of Edinburgh*, **63**, 85–113.

VEALE, C. & PARNELL, J. 1996. Metal–organic interactions in the Dinantian Solway Basin: inferences for oil migration studies. *In*: STROGEN, P., SOMERVILLE, I. D. & JONES, G. LL. (eds) *Recent Advances in Lower Carboniferous Geology*. Geological Society, London, Special Publications, **107**, 51–63.

WALKER, T. R. 1976. Diagenetic origin of continental red beds. *In*: FALKE, H. (ed.) *The Continental Permian in Central, West and South Europe*. NATO Advanced Study Institute Series C, Mathematics & Physical Sciences, 22.

WANG, W. H. 1992. Origin of reddening and secondary porosity in Carboniferous sandstones, Northern Ireland. *In*: PARNELL, J. (ed.) *Basins on the Atlantic Seaboard: Petroleum Geology, Sedimentology and Basin Evolution*. Geological Society, London, Special Publications, **62**, 243–254.

WILSON, G. V. 1921. The Lead, Zinc, Copper and Nickel Ores of Scotland. Mineral Resources Memoir of the Geological Survey, **17**.

The presence of sulphate-reducing bacteria in live drilling muds, core materials and reservoir formation brine from new oilfields

CAROLINE MCGOVERN-TRAA[1], JYH-YIH LEU[1], W. ALLAN HAMILTON[1], IAIN S. C. SPARK[2] & IAN T. M. PATEY[2]

[1] *Department of Molecular and Cell Biology, Institute of Medical Sciences, University of Aberdeen Aberdeen, AB25 2ZD, UK*

[2] *Corex (UK) Ltd, Units B1–B3, Airport Industrial Park, Howe Moss Drive, Dyce, Aberdeen AB2 0GL, UK*

Abstract: The sulphate-reducing bacteria (SRB) produce the toxic gas hydrogen sulphide (H_2S) and cause numerous problems in the oil and gas industry. It has been suggested that these organisms are either introduced to the formation by drilling or water injection or that they may be indigenous. Little is known of the activity and survival mechanisms of the SRB under reservoir conditions. In this study a number of samples from hydrocarbon reservoirs were inoculated into sulphate-rich growth medium to determine the presence and activity of the SRB. The samples studied included freshly drilled core material and live drilling muds. The majority of samples analysed contained SRB which produced H_2S to varying degrees following different incubation times. Organisms capable of producing H_2S at temperatures up to 95°C were detected. Often long incubation periods, in excess of 6 months, were necessary before H_2S was produced. Many of the blank mud samples analysed (e.g. prior to use in the well) were already contaminated with sulphidogenic bacteria. Some of the cultures obtained to date have been studied in reservoir conditions core flood tests. It is hoped that the use of these laboratory tests will highlight any potentially damaging reactions prior to their occurrence in hydrocarbon formations.

Microorganisms have many varied roles in the oil and gas industry. Such roles include the contamination of crude, diesel and lubricating oils in pipelines and storage systems. The production platforms themselves, steel or concrete, may be subjected to corrosion or biofouling by different marine bacteria. Drilling mud may cause pollution problems or become biodegraded, thus losing its efficacy (Freeman 1983). In the actual formation, bacteria may be utilized for microbial enhanced oil recovery (MEOR). On the other hand, bacteria in hydrocarbon formations can contribute to reservoir souring and other formation damage mechanisms such as polymer plugging (Fig. 1).

Sulphate-reducing bacteria (SRB) are considered to be the major contributors to the biogenic souring of hydrocarbon reservoirs. The SRB create numerous problems in the oil industry (Rosnes *et al.* 1991*a*) including:

- corrosion of pipelines and production facilities
- reduction in value of H_2S-rich oil and gas
- precipitation of sulphides causing formation damage
- health and environmental concern.

Rosnes *et al.* (1991*b*) proposed three theories as to the origin of SRB in hydrocarbon reservoirs. The organisms may be either introduced by the drilling process or following seawater injection, or they may be indigenous to the formation, being laid down with the original sediments and/or introduced during geological time.

Aims

The aims of this research project are three-fold:

- To test a range of hydrocarbon reservoir samples for the presence and activity of SRB.
- To develop microbial, reservoir condition core flood tests to study these organisms.
- To determine the effect of different production chemicals on the growth and activity of SRB and other bacteria isolated from hydrocarbon reservoirs.

Methods

A range of samples from hydrocarbon reservoirs have been studied in the course of this project including:

- freshly drilled core material
- live drilling muds
- completion chemicals
- formation/aquifer brine samples
- crude oil.

From Meadows, N. S., Trueblood, S. P., Hardman, M. & Cowan, G. (eds), 1997, *Petroleum Geology of the Irish Sea and Adjacent Areas,* Geological Society Special Publication No. 124, pp. 229–236.

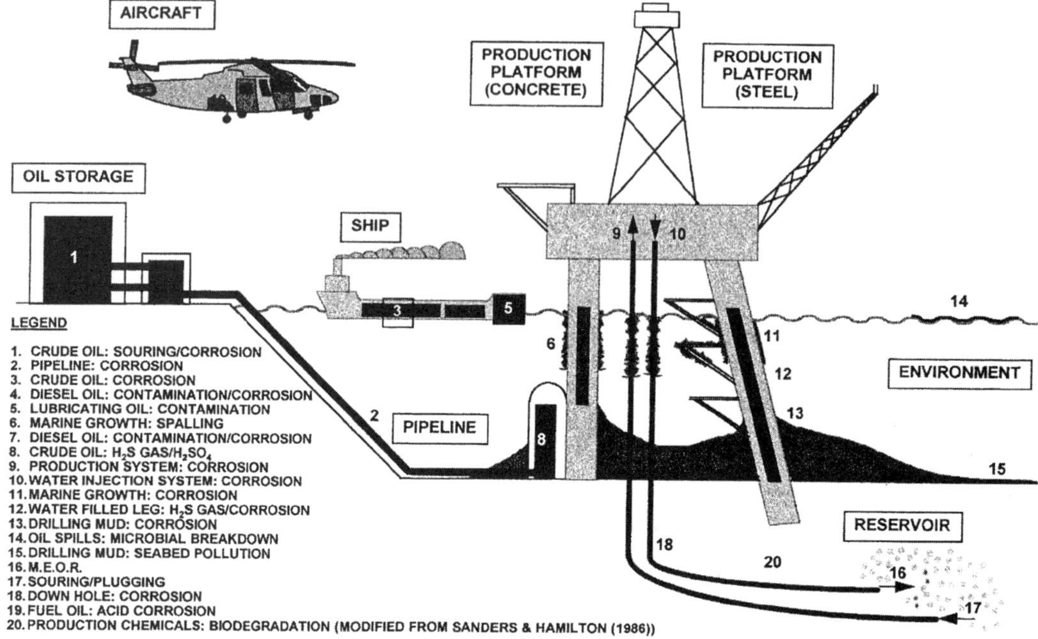

Fig. 1. Microorganisms and the oil industry.

This paper will discuss the main findings from the first three sample types only.

Core material was obtained from a new field prior to any production or injection. The core was sealed in a fibreglass sleeve immediately on arrival at the surface. This was then transported to Aberdeen where two sections of rock were removed and immersed in anaerobic jars filled with light mineral oil, which had been filtered, depolarized and degassed prior to use. The jars were then filled with Argon gas and sealed. In the laboratory the core was removed from each of the anaerobic jars and the outer surface of the rock, which was visibly infiltrated with drilling mud solids, was pared away using a hammer and chisel. The inner segment of rock was then separated from the outer area and small cubes of rock from each section (2 cm^2) were immersed in solvent mixture in preparation for scanning electron microscopy (SEM) analysis. The remainder of the rock was gently crushed in a sterile mortar and pestle and stored in nitrogen-flushed sterile plastic containers ready for use. Approximately 2 g of core were inoculated into 50 ml volumes of sulphate-rich growth medium suitable for the cultivation of SRB (Widdel & Pfennig 1981). Enrichment cultures were set up according to the scheme in Table 1.

Drilling mud and completion fluid samples were taken in sterile plastic bottles (250 ml), flushed with nitrogen prior to use. Samples were taken from the mud pits at regular depth intervals, filled to the top of the containers and transported immediately to the laboratory for inoculation into growth

Table 1. *Test parameters for the inoculation of two core samples into sulphate-rich growth medium for the enhancement of SRB*

Parameters tested	Variables
core samples	core 1 (3504 feet)
	core 2 (3544 feet)
Core area	inner
	outer
NaCl concentration	0.2% NaCl (fresh water environment)
	2.0% NaCl (marine environment)
Carbon substrate	lactate
	acetate
	propionate
	butyrate
	(benzoate + hexadecane)
Incubation temperatures	30°C
	55°C
	70°C

medium. Approximately 2 ml of mud was inoculated into each 50 ml volume of growth medium using a sterile syringe.

Samples of growth medium were removed from the enrichment cultures at regular intervals over a period of 6 months to test for the production of H_2S by the methylene blue colorimetric assay (Cline 1969).

Results

Core samples

Samples of core material taken from freshly drilled core were studied using SEM. Figure 2 shows an SEM micrograph of the inner section of rock from a depth of 3504 feet. The sample shows no drilling mud solids invasion. The sandstone quartz rock

Fig. 2. Scanning Electron micrograph of the inner section of core sample 1 (depth 3504 feet) revealing the presence of copious amounts of biofilm material (×300 magnification) (scale bar equals 33.3 μm).

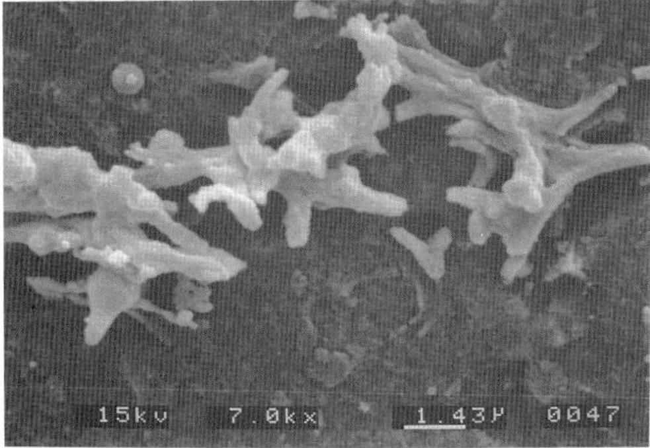

Fig. 3. Scanning electron micrograph of the inner section of core sample 1 (depth 3504 feet) showing individual cells entrapped within a cellular/polymer matrix. Rod-shaped cells, some with a central spore, were evident (×7000 magnification) (scale bar equals 1.43 μm).

Table 2. *The production of H_2S from core sample enrichment cultures incubated for 6 months at 30°C in Widdels Medium with a range of carbon substrates and either 2% or 0.2% NaCl concentration*

		Core 1 (3504 feet)		Core 2 (3544 feet)	
Culture	Substrate	Inner	Outer	Inner	Outer
2.0% NaCl	lactate				
	acetate				
	propionate			I	
	butyrate				
	hex/ben*				O
0.2% NaCl	lactate			I,O	I,O
	acetate				O
	propionate				
	butyrate				O
	hex/ben*	I			

I,O – H_2S produced from both inner & outer core samples.
O – H_2S produced from outer core only.
I – H_2S produced from inner core only.
* Hexadecane/benzoate.

grains appear to be coated in 'fluffy' polymeric biofilm material which can be distinguished as individual bacterial cells at a higher magnification (Fig. 3).

Core samples were tested for H_2S production over a 6 month incubation period, Tables 2 and 3 show qualitatively which enrichment cultures produced H_2S following 6 months incubation at 30°C and 55°C, respectively. Differences in the

Table 3. *The production of H_2S from core sample enrichment cultures incubated at 55°C in Widdels Medium with a range of carbon substrates and either 2% or 0.2% NaCl concentration*

		Core 1 (3504 feet)		Core 2 (3544 feet)	
Culture	Substrate	Inner	Outer	Inner	Outer
2.0%	lactate		O		
	acetate	I,O	I,O		
	propionate	I			
	butyrate	I,O	I,O	I,O	I,O
	hex/ben*	I,O	I,O		O
0.2%	lactate	I,O	I,O		
	acetate	I,O	I,O		O
	propionate				
	butyrate		O		
	hex/ben*		O		

I,O – H_2S produced from both inner & outer core samples.
O – H_2S produced from outer core only.
I – H_2S produced from inner core only.
* Hexadecane/benzoate.

amount and rate of H_2S production were apparent in the different cultures. Greater amounts of H_2S were detected in the core 2 enrichment cultures at 30°C, whilst more H_2S was produced in the core 1 enrichments at 55°C.

Samples of culture fluid were analysed using both light and scanning electron microscopy. Figure 4 shows a sample of culture broth from one acetate enrichment culture growing at 55°C. The same spore-forming rod-shaped bacteria were visible as were noted in the original rock samples (Fig. 3).

Core flood tests. The mixed populations of bacteria obtained during this study are now being used in core flood tests to determine what sort of reactions occur at reservoir conditions of pressure and temperature. Culture H12, an enrichment culture from core 1 growing in sulphate-rich growth medium with 0.2% NaCl concentration and the carbon substrate lactate, has been used in core flood tests to date. This bacterial consortium produced H_2S in bottle tests at 55°C. When this culture was used in core flood tests at reservoir pressure and temperature however, no H_2S was detected from the effluent stream, whereas copious amounts of polymeric material were evident, often coating the bacterial cell surface. Figures 5 and 6 show culture H12 prior to and following a shut in period of 10 days in a static reservoir conditions core flood test. The polymeric material evident following the core flood test caused the cells to aggregate. Such aggregates could block pore throats if permitted to migrate through the reservoir. The core plug was off-loaded at the end of the core flood test and analysed using SEM. Gypsum crystals were

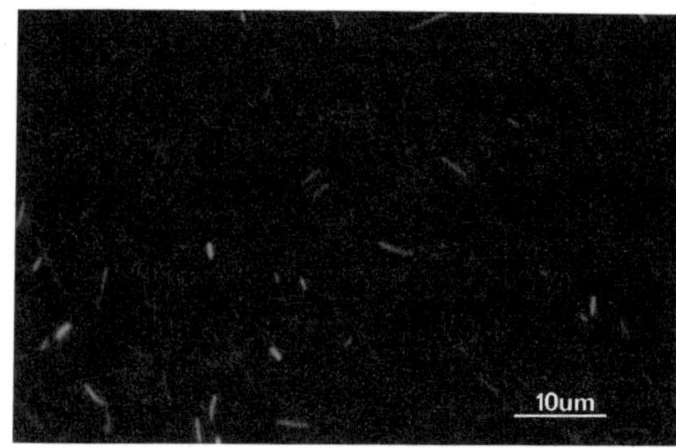

Fig. 5. Light micrograph of the stable enrichment culture H12 prior to use in a static reservoir conditions core flood test. This culture was isolated from the outer area of core 1 (depth 3504 feet) grown in Pfennigs/acetate medium (0.2% NaCl) at 55°C. Cells were stained using acridine orange (×1000 magnification) (scale bar equals 10 μm).

apparent in the rock caused by an incompatibility reaction between the bacterial growth medium and the simulated sulphate-rich formation brine (Fig. 7). This depletion in soluble sulphate probably accounts for the lack of H_2S production by the SRB during the core flood test.

Drilling mud samples. To date, three sets of drilling muds from different oil companies have been studied. These muds were from three different reservoirs of varying depth and production history (Table 4).

The samples from Company A were obtained from a reservoir with bottom-hole temperature of

Fig. 4. Scanning electron micrograph of a stable enrichment culture isolated from the outer section of core 2 in Pfennigs/acetate medium (0.2% NaCl) at 55°C. Many rod-shaped bacteria, some containing central spores, were evident (×4400 magnification) (scale bar equals 2.27 μm).

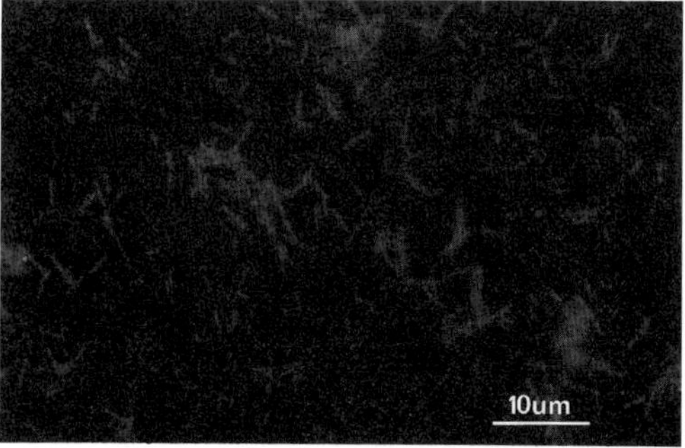

Fig. 6. Light micrograph of the stable enrichment culture H12 following a shut-in period of 10 days in a static reservoir conditions core flood test. The culture was isolated from the outer area of core 1 (depth 3504 feet) grown in Pfennigs/acetate medium (0.2% NaCl) at 55°C. Cells were stained using acridine orange (× 1000 magnification) (scale bar equals 10 μm).

Fig. 7. Scanning electron micrograph of the core plug used in a static reservoir conditions core flood test with culture H12 isolated from the outer section of core 1 (depth 3504 feet). Gypsum crystals ($CaSO_4$) were evident totally covering the rock surface. (×616 magnification) (scale bar equals 50 μm).

Table 5. *The production of H_2S from drilling muds from company A following 6 months incubation. The muds were inoculated into Widdels Medium for the enrichment of SRB*

Drilling mud sample used	Depth (feet)	Incubation temperature		
		30°C	55°C	70°C
Seawater/ guar gum sweeps	A1 (Blank)	med	0	0
	A2 (590 ft)	high	0	0
	A3 (1000 ft)	high	0	0
	A4 (1550 ft)	high	0	0
Salt saturated polymer	A5 (2000 ft)	0	0	0
	A6 (2500 ft)	0	0	0
	A7 (2950 ft)	0	0	0

H_2S levels detected: 0, no H_2S; low <10 ppm H_2S; med, 10–100 ppm H_2S; high, >100 ppm H_2S.

28°C. Interestingly, only mesophilic SRB (mSRB) were detected in these samples, e.g. organisms that grow at about 30°C (Table 5). No H_2S was detected from salt-saturated polymer samples used to drill the bottom 1500 feet of this well. It is surprising to note that in all of the drilling muds studied to date, all of the blank samples were also found to contain H_2S-producing bacteria.

The results from Company B show H_2S production from the water based muds used in the vertical section of the well (Table 6). The greatest amounts of H_2S were produced at 30°C, although H_2S was also detected at 55°C and 70°C. The blank samples again contained H_2S-producing bacteria, although less H_2S was detected in these enrichments. The oil based muds tested showed no H_2S production after 2 months incubation.

Two sets of muds were used by Company C in the well studied. Petrofree mud was used down to about 10 000 feet and a water based mud to total depth (TD) of 13 000 feet. The greatest H_2S levels were detected at 55°C, lower amounts were obtained at 30°C and no H_2S detected at 70°C (Table 7). The petrofree mud was broken down extensively in culture medium with 0.2% NaCl, whilst little biodegradation was apparent in culture broth containing 2% NaCl. The greatest levels of H_2S were usually detected in enrichment cultures with the greatest amount of mud biodegradation.

At 70°C, some mud breakdown was visible from culture C4 (depth 8500 feet) in medium with 0.2% NaCl, however, no H_2S was detected in these samples. Organisms coexisting with the SRB may be responsible for the biodegradation of drilling mud components, producing metabolic byproducts which the SRB can then metabolize. Figure 8

Table 4. *Drilling mud samples from three different oil companies which were incubated into sulphate-rich growth medium for the enrichment of SRB*

Company	Bottom hole depth (ft)	Well type	Drilling muds used	
A	3000	vertical	0–1500 ft	seawater/guar gum sweeps
			1500–3000 ft	salt saturated polymer
B	9500 * (TVD 4500)	horizontal	0–3500 ft	build mud
			3500–9500 ft	oil-based mud
C	13000	vertical	0–10 500 ft	pseudo oil based mud (petrofree)
			10 500–13 000 ft	water-based mud

*TVD, true vertical depth.

Table 6. *The production of H_2S from drilling muds from company B following 6 months incubation. The muds were inoculated into Widdels Medium for the enrichment of SRB*

Drilling mud sample used	Depth (feet)	Incubation temperature		
		30°C	55°C	70°C
Build mud	B1 (Blank)	high	low	low
	B2 (2300 ft)	high	med	med
	B3 (3400 ft)	high	med	med
Oil-based mud	B4 (3500 ft)	0	0	0
	B5 (4500 ft)	0	0	0
	B6 (5700 ft)	0	0	0
	B7 (8500 ft)	0	0	0
	B8 (9500 ft)	0	0	0

H_2S levels detected: 0, no H_2S; low <10 ppm H_2S; med, 10–100 ppm H_2S; high, >100 ppm H_2S.

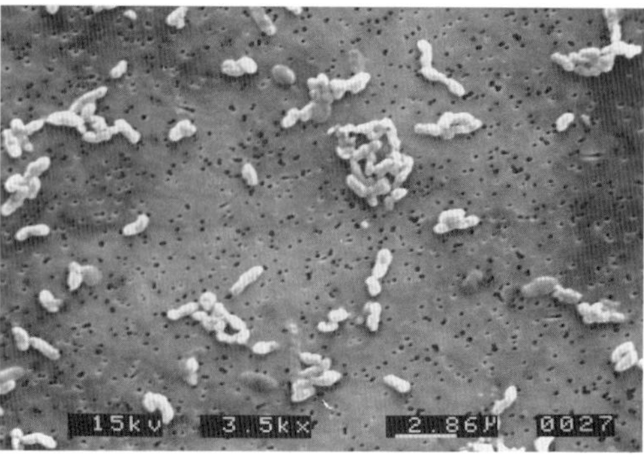

Fig. 8. Scanning electron micrograph from the enrichment culture of the pseudo oil based drilling mud C1 (depth 2620 feet) inoculated into Widdels medium/acetate (0.2% NaCl) and incubated at 55°C for 6 months. This micrograph shows short rod-shaped bacterial cells less than 2 μm in length. (×3500 magnification) (scale bar equals 2.86 μm).

shows bacterial cells isolated from one petrofree enrichment culture at 55°C.

The water based muds used by Company C produced extremely high levels of H_2S in both 30°C and 55°C enrichment cultures (Table 8). Little H_2S was detected at 70°C. An extra incubation temperature of 95°C was used for these samples to mimic the bottom-hole temperature since the reservoir was 13 000 feet deep. The cultures set up at 95°C included some sulphur-based enrichments. Some H_2S was detected in the sulphur containing enrichment cultures of sample C5 incubated at 95°C. This drilling mud sample occurred at the point of mud change over and was found to be a rich source of bacteria. Although it has proved extremely difficult to grow these hyperthermophilic organisms in the laboratory, archaebacterial DNA has been detected from the original drilling mud which may suggest that these sulphur reducing bacteria may in fact be archaebacteria (C. Devine, pers. comm.). Some archaebacterial species produce H_2S from sulphur rather than sulphate.

A number of pure SRB cultures have been obtained from several of the drilling mud samples studied to date. These cultures have been characterized using molecular biology techniques. Two of these isolates are thought to be novel

Table 7. *The production of H_2S from pseudo oil based drilling mud from company C following 6 months incubation. The mud samples were inoculated into Widdels Medium for the enrichment of SRB*

Mud sample Depth (ft)	0.2% NaCl 30°C	2% NaCl 30°C	0.2% NaCl 55°C	2% NaCl 55°C	0.2% NaCl 70°C	2% NaCl 70°C
C1 (2600 ft)	0	0	**High**	**High**	0	0
C2 (4500 ft)	**Low**	0	**High**	0	0	0
C3 (6500 ft)	**High**	High	**High**	Med	0	0
C4 (8500 ft)	**High**	0	**Med**	Low	**0**	0

H_2S levels detected: 0, no H_2S; low <10 ppm H_2S; med, 10–100 ppm H_2S; high, >100 ppm H_2S.
Characters in bold indicate that biodegradation of mud took place. For all other characters, no biodegradation took place.

Table 8. *The production of H$_2$S from water-based muds from company C following 6 months incubation. The mud samples were inoculated into Widdels Medium for the enrichment of SRB*

Mud sample Depth (ft)	0.2% NaCl 30°C	2% NaCl 30°C	0.2% NaCl 55°C	2% NaCl 55°C	0.2% NaCl 70°C	2% NaCl 70°C	2% NaCl 95°C*
C5 (10500ft)	high	high	high	high	low	0	low
C6 (11500ft)	high	high	high	low	0	0	NT
C7 (13000ft)	high	high	high	low	0	0	NT

H$_2$S levels detected: 0, no H$_2$S; low <10 ppm H$_2$S; med, 10–100 ppm H$_2$S; high, >100 ppm H$_2$S.
NT, not tested.
* Sample C5 inoculated into a range of culture media.

species of SRB. Figure 9 shows a light micrograph of one of these SRB pure cultures. (J-Y. Leu, pers. comm.)

Completion chemicals. Completion chemicals used in a clean up sequence were analysed for one oil company (Table 9). One of the chemicals, a guar gum pill was found to be extremely H$_2$S-rich on arrival at the laboratory. Figure 10 shows a phase contrast light micrograph of this sample. The sample was found to be a rich source of microorganisms including H$_2$S-producing bacteria.

Future work

It is hoped in the second phase of this project to further develop the microbial core flood tests. In this way production chemicals can be screened prior to use in the reservoir thus preventing many mechanisms of microbial formation damage.

Table 9. *Completion chemicals incubated into Widdels Medium for the enrichment of SRB*

Code	Sample
1	Mud 9800 ft
2	Baracarb OBM 8400 ft
3	Dirt Magnet In
4	Claustic Pill Clean Up In
5	Guar Gum Pill In Last
6	Baraclean In
7	Guar Gum Baraclean In
8	Caustic Pill Out
9	Dirt Magnet Out
10	HI VIS Baraclean Out
11	Baraclean Out
12	HI VIS Out First of 3
13	HI VIS Out Last of 3

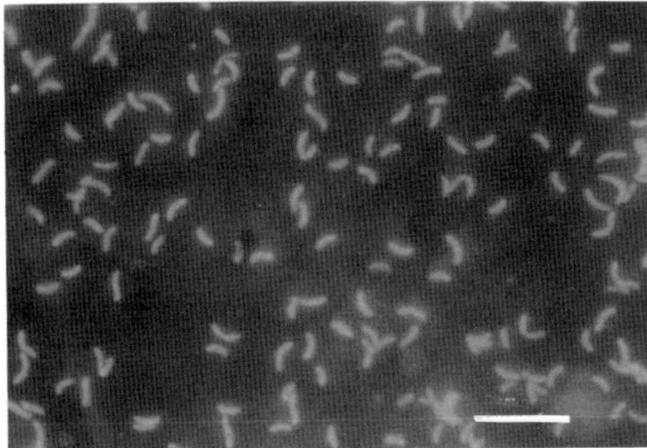

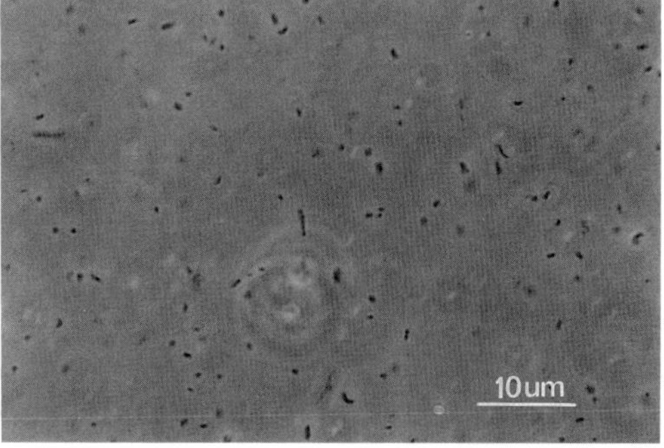

Fig. 9. Light micrograph of the pure SRB culture EM2 isolated from the water based drilling mud C5 from company C. The micrograph shows vibrio-shaped bacterial cells stained with the fluorescent stain acridine orange (×1000 magnification) (scale bar equals 10 μm).

Fig. 10. Light micrograph of guar gum containing completion chemical showing a mixed bacterial population. The cells were visualized using phase contrast microscopy (×1000 magnification) (scale bar equals 10 μm).

The authors would like to thank Douglas Field, Andy Brickell (BHP); Dave Dawson, Alan Leiper (Chevron (UK) Ltd); Liz Aston, Chris Carr (Ranger Oil (UK) Ltd and Fadel Hreeb (Waha Oil Company, Libya) for supplying the materials and their help throughout the project.

References

CLINE, D. J. 1969. Spectrophotometric determination of hydrogen sulphide in natural waters. *Limnology and Oceanography,* **14,** 454–458.

FREEMAN, H. B. 1983. Microbial contamination of drilling fluids and its implications on drilling. *Petroleum Society of Canadian Institute Mining,* 34th Annual Technical Meeting, Paper Number 83-34-22.

ROSNES, J. T., GRAVE, A. & LIEN, T. 1991*a.* Activity of sulphate-reducing bacteria under simulated reservoir conditions. *Society of Petroleum Engineers Production Engineering,* May, 217–220.

——, TORSVIK, T. & LIEN, T. 1991*b.* Spore-forming thermophilic sulphate-reducing bacteria isolated from North Sea oil field waters. *Applied and Environmental Microbiology,* **57,** 2302–2307.

SAUNDERS, P. F. & HAMILTON, W. A. 1986. Biologically induced corrosion. *National Association of Corrosion Engineers,* **8,** DEXTER, S. C. (ed.), 47–68.

WIDDEL, F. & PFENNIG, N. 1981. Studies on dissimilatory sulphate-reducing bacteria that decompose fatty acids. *Archives of Microbiology,* **129,** 395–400.

Clastic sabkhas and diachroneity at the top of the Sherwood Sandstone Group: East Irish Sea Basin

JILLIAN THOMPSON[1] & NEIL S. MEADOWS[2]

Geochem Group Limited, Chester Street, Chester CH4 8RD, UK

[1] *Present address: Robertson Research, Tyn-y-Coed, Llandudno, North Wales LL30 1SA, UK*

[2] *Present address: Redrock Associates Limited, 38 Queens Drive, Prenton, Wirral L43 0RP, UK*

Abstract: The identification of early evaporite sulphate cements with a marine isotopic signature within the Ormskirk Sandstone Formation supports previous models of diachroneity at the contact of the Sherwood Sandstone Group and the Mercia Mudstone Group in the East Irish Sea Basin. Basin-wide correlations suggest that, at the onset of transgression, a range of sandflat and aeolian dune environments co-existed on the margins of the basin with marginal marine environments toward the basin centre. One of the major implications of this model is that the high quality aeolian sandstone reservoirs encountered near the top of the Ormskirk Sandstone in the southern parts of the East Irish Sea Basin have no reservoir correlatives in large areas of the basin but can be predicted on the margins.

Poor biostratigraphical control combined with historically confused stratigraphical nomenclature within the Sherwood Sandstone Group (SSG) have obstructed the recognition of sedimentological and palaeoenvironmental relationships within the East Irish Sea Basin (EISB) and its margins. Despite an increased understanding of the sedimentary environments within the SSG over the past two decades, little attention has been given to relating sedimentological and stratigraphical relationships on a regional scale. This paper offers a sedimentological model for the top of the Ormskirk Sandstone Formation (OSF) which assumes, in very broad terms, the gradual encroachment of marine conditions southwards across the basin at the onset of Mercia Mudstone Group (MMG) deposition. The key to the model was the discovery of early marine evaporite cements within cored sediments from British Gas well 110/8a-5 (Fig. 1), which effectively invoke marginal marine conditions penecontemporaneous with deposition at the top of the OSF. This concept is taken further by integrating these observations with broad-scale facies relationships across large parts of the EISB.

Stratigraphic and sedimentological background

Sherwood Sandstone Group

The clastic sediments of the SSG have long been recognized as representing continental 'red bed' environments comprising mainly the deposits of a major braided fluvial channel system with subordinate aeolian and sandflat deposits (Colter & Barr 1975; Bushell 1986). Although the precise extent of the river system is not known, its source is assumed to have been the Hercynian massifs of southwest England and northern France (Fitch *et al.* 1996; Audley-Charles 1970). The fluvial fairway appears to have extended northwards through the Permo-Triassic basins of the western UK, entering the Irish Sea via the northwest outlet of the Cheshire Basin. Beyond the confines of the channel system, a semi-arid to arid climate, combined with an apparently seasonal northeasterly trade wind belt system (Parrish & Curtis 1982), had a significant effect on sedimentation. The characteristics of the various depositional systems are summarized here in terms of four major facies associations represented by fluvial channel, sheetflood, aeolian and playa lake deposits. A more in-depth synthesis of these facies is given in Meadows & Beach (1993).

Fluvial channel facies. Examples of both major and minor fluvial channel deposits exhibit similar sedimentological characteristics, being dominated by intervals of erosively based, cross-stratified, fine to medium-grained sandstones. These form compound channel-fill sequences which commonly fine upwards in grain size and locally pass vertically into channel abandonment fines. Stacking of successively deposited major channel-fill deposits yields multistorey channel-fill sequences with preserved thicknesses locally in excess of 20 m. Minor (probably ephemeral) channel deposits are commonly recorded as single channel fills that are

From Meadows, N. S., Trueblood, S. P., Hardman, M. & Cowan, G. (eds), 1997, *Petroleum Geology of the Irish Sea and Adjacent Areas,* Geological Society Special Publication No. 124, pp. 237–251.

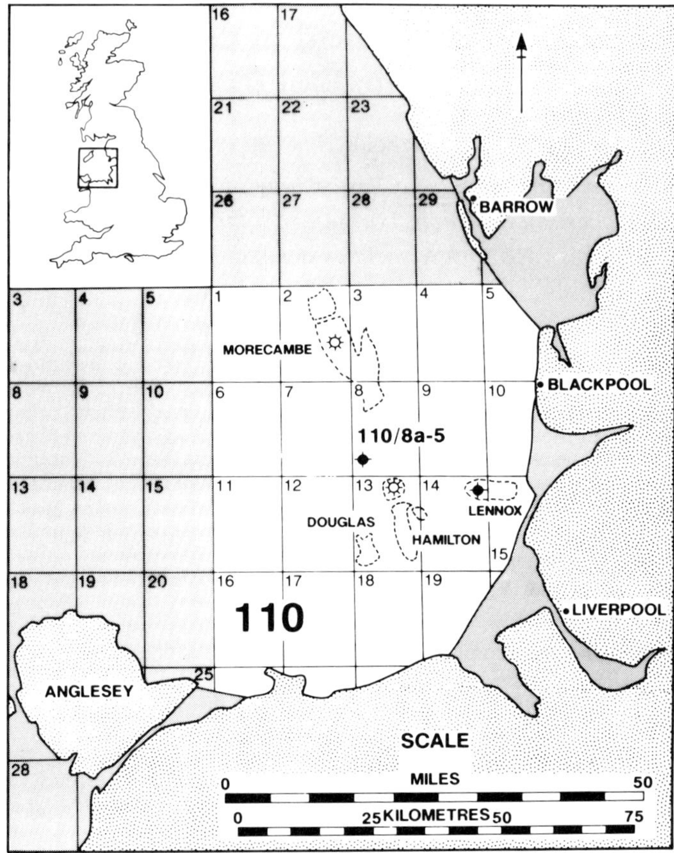

Fig. 1. Quadrant 110, East Irish Sea Basin showing the location of 110/8a-5. Also shown are wells 110/2-6, 110/13-5 and 110-14-3.

often closely interdigitated with a range of non-channelised sandflat deposits.

Sheetflood facies. Sheetflood deposits are depositionally and lithologically varied, typically comprising tabular sheets of irregularly laminated argillaceous sandstones with heterogeneous and variable grain size, sorting and porosity distributions. Although frequently disrupted by soft-sediment deformation and bioturbation, primary depositional structures include plane-bed and current ripple lamination, whilst cross-stratified sets with erosive bases are locally observed. The range of structures preserved is consistent with successive flooding over low-relief sandflats, although subaerial modification by aeolian processes is also claimed to have played an important role in their development (see Herries 1992 and this volume).

Aeolian facies. Aeolian sediments are represented by dune and sandsheet deposits and generally com-prise fine to coarse-grained sandstones with distinctive bimodal sorting and styles of lamination. Dune sets rarely attain preserved thicknesses of more than a metre, although they locally occur as composite sequences up to tens of metres in thickness. Sandsheet deposits are laterally persistent and vary greatly in recorded thickness, from millimetre-scale up to metre-scale.

Playa facies. Playa lake deposits comprise essentially non-reservoir siltstone and claystone-dominated sequences representing the settling of fines from suspension within shallow standing bodies of water. Although similar in character to fluvial abandonment deposits, they are differentiated by their association with non-channelized sandflat strata and typically display greater lateral continuity on a reservoir scale. Internal structure tends to be blocky, although thin sandstone laminae may display current or wave rippling and desiccation cracks are locally abundant. Peripherally developed playa-margin sediments represent the transition between the deposits of playa lakes and wet (sheetflood dominated) sandflat areas and exhibit the characteristics of both environments.

The basis of stratigraphic nomenclature for the SSG was established nationally by Warrington *et al.* (1980), and by Jackson *et al.* (1987) and Jackson & Mullholland (1993) in detailed evaluations of the EISB. In the study area, the SSG was divided into upper and lower formations. The lower (fluvially dominated) is termed the St Bees Sandstone Formation by analogy with onshore fluvial sandstones at St Bees Head, west Cumbria. The upper, comprising mixed fluvial, aeolian, sheetflood and playa sediments, has been locally referred to as the Lower Keuper Sandstone (Colter & Barr 1975), the Helsby Sandstone Formation (Warrington *et al.* 1980) and the Ormskirk Sandstone Formation (Jackson *et al.* 1987). The Lower Keuper Sandstone, originally a subdivision of the German Trias, was shown to be 'Bunter' in age compared with Germany (Warrington 1967) and hence inappropriate for the northwestern UK. The use of Helsby Sandstone (Warrington *et al.* 1980) was based upon work in north Cheshire where three distinct members were recognized (Thompson 1969, 1970*a,b*) on facies-related criteria; an upper Frodsham (aeolian-dominated), middle Delamere (fluvially dominated) and a lower Thurstaston Member (mixed aeolian and fluvial). This scheme was subsequently applied to the reservoir sequence of the Morecambe Field (Colter & Barr 1975; Ebbern 1981; Bushell 1986), although it became apparent that facies associations offshore were often distinct from onshore. As a compromise, Jackson *et al.* (1987) proposed the term Ormskirk Sandstone Formation and so helpfully restricted the designation of members to appropriate basins, thus dispensing with formal and rigorous subdivision within the upper SSG of the EISB.

Biostratigraphical data are extremely sparse for the SSG in the EISB, and of variable quality and abundance within the overlying MMG where diagnostic evidence, mainly spores and pollen, is limited for many sequences. Following a reassessment of palynological assemblages within the Cheshire Basin and the west Lancashire–East Irish Sea Basin, an Anisian age (Middle Triassic) has been proposed (Fig. 2) for all sequences from the basal OSF to a level above the Northwich Halite in the MMG (Benton *et al.* 1994). Previously, sequences within the Ormskirk Sandstone and its transition into the MMG within these areas had been assigned to the late Scythian. These data make it clear that these strata and those immediately overlying belong to an equivalent of the Röt transgression of south Germany as first suggested by Ireland *et al.* (1978) and confirmed by Pollard (1981).

Mercia Mudstone Group

The basal part of the MMG onshore, the Tarporley Siltstone Formation and its equivalents (formerly the 'Keuper' Waterstones), is represented in west Lancashire and northern Cheshire by intertidal to supratidal (estuarine) mudstones and siltstones (Ireland *et al.* 1978), and on the Fylde coast and most parts of the East Irish Sea Basin by sequences of interbedded mudstones and laterally persistent evaporites (Warrington *et al.* 1980; Wilson 1990). Higher parts of the MMG comprise mudstones and fine clastic units with abundant sulphate and dolomite horizons, locally interspersed with thick, regionally extensive halites containing variable amounts of mudstone (Wilson 1990).

It has been long agreed that a substantial part of the MMG was deposited in shallow bodies of water in which current activity was negligible and subaerial exposure was common (Warrington 1981). An aeolian 'loessic' origin has also been advocated for blocky mudstone sequences within parts of the Cheshire Basin (Taylor *et al.* 1963; Arthurton 1980). Connection with a marine source, at least intermittently, is evidenced by sparse palaeontological, sedimentological and geochemical data. Within the Cheshire Basin, for example, acritarchs, indicative of marine incursion, exhibit a sparse though widespread occurrence, although some authors suggest this to reflect an allochthonous source, analagous to the present day Ranns of Kutch in India (Glennie & Evans 1976) where marine microplankton have been swept inland a distance of 150 km by wind and high tides. Sedimentological evidence for early marine conditions is provided by limited trace fossil data. The identification of *Skolithos-Glossifungites* inchnofacies in the Tarporley Siltstone Formation of northern Cheshire (Ireland *et al.* 1978; Pollard 1981) led these authors to infer intertidal conditions at the onset of MMG deposition. More conclusive evidence is provided by geochemical data. Published sulphur isotope analyses (Naylor *et al.* 1989) and bromine data (Haslam *et al.* 1950; Tucker 1981) indicate that most of the lower Cheshire evaporites have a marine source. It thus appears that halite and gypsum deposits were precipitated from hypersaline waters during episodes of intense evaporation and steady reflux from marine source waters. At other times water salinity is likely to have varied both spatially and temporarily in response to fluctuating rates of marine versus freshwater (fluvial) input and the intensity of evaporation.

Confusion regarding stratigraphical terminology also afflicts the MMG and its contact with the Ormskirk Sandstone. Within the Morecambe Field, for example, confusion between old and new

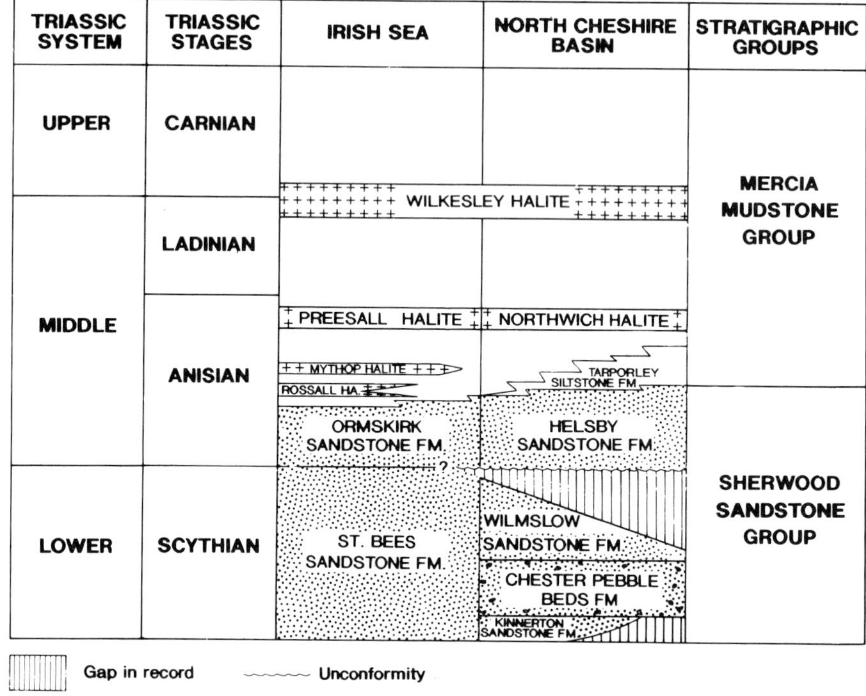

Fig. 2. Simplified stratigraphical breakdown of the Sherwood Sandstone Group and Mercia Mudstone Group in the Irish Sea and North Cheshire Basins (adapted from Benton *et al.* 1994 and Jackson & Mulholland 1993).

stratigraphic terms led Ebbern (1981) and Bushell (1986) to place the 'Keuper Waterstones' erroneously within the SSG.

Previous models of diachroneity

Insights into the nature and origin of the SSG–MMG boundary have varied over the past few decades. Thompson (1970a) concluded that the base of the Tarporley Siltstone within the northern part of the Cheshire Basin, although locally approximating to a time plane, must be diachronous to at least some degree since it represents the spread of intertidal, lagoonal and probable marine conditions across the basin from north to south. He went as far as to suggest that the Frodsham Member formed part of a coastal dune system which migrated diachronously southwards parallel to the advance of the Waterstones (Thompson 1969). Diachroneity at this interface was investigated palynologically by Warrington (1970) who, across Northern Ireland, northwest England and the Midlands, recognized a marked southwards younging of the basal MMG (Tarporley Siltstone and its equivalents). More recently, Warrington (*in* Wilson & Evans 1990 and

Benton *et al.* 1994) noted a correlation between the Bromsgrove Sandstone Formation (SSG) at Stratford-upon-Avon with the Kirkham Mudstone Formation (MMG) in west Lancashire, both dated as Anisian.

Clastic sabkhas at the top of the Ormskirk Sandstone Formation

Examination of cores from well 110/8a-5 (Fig. 1) has revealed early evaporite precipitates in the upper OSF in quantities previously undetected by routine analysis. The core (230 ft/82 m cored interval) is located in the reservoir sequence at the top of the formation. It comprises facies consistent with those described above and represents the full range of fluvial–aeolian facies associations encountered elsewhere in the EISB in equivalent successions (Meadows & Beach 1993). Early evaporite cement phases are restricted to the uppermost 90 ft (30 m) of the cored Ormskirk Sandstone interval. Here they are first discussed in sedimentological and diagenetic terms and later with respect to isotopic and pore-fluid compositions.

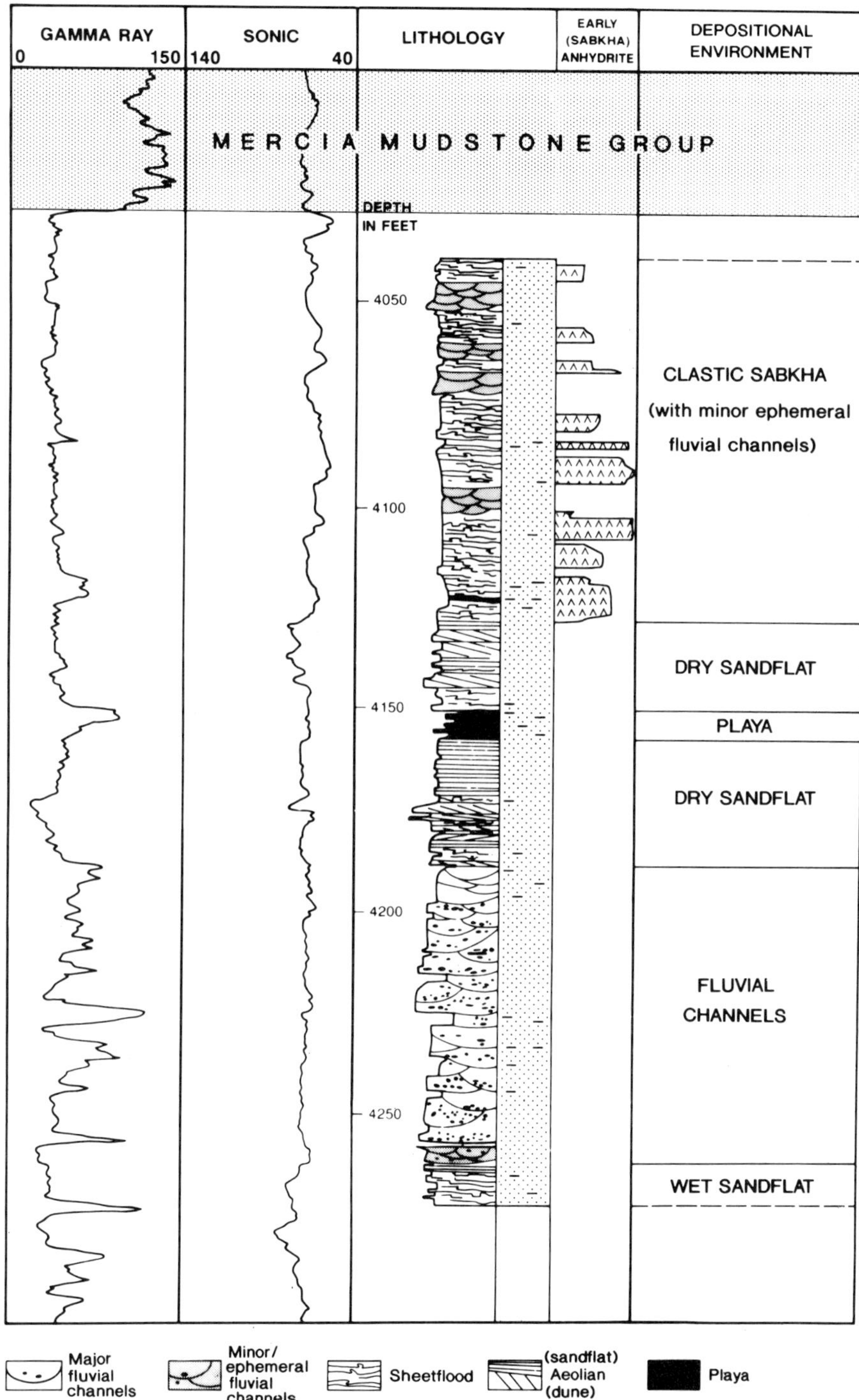

Fig. 3. Schematic facies relationships and anhydrite distribution within the cored interval of well 110/8a-5 plotted against sonic log and gamma ray data (supplied by British Gas Exploration and Production Limited).

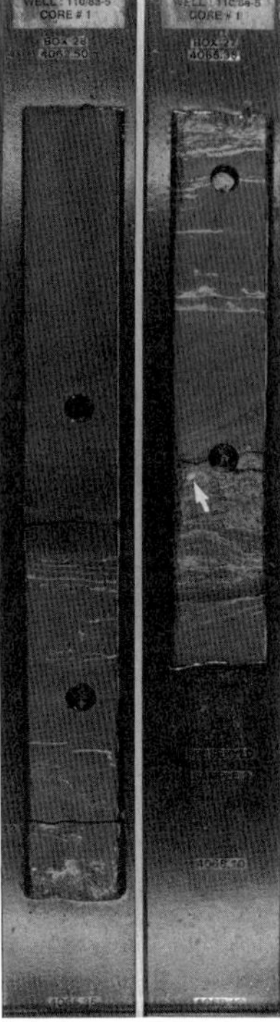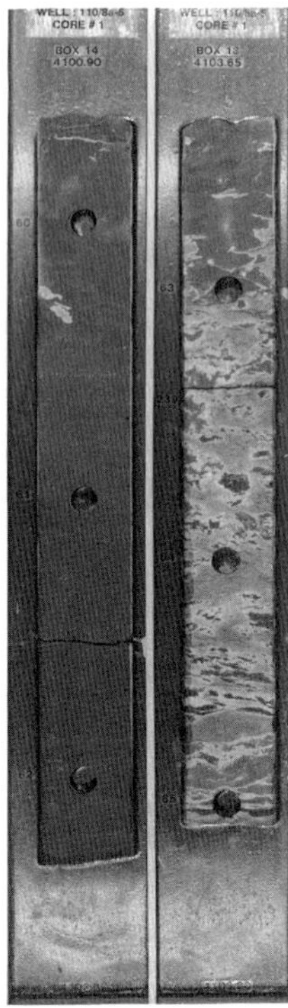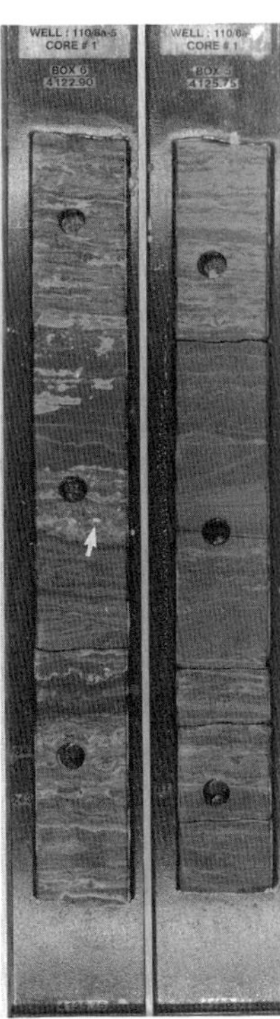

Fig. 4. Representative core photographs from 110/8a-5 showing distinctive anhydrite cement fabrics (white irregular laminae) and rare nodules (arrowed).

Sedimentary and diagenetic features

Facies relationships within the core are summarized in Fig. 3. Major fluvial channel deposits dominate the lower part of this interval, mainly comprising fine to medium-grained, cross-stratified and pebbly intraclastic sandstones with locally interspersed channel abandonment fines. Above this, a range of non-channelized sandflat, playa and aeolian dune strata are represented together with minor or ephemeral fluvial channel sandstones. Dry aeolian sandsheet and dune deposits are more prevalent at the base of this sandflat interval and diminish in abundance towards the very top of the sequence.

Early evaporite cements are generally sediment-supported and dominated by anhydrite with sub-ordinate dolomite. The host sediments comprise sheetflood and playa margin facies, representing the products of tractional sheetflow, settling of suspended fines and varying degrees of reshaping by aeolian processes at the sediment surface or by soft-sediment deformation. Over time, a high and continuously rising water table is likely to have favoured deposition and preservation of such sequences. Conspicuous anhydrite cements (Fig. 4) are concentrated into millimetre- and centimetre-scale laminae or, more rarely, discrete nodules up to 1 cm in diameter. The latter are locally reworked into the bases of cross-bed sets associated with large sheetfloods or ephemeral fluvial channels, i.e. sequences otherwise devoid of early evaporite cements. Soft-sediment deformation is ubiquitous within the laminar-cemented horizons where it is manifest as gentle contortions and doming or more intense disturbances characterized by partially homogenized fabrics.

Two conclusions are drawn from the cement fabrics. Firstly, the identification of reworked anhydrite intraclasts within intervening sequences provides direct evidence for evaporite precipitation during or very soon after sediment accumulation. Secondly, restriction of the laminar and nodular cements to wavy-laminated sandflat and playa-margin deposits suggests a strong genetic link between these sequences and evaporite precipitation. In thin section anhydrite comprises felted and subparallel arrangements of laths (Fig. 5a), either surrounding grains or concentrated into nodules, and in essence shows textures typical of early evaporitic origin. It is closely associated in

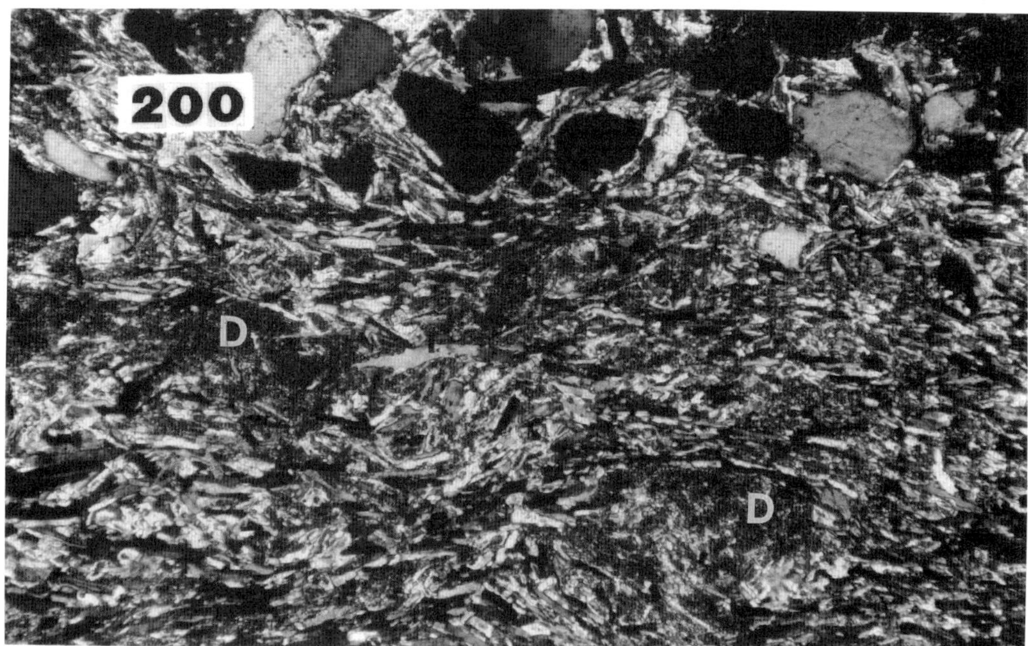

(a)

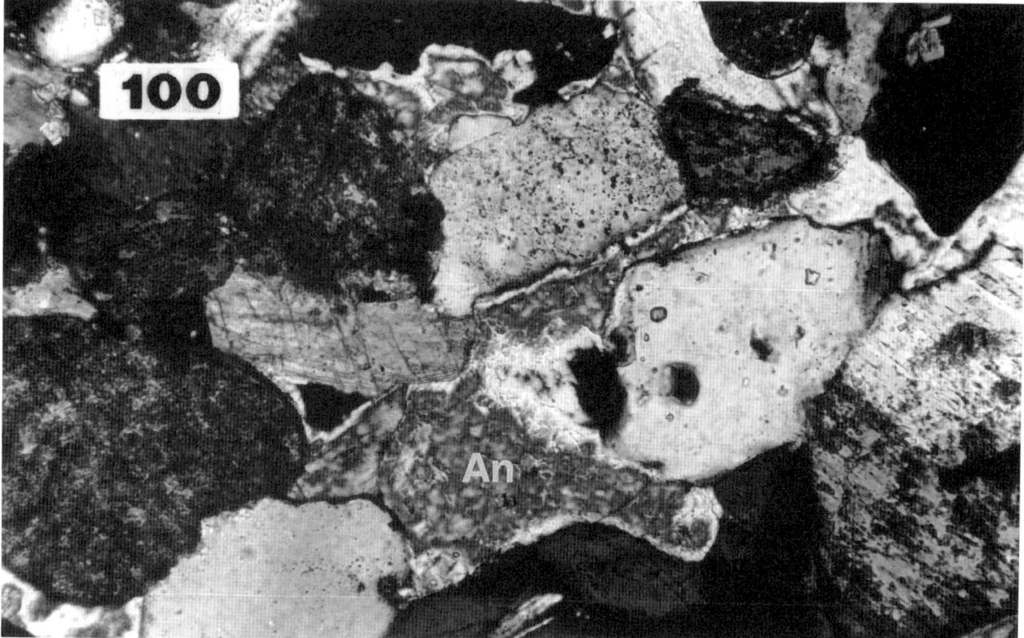

(b)

Fig. 5. Thin section photomicrographs from clastic sabka sequences in 110/8a-5. (a) Early evaporite cement composed of felted subparallel laths of anhydrite with associated sucrosic dolomite (D). (b) Poikilotopic anhydrite (An) represents a diagenetically late cement. Scale bar in microns, XPL.

places with early diagenetic microcrystalline dolomite which displays a sucrosic texture and locally forms micronodules and hollow circular rims. A later form of poikilotopic anhydrite is also observed (Fig. 5b), but is morphologically quite distinct from the earlier form (see Greenwood & Habesch, this volume, for an in-depth diagenetic synthesis).

Sulphur isotope data

To establish the sulphate source for the early evaporite cements, the sulphur isotope compositions of selected anhydrite-cemented samples were analysed and compared with published Triassic seawater compositions. The examples chosen contained either early or late phases or both, so that their signatures could be compared. The results (Table 1) reveal a range of $\delta^{34}S$ values between +18.6‰ and +21.4‰ with a mean of +20.0‰, there being no apparent difference between early and late anhydrite or mixtures of the two. The narrow range of $\delta^{34}S$ values encountered shows close agreement with the Triassic seawater composition of +19.48‰ quoted by Veizer *et al.* (1980) and similar anhydrite $\delta^{34}S$ values of +18.4‰ to +20.8‰ obtained by Naylor *et al.* (1989) from the MMG of the Cheshire Basin. The latter authors proposed a marine source for the sulphate, assuming a Triassic seawater value of +16.5±1.5‰ (Claypool *et al.* 1980) and an enrichment of +1.5‰ in precipitated anhydrite relative to aqueous sulphate. The source of evaporitic sulphate within well 110/8a-5 thus appears to be Triassic seawater, with little or no dilution from a continental meteoric component (for which signatures are likely to have been lighter and much more variable). Similar values for both early and late anhydrite cement fabrics are inferred to reflect a simple dissolution–reprecipitation of the early cement phase without significant fractionation (Greenwood & Habesch, this volume).

Table 1. *Table of sulphur isotope compositions within anhydrite cements*

Sample depth (ft)	$\delta^{34}S_{CDT}$ (‰)	Primary (P) and Late (L) anhydrite cements
4043.40	+21.4	L
4085.20	+19.6	P
4091.20	+21.2	P
4093.30	+20.6	P & L
4105.30	+18.8	P
4106.30	+19.7	P & L
4160.90	+18.6	L

The clastic sabkha model

The evidence suggests that early evaporite cements within the Ormskirk Sandstone are facies-specific and that the origin of the sulphate was Triassic seawater. A model of evaporite precipitation is proposed which invokes clastic sabkha development marginal to marine environments of the Mercia Mudstone and provides a basis for sedimentological and palaeogeographical correlation throughout large parts of the EISB.

Early cementation is likely to have been restricted to the low relief and laterally extensive sandflat areas bordering the Mercia Mudstone shoreline. The combined effects of a high water table and intense evaporation from the sediment surface are likely to have encouraged saturation of saline pore fluids above the water table and ultimately the interstitial and displacive growth of evaporite minerals, both as layers and nodules, within near-surface sediments (Fig. 6). Soft-sediment deformation was probably initiated by early cement growth and surface desiccation processes, and was later emphasized by a combination of burial compaction, dewatering and localized evaporite dissolution. Ponded bodies of saline water, effectively forming coastal playas or salinas, also appear to have developed locally, possibly subject to recharging by washover processes. These are interpreted to have been responsible for the localized deposition and precipitation of sulphate-rich playa-margin facies.

Facies correlation at the top of the Ormskirk Sandstone Formation

The identification of marginal marine clastic sabkhas within the south–central part of the EISB has profound implications for sedimentological and palaeogeographical models for the top OSF. Though it has been recognized for some time that facies distributions vary with respect to their location in the basin, few accounts explain the variabilities. Allowing for local differences, it is clear for example that aeolian facies are more prevalent close to the basin margins which represent areas of relative elevation and lower water table levels. A re-evaluation of sedimentological environments and correlations across large parts of the EISB is required to account for broad-scale facies characteristics and their relationship with clastic sabkha deposits in 110/8a-5. To achieve this, representative wells have been selected from across the basin that, together with 110/8a-5, represent a crude northwest to southeast traverse from basin centre to margin (Fig. 1). Core from the OSF has been analysed from each of these wells, including

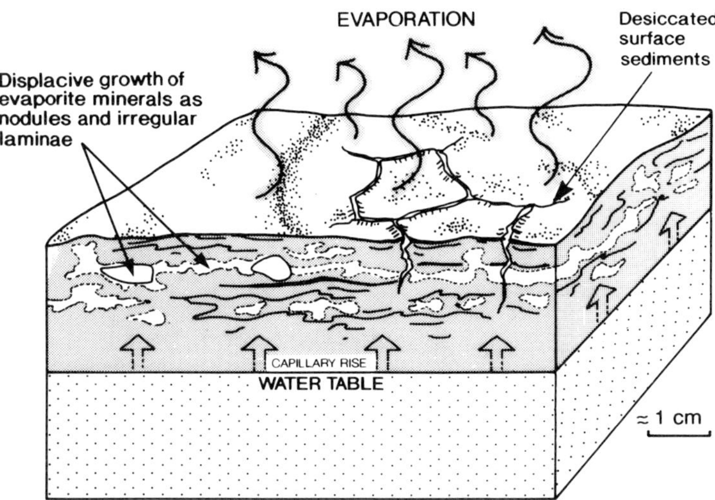

Fig. 6. Genetic model for early evaporite precipitation in near-surface sabkha sediments.

110/2-6 (Morecambe Field), 110/13-5 (BHP) and 110/14-3 (BG). The thickness of the OSF varies across the traverse, being greatest in the basin centre (800 ft/244 m in 110/2-6) and relatively thin towards the margins (500 ft/152 m in 110/13-5 and 515 ft/157 m in 110/14-3).

A correlation of these successions is presented in Fig. 7 together with a generalized facies breakdown. The entire OSF interval is displayed for 110/14-3, although its upper sections only are presented in the other wells. A major obstacle to correlation is establishing a reliable time plane throughout these continental sequences, given the poor biostratigraphic control and the absence of regional sedimentological markers. More importantly, the contact between the SSG and the MMG, for the reasons outlined above, no longer approximates to a time plane for facies correlation on a basinwide scale. As a compromise, therefore, it was decided to align the successions along the top of a well defined major fluvial channel sequence recognized in each well. The orientation of the wells is approximately parallel to the regional south to north palaeoflow within this part of the basin and it seems reasonable to suppose that the fluvial sandstones represent, in broad terms, part of the same fluvial system. Given the inherent mobility of channel systems, however, the existence of a time plane at the top of major fluvial packages is a gross oversimplification since cessation of fluvial activity on a basinwide scale is likely to have taken place at different times, with some channels remaining active while others were abandoned. Weakly

supporting the correlation is the presence of a potentially correlable playa sequence (Fig. 7) within 110/2-6 and 110/8a-5 which may thin towards 110/13-5 and pinch out at the basin margin. However, there is no independent evidence to suggest that the playa sequences are anything other than coincidental.

Well correlation according to Fig. 7 reveals several interesting trends. Aeolian facies are abundant above the channel sandstones in wells 110/2-6 and 110/8a-5 and are replaced upwards by progressively wetter sandflat deposits dominated by a range of sheetflood, playa and ephemeral fluvial channel deposits. A similar 'wetting'-upward profile is evident in 110/13-5, but is less distinct due to a greater abundance of aeolian sandstones. In 110/14-3 the upper sandflat interval is not only much thicker but is almost completely dominated by aeolian dune and sandsheet deposits. Facies not only become more aeolian towards the basin margins, as previously noted, but form part of a much thicker sandflat sequence (relative to the top fluvial channel datum).

Significantly, the correlation proposed in Fig. 7 also suggests that reservoir sequences at the top of the OSF pass laterally northwards and northwestwards into contemporaneous argillaceous and saliferous beds in the lower part of the MMG. Sulphate cement accumulations within the clastic sabkhas of 110/8a-5 may therefore represent the landward (up-dip) equivalents of early sulphate or halite accumulations within the basal part of the MMG.

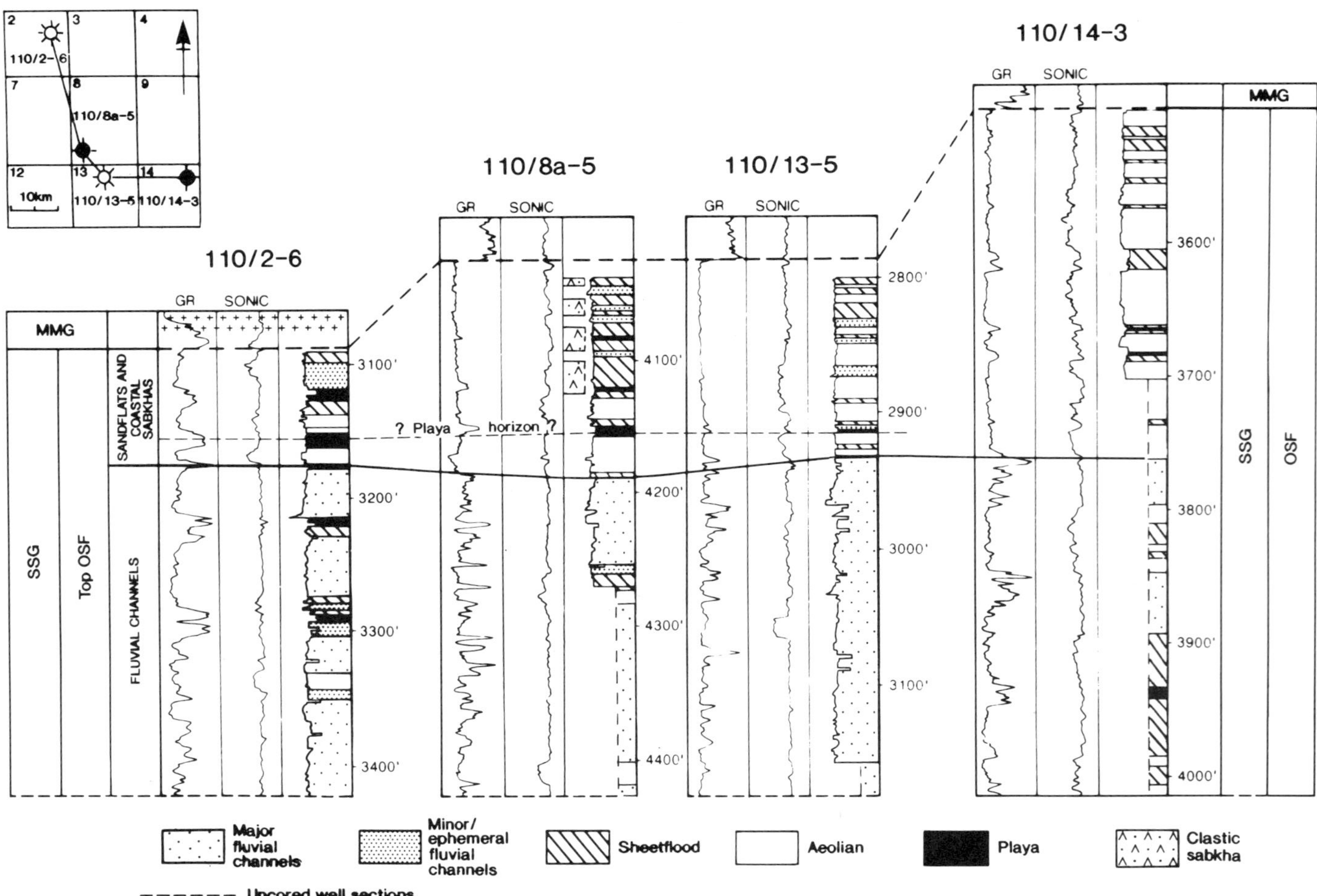

Fig. 7. A correlation of 110/8a-5 with representative well sections throughout the East Irish Sea Basin. Note the southwards and southeastwards thickening of the sandflat package at the top of the Sherwood Sandstone Group.

Implications of diachroneity at the top of the Ormskirk Sandstone Formation

Figure 8 summarizes schematically the facies correlations suggested above and attempts to place the distribution of halite members within the MMG in the same context. Regional mapping and correlation have suggested that progressively younger halite sequences within the MMG thin markedly towards the basin margins (Wilson 1990) and successively overlap each other, progressively overstepping the SSG. In a diachronous model, therefore, some of the oldest halite and mudstone sequences within the EISB are inferred to represent the age-equivalents of a southwards-thickening wedge of continental sandflat sediments that forms the top of the Ormskirk Sandstone. The Preesall Halite is possibly the first major halite sequence to extend continuously beyond the confines of the

EISB and into the Cheshire Basin (Warrington & Ivimey-Cook 1992), where it is slightly thinner and known locally as the Northwich Halite (Evans *et al.* 1968). It has been assigned a late Anisian age by Warrington (*in* Wilson & Evans 1990 and Benton *et al.* 1994). The equivalent of the overlying Wilkesley Halite (earliest Carnian) persists southeastwards beyond the Cheshire Basin where it is represented by the Stafford Halite (Warrington *et al.* 1980).

A major implication of this model is that aeolian-dominated sequences encountered at the top of the Ormskirk Sandstone in the southern part of the EISB and along its margins have no contemporary reservoir equivalents in large parts of the basin centre, since Sherwood Sandstone deposition had already ceased in many basinal areas when these sands were being deposited. This does not imply an absence of good (high reservoir quality) aeolian

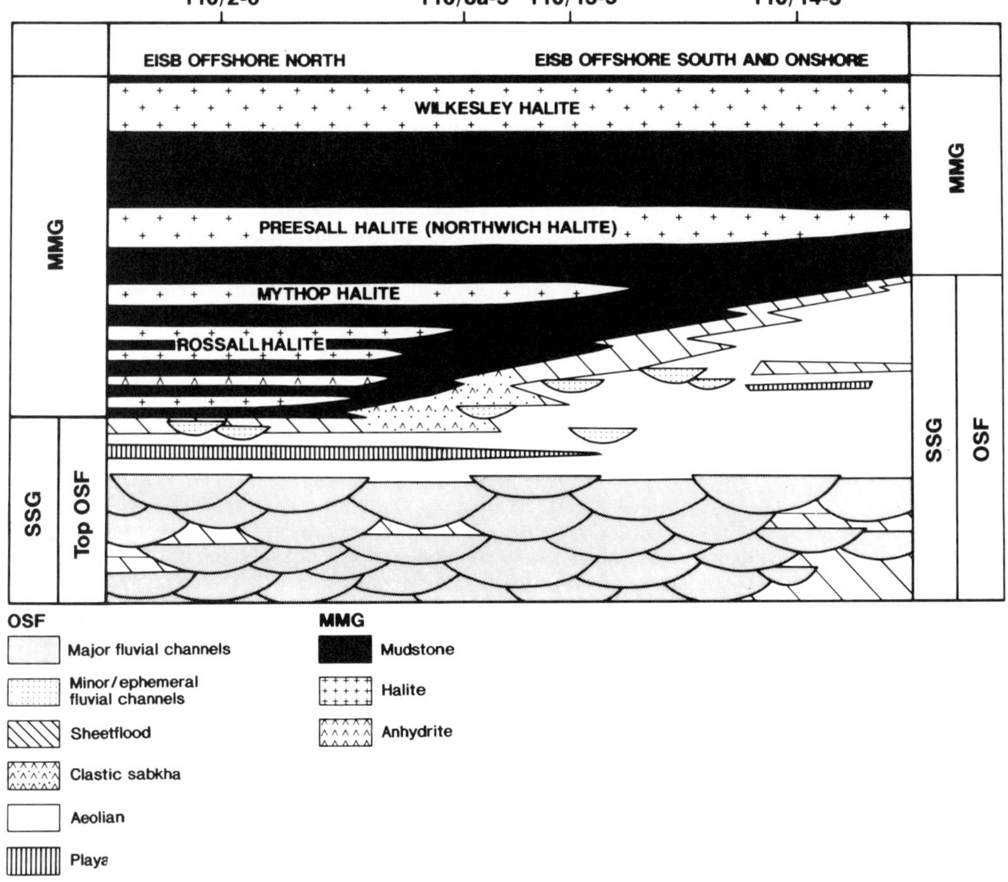

Fig. 8. Schematic facies correlation at the inferred diachronous contact between the Sherwood Sandstone Group and Mercia Mudstone Group.

reservoirs in the basin centre, but rather that they are distributed elsewhere in the reservoir sequence and tend not to be within the uppermost reservoir interval.

Another important aspect of the model is the potential to review the SSG–MMG boundary in terms of a sequence stratigraphic framework. The effects of base level-controlled accommodation space on sedimentary response are well documented for humid systems (Posamentier & Vail 1988) and the fundamental principles are equally relevant when applied to arid depositional systems. The whole transition from fluvial to sandflat environments at the top of the OSF may represent the systematic rise in water table and decrease in fluvial amalgamation known to reflect base level rise, culminating in diachronous marine flooding (MMG) and the local development of coastal sabkhas in juxatoposed OSF sediments. The coastal evaporites in 110/8a-5 may in theory represent the up-dip expression of a discrete evaporite horizon within the MMG and, if repeated regionally, this phenomenon could provide a potential means of correlating marine evaporate layers in the MMG with sabkha sequences in the top SSG. However, although the presence of early sulphate cements is not unique, their abundance in 110/8a-5 is exceptional (probably a preservational effect). The basis for this kind of systematic sequence analysis is thus tentative and beyond the scope of this paper.

Palaeogeographic reconstructions

A series of palaeogeographies (Figs 9a–c) are proposed for the upper part of the Ormskirk Sandstone based on the model outlined above and the broad-scale facies distributions recognized across the basin.

In the first of these (Fig. 9a), major braided channel systems were apparently active across the whole of this part of the basin, although their influence diminished eastwards towards the basin margin (110/14-3). They flowed northwards or northwestwards, parallel to the regional palaeoslope, and deposited thick sequences of stacked channel-fill sandstones at a time when fluvial gradients were high. Channel migration and avulsion appear to have restricted the preservation of within-channel abandonment and any adjacent non-channelized deposits.

Major fluvial systems were then abandoned wholesale in this part of the basin, the precise mechanism of fluvial shut-off being unclear but possibly related to regional base level rise as noted above. Alternatively, it could relate to more remote climatic or tectonic changes taking place in the fluvial hinterland. Sandflat and aeolian dune systems were subsequently established across the

basin (Fig. 9b), with 'drier' aeolian facies concentrated towards the margins where water tables were comparatively low and 'wetter' sheet-flood and playa facies more common in the centre where water tables were higher but subject to episodic fluctuations. Channel sandstones were still deposited locally, but were representative of minor or ephemeral fluvial systems with highly variable discharge patterns. Although it is possible that oceanic base level rise controlled sedimentation in a very broad sense, climatic fluctuations are also interpreted to have contributed to correlable facies variation across the basin. For instance, a regionally dry episode, represented by abundant aeolian dune and sandsheet deposits, is suggested in sequences immediately overlying the major fluvial channel system, whilst a regional pluvial episode may account for laterally extensive playa lake sequences (Fig. 7).

Figure 9c illustrates the first effects of marine transgression (Röt equivalent) and MMG deposition in the centre of the EISB. Whilst muds and halites were deposited in the vicinity of 110/2-6, clastic coastal sabkhas were developing in the upper OSF near well 110/8a-5. Seawater sulphates were drawn up into sediments of the arid coastal plain and precipitated locally as laminar and nodular anhydrites. Shallow saline ponds accumulated locally. Dry sandflat and aeolian dune environments persisted along the basin margins but were pushed back peripherally as the transgression continued. Broadly analogous environments are presently developing along parts of the Saudi Arabian coastline in areas where strong offshore winds carry an abundant sand supply to the coast and effectively prevent classic carbonate sabkha development (Fryberger et al. 1983). Offshore trade winds may have operated in a similar way within the EISB during simultaneous deposition of the OSF and its offshore equivalents in the MMG.

Discussion

A model of basin evolution is proposed for the EISB which invokes the contemporaneous deposition of the top OSF and the basal MMG during the Anisian. At this time, marine, locally hypersaline environments encroached southwards across the basin, gradually replacing continental clastic systems with finer-grained muds and evaporite accumulations. At the margins of this transgressive system, a flat, extensive coastal plain was developed which locally supported coastal clastic sabkhas and early evaporite precipitation from marine pore waters. Given the inherent ephemeral nature of early evaporite cements, their abundance within 110/8a-5 may simply represent a freak preservational phenomenon. Cement precipitation

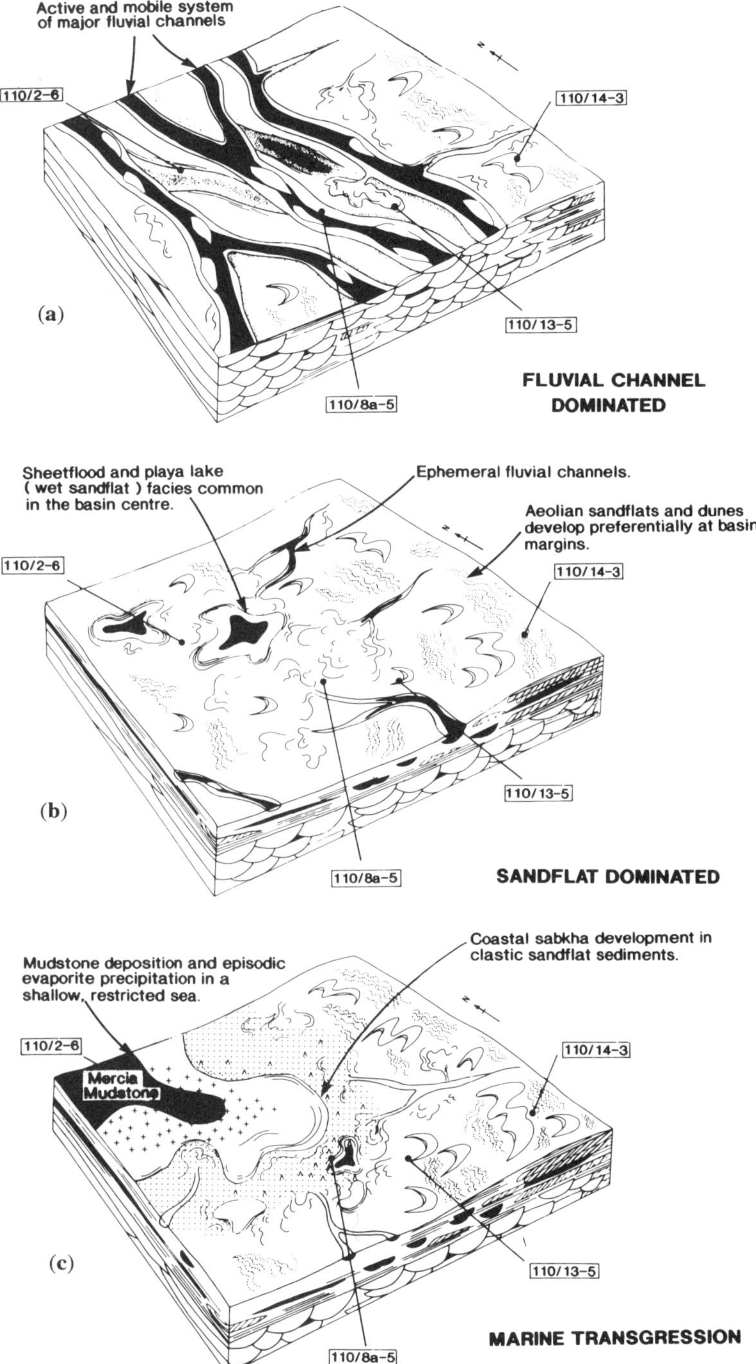

Fig. 9. Schematic palaeogeographies based on well correlations for the upper Ormskirk Sandstone Formation and lower parts of the Mercia Mudstone Group in the vicinity of well 110/8a-5. (**a**) Active fluvial channel activity across the entire study area. (**b**) Cessation of major fluvial channel activity and extensive sandflat development (dry sandflats preferentially developed towards the basin margins). (**c**) Onset of marine transgression and Mercia Mudstone deposition. Early evaporite precipitation within coastal sabkhas.

may have been widespread, taking place within all sandflat sediments characterized by net evaporitic loss and with access to marine pore fluids.

At the onset of transgression, coastal geomorphology may have been quite complex and subject to constant shifting in response to subtle relative sea-level changes. Early sulphate and halite accumulations may have thus had an erratic distribution, making them an unreliable tool for correlation in the lower part of the Mercia Mudstone succession. As the transgression advanced, much more uniform mudstone and evaporite cycles were deposited, incorporating discrete and laterally correlable halite sequences, each representing deposition in shallow-water bodies of regionally uniform depth. Gradual reduction of topographic relief between the Irish Sea and Cheshire Basins appears likely to have accompanied the transgression and the associated deposition of marine sequences.

As the transgression proceeded southwards, expansive intertidal plains replaced continental sedimentation across the Cheshire Basin (Tarporley Siltstone Formation). Overlying halite successions are demonstrably correlable with several of the EISB halites (most notably the Preesall/Northwich Halite), suggesting that the two basins were cojoined at this time with little relief to distinguish them.

The authors express their thanks to BG Exploration and Production and BHP Petroleum whose personnel made available the core materials and wireline data and kindly gave us permission to publish the data. A special thanks goes to Martin Hardman of BG with whom we had many profitable discussions. Thanks are also due to D. B. Thompson and S. Flint for critically reviewing the manuscript. Stable isotope analyses were performed at the Scottish Universities Research and Reactor Centre. We kindly acknowledge the assistance of colleagues at Geochem Group Limited, principally Claire Harris and Sharon Hindley for draughting and Hywel Jones for photography.

References

ARTHURTON, R. S. 1980. Rhythmic sedimentary sequences in the Triassic Keuper Marl (Mercia Mudstone Group) of Cheshire, northwest England. *Geological Journal*, **15**, 43–58.

AUDLEY-CHARLES, M. G. 1970. Triassic palaeogeography of the British Isles. *Quarterly Journal of the Geological Society, London*, **126**, 49–89.

BENTON, M. J., WARRINGTON, G., NEWELL, A. J. & SPENCER, P. S. 1994. A review of the British Middle Triassic tetrapod assemblages. *In:* FRASER, N. C. & SUES, H. D. (eds) *In the Shadow of the Dinosaurs: Early Mesozoic Tetrapods*. Cambridge University Press, New York, 131–160.

BUSHELL, T. P. 1986. Reservoir geology of the Morecambe Field. *In:* BROOKS, J., GEOFF, J. C. & VAN HOORN, B. (eds) *Habitat of Palaeozoic Gas in NW Europe*. The Geological Society, London, Special Publication, **23**, 189–208.

COLTER, V. S. & BARR, K. W. 1975. Recent developments in the geology of the Irish Sea and Cheshire Basins. *In:* WOODLAND, A. W. (ed.) *Petroleum and the Continental Shelf of North West Europe, Volume 1: Geology*. Applied Science, London, 61–75.

CLAYPOOL, G. E., HOLSER, W. T., KAPLAN, I. R., SAKAI, H. & ZAK, I. 1980. The age curves of sulphur and oxygen isotopes in marine sulphate and their mutual interpretation. *Chemical Geology*, **28**, 199–260.

EBBERN, J. 1981. The geology of the Morecambe Gas Field. *In:* IILLING, L. V. & HOBSON, G. D. (eds) *Petroleum Geology of the Continental Shelf of North West Europe*. Heyden, London, 485–493.

EVANS, D. J., REES, J. G. & HOLLOWAY, S. 1993. The Permian to Jurassic stratigraphy and structural evolution of the central Cheshire Basin. *Journal of the Geological Society, London*, **150**, 857–870.

EVANS, W. B., WILSON, A. A., TAYLOR, B. J. & PRICE, D. 1968. *Geology of the Country around Macclesfield, Congleton, Crewe and Middlewich*. Memoirs of the Geological Survey, UK, 328 pp.

FITCH, F. J., MILLER, J. A. & THOMPSON, D. B. 1966. The palaeogeographic significance of isotopic age determinations on detrital micas from the Triassic of the Stockport–Macclesfield district, Cheshire. *Palaeogeography, Palaeoclimatology, Palaeoecology*, **2**, 281–313.

FRYBERGER, S. G., AL-SARI, A. M. & CLISHAM, T. J. 1983. Eolian dune, interdune, sand sheet and siliciclastic sabkha sediments of an offshore prograding sand sea, Dhahran, Saudi Arabia. *Bulletin of the American Association of Petroleum Geologists*, **67**, 280–312.

GLENNIE, K. W. & EVANS, G. 1976. A reconnaissance of the recent sediments of the Ranns of Kutch, India. *Sedimentology*, **23**, 625–647.

GREENWOOD, P. J. & HABESCH, S. M. 1997. Diagenesis of the Sherwood Sandstone Group in the southern East Irish Sea Basin (Blocks 110/13, 110/14 and 110/15): constraints from preliminary isotopic and fluid inclusion studies. *This volume*.

HASLAM, J., ALLBERRY, E. C. & MOSES, G. 1950. The bromine content of the Cheshire salt deposit and some borehole and other brines. *Analyst*, **75**, 352–356.

HERRIES, R. D. 1992. *Sedimentology of Continental Erg–Margin Interactions*. PhD thesis, University of Aberdeen, 228 pp.

—— & COWAN, G. 1997. Challenging the 'sheetflood' myth: the role of water-table-controlled sabkha deposits in redefining the depositional model for the Ormskirk Sandstone Formation (Lower Triassic), East Irish Sea Basin. *This volume*.

HOLLAND, T. H. 1912. The origin of desert salt deposits. *Proceedings of the Liverpool Geological Society*, **11**, 227–250.

HOLLOWAY, S. 1985. Triassic Sherwood Sandstone Group (excluding the Kinnerton Sandstone Formation and the Lenton Sandstone Formation). *In*: WHITTAKER, A. (ed.) *Atlas of Onshore Sedimentary Basins in England and Wales: Post-Carboniferous tectonics and stratigraphy.* The Geological Society, London, 31–34.

IRELAND, R. J., POLLARD, J. E., STEEL, R. J. & THOMPSON, D. B. 1978. Intertidal sediments and trace fossils from the Waterstones (Scythian–Anisian?) at Daresbury, Cheshire. *Proceedings of the Yorkshire Geological Society*, **41**, 399–436.

JACKSON, D. I. & MULHOLLAND, P. 1993. Tectonic and stratigraphic aspects of the East Irish Sea Basin and adjacent areas: contrasts in their post-Carboniferous structural styles. *In*: PARKER, J. R., *Petroleum Geology of Northwest Europe: Proceedings of the 4th Conference.* The Geological Society, London, 791–808.

——, ——, JONES, S. M. & WARRINGTON, G. 1987. The geological framework of the East Irish Sea Basin. *In*: BROOKS, J. & GLENNIE, K. W. (eds) *Petroleum geology of North West Europe*, Vol. 1. Graham and Trotman, London, 191–203.

MEADOWS, N. S. & BEACH, A. 1993. Stuctural and climatic controls on facies distribution in a mixed fluvial and aeolian reservoir: the Triassic Sherwood Sandstone in the Irish Sea. *In*: NORTH, C. P. & PROSSER, J. (eds) *Characterization of Fluvial and Aeolian Reservoirs.* Geological Society, London, Special Publication, **73**, 247–264.

NAYLOR, H., TURNER, P., VAUGHAN, D. J., BOYCE, A. J. & FALLICK, A. E. 1989. Genetic studies of red bed mineralization in the Triassic of the Cheshire Basin, Northwest England. *Journal of the Geological Society, London*, **146**, 685–700.

PARRISH, J. T. & CURTIS, R. L. 1982. Atmospheric circulation, upwelling and organic-rich rocks in the Mesozoic and Cenozoic eras. *Palaeogeography, Palaeoclimatology, Palaeoecology*, **40**, 67–101.

POLLARD, J. E. 1981. A comparison between the Triassic trace fossils of Cheshire and south Germany. *Palaeontology*, **24**, 555–588.

POSAMENTIER, H. W. & VAIL, P. R. 1988. Eustatic controls on clastic deposition, II, sequence and systems tract models. *In*: WILGUS, C. K., HASTINGS, B. S., KENDALL, G. St C., POSAMENTIER, H. W., ROSS, C. A. & VAN WAGONER, J. C. (eds) *Sea Level Changes: An Integrated Approach.* Society of Economic Palaeontologists and Mineralogists, Special Publication, **42**, 125–154.

TAYLOR, B. J., PRICE, R. H. & TROTTER, F. M. 1963. *Geology of the Country around Stockport and Knutsford.* Geological Survey of Great Britain Memoir, England and Wales, Sheet 98.

THOMPSON, D. B. 1969. Dome-shaped aeolian dunes in the Frodsham Member of the so-called 'Keuper' Sandstone Formation (Scythian–?Anisian: Triassic) at Frodsham, Cheshire (England). *Sedimentary Geology*, **3**, 263–289.

—— 1970a. The stratigraphy of the so-called Keuper Sandstone Formation (Scythan–?Anisian) in the Permo-Triassic Cheshire Basin. *Quarterly Journal of the Geological Society of London*, **126**, 151–181.

—— 1970b. Sedimentation of the Triassic (Scythian) Red Pebbly Sandstones in the Cheshire Basin and its margins. *Geological Journal*, **7**, 183–216.

TUCKER, M. E. 1978. The marginal Triassic rocks from south Wales: shore zone clastics, evaporites and carbonates. *In*: MATTER, A. & TUCKER, M. E. (eds) *Modern and Ancient Lake Sediments.* The International Association of Sedimentologists, Special Publication, **2**, 205–224.

TUCKER, R. M. 1981. *The Origin, Geochemistry and Sedimentology of the Triassic Cheshire Halite.* PhD Thesis, University of Newcastle-upon-Tyne.

VEIZER, J., HOLSER, W. T. & WILGUS, C. K. 1980. Correlation of $^{13}C/^{12}C$ and $^{34}S/^{32}S$ secular variations. *Geochimica et Cosmochimica Acta*, **44**, 579–587.

WARRINGTON, G. 1967. Correlation of the Keuper series of the Triassic by miospores. *Nature*, **214**, 1323–1324.

—— 1970. The 'Keuper' Series of the British Trias in the northern Irish Sea and neighbouring areas. *Nature*, **226**, 254–256.

—— 1981. The indigenous micropalaeontology of British Triassic shelf sea deposits. *In*: NEALE, J. W. & BRASIER, M. D. (eds) *Microfossils from Recent and Fossil Shelf Seas.* The British Micropalaeontological Society/Ellis Horwood, Chichester, 61–70.

—— & IVIMEY-COOK, M. L. 1992. Triassic. *In*: COPE, J. C. W., INGHAM, J. K. & CRAWSON, P. F. (eds) *Atlas of Palaeogeography and Lithofacies.* Geological Society, London, Memoir No. 13, 97–106.

——, AUDLEY-CHARLES, M. G., ELLIOT, R. E., EVANS, W. B., IVIMEY-COOK, H. C., KENT, P. E., ROBINSON, P. L., SHOTTON, F. W. & TAYLOR, F. M. 1980. A correlation of Triassic rocks in the British Isles. *Special Report of the Geological Society, London*, No. 13, 78 pp.

WILSON, A. A. 1990. The Mercia Mudstone Group (Trias) of the East Irish Sea Basin. *Proceedings of the Yorkshire Geological Society*, **48**, 1–22.

—— & EVANS, W. B. 1990. *Geology of the Country around Blackpool.* *Memoir of the British Geological Survey*, sheet 66 (England and Wales).

Challenging the 'sheetflood' myth: the role of water-table-controlled sabkha deposits in redefining the depositional model for the Ormskirk Sandstone Formation (Lower Triassic), East Irish Sea Basin

ROBERT D. HERRIES[1,2] & GREIG COWAN[3]

[1] *Department of Geology and Petroleum Geology, University of Aberdeen, Aberdeen AB9 2UE, UK*

[2] *Present address: Amerada Hess Ltd., 33 Grosvenor Place, London SW1X 7HY, UK*

[3] *British Gas plc, 100 Thames Valley Park Drive, Reading, Berkshire RG6 1PT, UK*

Abstract: Two distinct units can be recognized in the Lower Triassic Ormskirk Sandstone Formation in the East Irish Sea Basin. Perennial fluvial channel, lacustrine and minor aeolian sandstones dominate the upper part, but the lower is characterized by thick (tens of metres) enigmatic wavy bedded deposits interbedded with aeolian and minor fluvial channel sandstones. Previous interpretations that invoked an overbank 'sheetflood' or 'sandflat' origin for the wavy bedding are difficult to reconcile with the sedimentary structures, which are more consistent with algal and evaporitic processes in modern sabkhas. These deposits contain widespread (km) 'drying upward' patterns 2–10 m thick, reflecting a sabkha to more aeolian dominated setting. Such thicknesses are consistent with an origin related to 23 000 year Milankovitch cyclicity.

The lower, wavy bedded portion of the Ormskirk Sandstone Formation is regionally truncated by fluvial channel-belt sandstones. Their extent reflects a combination of high sediment supply, uniform subsidence and low basinal relief. Lacustrine units near the top of the formation can be correlated regionally, with one passing southward into a major aeolian unit that thickens to 60 m in the south of the basin. This transition demonstrates the diachroneity of the Mercia Mudstone Group transgression. The key to correlating the Ormskirk Sandstone Formation lies in identifying intervals dominated by fluvial, lacustrine or sabkha-aeolian deposits. Correlation of water-table-controlled 'drying upward' patterns within sabkha–aeolian deposits allows finer scale sub-division, and provides an estimate for the degree of fluvial truncation at the top of the unit.

Between one-third to a half of the Lower Triassic Ormskirk Sandstone Formation of the East Irish Sea Basin (Fig. 1) consists of flat-lying, wavy bedded deposits long thought to have a sheetflood origin (Colter & Ebbern 1978). They are interbedded on a millimetre to metre scale with aeolian sandstones and, on a larger scale (metres to tens of metres), with major fluvial channel sandstones. Contrary to previous interpretations, they share few of the features typically associated with sheetflood processes, instead consisting of wavy and irregular strata previously attributed to soft sediment deformation and bioturbation (Meadows & Beach 1993*a,b*). However, most, if not all of the features associated with the wavy and irregular stratification can be explained by analogy with modern, aeolian-dominated siliciclastic sabkhas. The aim of this paper is to demonstrate the significance of these sabkha deposits in re-defining previous depositional models, and to show how widespread but consistent, changes in surface dampness produce correlatable drying upward depositional patterns. Although these, along with some of the widespread fluvial depositional units can be used as a basis for regional correlation, low well density in some areas currently precludes numbering them as 'cycles' similar to those of the Permian Rotliegendes in the Southern North Sea.

Historical background

Much of our understanding of the Lower Triassic Ormskirk Sandstone Formation in the East Irish Sea Basin has come from the hydrocarbon industry, following the discovery of the South Morecambe Field in 1974. Colter & Ebbern (1978) were the first of several industry authors (e.g. Ebbern 1981; Bushell 1986) to describe the reservoir characteristics of the field. All proposed a wholly fluvial origin for the sequence, which was interpreted as a series of ribbon-like channel sandstones encased within less permeable overbank 'sheetflood' deposits. From a reservoir development viewpoint (Bushell 1986), the emphasis was on reservoir

From Meadows, N. S., Trueblood, S. P., Hardman, M. & Cowan, G. (eds), 1997, *Petroleum Geology of the Irish Sea and Adjacent Areas*, Geological Society Special Publication No. 124, pp. 253–276.

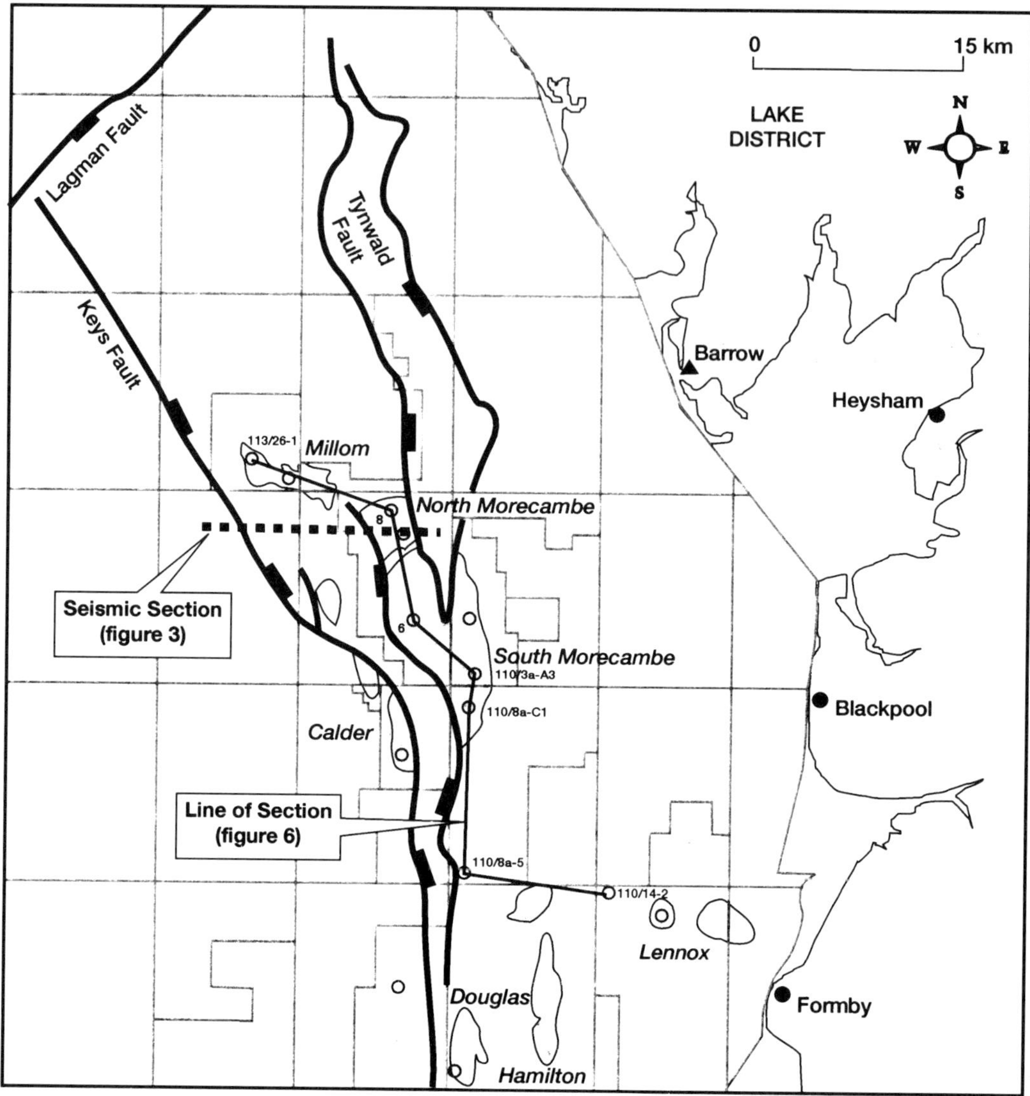

Fig. 1. Location map of East Irish Sea Basin showing the wells used to construct the correlation panel in Fig. 6.

drainage, by intercepting as many channel sandstones as possible within a single well. Despite the wholly fluvial interpretations, the presence of well rounded 'millet seed' grains was noted by these authors and attributed to fluvial reworking of aeolian sands. Woodward & Curtis (1987) were the first to document aeolian dune and sandsheet deposits offshore, though these had been previously identified in the neighbouring Cheshire Basin by Thompson (1969, 1970a,b).

It was not until the early 1990s that the full extent of the aeolian deposits was documented (Stuart & Cowan 1991; Herries 1992; Cowan 1993; Meadows & Beach 1993a,b). By this time, the fluvial emphasis on field development had shifted, following the recognition that the high permeability aeolian sandstones provided most of the flow into the wellbore (Cowan 1993, fig. 13). This observation, and the recognition that some of the depositional associations were extremely widespread led to a change in development emphasis from reservoir drainage to reservoir productivity. The assumption that the 'sheetflood' sandstones were genetically linked to the major fluvial channels was questioned by Herries (1992), who likened many of the structures in the 'sheetflood' deposits to those described from modern aeolian dominated siliciclastic sabkhas (e.g. Fryberger *et al.* 1983). Support for this interpretation was provided by T. M. Goodall (pers. comm. 1992), at that time undertaking PhD research on Sabkhat Matti in the United Arab Emirates (Goodall 1995).

A detailed dipmeter study of the same area (Cowan *et al.* 1993) revealed that major facies changes in the upper half of the Ormskirk Sandstone Formation were accompanied by subtle changes in structural dip, perhaps indicating some causal relationship. On the basis of seismic character, and the notion that fluvial palaeoflow was parallel to elongate N–S depositional thickening trends associated with basin margin faults (notably the Keys, Sigurd, Lagman and Formby Point faults), a model that predicted aeolian sand distribution was proposed by Meadows & Beach

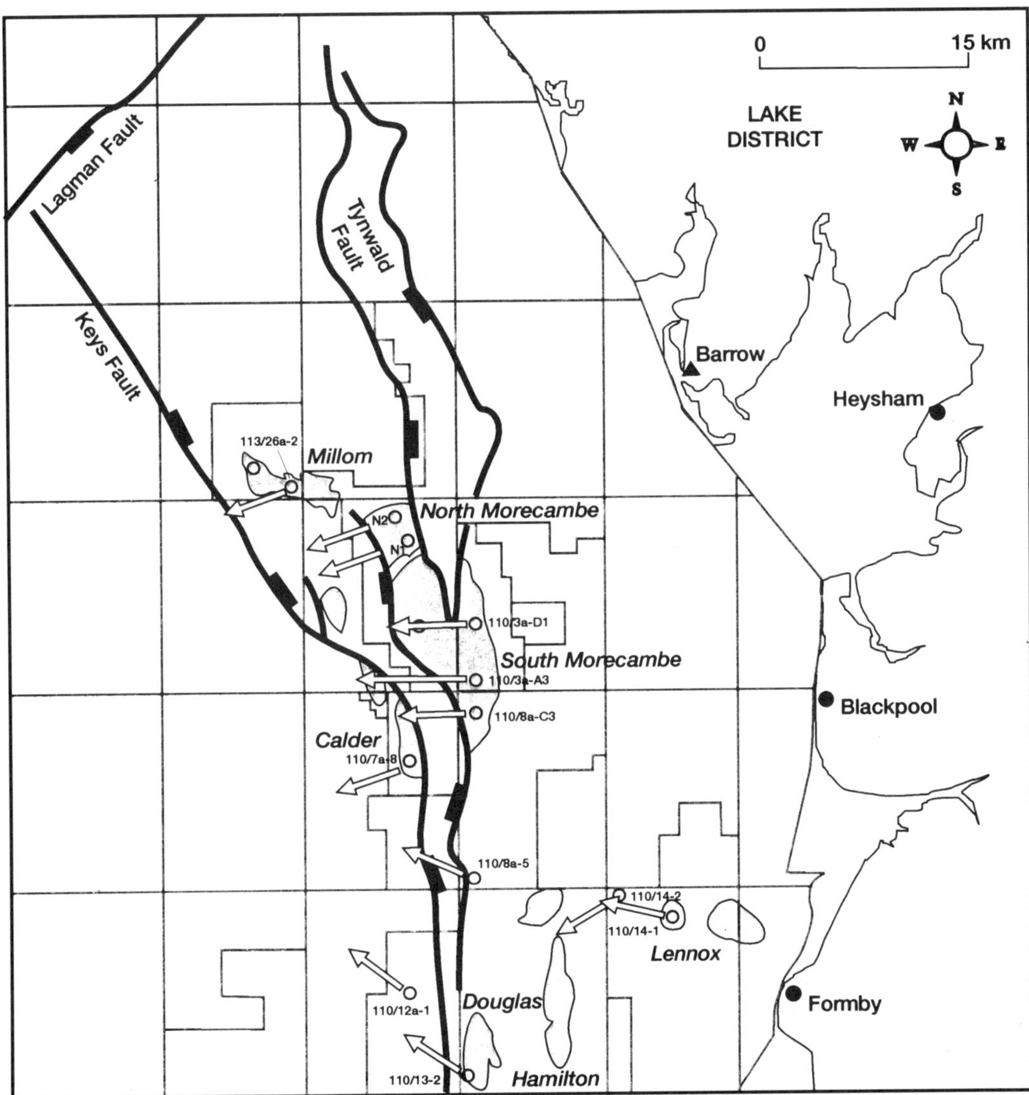

Fig. 2. Palaeocurrent vectors for cross-stratified fluvial sandstones. Consistent westward fluvial transport directions east and west of the Keys Fault are difficult to reconcile with previous models that invoked northward flows along the fault. Arrow tails indicate well locations.

(1993*b*). This contradicted dipmeter interpretations from the Morecambe fields, which indicated westerly fluvial palaeoflow (Cowan 1993).

Figure 2 shows new data indicating that westerly fluvial palaeocurrent vectors are not confined to the Morecambe fields, but occur on both the hanging wall and footwall of the Keys Fault as well as in the more southerly part of basin. This, and the presence of strong, continuous seismic layering (Fig. 3) is difficult to reconcile with Meadows & Beach's (1993*b*) model, which predicted major northward flowing fluvial systems along the hangingwalls of the basin bounding faults, with coeval aeolian or sabkha activity located on growth related rollovers to these faults. This paper will demonstrate that with one notable exception, the thick accumulations of fluvial and aeolian-dominated sandstones are mutually exclusive and form laterally extensive, if not basin-wide, correlatable units.

Depositional associations

Descriptions of the sedimentary facies encountered in the Ormskirk Sandstone Formation can be found in Bushell (1986), Herries (1992) and Meadows & Beach (1993*b*). The following summarizes the key sedimentary features of the facies associations, broadly grouping them into: (i) single and multi-storey fluvial channel-belt, (ii) lacustrine, playa and lake margin, and (iii) aeolian dominated sabkha associations. During deposition of the Ormskirk Sandstone Formation, each association has, at one time or another, extended over large areas of the basin, and it is this which is responsible for the seismic scale layering of Fig. 3. In the northern and central areas (around blocks 110/2 & 3) the upper half of the Ormskirk Sandstone Formation is dominated by fluvial and lacustrine facies associations. However, to the south, the upper part of this interval contains an increasing proportion of aeolian sandstone (blocks 110/13, 14, 15) as discussed by Thompson & Meadows (this volume).

Fluvial channel-belt association

Description. Multistorey fluvial units are dominated by well organized cosets of trough cross-stratified, fine to medium grained sandstone, forming distinct units up to 30 m thick (Fig. 4a). Stacking of such units with minor intervening aeolian sandstones produces fluvially dominated sandstones up to 80 m in thickness. Individual cross-sets are centimetres to a metre or so in thickness. Sets of low angle and horizontal strata are confined to finer grained lithologies, forming up to 15% of the overall thickness. Abandonment fines are occasionally preserved, and whilst extra-formational pebbles are extremely rare, mudstone intraclast lags up to 30 cm thick are developed locally.

The correlatable extent of these fluvial units is proportional to their thicknesses, so that a 5 m thick sandbody may extend a few kilometres transverse to palaeoflow, whereas the thickest fluvial sand-bodies occur basin-wide. Smaller, single storey fluvial units 1–2 m thick occur isolated within intervals dominated by wavy strata (Fig. 4b), and are usually restricted to single wells. They too contain stacked bedforms, but preserve greater proportions of low angle and finer grained channel abandonment lithologies.

Interpretation. Attempting to identify fluvial channel type from core or even a single channel

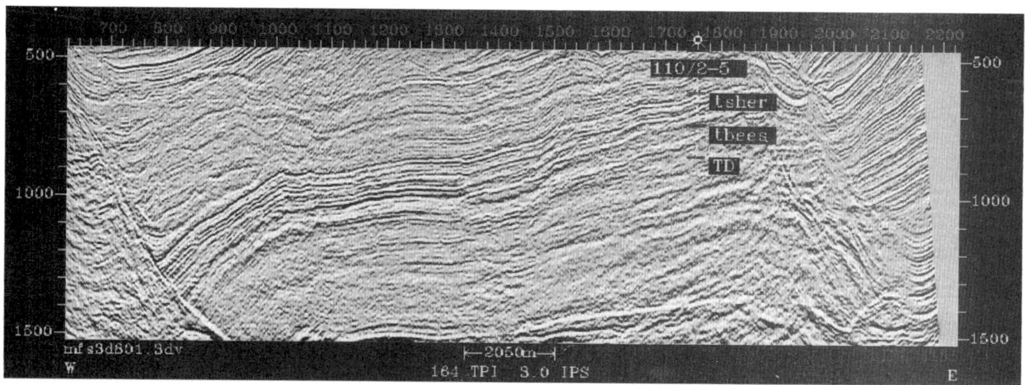

Fig. 3. Seismic section through the North Morecambe Field showing seismic-scale parallel reflectors in the Ormskirk Sandstone Formation. Note the minimal thickening into the Keys Fault, to the left of the figure. Location of line shown in Fig. 1.

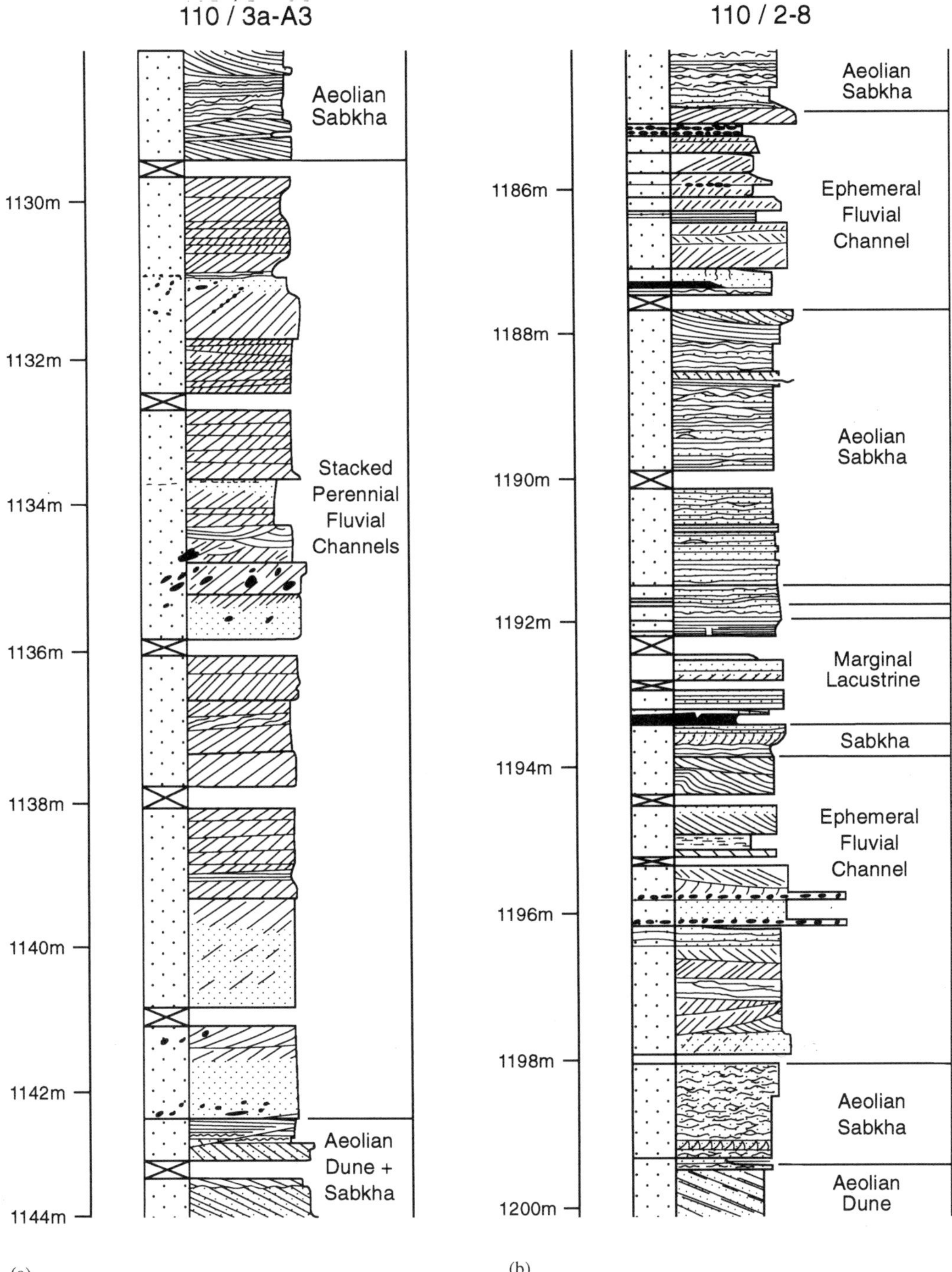

Fig. 4. (a) Core log of stacked perennial fluvial channel sandstones overlying aeolian deposits in well 110/3a-A3. Grain sizes range from very fine to coarse; scale in metres. (b) Core log showing ephemeral fluvial channel sandstones interbedded with aeolian sabkha deposits in well 110/2-8. Sand grade ranges from very fine to coarse; scale in metres.

cross-section is a difficult process, given that modern channels of 'braided' and 'meandering' aspect contain both channel-fill and laterally accreted elements (see Bridge 1985 for critical discussion). Additionally, it is quite conceivable that under low rates of accommodation, enhanced fluvial reworking by a high sinuosity channel would preserve only the basal sandy channel portion, thus preserving a sand-dominated deposit that might traditionally be interpreted as having a braided river origin.

The volumetrically minor single storey sandbodies (1–2 m thick) that occur interbedded with sabkha deposits are interpreted to represent either isolated, ribbon- to sheet-like channels (*sensu* Friend *et al.* 1979), or individual channel-fills forming parts of amalgamated sheets only a single channel thick. The higher proportion of preserved fines is simply due to the preservation of channel abandonment deposits as well as those resulting from bedform migration and does not necessarily reflect higher channel sinuosities. These single channel units may have been ephemeral, perhaps flowing over a single wet season.

In the Ormskirk Sandstone Formation, the scarcity of muddy overbank material, the well organized nature of the cosets, and the high ratio of sandy bedform to horizontally stratified sandstone and abandonment deposits suggest permanent flow, probably in large braided rivers. The absence of fine grained overbank material cannot be accounted for by total reworking as mudstone intraclasts, since these are generally rare (though see Cowan *et al.* 1993, fig. 14). The scarcity of overbank material may have driven earlier studies to invoke an overbank origin for the interbedded aeolian sabkha deposits.

The dimensions of some fluvially dominated units (up to 80 m thick and tens of kilometres in extent) are an order of magnitude greater than other ancient fluvial channel belts (e.g. Fielding & Crane 1987). These deposits are thought to reflect the vertical and lateral amalgamation of several multistorey channel belts associated with a permanent westwardly flowing river system (Figs 2 & 5). Their thickness reflects generation of considerable accommodation space which, at this time, is thought to have been driven by regional subsidence in response to thermal sag (Quirk & Kimble 1996, this volume). As might be expected in such a tectonic setting, these fluvial systems at times extended basin-wide (Fig. 6), and may have been climatically controlled.

Lacustrine, playa and lake margin association

Description. Mixed associations of mudstones, planar and rippled sandstones, and heterolithic sandstones and mudstones occur mainly in the uppermost 50 m of the Ormskirk Sandstone Formation. Individually, they rarely exceed 2 m in thickness, yet are correlatable over large areas of the basin (Fig. 6). Lower in the sequence, they may be equally thick, but are of more limited extent.

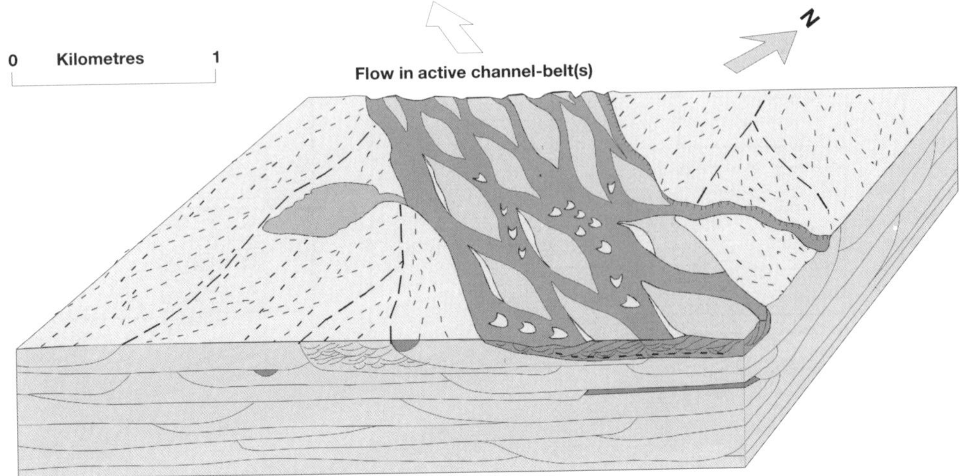

Fig. 5. Schematic model for perennial fluvial channel-belt deposition of Fig. 4a. Amalgamated deposits such as these dominate a unit approximately 60 m thick, which can be correlated across the study area from the Millom Field in the north to block 110/14 in the south (Fig. 6).

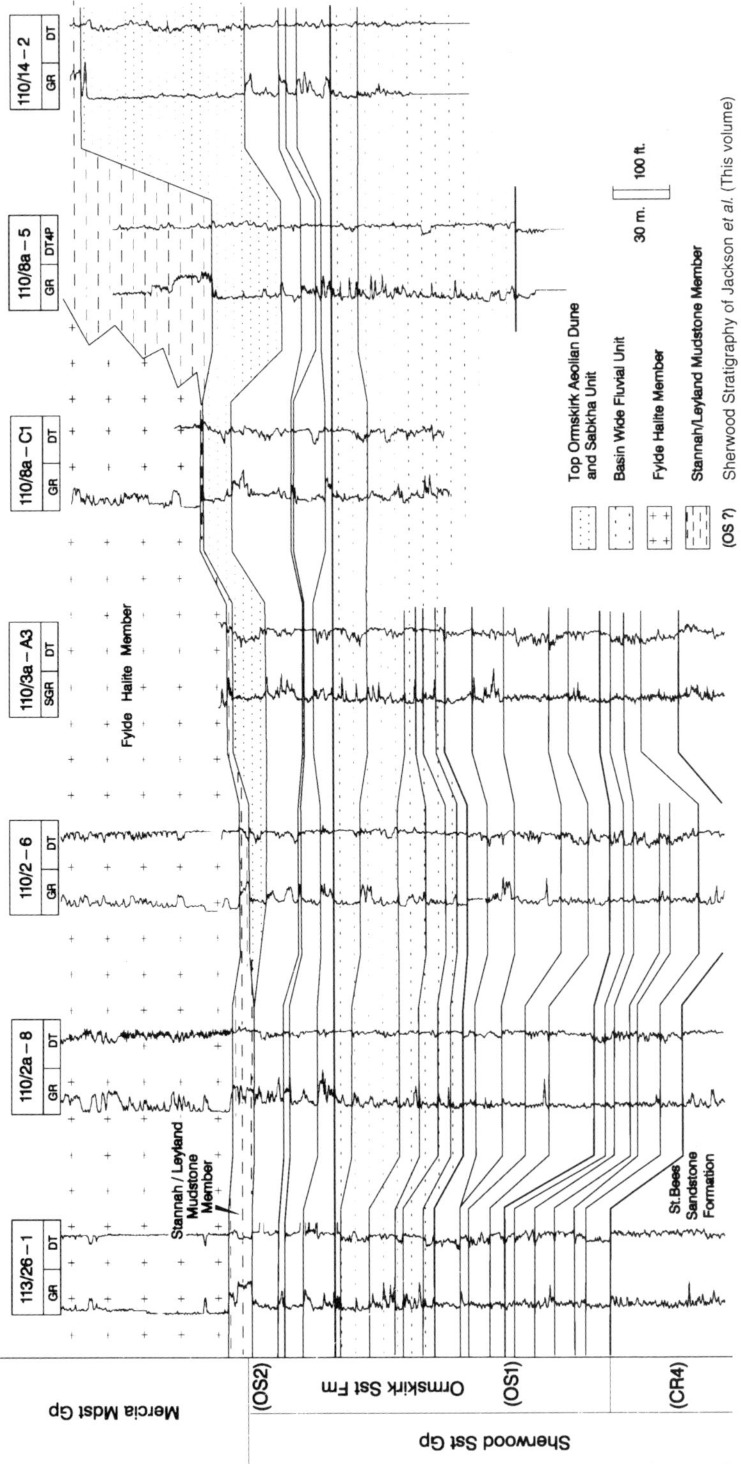

Fig. 6. North–south correlation of the Ormskirk Sandstone Formation using top of regionally extensive, stacked fluvial sandstones referred to in Fig. 5 as datum. The overlying aeolian dune unit extends as far north as South Morecambe (well 110/3a-A3) and appears to correlate with the lower part of the Stannah Mudstone Member. The mudstone thins to the south, where the overlying Fylde Halite Member directly overlies the Ormskirk Sandstone Formation in well 110/8a-C1. Note the regional extent of some of the higher lacustrine mudstone horizons.

Vertically, individual units are loosely arranged in fining- then coarsening-upward packages each a metre or so in thickness. The central mudstones are apparently homogeneous or may be finely laminated and possibly varved. Coalesced carbonate nodules of possible pedogenic origin are sometimes present, along with sand-filled sub-vertical cracks more often attributable to water escape than desiccation. The disruption that results from this cracking can be extreme, with cracks frequently linking thin (1–2 cm) sandy interbeds. This results in a heterolithic deposit consisting of mudstone 'blocks' suspended in a sandstone matrix. The heterolithic facies is transitional to sandier lithologies consisting of thin (millimetre to centimetre scale) structureless mudstone drapes separating slightly wavy laminated sandstones; the latter are distinguished from the damp sabkha deposits described below, by having discrete mudstone laminae.

These vertical transitional facies relationships also occur laterally. In a 12 km N–S traverse across the Morecambe fields, muddier deposits overlying fluvial channel belt sandstones pass southward into heterolithic and argillaceous wavy bedded sandstones of possible damp sabkha origin (Herries 1992).

Interpretation. The argillaceous nature and lateral extent of these units indicates widespread lacustrine deposition. The symmetrical fining- then coarsening-upward profiles reflect a transgressive-regressive lacustrine cycle. Fine lamination results from perennial lacustrine deposition with possible seasonal varving. The vertical and lateral relations of the more heterolithic sandstone/mudstone facies suggest that it represents more marginal lacustrine deposition, where periods of sheetflooding were followed by ponding and, at times, desiccation.

The distinction of these sheetflood-ponded units from the damp sabkha deposits described in the next section is quite critical, since it is only the former that contain evidence for subaqueous deposition. This distinction is highlighted by the observation that lacustrine units invariably overlie fluvial channel deposits rather than sabkha sandstones, suggesting a genetic relationship with the fluvial systems. These marginal lacustrine units may be considered as 'playa' deposits, in the sense that they were ephemerally ponded, and may reflect deposition in truly isolated lakes, or as periodically exposed mudflats and sandflats (*sensu* Hardie *et al.* 1978) fringing the more permanent lakes. Pedogenic and desiccation features associated with the lake centre mudstones indicate that longer term lake level falls also occurred, but were not associated with progradation of the sandier lake margin deposits.

Aeolian dominated sabkha association

Description. A variety of wavy and irregularly stratified fine to medium grained sandstones are interbedded with aeolian sandstones on a millimetre to metres scale. The association contains only rare fluvial or lacustrine interbeds and dominates sections up to 50 m thick in the lower half of the Ormskirk Sandstone Formation. These 'wavy bedded' units are characterized by undulose, wavy, crinkly, and irregular stratification defined by preferentially cemented, thin, very fine grained and silty laminae a few millimetres to 1–2 cm apart.

Two broad categories of wavy strata are identified on the basis of the degree of continuity or disruption of the finer grained wavy and crinkly laminae across the width of the core. The first consist of increasingly disrupted finer grained laminae that are usually associated with patches and lenses of structureless, coarser grained, and more porous 'aeolian type' sandstone 1–2 cm thick. The disrupted finer laminae rarely extend the full 10 cm width of a vertical core (Fig. 7a). They are also frequently interbedded with horizontal and low-angle aeolian wind ripple and, to a lesser degree, sandflow cross-strata (Fig 7b). A second, more continuously stratified type (Figs 7a & 8a), contains less evidence for aeolian activity, and occasionally preserves 1–2 cm thick sets of vague and slightly irregular cross-stratification, attributed here to upwind accretion of climbing adhesion ripples (e.g. Hunter 1981). Rare vertical and horizontal burrows up to 3 mm in diameter and 15 mm in length (Fig. 8b), and scarce isolated subaqueous rippled horizons are associated with continuously stratified facies. No discrete mudstone drapes occur in this facies, and other evidence for aqueous deposition such as intraclast horizons and climbing ripple cross-stratification is extremely rare.

The key feature that allows the cleaner, interbedded sandstones to be described as aeolian relies on identifying wind ripple strata. These have an extremely regular 'varve-like' appearance (Fig. 7b), first described by Rim (1951), and documented more fully by Hunter (1977). The stratification reflects the bimodal grain-size distribution, but because each wind ripple stratum may be only a few grains thick, the characteristic inverse grading is not always seen; instead, it is the sharp grain-size decrease marking the base of each stratum that is most distinctive (and distinguishes them from aqueous upper plane bed deposits).

Units of horizontal wind ripple strata range from several centimetres to as little as 1–2 mm in thickness. As the thickness of these units increases, the strata become gently inclined to around 15°, and individual sets 10–20 cm thick may stack as

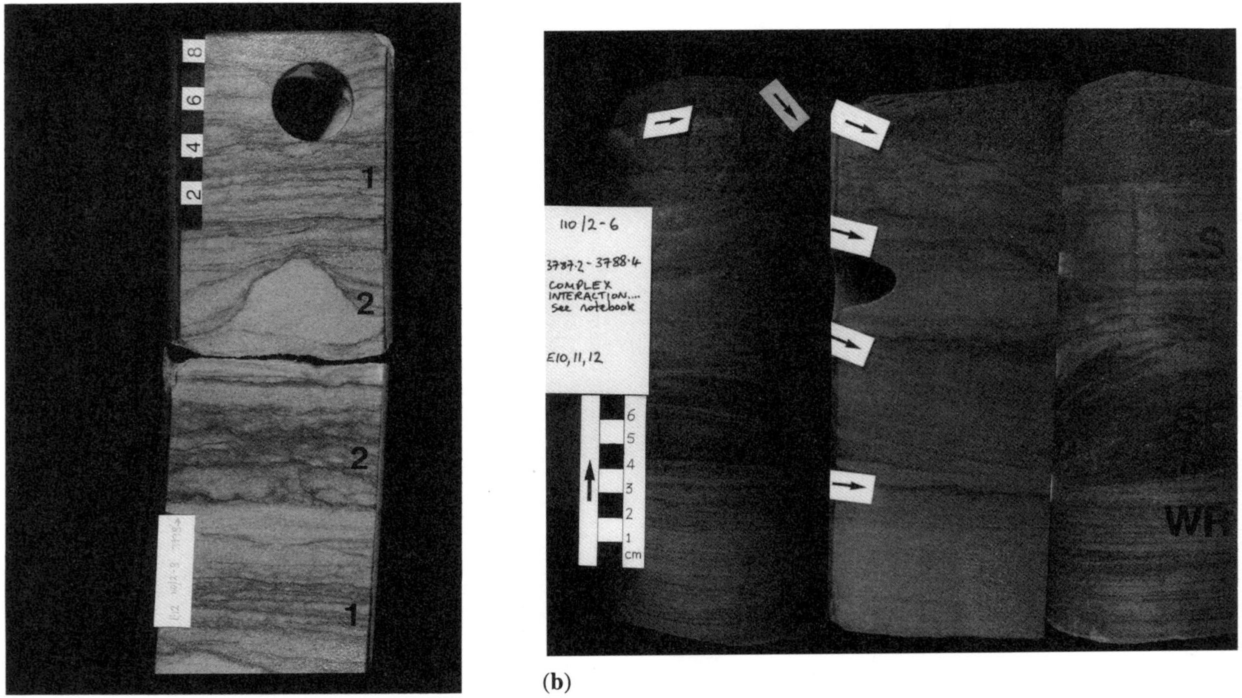

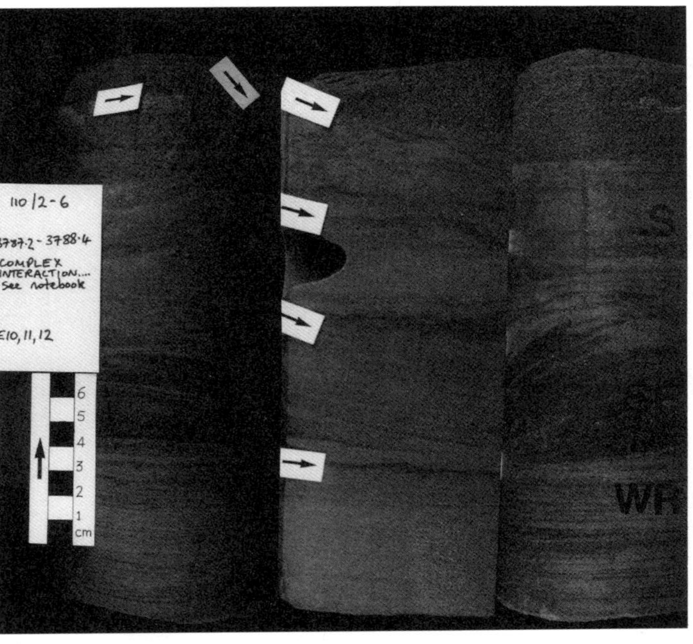

Fig. 7. (a) Core photograph from well 110/2-8 (1219 m) illustrating variations in stratal continuity between the more continuous *damp sabkha* facies (1) and more disrupted *aeolian sabkha* (2) stratification styles in poorly sorted fine to medium grained sandstone. The distinctive 'sand patch' contains subtle convex-up internal cross-strata which, in combination with the thinning of the onlapping strata, is consistent with aeolian sand infilling an actively growing pustule or ridge. In the Ormskirk Sandstone Formation, these features are associated with the more disrupted aeolian sabkha strata. Scale bar in centimetres. (b) 360° view of slabbed core from well 110/2-6 (1155 m) illustrating the complex small-scale interactions that occur between aeolian and sabkha facies. The arrows highlight irregular silty horizons that result from deflation of fine to medium grained wind ripple sandsheet (WR), aeolian dune sandflow (SF) and wavy stratified sabkha sands (S) to a level determined by the rising water-table. Note that the deflation surfaces sometimes retain some minor erosive topography. Scale bar in centimetres.

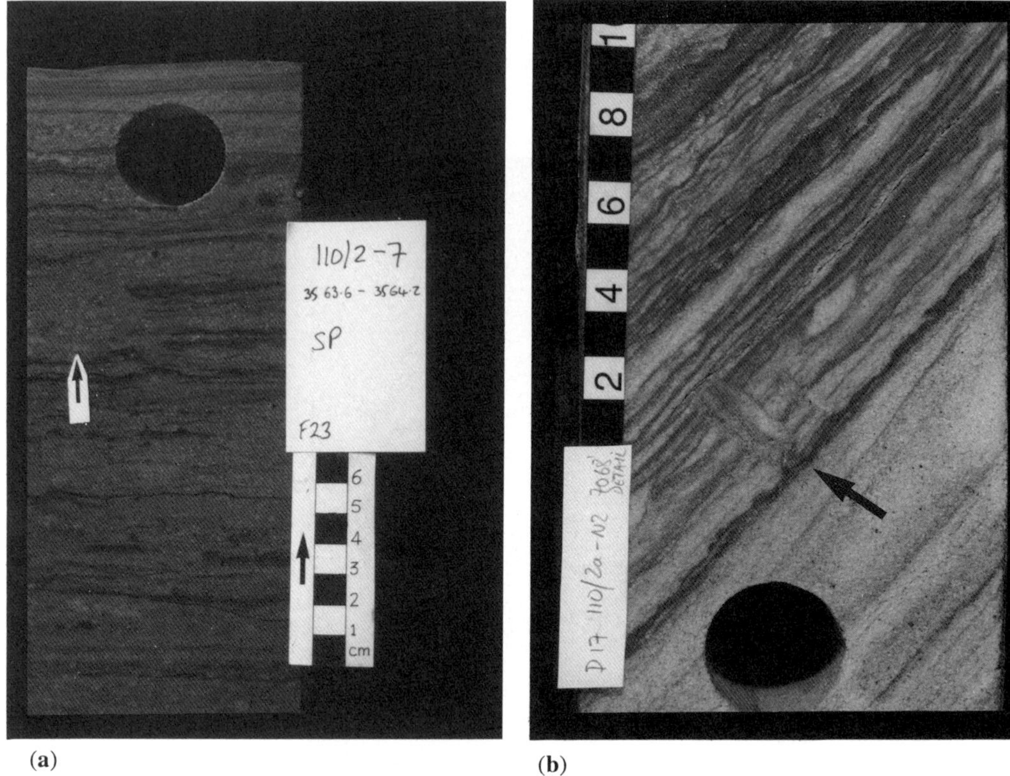

(a)   **(b)**

Fig. 8. (**a**) Core photograph from well 110/2a-7 (1087 m) showing continuous wavy and crinkly silty strata in poorly sorted, fine grained damp sabkha sandstone. The patch of structureless sandstone (arrowed) may have resulted from burrowing, or from the infill of microtopography sculpted by wind or water. Scale bar in centimetres. (**b**) Core from deviated well 110/2a-N2 (2155 m) showing small vertical (?insect) burrow (arrowed) in poorly sorted, silty to fine grained, damp sabkha facies. These overlie a surface truncating well sorted, medium grained, low-angle aeolian sandstones. Scale bar in centimetres.

cosets up to 1 m in thickness. Bounding surfaces separating individual sets may be planar or more commonly slightly irregular, consisting of a millimetre or two of wavy stratified facies. A third type of aeolian unit is characterized by the presence of sets of higher angle (up to around 25°) avalanche sandflow cross-strata, ranging from 2 cm to an average of around 0.5 m, though exceptionally up to 2 m in thickness.

Interpretation. The thickness, regional extent, and only minor interbedded fluvial channel sandstones of this association make an overbank origin unlikely. Whilst interpretation of the aeolian sandstones is relatively straightforward, there is clearly a polygenetic origin for the much of the wavy bedding, with evidence for algal binding, aeolian, adhesion, biological and evaporitic processes. Some of the deformation appears to have had surface expression, and this forms the key distinction between it and the more continuously stratified type

of wavy bedding. Some surface features may be erosive (Fig. 8a), but more often they can be shown to have developed prior to, and during the deposition of the blanketing strata (Fig. 7a). Similar surface features occur in modern aeolian dominated sabkhas as a result of growth, breakage and aeolian infill of surface micro-topography generated by evaporitic and algal domes and ridges (Fig. 9). Other larger scale deformation may be attributed to the subsurface growth of evaporitic cements (Goodall 1995). Overall, the observations are consistent with incremental aggradation of strata on an irregular, evaporite and algally bound sabkha surface, periodically inundated by aeolian sand. The interpretation is supported by the similarities to sediments trenched in modern sabkhas (e.g. Fryberger *et al.* 1983; Smoot & Castens-Seidell 1994; Goodall 1995).

Flat-lying wind ripple strata probably result from localized sandsheet deposition within an aeolian sabkha setting in conditions where sand supply was

Fig. 9. Pustular structures consisting of a weakly cemented silty crust capable of trapping wind blown sand may have dominantly (a) evaporitic or (b) algal crust origins. The expansion of the surface may be due to evaporite growth, or trapped gas generated by the decay of buried algal material. Scale bars in (a) 10 cm and (b) 25 cm long. Photographs from Sabkhat Matti, United Arab Emirates, courtesy of Tim Goodall.

insufficient to promote slipfaced dune development (Fig. 10a). Stacked low-angle wind ripple sets may represent the basal parts of successive dunes with well developed wind ripple aprons (e.g. Rubin & Hunter 1985). However, the lack of upward steepening and only rare slipface preservation suggest that these may represent the zibar deposits. Zibar are broad, dome-shaped, regularly arranged dunes (but not necessarily as coarse grained as Nielson & Kocurek (1986) have described) that are characteristic of modern sabkha to dunefield transitions such as those seen in Sabkhat Matti, Abu Dhabi (Goodall 1995) (Fig. 10b).

Aeolian dune deposits are characterized by sand-flow cross-strata resulting from dune slipface migration (Fig. 11). Stacked cross-sets that are separated by asymptotic wind ripple strata may either represent the deposits of compound dunes (draa), or successive migration of simple bedforms with dry interdunes. Their distinction rests on whether or not the intervening bounding surfaces were originally horizontal (successive migration of simple dunes), or whether they dip less than about $10°$ in a direction lying within $90°$ of the foreset azimuth (superimposed dunes migrating obliquely down the draa lee slope). The distinction between the dune and draa deposits in core is likely to be equivocal.

Stacked aeolian dunes that are separated by thin sets of irregular strata (Fig. 7b) indicate that the bedforms were deflated by succeeding interdunes to the water-table-controlled vadose zone. In such settings, their lower portions and intervening damp/evaporitic interdunes can only be preserved by a rising water-table (Fig. 11). Thus, there is no way of telling whether stacked dune cross-sets represent successive dunes or, for example, every thousandth dune that was preserved. Uniform dune and interdune thickness might favour the former, but thicker interdune deposits may represent amalgamated interdunes (Simpson & Loope 1985) or widespread, dune-free, damp or aeolian sabkha development.

Discussion: sabkhas versus sheetfloods

Wavy and crinkly stratified facies of the Ormskirk Sandstone Formation have been previously termed 'sheetflood' (Bushell 1986; Cowan 1993) and 'sandflat' (Meadows & Beach 1993a,b). Both terms are inappropriate; 'sheetflood' because there is no evidence that repeated sheetflood events and 'soft sediment deformation and bioturbation' (Meadows & Beach 1993a) were responsible for either depositing or modifying these sandstones, and 'sandflat', since this is already in common usage for a narrow, very low gradient, sheetflood prone, fringing sand plain at the distal reach of an alluvial fan (Hardie *et al.* 1978; Hubert & Hyde

1982). Both terms previously used to describe these deposits imply sheetflood processes, yet features indicative of waning flow are absent. There are clear similarities between this association and illustrations of trenches through modern sabkhas (Fryberger *et al.* 1983, Smoot & Castens-Seidell 1994, Goodall 1995).

Unfortunately, many geologists associate sabkhas with evaporites, and indeed, Handford's (1981) definition of aeolian sabkhas requires their formation in evaporitic conditions. Apart from the possible salt ridge structures, direct evidence for evaporitic conditions is absent (though see Thompson & Meadows, this volume). This is probably because relatively fresh groundwater is evaporitically concentrated in the vadose zone, but as the water-table rises, the evaporites are dissolved (e.g. Smoot & Castens-Seidell 1994). In the case of the Ormskirk Sandstone Formation, it may be more appropriate to adopt Kinsman's (1969) definition of sabkhas as 'equilibrium geomorphic surfaces whose level is controlled by the local groundwater-table'.

The broad subdivision of wavy strata based on their relative disruption is thought to reflect the relative dampness of the depositional surface, the least continuous strata probably reflecting more evaporitic and aeolian influence than more continuously stratified siltier sandstones. This distinction is important both for correlation of water-table-controlled patterns of deposition and also in terms of reservoir properties, particularly vertical permeability, which is lower in the more continuously stratified facies. In order to distinguish the two in subsequent discussions, the less evaporitically influenced, more continuously stratified facies are termed *damp sabkha*, and the more disrupted strata containing a higher proportion of wind ripple strata, *aeolian sabkha.*

Wilson's sand saturation concept

Given the variable aeolian development seen in core and the demonstrable effect of the water-table on deposition, the sand saturation concept proposed by Wilson (1971), rekindled by D. B. Loope (pers. comm. 1990) and referred to by Loope & Simpson (1991) is a useful tool for interpreting these deposits (Fig. 12). The concept empirically relates the actual amount of sand transported to the potential sand-carrying capacity of the wind and recognizes three distinct conditions: (i) Under full *sand saturation*, dunes with dry, sandy interdunes aggrade by bedform climb, in a manner analogous to subaqueous climbing ripples. (ii) *Metasaturation* of the wind (moderate sand saturation) also allows dunes to form and migrate, but the interdunes are deflationary to a level controlled by the water-table (or bedrock) because there is insufficient sand

Fig. 10. (a) Fine grained rippled aeolian sandsheet prograding over more irregular, water-table-controlled medium to coarse grained sabkha surface. Shovel for scale, approximately 1 m long. (b) Fine to medium grained zibar (low amplitude, slipfaceless dune) prograding over medium to coarse grained, wind rippled sandsheet. Vehicle for scale. Both photographs from Sabkhat Matti, United Arab Emirates, courtesy of Tim Goodall.

Fig. 11. Aeolian dune consisting of fine grained sand advancing over water-table-controlled fine to medium grained sandy interdune sabkha. This situation represents metasaturation of the wind, where the lowermost portion of the dune will only be preserved if the water-table in the deflationary interdune rises. Human for scale. Photograph from Sabkhat Matti, United Arab Emirates, courtesy of Tim Goodall.

available to deposit sand in the interdunes (Fig. 11). Dunes cannot aggrade by bedform climb except in the unlikely event of sediment supply exactly matching the rising water-table. Preserved 'slivers' of lowermost dune toesets reflect capture by a rising water-table. The resulting deposits have been termed 'amalgamated interdunes' in an elegant paper by Simpson & Loope (1985). (iii) *Undersaturation* produces purely deflationary 'Stokes' surfaces. Insufficient sand is available for dune formation but can support localized sandsheet development (Fig. 10a). Preservation of these deposits requires water-table rise to prevent their subsequent deflation (Figs 11 & 12).

At modern erg-margins, there is a downwind increase in sand saturation of the wind (Fig. 12), which is commonly manifested in a transition from deflated areas to sandsheets and zibar (Fig. 10), small simple dunes (Fig. 11) and finally compound dunes. In time, these zones will migrate downwind under net deflation, or upwind in relatively saturated wind conditions. Thus, if the regional control on sand supply varied, discrete intervals reflecting these changes should be preserved in the rock record, given relatively constant subsidence.

Drying-upward depositional patterns

Character and lateral extent

Within the aeolian and sabkha dominated sections of the Ormskirk Sandstone Formation, a repeated arrangement of facies, 2–10 m thick is consistently developed (Fig. 13). A conceptual model based on Wilson's (1971), sand saturation concept illustrating these observations is shown in Fig. 14. Overlying the basal surface (Fig. 14a) that truncates the underlying unit are damp sabkha deposits, sometimes associated with isolated lacustrine and fluvial channel sandstones (Figs 14b and 15). These strongly water-table influenced deposits grade up into aeolian sabkha sandstones (Fig. 14c) with a matching increase in both the frequency and thickness of interbedded dry aeolian deposits. Initially these take the form of dry aeolian sandsheets and zibar (e.g. Fig. 10), finally culminating in slipfaced dunes (Figs 11 and 14c). However, even where aeolian dunes are dominant, the horizontal bounding surfaces separating them are generally wavy and slightly silty or overlain by thin sabkha deposits, rather than planar and smooth. This

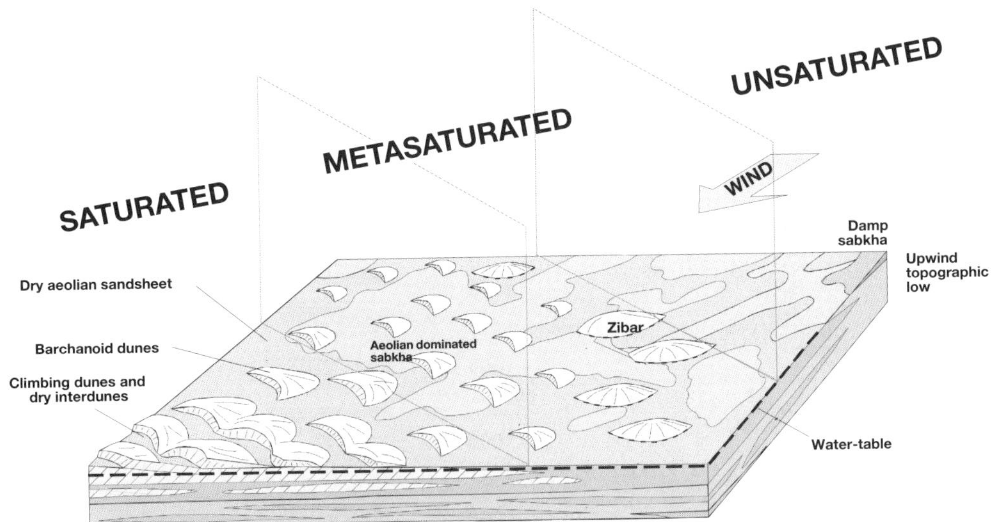

Fig. 12. Model illustrating the sand saturation concept of Wilson (1971). Sand saturation is a measure of the actual versus the potential amount of sand transported by the wind. The lowermost portions of dunes that form in metasaturated conditions will only be preserved if the water-table rises. Under fully saturated conditions, dunes with dry, sandy interdunes may aggrade by bedform climb.

suggests that sand saturation rarely exceeded the metasaturated state required to form compound dunes or draa, and that interdunal deflation extended to a level controlled by the water-table. Throughout the development of these units, there is no requirement for a fluvial sand supply. The frequent thin aeolian interbeds suggest that aeolian activity transported the bulk of the sediment, perhaps from dried up fluvial sands around the basin margins.

Overall, these vertically ordered facies patterns form units that can be correlated for at least 20 km in the Morecambe–Millom fields area (Fig. 6). In a vertical sense, these patterns appear to reflect the transition from under- through meta- to fully sand saturated conditions. No lateral sabkha–zibar–dune transition can be demonstrated, but the conceptual model illustrated in Fig. 14 implies diachroneity, though this may have been geologically instantaneous. The widespread nature of these drying upward units suggests that they developed in response to basin-wide changes in the balance between sand saturation and the water-table level, and there are a variety of mechanisms that could affect this.

Origin and preservation

There are two, seemingly conflicting elements of the drying upward patterns that require explanation. The first is the long-term process that produced the

overall drying upward pattern and its abrupt truncation; superimposed on this is a second, shorter term mechanism that allowed deflation and sabkha deposition to occur intermittently throughout the longer term drying upward event.

A number of combinations of wind strength, sand availability, water-table level and subsidence could account for the origin and preservation of these depositional patterns. These processes are interdependent, so in order to demonstrate the range of possibilities, two end-members are considered. The first requires intermittently increasing subsidence of a constant sabkha/dune environment simply to preserve progressively greater proportions of stacked dune sandstone. The second, favoured option requires uniform subsidence to preserve an evolving sabkha to dunefield environment developed in response to increasing sand saturation of the wind.

(i) Subsidence driven preservation. The Ormskirk Sandstone Formation exhibits only subtle thickness variations across the East Irish Sea Basin (Figs 3 and 6). Combined with the negligible influence of the major basin bounding faults on fluvial palaeoflow (Fig. 2) this is consistent with deposition during a thermal sag phase (Quirk & Kimble, this volume). The variable subsidence driven scenario requires intermittent but increasing rates of subsidence. Approximately 1450 m of Sherwood Sandstone Group red beds, the uppermost 250 m

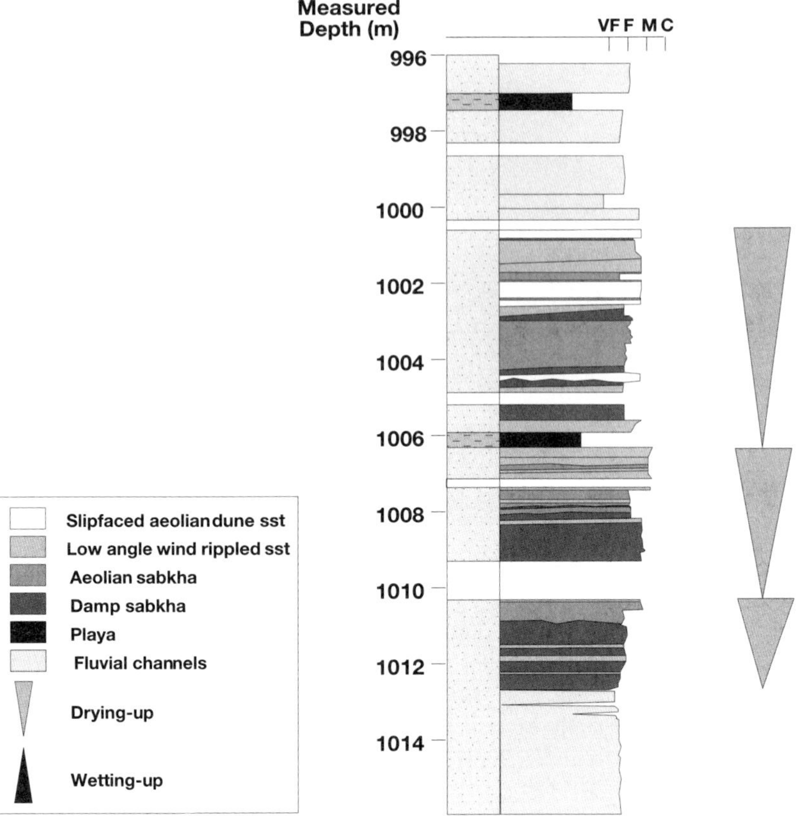

Fig. 13. Drying-upward depositional patterns seen in core, well 110/3a-A3, scale in metres.

consisting of Ormskirk Sst, were preserved in the East Irish Sea Basin (Jackson & Mulholland 1993). These deposits are sandwiched between the Zechstein equivalent Upper Permian Manchester Marls and the overlying latest Scythian–Anisian Mercia Mudstone Group (Warrington *et al.* 1980). Assuming that Sherwood Sst Gp deposition spanned the Scythian, estimated by most authors to have had a duration of 5 Ma (see Gradstein *et al.* 1994), gives an average subsidence rate of 0.29 m per thousand years.

It is difficult to account for periodically increasing subsidence necessary to generate a drying upward pattern. Nevertheless, a water-table rise that matches the calculated subsidence rate is certainly capable of preserving the observed thicknesses of aeolian dune cross-strata. A 30 cm cross-set preserved over 1000 years requires that the bedform migrates less than a single wavelength (the reconstitution time) in that period. By modern analogy, only large, slow-moving, draa-scale bedforms have reconstitution times similar to, or

greater than this, with typical wavelengths for transverse draa being 0.8–1.2 km (Lancaster 1988). Unfortunately, there is little evidence for compound dunes (stacked cross-strata separated by low-angle bounding surfaces) in the Ormskirk Sandstone Formation. It is far more likely that the single cross-sets are the product of simple barchanoid dunes. However, these migrate several metres per year and are unlikely to generate the required thicknesses of cross-strata if regional subsidence were the only control. Moreover, this scenario fails to account for the sand-trapping effect of the rising water-table. Cohesion resulting from increasing intergranular moisture content significantly increases the wind threshold velocity required to entrain sand (Hotta 1985 *in* Havholm & Kocurek 1994). The effect of a high water table would be to curtail aeolian dune development by reducing the sand saturation of the wind.

(ii) Sand supply dominant. Short-term, meteorologically/seasonally-induced water-table fluctu-

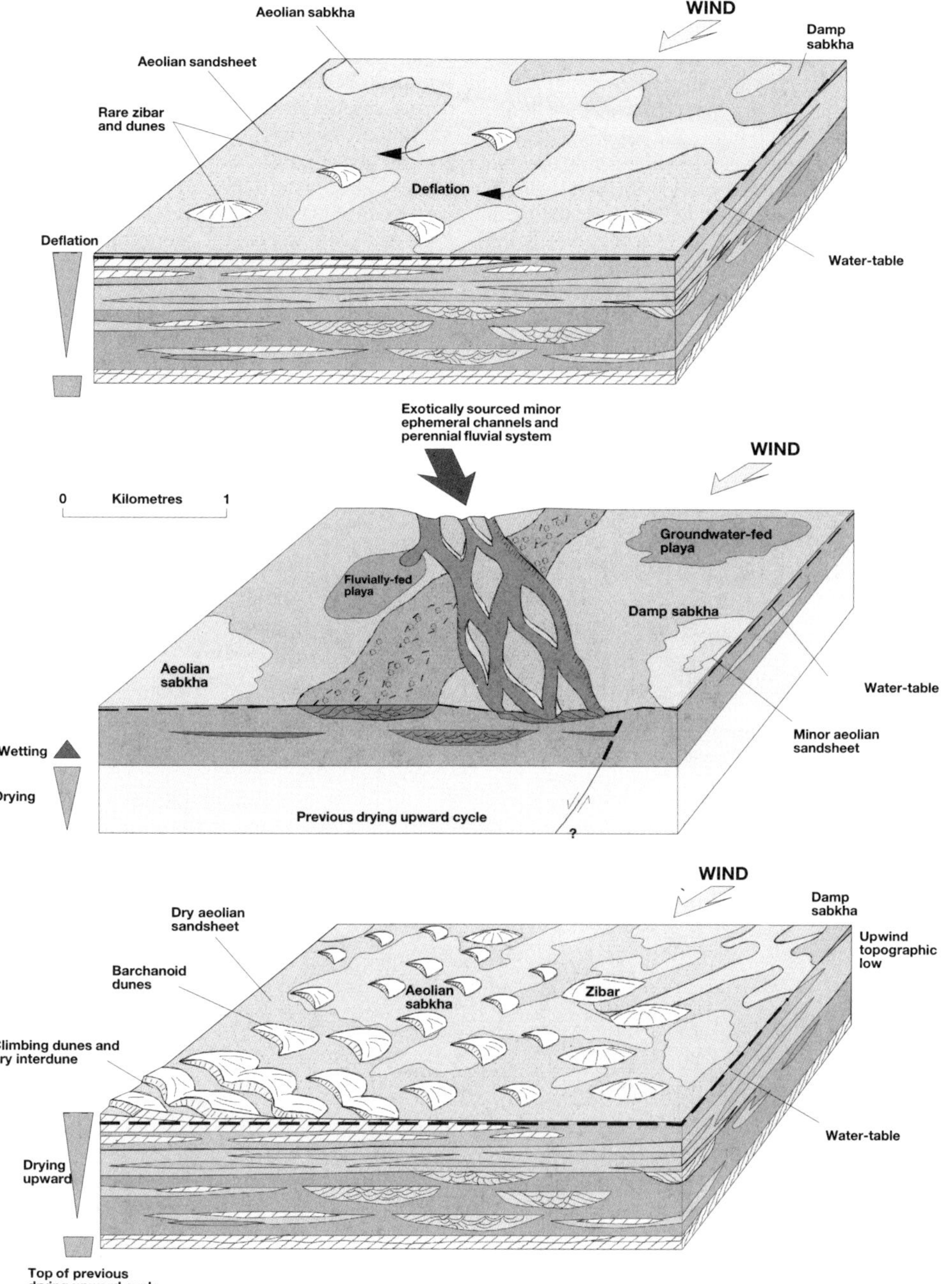

Fig. 14. Block diagrams illustrating the development of a drying-upward pattern of deposition resulting from increasingly sand saturated conditions: (a) deflation of previous drying upward deposits to a level determined by the water-table (a 'Stokes' surface); (b) fluvial channels sourced in wetter highlands enter topographically lower areas of the damp sabkha; the rising water-table allows sediment aggradation; (c) increasing sand saturation leads to a downwind and vertical transition from damp sabkha to aeolian sabkha and dune development. Ultimately, climbing dunes and dry interdunes produce dry sand seas (ergs).

ations superimposed on longer-term, climatically controlled water-table movements can more readily explain the expected changes in sand saturation, preservation of aeolian dune sands, and intermittent deflation to a level controlled by the water-table.

Overall drying upward patterns reflect a long-term increase in relative sand saturation and aeolian deposition. This may result from increasing upwind sand availability sourced by dried up river beds or long term water-table fall. In upwind sediment source areas, a falling water-table relative to the sediment surface releases dry sediment for deflation and transport to sink areas. Conversely, as deflation or recharge causes the water-table to approach the sediment surface, moisture will bind the sediment, restricting its availability. The result is a fall in sand saturation, progressive downwind deflation, and abrupt truncation of previously drying-upward units (Fig. 15). Abrupt terminations are an inherent part of this model and not due to asymmetry of, for example, the climatic control of relative water-table movement.

Superimposed on these long-term patterns are short-lived water-table fluctuations relative to the sediment surface. Greater than normal deflation during large storms or windier years resulted in interdunal sabkha deposition in even the most aeolian prone portions of the drying upward units. More rapid water-table rises prior to the next deflationary event allowed preservation of aeolian dune cross-strata. Overall, more time is likely to be stored in the deflationary hiatuses than in the preserved rocks, a point highlighted by Havholm & Kocurek (1994) in the partially analogous Middle Jurassic Page Sandstone of Arizona and Utah.

Climatic control

The sand supply dominated model appears to require distinct seasonality, in order to account for

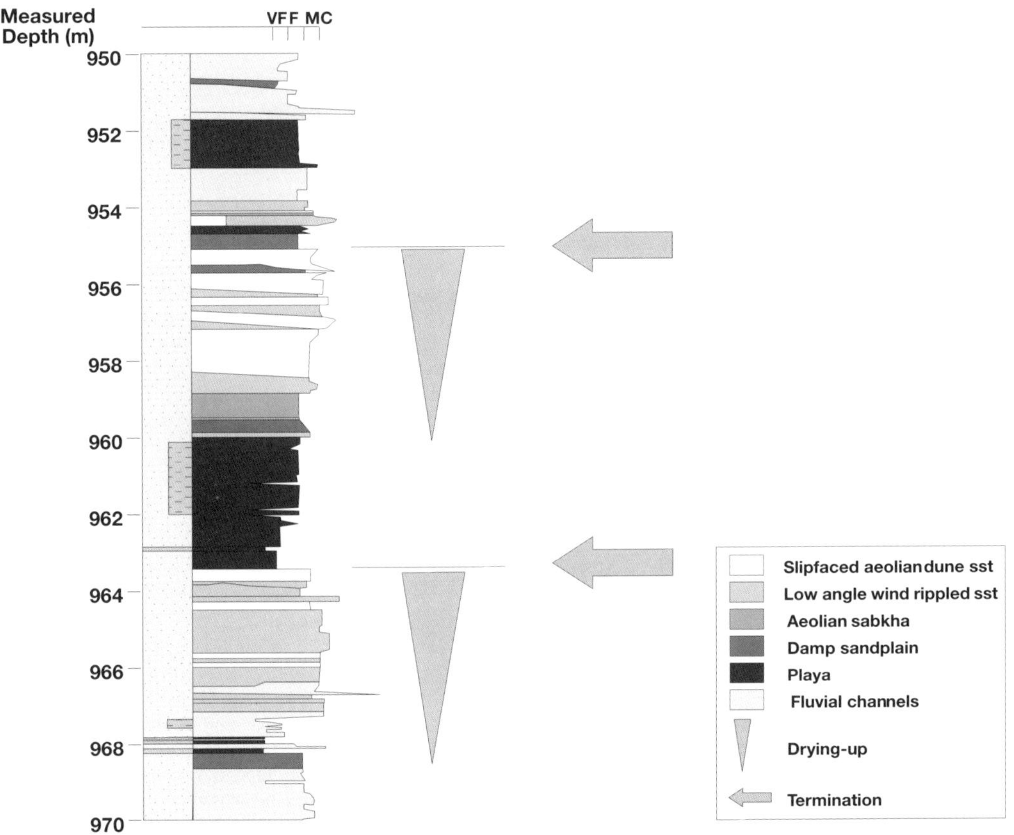

Fig. 15. Core log illustrating abrupt terminations of drying-upward depositional units and subsequent playa lake development in well 110/2-6, scale in metres.

the water-table fluctuations, plus longer term, possibly cyclic, changes in water-table level to generate the drying upward patterns. At the start of the Lower Triassic, Britain lay between 15° and 20° north of the equator (Dubiel *et al.* 1991). In contrast to present-day zonal climates, Triassic climatic belts are unlikely to have paralleled latitude (Dubiel *et al.* 1991). The Pangaean continent straddled the equator at this time, resulting in maximum monsoonal circulation (Parrish 1985), and tropical and sub-tropical belts that extended as far as 40° north and south of the equator in some regions (Kutzbach 1994). Climatic modelling suggests that interior lowlands underwent extreme seasonal temperature variations with seasonal precipitation followed by extreme heat and humidity (Wilson *et al.* 1994). This would be expected to produce the sizeable changes in water-table level invoked above.

Over the past 150 000 years, significant changes in monsoonal circulation patterns have been related to the 23 000 year precession cycle (Kutzbach 1981). Modelling these effects on Pangaean climates predicts that changes in fluvial runoff of 25% above and below the Triassic average may have occurred (Kutzbach 1994). As a mechanism for generating the drying upward units, the precession cycle is particularly attractive; 23 000 years should be represented by approximately 8 m of preserved sediment, which is within the 2–10 m range of thicknesses observed for the drying upward depositional units. Support for a climatic control comes from evidence for changing wind directions preserved in drying upward units of the Millom Field. Lowermost dunes indicate westerly directed winds, but in some units, there is a gradual upward clockwise rotation of 90° to more northerly directed winds.

Regional depositional patterns and correlation

The lower third to half of the Ormskirk Sandstone Formation is dominated by sabkha and aeolian deposits in which drying upward patterns are developed. Minor, probably ephemeral, fluvial and lacustrine deposits form the bases of many drying upward cycles. The sharp gamma peaks associated with fluvial abandonment fines are unlikely to be correlatable. A basin-wide fluvial erosion surface (OS1 surface of Jackson & Johnson 1996) is present mid-way up the Ormskirk Sandstone Formation and is correlatable from the Millom Field (113/26a-1) in the north to block 110/14 in the south (Fig. 6). Erosive truncation of the underlying sabkha and aeolian dominated sequence can be demonstrated by correlating drying upward units (Herries 1992). This indicates 10 m of erosion over

the 6.3 km distance between 110/2-6 and 110/3a-A3 (Fig. 16). Cowan *et al.* (1993) identified this level as the position where a subtle structural rotation, identified by dip azimuth vector plots occurred. In well 113/26a-2, where borehole image data are available, changes in fluvial palaeoflow occur above and below this surface, possibly indicating a degree of tectonic control.

The erosion surface is overlain by the fluvial and lacustrine dominated upper Ormskirk Sandstone Formation. The 30 m thick fluvial channel-belt unit immediately overlying this surface reflects a period of widespread fluvial dominance, and marks a major change in the Ormskirk Sandstone Formation. The increased fluvial runoff may have been climatically driven, but its basin-wide extent is probably a function of low relief generated by uniform, basin-wide subsidence, and high sediment supply relative to accommodation rate, allowing amalgamation of successive channel-belts. The Kayenta Formation of Colorado/Utah (e.g. Bromley 1991), and the Westwater Canyon Member of the Morrison Formation in northeastern New Mexico (e.g. Cowan 1991) are similarly high sand:gross fluvial systems dominated by amalgamated channel belt deposits that possibly formed under similar conditions. Other fluvially dominated units occurring higher in the sequence may also be correlatable. Thin (2–5 m), interbedded sabkha to aeolian intervals may be fluvially eroded remnants of drying upward units, but can sometimes be traced to adjacent wells.

Extensive lacustrine units in the uppermost 50 m of the formation allow confident correlation, and are better developed north of South Morecambe (Fig. 6). One particular mudstone unit, informally termed the 'hundred foot playa', lies approximately 30 m beneath the basal Mercia Mudstone Group and forms a basin-wide marker horizon. The increasingly argillaceous nature of the uppermost fluvial deposits, and the development of lacustrine units at this level may reflect a rising water-table, possibly in response to eustatic sea level rise during the Triassic (Parrish *et al.* 1982). This culminated with the drowning of the basin by the Mercia Mudstone Group.

Despite the evidence for a rising water-table near the top of the Ormskirk Sandstone Formation, the best developed aeolian dune (and probable draa) sandstones occur here (Figs 6 and 17). Bounding surfaces separating sets of cross-strata are planar rather than wavy, suggesting deflation of dry interdunes under sand saturated conditions. This is surprising, given the sand transport-limiting effects of a presumably elevated water-table. It is possible that sand supply was enhanced at this time by a change in wind patterns in response to the proximal lake shoreline. Regional correlation indicates that

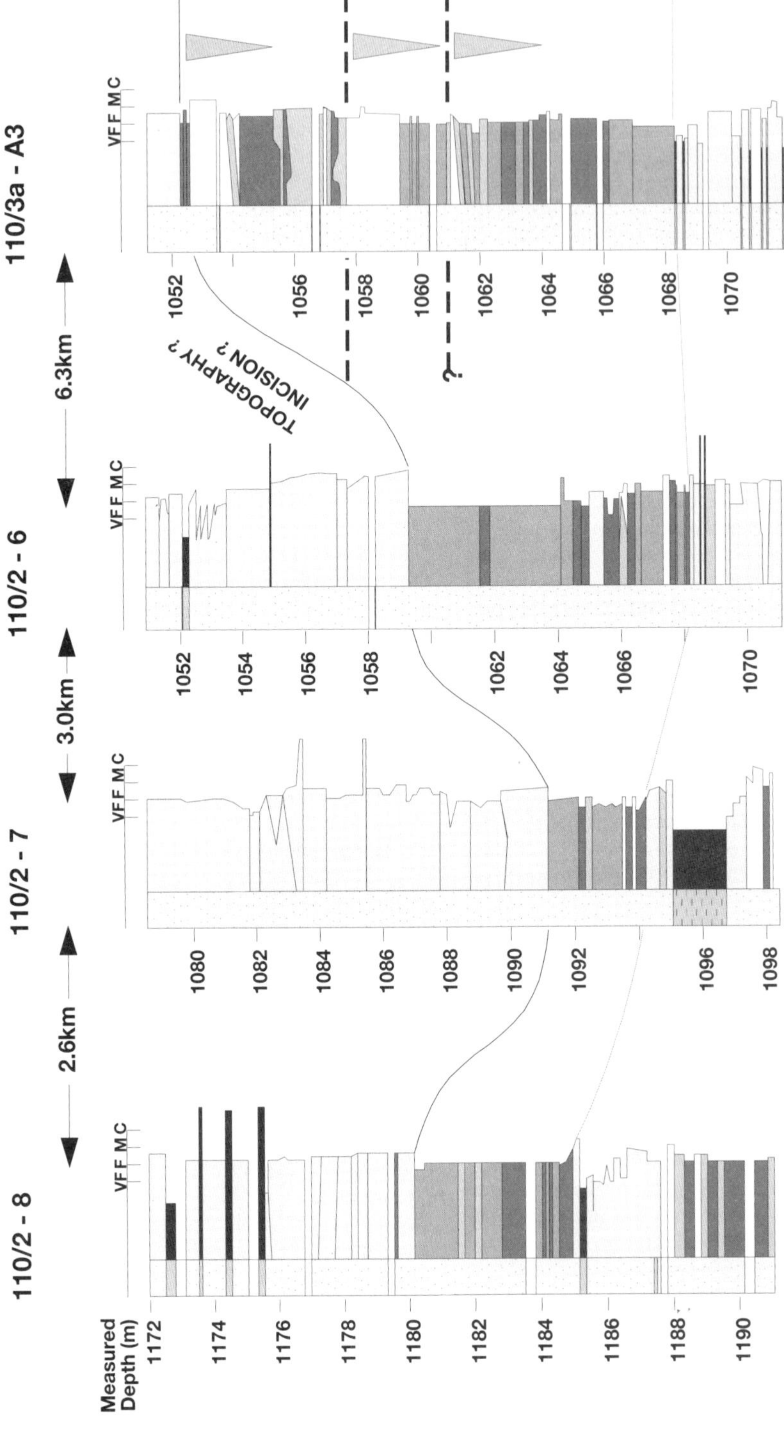

Fig. 16. Correlation panel illustrating widespread incision by a major perennial fluvial system of the underlying sabkha deposits across the North and South Morecambe fields. The base of the overlying Fylde Halite Member is used as a datum, but the degree of fluvial incision can be determined by the apparent removal of the upper two drying-upward units seen in well 110/3a-A3. The lowermost correlation is a playa unit, which marks the base of this thick, extensive sabkha package. Key to symbols in Fig. 15, scale in metres.

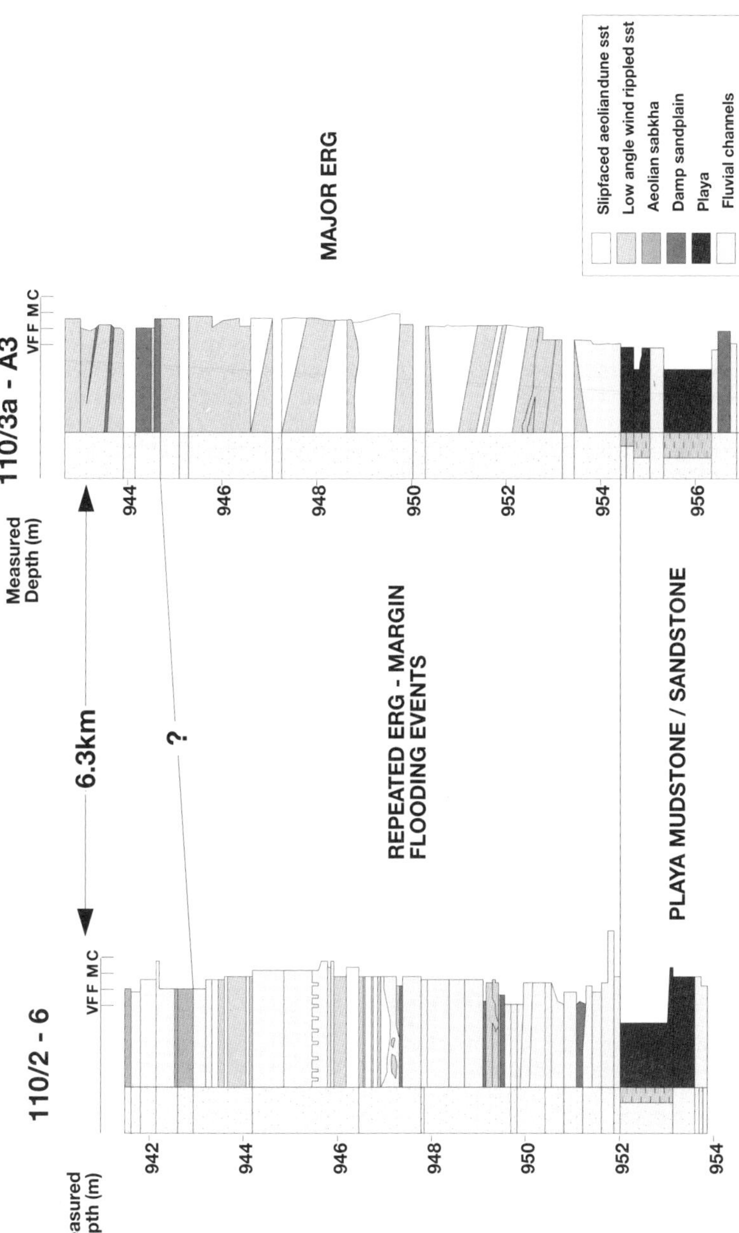

Fig. 17. Core log from top of Ormskirk Sandstone Formation in well 110/3a-A3 showing northernmost extension of the aeolian tongue passing into a more fluvially influenced setting in well 110/2-6 to the north. This rare example of apparent fluvial-aeolian coexistence is constrained by lacustrine units above and below. The aeolian deposits probably formed under sufficiently sand saturated conditions to be unaffected by the adjacent water-table rises, which were unable to affect the dry interdunes, but to the north, were able to periodically inundate the dunefield. Preservation was a consequence of the long-term, ?subsidence-driven rise in water-table lagging behind the aggradation of the dry, sandy dune-interdune system. Scale in metres.

the aeolian sandstone pinches out between South and North Morecambe, where it is laterally equivalent to a sabkha and lacustrine dominated interval. Southward, it thickens to over 60 m, forming the main reservoir for the Hamilton, Douglas and Lennox accumulations. Thus the apparent northward equivalence to sabkha sandstones and lacustrine mudstones is a function of diachroneity of the basal Mercia Mudstone Group, highlighted by the southward thinning of the Stannah Mudstone Member (Jackson *et al.* 1996) (Fig. 6). It is only at this level that coeval fluvial and aeolian interaction can be demonstrated in a part of the sequence that shows no evidence for a dominant water-table control (Fig. 17).

Conclusions

1. Wavy stratified silty sandstones, previously ascribed a sheetflood origin, are reinterpreted as siliciclastic sabkha deposits. They do not originate from overbank flows but aggrade under a rising water table as windblown sand and silt adheres to the damp substrate or is trapped by algal and evaporitic microtopography.
2. Repeated depositional patterns 2–10 m thick contain upward decreasing proportions of damp and aeolian sabkha, matched by increasing amounts of dry aeolian deposits. These 'drying upward' units, probably result from increasing sand saturation of the wind, and are prevalent in the lower half of the Ormskirk Sandstone Formation.
3. Their regional extent (20 km in the Morecambe–Millom area) suggests an allocyclic control, probably climatic. The 23 000 year precession cycle is favoured because of the close correspondence between predicted and actual thickness of drying upward units, and the strong influence exerted by the precession cycle on

recent (150 000 years) monsoonal circulation.
4. Preservation of asymmetrical drying upward units rather than drying–wetting patterns is expected, even if the climatic control is cyclical. Gentle relative water-table rise would be sufficient to reduce sand availability, curtail aeolian sand transport, and cause an abrupt return to water-table-controlled sabkha conditions prior to the next drying upward depositional event.
5. Approximately midway through the Ormskirk Sandstone Formation, a major, basin-wide fluvial unit truncates the underlying sabkha and aeolian sandstones possibly in response to a wetter climate. Its regional extent is a combined function of low relief, uniform subsidence across the basin, high sediment supply and low accommodation rates, resulting in amalgamation of successive sandbodies.
6. Near the top of the Ormskirk Sandstone Formation, extensive lacustrine deposits in the North Morecambe area pass southward into aeolian sandstones over 60 m thick. These appear to reflect erg development under fully sand saturated conditions, and are unlike those seen lower in the formation. The coeval mudstones to the north are evidence for the diachroneity of the overlying Mercia Mudstone Group.

We would like to thank the directors of British Gas Exploration and Production Ltd for permission to publish this work. Fina are acknowledged for generous funding of RDH's PhD research, which was improved by discussion of modern analogues with T. Goodall (Aberdeen University). Correlation of the Millom area was carried out whilst RDH was employed by PMGeos Ltd, Aberdeen. Z&S Geology carried out the dipmeter interpretation. The clarity of the text and figures has been improved by the comments of D. Quirk and an anonymous referee, and we are especially grateful for the stimulus of J. Fuller, J. Smith and J. Turner.

References

BRIDGE, J. S. 1985. Paleochannel patterns inferred from alluvial deposits: a critical evaluation. *Journal of Sedimentary Petrology*, **55**, 579–589.

BROMLEY, M. H. 1991. Architectural features of the Kayenta Formation (Lower Jurassic), Colorado Plateau, U.S.A: relationship to salt tectonics in the Paradox Basin. *Sedimentary Geology*, **73**, 77–99.

BUSHELL, T. P. 1986. Reservoir geology of the Morecambe Field. *In*: BROOKS, J., GOFF, J. C. & VAN HOORN, B. (eds) *Habitat of Palaeozoic Gas in Northwest Europe*. Geological Society, London, Special Publication, **23**, 189–207.

COLTER, V. S. & EBBERN, J. 1978. The petrography and reservoir properties of some Triassic sandstones of the Northern Irish Sea Basin. *Quarterly Journal of the Geological Society, London*, **135**, 57–62.

COWAN E. J. 1991. The large scale architecture of the fluvial Westwater Canyon Member of the Morrison Formation (Upper Jurassic), San Juan Basin, New Mexico. *In*: MIALL, A. D. & TYLER, N. *The Three Dimensional Facies Architecture of Terrigenous Clastic Sediments and its Implications for Hydrocarbon Discovery and Recovery*. Society of Economic Paleontologists and Mineralogists, Concepts in Sedimentology, **3**, 80–93.

COWAN, G. 1993. Identification and significance of aeolian deposits within the dominantly fluvial Sherwood Sandstone Group of the East Irish Sea Basin, U.K. *In*: NORTH, C. P. & PROSSER, D. J. (eds) *Characterization of Fluvial and Aeolian Reservoirs*. Geological Society, London, Special Publication, **73**, 231–245.

——, OTTESEN, C. & STUART, I. A. 1993. The use of dipmeter logs in the structural interpretation and palaeocurrent analysis of the Morecambe Fields, East Irish Sea Basin. *In*: PARKER, J. R. (ed.) *Petroleum Geology of Northwest Europe: Proceedings of the 4th Conference*. Geological Society, London, 867–882.

DUBIEL, R. F., PARRISH, J. T., PARRISH, J. M. & GOOD, S. C. 1991. The Pangaean Megamonsoon – evidence from the Upper Triassic Chinle Formation, Colorado Plateau. *Palaios*, **6**, 347–370.

EBBERN, J. 1981. The geology of the Morecambe Gas Field. *In*: ILLING, L. V. & HOBSON, G. D. (eds) *Petroleum Geology of the Continental Shelf of North-West Europe*. Institute of Petroleum, 85–493.

FIELDING, C. R. & CRANE, R. C. 1987. An application of stratistical modelling to the prediction of hydrocarbon recovery factors in fluvial reservoir sequences. *In*: ETHRIDGE, F. G., FLORES, R. M. & HARVEY, M. D. (eds) *Recent Developments in Fluvial Sedimentology*. Society of Economic Petrologists and Mineralogists, Special Publication, **39**, 321–327.

FRIEND, P. F., SLATER, M. J. & WILLIAMS, R. C. 1979. Vertical and lateral building of river sandstone bodies, Ebro Basin, Spain. *Journal of the Geological Society, London*, **136**, 39–46.

FRYBERGER, S. G., AL-SARI, A. M. & CLISHAM, T. J. 1983. Eolian dune, interdune, sand sheet, and siliciclastic sabkha sediments of an offshore prograding sand sea, Dhahran area, Saudi Arabia. *American Association of Petroleum Geologists Bulletin*, **67**, 280–312.

GOODALL, T. M. 1995. *The Quaternary geology and geomorphology of Sabkhat Matti, U.A.E: a modern analogue for the Permian desert facies of the Southern North Sea*. PhD Thesis, University of Aberdeen.

GRADSTEIN, F. M., AGTERBERG, F. P., OGG, J. G., HARDENBOL, J., VAN VEEN, P., THIERRY, J. & HUAHG, Z. 1994. A Mesozoic time scale. *Journal of Geophysical Research*, **99**, B12, 24051–24074.

HANDFORD, C. R. 1981. A process-sedimentary framework for characterizing recent and ancient sabkhas. *Sedimentary Geology*, **30**, 255–265.

HARDIE, L. A., SMOOT, J. P. & EUGSTER, H. P. 1978. Saline lakes and their deposits: a sedimentological approach. *In*: MATTER, A. & TUCKER, M. E. (eds) *Modern and Ancient Lake Sediments*. International Association of Sedimentologists, Special Publication, 2, 7–41.

HAVHOLM, K. G. & KOCUREK, G. 1994. Factors controlling aeolian sequence stratigraphy: clues from super bounding surface features in the Middle Jurassic Page Sandstone. *Sedimentology*, **41**, 913–934.

HERRIES, R. D. 1992. *Sedimentology of continental erg-margin interactions*. PhD Thesis, University of Aberdeen.

HOTTA, S., KUBOTA, S., KATORI, S. & HORIKAWA, K. 1985. Sand transport by wind on a wet sand surface. *In*: EDGE, B. L. (ed.) *Proceedings, 19th Coastal Engineering Conference, Houston*. American Society of Civil Engineers, New York, 1265–1281.

HUBERT, J. F. & HYDE, M. G. 1982. Sheet-flow deposits of graded beds and mudstones on an alluvial sandflat-playa system: Upper Triassic Blomidon redbeds, St Mary's Bay, Nova Scotia. *Sedimentology*, **29**, 457–474.

HUNTER, R. E. 1977. Basic types of stratification in small eolian dunes. *Sedimentology*, **24**, 361–387.

—— 1981. Stratification styles in eolian sandstones: some Pennsylvanian to Jurassic examples from the Western Interior USA. *In*: ETHRIDGE, F. G. & FLORES, R. M. (eds) *Recent and Ancient Non Marine Depositional Environments: Models For Exploration*. Society of Economic Petrologists and Mineralogists, Special Publication, **31**, 315–329.

JACKSON, D. I. & JOHNSON, H. 1996. Lithostratigraphic nomenclature of the Triassic, Permian and Carboniferous of the UK offshore East Irish Sea Basin. British Geological Survey, Nottingham.

—— & MULHOLLAND, P. 1993. Tectonic and stratigraphic aspects of the East Irish Sea Basin and adjacent areas: contrasts in their post-Carboniferous structural styles. *In*: PARKER, J. R. (ed.) *Petroleum Geology of Northwest Europe: Proceedings of the 4th Conference*, Geological Society, London, 791–808.

——, JOHNSON, H. & SMITH, N. J. P. 1997. Stratigraphical relationships and a mixed lithostratigraphical nomenclature for the Carboniferous, Permian and Triassic rocks of the offshore East Irish Sea Basin. *This volume*.

KINSMAN, D. J. 1969. Modes of formation, sedimentary associations, and diagnostic features of shallow-water supratidal evaporites. *American Association of Petroleum Geologists Bulletin*, **53**, 830–840.

KUTZBACH, J. E. 1981. Monsoon climate of the early Holocene: climatic experiment using the earth's orbital parameters for 9000 years ago. *Science*, **214**, 59–61.

—— 1994. Idealized Pangean climates. Sensitivity to orbital change. *In*: KLEIN, G. D. (ed.) *Pangea: Paleoclimate, Tectonics, and Sedimentation During Accretion, Zenith, and Breakup of a Supercontinent*. Geological Society of America Special Paper, **288**, 41–55.

LANCASTER, N. 1988. The development of large aeolian bedforms. *Sedimentary Geology*, **55**, 69–89.

LOOPE, D. B. & SIMPSON, E. L. 1991. Significance of thin sets of aeolian cross-strata, Jurassic of Eastern Utah (USA). *In*: NORTH, C. P. & PROSSER, D. J. (eds) *Abstracts for the British Sedimentological Research Group Conference on Characterization of Fluvial and Aeolian Reservoirs*. University of Aberdeen, Scotland, 24–27 March 1991.

MEADOWS, N. S. & BEACH, A. 1993*a*. Controls on reservoir quality in the Triassic Sherwood Sandstone of the Irish Sea. *In*: PARKER, J. R. (ed.) *Petroleum Geology of Northwest Europe: Proceedings of the 4th Conference*. Geological Society, London, 823–833.

—— & —— 1993*b*. Structural and climatic controls on facies distribution in a mixed fluvial and aeolian reservoir: the Triassic Sherwood Sandstone in the Irish Sea. *In*: NORTH, C. P. & PROSSER, D. J. (eds) *Characterization of Fluvial and Aeolian Reservoirs*.

Geological Society, London, Special Publication, **73**, 247–264.

NIELSON, J. & KOCUREK, G. 1986. Climbing zibars of the Algodones. *Sedimentary Geology*, **48**, 1–15.

PARRISH, J. T. 1985. Latitudinal distribution of land and shelf and absorbed radiation during the Phanerozoic: *United States Geological Survey, Open-File Report*, No. 85–31.

——, ZEIGLER, A. M. & SCOTESE, C. R. 1982. Rainfall patterns and the distinction of coals and evaporites in the Mesozoic and Cenozoic. *Palaeogeography, Palaeoclimatology and Palaeoecology*, **40**, 67–101.

QUIRK, D. & KIMBLE, G. 1997. Structural evolution of the Isle of Man. *This volume.*

RIM, M. 1951. The influence of geophysical processes on the stratification of sandy soils. *Journal of Soil Science*, **2**, 2–5.

RUBIN, D. M. & HUNTER, R. E. 1985. Why the deposits of longitudinal dunes are rarely recognised in the geologic record. *Sedimentology*, **32**, 147–157.

SIMPSON, E. L. & LOOPE, D. B. 1985. Amalgamated interdune deposits in the White Sands, New Mexaco. *Journal of Sedimentary Petrology*, **55**, 361–365.

SMOOT, J. P. & CASTENS-SEIDELL, B. 1994. Sedimentary features produced by efflorescent salt crusts, Saline Valley and Death Valley, California. *In*: RENAUT, R. W. & LAST, W. M. (eds) *Sedimentology and Geochemistry of Modern and Ancient Saline Lakes*. Society of Economic Paleontologists and Mineralogists, Special Publication, **50**, 73–90.

STUART, I. A. & COWAN, G. 1991. The South Morecambe Field, Blocks 110/2a, 110/3a, 110/8a, UK East Irish Sea. *In*: ABBOTTS, I. L. (ed.) *United Kingdom Oil and Gas Fields, 25 Years Commemorative Volume*, Geological Society, London, Memoir, **14**, 527–541.

THOMPSON, D. B. 1969. Dome-shaped aeolian dunes in the Frodsham Member of the so-called "Keuper Sandstone Formation" (Scythian–?Anisian: Triassic) at Frodsham, Cheshire (England). *Sedimentary Geology*, **3**, 263–289.

—— 1970a. The stratigraphy of the so-called Keuper Sandstone Formation (Scythian–?Anisian) in the Permo-Triassic Cheshire Basin. *Quarterly Journal of the Geological Society, London*, **126**, 151–181.

—— 1970b. Sedimentation of the Triassic (Scythian) Red Pebbly Sandstones in the Cheshire Basin and its margins. *Geological Journal*, **7**, 183–216.

THOMPSON, J. & MEADOWS, N. S. 1997. Clastic sabkhas and diachroneity at the top of the Sherwood Sandstone Group; East Irish Sea Basin. *This volume.*

WARRINGTON, G., AUDLEY-CHARLES, M. G., ELLIOT, R. E. *ET AL.* 1980. *A Correlation of the Triassic Rocks of the British Isles*. Geological Society, London, Special Report, **13**.

WILSON, I. G. 1971. Desert sandflow basins and a model for the development of ergs. *Geographical Journal*, **137**, 180–199.

WILSON, K. M., POLLARD, D., HAY, W. H., THOMPSON, S. L. & WOLD, C. N. 1994. General circulation model simulations of Triassic climates: Preliminary results. *In:* KLEIN, G. D. (ed.) *Pangea: Paleoclimate, Tectonics, and Sedimentation During Accretion, Zenith, and Breakup of a Supercontinent*. Geological Society of America Special Paper, **288**, 91–116.

WOODWARD, K. & CURTIS, C. D. 1987. Predictive modelling of the distribution of production constraining illites – Morecambe Gas Field, Irish Sea, Offshore UK. *In:* BROOKS, J. & GLENNIE, K. (eds) *Petroleum Geology of North West Europe*. Graham & Trotman, London, 205–215.

Early Dinantian evaporites of the Easton-1 well, Solway Basin, onshore, Cumbria, England

JIM WARD

Edinburgh Oil & Gas plc, 10 Coates Crescent, Edinburgh EH3 7AL, Scotland

Abstract: An early Viséan, anhydrite-rich, carbonate-clastic succession, of vertical thickness 3784 feet (no base) was found in 1990 at the Easton-1 well in the Solway Basin, onshore Cumbria. The succession is stratigraphically located in the Lower Border Group where only traces of evaporites had been found at outcrop. It is proposed to name the succession the 'Easton Anhydrite Facies of the Lower Border Group', abbreviated to 'EA Beds'.

The EA Beds are divided into three units. Unit 1 at the top, about 400 feet thick, contains two sequences of upwardly thickening anhydrite beds within mixed lithologies. Unit 2, about 2100 feet thick, consists of nine cyclic sequences of limestones, shales and sandstones with numerous interbeds of anhydrite. Overall similarity of several of the sequences suggests similar origins. Unit 2 sequences are grouped into three megacycles, two of approximately 750 feet and one 590 feet thick. Deposition is interpreted to have taken place in arid or partly arid climatic conditions, within a shallow and rapidly, but uniformly subsiding basin with glacio-eustatically controlled rise and fall of sea level. Many of the Unit 2 anhydrite beds are interpreted to be of subaqueous origin. Unit 3, about 1500 feet thick, displays complex patterns of cyclicity with sedimentation apparently responding directly to solar insolation under orbitally induced control within Milankovitch periodicities of 100 000 and 400 000 years. Seismic interpretation indicates that the regional extent of the EA Beds is greater than 1000 km^2.

This paper describes the thickest vertical succession (3784 feet) of Carboniferous evaporitic strata found in Europe. It was discovered at Easton-1 in 1990, the only well to have penetrated a significant thickness of the Viséan-aged Lower Border Group, where no substantial amounts of evaporites had previously been recorded. Minor references to these evaporites have been made by Chadwick *et al.* (1993, 1995). The well was drilled by Edinburgh Oil & Gas plc and partners near the hamlet of Easton 16 km north of Carlisle in Cumbria (co-ordinates 344 120, 571 705) (Fig. 1), on the crest of a Variscan inversion anticline mapped on seismic. From surface, the Upper, Middle and Lower Border Groups were penetrated, the objective being gas-bearing sandstones (non-commercial gas was found in the Lower Border Group). At depth 3458 feet the top of an evaporitic succession was encountered, corresponding to a prominent seismic package of reflectors at about 600 ms (TWT) below ground level (Fig. 2) interpreted as at a stratigraphic level approximating to the top of the Bewcastle Beds of the Lower Border Group (Fig. 3). At outcrop, minor indications of former evaporites had been reported in that group (Leeder, 1974a, 1975a, 1975b). The succession continued down through the equivalent of the Lynebank Beds into underlying strata to the total depth of the well at 7415 feet (TVD 7218 feet). The only evaporitic mineral found was anhydrite. The age of the succession, as determined by limited palynology is early Viséan. Following a suggestion by J. Collinson it is proposed that the succession be named the 'Easton Anhydrite Facies of the Lower Border Group' (abbreviated to 'EA Beds').

Previous Work

Numerous studies of the geology of the Solway–Northumberland Basin are comprehensively referenced by Chadwick *et al.* (1995) within a succinct account of the structure and evolution of the region. Work by Day (1970) and Leeder (1974a, 1975a, 1975b) have been the main sources for the present study. Other work of importance in the region includes Craig (1956), Craig & Nairn (1956), Nairn (1956), Lumsden & Wilson (1961), Lumsden *et al.* (1967), Deegan (1973), Johnson (1984), Leeder & Strudwick (1987), Leeder (1988), Kimbell *et al.* (1989), Smith & Holliday (1991), Chadwick *et al.* (1993) and Maguire *et al.* (1996).

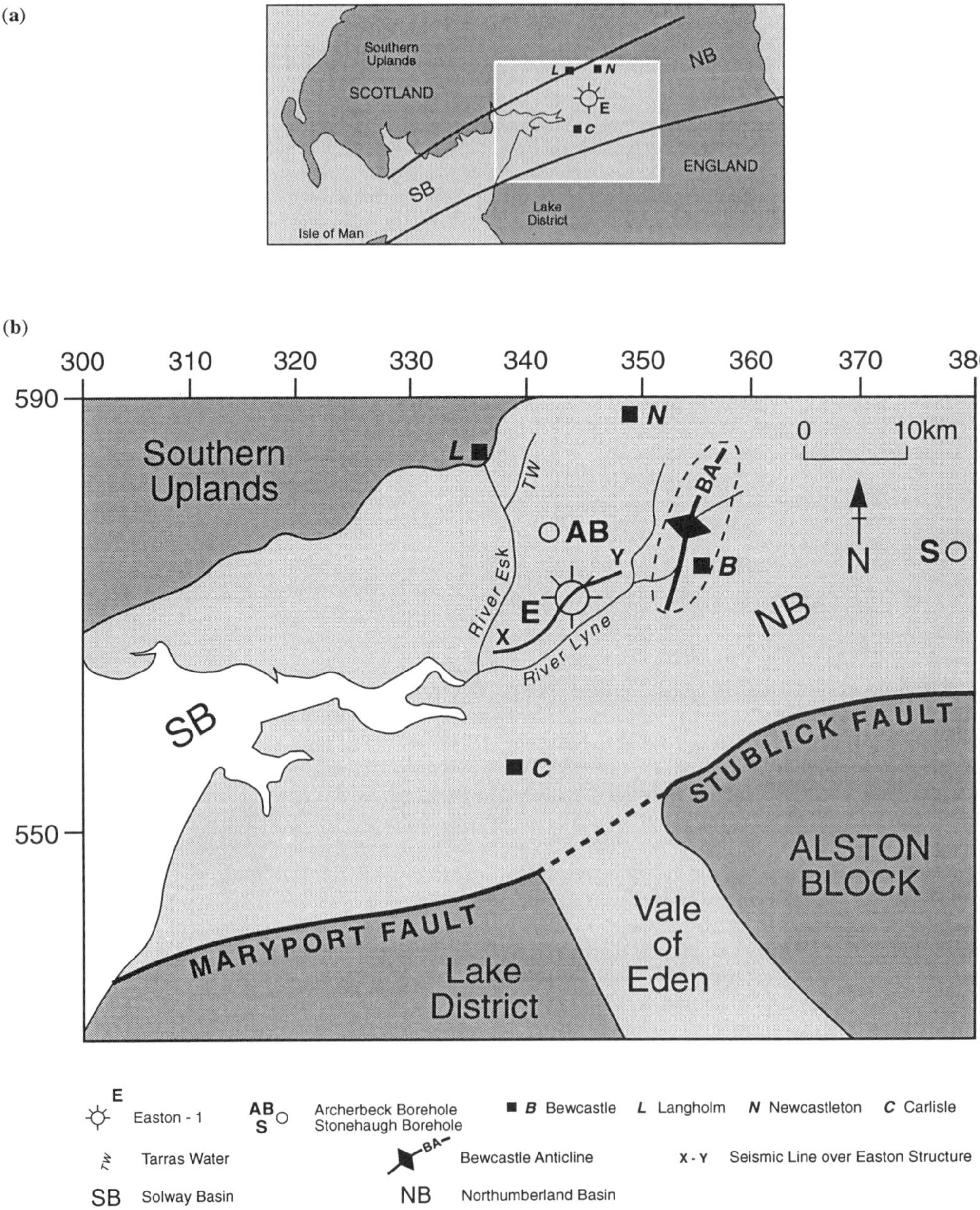

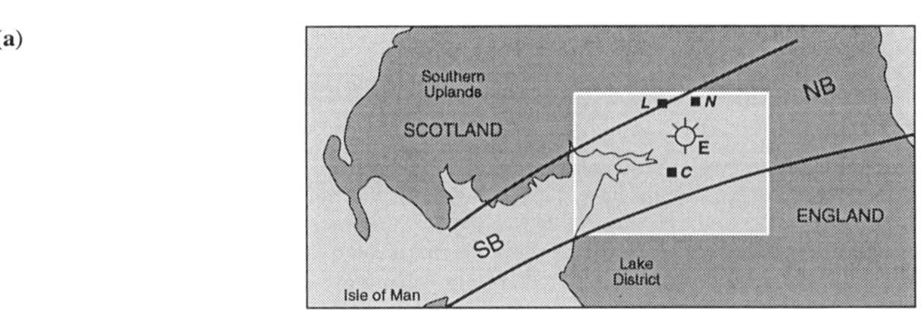

Fig. 1. (**a**) Regional location map. (**b**) Enlargement of main area referred to in text.

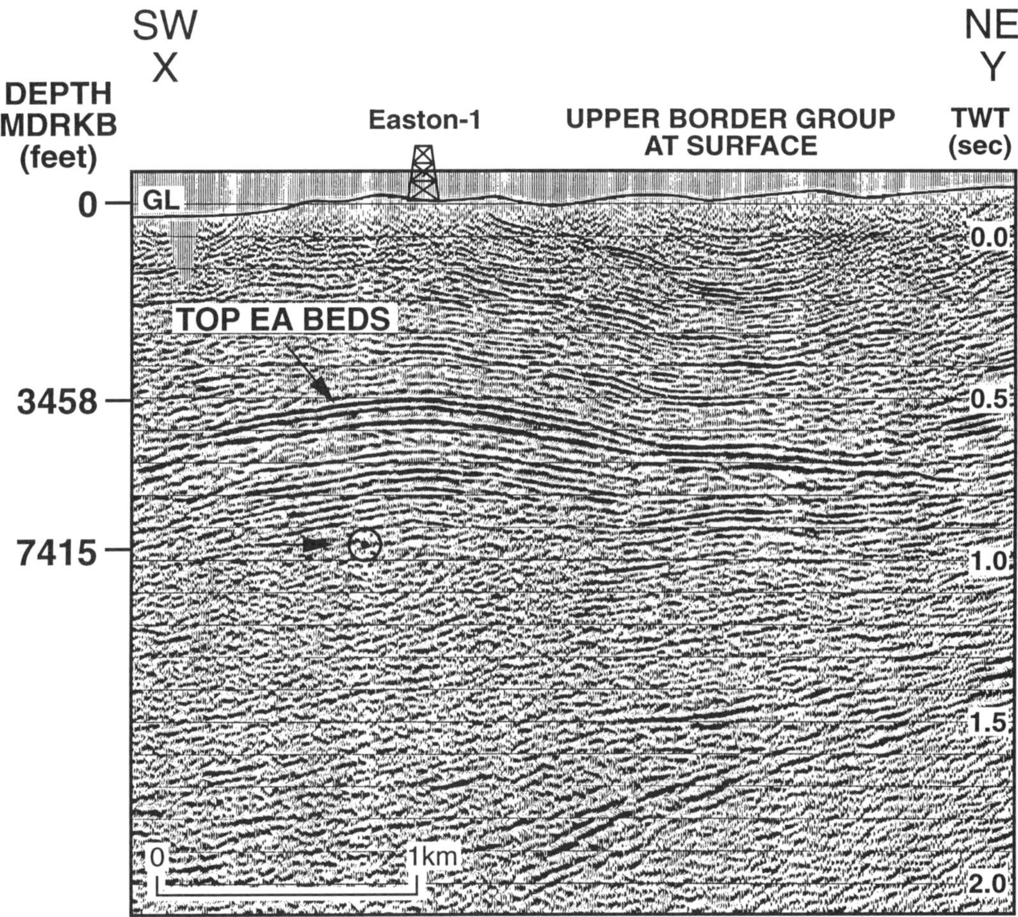

Fig. 2. Seismic line XY showing Easton-1 well location. Top of the EA Beds is approximately 0.6 s (TWT) below ground level (GL). Total depth of the well is 7415 feet, mean depth below rig kelly bushing (MDRKB) equivalent to 1.07 s (TWT) below GL.

Regional geology

Solway Basin: geological summary: from Lower Border Group to Upper Border Group (Figs 1, 3a, 3b)

The elongate northeast–southwest trending Solway Basin transects the Irish Sea north of the Ramsay–Whitehaven Ridge. The basin extends to the southwest as the Peel Basin and onshore northeast-wards to merge with the Northumberland Basin to the east of the Bewcastle anticline, forming the composite Solway-Northumberland Basin (Fig. 1).

Subsidence started in the late Devonian or Courceyan by crustal extension in the area coincident with the Iapetus Suture zone (Leeder 1982). Lower Palaeozoic rocks of the Southern Uplands to the north and the Lake District and Alston block to the south form the basin boundaries. Earliest sediments overlying lavas on the basin margins are believed to be of Courceyan age. These and most subsequent sediments are of predominantly shallow water origin indicating that sedimentation kept pace with subsidence, throughout the early Dinantian. Yoredale cyclicity of limestones, shales and sand-stones is recurrent in this area in the Dinantian (Day 1970; Leeder 1975a; Leeder & Strudwick 1987). Dinantian sediments at outcrop are of two main types: shallow marine carbonates which predom-inate in the southwest interdigitating with fluvio-deltaic clastics that episodically prograded from the northeast. Regionally the greatest thicknesses of Dinantian sediments are found in the Border Groups (Figs 3a, 3b).

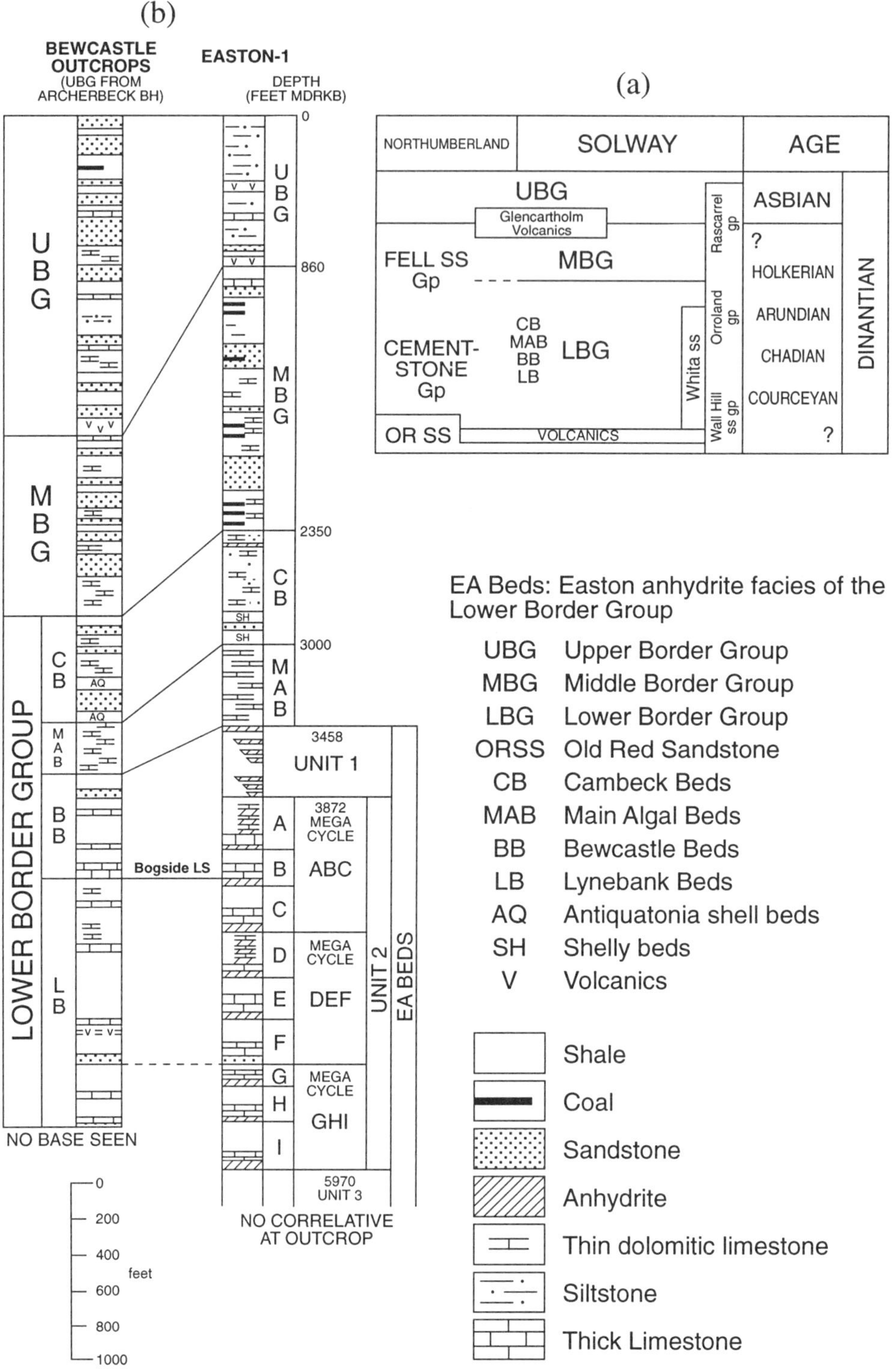

Fig. 3. (**a**) Regional stratigraphic correlation of the Border Group (Courceyan to Asbian) in the Solway and Northumberland Basins. (**b**) Generalized lithostratigraphic correlation of the Bewcastle outcrop and Archerbeck BH strata with Easton-1, from the surface to the base Unit 2 of the EA Beds. Note that Unit 3 is shown in Figs 8, 9 &10.

The Lower Border Group (LBG), of Courceyan to probable Holkerian age, consists mainly of limestones, shales, sandstones and the recently discovered evaporites. The overlying Middle Border Group (MBG), of Holkerian to Asbian age contains thick deltaic sandstones, exposed on the Bewcastle and Larriston Fells and laterally equivalent to the Fell Sandstones of Northumberland. The MBG becomes increasingly more marine to the southwest onto the north margins of the Lake District in Cumbria. In the northeast of the Solway Basin, tuffaceous beds of the Glencartholm Volcanics mark the approximate boundary with the overlying Upper Border Group (UBG) of early Asbian age consisting of cyclic sequences of thin limestones, shales, sandstones and coals, thought to have been deposited in a tidally influenced environment (Chadwick *et al.*, 1995). Coals at the top of those sequences are found to thicken towards Northumberland where they are known as the Scremerston Coal Group. Surface rocks, below drift at Easton-1 are believed to be from the UBG, and therefore no stratigraphically higher rocks are described here.

Lower Border Group Geology

Some details of this complex group within which the EA Beds are located are given below. Figures 3a and 3b summarize the regional stratigraphy and lithologies of this group.

The LBG (equivalent to the Cementstone Group of Northumberland and Berwickshire) is found at outcrop from the north Solway shore to the Langholm and Bewcastle districts. Based on seismic interpretation the group is of great thickness and may reach 3000–4000 m north of the Maryport–Stublick fault systems. The group is of varied lithofacies at outcrop. Craig (1956), Craig & Nairn (1956), Deegan (1973) and Maguire *et al.* (1996) have described LBG sediments of the north Solway shore. There, in the southwest, the succession is of mainly terrestrial clastics, including boulder conglomerates, deposited under tectonically active conditions. At Kirkbean further to the northeast along the shore, LBG sediments are shales and limestones with several sandstones, deposited on a more stable shallow marine shelf. These strata are correlated with the LBG of Langholm and Bewcastle. At Langholm the 1000 feet thick Whita Sandstone of mainly fluvial origin (Nairn 1956; Leeder 1974*a*; Lumsden *et al.* 1967), interfingers with, and is overlain by, shallow water limestones, shales and sandstones. Lumsden *et al.* (1967) and Day (1970) following Garwood (1931) divided the LBG from bottom up into the Lynebank, Bewcastle, Main Algal and Cambeck Beds. Biostratigraphical markers were used as

boundaries between the sub-divisions, but there is uncertainty and some disagreement on the correlation of these markers, not all of which are of regional extent. The ages of the divisions of the LBG are also still under discussion (see Mahdi & Butterworth 1994).

Day (1970) drew attention to Yoredale-type cyclicity in the Lynebank Beds (1400 feet at outcrop), comprising upward cycles of limestones, shales and minor sandstones. Some of the limestones reach thicknesses of 30–40 feet. The Bewcastle Beds (680 feet) have at their base the Bogside Limestone, 50–60 feet thick, which is a useful stratigraphic marker. The Main Algal Beds (280 feet) contain some fourteen, predominantly algal, limestones interbedded with shales and sandstones. The Cambeck Beds (600 feet) have a distinctly shelly fauna, two zones in particular near the base being packed with *Antiquatonia* shells.

Leeder (1974*a*, 1975*a*) carried out detailed studies of both the LBG limestones and the clastics of the Solway and Northumberland Basins. The environment was one where carbonates alternated with fluvio-deltaic clastics of the Whita, Bewcastle and Lynebank delta systems.

Previous records of Dinantian evaporites in the Solway Basin

Llewellyn *et al.* (1968) found minor amounts of anhydrite at a borehole at Egremont in west Cumbria. The anhydrite was within the Seventh Limestone Group stratigraphically equivalent to the MBG but of more marine facies. They suggested the possible existence of widespread evaporites in the Dinantian of the Solway Basin.

No LBG evaporites have been found at outcrop; long exposure to weathering agents having removed any from the near-surface. There are, however, indications of the former presence of evaporites in both the Bewcastle and Langholm areas.

Day (1970) recorded breccias of unknown origin as a particular feature of the Lynebank Beds, but found fewer in the Bewcastle Beds. These are now interpreted as likely to be dissolution breccias, relics of former evaporites (Chadwick *et al.* 1993). At Banks Quarry (5558 7490), near Bewcastle (Day 1970), for example, (Fig. 4a) a 60 cm thick breccia is well exposed, consisting of loose, angular clasts of dolomitic limestone up to 10 cm long, with minor claystone matrix; some imbrication is present. Limestone beds above are flat-lying and not disturbed. Although the underlying beds at this location are not exposed, elsewhere, for example, in the River Lyne, breccias are found to be sandwiched between parallel, undisturbed beds and have sharp flat contacts with the other lithofacies. Breccias in the district may be either clast or matrix supported, friable or well-cemented; some contain both shale and limestone clasts.

(a)

(b)

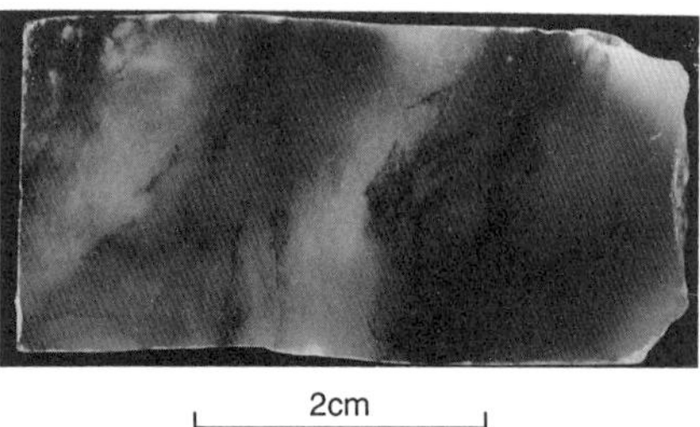

Fig. 4. (a) Breccia near top of Lynebank Beds, Banks Quarry (5558, 7490), Bewcastle. Breccia is consists of loosely packed dolomitic limestone clasts up to 10 cm long, very angular, with minor claystone; some imbrication is visible. Interpreted as a solution collapse breccia after anhydrite. (b) Polished slab surface of sidewall rotary core of anhydrite bed in Unit 1. Anhydrite is massive, mottled grey and white, and very pure.

Previously unrecorded brecciated dolomitic limestones occur as three beds between 15 and 30 cm thick separated by dark grey mudstones at a newly exposed cliff section in the Tarras Water (upstream of the road bridge, NY386 815), near Langholm. At that location, also, swallow tail-shaped crystal casts up to 15 cm long, thought to have been of gypsum, occur on a surface of a ferroan dolomite bed. These beds may correlate with strata higher in the LBG than the Bewcastle Beds (Mahdi & Butterworth 1994).

Halite casts on the undersides of sandstone beds were recorded in the Wall Hill Sandstone Gp (LBG) on the north Solway shore (Deegan 1973) and by Leeder (1974a) in the LBG of the Langholm area, in both cases interpreted as having formed in playa lake environments. No halite was found in either the Easton well cuttings or recorded on the logs; thin salt beds would have been susceptible to wash-out by the water-based drilling fluid. Leeder (1975b) described abundant calcite pseudomorphs of gypsum in LBG algal stromatolites. His palaeo-geographic map (1975a) shows evaporites as conjecturally located in the centre of the Northumberland Basin.

Geology of the Easton-1 well

Stratigraphy of Easton-1

Interpretation of the EA Beds has been largely based on wireline logs from Easton-1: gamma ray, neutron-density, resistivity and sonic logs. The Formation Micro-Scanner (FMS) log was recorded through the EA Beds to 6000 feet. A computerized lithological interpretation log : the Schlumberger ELAN log was particularly useful in revealing apparent cyclicity in the EA Beds. The ELAN log was programmed to classify lithologies as either sandstone, limestone, shale or anhydrite. Dolomite, although present in the succession, was not included in the computer model. This omission is not thought to adversely affect interpretation. Cuttings analysis showed dolomite to be associated with most of the thin limestones but with fewer of the thicker limestones (c. 50 feet). Dolomite was also associated with some anhydrites and as cement in sandstones. These lithologies were well con-strained by cuttings to log correlations. No facies analysis of cuttings or sidewall cores has yet been made. Detailed petrology was carried out on seven thin-sectioned sandstones and on some shales.

Due to the apparent uniqueness of the strati-graphy of the EA Beds in the Carboniferous, the complete succession is shown in Figs 6–9. The lithology columns are based on the ELAN computerized interpretation. Depths in the text are drill depths in feet below rig kelly bushing except where mentioned as true vertical. The well had a deviation of between 10° and 20° in the EA Beds.

Up-hole section from UBG to the Bewcastle Beds of the LBG (0–3458 feet) (Fig. 3b)

The following summary of the lithostratigraphy of the up-hole section of the well provides some evidence for tentatively assigning the top of the EA Beds to a level close to the Bewcastle Beds of the LBG. The uphole stratigraphy of Easton-1 differs considerably from that at outcrop, as illustrated by Fig. 3b. Lack of palaeontological control at the well means that boundaries between the three Border Groups or those of the divisions of the LBG are uncertain. The UBG and MBG correlations are based on lithological comparison with outcrop geology and with the Archerbeck and Stonehaugh boreholes (Lumsden & Wilson 1961; Smith & Holliday 1991). The up-hole section down to approximately 2350 feet is interpreted as of UBG underlain by MBG strata. Seismic interpretation indicates that some sections of those beds may be repeated by reverse faulting. Tuffaceous clays with base at 860 feet are the probable Glencartholm Volcanics at the base of the UBG. The boundary between the MBG and LBG may be gradational as no distinctive change is seen on well logs. Massive sandstones thought to be of the MBG terminated at 2135 feet and numerous thin coals at 2300 feet. Coals are only rarely recorded in the LBG. From about 2350 feet downwards, the limestones became thicker: up to 10 feet thick; they were oolitic and dolomitic suggesting the Cambeck Beds of the LBG had been reached. Well log interpretation showed two thin beds of anhydrite at 2370 and 2402 feet, the only evaporites to be recorded in the Cambeck Beds. Shelly beds that occurred at 2850–3000 feet may possibly be the *Antiquatonia* beds at the base of the Cambeck Beds. No positive identification of the Main Algal Beds was possible from wellsite cuttings descriptions, but limestones between 3000 and 3458 feet were variously algal, oolitic, pelloidal, shelly, crinoidal and serpulid as found in those beds at outcrop.

Easton Anhydrite Facies of the LBG (EA Beds): 3458–7415 feet

A very significant change of lithology occurred at 3458 feet with the incoming of numerous beds of anhydrite. This succession of anhydrite interbedded with limestones, shales, siltstones and sandstones continued down to the total depth of the well at 7415 feet. The vertical thickness of the succession drilled at Easton is 3784 feet. (Seismic inter-

pretation indicates the presence of several thousand feet of conformable strata below the succession in some parts of the basin.)

Absence of biostratigraphic control at Easton precludes exact correlation of the EA Beds. However, a useful lithostratigraphic marker in the LBG is the massive Bogside Limestone, about 50 or 60 feet thick at outcrop and found at the base of the Bewcastle Beds. No limestone at any stratigraphically higher level in the Border Groups attains this thickness at outcrop. At Easton-1 the first limestone of such size in downward succession in the well was found with base at 4314 feet above a thin anhydrite bed. It is considered to be the probable correlative of the Bogside Limestone, based on comparable thickness. The base of this limestone at outcrop and its supposed correlative at Easton-1 form the datum for Fig. 3b. Anhydrite is found in association with this limestone and adjacent lithologies in the well. The overlying evaporitic beds are therefore interpreted as correlatives of the Bewcastle Beds but possibly also part of the Main Algal Beds. The underlying strata for the next 1400 feet or so should then be equivalent to the Lynebank Beds of the outcrop, but uncertainty remains as to the precise stratigraphic position of the evaporitic succession.

Anhydrite Lithofacies. Anhydrite was identified by neutron-density and gamma ray logs. The majority of the anhydrite beds had densities very close to 2.95 g/cm^3 indicating great purity. Corresponding gamma ray log values were 12–15 API units. Logs show many of the beds to have sharp boundaries with associated lithologies, whether carbonates, shales or sandstones. There are about 120 discrete beds ranging in thickness from a foot to over twenty feet; thicker beds tend to be nearer the top of the EA Beds. Cumulative thickness measured from well logs comes to about 320 feet. However, the total amount of anhydrite would be greater than this since from cuttings description it occurred also as nodules or veins within mudstones, shales and limestones. Five thin sections from sidewall cores of sandstones in the EA Beds had proportions of anhydrite varying from 0.5 to 5% present as cement.

Anhydrite in cuttings samples was grey or white in colour and massive One sample was obtained by sidewall rotary coring at 3750 feet in the centre of a 16 foot thick bed. Figure 4b shows a photograph of the polished surface of the anhydrite (orientation of the surface unknown). The anhydrite has a mottled white to grey colour. It is apparently very pure with no other minerals visible in the hand specimen. Hairlike, dark grey near-horizontal stylolites are present on the outer surface (not shown) parallel to the long side of the core which would have been cut approximately horizontally into the anhydrite bed at the well.

The purity of the anhydrite, the thickness of the beds and sharp contacts with adjacent strata would favour a subaqueous rather than a sabkha origin for the majority of the beds (Warren & Kendall 1985). Further support for a subaqueous origin comes from the FMS log which shows images of banding and lamination within many of the thicker anhydrite beds (Fig. 5). It is therefore probable that many of the purer anhydrite beds were precipitated as gypsum from salinas (Kendall 1992).

Stratigraphy and facies association of the EA Beds (3458–7415 feet)

The EA Beds have been divided into three units from top to bottom in the Easton well, based almost entirely on log character, in particular that of the ELAN. It is believed these divisions approximate to real stratigraphic boundaries within the EA Beds. The units are described in terms of lithology and cyclicity below. Palaeoclimate, palaeogeography, depositional environments and the origin of cyclicity in the EA Beds are then considered.

EA Beds: Unit 1 (3458–3872 feet). The top of the EA Beds is taken to be at the topmost anhydrite bed found at 3458–3465 feet, (Fig. 6). Two successions from depths 3531–3693 feet and from 3741–3849 feet contain generally upward thickening beds of anhydrite interbedded with limestone, some dolomite, sandstone and shale. The two uppermost beds of anhydrite in each of these sequences are also the thickest found in the well: at 3531-3557 feet (26 feet) and 3741–3756 feet (16 feet). These two successions contain about 20% and 24% anhydrite, respectively. Limestones at the top of Unit 1 were dolomitic.

EA Beds: Unit 2 (3872–5970 feet). These beds consist of nine large cyclic sequences of limestones, shales, sandstones and anhydrite, eight being between 220 and 270 feet thick and one 82 feet thick. The sequences have been labelled A to I from top to bottom, (Fig. 7) and arranged into three vertical groups ABC, DEF and GHI so that in Fig. 7 the bases of the thickest parts of the limestones in B, E and H are all approximately level. The bases of the C, F and I limestones then also become nearly level and the complexity and large scale of cyclicity is apparent.

Features common to sequences B, C, E, F, H and I result in overall similarity, suggesting they were deposited under broadly similar conditions. Each sequence contains a limestone some 50–80 feet thick (limestone 50% or more) overlain by a shaly

section. The shale is succeeded upwards by variably sized, stacked, small cycles of limestone, shale and usually a sandstone; the small cycles terminate near the base of the next thick limestone.

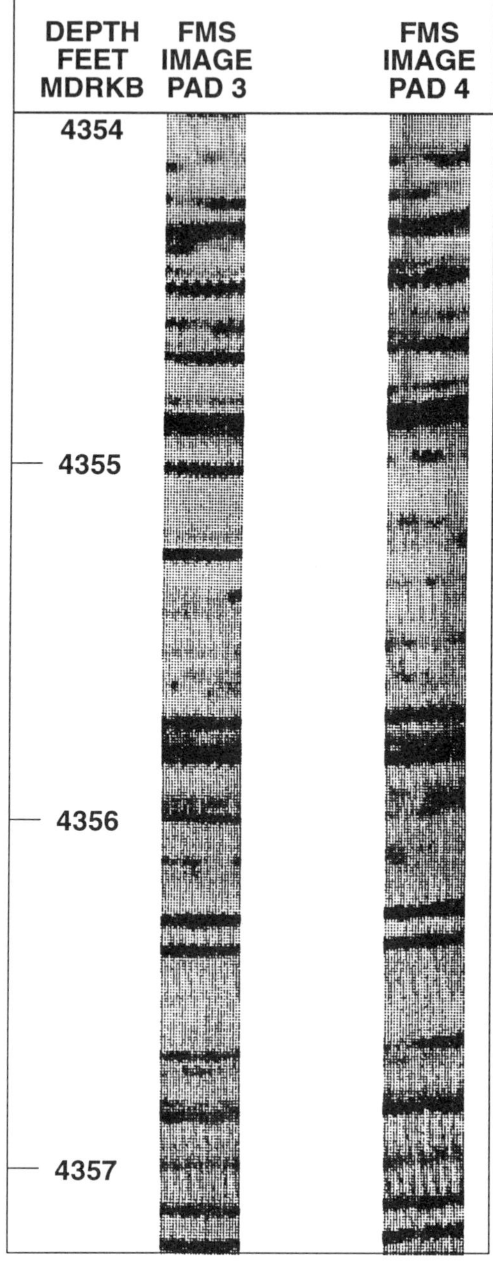

Fig. 5. Formation Microscanner (Schlumberger) (FMS) images of anhydrite, depth 4354–4357.25 feet, at base of Sequence B, Unit 2. Images show dark lineation and banding, interpreted as indicative off subaqueous origin.

Numerous beds of anhydrite are found in recurrently similar positions within the sequences: below all the main limestones, within most of their lower parts and immediately above several limestones. The smaller scale sedimentary cycles within and above the main shaly sections contain many anhydrite beds.

Sequences B, E, H and I have thick limestones that shale upwards. Most shaly sections contain at least one thin bed of nearly pure shale, 5–10 feet

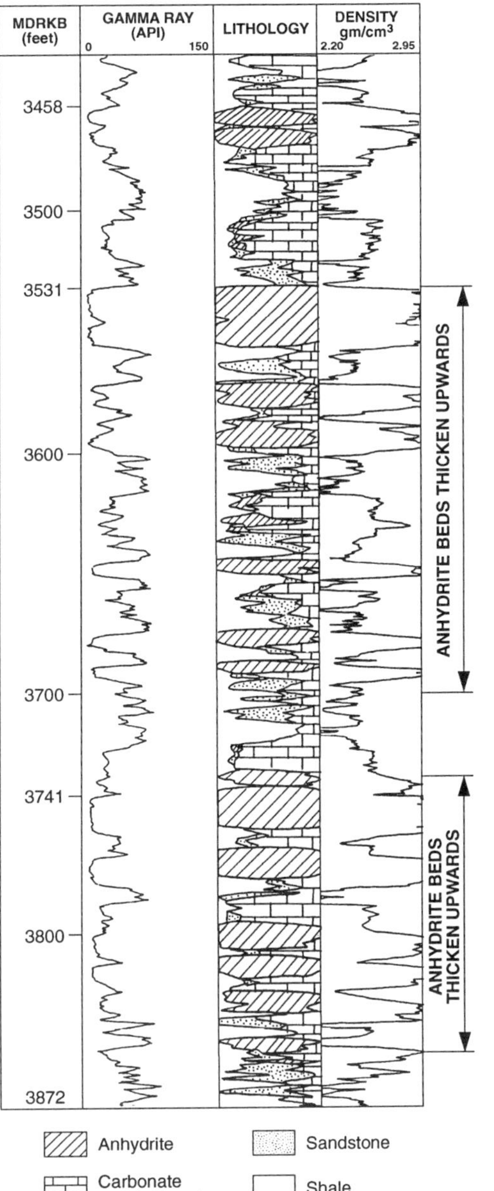

Fig. 6. Unit 1; 3458–3872 feet. Top anhydrite at 3458 feet. The thickest anhydrites in the EA Beds are at 3531–3557 feet (26 feet) and 3741–3757 feet (16 feet) at the tops of two sequences of upwardly thickening anhydrite beds.

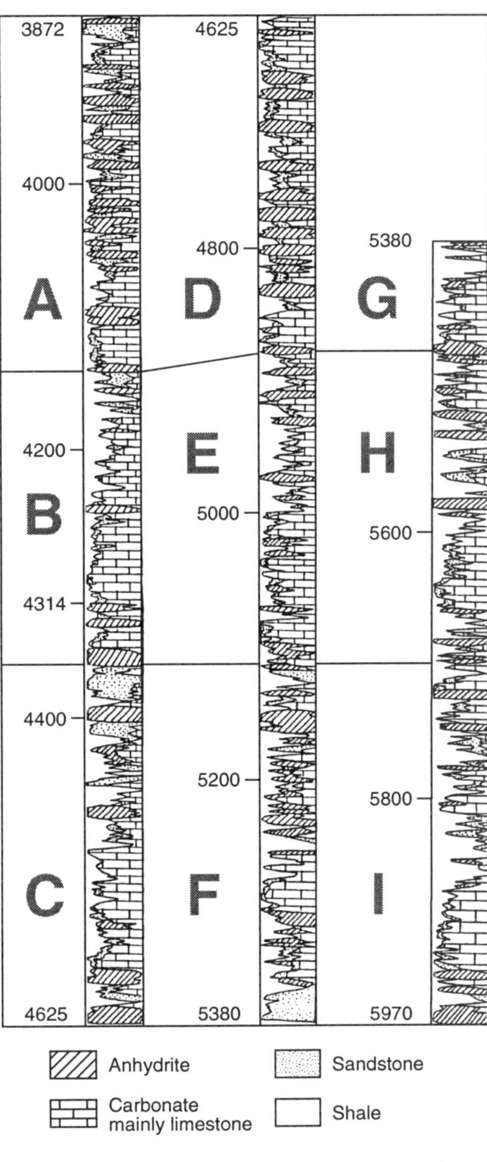

Fig. 7. EA Beds, Unit 2; 3872–5970 feet. Detailed lithological logs based on ELAN log of Easton-1. Megacycles ABC, DEF, GHI occur in direct overlying succession. Sequences B, C, E, F, H and I display many similarities. A and D are alike; G may be a truncated sequence. Limestone in B, with base at 4314 feet is interpreted as the Bogside Limestone equivalent, at base Bewcastle Beds.

thick; some have more than one of these beds. In B and E, limestones become shalier gradually, followed by an upward decrease in shale with beds of anhydrite overlying thin limestones and sands. By contrast, in H and I above the shaly section, anhydrite underlies thinner limestones that form small shaling upwards cycles. Sandstone beds not associated with anhydrite occur in the shaly sections of H and I. C and F each have shaly sections with interbeds distributed more erratically than in the others. Both have high proportions of sandstone and anhydrite and in C, thick anhydrites directly overlie thick sands. Sequences C and D terminate at thick anhydrites lying near the base of the next major limestone. A thick anhydrite also lies at the base of Unit 2.

Sequence A and D are mutually alike and although of approximately similar thicknesses to the other sequences differ from them significantly in containing thinner basal limestones and many anhydrites interbedded with limestones, sandstones and minor shales. Sequence A has 15 anhydrite beds interbedded with dolomitic limestones, dolomitic sands and minor shales. Sequence D has 13 anhydrite beds, but interbedded sandstones are thinner than in A. In both sequences, sandstones tend to be overlain by anhydrites.

The 82 feet thick section labelled sequence G is problematical. It may be an incomplete sequence truncated by a hiatus; alternatively it may represent the uppermost beds of an anomalously thick sequence H. But since the limestone at the base of G is more comparable in thickness to the limestones near the bases rather than the tops of the other sequences the former interpretation is preferred.

Sequence B has near its base the supposed correlative of the Bogside Limestone, as inferred above. The type section of the Bewcastle Beds described by Day (1970), Leeder (1975a) at Ashy Cleugh, near Bewcastle has the Bogside Limestone at its base passing upwards into shale, sandstone and then a series of smaller, though similar cycles. Anhydrite is of course absent at the outcrop. The basal few hundred feet of the outcrop section is broadly comparable to sequence B. Higher in the outcrop of the Bewcastle Beds, collapsed shaly strata are possibly correlatives of the anhydrite rich sequence A. At outcrop however, the proportion of sandstone in the Bewcastle Beds is greater than found in sequence B. Provided the correlation is valid then the sequences A and B would be within the Bewcastle Beds. The C, E, F, G and H sequences should then be correlatives of the Lynebank Beds; sequence I, may belong within deeper strata not exposed at outcrop.

The thinner limestones of the Unit 2 sequences were dolomitic; the thicker limestones less so, or not at all. Algal limestones were recorded throughout the sequences. Serpulids, shells and crinoids were associated with the thick limestones. Ostracods occurred in some shaly sections.

Megacycles. Strong similarities between the pairs of sequences: A and D, B and E, C and F result in overall similarity between the vertical successions ABC and DEF. These are therefore considered to be two large-scale composite cycles or megacycles. GHI is interpreted as another megacycle consisting of two lower sequences overlain by the anomalous G sequence. The two lower sequences of each of the three megacycles then are broadly comparable while the upper sequences A and D are mutually similar though unlike all the rest. ABC and DEF are 753 and 755 feet thick, respectively. GHI is 590 feet thick. The megacycles are sufficiently alike that Fig. 7 could perhaps be mistaken for a stratigraphic section or fence diagram through the same formations in three different boreholes, rather than three megacycles lying directly one above the other in the same borehole.

EA Beds: Unit 3 (5970–7415 feet). From the base of Sequence I of Unit 2 at 5970 feet until total depth at 7415 feet the succession becomes markedly different (Figs 8, 9 & 10). Lithologies are similar to those above, but the limestones become thinner and are predominantly interbedded with anhydrite forming composite limestone-anhydrite beds that contain only minor amounts of shale. The anhydrite may underlie, overlie or be interbedded within the limestone. These composite beds are mainly between 8 and 20 feet thick, separated by shaly sections. The proportion of shales or claystones in the succession increases downhole; sandstones are numerous but thin. Unit 3 is divided into two parts A and B. Many of the limestones in Unit 3 were dolomitic, algal and oolitic, as recorded from wellsite cuttings descriptions.

EA Beds: Unit 3A (5970–6542 feet) (Fig. 8). Unit 3A has been divided into five cycles. The uppermost, from 5970 to 6140 feet consists of three complex beds of mainly limestone and anhydrite each about 40 feet thick separated by shalier facies and overlain by a 40 foot shaly section. Anhydrite lenses thicken upwards resembling the Unit 1 sequences, but on a smaller scale. From 6140 to 6542 feet, composite limestone-anhydrite beds, 14–20 feet thick also tend to occur in distinct groups of three, with thin shales between the beds, each group separated from the next by a shaly section containing interbeds of thin sandstone and limestone. The trend is clearly cyclic and particularly well developed from 6200 to 6542 feet where there are three similar sequences of about

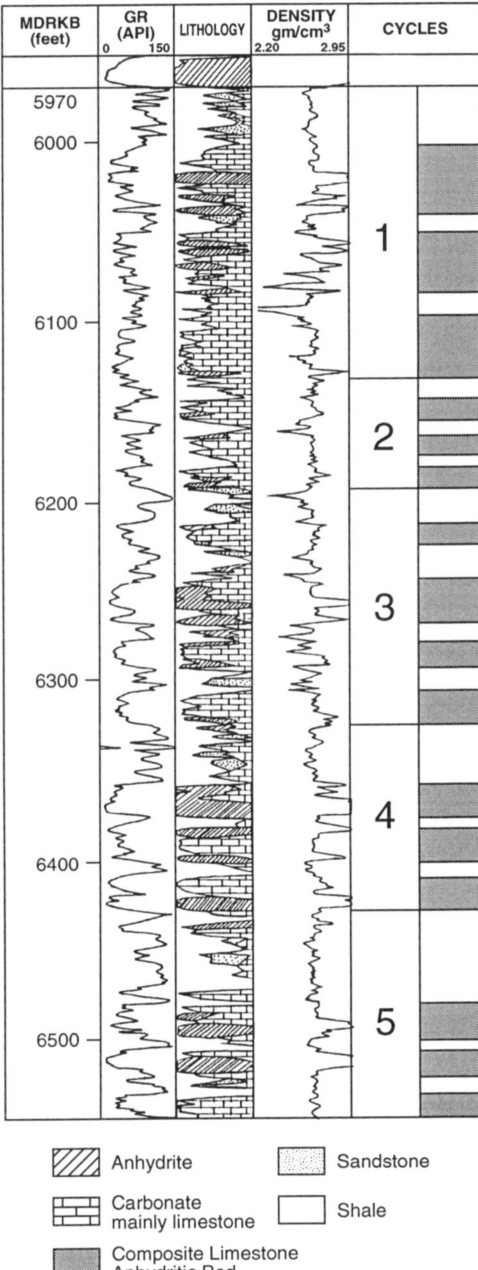

Fig. 8. EA Beds, Unit 3A; 5970–6542 feet. Cyclic sequences have mainly three composite limestone-anhydrite beds overlain by shale with interbeds of sandstone and carbonates. (Thick anhydrite at base of Unit 2 is shown above 5970 feet.

100–110 feet in thickness. In each of these the proportion of anhydrite increases upwards from the lowest to the highest of the three composite beds.

EA Beds: Unit 3B (6542–7415 feet) (Fig. 9). Below 6550 feet the shale increases to between 50% and 70% of the succession. Above the composite beds the thick shaly section contains interbeds of limestone and dolomitic sandstones; the lowermost few feet of the shaly section tends to be 100% shale. Cuttings descriptions record several thin coals.

From 6700 to 6857 feet four composite limestone-anhydrite beds about 8 feet thick are cyclic at about 50 foot intervals enclosing three shaly sections with interbeds of minor sandstones and limestones of the type found in Unit 3A. Another four cyclical composite beds up to 14 feet thick are found from 6920 to 7046 feet, at approximately 40 foot intervals, enclosing shaly sections with interbeds as above. Anhydrite tends to be at the base of the limestone and a shaly section occurs above it. Interbeds of sandstones and thin carbonates in the shaly section vary with each set of cycles.

At significant stratigraphical positions relative to the 50 and 40 foot cycles occur three cycles of a particular type (Fig. 9). In these cycles the lithologies of the upper half are nearly the same as those of the lower half, but in reversed order. They have the general form ABCDCBA where D is a thin bed that resembles a centre of symmetry. The cycles are found at 6610–6652 feet, i.e. directly above the 50 foot cycles; at 7093–7176 feet, i.e. directly below the 40 foot cycles and at 6856–6920, in between the two sets of cycles.

Close to total depth, a thick anhydrite bed (7341–7348 feet) is present within a large composite limestone-anhydrite bed. One of the few coals recorded by the density log occurs within this unit at 7350 feet.

Interpretation: origins of the EA Beds

The origins of evaporites are known to be very diverse, as described by Schreiber (1988) Kendall & Warren (1988) and Handford (1988). The varied lithological associations of the EA Beds suggests that they too have diverse origins. It is not always possible, however, to determine the depositional environment of a particular evaporite, even where good subsurface control is available (Kendall 1988) and it is probable that only partial interpretation of the origins of the EA Beds can be made with the available data.

Factors interpreted as having controlled the deposition and cyclicity of the EA Beds are discussed below. Unit 2 is considered in most detail since

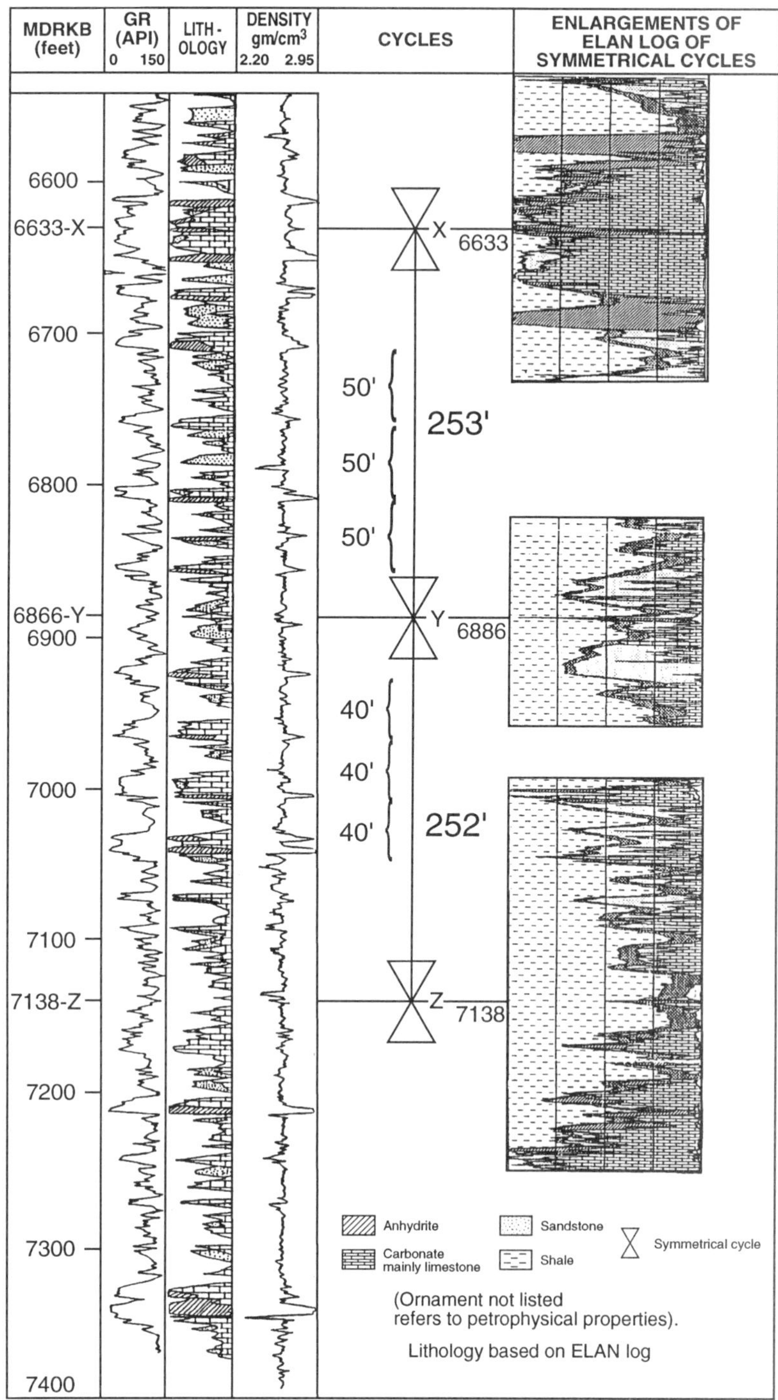

Fig. 9. Unit 3B: 6542–7415 feet. Cycles containing composite limestone-anhydritie beds overlain by shaly section with interbeds of sandstone and carbonates are 40–50 feet thick. Reversed or symmetrical cycles occur with centres at X, Y and Z. Three consecutive 50 foot cycles occur between X and Y; three 40 foot cycles between Y and Z. XY and YZ are approximately 250 feet. XYZ is a pseudo-symmetrical megacycle. RHS shows ELAN log sections of symmetrical cycles with centres at X, Y and Z.

it can be related to outcropping rocks that have been described by earlier workers. Only brief comments on Units 1 and 3 are possible due to the lack of correlatives at outcrop and the limited database used here. Aspects of Unit 3 cyclicity are discussed in a later section.

Causes of cyclicity

Subsurface preservation of thick successions with complex cyclicity on the scale of the Unit 2 sequences and megacycles, must have entailed rapid subsidence and burial. This comment also applies to Unit 3. Early Dinantian subsidence in the Solway Northumberland Basin is known to have been rapid (Taylor *et al.* 1971). Chadwick *et al.* (1993) have suggested that subsidence in the Solway–Northumberland Basin was possibly more rapid than in any other contemporaneous British basin. Since sediments are predominantly of shallow water facies (Taylor *et al.* 1971), sedimentation is inferred to have kept pace with subsidence and therefore burial was also rapid.

The recurrence in Unit 2 of certain lithological associations, such as the major limestones, or the thick beds of anhydrite at nearly equal intervals suggests that subsidence was continuous and took place at a near constant rate (Schwarzacher 1993). Kimbell *et al.* (1989), have proposed that the synextensional phase in the Northumberland Trough comprised essentially one continuous period of subsidence resulting from basin-margin faulting. Had subsidence been pulsed by irregular fault movements of varying magnitude, the recurrence of such similar and comparably large-sized sequences would hardly seem possible.

The Bewcastle and Lynebank delta systems were not large. Leeder (1974*a*) compared the Bewcastle delta with the Guadalupe delta of the Texas Gulf Coast. Few sandstones at outcrop are much thicker than about 10–20 feet. It is improbable that in those deltas, autocyclic processes such as lobe avulsion could have generated cycles on a scale of 250 feet, particularly where there is no evidence that these cycles were disrupted by synsedimentary faulting; but autocyclicity could account for the numerous, small (20–30 feet thick) cycles within the shaly sections of those sequences.

Combined with uniform rates of subsidence, ordered rise and fall of sea level can account for large-scale cyclicity. The nearly equal spacing of the lithologies in the Unit 2 sequences might be attributable to glacially influenced eustasy resulting from fourth or fifth order Milankovitch periodicities. Ramsbottom (1973) proposed eustatic control of the Yoredale cycles in northern England. In response to sea level rise and fall, limestones were deposited during transgression, overlain by deeper water shale and succeeded by regressive clastics and coal. The Unit 2 sequences resemble such cycles in overall disposition of lithologies, but are on a larger scale and contain anhydrite, the latter most probably reflecting different climatic depositional conditions, as discussed below.

Palaeoclimate and palaeogeography

Both palaeoclimate and palaeogeography are likely to have influenced deposition of the EA Beds as follows.

(i) Palaeoclimate. In early Carboniferous times the British Isles area was situated within a few degrees of the Equator (Scotese *et al.* 1979; Smith *et al.* 1981). Evidence for climatic aridity in those times has been found by several workers. Besly (1987), following the work of Leeder (1974*a*, 1975*a*), Cameron & Stephenson (1985), inferred that climatic conditions in the Lower Carboniferous of northern Britain were seasonally arid. Scott (1986) inferred aridity from evidence of evaporites in the Cementstone Group (Tournaisian) of northeast England and southern Scotland; LBG conditions were hot and arid according to Leeder *et al.* (1989). Wright (1990), suggested that a semi-arid climate prevailed in southern Britain from the Tournaisian to the mid-Viséan, a period lasting some 20 million years. Smith & Holliday (1991) found some evidence for arid climatic conditions extending into MBG times (Holkerian) at the Stonehaugh BH, 20 km east of Easton. The unprecedented thickness of the EA Beds, however, is the strongest evidence so far in support of the view that in north Britain aridity, whether seasonal or episodic, persisted from at least the Chadian to the Arundian and possibly into the Holkerian, an interval of the order of three million years. But outcrop evidence of the former presence of vegetation during LBG times, such as plant fragments and rootlet horizons (Leeder 1974*a*) suggests that aridity may not have been continuous.

(ii) Palaeogeography. In LBG times, the Solway–Northumberland Basin was elongate, stretching some 150 to 200 km northeast of the Isle of Man, its waters constrained within actively subsiding fault-bounded, interlinked graben or half-graben. Access to open sea may have been only to the southwest, while a shoreline lay to the northeast, the direction from which deltaic clastics prograded. Absence of deep water sediments suggests a subaqueous environment frequently resembling a shallow marine lagoon on an inner platform area. Carbonates are of mainly low-energy facies; there is no evidence for strong tidal currents (Leeder 1975*a*).

Such a setting in arid conditions might frequently have led to restricted marine circulation, hyper-salinity and gypsum precipitation. The numerous beds of anhydrite of probable subaqueous origin point to the likelihood that hypersalinity was a common occurrence. Fluctuations in sea level may have profoundly affected evaporite generation: a fall leading to restricted marine influx, hyper-salinity and gypsum precipitation; a rise causing salinity dilution leading to carbonate production.

Depositional setting of megacycles ABC, DEF, GHI

Depositional settings suggested for the lower and middle sequences B, C, E, F, H and I of the megacycles are as follows: uniform rate of subsidence in equilibrium with sedimentation, regular rise and fall of sea level and relatively comparable climatic conditions. The anhydrite-rich A and D sequences may have formed in more arid conditions than the above six.

Similar lithologies and thicknesses in the sequences suggest similar rates of sedimentation, overall sequence thickness therefore being pro-portional to depositional time (Swarzacher 1993). The ABC and DEF megacycles would then re-present two time-equivalent geological episodes each divided into three shorter intervals, in which the upper sequence possibly records an interval of greater aridity. The justification for calling GHI a megacycle rests on it having two lower sequences comparable, in particular, to the B and E sequences. If the apparently anomalous sandstone at the base of sequence F were an erosional remnant after truncation of a once-larger sequence G, then GHI might represent the first of three successive time-equivalent megacycles.

Relationship of the Unit 1 and 2 sequences to the outcrops

The Lynebank and Bewcastle Beds at outcrop contain fewer breccias than there are beds of anhydrite at Easton. Anhydrite-rich successions such as Unit 1, or sequences A and D of Unit 2, at outcrop or near-surface would have been subject to rapid dissolution and erosion leading to collapse and disintegration of adjacent strata. Few traces of evaporites would remain. On the other hand suc-cessions that originally contained few anhydrite beds might be better preserved at surface. The most complete outcrop correlative of the Unit 2 sequences is that of the Bewcastle Beds at Ashy Cleugh near Bewcastle. No breccias are found in the lower two hundred feet of the beds. The equivalent at Easton, interpreted as sequence B, is

in fact the one that contains least anhydrite of all the Unit 2 sequences.

Fewer relics of evaporites at outcrop compared to the subsurface may also reflect differences in depositional environments between Easton and Bewcastle. Deltaic influences may have been more significant at Bewcastle than Easton with consequent effects on evaporite deposition. Sandstones are thinner at Easton than Bewcastle. Some 11 km north of Bewcastle, contemporary strata in the Newcastleton area may be analogous to Easton, outcrops there also being deficient in sandstones of delta distributary origin. Instead, the area was a mud-dominated embayment, where brackish water conditions were indicated by the presence of monotypic fauna, (Leeder 1974a). In the subsiding Solway–Northumberland half-graben, sands may have tended to migrate south down the hanging wall dip-slope, to flow parallel to the controlling Maryport–Stublick fault system, leaving Easton and Newcastleton relatively distal to the Bewcastle Delta system, and generally by-passed by major delta channels.

In a shallow basin with a tendency towards raised salinity, areas supplied by fresh water from interdistributary deltaic channels could be predicted to preferentially support vegetation. Fewer salinas would develop in the delta, resulting in fewer evaporites. Ample outcrop evidence for the pres-ence of vegetation in the Bewcastle and Lynebank Beds is found in the form of rootlet horizons, plant debris and, rarely, thin coals. Leeder (1974a) described the occurrence of rootlet horizons at the tops of coarsening upwards sequences representing the infill of hypersaline interdistributary bays, lagoons and lakes, and of similar sequences representing the outgrowth of the delta complex in a marine environment. Rootlets occur, too, within the sandstones of fining-upward sequences formed within coastal lakes, or in playas, where also, large pseudomorphs of evaporites are recorded on the underside of sandstones. Rootlets are found abun-dantly in fining-up sequences attributable to environments of back swamps, levees and channel fills. These observations might indicate that climatic conditions were not continuously arid during deposition of the LBG, but the lack of thick coals may indicate that humid periods were short-lived. Fresh water prevented high salinites from developing proximal to the delta, whereas more distal areas had waters of varying degrees of salinity. Depending on local conditions, the for-mation of variable cycle types can be envisaged, from the typical Yoredale-type cycle: limestone, shale, sandstone, seatearth, coal; to cycles without sandstone, or without coal; or with coal overlain by anhydrite where a fresh water lagoon later became a salina. In sequence C of Unit 2, for example, coal in drill cuttings was associated with the uppermost thick sandstone that lies below a 10 foot anhydrite bed. At Bewcastle, then, fewer evaporites may have formed within small cycles due to the influence of fresh water from prograding deltas. Where breccias are found at outcrop they may represent the presence of former evaporites deposited from large-scale, extensive salinas, possibly formed during regional lowering of sea level.

In summary, different depositional conditions may have periodically prevailed at Easton and Bewcastle during deposition of Unit 2. Fresh deltaic waters at times inhibited evaporite formation at Bewcastle; more distal waters subject to hypersalinity at Easton resulted in gypsum precipitation.

Depositional environments of Unit 1 and Unit 3

Interpretation of these units is hindered by lack of detailed petrography and the absence of equivalent outcrops.

Unit 1 has no recognizable correlative at outcrop, either due to non-deposition, or to weathering and dissolution. Sequences of upwardly thickening beds of anhydrite within mixed clastic and carbonate successions, as found in Unit 1, possibly record progressive increase in aridity. Log analysis shows the thick anhydrites (up to 26 feet) in these sequences to be mineralogically pure and therefore likely to have been formed in subaqueous environments, inferred to be salinas.

No sediments equivalent to Unit 3 are exposed at surface in the Bewcastle area. The trend of upwardly increasing proportions of anhydrite from the lowest to highest composite bed in each of the Unit 3A cycles might, as suggested for Unit 1, show increasing aridity. Simplistically, Unit 3A containing relatively more anhydrite than Unit 3B might also have been deposited in more arid conditions than the latter.

Four thin coals were recorded in drill cuttings within Unit 3B. One coal at about 7350 feet underlies a 6 feet thick anhydrite. Light grey, soft claystones, some of which were carbonaceous, were also recorded. These sediments might be analogous to the shallow late Oligocene lacustrine systems of the Ebro Basin of Spain when carbonates, evaporites and thin coals formed in conditions that varied from arid to semi-arid (Cabrera & Saez, 1987).

Summary of inferred environments of deposition of the EA Beds

The mineralogically pure anhydrites of the EA Beds have been interpreted as of subaqueous origin, deposited originally as gypsum from salinas. Thick anhydrites of Unit 1 and 2 may have formed in regionally extensive salinas, whereas those associated with small depositional cycles including those of Unit 3B are suggested as as having been formed by deltaic or lacustrine processes. Some indications that environments, such as sabkhas, may have been represented in the EA Beds comes from the presence in drill cuttings of nodules and bands of anhydrite associated with limestones, sandstones and mudstones and as cement within sandstones; extensive shallow evaporitic settings frequently consist of a mosaic of sabkha and salina environments. Both the outcrop records of extensive pseudomorphing of gypsum within LBG stromatolites (Leeder 1975a) interpreted as of intertidal origin, and the occurrence of halite molds on some LBG sandstones formed in playa lakes (Leeder 1974a), would suggest those environments should be represented in the EA Beds. At present, however, most of the available evidence points to the anhydrites at Easton as being of mainly subaqueous origin.

Controls on Unit 3 cycles (Fig. 10)

Reversed cycles of the type ABCDCBA, though uncommon (Kendall 1988) are found particularly in evaporite settings. Krumbein & Sloss (1963) described this type of cycle as the ideal evaporitic cycle. Matthews (1977) recorded many such cycles in the Lucas Formation of the Detroit River Group, Devonian, Michigan and described them as reversed sequences. Cyclic fluctuations in the salinity of the basinal waters resulted in precipitation of different evaporitic minerals. These recurrent changes of salinity gave rise to reversed, or symmetrical cycles of evaporites. Kauffmann (1977) described symmetrical shallowing and deepening cycles from the Cretaceous western interior of the USA. Such cycles can occur when basinal subsidence is so rapid that shoaling upward sequences on tidal flats are prevented from developing. It is proposed here that the reversed or symmetrical cycles of Unit 3B can be accounted for by a similar process of alternating shallowing and deepening conditions during a period of rapid subsidence. The presence of evaporite minerals in such cycles reflects a tendency for the shallowing phase to be hypersaline due to climatic aridity.

Symmetrical cycles at Easton are only readily apparent from the ELAN log and to a lesser extent the gamma ray (Fig. 10). The thin beds at the centres of the three cycles: at 6633 feet, point X; at 6886 feet, point Y and at 7138 feet, point Z each have the appearance of a centre of symmetry, though the symmetry is not exact, nor are the

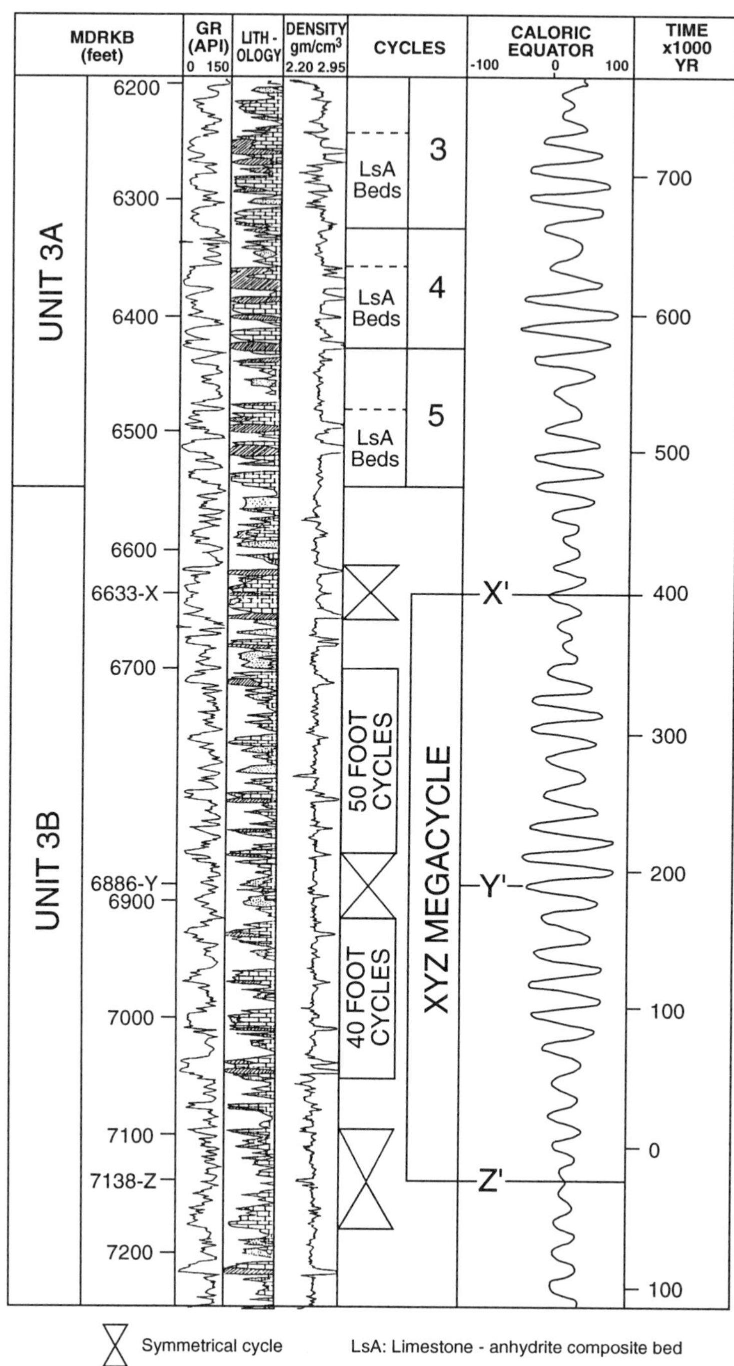

Fig. 10. Comparison of lower unit 3A and Unit 3B cycles with the caloric equator curve of Berger (1978). Limestone-anhydrite composite beds of cycles 3, 4 and 5 of Unit 3A approximately match high amplitude caloric equator cycles. Centre points X and Z of symmetrical cycles in Unit 3B correspond to X' and Z', midpoints of low amplitude Berger cycles. Pseudo-symmetrical megacycle XYZ has a duration of 425 000 years; cycles 3, 4 and 5 are approximately 100 000 years.

lithologies on either side exactly similar. Centre points of symmetry X, Y and Z are such that XY and YZ are almost the same thickness (253 and 252 feet, respectively). Thus point Y is the centre of a 500 foot thick succession that displays an overall approximation to symmetry. It can therefore be described as a pseudo-symmetrical megacycle. It is difficult to envisage a sedimentary mechanism that could generate such cyclicity and no analogues of this type of succession are known elsewhere in the Lower Carboniferous of Britain.

Borer & Harris (1991) in a study of the Yates Formation (Permian Basin, USA) were able to relate lithological cycles containing bundles of mainly three clastic beds to the caloric equator curve of Berger (1978). Composite gamma ray logs of such cycles could be matched with high amplitude cycles of the Berger curve. This allowed estimates for the duration of both large and small-scale component cycles to be made. Resemblance between the patterns of cyclicity of the three lower cycles of Unit 3A: 3, 4 and 5, to the Yates cycles suggested the possibility of a corresponding connection between the former and the caloric equator curve. The lower three cycles 3, 4 and 5 of the Unit 3A each contain groups of three composite limestone-anhydrite beds; they lie directly above the XYZ megacycle. The ELAN log of cycles 3, 4 and 5 was matched against the high amplitude cycles of the caloric equator curve for latitude 10° north (the early Dinantian Solway Basin lay close to the equator). Figure 10 shows the match to be reasonably good. The points X and Z on the lithology log are then found to line up with the points of symmetry of the low amplitude cycles of the Berger curve, points X′ and Z′ (Fig. 10), the lithological log apparently tracking the variation in the caloric equator. The possibility that the arrangement of the various components of Fig. 10 could have happened by coincidence appears unlikely. The implication has to be that sedimentary processes are responding directly to orbitally induced control.

Berger's theoretical caloric equator curves were calculated for the period of the last one million years before present. Figure 10 shows them to perhaps be applicable to geology of about 350 million years ago. Provided the correlation is valid, since the Berger curve has a linear time scale, deposition times for the Unit 3A and 3B beds can be read directly. The duration of the XYZ megacycle is 425 000 years; cycles 3, 4 and 5 of Unit 3A are each of the order of 100 000 years, both periodicities being within the scale of Milankovitch orbitally induced cycles. Sedimentation rate is about 1.18 feet per 1000 years (36 cm per 1000 years) and the time scale for the deposition of Unit 3 would be about a million years.

Regional extent of the EA Beds

The package of seismic reflectors which at Easton marks the top of the EA Beds can be mapped in the onshore Solway Basin over an area of some 600 km^2 extending from west Cumbria to the flanks of the Bewcastle anticline in the east, Fig. 11. The reflectors fade at the basin margins. To the east of the Bewcastle anticline similar reflectors visible on seismic lines over an area of 1000 km^2 can be correlated to the Easton area as corroborated by outcrop to subsurface data correlation. Provided that the same seismic reflector over all those areas represents evaporites then the evaporitic seaway from which the Unit 1 anhydrites were deposited would have had an area of around 1600 km^{-2} not including the Solway–Irish Sea to the west (Fig. 11).

The nearest Dinantian evaporites of significant thickness are those recorded in the Belfast Harbour Borehole in Northern Ireland where 655 feet of interbedded anhydrites, shales and clastics with no base seen have been described by Smith (1986). If these evaporites were of similar age to the EA Beds, and if the North Channel that had breached the Longford–Down massif in the Devonian (Simon & Bluck 1984) remained open till the Dinantian then an evaporitic seaway may have connected the Midland Valley basins to those of the Solway–Northumberland trend. The two areas at times may even have been linked via Berwickshire where Scott (1986) recorded evaporites.

The occurrence of early Dinantian evaporites as far afield as northwest Ireland (West et al. 1968; Sevastopulo 1981), in boreholes in the Midlands of

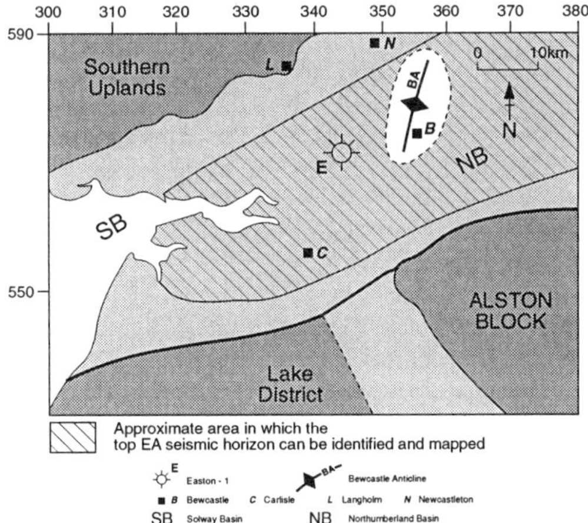

Fig. 11. Hatched area represents the possible regional extent (over 1000 sq km) of evaporite deposition at top of Unit 1 of the EA Beds.

England at Brimmington, Eyam (Strank 1985) and Hathern (Llewellyn & Stabbins 1970) and possibly in the southern Irish Sea (Grayson & Oldham 1987) points to the likelihood that deposition of evaporites was regionally widespread in contemporaneously subsiding early Dinantian basins.

Conclusions

The Easton Anhydrite Facies of the LBG (early Viséan) is a significant discovery in British Lower Carboniferous stratigraphy. The exceptional thickness of the succession (at least 3784 feet) provides strong evidence for the persistence of arid or semi-arid climatic conditions throughout much of LBG time (Chadian to Arundian) in north Britain. The complex and large-scale cyclicity of the Unit 2 sequences and their arrangement in megacycles is attributed to equilibrium between rate of subsidence and sedimentation, with ordered sea level oscillations of possible glacio-eustatic origin, and regular changes in climatic aridity. Unit 3 cyclicity is related to variations in solar insolation, cycle periodicities falling within Milankovitch ranges of 100 000 and 400 000 years. Depositional time of Unit 3 is estimated to be about one million years. The regional extent of the evaporitic sea from which the Unit 1 anhydrites were deposited may have been well in excess of 1000 km^2.

The author thanks John Collinson and an anonymous referee for their constructive criticisms of an earlier draft of this paper. Neil Meadows is thanked for much help and advice. Regional seismic mapping was carried out by A. Chan (Edinburgh Oil & Gas plc), palynology by S. A. Brindley and M. Crump (Geochem) and petrography by G Blackbourne. J. Smith (Schlumberger) gave advice on log interpretation. Useful discussions on this paper were held with D. Clark-Lowes. Alf Bissett of Edinburgh Oil & Gas is thanked for allowing me time to write this paper. My partner Barbara Thorpe is praised for enthusiasm, particularly for field trips. Jane Cochrane is thanked for typing the manuscript.

References

BERGER, A., 1978. Long-term variations of caloric insolation resulting from the Earth's orbital elements. *Quaternary Research*, **9**, 139–167.

BESLEY, B. M. 1987. Sedimentological evidence for Carboniferous and Early Permian palaeoclimates of Europe. *Annales de la Societe Geologique du Nord*, **151**, 131–143.

BORER, J. M. & HARRIS, P. M. 1991. Lithofacies and cyclicity of the Yates Formation, Permian Basin: implications for reservoir heterogeneity. *The American Association of Petroleum Geologists Bulletin*, **75**, 726–779.

CABRERA, L. L. & SAEZ, A. 1987. Coal deposition in carbonate-rich shallow lacustrine systems: the Calef and Mequinenza sequences (Oligocene, eastern Ebro Basin, NE Spain). *Journal of the Geological Society, London*, **144**, 451–461.

CAMERON, I. B. & STEPHENSON, D. 1985. *British Regional Geology: The Midland Valley of Scotland*. HMSO, London.

CHADWICK, R. A., HOLLIDAY, D. W., HOLLOWAY, S. & HULBERT, A. G. 1993. The evolution and hydrocarbon potential of the Northumberland–Solway Basin. *In*: PARKER, J. R. (ed.) *Petroleum Geology of Northwest Europe: Proceedings of the 4th Conference*. Geological Society, London, 717–726.

——, ——, —— & —— BGS 1995. The structure and evolution of the Northumberland–Solway Basin and adjacent areas. Contributor: LAWRENCE, D. J. D., *British Geological Survey, subsurface memoir*, HMSO, London.

CRAIG, C. Y. 1956. The Lower Carboniferous outlier of Kirkbean, Kirkcudbrightshire. *Transactions of the Geological Society of Glasgow*, **22**, 113–132.

CRAIG, G. Y. & NAIRN, A. E. M. 1956. The Lower Carboniferous outliers of the Colvend and Rerrick shores, Kirkcudbrightshire. *Geol. Mag.*, **93**, 249–256.

DAY, J. B. W. 1970. Geology of the country around Bewcastle. *Memoir of the Geological Survey of Great Britain*, Sheet **12** (England and Wales).

DEEGAN, C. E. 1973. Tectonic control of sedimentation at the margin of a Carboniferous depositional basin in Kirkcudbrightshire. *Scottish Journal of Geology*, **9**, 1–28.

GARWOOD, E. J. 1931. The Tuedian Beds of Northern Cumberland and Roxburghshire, east of the Liddel Water. *Quarterly Journal of the Geological Society, London*, **87**, 97–159.

GRAYSON, R. F. & OLDHAM, L. 1987. A new structural framework for the northern British Dinantian as a basis for oil, gas and mineral exploration. *In*: MILLER, J., ADAMS, A. E. & WRIGHT, V. P. (eds). *European Dinantian Environments*. Wiley, Chichester.

HANDFORD, C. R. 1988. Depositional interaction of siliciclastics and marginal marine evaporites. *In*: SCHREIBER, B. C. (ed.) *Evaporites and Hydrocarbons*. Columbia University Press, New York, 139–181.

JOHNSON, G. A. L. 1984. Subsidence and sedimentation in the Northumberland Trough. *Proceedings of the Yorkshire Geological Society*, **45**, 71–83.

KAUFFMAN, E. 1977. Geological and biological overview: western interior Cretaceous Basin. *Mountain Geology*, **14**, 75–99.

KENDALL, A. C. 1988. Aspects of Evaporite Basin Stratigraphy. *In*: SCHREIBER, B. C. (ed.) *Evaporites and Hydrocarbons*. Columbia University Press, New York, 11–65.

—— 1992. Evaporites. *In*: WALKER, R. G. & JAMES, N. P. (eds) *Facies Models: Response to Sea*

Level Changes. Geological Association of Canada.

KENDALL, C. G. St. C. & WARREN, J. K. 1988. Peritidal evaporites and their sedimentary assemblages. *In*: SCHREIBER, B. C. (ed.) *Evaporites and Hydrocarbons*. Columbia University Press, New York, 66–137.

KIMBELL, G. S., CHADWICK, R. A., HOLLIDAY, D. W. & WERNGREN, O. C. 1989. The structure and evolution of the Northumberland Trough from new seismic reflection data and its bearing on modes of continental extension. *Journal of the Geological Society, London*, **146**, 775–787.

KRUMBEIN, W. C. & SLOSS, L. L. 1963. *Stratigraphy and Sedimentation*, 2 edn. Freeman, San Francisco, 660 pp.

LEEDER, M. R. 1974a. Lower Border Group (Tournaisian) fluvio-deltaic sedimentation and palaeography of the Northumberland Basin. *Proceedings of the Yorkshire Geological Society*, **40**, 129–180.

—— 1974b. The origin of the Northumberland Basin. *Scottish Journal of Geology*, **10**, 283–296.

—— 1975a. Lower Border Group (Tournaisian) limestones from the Northumberland Basin. *Scottish Journal of Geology*, **11**, 151–167.

—— 1975b. Lower Border Group (Tournaisian) stromatolites from the Northumberland Basin. *Scottish Journal of Geology*, **11**, 207–226.

—— 1982. Upper Palaeozoic basins of the British Isles – Caldonide inheritance versus Hercynian plate margin processes. *Journal of the Geological Society, London*, **139**, 479–491.

—— 1988. Recent developments in Carboniferous geology: a critical review with implications for the British Isles and N.W. Europe. *Proceedings of the Geologists' Association*, **99**, 73–100.

LEEDER, M. R. & STRUDWICK, A. E. 1987. Delta-marine interactions: a discussion of sedimentary models for Yoredale-type cyclicity in the Dinantian of northern England. *In*: MILLER, J., ADAMS, A. E. & WRIGHT, V. P. (eds) European Dinantian Environments. Wiley, Chichester, 115–130.

——, FAIRHEAD, D., LEE, A., STUART, G., CLEMMY, H., AL-HADDEN, B., & GREEN, C. 1989. Sedimentary and tectonic evolution of the Northumberland Basin. *In*: ARTHURTON, R. S., GUTTERIDGE, P., & NOLAN, S. C. (eds) *The Role of Tectonics in Devonian and Carboniferous Sedimentation in the British Isles*. Yorkshire Geological Society, Leeds, 207–223

LLEWELLYN, P. G. & STABBINS, R. 1970. The Hathern Anhydrite Series, Lower Carboniferous, Leicestershire, England. *Transactions of the Institution of Mining and Metallurgy, (Sec.B: Appl. Earth Sci.)*, **793**, B1–15.

——, MAHMOUD, S. A. & STABBINS, R. 1968. Nodular anhydrite in Carboniferous Limestones, west Cumberland. *Transactions of the Institution of Mining and Metallurgy*, **B18**.

LUMSDEN, G. I. & WILSON, R. B. 1961. The stratigraphy of the Archerbeck Borehole, Canonbie, Dumfriesshire. *Bulletin of the Geological Survey of Great Britain*, No. 18, 1–89.

——, TULLOCH, W., HOWELLS, M. F. & DAVIES, A. 1967. The geology of the neighbourhood of Langholm. *Memoir of the Geological Survey of Great Britain*, Sheet 10 (Scotland).

MAGUIRE, K., THOMPSON, J. & GOWLAND, S. 1996. Dinantian depositional environments along the northern margin of the Solway Basin, UK. *In*: STOGEN, P., SOMMERVILLE, I. D. & JONES, G. L. L. (eds) *Recent Advances in Lower Carboniferous Geology*. Geological Society, London, Special Publications, **107**, 183–206.

MAHDI, S. A. & BUTTERWORTH, M. A. 1994. Palynology of the Dinantian Lower Border Group of the Solway Basin. *Proceedings of the Yorkshire Geological Society*, **50**, 157–171.

MATTHEWS, R. D. 1977. Evaporite cycles and lithofacies in Lucan Formation, Detroit River Group, Devonian, Midland, Michigan. *In*: FISHER, J. H. (ed.) *Reefs and Evaporites; Concepts and Depositional Models*, Studies in Geology No. 5. American Association of Petroleum Geologists, Tulsa, 73–91.

NAIRN, A. E. M. 1956. The Lower Carboniferous rocks between the rivers Esk and Annan, Dumfriesshire. *Transactions of the Geological Society of Glasgow*, **22**, 80–93.

RAMSBOTTOM, W. H. C. 1973. Transgression and regression in the Dinantian: a new synthesis of British Dinantian stratigraphy. *Proceedings of the Yorkshire Geological Society*, **39**, 567–607.

SCHREIBER, B. C. 1988. Subaqueous evaporite deposition. *In*: SCHREIBER, B. C. (ed.) *Evaporites and Hydrocarbons*. Columbia University Press, New York, 182–255.

SCHWARZACHER, W. 1993. *Cyclostratigraphy and the Milankovitch Theory*. Elsevier, Amsterdam, 219 pp.

SCOTESE, C. R., BAMBACH, R. K., BARTON, C., VAN DER VOO, R. & ZIEGLER, A. M. 1979. Palaeozoic base maps. *Journal of Geology, Chicago*, **87**, 217–278.

SCOTT, W. B. 1986. Nodular carbonates in the Lower Carboniferous Cementstone Group of the Tweed Embayment, Berwickshire: evidence for a former sulphate evaporite facies. *Scottish Journal of Geology*, **22**, 325–345.

SEVASTOPULO, G. D. 1981. Lower Carboniferous. *In*: HOLLAND, C. H. (ed.) *A Geology of Ireland*. Scottish Academic Press, 147–171.

SIMON, J. B. & BLUCK, B. J. 1984. Palaeodrainage of the southern margin of the Caledonian mountain chain in the northern British Isles. *Transactions of the Royal Society of Edinburgh: Earth Sciences*, **73**, 11–15.

SMITH, A. G., HURLEY, A. M. & BRIDEN, J. C. 1981. *Phanerozoic Palaeocontinental World Maps*. Cambridge University Press, p. 102.

SMITH, R. A. 1986. Permo-Triassic and Dinantian rocks of the Belfast Harbour Borehole. *Report BGS*, **18**, No.6, 1–13.

SMITH, S. A. & HOLLIDAY, D. W. 1991. The sedimentology of the Middle and Upper Border Groups (Viséan) in the Stonehaugh Borehole, Northumberland. *Proceedings of the Yorkshire Geological Society*, **48**, 435–446.

STRANK, A. R. E. 1985. The Dinantian biostratigraphy of a deep borehole near Eyam, Derbyshire. *Geological Journal*, **20**, 227–237.

TAYLOR, B. J., BURGESS, I. C., LAND, D. H., MILLS, D. A. C., SMITH, D. B. & WARREN, P. T. 1971. *British Regional Geology: Northern England*, 4th edn. HMSO for Institute of Geological Sciences, London.

WARREN, J. K. 1991. Sulfate dominated sea-marginal and platform evaporative settings: sabkhas and salinas, mudflats and salterns. *In*: MELVIN, J. L. (ed.) *Evaporites, Petroleum and Mineral Resources*. Developments in Sedimentology, **50**, Elsevier, Amsterdam, 555 pp.

—— & KENDALL, C. G. St. C. 1985. Comparison of sequencies formed in Marine Sabkha (subaerial) and saline (subaqueous) settings – Modern and ancient. *Bulletin of American Association of Petroleum Geologists*, **69**, 1013–1023.

WEST, N., BRANDON, A. & SMITH, M. A. 1968. A tidal flat evaporitic facies in the Viséan of Ireland. *Journal of Sedimentary Petrology*, **38**, 1079–1093.

WRIGHT, V. P. 1990. Equatorial aridity and climatic oscillations during the early Carboniferous, Southern Britian. *Journal of the Geological Society*, **147**, 359–363.

Fault analysis of the Cheshire Basin, NW England

R. A. CHADWICK

*Petroleum Geology and Basin Analysis Group, British Geological Survey, Keyworth,
Nottinghamshire NG12 5GG, UK*

Abstract: The Cheshire Basin forms a segment of the N–S trending English Channel–East Irish
Sea rift. The southern part of the basin has classic half-graben geometry, with the Wem–
Bridgmere–Red Rock Fault system forming its southeast margin, whereas its northern part is
much more symmetrical with much smaller basin margin faults. Fault strikes also show
systematic differences between the southern and northern parts of the basin. In the south,
dominant NE–SW (Caledonoid) fault-strikes indicate a strong degree of basement control,
particularly at the faulted southeast margin. Farther north, numerous faults with N–S strikes
constitute the majority of faults within the basin. They are interpreted to have formed perpendicular to the initial basin-forming extension (roughly E–W), and suggest a much weaker
basement influence. Fault displacement analysis also reveals a N–S split, with two distinct fault
populations showing different fractal (power-law) distributions. In the south, extension was
concentrated on the large, dominantly oblique-slip basin margin faults, whereas in the north,
extension was distributed on numerous small to medium-sized dip-slip faults. The two fault
populations are attributed to a variable degree of basement control. Major reactivated Caledonian
basement structures controlled basin faulting in the southern part of the basin, whereas farther
north the basin faults developed more-or-less independently of a less faulted, more isotropic
basement. Estimates of extension magnitude, incorporating the 'hidden' contributions of small
faults, indicate local basin extension factors in the range 1.07–1.16 (7–16%). However, because
of difficulties in measuring the true heaves of the basin margin faults, these values are likely to
be minimum estimates.

The Cheshire Basin forms part of a complex N–S
trending rift system, initiated in Permian times,
which extends for more than 400 km, from the
English Channel Basin in the south to the East Irish
Sea Basin in the north (Fig. 1). The basins of the rift
system are bounded by major, syn-depositional
normal faults, which controlled basin development.
The Cheshire Basin marked a particularly rapidly
subsiding segment of the rift system, with a
preserved Permo-Triassic and Lower Jurassic
succession locally more than 4000 m thick.

The rift system is particularly interesting because
it formed upon basement rocks of widely differing
type and structural fabric. This resulted in its
constituent basins showing markedly different
structural geometries along its length, as exemplified by their margin faults showing differing
trends characteristic of the underlying basement
(Chadwick & Evans 1995). This paper examines
the faults of the Cheshire Basin in detail, in order to
elucidate aspects of its extensional development.
Firstly, fault trends will be used to estimate the
initial basin-forming extension direction. Then, the
importance of very small faults, beneath the limit
of seismic resolution will be assessed from
fault displacement analysis. Results will be used to

obtain estimates of extension magnitudes and to
evaluate the role of basement fault reactivation.

Summary of basin development

Pre-Permian basement

The basement rocks which underlie the Permo-
Triassic and Carboniferous rocks of the Cheshire
Basin are thought to comprise principally Lower
Palaeozoic strata of the Welsh Caledonides; such
rocks have been proved beneath Carboniferous
cover by boreholes within the basin (Plant *et al.*
1996), as well as several shallow provings in
northeast Wales to the west of the basin. Southwest
of the Cheshire Basin, Lower Palaeozoic strata crop
out widely, with local Proterozoic inliers. Structurally the basement rocks lie close to the southeast
margin of the Welsh Caledonides, and are cut by
many NE-trending Caledonian faults (Fig. 1 inset).

The culminating phase of the Caledonian
Orogeny, termed the 'Acadian' events (Soper *et al.*
1987), took place in early to mid Devonian times
with large transpressive displacements (commonly
down-to-the-east) on many of the major basement
structures. These include the Bala Fault, the Clun

From Meadows, N. S., Trueblood, S. P., Hardman, M. & Cowan, G. (eds), 1997, *Petroleum Geology of the
Irish Sea and Adjacent Areas,* Geological Society Special Publication No. 124, pp. 297–313.

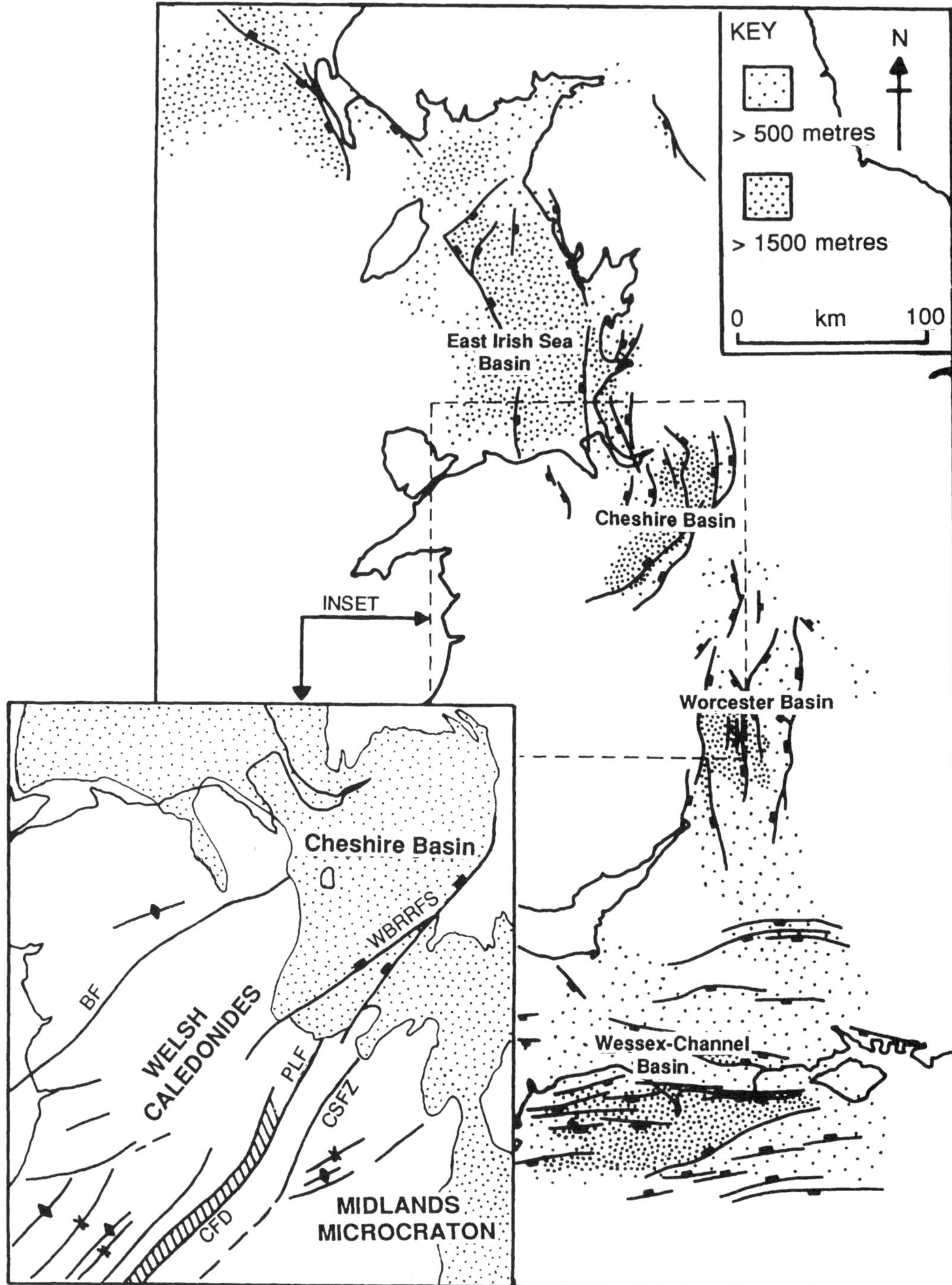

Fig. 1. The Permo-Triassic rift basins of southern Britain (stipple ornament denotes thickness of preserved Permo-Triassic strata). Inset shows Cheshire Basin in relation to outcropping basement structures. BF, Bala Fault; PLF, Pontesford–Linley Fault; CFD, Clun Forest Disturbance; CSFZ, Church Stretton Fault Zone; WBRRFS, Wem–Bridgemere–Red Rock Fault System.

Forest Disturbance, the Church Stretton Fault and the Pontesford–Linley Fault (Fig. 1). The Church Stretton Fault is imaged as a west-dipping thrust on a seismic profile (Smith 1987) and separates, at the surface, a zone of steep dips in its hanging wall block from relatively undeformed rocks in its footwall block. The Pontesford–Linley Lineament is a prominent feature on images of gravity and magnetic fields (Lee *et al.* 1990) and has a prolonged history of fault reactivation, possibly from Proterozoic times onwards (Wills 1978; Woodcock 1984; Soper *et al.* 1987). The Clun Forest, Pontesford–Linley and Church Stretton structures together form the northwest boundary of the Midlands Microcraton which, in Caledonian times, remained stable and relatively undeformed (Soper *et al.* 1987) and is believed to constitute an important rigid block in the basement structural mosaic.

There is ample evidence that reactivation of these Caledonian basement structures influenced both extensional basin development in the early Carboniferous (Plant *et al.* 1996) and also subsequent Variscan basin inversion in latest Carboniferous times. Thus, seismic reflection data (Evans *et al.* 1993; Chadwick & Evans 1995) show evidence of Variscan reverse faults in the basement along the southeast margin of the Cheshire Basin. Related Variscan structures can also be seen at outcrop in the area, ranging in size from the Potteries Syncline to the numerous small thrusts in coal workings of the North Staffordshire Coalfield (Taylor *et al.* 1963; Evans *et al.* 1968, Rees & Wilson in press). This zone of reverse faulting and folding lies roughly along the northeasterly continuation of the Clun Forest Disturbance and the Pontesford–Linley Lineament and is interpreted as being associated with Variscan transpressive reactivation of these, or related Caledonian basement structures.

Farther south, seismic reflection data (Plant *et al.* 1996), and outcrop information (BGS 1990), show that Carboniferous strata pinch out eastwards beneath the Permo-Triassic rocks of the Cheshire Basin such that Lower Palaeozoic rocks form the pre-Permian subcrop on the hanging wall block of the Wem Fault. In contrast, east of the Wem Fault, on its footwall-block, outcrops to the south indicate that quite thick Carboniferous strata are preserved beneath the Permo-Triassic cover (BGS 1990). This structural configuration again appears to be the consequence of basement fault reactivation, Variscan reversal of the structural precursor of the Wem Fault leading to uplift and erosion of its hanging wall block.

There are therefore, numerous instances of Caledonian basement structures being subject to repeated reactivation. This is particularly noticeable at the northwest margin of the Midlands Microcraton, along a line that was subsequently to become the southeast margin of the Cheshire Basin.

Permian to Cenozoic basin evolution

Variscan continental collision led to final suturing and consolidation of the Pangaean supercontinent, accompanied by regional uplift. During Permian times progressive peneplanation was accompanied by development of regional sag basins in the North Sea region. In late Permian, and particularly, early Triassic times, the northwest European region formed an isthmus between the rapidly developing Arctic–North Atlantic rift system to the north and the Tethys–Central Atlantic–Gulf of Mexico rift-wrench system to the south (Coward 1995). Regional extension, directed roughly E–W to NE–SW, became established as the dominant tectonic process and the Cheshire Basin formed as part of the English Channel–East Irish Sea rift. Transtensional reactivation of pre-existing basement structures played an important part in controlling the evolution of all the basins along the rift (Chadwick & Evans 1995). This is exemplified by the southeast margin of the Cheshire Basin (Fig. 1) which lies along strike from the important basement structures forming the northwest margin of the Midlands Microcraton (the Clun Forest, Pontesford–Linley and Church Stretton structures). As described above, these Caledonian features were reactivated in Variscan times (in transpressional mode) and there is abundant seismic reflection evidence (Evans *et al.* 1993, Chadwick & Evans 1995) that they were reactivated again in Permo-Triassic times (in transtensional mode), to form the basin-controlling Wem–Bridgemere–Red Rock Fault System (WBRRFS).

Because strata younger than early Jurassic are no longer preserved, the post-Triassic structural evolution of the basin is, to some extent, conjectural. Constraints on this period of basin development are provided by assessment of regional tectonics and comparison with neighbouring basins where more complete sequences are preserved.

Prior to the onset of North Atlantic seafloor spreading in mid-Cretaceous times, the post-Triassic development of the Cheshire Basin was characterized by episodes of crustal extension. It is likely that the principal post-Triassic extensional structures largely followed the template established in Permo-Triassic times. This is exemplified by the Lias sequence preserved in the centre of the basin which is stratigraphically very thick compared to equivalent strata elsewhere in southern Britain, suggestive of continued movement on the WBRRFS (Evans *et al.* 1993).

Comparison with more complete sequences in basins to the south and east (e.g. Whittaker 1985) indicates that extension occurred principally in the early Jurassic and again in late Jurassic to early Cretaceous times. For much of this period the region probably lay within a depositional regime, with thick sedimentary sequences laid down in the fault-controlled extensional basins. It is likely that by early Cretaceous times however, extension was accompanied by considerable erosion, particularly of the block areas, which led to development of the widespread late-Cimmerian unconformity (Rawson & Riley 1982, Whittaker 1985).

By mid-Cretaceous times extension had effectively ceased (e.g. Whittaker 1985), as seafloor spreading propagated northwards into the North Atlantic region. Post-extensional regional shelf subsidence became established, with much diminished structural demarcation between blocks and basins and deposition of a relatively uniform Upper Cretaceous sequence (the Chalk). Latest Cretaceous or earliest Cenozoic times probably marked the maximum post-Variscan burial of the basin (e.g. Lewis *et al.* 1992).

Regional uplift, commencing in latest Cretaceous or early Cenozoic times (e.g. Lewis *et al.* 1992; Brodie & White 1994), triggered a period of erosion which has probably continued to the present-day. Superimposed upon this regional uplift, were more localized basin inversions associated with minor crustal shortening, reversal of earlier normal faults and minor folding. Depth of burial studies (Plant *et al.* 1996) indicate that the Cheshire Basin may have suffered significant structural inversion with considerable reversal of the WBRRFS at its southeast margin.

Present-day structure

The Cheshire Basin is roughly elliptical in plan (Fig. 2), approximately 105 km long and typically 30–50 km wide with a long axis trending NE–SW and flanked to the east and west by Carboniferous and Lower Palaeozoic basement rocks. Knowledge of the basin subsurface structure is based upon the interpretation of several thousand line kilometres of seismic reflection data, details of which are given in Plant *et al.* (1996). In general terms the basin is deepest close to the WBRRFS which forms its southeast margin (Fig. 2). In contrast the western margin of the basin is relatively unfaulted, forming a feather-edge characterized by depositional onlap. Details of basin structure are illustrated by a series of cross-sections (Fig. 3). In the north (Fig. 3a), the basin is roughly symmetrical, its eastern margin not being marked by a major structure; the Red Rock Fault here having quite a modest throw (not more than a few hundred metres). The central part of the Cheshire Basin (Fig. 3b) is also roughly symmetrical, but with a slight eastward-deepening

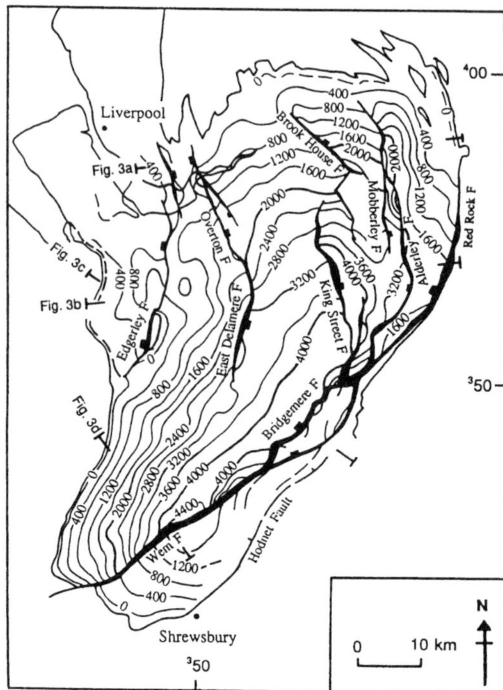

Fig. 2. Simplified depth-contours on the base Permo-Triassic surface, Cheshire Basin (and location of Fig. 3 cross-sections).

asymmetry; the Red Rock Fault which marks the eastern margin of the basin having a normal throw of at least 1800 m hereabouts. True half-graben asymmetry is developed in the southern part of the basin (Figs 3c and d), where the Wem and Bridgemere faults, forming the southeast margin of the basin typically have present-day normal throws (at the base Permo-Triassic level) of some 2500 m, down-to-the-west.

Principal basin-controlling faults

The large majority of the basin faults (Fig. 4) are small to medium, sub-planar, normal faults. Fault-dips are moderate to steep (mostly in the range 55–80°), being generally somewhat steeper in the south of the basin. Most of these smaller faults penetrate to the surface and have throws typically in the range 100–500 m, which tend to decrease upwards, indicating both Permo-Triassic and probable post-Triassic syn-depositional movements. A few significantly larger normal faults are present. These are sub-planar, and of considerable length (several tens of kilometers), with dips in the range 45–75°. Throws at base Permo-Triassic level are typically in the range 500–2500 m, decreasing

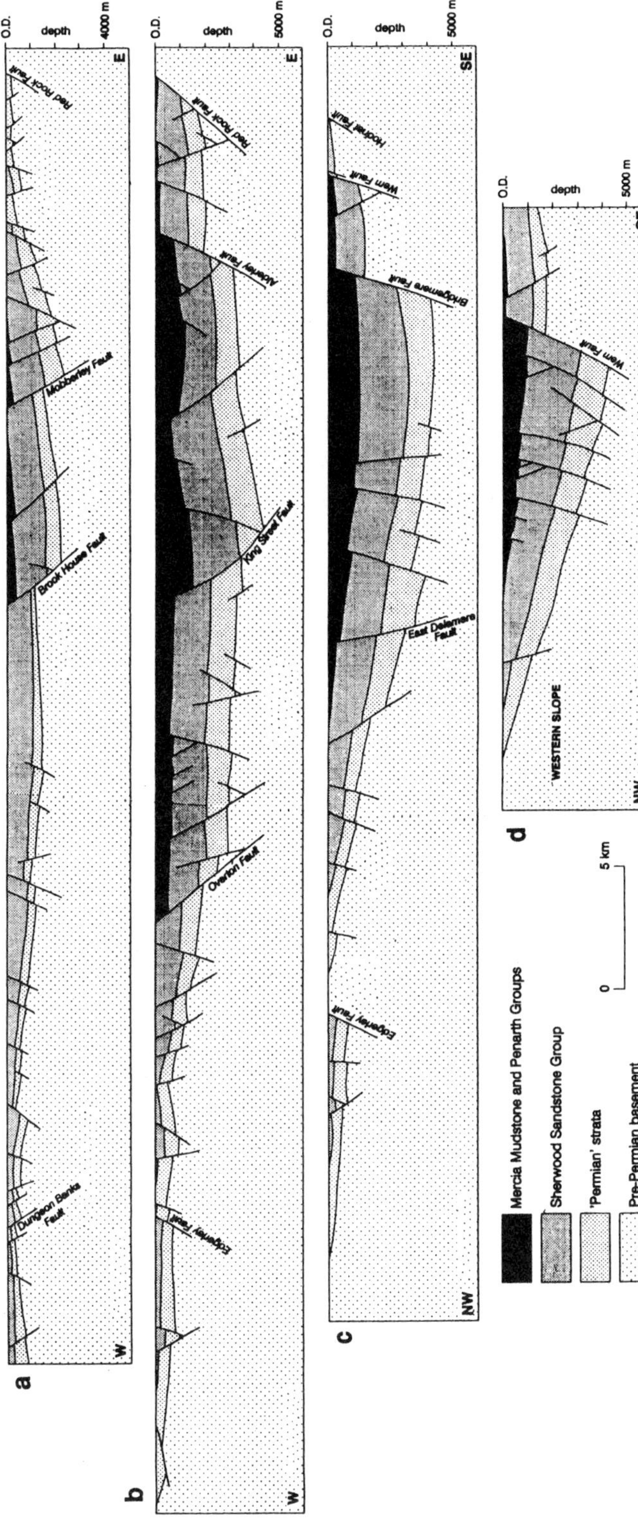

Fig. 3. True-scale cross-sections through the Cheshire Basin (see Fig. 2 for locations).

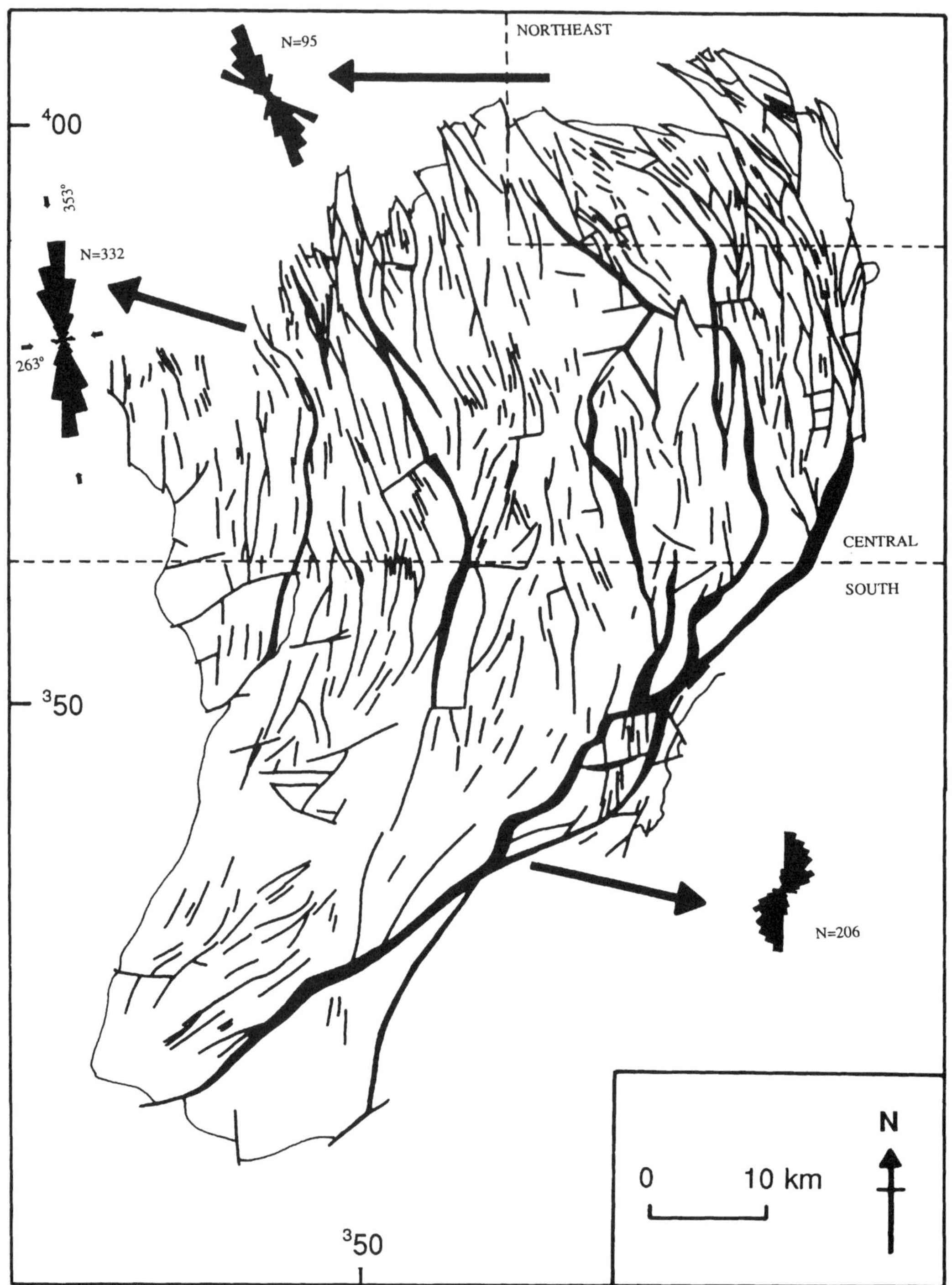

Fig. 4. Fault map at base Permo-Triassic surface, showing regional analysis of fault-strikes.

upwards, and again indicating syn-depositional basin-controlling displacements. The largest basin-controlling normal faults are described briefly below.

The **WBRRFS** is over 110 km long, forms the eastern margin of the Cheshire Basin and was the dominant influence on basin development (Figs 2 and 3). The fault system comprises an anastomosing network of large down-west normal faults, trending NE–SW in the south, bending round to N–S in the north. As described above, it is likely that the reactivation of older basement structures played an important part in determining the location and geometry of the fault system. The cumulative throw on the WBRRFS is greatest in its southern–central section where it is locally in excess of 4000 m at the base of the Permo-Triassic succession (Figs 3c & d). Throws on individual fault strands (e.g. the Wem and Bridgemere faults) are locally in excess of 2500 m. The throw on the fault system gradually diminishes northwards, as the Bridgemere Fault dies out and the Wem Fault passes into the Red Rock Fault (Figs 3a & b). Where it marks the eastern margin of the basin the observable throw on the Red Rock Fault is typically 1000–2000 m, but this decreases to a few hundred metres, or less, in the northeast of the basin. It should be stressed however, that because Carboniferous rocks crop out on the footwall-block of the Red Rock Fault, along much of its length, only minimum estimates of its true throw can be made; its real throw may be considerably larger. The constituent faults of the WBRRFS have planar or sub-planar fault-surfaces, which show a systematic variation in dip along the length of the fault system. The Red Rock Fault has rather moderate westerly dips, typically in the range 45–60° (Fig. 3b). Farther south the Wem and Bridgemere faults are significantly steeper with northwesterly dips in the range 60–70°. Locally, close to its junction with the Wem Fault, the Bridgemere Fault is very steep (up to 80°), and has the appearance on seismic data of a negative flower-structure (*sensu* Harding 1985). Small transfer faults associated with the WBRRFS trend roughly E–W and appear to cause local segmentation of the fault system.

All preserved stratigraphical units thicken dramatically westwards across the WBRRFS (Fig. 3), resulting in a marked upward decrease of displacement and indicating that the WBRRFS was active throughout the extensional phases of basin evolution, imposing on the basin its dominant easterly-deepening asymmetry. Depth of burial studies (Plant *et al.* 1996), suggest that during Cenozoic basin inversion the hanging wall blocks of the WBRRFS may have suffered differential uplift of up to 1000 m relative to the footwall blocks. By implication, associated reversal of the

WBRRFS would also have approached 1000 m locally. If this were the case, prior to reversal, the cumulative down-to-the-west normal throw on the WBRRFS may in places have approached 5000 m.

The **Hodnet Fault** is a large normal fault to the east of the WBRRFS (Fig. 2). It appears to splay southwards from the Wem Fault separating the Cheshire Basin from the Stafford Basin to the southeast. Sparse seismic coverage means that the fault is poorly understood, but a significant down-west gravity gradient (Plant *et al.* 1996) suggests a considerable westerly downthrow.

The **Alderley Fault** is approximately 30 km long, lying in the hanging wall block of the WBRRFS, splaying north from the Wem Fault and making an important contribution to the cumulative down-west throw at the basin margin hereabouts (Fig. 3b). The sub-surface nature of this fault is poorly understood but its throw, at base Permo-Triassic level, is believed to be of the order of several hundred metres, perhaps locally in excess of 1000 m (Plant *et al.* 1996).

Three large down-east normal faults, the Brook House, King Street and Mobberley faults, lie in the hanging wall of the WBRRFS, and link with it at depth in an antithetic relationship.

The NW–SE trending **Brook House Fault** is about 15 km long and constitutes the major structure in the northern part of the Cheshire Basin. Preserved strata thicken eastwards across the fault, notably the Permian sequence (Fig. 3a), indicating that the fault was an important structure early in basin evolution. The throw on the fault locally exceeds 1000 m, decreasing upwards and also northwards.

The **King Street Fault** is a 30 km long, N–S trending structure linking to the Brook House Fault via sub-vertical, possible transfer, faults. The fault is typically planar and quite steeply-dipping (~70°), but in parts a less-steep, somewhat listric geometry is developed (Fig. 3b). Throws on the King Street Fault approach 1000 m at the base of the Permo-Triassic sequence, but diminish north and south. Eastwards stratigraphical thickening is evident, particularly in the Sherwood Sandstone Group. In contrast to the Brook House Fault, thickening of the Permian sequence is relatively minor.

The **Mobberley Fault** is a 30 km long, N–S trending structure (Fig. 3a) lying to the east of the King Street and Brook House faults. Similar to these faults, it has an easterly downthrow locally approaching 1000 m, largely syn-depositional, with commensurate eastward thickening of stratigraphical units.

The **Overton** and **East Delamere** faults (Figs 3b & c) form a major N–S intra-basin structure 70 km long. The faults have planar geometry and moderate (~50–60°) easterly dips, steepening somewhat in the south. Easterly downthrows locally exceed 1000 m, at base Permo-Triassic level. Only minor eastward thickening of preserved strata is observed across the faults, perhaps indicating that the bulk of displacement took place after deposition of the Sherwood Sandstone Group.

The principal fault in the western part of the basin is the 60 km long, N–S trending **Edgerley Fault** (Figs 3b & c). This fault is unusual for the Cheshire Basin in that it displays a markedly curved 'listric' profile, with a rollover developed in its hangingwall-block (Plant *et al.* 1996). Throws on the fault are down-west, and locally approach 1000 m at the western edge of the Milton Green inlier, where basement rocks (Carboniferous strata) crop out in the footwall block.

In addition to normal faults, which are in the overwhelming majority, a small number of reverse faults have been observed. These are generally very steep features with small reverse throws, which appear to have formed by strongly oblique-reverse displacements in a thoroughly lithified ('brittle') basin-fill. A possible larger reverse fault is visible near the eastern margin of the basin. All of the reverse faults are interpreted to be associated with Cenozoic basin inversion (Plant *et al.* 1996).

Having summarized the evolution and present structure of the Cheshire basin, the remainder of the paper will describe various types of fault analysis aimed at elucidating the nature of the extensional processes which controlled basin development.

Derivation of extension direction

The first step in estimating the extension direction was to analyse fault orientations. Strikes of all faults cutting the base Permo-Triassic surface (Fig. 4) were measured. Very long faults were divided into individual segments of uniform strike. In total, 633 faults or fault segments were measured. Examination of the fault map reveals a systematic geographical variation of fault-strike, illustrated clearly by rose diagrams.

In the southern area trends are dominantly NNE to NE, some 71% of fault-strikes lying between 0° and 060° (Fig. 4). These trends are roughly parallel to the Wem and Bridgemere faults which form the southeastern margin of the basin hereabouts. A strong element of Caledonian basement control is indicated, consistent with the basement fault reactivation seen at the basin margin (see above). Statistically however, the observed trends are somewhat unusual, in that they do not have a normal distribution about a mean direction. The commonest (modal) fault-strike (16%) is 0–10°. This lies at the extreme limit of the dominant range and gives a strongly skewed overall distribution. It

is suspected that this is caused by poorly constrained fault-strikes in this area, due principally to the rather sparse seismic coverage hereabouts (Plant *et al.* 1996).

The central area contains the majority of the faults in the basin. Faults are present on all scales ranging from very large structures at the basin margin, through major intra-basin faults to small structures of limited lateral extent. Fault-strikes here have a well-defined distribution. The dominant range of fault-strikes is quite narrow, 76% of faults falling between 330° and 010°, with a modal direction (26%) of 350–360°. The single dominant direction is estimated at 353°. The roughly uniform distribution of fault-strikes either side of this direction indicates that the dataset is well constrained. Whilst it is possible that the trends of the major basin-controlling faults were determined by underlying basement structure, it is most unlikely that the multitude of intermediate and small intra-basin faults were directly controlled in this way. Instead these are interpreted to be dip-slip normal faults, which formed perpendicular to the dominant extension direction which is estimated therefore at 083–263°. A small but significant peak occurs in the fault distribution with a trend of 265–270°, perpendicular to the dominant fault trend. This corresponds to several small faults, which cross-cut the dominant N–S trending faults, sometimes offsetting them, and commonly partitioning adjacent areas of differing structural style. These faults are interpreted as transfer faults which, ideally, form parallel to the extension direction. This transfer fault trend is, therefore, consistent with the extension direction as derived from the dip-slip faults.

In the northeastern part of the basin a dominant NNW–SSE trend is evident, 60% of fault-strikes lying between 310 and 350°, with a modal direction (19%) of 330–340°. A subsidiary peak of 290–300° relates principally to poorly constrained small faults near the northern edge of the basin and may be an artefact. The dominant fault trend is somewhat enigmatic, being neither parallel to the main Caledonoid (NE–SW) trend, nor precisely perpendicular to the extension direction assumed for the central part of the basin (see above). One possibility is a gradual change in regional extension direction. However, the fact that the intra-basin faults of the East Irish Sea Basin (as distinct from its major basin margin faults which have trends characteristic of basement control) show dominant N–S trends (BGS 1994), very similar to the faults of the central Cheshire Basin, suggest that this is not the case. Perhaps more likely is an increased element of basement control in this area. This is consistent with NNW–SSE trending faults in the (Carboniferous) basement outcrop to the north of the basin, which are believed to date from Carboniferous times or earlier (Kirby *et al.* in press).

It is concluded that some of the fault trends in the Cheshire Basin indicate a degree of basement control, particularly at the southeast basin margin and, more generally, in the southern and possibly the far northeastern parts of the basin. Most of the faults however, particularly small intra-basin structures, seem to have formed independently of basement influence, developing perpendicular to the extension direction, which is estimated at 083–263°. This is in substantial agreement with the regional Permo-Triassic extension direction in southern Britain, estimated by Chadwick & Evans (1995) using different structural criteria, at 075–255°. It is important to stress that the extension direction may not have remained the same throughout basin evolution. It is likely that the dominant fault trend reflects the early extension direction which initiated basin development and forged the fundamental fault architecture of the basin. Regional tectonic considerations (e.g. Coward 1995) indicate that roughly E–W to NE–SW directed extension was probably dominant throughout the Permo-Triassic and early Jurassic phases of basin development, and accounted for most of the basin extension. Subsequent, possibly different, extension directions may not have contributed to significant new faulting, but rather just reactivated existing faults, albeit in oblique-slip modes.

Measurement of extension magnitude from observed fault displacements

The heave of a fault is defined as the horizontal displacement of a reference horizon across that fault, measured perpendicular to the fault strike. For a dip-slip normal fault therefore, the heave is equal to the true horizontal displacement across the fault, and provides a measure of the extension on the fault. In the more general case of an oblique-slip fault, the apparent heave can be defined as the horizontal displacement of a reference horizon across the fault measured in a plane parallel to the extension direction. Thus by summing the apparent heaves of faults which cut a cross-section parallel to the extension direction, a measure of the total extension can be obtained.

Twenty-nine equally spaced transects were chosen, parallel to the assumed extension direction (083–263°), and crossing the basin so as to encompass the Permo-Triassic outcrop and the basin-margin faults (Fig. 5). Taking each transect in turn, the apparent heaves on all mapped (i.e. seismically resolved) faults intersected by the transect were measured and summed to give the total apparent heave (Table 1). The transects vary in

length from 21.2 to 65.2 km with total apparent heave values in the range 1.56–5.98 km. The total apparent heaves on transects 7–18 are minimum values, because, on these transects, the eastern basin-margin includes faults (principally the Red Rock Fault) which have pre-Permian basement rocks at outcrop in their footwall-blocks. Thus, the full displacement of these faults at base Permo-Triassic level cannot be seen; observed heaves being minimum estimates. Nevertheless, the total apparent heave on each transect gives a first order estimate of the extension along that transect. The crustal extension factor γ can also be computed for each transect where:

$$\gamma = \frac{\text{present (extended) transect length}}{\text{pre-extension transect length}}.$$

Extension factors based upon the measured apparent heaves of seismically resolved faults are in the range 1.05–1.14 (Table 1).

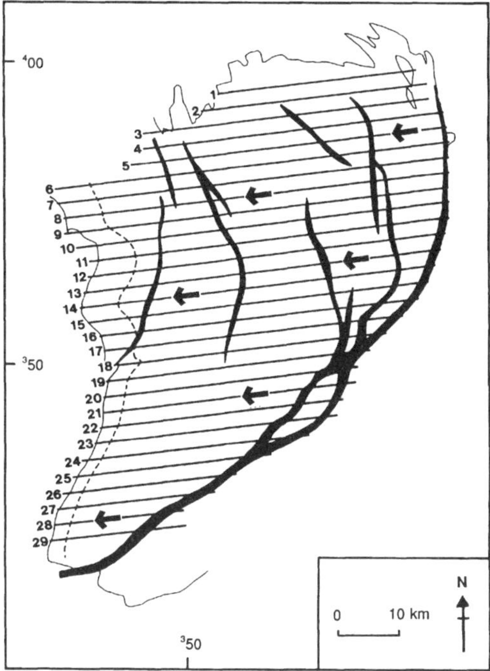

Fig. 5. Extension-parallel transects of the Cheshire Basin. Arrows denote dominant extension direction (083–263°), producing near dip-slip displacements on the eastern margin and intra-basin faults, but sinistral transtension on the WBRRFS at the southeast basin margin. Dashed line denotes pre-extension position of present-day base Permo-Triassic outcrop (notional pin-line immediately to east of WBRRFS).

Table 1. *Measured apparent heaves and extension factors from resolved faults*

Transect	Transect length (km)	Measured total heave (km) on resolved faults	Extension factor from resolved faults
1	32.6	1.56	1.05
2	37.3	2.02	1.06
3	47.4	3.70	1.08
4	49.4	3.87	1.08
5	52.2	4.12	1.09
6	65.2	5.10	1.08
7	64.7	4.31	1.07
8	63.6	4.13	1.07
9	63.3	4.44	1.08
10	61.4	5.54	1.10
11	59.4	4.53	1.08
12	58.8	5.01	1.09
13	58.8	5.70	1.11
14	58.4	5.37	1.10
15	55.7	5.98	1.12
16	52.3	5.83	1.13
17	49.0	4.57	1.10
18	44.1	3.85	1.10
19	43.5	3.41	1.09
20	41.7	2.95	1.08
21	41.7	2.85	1.07
22	40.4	2.53	1.07
23	40.2	2.65	1.07
24	34.8	2.63	1.08
25	29.5	2.61	1.10
26	27.2	2.62	1.11
27	24.8	2.69	1.12
28	21.2	2.58	1.14
29	22.1	2.44	1.12

Estimation of extension magnitude from all faults by fractal extrapolation

Background

Only in certain rare circumstances do observed fault heaves give an accurate estimate of true extension. This is principally because the hanging wall blocks of observed faults themselves extend by modes of strain beneath the limit of observational (generally seismic) resolution (e.g. White *et al.* 1986). Thus, extension estimates based on the sum of observed fault heaves may be significantly deficient, because the contribution to extension of small faults, beneath the limit of resolution, is neglected. There are many instances of extension estimates based on the restoration of (observed) faults being significantly less than estimates of extension from subsidence analysis and/or observed amounts of crustal thinning; for example the basins of the North Sea (Sclater & Shorey 1989; Ziegler & Van Hoorn 1989). One explanation for

this discrepancy is that small faults, beneath the seismic resolution threshold, can together account for substantial amounts of 'hidden' extension (e.g. Walsh *et al.* 1991).

Several recent studies have provided evidence in support of the contention that fault populations within a basin are self-similar across a wide range of scales (e.g. Kakimi 1980; Marret & Allmendinger 1992; Walsh & Watterson 1992). In other words, the relative numbers of large and small faults remain the same, irrespective of size. This scale-invariance is known as a fractal, or power-law, size distribution. Thus, the 'hidden' extension due to small faults can be addressed statistically by applying fractal scaling laws, which quantify scale-invariant phenomena.

For a set of faults, heaves and extension are related as follows:

$$E_{\text{total}} = \Sigma \Delta h_{\text{resolved}} + \Sigma \Delta h_{\text{hidden}}$$

where E_{total} is the total extension on a set of faults Δh is the heave on each fault

A major disadvantage of analysing measured fault heave data however, is that the heaves of small faults, whose distribution critically affects the prediction of hidden faults, are extremely difficult to measure accurately from seismic reflection data. Seismic data are, however, much more amenable to the accurate measurement of small fault throws. In the analysis described below therefore, the distribution of fault throws is the principal analytical tool. Thus:

$$T_{\text{total}} = \Sigma t_{\text{resolved}} + \Sigma t_{\text{hidden}}$$

where T_{total} is the total throw on a set of faults t is the throw on each fault.

Following Marrett & Allmendinger (1992), for a fault population that follows a fractal size distribution, the cumulative number of faults (N) with throw greater than or equal to t may be written:

$$N = t^{-c} \quad \text{i.e.} \quad \log_{10} N = -c.\log_{10} t$$

where c is the power-law exponent which characterizes the relative number of large and small faults.

Thus, for a fractal distribution, a plot of log (N) against log (t) will be a straight line of gradient $-c$.

In a fault population exhibiting power-law or fractal behaviour, the total throw due to small faults beneath the limits of detection can be quantified, provided the number of larger faults is known. Thus, the total throw due to displacements on all faults (T_{total}) can be estimated from the N largest faults (e.g. the resolved faults):

$$T_{\text{total}} = t_1 + t_2 + t_3 + \ldots\ldots + t_N + \Sigma t_{\text{hidden}}.$$

For 1–D sampling of the fault population, the total 'hidden' throw of small faults can be determined by extrapolating the fractal size distribution (Marrett & Allmendinger 1992):

$$\Sigma t_{hidden} = T_{MIN} \left\{ \frac{c}{1-c} \right\} (N + 1) \left\{ \frac{N}{(N+1)} \right\}^{\frac{1}{c}} \quad (1)$$

where Σt_{hidden} = total throw on faults of throw $< T_{MIN}$ (i.e. the total 'hidden' throw)

T_{MIN} = minimum throw of fully sampled faults exhibiting fractal behaviour

N = number of faults exhibiting fractal behaviour with throws $\geq T_{MIN}$

$-c$ = gradient of linear segment on log–log plot.

For this expression to be useful, the fault population must be such that c is less than unity. This allows Σt_{hidden} to converge, with the implication that the largest faults in the population account for most of the cumulative throw.

Fault analysis

Fault throw analysis was carried out on the extension-parallel transects (Fig. 5), which provide multiple 1-D sampling of the fault population at the base of the Permo-Triassic sequence. It is important to stress that this 1-D sampling procedure analyses the displacement statistics of the faults *in the planes of the transects*, which differ from the statistics of the faulted surface as a whole (this would be provided by a 2-D sampling procedure, as derived from a fault map). Thus, for example, with multiple 1-D sampling, a single long fault may intersect several adjacent transects and give several different measured throws at each fault/transect intersection. In the following discussions therefore, the term 'fault' strictly means 'fault-intersection'. Taking each transect in turn, the throw of every intersecting fault was measured and the cumulative number of faults with throws greater than each throw value computed.

Figure 6 illustrates a plot of log(N) against log(t) for all of the measured faults, on all the transects. Fractal size distribution is characterized by points which lie on a straight line. Thus, although limited segments of the throw distribution plot do show approximately linear trends, the fault dataset as a whole does not, having a generally curved, convex-upwards shape. The whole fault population of the Cheshire Basin does not therefore fall on a single fractal trend.

To examine more closely this apparent departure from ideal fractal behaviour, a more detailed

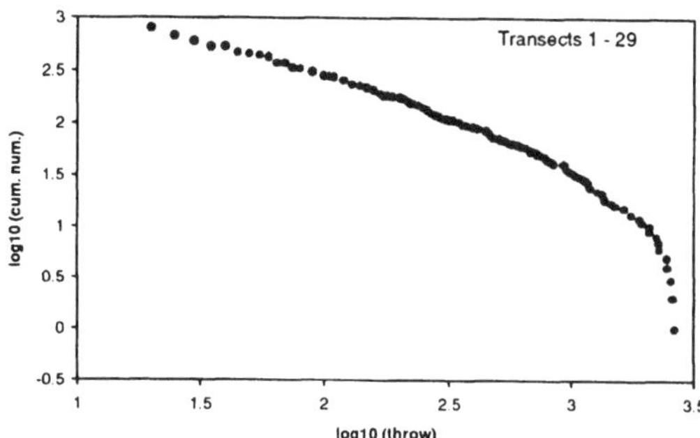

Fig. 6. Fault throw distribution in the Cheshire Basin – all transects.

analysis of the fault throws was carried out by examining each transect in turn. Throw distribution plots for individual transects are considerably more 'noisy' than for the whole fault set (because many fewer faults were measured, interpretation, mapping and sampling deficiencies were more significant). Nevertheless, the individual fault transect data differ significantly from the full fault dataset in that clear linear segments are evident for most transects between certain throw limits. Sample throw distribution log–log plots are illustrated in Figs 7 and 8 with critical plot parameters presented in Table 2.

Before examining the data from individual transects in detail, it is instructive to discuss briefly some general aspects.

Large faults (with throws typically in the range 500–1000 m) commonly depart significantly from a linear fractal trend. This is likely to be for two main reasons. Firstly, observed throws on some of the faults (for example the Red Rock Fault), may be significantly less than their true original throws, due both to the lack of preserved Permo-Triassic strata on their footwall-block and also to the effects of subsequent Cenozoic reversal. If it were possible to correct present-day observed throws to true original throws, the result would be to move the plotted points to the right, closer to a fractal trend. Secondly, the fractal trends predict the existence of rare but very large faults (throw ~10000 m). Such faults are physically difficult to develop, displacement tending to transfer to new faults when the throw has reached a certain critical value (dependent on the dip of the fault and the physical properties of the faulted medium). The several large faults which form the eastern margin of the Cheshire Basin may be the equivalent of a single very large fault which was unable to develop. It thus appears that very large faults depart from ideal fractal behaviour, because of mechanical difficulties connected with their formation.

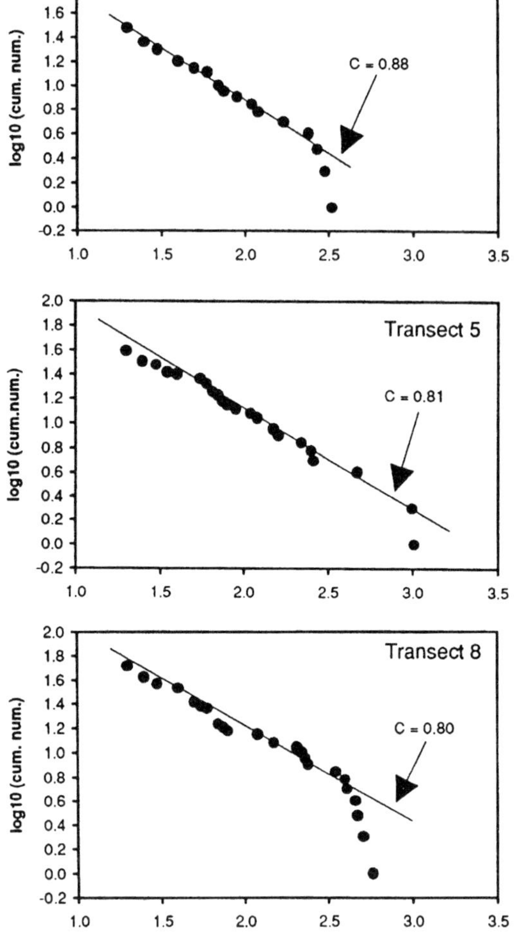

Fig. 7. Fault throw distributions typical of the northern part of the Cheshire Basin (transects 2, 5 and 8).

Turning now to more detailed aspects of the transect fault data, an intriguing systematic geographical variation in fault displacement distribution is revealed. Fault distribution plots from the northern part of the basin, exemplified by transects 2, 5 and 8 (Fig. 7), have linear segments with markedly higher gradients (typical values of $c. >0.80$) than from the southern part of the basin (typical values of $c. <0.6$), exemplified by transects 15, 16 and 22 (Fig. 8). Thus, in the northern part of the basin, the fault population comprises principally small and medium-sized faults, whereas farther south very large faults are present with, relatively speaking, fewer medium-sized and small faults.

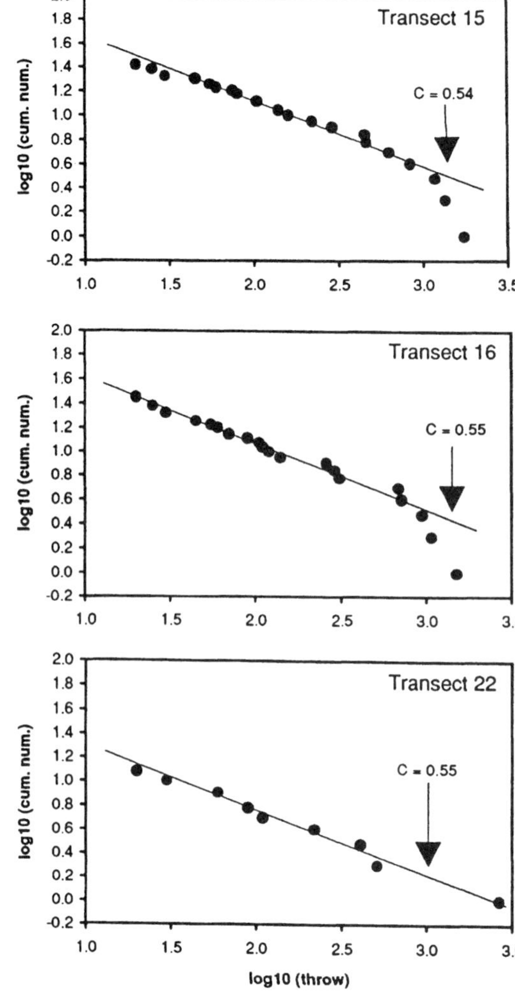

Fig. 8. Fault throw distributions typical of the southern part of the Cheshire Basin (transects 15, 16 and 22).

Small faults (with throws typically in the range 20–50 m) also depart somewhat from a linear fractal trend, though less markedly than the large faults. This is interpreted as an artefact due to inadequate sampling, rather than a fundamental property of the faults themselves, which are assumed to lie on a fractal trend. Small faults are liable to be undersampled for two main reasons. Firstly they are generally of limited length (e.g. Walsh & Watterson 1987), and as such are prone to remain undetected, particularly with a well-spaced seismic grid. Secondly, because small faults are close to the limit of seismic resolution, a significant proportion may have been missed by the interpreter, particularly in areas of complex structure and poor quality data.

Table 2. *Computation of 'hidden' throws, 'hidden' heaves and total heaves and extension factors for all faults (c = −gradient of linear segment, N = cumulative number of faults of throw >T_{MIN})*

Transect	Transect length (km)	T_{MIN} (m)	c	N	Computed total throw (km) on faults of throw $<T_{MIN}$	Computed total heave (km) on faults of throw $<T_{MIN}$	Measured total heave (km) on faults of throw $\geq T_{MIN}$	Total heave (km) on all faults	Extension factor from all faults
1	32.6	40	1.32	13.3	*****	*****	*****	*****	*****
2	37.3	30	0.88	21.8	4.767	2.142	1.79	3.93	1.12
3	47.4	40	0.79	35.0	5.227	2.349	3.57	5.92	1.14
4	49.4	40	0.80	28.1	4.457	2.003	3.64	5.64	1.13
5	52.2	55	0.81	22.1	5.128	2.316	3.68	6.00	1.13
6	65.2	50	0.80	25.4	5.031	2.272	4.30	6.57	1.11
7	64.7	100	0.81	16.3	6.853	3.094	3.37	6.46	1.11
8	63.6	40	0.80	35.8	5.689	2.557	3.73	6.29	1.11
9	63.3	35	0.81	44.0	6.531	2.935	4.25	7.19	1.13
10	61.4	60	0.82	30.0	8.141	3.676	4.87	8.55	1.16
11	59.4	90	0.81	21.7	8.238	3.737	3.95	7.69	1.15
12	58.8	130	0.82	12.5	7.279	3.317	4.18	7.50	1.15
13	58.8	55	0.77	24.0	4.366	1.971	5.44	7.41	1.14
14	58.4	70	0.68	16.8	2.432	1.098	5.00	6.10	1.12
15	55.7	45	0.54	20.7	1.050	0.474	5.84	6.31	1.13
16	52.3	20	0.55	28.4	0.675	0.303	5.83	6.13	1.13
17	49.0	55	0.64	18.1	1.717	0.775	4.31	5.09	1.12
18	44.1	50	0.57	14.5	0.914	0.413	3.73	4.14	1.10
19	43.5	20	0.61	19.0	0.575	0.258	3.41	3.67	1.09
20	41.7	20	0.67	24.5	0.975	0.438	2.95	3.39	1.09
21	41.7	20	0.64	20.6	0.713	0.321	2.85	3.17	1.08
22	40.4	30	0.55	11.4	0.390	0.175	2.51	2.69	1.07
23	40.2	30	0.52	8.3	0.243	0.109	2.55	2.66	1.07
24	34.8	25	0.51	10.0	0.237	0.107	2.62	2.73	1.09
25	29.5	25	0.49	8.9	0.191	0.086	2.55	2.64	1.10
26	27.2	20	0.47	10.5	0.168	0.076	2.62	2.70	1.11
27	24.8	20	0.56	7.7	0.178	0.080	2.69	2.77	1.13
28	21.2	20	0.57	7.2	0.173	0.078	2.58	2.66	1.14
29	22.1	20	0.56	8.8	0.206	0.093	2.44	2.53	1.13

*****, hidden displacements not computed for values of c greater than unity.

Recalling the geographical variations in the degree of basement fault reactivation indicated by the fault strikes, it may be that the fault displacement populations are revealing the same effect. Thus, in the south of the basin the fault distribution is determined by the strong influence of large, reactivated basement structures, which promotes the development of large, basin-controlling normal faults. Farther north, the more equitable distribution of fault throws perhaps reflects the effect of extension on more isotropic basement rocks.

Examination of the transect throw-frequency plots revealed that the minimum throw threshold for fractal behaviour (T_{MIN}) was typically in the range 20–50 m, though with a few exceptions. The cumulative number of faults (N) in each transect with throws $\geq T_{MIN}$, is given in Table 2 (the actual number of faults corresponding to T_{MIN} does not usually fall precisely on the linear fractal trend, instead the corresponding interpolated value on the

fractal trend was taken, thus N is not necessarily an integer). By substituting for T_{MIN}, c and N in Equation 1 (above), the total 'hidden' throw (Σt_{hidden}) on each transect was obtained (Table 2). Hidden throws for Transect 1 at the northern edge of the basin, were not computed because its value of c was markedly anomalous, being greater than unity. It may be that the fault parameters measured on this transect, where seismic data are relatively sparse, and fault throws poorly constrained due to removal of Permo-Triassic strata on footwall blocks, are unreliable.

For the practical reasons outlined above, the fault population study was directed at fault throws. To obtain estimates of 'hidden' extension, it was necessary to obtain the total heave of the small faults. To this end, a detailed study of the geometry of all clearly resolved faults seen on the transects was carried out (Fig. 9). The apparent dips of these faults show considerable scatter but generally lie in

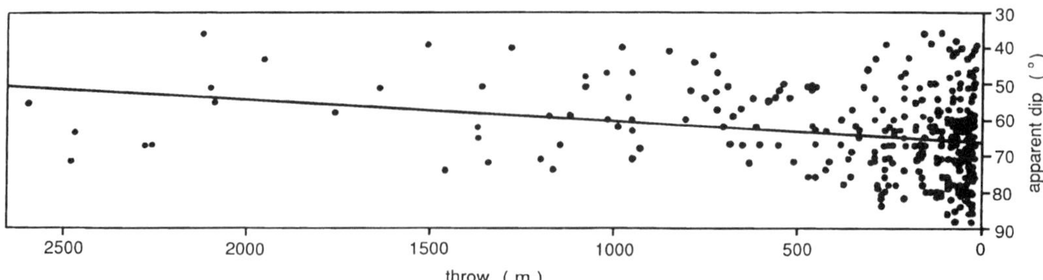

Fig. 9. Fault throw plotted against apparent dip for well-resolved faults intersecting the transects.

the range 40–80°, and define a broad linear trend, apparent dips decreasing somewhat with throw (as with the foregoing fault population analysis, this procedure merely examines the geometrical properties of the faults *in the plane of the transects*, there is no correction for obliquity, so fault geometries are apparent, rather than real). A least-squares best-fit line through the data has a gradient of $-0.00605°m^{-1}$, and an intercept at zero throw of 65.9°. From this, a mean apparent dip (α) was obtained for all faults of throw $<T_{MIN}$ (α varies only slightly with T_{MIN}, in the range 65.5–65.8°). Though covering a considerable range of scatter, it is considered that α reasonably represents the apparent geometry of an 'average' small fault intersecting the transects. Thus the total 'hidden' heave of these faults ($\Sigma\Delta h_{<T_{MIN}}$) was calculated for each transect as follows:

$$\Sigma\Delta h_{<T_{MIN}} = \frac{\Sigma t_{<T_{MIN}}}{\tan \alpha} .$$

Computed total heaves on faults of throw $<T_{MIN}$ are given for each transect in Table 2. Values are in the range 0.08–3.74 km, the largest 'hidden' extensions being in the north of the basin where small faults are relatively much more abundant.

In order to derive the total extension on each transect, the total observed heave due to faults of throw $\geq T_{MIN}$ was measured by discarding observed heaves on smaller faults (Table 2).This value was then added to the total 'hidden' heave to obtain the total heave (extension) on all faults on each transect (Table 2).

Discussion

By incorporating the contribution of small faults, the total heave obtained for each transect is in general considerably greater than the heave derived from only the seismically resolved faults (Fig. 10a). Values of heave (extension) range from less than

3 km in the south, to more than 5 km farther north, and more than 7 km in the central part of the basin. It is noticeable that the computed total heaves are only slightly larger than the observed heaves in the south of the basin, where the contribution of small faults is minimal.

Summing the total heaves of all faults on transects 2 to 29 gives a basinwide total heave of 138.50 km. The corresponding total heave on faults of throw $<T_{MIN}$ is 37.25 km. Thus, 'hidden' faults account for some 27% of the total fault-related extension in the basin. This figure falls within the range of 25–60% quoted by Marret & Allmendinger (1992) and is comparable to the 40% estimated by Walsh *et al.* (1991) in two independent studies of faults in the Viking Graben.

These basinwide figures comprise two distinct components. Considering only the northern part of the basin (transects 2–13), the total heave on all faults is 79.14 km and on 'hidden' faults 32.37 km (Table 2). In the south of the basin (transects 14–29) the total heave on all faults is 59.36 km, and on 'hidden' faults is 4.88 km. Thus, in the northern part of the basin, 'hidden' faults account for some 41% of the total extension, whereas in the south they contribute only 8% of the total extension.

Computed values of total extension on the transects are typically >6 km in the north of the basin (Fig. 10b), falling rapidly between transects 16 and 20 to <3 km in the south of the basin. This may in part be an artificial variation, because large faults in the south, lying to the east of the WBRRFS (e.g. the Hodnet Fault) have not been incorporated into the measurements due to lack of information. This is supported by the extension factors (Fig. 10b) which display rather more uniform behaviour across the basin. Nevertheless, increasing values of extension factor (1.10–1.14) in the extreme south of the basin (transects 27–29) may also be an artefact, related to the decreasing transect lengths; cumulative heaves by contrast remain rather consistent around 2.6 km on transects 22–29 (Fig. 10a). It is also reiterated that these estimations of extension

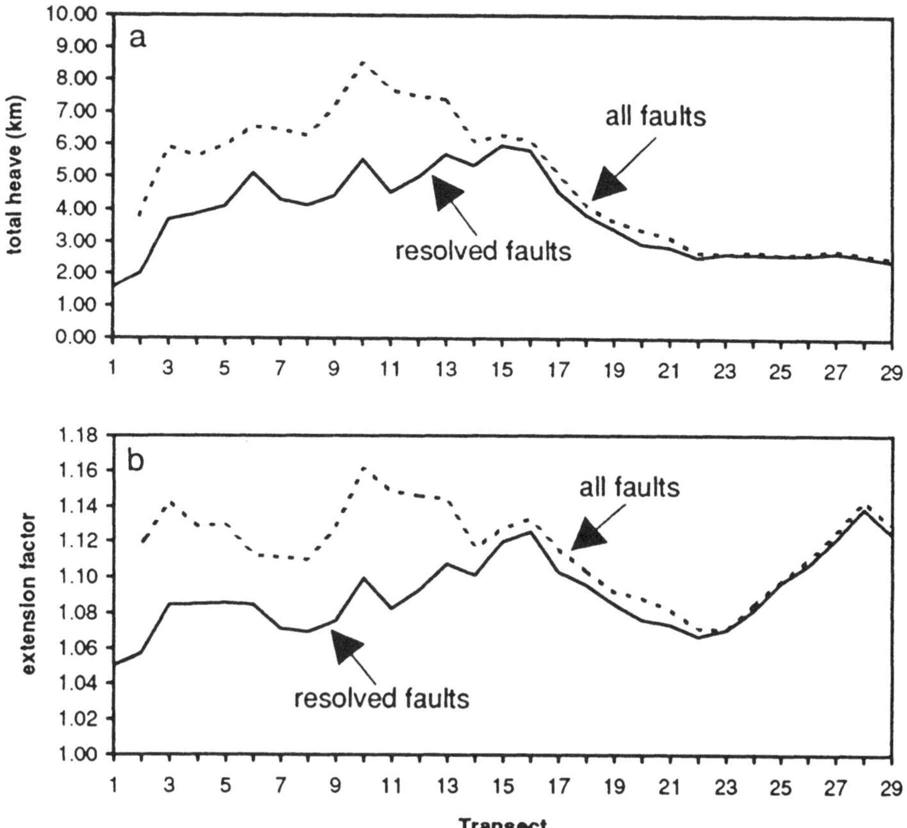

Fig. 10. Amounts of extension for each transect, measured from resolved faults (solid line) and computed for all faults by fractal extrapolation (dashed line) (**a**) heaves (**b**) extension factor.

and extension factor may be significant under-estimates; the lack of preserved Permo-Triassic strata on the footwall block of the WBRRFS precludes measurement of the true fault heave and the WBRRFS may have suffered significant reversal, with consequent decrease of heave during Cenozoic basin inversion. These provisos notwithstanding, it is concluded that extension factors are indeed highest in the central part of the basin, with values in excess of 1.15, declining somewhat to both north and south. The fact that the amount of extension at both the northern and southern edges of the basin is still well above zero, suggests that the present basin edges do not correspond to its original depositional limits. It is likely that the present-day basin limits owe much to structural exhumation and that the original depositional area of the basin spread much farther to the north and south.

By restoring the full computed extensions along each (extension-parallel) transect, it is possible to reconstruct the configuration of the basin prior to extension (Fig. 5). It is evident from this how the basin extended and the different ways in which this extension was related to displacements on the main basin-controlling faults. Thus, the fact that the dominant N–S trending faults (of the central and northern parts of the basin) form a distinct fractal population with trends perpendicular to the exten-sion direction supports the contention that they formed as dip-slip normal faults whose geometry was principally controlled by the extension process rather than by the underlying basement. This contrasts to the NE–SW trending faults farther south whose geometry was effectively pre-ordained by underlying basement trends, and which as a consequence, suffered strongly oblique (sinistral transtensional) displacements. It is notable that the faults forming the southeast basin margin locally have the appearance on seismic data of negative flower-structures, diagnostic of strongly oblique displacements (Chadwick & Evans 1995).

Conclusions

It is concluded that the extensional structural architecture of the Cheshire Basin resulted from a roughly E–W tensional stress field which reactivated, to a variable degree, pre-existing structures in a geographically variable basement.

Analysis both of fault strikes and fault displacements indicates two distinct fault populations, with a significantly greater degree of basement control being evident in the faulting in the south of the basin than farther north. This is ascribed to differing modes of basement extension. Thus, in the south of the basin, strong basement control was associated with the development of large basin-margin normal faults and marked basin asymmetry. The basin margin faults formed parallel to the major reactivated basement structures which form the northwest margin of the Midlands Microcraton. They trend NE–SW and thus suffered dominantly oblique-slip during basin extension. In the central and northern parts of the basin (except the extreme northeast) much weaker basement control is evident. This is perhaps due to a lack of large basement structures giving an effectively more 'isotropic' basement type. Here extension was distributed more equitably across a range of fault sizes. Faults developed here were essentially 'new', formed perpendicular to the extension direction and thus suffered dominantly dip-slip displacements.

Analysis of fault displacement populations demonstrates the importance of 'hidden' displacements on small faults in the estimation of total basin extension. Attention is again drawn to the contrast between large 'hidden' displacements in the central and northern parts of the basin, interpreted as having formed on relatively isotropic basement, and much smaller 'hidden' displacements in the south, where nearly all of the extension was concentrated on large faults overlying major reactivated basement structures.

The analysis presented here is focused on the faults at the base of the Permo-Triassic sequence in the Cheshire Basin. The faulting behaviour described therefore is essentially characteristic of consolidated basement rocks (Carboniferous and older). It is uncertain to what extent the observations noted here have wider applicability. The presence of more than one fault population in a basin (i.e. two or more fractal trends) seems likely to occur more widely. Similarly it may be the general rule that small intra-basin faults form a different population than the larger basin-margin faults. Whether the different populations will be geographically distinct, as in the Cheshire Basin, or spatially superimposed seems likely to vary from basin to basin. The specific conclusion that areas which extend by the reactivation of large basement structures will in general have fewer small faults and, by implication fractures and joints (assuming fractal behaviour), than areas extending upon more isotropic basement, is of considerable potential importance in the study of subsurface fluid movements, with relevance both to subsurface mineral exploration, and also to waste disposal.

This work is based upon interpretations carried out by the author and colleagues D. J. Evans and N. J. P. Smith during the BGS Cheshire Basin multidisciplinary R & D project under the leadership of Dr J. A. Plant. Thanks are given to J. G. Rees and R. P. Barnes for thorough and thoughtful reviews of the initial draft. The work is published with permission of the Director, British Geological Survey (NERC).

References

BGS. 1990. *Mid-Wales and the Marches*, 1:250000 series, Solid Geology.

—— 1994. *East Irish Sea*, 1:250000 series, Special Sheet Edition, Solid Geology.

BRODIE, J. & WHITE, N. 1994. Sedimentary basin inversion caused by igneous underplating: Northwest European continental shelf. *Geology*, **22**, 147–150.

CHADWICK, R. A. & EVANS, D. J. 1995. The timing and direction of Permo-Triassic extension in southern Britain. *In*: BOLDY, S. A. R. (ed.) *Permian and Triassic Rifting in Northwest Europe*. Geological Society, London, Special Publication, **91**, 161–192.

COWARD, M. P. 1995. Structural and tectonic setting of the Permo-Triassic basins of northwest Europe. *In*: BOLDY, S. A. R. (ed.) *Permian and Triassic Rifting in Northwest Europe*. Geological Society, London, Special Publication, **91**, 7–39.

EVANS, W. B., WILSON, A. A., TAYLOR, B. J. & PRICE, D. 1968. *Geology of the Country around Macclesfield, Congleton, Crewe and Middlewich*. Memoir of the Geological Survey, Great Britain, Sheet 110.

EVANS, D. J., REES, J. G. & HOLLOWAY, S. 1993. The Permian to Jurassic stratigraphy and structural evolution of the Cheshire Basin. *Journal of the Geological Society, London*, **150**, 857–870.

HARDING, T. P. 1985. Seismic characteristics and identification of negative flower-structures, positive flower-structures and positive structural inversion. *Bulletin of the American Association of Petroleum Geologists*, **69**, 582–600.

KAKIMI, T. 1980. Magnitude–frequency relation for displacement of minor faults and its significance in crustal deformation. *Bulletin of the Geological Survey of Japan*, **31**, 467–487.

KIRBY, G. A., AITKENHEAD, N., BAILY, H. E., BIRCH, B., CHADWICK, R. A., EVANS, D. J., HOLLIDAY, D. W., HOLLOWAY, S., HULBERT, A. G., PHARAOH, T. C. & SMITH, N. J. P. (in press). The structure and evolution of the Craven Basin and adjacent areas.

Subsurface Memoir of the British Geological Survey.

LEE, M. K, PHARAOH, T. C. & SOPER, N. J. 1990. Structural trends in central Britain from images of gravity and aeromagnetic fields. *Journal of the Geological Society, London,* **147**, 241–258.

LEWIS, C. L. E., GREEN, P. F., CARTER, A. & HURFORD, A. J. 1992. Elevated K/T palaeotemperatures throughout Northwest England: three kilometres of Tertiary erosion? *Earth and Planetary Science Letters,* **112**, 131–145.

MARRETT, R. & ALLMENDINGER, R. W. 1992. Amount of extension on 'small' faults: an example from the Viking Graben. *Geology,* **20**, 47–50.

PHARAOH, T. C., MERRIMAN, R. J., WEBB, P. C. & BECKINSALE, R. D. 1987. The concealed Caledonides of eastern England: preliminary results of a multidisciplinary study. *Proceedings of the Yorkshire Geological Society,* **46**, 355–369.

PLANT, J. A., JONES, D. G. & HASLAM, H. W. (eds) 1996. *Basin evolution, fluid movement and mineral resources in a Permo-Triassic rift setting: the Cheshire Basin.* BGS Technical Report WP/96/14R, 263 pp.

RAWSON, P. F. & RILEY, L. A. 1982. Latest Jurassic–early Cretaceous events and the 'Late-Cimmerian Unconformity' in the North Sea area. *Bulletin of the American Association of Petroleum Geologists,* **66**, 2628–2648.

REES, J. G. & WILSON, A. A. (in press). *Geology of the Country around Stoke-on-Trent.* Memoir of the British Geological Survey, Sheet 123.

SCLATER, J. G. & SHOREY, M. D. 1989. Mid-Jurassic through mid-Cretaceous extension in the Central Graben of the North Sea – Part 2: Estimates from faulting based on a seismic reflection line. *Basin Research,* **1**, 201–215.

SMITH, N. J. P. 1987. The deep geology of central England: the prospectivity of the Palaeozoic rocks. *In:* BROOKS, J. & GLENNIE, K. W. (eds) *Petroleum Geology of North West Europe.* Graham & Trotman, London, 217–224.

SOPER, N. J., WEBB, B. C. & WOODCOCK, N. H. 1987. Late Caledonian (Acadian) transpression in northwest England: timing, geometry and geotectonic significance. *Proceedings of the Yorkshire Geological Society,* **46**, 175–192.

TAYLOR, B. J., PRICE, R. H. & TROTTER, F. M. 1963. *Geology of the Country around Stockport and Knutsford.* Memoir of the Geological Survey, Great Britain, HMSO, London.

WALSH, J. J. & WATTERSON, J. 1987. Analysis of the relationship between displacements and dimensions of faults. *Journal of Structural Geology,* **10**, 239–247.

—— & WATTERSON, J. 1992. Populations of faults and fault displacements and their effects on estimates of fault-related regional extension. *Journal of Structural Geology,* **14**, 701–712.

——, WATTERSON, J. & YIELDING, G. 1991. The importance of small-scale faulting in regional extension. *Nature,* **351**, 391–393.

WHITE, N. J., JACKSON, J. A. & McKENZIE, D. P. 1986. The relationship between the geometry of normal faults and that of the sedimentary layers in their hanging-walls. *Journal of Structural Geology,* **8**, 897–910.

WHITTAKER, A. (ed.) 1985. *Atlas of Onshore Sedimentary Basins in England and Wales: Post-Carboniferous Tectonics and Stratigraphy.* Blackie, Glasgow.

WILLS, L. J. 1978. *A Palaeogeological Map of the Lower Palaeozoic Floor Below the Cover of Upper Devonian, Carboniferous and Later Formations.* Geological Society of London Memoir No. 8.

WOODCOCK, N. H. 1984. The Pontesford Lineament, Welsh Borderland. *Journal of the Geological Society, London,* **141**, 1001–1014.

ZIEGLER, P. A. & VAN HOORN, B. 1989. Evolution of the North Sea rift system. *In:* TANKARD, A. J. & BALKWILL, H. R. (eds) *Extensional Tectonics and Stratigraphy of the North Atlantic margins.* American Association of Petroleum Geologists Memoir, **46**, 471–500.

Characteristics of fault zones in sandstones from NW England: application to fault transmissibility

ALASTAIR BEACH, J. LAWSON BROWN, ALASTAIR I. WELBON,
JEAN E. MCCALLUM, PAUL BROCKBANK & STEVEN KNOTT
*Alastair Beach Associates Ltd, 11 Royal Exchange Square,
Glasgow G1 3AJ, UK*

Abstract: Faults within outcrops of Permo-Triassic sandstones from NW England have been analysed, and their throws and fault zone thicknesses measured. These data were collected in order to provide a basis for estimating fault zone transmissibility. Logarithmic plots of fault throw versus fault zone thickness from the same localities show a positive correlation, and this relationship can be used to estimate fault zone thicknesses from fault throws derived from seismic data within similar successions. By combining fault zone thickness with laboratory-derived measurements of fault zone permeability, we show that fault transmissibilities can be estimated for input into reservoir simulation. These results can be applied to the assessment of seal potential and transmissibility in offshore oil and gas fields reservoired in Permo-Triassic sandstones. Layer thickness has a control on the fault populations, giving rise to separate population slopes for faults contained within a distinctive layer compared with those that extend outwith the layer. If layer thickness effects are not taken into account, extrapolation of a single power law trend may lead to an overestimation of the number of small faults in a population.

Faults in high porosity sandstones have become a recent focus of interest because of their role in compartmentalization of hydrocarbon reservoirs (Pittman 1981; Aydin & Johnson 1983; Underhill & Woodcock 1987; Edwards *et al.* 1994; Knott 1994; Fowles & Burley 1994; Antonellini & Aydin 1994) and aquifers (Randolph & Johnson 1989). The majority of these studies have emphasized the petrographical and petrophysical properties of fault zones and in particular granulation seams, which are small-scale structures characteristic of faulted, high porosity sandstones. The present study explores the relationship between fault displacement and fault zone thickness in an attempt to apply the results to estimating fault transmissibility in the subsurface. Outcrops of faulted Permo-Triassic sandstones in the NW England region (Fig. 1) provide an opportunity to analyse these parameters in detail.

Fault analysis

Fault analysis (e.g. Walsh & Watterson 1988; Yielding *et al.* 1992) was used as the main approach for the quantification of the two parameters studied – throw and thickness. Analysis was carried out by plotting these parameters on cumulative number graphs in logarithmic space (population graph, Yielding *et al.* 1992). The reason this type of graph is employed is that parameters such as fault throw, measured from the same fault population, plot as a straight line if the fault population grows in line with a power law relationship – i.e. they are said to belong to a fractal population. The straight line on the graph therefore characterises the population in a quantifiable form – the slope of the straight line gives the relative numbers of small faults and large faults in the population. Steep slopes indicate large numbers of small faults for a given number of large faults. Shallower slopes, conversely, imply smaller numbers of small faults for the given number of large faults.

Faults in nature very often do not follow exactly the rules of the fractal concept, but statistically, fault populations are self similar over a range of scales (Yielding *et al.* 1992; Gaultier & Lake 1993). When plotted on population graphs, natural fault datasets often deviate from the ideal straight line. Where the limit of resolution is reached, for example in seismic data, not all faults below this limit are imaged, so the data collected are only a restricted sample of the full population. This results in a deviation of the curve away from a straight line at the left hand side of the plot – the slope noticeably shallows (see later figures). Also, in a restricted sampling area, such as a depth structure map of a field, not all the larger faults in the total population will be measured and this will also result in a deviation away from a full characterization of the fault population (Yielding *et al.* 1992).

From Meadows, N. S., Trueblood, S. P., Hardman, M. & Cowan, G. (eds), 1997, *Petroleum Geology of the Irish Sea and Adjacent Areas,* Geological Society Special Publication No. 124, pp. 315–324.

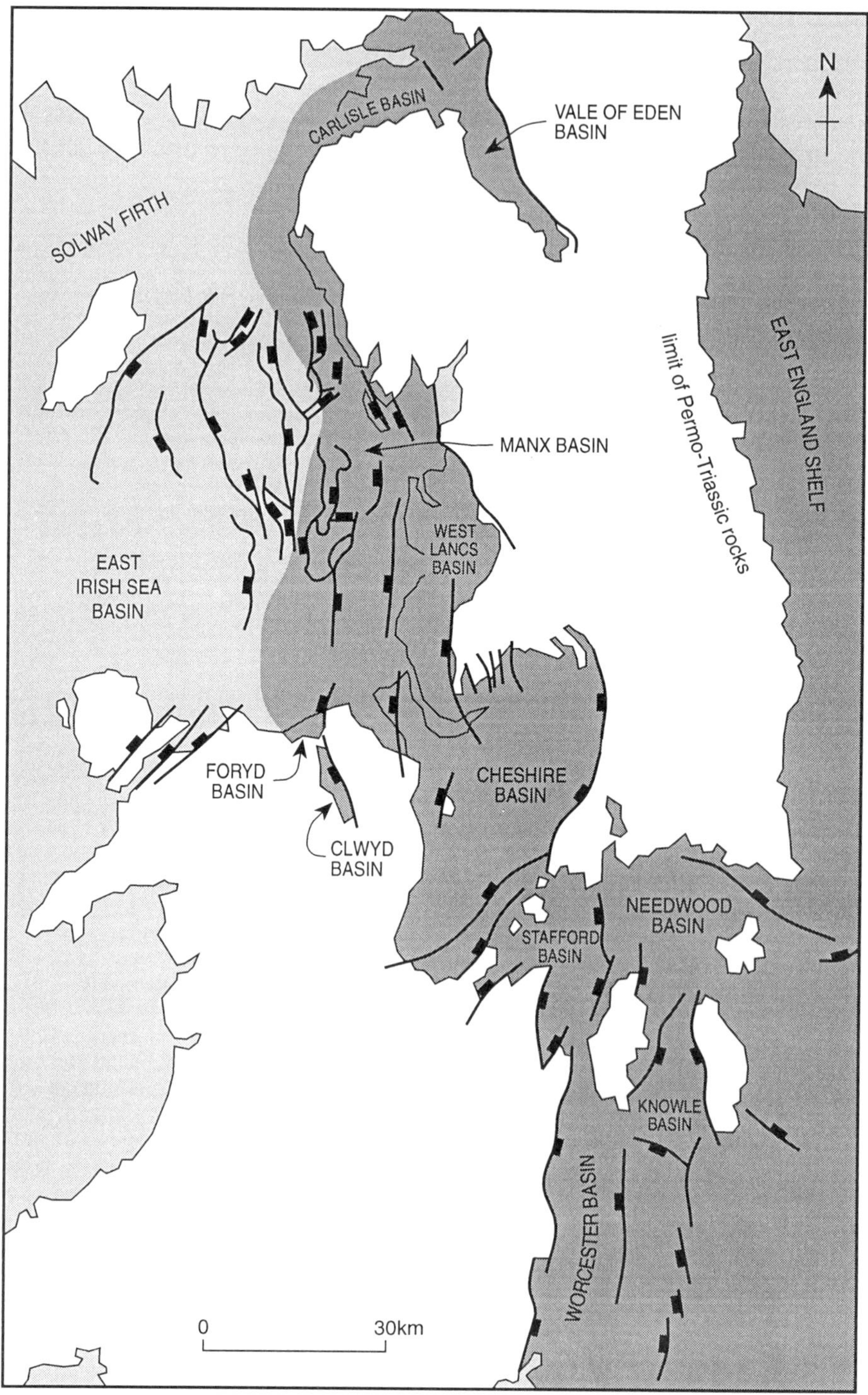

Fig. 1. Simplified geological map of the Triassic and Jurassic basins of NW England and Wales (based on Evans *et al.* 1993).

Outcrops

Outcrops provide the best data for quantitatively analysing faults that are likely to be important in the structural compartmentalization of reservoirs (e.g. Edwards *et al.* 1994). Cores generally provide samples that only give a restricted range of fault sizes. Often fault zones are deliberately avoided in cored sections, for obvious reasons. Furthermore, fault zones, when they are cored, often appear as zones of rubble and rarely provide good *in situ* data on fault zone characteristics such as fault zone thickness. Well images such as bore hole scanners and high resolution resistivity tools can provide excellent images of fault zone thicknesses but only over a limited range of sizes. There is also a limit to the resolution of these tools which is particularly apparent when small, closely spaced faults are not clearly imaged.

In contrast, outcrops, particularly the area of NW England, provide an ideal natural laboratory for the collection and quantification of fault characteristics at reservoir scale. Faults over a range of throws are exposed. The range of relative fault densities, i.e. the relative number of small and large faults in an area, is sufficiently large to cover all degrees of compartmentalization from heavily faulted to relatively unfaulted. The Permo-Triassic lithologies outcropping in NW England have long been recognized as being analogous to the subsurface reservoirs currently producing hydrocarbons in the East Irish Sea Basin.

Fault growth

One of the prime aims of this study was to quantify fault zone thickening. The most commonly observed class of structures associated with fault zone thickening in high porosity reservoir quality sandstones, are granulation seams and slip surfaces (Aydin & Johnson 1983; Underhill & Woodcock 1987; Knott 1994; Fowles & Burley 1994). The incremental growth of granulation seams is due to initial collapse of pores and fracturing of cement during the early stages of deformation followed by grain size reduction through crushing and fracturing. Because of the reduced grain size within the fault zone, point contacts increase and so friction increases. Therefore a greater shear stress is required to fracture the fault rock compared with the undeformed host rock. The fault zone thus thickens outwards into the undeformed sandstone (Aydin & Johnson 1983).

This process continues until the individual granulation seams anastomose and form a wider zone of composite granulation seams. In some instances stress is concentrated within the rigid zones of granulation seams, particularly when they become relatively thick (say 0.2–0.3 m), which at high shear stress values may eventually cause a slip surface to form. The slip surface is a discrete slip plane within the sandstone which, in outcrop, is usually highly polished and striated with very low permeability (Antonellini & Aydin 1994).

In order to examine these structures, we have collected and analysed data from two areas – Cheshire (Cheshire Basin) and Cumbria (Vale of Eden). These are described in the next sections.

Cheshire

The Cheshire Basin (Fig. 2) formed in the hanging wall to the Wem-Red Rock fault zone during Early Permian rifting (Evans *et al.* 1993). The basin was modified by Early Triassic and Jurassic normal faulting and underwent significant uplift (up to 1.5 km) during the Tertiary (Green *et al.,* this volume). This is similar to the sequence of events recorded in the Cumbrian Permo-Triassic basins. The Cheshire Basin extends offshore to the East Irish Sea Basin (Fig. 2). Recent work on the tectonics and stratigraphy of the basin include Evans *et al.* (1993) and Williams & Eaton (1993); whilst in Naylor *et al.* (1989), Metcalfe *et al.* (1993), Rowe *et al.* (1993) and Scrivener *et al.* (1993) there are discussions on the nature of mineralisation and faulting.

Faults were examined in the Cheshire Basin at four localities – Heathfield, Thurstaston Hill, Hilbre Island and Helsby Hill. The outcrops are found in the Triassic Sherwood Sandstone Group (Chester Pebble Beds Formation; Wilmslow Sandstone Formation and Helsby Sandstone Formation) (Fig. 3).

Outcrops

The road cut at Heathfield is located on the eastern side of the Wirral Peninsula (Fig. 2). Exposures are within the Triassic Bunter Pebble Beds (Chester Pebble Beds Formation) (Fig. 3) and consist of interbedded fluvially deposited sandstone, siltstone and mudstone. White (weathered) and red iron oxide stained sandstones are offset by several faults in the road cuttings in this area. The faults have displacements of a few tens of millimetres to a metre or so, and the fault zones show local silt and sand smear.

Thurstaston Hill lies 7 km southwest of Birkenhead on the western side of the Wirral Peninsula (Fig. 2). Outcrops of Upper Mottled Sandstone (Wilmslow Sandstone Formation) and Keuper Sandstone Basement Beds (Helsby Sandstone Formation) containing faults, granulation

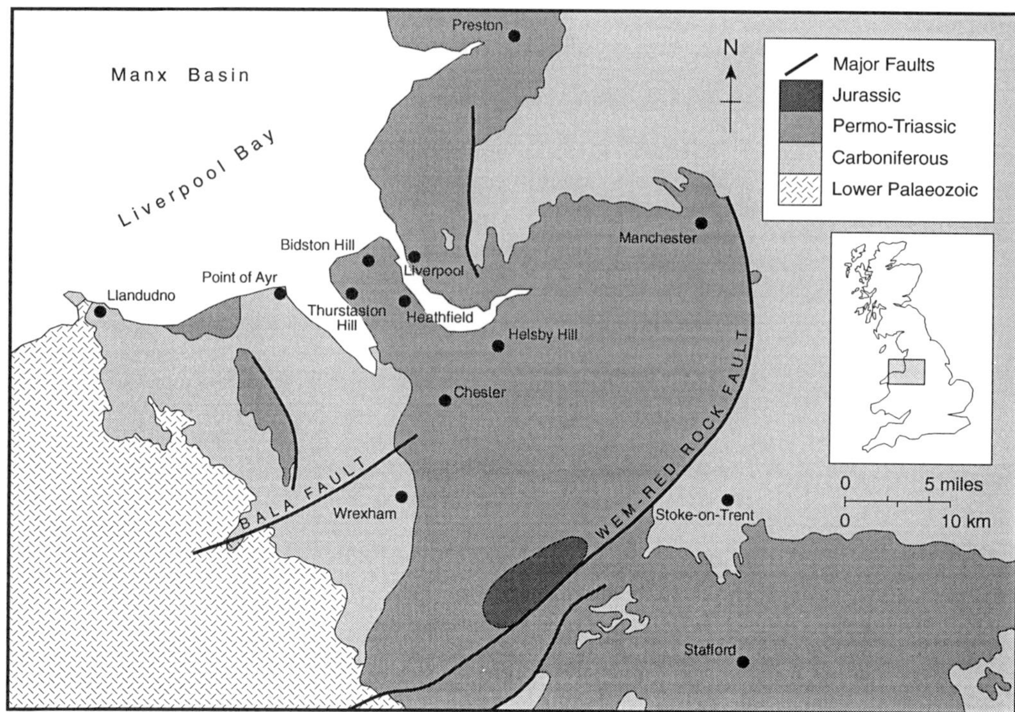

Fig. 2. Geological map of the Cheshire area.

seams and slip surfaces occur on the flat heathland of Thurstaston Common.

Hilbre Island is the largest of three islands found on the western side of the Wirral Peninsula (Fig. 2). The rocks exposed here consist of Triassic Pebble Beds (Chester Pebble Beds Formation), with interbedded siltstone, sandstone and shale. Several fault zones cross the island and are well exposed along the coastline. The throw on the faults ranges from centimetres to metres and there are changes in fault zone thickness correlating with variations in lithology and fault throw.

Helsby Hill lies immediately above the village of Helsby, which is located 10 km northeast of the city of Chester (Fig. 2). On the northern side of the hill there is a well exposed cliff. The lower part of the cliff consists of Upper Mottled Sandstone (Wilmslow Sandstone Formation), with the Helsby Sandstone Formation comprising the upper part. A fault with several metres of displacement cuts through the cliff on the north side of the hill. Deformation bands are common but slip surfaces predominate. Slip surfaces are more common in the cleaner quartzose Upper Mottled Sandstone. At this locality the nature of the fault zone and the relationship between style of fault zone deformation, thickness and lithology is well exposed.

Data

Fault throw data from the Cheshire area have been plotted against cumulative number in logarithmic space (Fig. 4). A straight line central segment may be defined. The slope of the left hand segment of the plot shallows, which as previously described, is typical of cumulative number log–log plots. The shallowing is normally attributed to undersampling, but there may be an alternative explanation: a second fault population is present (see later discussion). Two straight line segments are also apparent on the cumulative number versus fault zone thickness plot (Fig. 5) although the central segment spans only one order of magnitude.

When the fault throw and corresponding fault zone thickness are plotted together in logarithmic space (Fig. 6), a positive correlation may be defined. Sandstone/sandstone juxtaposition points occur across most of the area covered by the data points, but sandstone/siltstone and sandstone/shale points tend to plot in the upper left part of this area.

Cumbria

The Permo-Triassic successions of Cumbria lie in three basins, the East Irish Sea Basin, the Solway

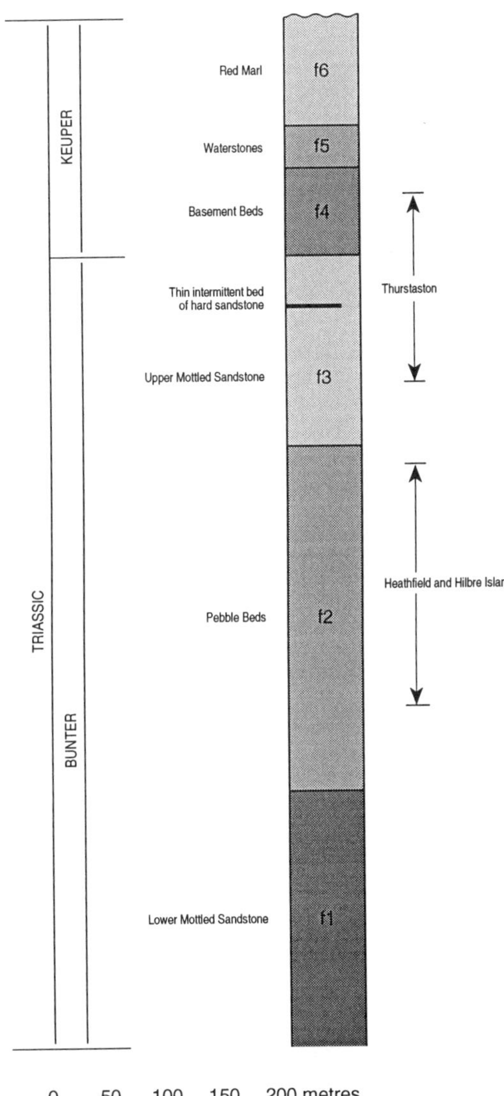

Fig. 3. Stratigraphy for localities in the Cheshire area (after BGS sheet 96, Liverpool).

Basin and the Vale of Eden Basin (Figs 1 & 7). Faults were analysed from outcrops in the Vale of Eden Basin (Fig. 7). This basin developed during Permo-Triassic extension as a NW–SE trending half graben in the hanging wall of the Pennine Fault located along the western margin of the Alston block of the Pennine High. Carboniferous rocks crop out in the Pennines to the east and Carboniferous and pre-Carboniferous rocks crop out in the Lake District Massif to the west. The area experienced significant uplift during Tertiary times (Green *et al.,* this volume).

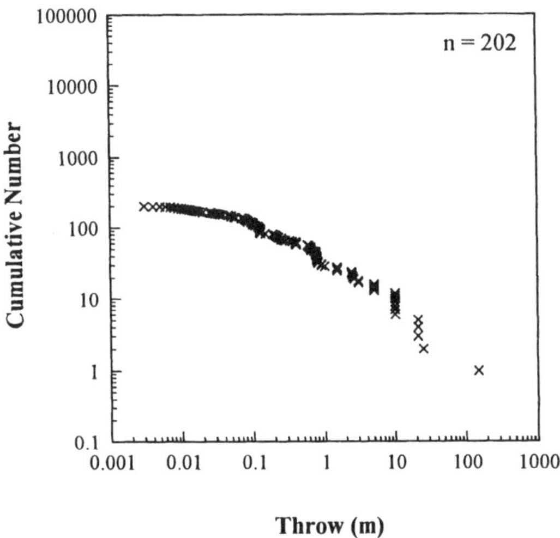

Fig. 4. Fault throw population graph from the Cheshire area.

The Permo-Triassic succession in the Vale of Eden (Fig. 8) consists of the Lower Permian Penrith Sandstone, the Upper Permian Eden Shales and the Triassic Sherwood Sandstone Group. Faults were examined at three localities in the southern and central Vale of Eden. These faults cut the Penrith Sandstone.

Outcrops

Lacy's Cave is located on the east side of the River Eden, approximately 8 km northeast of the town of

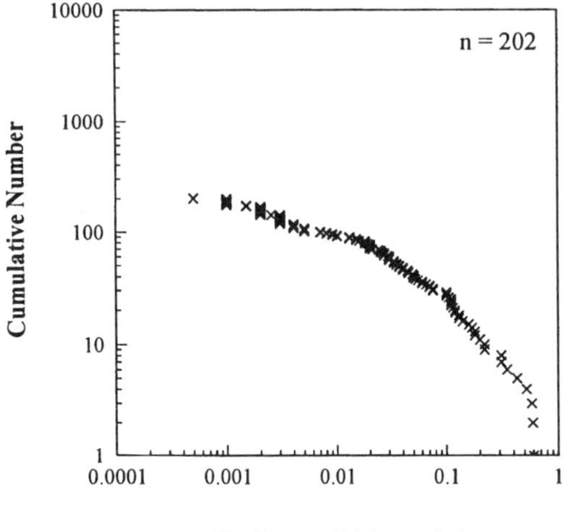

Fig. 5. Fault zone thickness population graph from the Cheshire area.

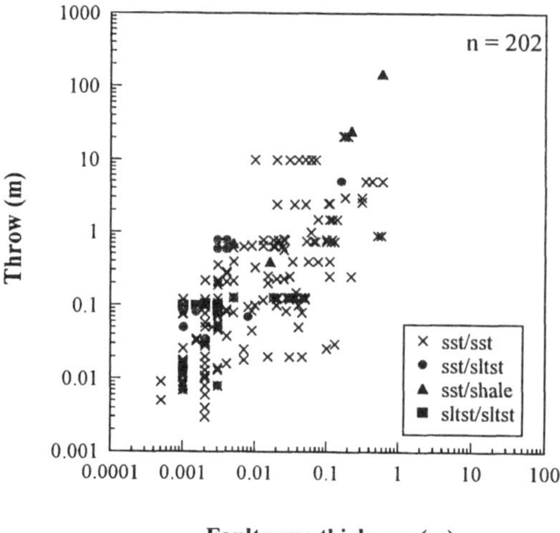

Fig. 6. Logarithmic throw–thickness plot of the Cheshire data.

Penrith, in the central Vale of Eden (Fig. 7). The cave exposures are within the upper part of the Penrith Sandstone, just beneath the evaporite beds of the Eden Shales (Fig. 8). Numerous granulation seams and slip surfaces are exposed in three-dimensions at this locality (see also Fowles & Burley 1994).

George Gill (Fig. 7) is located 3 km east–southeast of Appleby. The exposures studied are on the north side of a small gorge. The exposures are within Penrith Sandstone, present in aeolian facies. This locality is roughly 150 m away from a fault with 75 m throw, and is situated in its hanging wall. There are many anastomosing granulation seams and slip surfaces within a low relief cliff face. The conjugate granulation seams dip to the east and west and some have offsets of a few millimetres.

Belah Scar is situated in the southernmost Vale of Eden, on the southern bank of the River Belah, 2 km south of the village of Brough (Fig. 7). The exposed section is within the lower part of the Penrith Sandstone and consists of fluvially deposited sandstones and conglomerates with some aeolian facies. Exposed at this locality are several syn-sedimentary faults. A conglomerate bed in the upper part of the cliff face thickens across some of the faults.

Data

Fault throw data from the Cumbria outcrops have been plotted on a population graph in logarithmic space (Fig. 9). This plot shows the characteristic

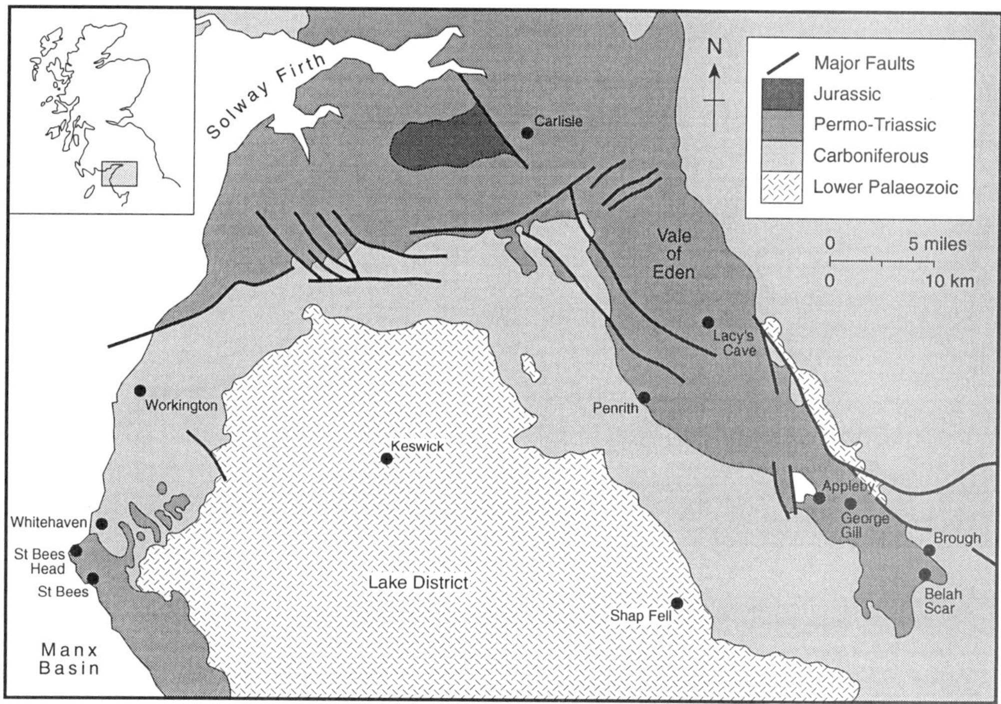

Fig. 7. Geological map of the Cumbria area.

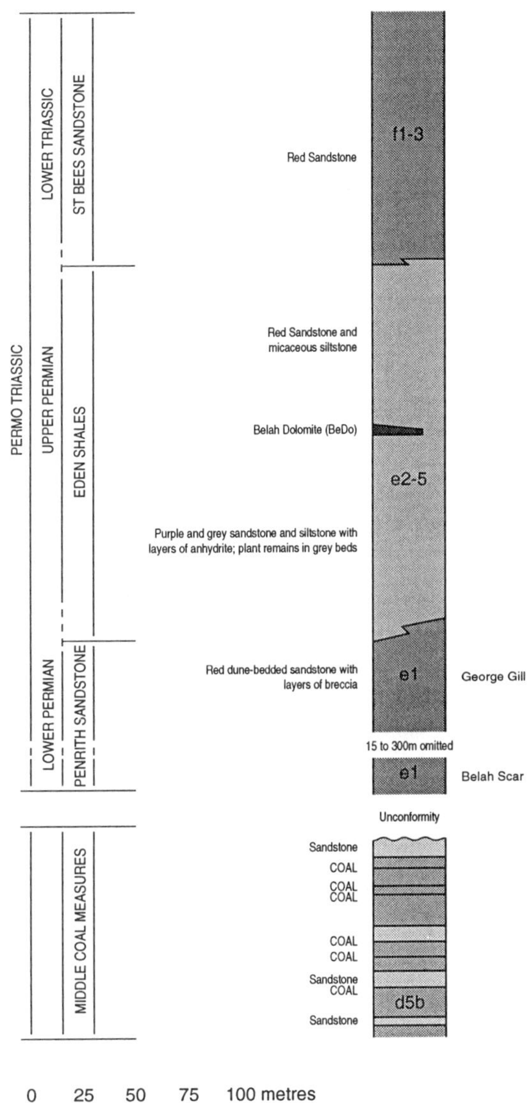

Fig. 8. Stratigraphy for localities in the Cumbria area (after BGS Sheet 31, Brough-under-Stainmore).

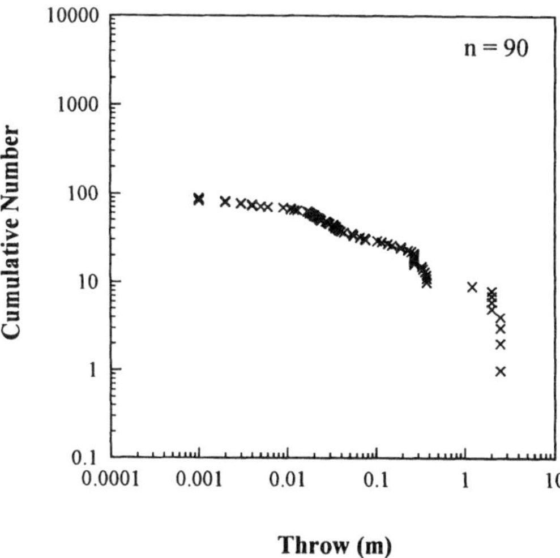

Fig. 9. Fault throw population graph from the Cumbria area.

central one spanning less than an order of magnitude.

Throw–thickness data for Cumbria show a broad positive correlation (Fig. 11), similar to the Cheshire throw–thickness data. Belah Scar data points, which come from argillaceous sandstones, plot mainly in the left hand part of the figure, whilst points from the aeolian sandstones of George Gill plot in the lower right.

shape for a cumulative number logarithmic plot, with right hand, central segment and left hand segments. As noted previously, the change in slope shown by the left and right hand portions of the plot are often attributed to undersampling, but as with the Cheshire throw population plot (Fig. 4), there may be an alternative explanation for the change in slope of the left hand segment, i.e. that two populations are present (see discussion). The fault zone thickness population plot for Cumbria (Fig. 10) also shows two straight line segments with the

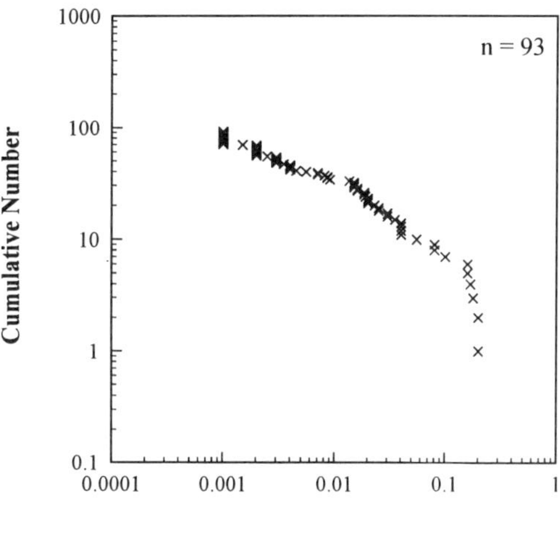

Fig. 10. Fault zone thickness population graph from the Cumbria area.

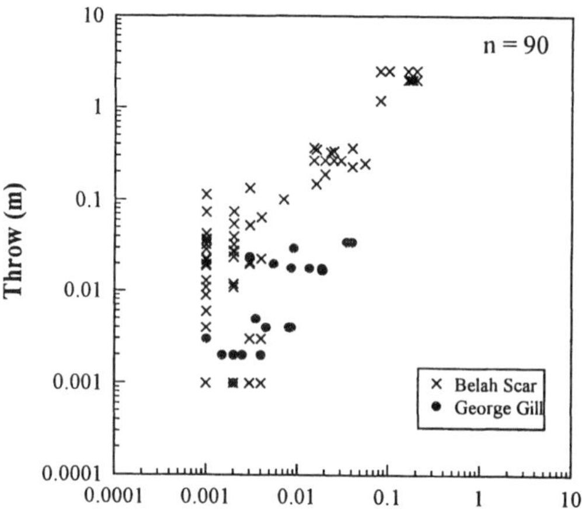

Fault zone thickness (m)

Fig. 11. Logarithmic throw–thickness plot of the Cumbria data.

Discussion

Previous studies (Knott 1994; Knott *et al.* 1995) of throw–thickness data have demonstrated a positive correlation with slope of approximately one in logarithmic space. From these studies the positive correlation has been related to fault zone thickening with increasing throw. The scatter of the data points about the straight line trend has been related in these studies to strain hardening and thicker fault zones in high porosity sandstones, and strain softening and thinner fault zones in argillaceous sedimentary rocks. The data presented here show broadly similar patterns. However, a more detailed analysis is required to understand in detail the relations between throw, fault zone thickness, host rock porosity and permeability, cementation effects in fault zones and lithological juxtaposition across faults.

A measure of fault transmissibility used in reservoir simulation is fault zone thickness multiplied by fault rock permeability (Fig. 12). On the basis of this measure of transmissibility, a transmissibility matrix (Fig. 13) has been developed which uses the throw–thickness relationship described above and laboratory-based measurements of fault rock permeability (e.g. Antonellini & Aydin 1994). Fault zone thickness can be measured directly from borehole images, or estimated from fault throw data on depth structure maps using throw–thickness relationships. Transmissibilities can be obtained from the matrix (Fig. 13) by choosing a likely fault zone permeability from laboratory-based measurements. Permeability is partly dependent on the

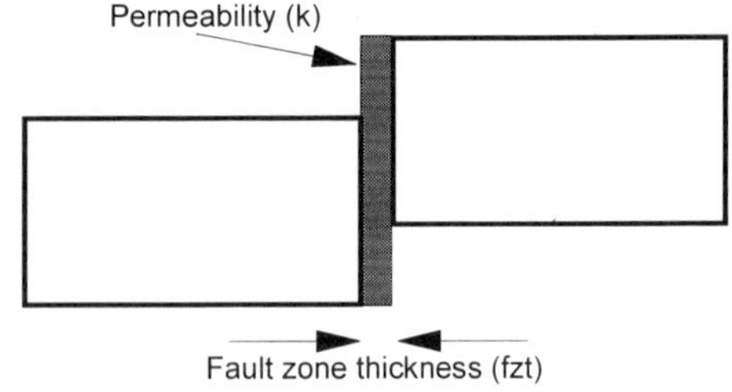

Fig. 12. A method for estimating fault transmissibility in reservoir simulation where transmissibility (T) = fault zone permeability (k) multiplied by fault zone thickness (*fzt*).

composition of the original host rock – argillaceous sandstones have different fault rock characteristics compared with high porosity, mature sandstones, for example (Antonellini & Aydin 1994).

As previously noted, the left hand parts of throw population curves obtained from depth structure maps are often affected by truncation due to undersampling of small fault throws below the limit of seismic resolution (Yielding *et al.* 1992). This is usually manifested by a reduction in slope occurring at the limit of resolution. In outcrop studies such as those reported here, the limit of resolution is less than 1 mm, and all small structures in each outcrop are measured. There should thus be no undersampling of data down to the lower limit of

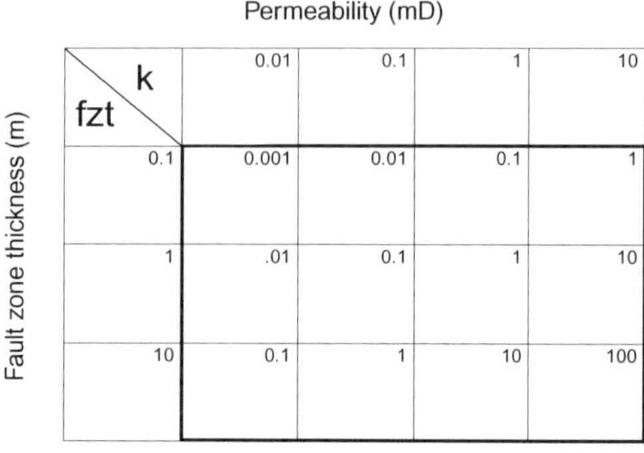

Fig. 13. Transmissibility matrix showing examples of fault zone thickness and fault zone permeability and the corresponding values of transmissibility. In practice, fault zone thickness can be estimated from throws on depth structure maps using the graphs in Figs 6 and 11, and fault zone permeability can be obtained from laboratory-based measurements (e.g. Antonellini & Aydin 1994).

measurement. Therefore, when population graphs from outcrops suggest that two straight line segments are present (e.g. Figs 4, 5, 9 & 10), an alternative explanation to truncation is required for the change in slope between the central and left hand segments.

One potential explanation is that grain size and layer thickness have affected fault growth (Knott *et al.* 1995). In the early stages of fault growth, ductile processes, such as grain boundary sliding and grain rolling, may be important (Knipe 1992). As fault growth continues and brittle processes such as grain and cement fracturing begin to dominate, the effect of grain size is no longer as important. After this, layer thickness may start to exert control. Figures 3 and 8 show the overall thicknesses of the main sedimentary units in the outcrop areas studied. Faults which grew entirely within one main unit might be expected to have a consistent power law relationship, particularly within broadly homogeneous sandstones. As faults grew outwith these, either upwards or downwards, then growth of the faults would be perturbed and therefore might follow a different growth law. Very small faults may be controlled by thinner stratigraphic units, or indeed individual beds. As an example, the Cumbria throw plot (Fig. 9) shows a change in slope between left hand and central segments at a throw of 0.015 m (1.5 cm) and a deterioration of slope and data gap between central and right hand segments at a throw of 0.25 m. Using throw–fault length relationships (e.g. Cowie & Scholz 1992), faults with throws of greater than 0.25 m are likely to have lengths in excess of 100 m. These faults may belong to a fault population that grew beyond the boundaries of the Penrith sandstone which is approximately 200 m thick in the Vale of Eden (Fig. 8). Faults with throws less than 0.25 m may belong to a population that grew entirely within the Penrith Sandstone. Stratigraphic subdivisions of the Penrith Sandstone may control the change in slope on the population graph at a throw of 0.015 m.

The effect of layer thickness on fault populations has also been proposed for the Permian Yellow Sands of the Newcastle area, England, and the Nubian Sandstone Formation of western Sinai, Egypt (Knott *et al.* 1995); and also for the brittle crust as a whole (Westaway 1994; Knott *et al.* 1995). The observation that layer thickness can affect fault populations is an important one if predictions are to be made about the number and density of faults in the subsurface. Extrapolation of straight line trends on some population graphs may overestimate the number of faults below the limit of resolution.

Conclusions

The main conclusions of this study are as follows:

1. Throw–thickness data from NW England, when plotted together in logarithmic space, show a positive correlation. The scatter of the data about this trend is probably due to variation in the host rock lithologies.
2. The throw–thickness relationship can be used to estimate fault zone thickness from fault throws and vice versa.
3. A value of fault transmissibility can be estimated from fault zone thickness (obtained either from well data, or from the throw–thickness plot) and fault rock permeability (obtained from laboratory-based measurements of fault rock samples).
4. Outcrop measurements of fault throw and thickness data indicate the presence of self similar populations which allow the quantification of these parameters for application to subsurface studies.
5. Layer thickness effects must also be taken into account before extrapolation is carried out below the limit of resolution, otherwise the number of small faults may be overestimated.

Many thanks to Michelle Smith, Martin Kidd and Graham Barr for their help in producing this paper.

References

Antonellini, M. & Aydin, A. 1994. Effect of faulting on fluid flow in porous sandstones: petrophysical properties. *Bulletin of the American Association of Petroleum Geologists,* **78**, 355–377.

Aydin, A. & Johnson, A. M. 1983. Analysis of faulting in porous sandstones. *Journal of Structural Geology,* **5**, 19–31.

Cowie, P. A. & Scholz, C. H. 1992. Displacement–length relationships for faults; data synthesis and discussion. *Journal of Structural Geology,* **14**, 1149–1156

Edwards, H. E., Becker, A. D. & Howell, J. A. 1994. compartmentalization of an aeolian sandstone by structural heterogeneities: Permo-Triassic Hopeman Sandstone, Moray Firth, Scotland. *In:* North, C. P. & Prosser, D. J. (Eds) *Characterization of Fluvial and Aeolian Reservoirs.* Geological Society, London, Special Publication, **73**, 339–365.

Evans, D. J., Rees, J. G. & Holloway, S. 1993. The Permian to Triassic stratigraphy and structural evolution of the central Cheshire Basin. *Journal of the Geological Society, London,* **150**, 857–870.

Fowles, J. & Burley, S. 1994. Textural and permeability characteristics of faulted, high porosity sandstones. *Marine and Petroleum Geology,* **11**, 608–623.

Gaultier, B. D. M. & Lake, S. D. 1993. Probabilistic modelling of faults below the limit of seismic resolution in Pelican field, North Sea, offshore

United Kingdom. *Bulletin of the American Association of Petroleum Geologists*, **77**, 761–777.

GREEN, P. F., DUDDY, I. R. & BRAY, R. J. Variation in thermal history styles around the Irish Sea and adjacent areas: implications for hydrocarbon occurrence and tectonic evolution. *This volume.*

KNIPE, R. J. 1992. Faulting processes and fault seal. *In:* LARSEN, L. M. (ed.) *Structural Modelling and its Application to Petroleum Geology.* Norwegian Petroleum Society (NPF), Special Publication, **1**, 325–342.

KNOTT, S. D. 1994. Fault zone thickness versus displacement: results from Permo-Triassic sandstone outcrops in NW England. *Journal of the Geological Society, London*, **151**, 17–25.

——, BEACH, A., BROCKBANK, P. J., BROWN, J. L., McCALLUM, J. E. & WELBON, A. I. 1995. Spatial and mechanical controls on normal fault populations. *Journal of Structural Geology*, **151**, 359–372.

METCALFE, R., ROCHELLE, C. SAVAGE, D. HOOKER, P. J. & GALE I. N. 1993. Hydrochemical modelling of red bed diagenesis and heavy metal mineralisation. *In:* PARNELL, J. RUFFELL, A. H. & MOLES, N. R. (eds) *Geofluids '93* extended abstracts.

NAYLOR, H., TURNER, P., VAUGHAN, D. J., BOYCE, A. J. & FALLICK, A. E. 1989. Genetic studies of red bed mineralisation in the Triassic of the Cheshire Basin, northwest England. *Journal of the Geological Society, London*, **146**, 685–700.

PITTMAN, E. D. 1981. Effect of fault-related granulation on porosity and permeability of quartz sandstones, Simpson Group (Ordovician), Oklahoma. *Bulletin of the American Association of Petroleum Geologists*, **65**, 2381–2387.

RANDOLPH, L. & JOHNSON, B. 1989. Influence of faults of moderate displacement on groundwater flow in the Hickory Sandstone aquifer in Central Texas. *Geological Society of America Abstracts with Programs*, **2**, A242.

ROWE, J., BURLEY, S., GAWTHORPE, R., COWAN, G. & HARDMAN, M. 1993. Palaeo-fluid flow in the East Irish Sea Basin and its margins. *In:* PARNELL, J. RUFFELL, A. H. & MOLES, N. R. (eds) *Geofluids '93* extended abstracts.

SCRIVENER, R. C., SHEPHERD, T. J. & WARRINGTON, G. 1993. Genetic relationship of base and precious metal mineralisation to Permo-Triassic basins in England. *In:* PARNELL, J. RUFFELL, A. H. & MOLES, N. R. (eds) *Geofluids '93* extended abstracts.

UNDERHILL, J. R. & WOODCOCK, N. H. 1987. Faulting mechanisms in high-porosity sandstones; New Red Sandstone, Arran, Scotland. *In:* JONES, M. E. & PRESTON, R. M. F. (eds) *Deformation of Sediments and Sedimentary Rocks.* Geological Society, London, Special Publication, **29**, 91–105.

WALSH, J. J. & WATTERSON, J. 1988. Analysis of the relationship between displacements and dimensions of faults. *Journal of Structural Geology*, **10**, 239–247.

WESTAWAY, R. 1994. Quantitative analysis of populations of small faults. *Journal of Structural Geology*, **16**, 1259–1273.

WILLIAMS, G. D. & EATON, G. 1993. Stratigraphic and structural analysis of the Late Palaeozoic–Mesozoic of NE Wales and Liverpool Bay: implications for hydrocarbon prospectivity. *Journal of the Geological Society, London*, **150**, 489–499.

YIELDING, G., WALSH, J. & WATTERSON, J. 1992. The prediction of small-scale faulting in reservoirs. *First Break*, **10**, 449–460.

Faulting and porosity modification in the Sherwood Sandstone at Alderley Edge, northeastern Cheshire: an exhumed example of fault-related diagenesis

JOHN ROWE[1,3] & STUART D. BURLEY[1,2]

[1] *Diagenesis Research Group, Department of Geology, University of Manchester M13 9PL, UK*

[2] *BG plc, Gas Research and Technology Centre, Ashby Road, Loughborough, LE11 2QU, UK*

[3] *Present address: Rock Deformation Research Group, Department of Earth Sciences, University of Leeds LS2 9JT, UK*

Abstract: The Alderley Horst is an exhumed, tilted structure close to the eastern margin of the Cheshire Basin, broadly analogous to the tilted fault blocks in the East Irish Sea Basin. Sediments of the Lower and Middle Triassic Sherwood Sandstone Group are exposed at the crest of the horst whilst its flanks are capped by sediments of the Mercia Mudstone Group. Extensive diagenesis and porosity modification have taken place in the Sherwood Sandstone across the Alderley Horst.

Diagenetic modification includes reduction of hematite grain coatings and quartz cementation, together with Cu–Pb–Zn–Fe sulphide and barite mineralization. The distribution of this cementation is spatially related to faults that cross the horst and its style and extent varies between individual faults and between fault sets of different orientations. Different spatial distributions are observed for quartz cements, barite, sulphides, ferroan calcite, reduction of hematite grain coatings and secondary ore concentrations. Sandstone units close to the base of the Mercia Mudstone Group are more intensely altered than lower units showing that hematite reduction and mineralization are concentrated at the crest of the horst.

Fault-related mineralization and porosity modification post-dates all of the fault movement as brecciation and deformation of barite and sulphide cements are not observed. The intensity of mineralization decreases at structurally lower levels on the WNW–ESE oriented faults, so precluding them as conduits for the vertical migration of mineralizing fluids. However, the horst-bounding Kirkleyditch Fault was probably an important migration pathway that enabled valving of fluids from dowthrown sources in the Mercia Mudstone Group, Manchester Marls and Carboniferous shales, although the possiblity of fluid migration up-dip from the south cannot be completely discounted. The mineralizing fluids were ponded at the crest of the horst where they were baffled by pre-existing structural barriers.

The hypothesis that basinal brines are the fluids involved in Mississippi Valley-type Pb–Zn and 'red bed'-type Cu–Pb–Zn mineral deposits has attracted much attention in recent years (e.g. Jackson & Beales 1967; Dunham 1970; Sverjenski 1984; Bethke & Marshak 1990). These saline mineralizing fluids are inferred to be either the result of deeply circulating meteoric waters or expulsion of formation waters from sediments buried in the basin. High salinities may result from either halite dissolution or migration of evaporitic brines (Land 1987; Walter *et al.* 1990). There is also commonly an association with hydrocarbons (e.g. Price *et al.* 1983; Montacer *et al.* 1987; Quing & Mountjoy 1994). The brines are generally considered to be expelled from the deeper parts of basins, perhaps episodically along hydraulic gradients, to either intra-basinal structural highs or to basin margins

where they mix with shallower fluids, cool and precipitate ore minerals (Cathles & Smith 1983; Bethke 1986).

The East Irish Sea Basin (EISB) together with its adjacent margins, including the Cheshire Basin, is a petroleum province containing proven reserves of oil and gas (e.g. Colter 1978; Bushell 1986; Stuart & Cowan, 1991; Stuart, 1993) associated with extensive and varied mineralizations that include sandstone-hosted Cu–Pb–Zn–Fe sulphides and fault-related diagenetic cementation (Clarke 1980; Carlon 1981; Carlon & Thompson 1981; Holmes *et al.* 1983; Naylor *et al.* 1989; Rowe *et al.* 1993). Petrographic studies of the mineral deposits suggest that they constitute an integral part of the diagenetic history of the basin and can be related temporally and spatially to burial diagenesis in its non-mineralized areas (Burley 1987; Rowe 1994). The

From Meadows, N. S., Trueblood, S. P., Hardman, M. & Cowan, G. (eds), 1997, *Petroleum Geology of the Irish Sea and Adjacent Areas*, Geological Society Special Publication No. 124, pp. 325–352.

basin thus provides an opportunity to investigate the relationship between migration of hydrocarbon-related brines and basin-margin mineralization (Fig. 1).

This paper addresses the type and extent of diagenesis and mineralization developed in Triassic sandstones at a structural high on the basin margin at Alderley Edge, northeast Cheshire. Sherwood Sandstone Group (SSG) sediments of the Alderley area are the lateral equivalents of the St Bees and Ormskirk Sandstone Formations of the EISB (Meadows & Beach 1993). The Ormskirk Sandstone Formation is the reservoir for the hydrocarbon accumulations in the EISB (Stuart & Cowan 1991; Greenwood & Habesch, this volume; Yaliz, this volume; Haig et al., this volume) and, as such, study of the palaeo-fluid flow regime in an equivalent sedimentary sequence in a similar structural setting should provide important information about the potential diagenetic influences upon reservoir quality. Particular emphasis is given hereafter to the extent of mineralization and distribution of porosity modification around the faults at Alderley Edge.

Geological setting

Stratigraphy

Alderley Edge is located close to the eastern boundary of the Cheshire Basin, an extensional half graben forming the northerly continuation of the Permo-Triassic rift system that extends through western England from the Worcester Graben to Cheshire and thence into the East Irish Sea Basin (Burley 1987; Jackson et al. 1987; Fig. 1).

In the Cheshire Basin, Carboniferous sediments are overlain unconformably by the Permian Collyhurst Sandstone, Manchester Marls and the Kinnerton Sandstone. Above this sequence are two Triassic units which dominate the basin fill (Fig. 2). The SSG is up to 1500 m thick and comprises a sequence of usually red-coloured, continental sandstones of mixed fluvial and aeolian origin. In the Cheshire Basin the SSG is divided into three main formations which, in ascending order, comprise the Chester Pebble Beds Formation, the Wilmslow Sandstone Formation and the Helsby Sandstone Formation (Warrington et al. 1980). Local names are used to define widely recognized conglomerate or sandstone members within the Helsby Sandstone Formation at Alderley Edge (Warrington & Thompson 1971; see Fig. 2).

The overlying Mercia Mudstone Group (MMG, up to 2000 m in thickness) comprises interbedded mudstones, siltstones and evaporites deposited in coastal sabkha and shallow epeiric sea environments. Uplift during the Cretaceous and Tertiary removed the post-Triassic cover at Alderley Edge and throughout most of the Cheshire Basin (Lewis et al. 1992; Chadwick & Rowley 1997) such that no Pre-Quaternary sediments younger than middle Liassic are preserved anywhere in the basin.

Structure

Alderley Edge is a 3 km-wide tilted horst, dipping to the SSW, bounded by the N–S oriented Alderley and Kirkleyditch faults, with normal displacements of up to 250 and 330 m, respectively (Figs 3 & 4; Thompson 1970). As a result, MMG sediments are directly juxtaposed against the SSG on the east and west of the horst.

Within the Alderley Horst, poorly exposed, smaller N–S-oriented, dip–slip normal faults (e.g. the Whitebarn Fault), crop out sub-parallel to the Alderley and Kirkleyditch faults (Fig. 3). Throws on these smaller faults are typically in the region of 10–20 m.

The horst is also crossed by subordinate WNW–ESE-oriented, predominantly dip–slip faults and associated splays with displacements of generally less than 10 m. Nearly all these faults downthrow to the NE and progressively down step the SSG sediments towards the Cheshire Plain. It is these WNW–ESE-oriented faults that are associated with the greatest concentration of hematite reduction, barite cementation and sulphide mineralization.

Mineralization

Colour leaching of the host sandstones, barite cementation and sulphide mineralization have been noted at Alderley Edge by many authors. Copper has been mined sporadically from the middle Bronze Age until the early parts of this century (Carlon 1979). The primary sulphide mineralization comprises bravoite, pyrite, chalcopyrite, sphalerite and galena (Ixer and Vaughan 1982; Holmes et al. 1983). This sulphide assemblage is associated with quartz and barite cements. Recent oxidizing groundwaters have altered much of the primary sulphide assemblage to a mixture of limonite–goethite, Cu–Pb–Zn carbonates, sulphates and Fe–Ni–Co hydroxides (Ixer and Vaughan 1982) that have been variably remobilised throughout the SSG sandstones adjacent to the WNW–ESE oriented faults. Most of the primary sulphide and barite mineralization is developed in the uppermost SSG towards the top of the horst structure (Warrington 1980; Fig. 4).

Many authors have noted the close spatial association of the primary mineralization and barite cementation to the faulting. Warrington (1965) produced the first published map of the Alderley Edge orefield showing the localization of sulphide

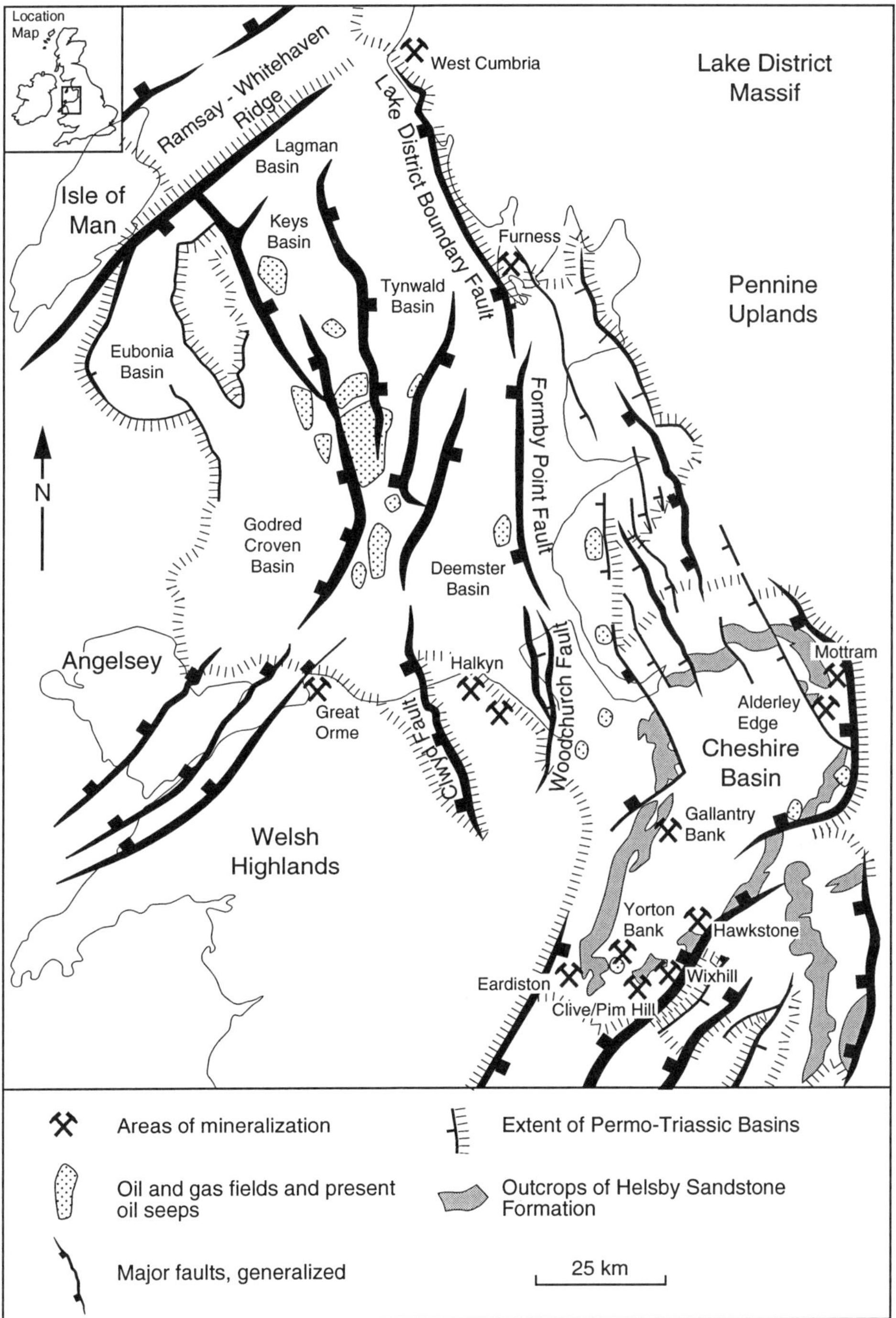

Fig. 1. Simplified structural map of the Cheshire Basin and the adjacent East Irish Sea Basin showing the location of Alderley Edge and related basin margin mineralizations in the context of the main hydrocarbon accumulations. Fault width approximates to displacement. The inset shows the context of the map within the UK.

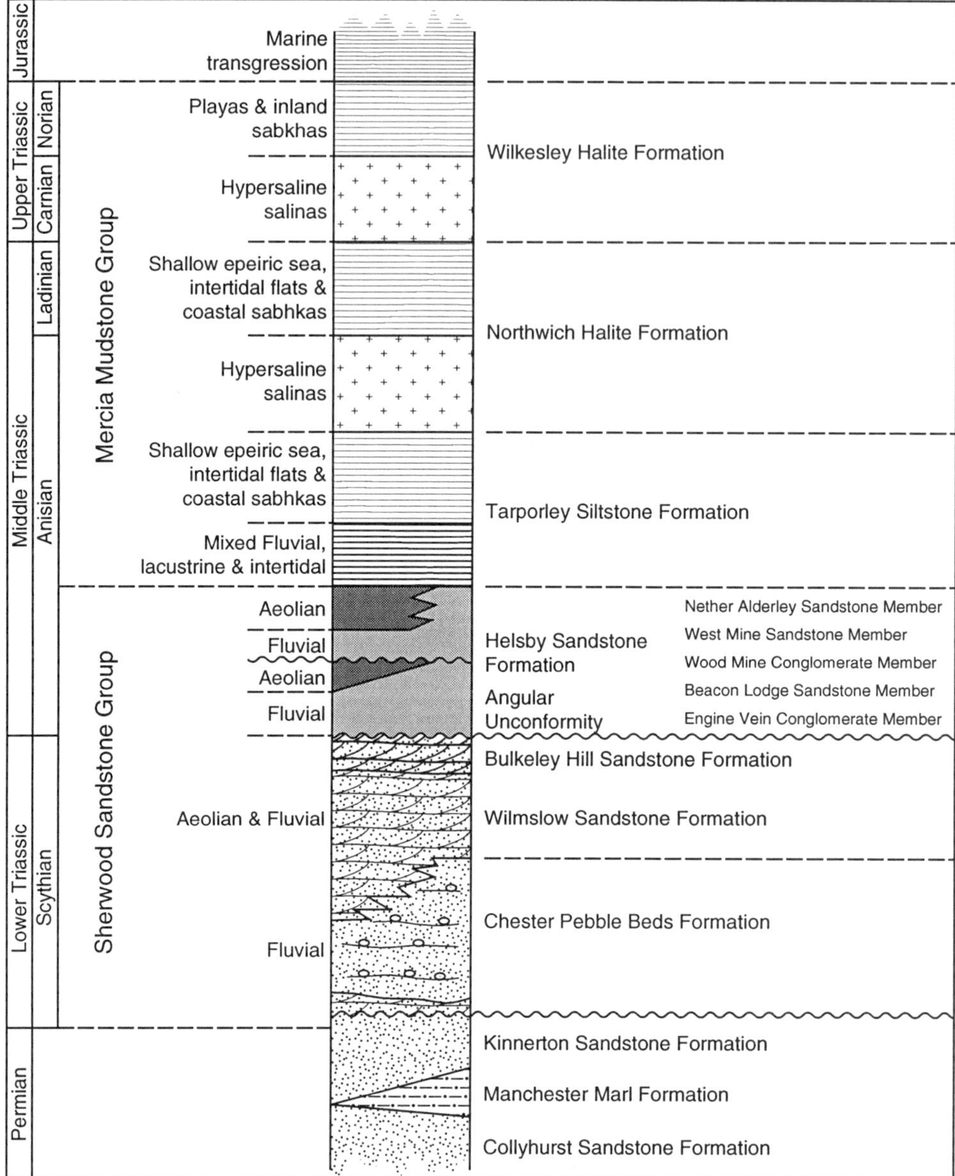

Fig. 2. The stratigraphy and interpreted depositional environments of the Permo-Triassic succession in the northeast of the Cheshire Basin (based on Evans *et al.* 1993; Thompson 1970, 1985, 1991; Warrington and Thompson 1971). The local sub-divisions of the Helsby Sandstone Formation at Alderley Edge are indicated.

and barite cements in the direct proximity of the faults and deduced that the faults were the main control on mineralization. Subsequently, Carlon (1979), Warrington (1980), Naylor *et al.* (1989) and Manning (1990) all advocated that warm, chloride-rich, saline brines migrated up the faults to produce the mineralization. Naylor *et al.* (1989) argued that

the mineralization took place at temperatures of between 60 and 150°C with the ore-forming sulphur being derived from Triassic evaporite brines. These brines were considered to have mixed with reducing waters, possibly in the presence of hydrocarbons, that were transported up the basin margin faults from source rocks in the Carbon-

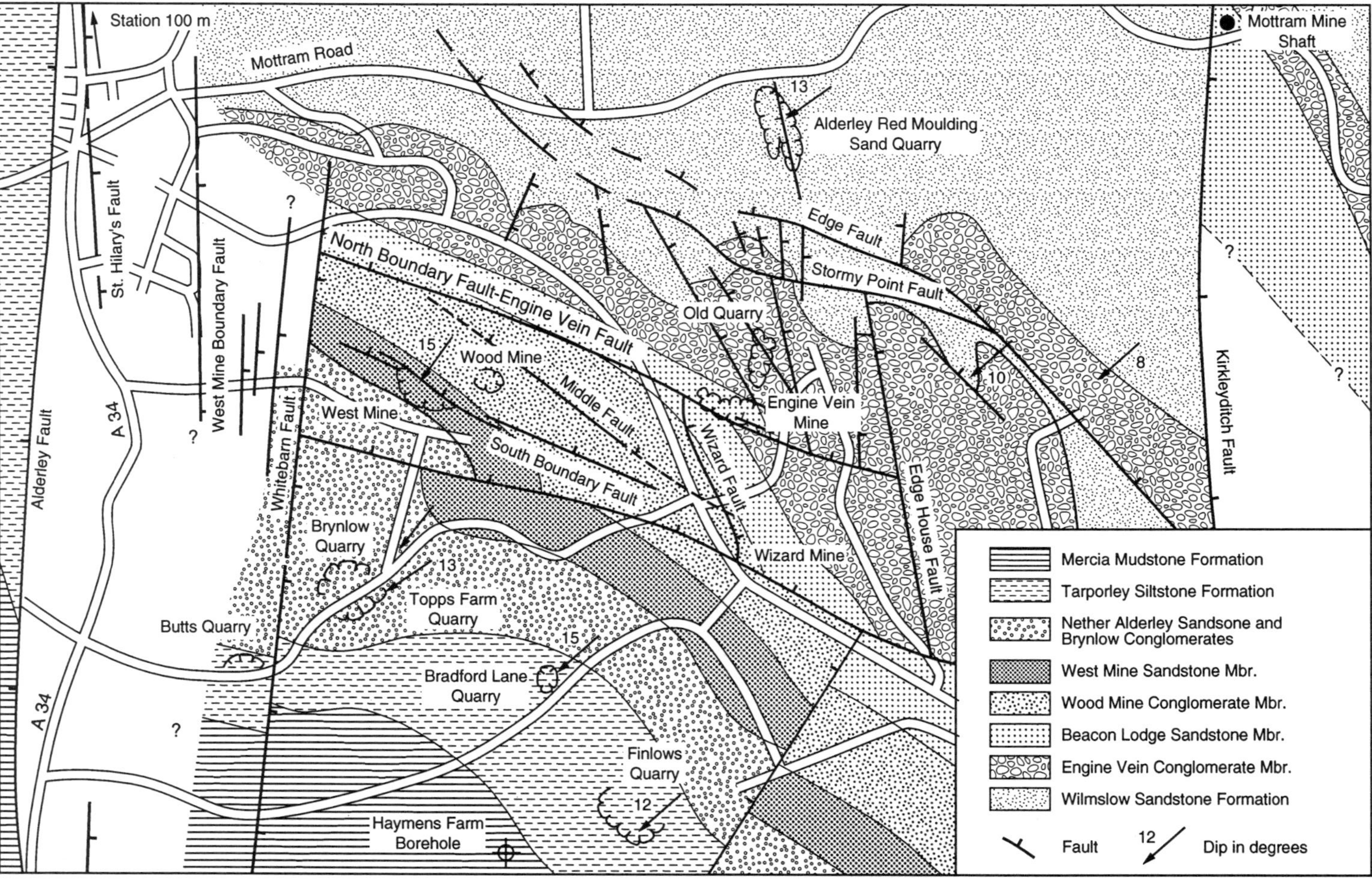

Fig. 3. Geological map of the Alderley Edge area detailing the location of faults and mine workings mentioned in the text (based on the BGS 1:10560 6″ maps of D. F. M. Trotter and Thompson 1970, 1991, with modifications and additions).

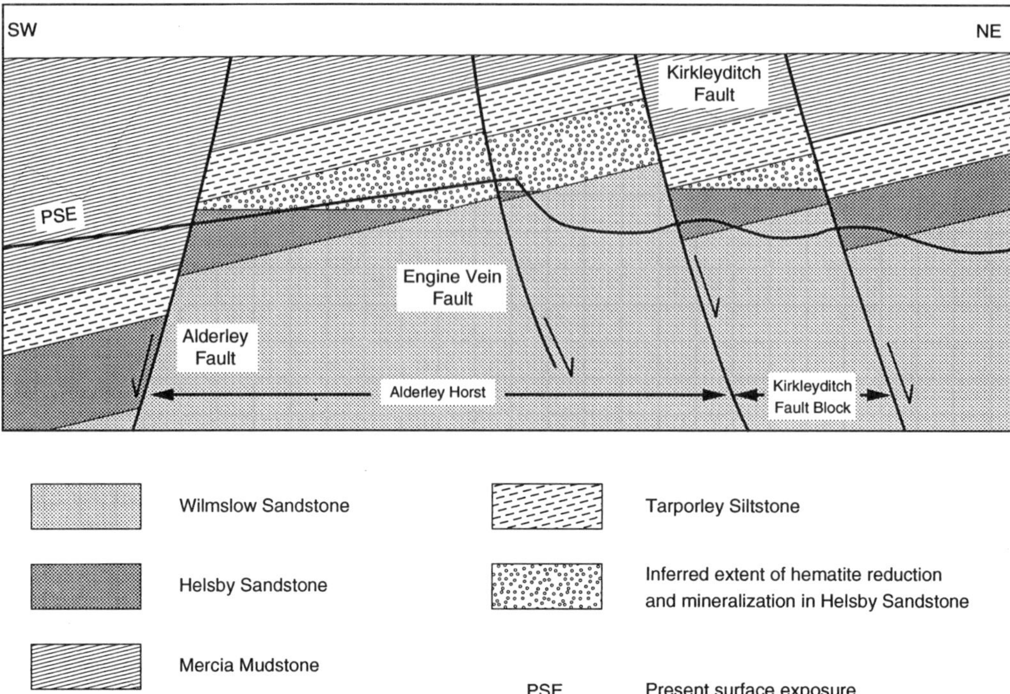

Fig. 4. Geological cross-section aligned SW–NE through the Alderley Horst illustrating the projected trap structure prior to present exhumation (based on Warrington, 1980). The Kirkleyditch Fault Block is shown mineralized to spill point, assuming fault-seal against the Tarporley Siltstones of the Mercia Mudstonre Group.

iferous. On the basis of hydrochemical modelling, Metcalfe *et al.* (1993) reasoned that the sulphide mineral assemblage at Alderley Edge could have been precipitated by the mixing of a relatively small volume of reducing metalliferous brine (>3% of the total fluid volume) with an oxidizing SSG formation water.

The present contribution examines representative faults of various orientations and displacements exposed at outcrop and in the mines at Alderley Edge and sets out to address which, if any, of the faults on the Alderley Horst were the major conduits for the mineralizing fluids.

Methodology

Data were generated from detailed outcrop and underground mine observations combined with the drilling of horizontal cores in the Wood and Engine Vein mines. Samples of mineralized and unmineralized as well as red and white coloured sandstones were collected from around Alderley Edge and in the Cheshire Basin generally for petrographic study. Details of sample localities are given in Burley (1987) and Rowe (1994).

Representative fault zones in Wood and Engine Vein mines were selected for detailed study. Drilling was undertaken normal to fault planes, in a continuous sequence where possible, to identify how cementation, diagenesis and mineralization are spatially and temporally related to fault displacement. Lithological and structural logs were constructed for the cored intervals. The number of cataclastic slip bands (CSBs: nomenclature of Fowles & Burley, 1994) and slip surfaces associated with each fault were plotted on a histogram against distance from the fault plane to investigate the distribution of deformation around fault planes.

Samples of mineralized sandstone selected from the cores were thin sectioned for petrographic study. Polished thin sections were examined and photographed on a Nikon Labophot-pol 2 polarizing microscope. Quantitative point-count analysis was performed on 24 unmineralized, red coloured sandstones from around the area of Alderley Edge and on 29 mineralized sandstone samples from Wood and Engine Vein mines, with 300 counts per slide, to determine grain size and sorting parameters, detrital and authigenic components as well as porosity amounts and types. All averages quoted in the text are arithmetic mean calculations. Selected representative samples of fault slip surfaces, CSBs and mineralized sandstone were examined in a JEOL 6400 scanning electron microscope (SEM) as broken surfaces in secondary electron mode and as polished thin sections in backscatter electron (BSE) and cathodoluminescence (CL) mode. Finely crystalline hydroxides and oxides were identified by XRD of <2 µm fraction separates of disaggregated sandstones. Fluid inclusion petrography and microthermometry of inclusions in barite cements were undertaken on a Fluid Inc. USGS gas-flow heating–cooling stage mounted on a Nikon Optiphot optical microscope.

In this paper, a fault damage zone is considered as the rock volume adjacent to the fault that exhibits macro- and micro-scale evidence of deformation as a result of displacement. The slip surface is the plane along which the majority of the strain has been accommodated and where displacement is at a maximum. The damage zone varies in width and symmetry about the slip surface as a function of the way strain is accommodated in the rock volume adjacent to the fault.

Representative core and hand specimen samples from the Engine Vein and Wood mines have been deposited with the Curator of Mineralogy at the Manchester Museum, Oxford Road, Manchester, UK.

Results

Faults and faulting at Alderley Edge

The main horst-bounding faults are poorly exposed; the Alderley Fault not at all and the Kirkleyditch Fault is only accessible in mine exposures in the village of Kirkleyditch to the east of Alderley (SJ785873; see Fig. 3). Some 30 m below the present surface at Kirkleyditch the main slip surface exposed in mine passages is extensively barite-cemented and slickensided, and juxtaposes sandstones of the Wilmslow Formation against sandstones presumed from regional mapping to belong to the Beacon Lodge Sandstone Member. At this locality displacement on the Kirkleyditch Fault, estimated from juxtaposed stratigraphies, appears to be in the order of 120 m. A 2 m wide zone of intense CSB development is associated with the hangingwall of the main fault whilst the footwall contains numerous small displacement faults and abundant CSBs. Both footwall and hangingwall sandstones are pervasively hematite reduced in the workings along and across the fault, and disseminated galena and copper carbonates are present close to the fault plane.

The smaller N–S oriented dip–slip normal faults within the horst are also generally poorly exposed, but the Whitebarn Fault and related N–S oriented faults are well exposed in West Mine (see Fig. 3). The Whitebarn Fault has a displacement of about 12 m, and a pronounced shale smear is present along the fault plane in the mine exposures. The slip surface is locally leached of hematite and patchily barite cemented. No metalliferous mineralization is present along this fault or the associated sub-parallel smaller displacement faults immediately to the west of the Whitebarn Fault.

By contrast, the WNW–ESE-oriented dip–slip faults, all downthrown to the NE, are characteristically intensely mineralized, associated with extensive barite cementation and red-colour leached at their highest structural levels. It is these faults that the miners have long exploited in their search for ore. Displacement on these faults varies from a few metres up to 50 m for the Edge Fault (see Fig. 3). At outcrop the faults are essentially straight, with numerous branch points and intersections. Small splay faults with displacements of <10 cm and angular divergences of up to 30° are commonly associated with the WNW–ESE faults. In the sandstone horizons, deformation adjacent to the main fault surface is accommodated in numerous CSBs, which cluster for short distances away from the main fault. Zones of intense CSBs development are usually restricted to within 10 m of the through-going fault. In mudstone horizons, the dips of the faults and CSBs decrease rapidly and flatten into low-angled thrusts. Minor bedding-parallel slip is developed at sandstone and mudstone contacts.

Detailed observations on the mineralized WNW–ESE-oriented faults

The Engine Vein Fault (Figs 5 and 6), which generally trends 115° and dips at 70° towards the NE, is well exposed at outcrop and in Engine Vein Mine, and has a total displacement of around 8 m where measured in Ring Shaft (see Fig. 5 for the locality). Some 50 m east of Bear Pit (also see Fig. 5) the Engine Vein Fault bends to trend 95° and abundant splays are developed in the footwall. For much of its length to the east of Bear Pit, the Engine Vein Fault is a composite of two dip–slip normal faults (5 m apart at the eastern end of the outcrop at the surface), defining a discrete fault zone, with associated damage zones of footwall and hangingwall cataclastic deformation up to 2.5 m in width. Most of the displacement is accommodated on the northerly of the slip surfaces. CSB development is most pronounced in the footwall; SE of the bend in the fault trace small displacement faults and CSBs are common up to 6 m away from the southerly

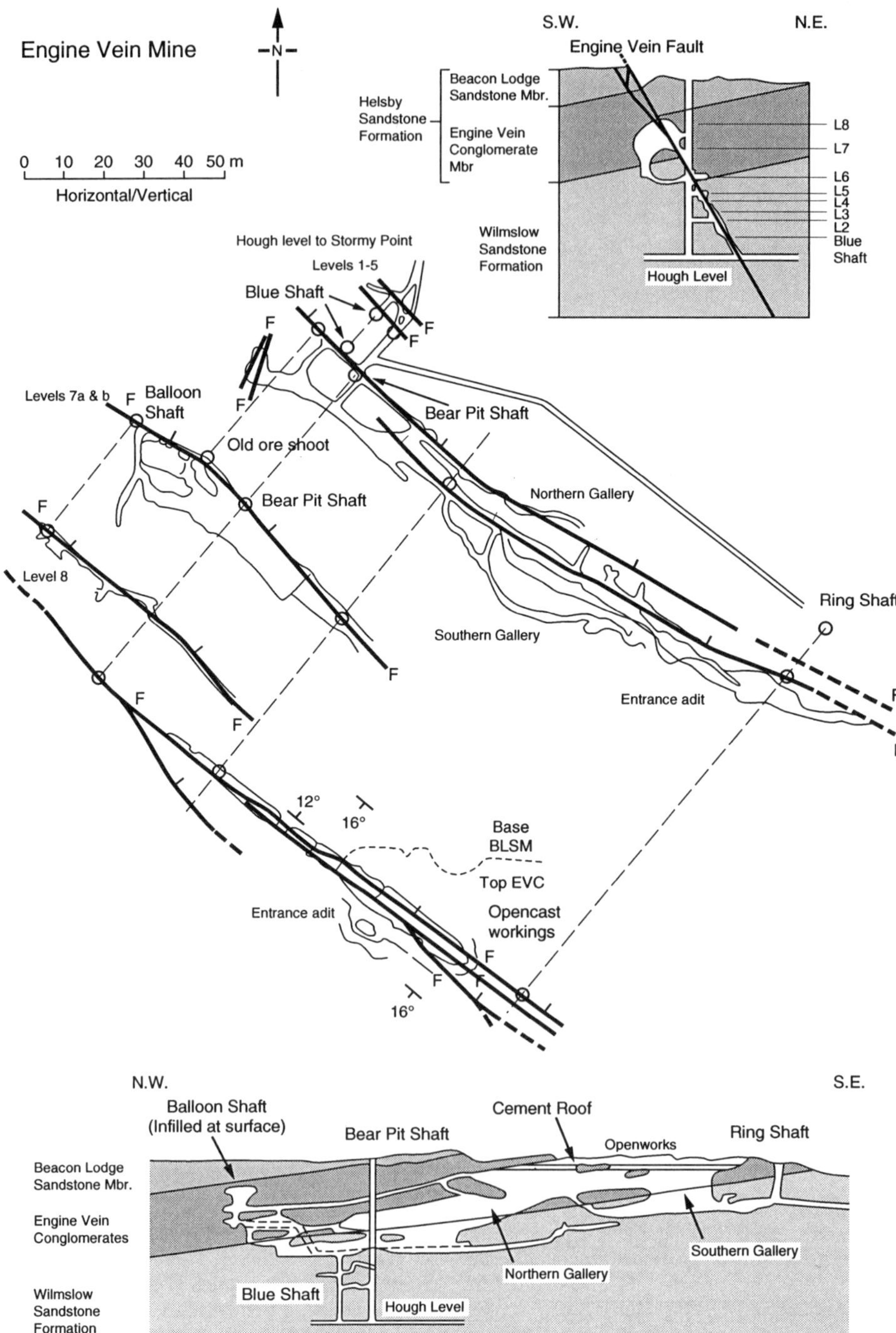

Fig. 5. Map and sections (dip and strike) of the Engine Vein Mine showing the relationship of the mine workings to stratigraphy and faults. Distribution of mine workings taken from original mine plans and the authors' surveys.

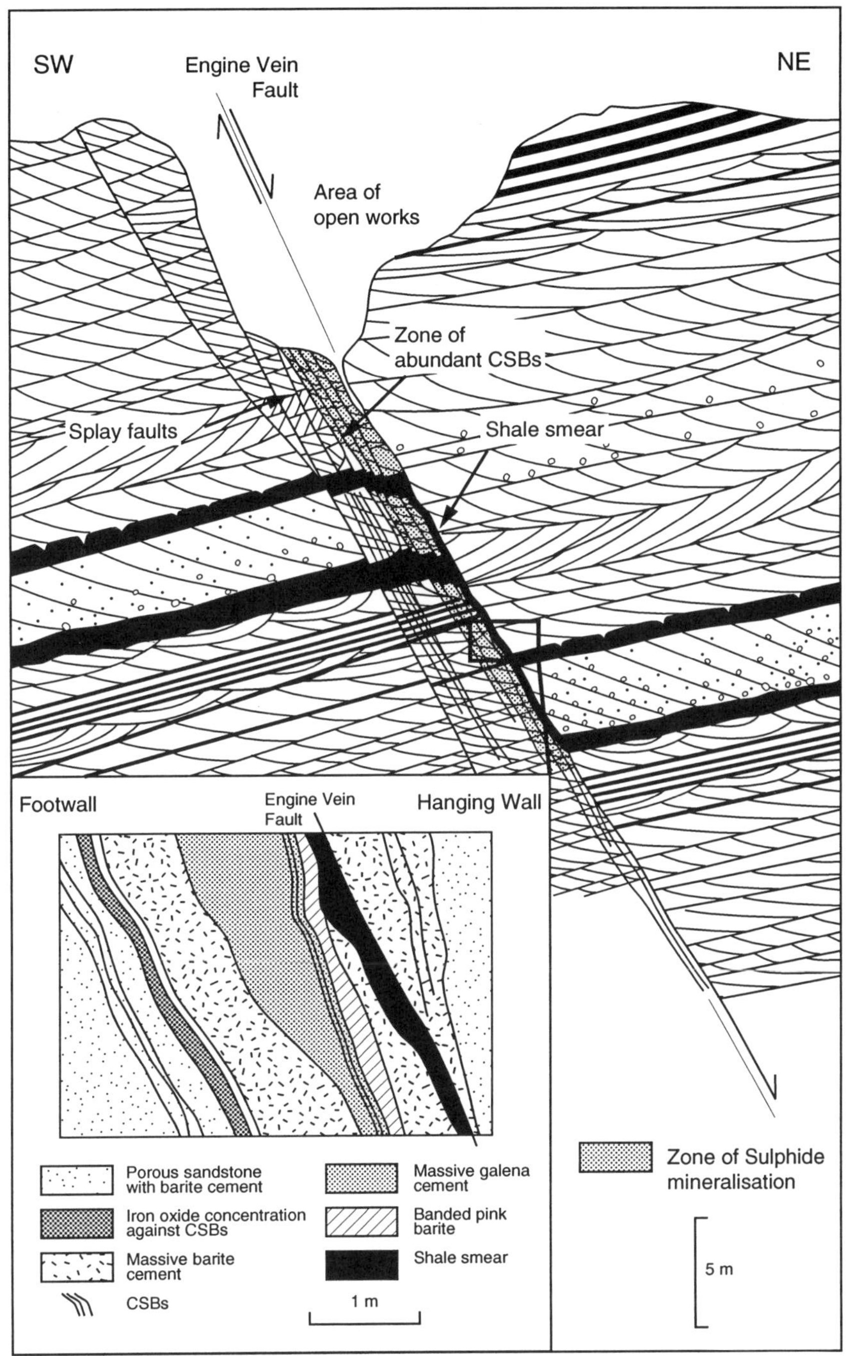

Fig. 6. Detailed cross-section through the Engine Vein Fault at the eastern end of the workings (vicinity of Ring Shaft) showing the development of shale smearing along the fault plane and the distribution of mineralization against the fault. (Sedimentological features taken from Thompson, 1970, his fig. 13.)

Table 1. *Distribution of mineralization and deformation around the WNW–ESE oriented faults studied in detail in Wood Mine*

	North Boundary Fault		Middle Fault		South Boundary Fault	
	Mineralization	Deformation	Mineralization	Deformation	Mineralization	Deformation
Footwall	Sulphides confined to within 3 m of slip surface; locally abundant Extensive secondary mineralization up to 7 m from the slip surface Hematite reduced	Uncommon CSB development up to 2 m from the slip surface	Sulphides and copper carbonate minerals concentrated within 5 m of the slip surfaces; large cavenous mine workings in footwall Hematite reduced	CSBs present within 5 m of, and increase in abundance towards, the slip surfaces	Sulphides within 1 m of the slip surface, locally abundant Hematite reduced	Largely undeformed
Slip surface	Poikilotopic sulphide cements: distribution varies laterally along strike of the slip surface Extensive barite cementation Hematite reduced	Throw of fault estimated at 5 m which produced a 1 m wide zone of intense cataclasis	Galena and copper carbonates abundant in zone between slip surfaces Abundant barite cement Hematite reduced	Twin slip surfaces 2 m apart with 0.5 throw visible across each Common CSBs with radiating barite cements: preferentially developed in CSB hanging wall	Intense development of galena and barite cements Hematite reduced	Twin slip surfaces, 3 m apart with total throw estimated at 4 m. Each surface characterized by ~0.5 m zone of intense cataclasis En-echelon, barite-filled, sigmoidal tension fractures developed in zone between slip surfaces
Hanging wall	Manganese hydroxide within 2 m of the slip surface and rare sulphides Sandstone leached of hematite and cemented by poikilotopic barite	CSBs abundant within 4 m of slip surface	Metalliferous mineralization is rare Hematite reduced	CSBs developed up to 3 m from the slip surfaces	No evidence of metalliferous mineralization Hematite reduced	Largely undeformed

main fault slip surface. Barite and sulphide mineralizations (especially galena) are concentrated in the fault zone in the Wilmslow and Helsby Sandstone Formations in areas of abundant CSB development, although no veining *sensu stricto* is observed in the remaining exposures of the Engine Vein in either the mine workings or surface exposures (although see Thompson 1970; his fig. 13). Pronounced hangingwall drag is developed in the fine grained at the top of the Engine Vein Conglomerate Member, and a continuous shale smear, 3–15 cm thick, is developed on the main fault surface for the full displacement distance of 8 m between the mudstone intervals present towards the base of the Engine Vein Conglomerates (see Fig. 6).

Surface exposures around Wood Mine are poor, but underground, three WNW–ESE fault surfaces and fault zones are well exposed (Fig. 7). Distributions of mineralization and deformation around the WNW–ESE faults studied in Wood Mine are summarized in Table 1. The North Boundary Fault (NBF) is the westwards extension of the Engine Vein Fault, and in Wood Mine comprises a single slickensided surface with a throw of approximately 7 m. The NBF zone and adjacent wall-rock sandstones were extensively cored during this study and a log illustrating the distributions of mineralization and deformation fabrics is presented as Fig. 8. Within ~1 m of the fault plane in the footwall is an area of intense cataclastic deformation, expressed as a zone of sub-parallel CSBs (Figs 8 and 10).

The Middle Fault (MF) is a small displacement (total throw approximately 1 m) normal fault that strikes NW–SE. It may have originated as a splay of the Engine Vein–North Boundary Fault system (see Fig. 3). In the mine it consists of two, sub-parallel, dip–slip normal faults (Fig. 7). Each fault has a displacement of approximately 0.5 m down-thrown to the NE and both displace a 30 cm thick pebbly sandstone unit (Fig. 9e). The distributions of

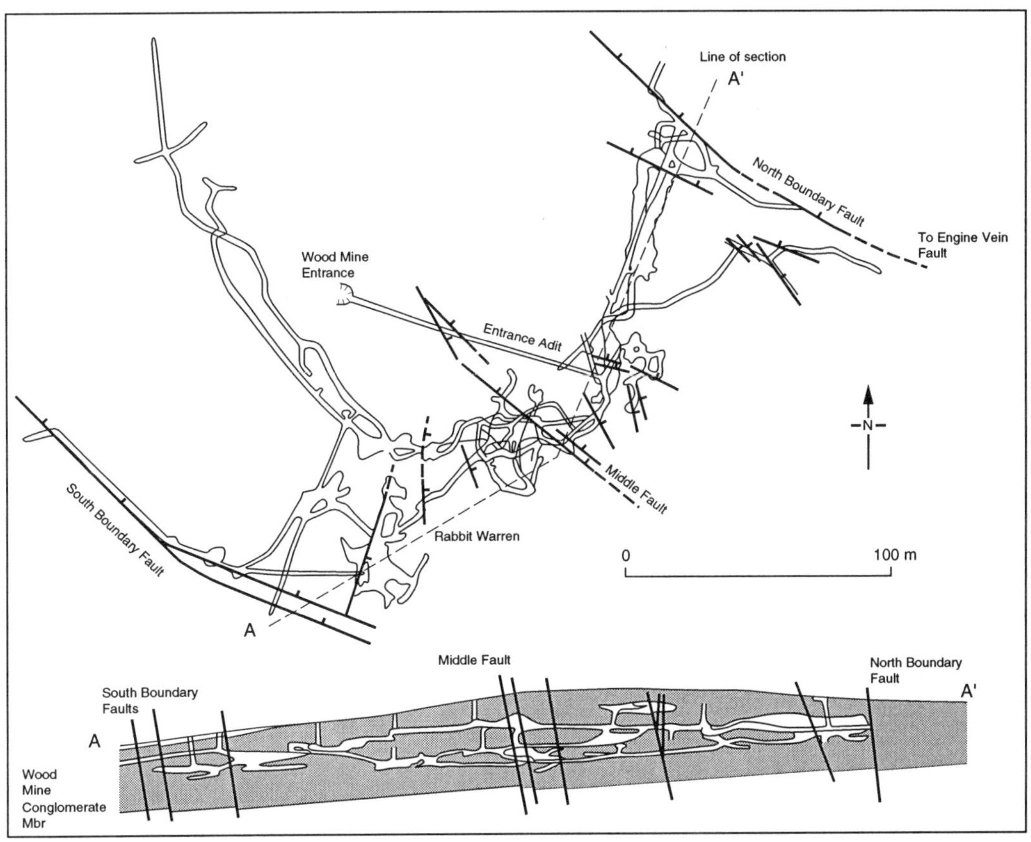

Fig. 7. Map and dip cross-section of Wood Mine (adapted from Warrington, 1965 with additions) showing the main faults in relation to the mine workings.

J. ROWE & S. D. BURLEY

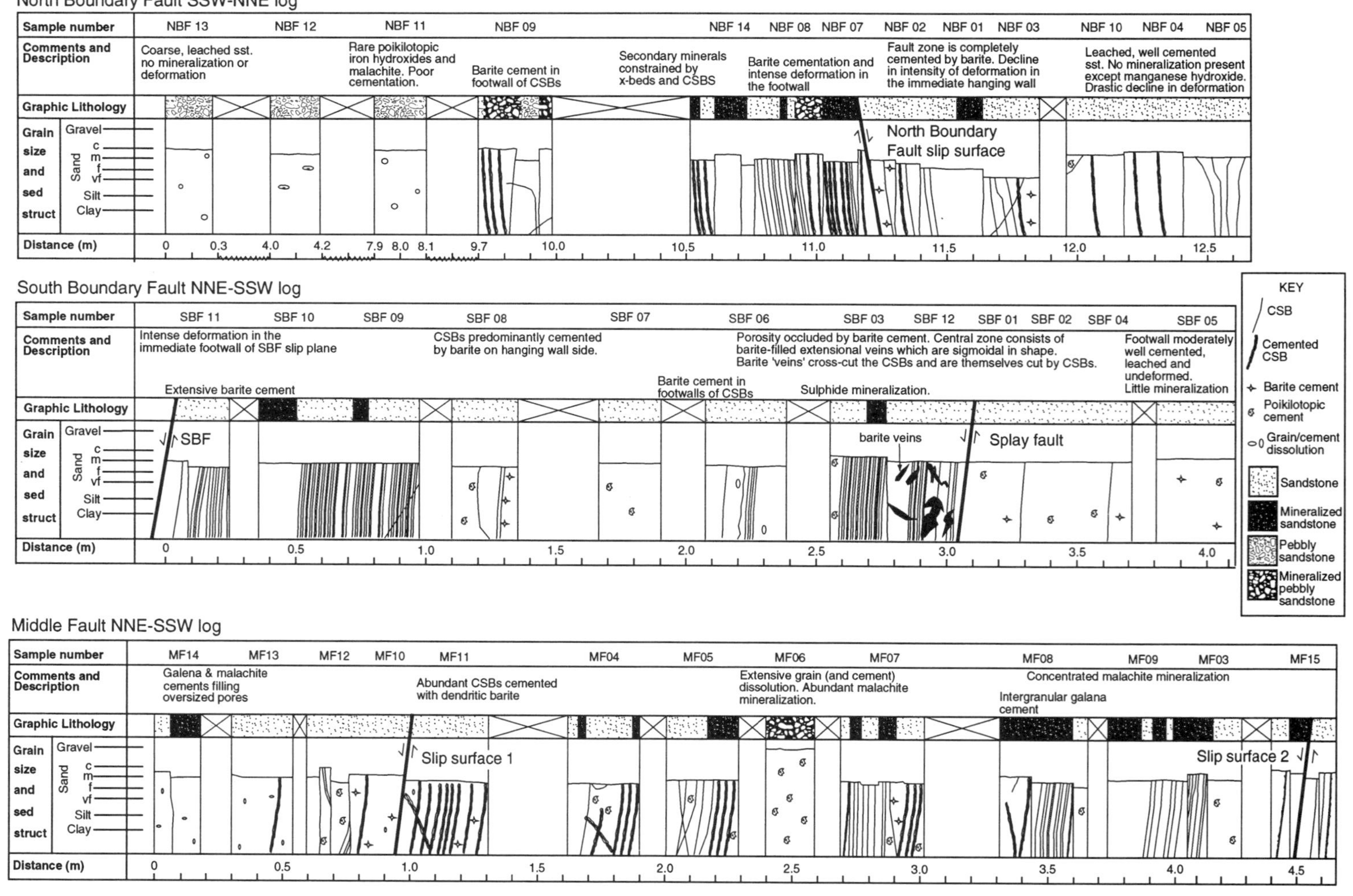

Fig. 8. Graphic logs drawn from the horizontal cores drilled through the North Boundary, Middle and south Boundary faults in Wood Mine. The distribution of structural features and mineralization are shown.

mineralization and deformation fabrics around the MF are illustrated as Figure 8. The more westerly of the slip surfaces includes large mudstone clasts which are deformed and smeared along the fault surface. Measurements of the abundance of CSBs in the cores taken from the MF zone reveal that they increase in number from the undeformed hangingwall towards the fault zone, with the concentration of CSBs being greatest between the faults, and decreasing away into the footwall (Figs 8 and 10). In the footwall, CSBs and mineralization can be seen for a distance of up to 5 m away from the fault plane.

The South Boundary Fault (SBF, see Figs 3 and 7) is a steeply inclined normal fault with a total displacement of 4 m that strikes 270°N. For most of its length it comprises a single normal fault surface (throw 2–3 m) with a complex zone of conjugate CSBs immediately in the footwall. This zone of cataclasis is around 1 m in width and displays intense deformation. CSBs associated with the SBF are both anastomising and intersecting. Again, they rapidly decrease in abundance away from the main fault surface. Graphic logs of the cored intervals and a histogram of CSB abundance are illustrated in Figs 8 and 10, respectively.

To the NW the SBF intersects a less steeply dipping fault that trends 305°. The area between the faults probably represents a breached transfer zone, and is a site of strong deformation. In the immediate hangingwall of the extension to the SBF is a zone of complex deformation including a series of *en echelon* barite-filled sigmoidal tension fractures (Fig. 9d). These barite-filled extensional fractures occur only in a 50 cm wide zone of extreme cataclastic deformation in the hangingwall of the SBF extension. The extensional fractures are up to 20 cm in length and 2 cm in width and are developed approximately normal to the plane of the SBF extension. This, along with similar barite-cemented extensional fractures in the Wizard Mine (see Fig. 3 for location) and the single vein in the Engine Vein Mine reported by Thompson (1970), are the only observed occurrences of true veining in the mines at Alderley Edge.

The effect of faults on porosity

Background porosity

The average total porosity measured from un-mineralized and undeformed samples of sediments from the Helsby Sandstone Formation is 17% of the sandstone volume ($n = 24$). Of this total, an average of 13% porosity is intergranular and 4% is grain dissolution porosity, principally resulting from the dissolution of detrital feldspar grains. Intergranular porosities of up to 27% are recorded from the better sorted sandstones of the Helsby Sandstone where no mineralization is developed.

Main slip surfaces

Slip surfaces of all faults examined in detail with throws from ~4 m (SBF) to ~330 m (Kirkleyditch Fault), are similar in nature. Displacement is characteristically confined to two or three through-going planes which are highly polished and preserve well defined slickensides. Brecciation is absent along slip surfaces except in areas of fault surface rugosity. Fault breccias are localized to extensional jogs, kinks and steps on the fault surface. Even where fault breccias are present, cementing minerals are not brecciated indicating that cementation post-dates faulting.

Sandstone porosities measured from samples taken across the main fault surface of the Kirkleyditch Fault have an average value of 0.8% of the sandstone volume ($n = 2$) and those from the Engine Vein Fault in Engine Vein Mine have no visible porosity ($n = 3$). Examination of these surfaces in the SEM shows the presence of comminuted quartz grain fragments 2–5 micron in size, cemented with sub-micron-sized quartz overgrowths and discrete quartz crystals.

Despite the sand-dominated nature of the Helsby Sandstone (approximately 85% sand), shale smears are a characteristic feature of many of the main faults on the horst. Laterally discontinuous mudstones, up to 1 m in thickness, are not uncommon in the lower part of the Engine Vein and the Wood Mine conglomerates. Shale smears are developed along most of the faults with <10 m displacement which intersect such mudstone horizons. The smears are usually thin (1–10 cm in width) but appear to be continuous between the displaced mudstones and are an important contributor to porosity and permeability reduction across the faults.

Cataclastic Slip Bands

Millimetre-wide CSBs are associated with all faults of metre-scale throw examined around the horst. In surface outcrops of unmineralized red bed sandstones, individual CSBs weather proud of their host and are a characteristic white colour. This is a result of the grain size reduction and communition (powdering) of the sandstone grains within the CSB. In areas of mineralization CSBs are also the sites of enhanced barite cementation which is commonly concentrated adjacent to individual CSBs.

Figures 8 and 10 show that each of the fault planes examined in Wood Mine is accompanied by a damage zone which is defined by the clustering of large numbers of CSBs adjacent to the main slip surface. This damage zone is typically in the order of 200–400 cm wide either side of the fault plane. The damage zones associated with the WNW–ESE faults, with throws of 10 m or less, are of similar width to those observed adjacent to the Kirkleyditch Fault, with a throw of up to 330 m. Hence, the width of the damage zone, in this instance, does not appear to be influenced by the total throw across the fault surface.

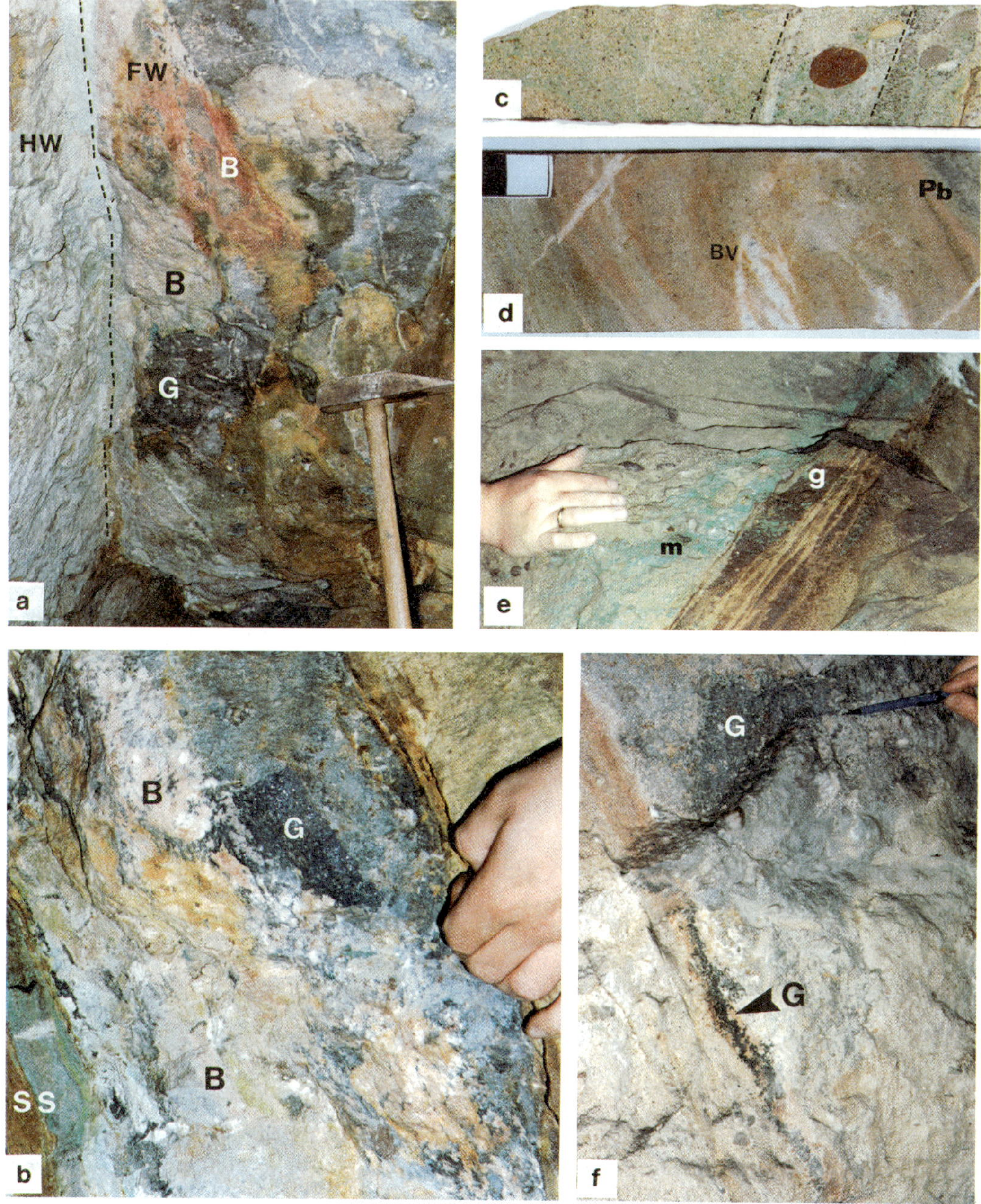

Granulation and porosity reduction produced by CSBs

In thin section, grains present within the CSBs have dramatically smaller average size than those in their adjacent sandstone hosts (Fig. 11a,b). The reduction in grain size from undeformed sandstone to CSB observed using an optical microscope is typically from coarse sand to silt. When viewed in the SEM, extensive sub-silt-sized grain communition is also apparent. SEM-CL examination of the CSBs reveals extensive grain fracturing. Sandstones within CSBs also display a dramatic reduction in sorting, from moderate in the undeformed sandstone, to very poor within the CSBs. Average total visible porosity measured from point counting along CSBs is 3% of the sandstone volume ($n = 12$). Porosity of the CSBs, and by implication, permeability and fluid transmissibility, are thus greatly reduced in comparison with their counterparts in the undeformed sandstone.

Porosity modification of the host sandstones

Nature of the cementation and mineralization

Virtually all of the diagenetic porosity modification and mineralization at Alderley Edge takes the form of disseminated pore-filling and grain-replacive cementation spatially associated with the WNW–ESE faults on the crest of the horst. More rarely, mineralization cements breccias associated with irregularities along fault planes. Veining is extremely rare; even the sulphide mineralization is not found in veins. This is particularly the case in Engine Vein Mine, which is completely misnamed, as barite and sulphide mineralization occur only as intergranular cements and replacements concentrated in the footwall of the Engine Vein Fault (Fig. 9a,b; although see Thompson 1970, his figure 13). The only observed occurrences of veining today are along the South Boundary Fault in a breached transfer zone and along the Wizard Fault in Wizard Mine. In both these cases barite occurs in extensional veins.

Diagenetic mineralogy

The diagenetic sequence for the Helsby Sandstone Formation sediments at the crest of the Alderley Horst, based upon thin section optical petrography, is shown in Fig. 12. The main sequence of burial cementation generally conforms to the order barite, copper sulphides, lead and zinc sulphides, followed by ferroan calcite cementation, although barite occurs in several generations, some of which post-date sulphide cementation. Each of the main cements is briefly described below.

Hematite grain coatings and associated clays. Hematite is present as coatings to detrital grains in the majority of the sandstones of both the Wilmslow Sandstone and the lowest Helsby Sandstone formations in the Alderley area. The hematite is mixed with illitic and smectitic clays that predominantly occur as tangential grain coatings. These minerals are considered to have precipitated during early diagenesis in an oxidising, neutral to alkaline geochemical environment (see Walker 1967 for the modern analogue and Burley 1984 and 1987 for the evidence in the Sherwood Sandstone). Hematite and the grain coating clays are typically a volumetrically minor component of these sandstones and constitute ≤1% of the total sandstone volume. By contrast, sandstones present at the mineralized crest of the Alderley Horst are, without exception, devoid of this grain coating hematite.

Fig. 9. Examples of faults and mineralization in core and mine workings. (**a**) The zone of mineralization associated with the Engine Vein Fault close to Ring Shaft, Engine Vein Mine showing the concentration in the footwall (FW) of galena (G) and barite (B) cements against the fault plane (dashed line). Hangingwall annotated HW. (**b**) Detail of the barite (B) and galena (G) cementation developed against a thin shale smear (SS). Engine Vein Mine close to Ring Shaft. (**c**) Section of core NBF07 showing the Wood Mine Conglomerate from the damage zone in the immediate footwall of the NBF. Galena cement is concentrated against a number of sub-parallel CSBs that cross the core (highlighted with dashed lines) whilst malachite cement is dispersed throughout the sandstone. Length of core is 22 cm. (**d**) Core SBF12 sampled from the immediate hangingwall of the splay from the South Boundary Fault. The sandstones are extensively cemented by barite. Discrete barite-filled extensional veins (BV) are present adjacent to the splay fault. Concentrations of galena cement (Pb) occur in discrete bands enclosed within the barite cement. The scale bar is graduated in centimetres. (**e**) Clustering of cataclastic slip bands and associated mineralization at the Middle Fault in Wood Mine (Slip surface 2 on Fig. 8c). Goethite (g) cementation results from the oxidation of a former carbonate cement whilst malachite mineralization (M) overprints the pebble band in the sandstone and the fault damage zone. (**f**) Detail of Wood Mine Conglomerate Member sandstones from between the two slip surfaces on the Middle Fault Zone illustrating barite and galena (G) cementation concentrated against CSBs. The pencil highlights an area where galena cementation is concentrated.

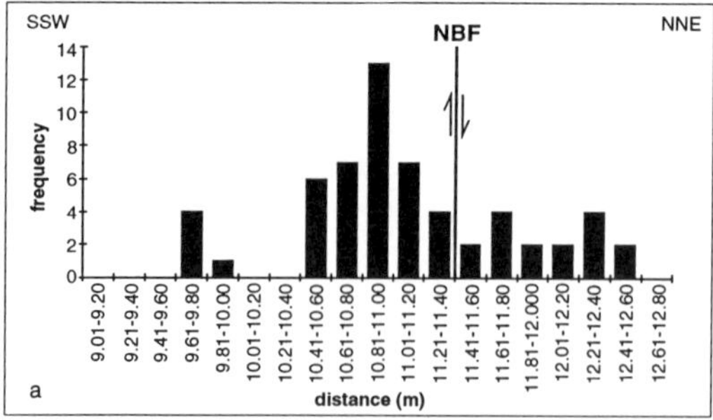

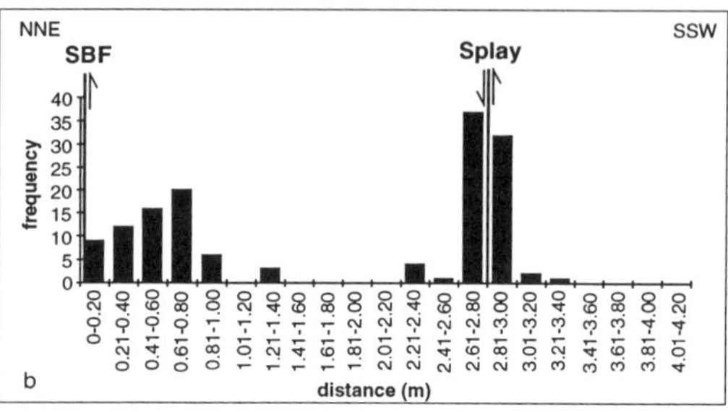

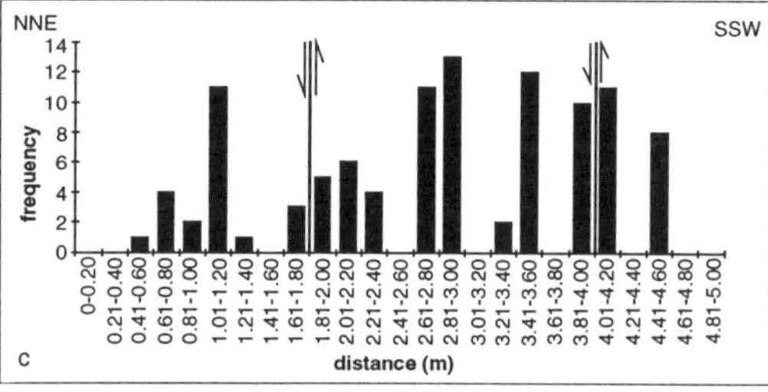

Fig. 10. Variations in the number of CSBs across the North Boundary, Middle and South Boundary faults in Wood Mine. Distances relate to the scale on the horizontal graphic logs through the faults shown on Fig. 8.

Grain coating illitic clays are still present in the mineralized sandstones, but the admixed hematite has been reduced.

Non-ferroan calcite cements. Non-ferroan calcite cements are absent from the mineralized sandstones. However, in red coloured Helsby Sandstones from around the horst and elsewhere in northern Cheshire, Lancashire and Cumbria, spheroidal and blocky non-ferroan calcite cements are common (Burley 1987; Rowe 1994; Strong and Pearce 1995). These cements exhibit displacive and replacive growth fabrics, enclose hematite grain pellicles and are considered to be pedogenic carbonates.

Quartz overgrowths. Authigenic quartz overgrowths represent a common, pervasive but volumetrically minor (<2% of the sandstone volume) authigenic component of the mineralized sandstones. The overgrowths are typically irregular in habit and have highly corroded outlines, suggestive of replacement by a later carbonate which has been subsequently dissolved. They are enclosed in barite and sulphide cements and probably formed during moderate burial.

Feldspar and early carbonate cement dissolution. Extensive grain dissolution porosity is developed in the areas of mineralization. Detrital feldspars are almost absent in the mineralized sandstones (<2% of the sandstone volume), being less common than in equivalent red coloured sandstones (up to 12% of the sandstone volume). Estimates of original detrital feldspar abundance indicate that 3–8% of the total present porosity has been produced by feldspar dissolution. Where present, feldspars typically show evidence of dissolution and commonly appear 'skeletal'. Large, oversized, secondary pores are considered to be the sites of former feldspar grains, now completely dissolved.

The timing of feldspar dissolution can be relatively constrained. The presence of open secondary pores containing relic feldspar grains suggests that dissolution post-dated the cessation of mechanical compaction. Many of the oversized pores are variously occluded by barite and sulphide cements so that dissolution preceded the main mineralizing event at Alderley. Grain dissolution in the Sherwood Sandstones is known to be associated with early red bed diagenesis (Burley 1984), but the enhanced volumes of feldspar dissolution in mineralized sandstones suggests a second period of grain dissolution associated with the mineralization (Fig. 12).

Dissolution of early calcite cements is also considered to have taken place, although the evidence for this is rather circumstantial, being based on the absence of non-ferroan calcite in the mineralized sandstones.

Sulphide cements. Samples of the primary sulphide mineral assemblage from the Engine Vein Mine are dominated by galena, sphalerite, bravoite, chalcopyrite and pyrite, although additional sulphide minerals are recorded from Alderley Edge (Ixer and Vaughan 1982). Galena is the most abundant sulphide present. Textural relationships indicate that the sulphide minerals were precipitated in both intergranular and grain dissolution porosity after cementation by quartz overgrowths, and are highly replacive with respect to both detrital grains and overgrowths (Fig. 11c,d). The sequence pyrite–bravoite–chalcopyrite–sphalerite–galena is typical, although not ubiquitous.

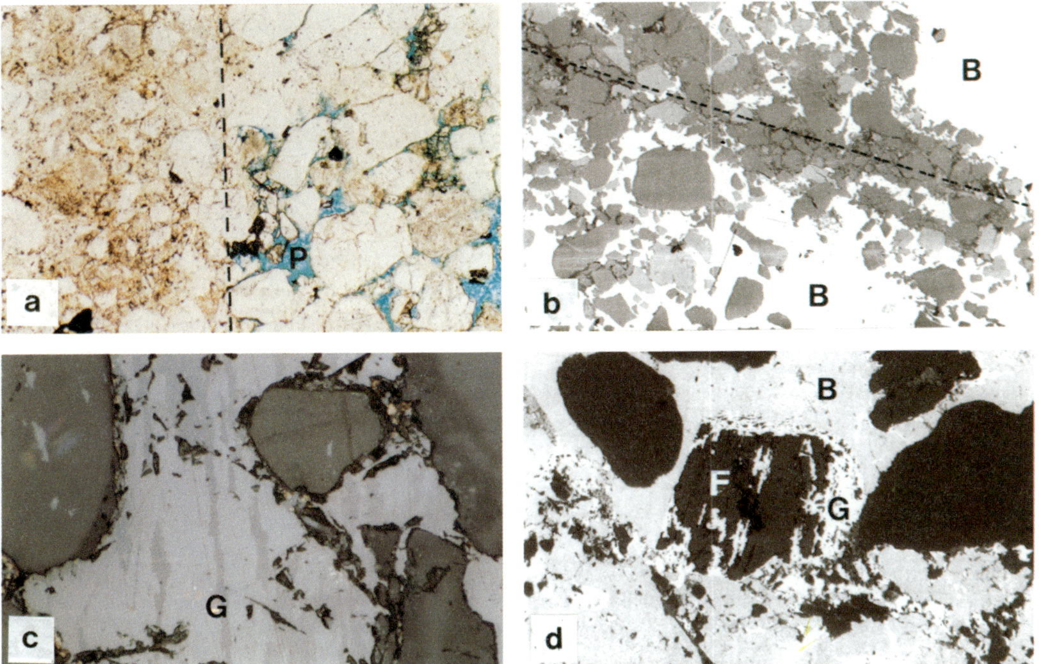

Fig. 11. Petrographic textures. (**a**) Zone of cataclasis (CSZ, boundary dashed) developed in a porous sandstone from the Engine Vein Conglomerates (porosity annotated P) illustrating the decrease in grain size and sorting associated with cataclasis. Optical photomicrograph in plane polarised light, 4 mm in width. (**b**) SEM-BSE image of an individual CSB (annotated with dashed line) from the SBF in Wood Mine which is partly cemented by barite cement (annotated B, high back-scatter coefficient). The cement is undeformed indicating the precipitation of barite post-dates cataclasis. Width of micrograph 2.5 mm (**c**) Extensive pore-filling and grain replacive galena cement (annotated G), from the footwall of the NBF in Wood Mine. The galena is shown in plane polarised, reflected light. Optical photomicrograph, 40 μm wide. (**d**) High contrast SEM-BSE image detailing authigenic galena (G, high back-scatter coefficient) which has partly replaced a K-feldspar grain (dark grain in the centre annotated F) and is in turn enclosed by later pore-filling barite cement. (annotated B). Middle Fault, Wood Mine, width of micrograph 600 μm.

The sulphide mineralization occurs exclusively as disseminated pore-filling intergranular cements and grain replacements in formerly porous sandstones (Figs 9, 11c,d). Veins of sulphide are not present. However, large areas of lensoidal sulphide mineralization are developed close to the faults where galena and pyrite are extremely replacive with respect to the host sandstone grain framework (cf. Thompson 1970; his fig. 13 and see our Figs 6s and 9). Sulphide mineralization is concentrated within ~5 m of the WNW–ESE faults, and is only infrequently observed further away. The intensity of sulphide mineralization declines dramatically with increasing distance from the mineralized faults (Fig. 9).

As a proportion of all the Helsby Sandstone Formation samples examined, sulphides constitute an average of only 2% of the total sandstone volume. However, where abundant close to the mineralized faults, these cements account for up to 38% of sandstone volume.

Barite. Barite is the dominant component of the mineralization assemblage, constituting an average of 18% of the volume of the mineralized Helsby Sandstone Formation sandstones studied ($n = 29$). It is present in the fault zones and occurs as a pervasive cement in the mineralized sandstones proximal to and distal from the mineralized faults, even occurring in sandstones that retain their red coloration.

Four discrete types of barite mineralization are recognized:

(i)   The most abundant occurrence is as widespread intergranular pore-filling cementation within the Helsby and Wilmslow Sandstone formations across the horst. This barite

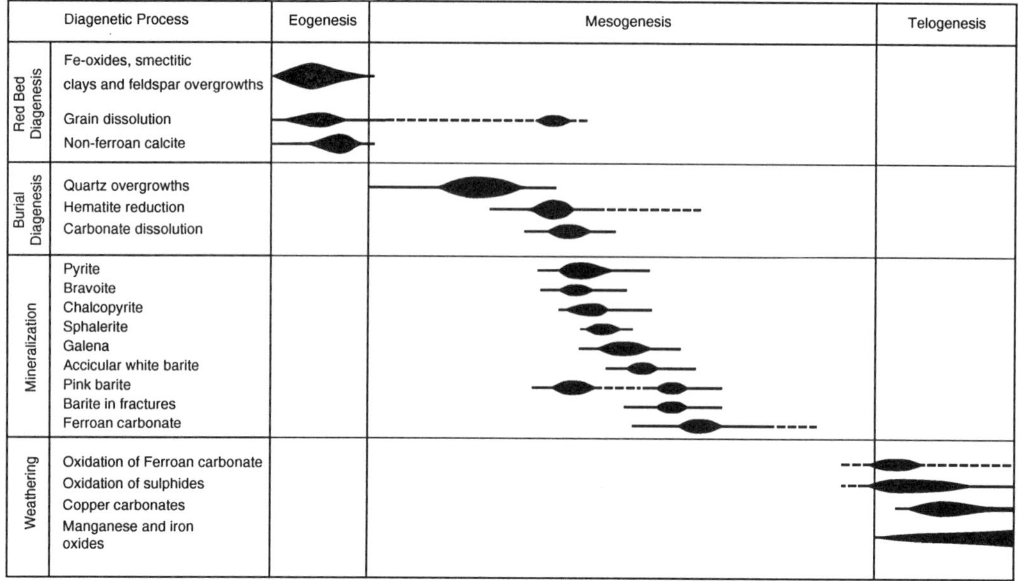

Fig. 12. Interpreted paragenetic sequence combining early diagenesis, burial diagenesis, mineralization and Recent weathering effects for the sediments of the Sherwood Sandstone Group at Alderley Edge.

occurs as large, white coloured lath shaped crystals (up to 2 cm in length) that form randomly-oriented aggregates clearly visible in hand specimen. These are not restricted to the immediate vicinity of faults and even extend beyond the hematite-leached sandstones. In thin section this type of barite cementation constitutes up to 16% of the sandstone volume and post-dates quartz overgrowth cementation.

(ii) Barite is also present as enhanced intergranular cementation centred on and increasing adjacent to CSBs, CSZs and fault slip surfaces, often in the footwall side of the fault zone (Fig. 9d). In these cemented zones the barite occurs as interlocking lath-shaped crystals, often totalling >30% of the sandstone volume, that are highly replacive to the host sandstones and their quartz overgrowths. In hand specimen this type of barite cement can be either pink or white coloured and superficially resembles veining. Careful observation and thin section study reveals that in all cases, however, this barite is always an intergranular cement, the appearance of veining being a combination of the replacive nature of the cementation and its abundance.

(iii) Barite is occasionally found as vein-filling crystals in sigmoidal extensional fractures (e.g. on the South Boundary and Wizard faults; see Fig. 9d). In this case the barite occurs as large euhedral crystal laths oriented at right angles to the vein walls.

(iv) Rarely, barite is present as euhedral, acicular, white-coloured crystals associated with intergranular galena cements and replacements close to the mineralized faults. These barite crystals in thin section form highly replacive poikilotopic cements that infill remaining intergranular and grain dissolution pore space not cemented by sulphide.

Hand specimen and thin section textural relationships indicate that most barite was precipitated after quartz overgrowths, hematite reduction and sulphide mineralization, and was subsequent to cataclastic deformation and faulting (Fig. 11). The vein-filling barites are no exception to this generalization, although the pink coloured barite cements were precipitated before sulphide mineralization.

Ferroan calcite cements. Large poikilotopic crystals of ferroan calcite occur as a minor pore-filling cement in both mineralized and unmineralized sandstones, rarely exceeding 2% of the volume. Textural relationships indicate that this generation of calcite cement post-dates quartz overgrowths, barite and sulphide mineralization.

Secondary copper minerals. Copper carbonate and silicate minerals (principally malachite, azurite and chrysocolla; Carlon 1979; Ixer & Vaughan 1982) are widely present as cements in the Helsby Sandstone Formation across the Alderley Horst. They are preferentially present in the most porous and permeable horizons of the sandstones (such as foresets of cross-beds) well beyond the areas where sulphides are present. The copper carbonates form pore filling microcrystalline and microsparry aggregates that coat detrital grains in the porous sandstones and form dense, microcrystalline anhedral masses within barite, galena and pyrite cements. Chrysocolla forms radiating acicular crystal aggregates 20–50 µm in diameter that also coat all detrital grains and cements with the exception of geothite and manganese oxides.

These copper minerals are clearly very late in the diagenetic sequence and are considered as oxidation products of a primary copper sulphide mineral assemblage, of which chalcopyrite is present as relics (Ixer and Vaughan 1982). They have a much wider distribution away from faults than that of the sulphide mineralization and they 'overprint' CSBs and fault zones (Fig. 9e).

Goethite–limonite and manganese hydroxide–oxides. A mixture of goethite and limonite (identified by XRD analysis) and a complex mixture of manganese oxides are widespread in several forms across the horst and typically amount to between 2 and 5% of the sandstone volume. The main occurrence of goethite–limonite is as a brown-coloured microcrystalline cement that in hand specimen gives the sandstone a distinctive 'spotty' appearance. It is intimately associated with copper carbonates and also 'overprints' sedimentological and structural features. The texture and distribution of this goethite–limonite mixture suggests that it may represent the oxidation product of the dissolution of a former ferroan carbonate cement.

Goethite and limonite also occur as common large, irregular leisegange rings and bands which overprint both sulphide and copper carbonate cements, but attain the greatest concentration adjacent to and within the cataclastic zones of the mineralized faults.

Manganese hydroxide–oxides are present as distinct areas of intense, black coloured cementation, best developed adjacent to sulphide cements. These appear as opaque, microcrystalline masses and grain coatings in thin section, and as with goethite and limonite, coat all other cements. They occur to an exceptional degree in the Wizard Mine for 0.5 km along and adjacent to the Wizard Fault.

Constraints on fluid types

Two-phase, aqueous fluid inclusions are common in the authigenic barite cements. Fluid inclusion homogenization temperatures (T_h), depression of freezing points (T_f), eutectic temperatures (T_e) and ice melting temperatures ($T_{m_{ice}}$) were measured from pore-filling barite cements collected from the

Table 2. *Petrographic and microthermometric data measured from fluid inclusions in barite cements*

Generation	Fluid inclusion assemblage	Long axis length (µm)	% gas vol.	T_f (°C)	T_e (°C)	$T_{m_{ice}}$ (°C)	Wt% NaCl equiv.	T_h (°C)
Primary	P1	5	10	−84	−46	−15	18.6	130
Primary	P1	6	15	−72	−42	−9	12.9	192
Primary	P1	4	10	−70	−34	−10	13.9	166
Primary	P1	7	10	−87	−43	−14	17.8	132
Primary	P1	4	25	−71	−42	−9	12.9	189
Secondary	S1	5	15	?	?	?	?	197
Secondary	S1	6	10	−73	−46	−15	18.6	119
Secondary	S1	4	15	−69	−53	−11	15.0	186
Secondary	S1	2	10	−87	−39	−13	16.9	125
Secondary	S1	4	20	−75	−50	−10	13.9	214
Secondary	S1	2	20	?	?	?	?	179
Secondary	S1	4	20	?	−55	−10	13.9	212
Secondary	S1	3	10	?	?	?	?	173
Secondary	S1	5	10	−73	−42	−9	12.9	214
Secondary	S1	3	20	−66	−38	−10	13.9	197
Secondary	S1	5	20	−75	−25	−12	16.0	232
Secondary	S1	8	10	−69	−49	−11	15.0	231
Secondary	S2	3	10	−71	−41	−11	15.0	231
Secondary	S2	5	15	−65	−37	−7	10.5	164

Engine Vein Fault and from extensional veins on the South Boundary Fault in Wood Mine. The results of measurements on two wafers are reported here and listed in Table 2.

Petrographic observations reveal that the primary inclusions occur aligned along growth zones and are cut by trails of secondary inclusions which occur in microfractures. Prior to microthermometry, individual fluid inclusion populations were characterized by consistent liquid to vapour ratios, typically with 10% vapour, indicating formation above 60°C. During heating however, all the inclusions leaked as the temperature was increased. As a result measured T_h values ranged from 119 to 232°C and are considered totally unreliable.

Cooling to between –65 and –89°C was required to freeze the inclusions. Eutectic temperatures varied between –34 and –55°C, with most T_e <45°C, although the small size of many of the inclusions made accurate measurement of T_e difficult. Final ice melting temperatures ranged from –9 to –15°C. The variation in freezing and melting behaviour is considered to be real and reflect compositional variation in the trapped fluids as leakage affects neither the salinity nor composition of fluids within inclusions. Eutectic temperatures of >–50°C are typical of complex chloride brines containing $NaCl–CaCl_2–MgCl_2$ species (quoted in Burley *et al.* 1989) whilst the presence of other dissolved anionic species (such as sulphate) will raise the temperature of the eutectic (Crawford 1981). The T_e behaviour is thus consistent with the presence in the inclusions of a complex chloride brine containing variable proportions of sulphate. Ice melting temperatures translate to a range of salinities of between 13 and 19 wt% in NaCl equivalents.

Distribution of hematite reduction, cements and mineralization

Relationship to faults

The sulphide mineralization, barite cementation and hematite reduction are spatially associated with the WNW–ESE faults, and to a lesser extent, with the N–S oriented faults. Sulphide mineralization, and also the remobilized secondary copper mineralization, tend to be concentrated in footwalls of WNW–ESE faults, although this relationship is not exclusive.

Both barite and sulphide cements are concentrated adjacent to the WNW–ESE faults and the associated CSBs, with the concentration of mineralization increasing dramatically in the close proximity of the faults. Abundant sulphide mineralistion is restricted to within 2 m of the fault planes, particularly where shale smears or clusters of CSBs are developed along or close to the fault. This relationship is displayed very well in the Engine Vein Mine, where, in the vicinity of Ring Shaft, the 'Engine Vein' mineralization is still exposed (Figs 5 and 9). Here massive, replacive galena and barite cements are concentrated against the Engine Vein Fault in a 1.2 m wide zone in the footwall. Neither the sulphide nor barite cements cross-cut small faults or the main Engine Vein Fault; they are always concentrated adjacent to it, giving rise to a banded, vein-like appearance to the mineralization (Figs 9a,b). A comparable relationship is seen in association with the Middle Fault in Wood Mine. Here many of the CSBs in both the footwall and hangingwall are extensively mineralized with pink barite cement. Galena and barite are clearly concentrated as zones of cementation, several centimetres wide, against the CSBs in the footwall (Fig. 9f).

The distribution of copper carbonate mineralization is more widespread than that of the sulphides. However, the copper carbonates are a secondary remobilization of former copper sulphides. It is probable that the original copper sulphides were concentrated around the WNW–ESE faults in the same way as the remaining sulphide cements are and that their present distribution is related to the movement of shallow groundwaters responsible for the oxidation of the sulphides.

Relationship to the structure of the Alderley Horst

The regional spatial distribution of sulphides, barite and hematite reduction across the Alderley Horst is shown in Fig. 13. Most of the mineralization is concentrated at the crest of the horst, although barite cementation and hematite reduction are developed along the Whitebarn and Kirkleyditch faults. Unfortunately, there is no exposure to document a link between the mineralization at Kirkleyditch and that associated with the crest of the Alderley Horst. Sulphide mineralization is obviously spatially associated with the dominant WNW–ESE faults and the Kirkleyditch Fault, but does not appear to be developed west of the Whitebarn Fault. Hematite reduction, quartz and barite cementation have a more widespread occurrence than sulphide mineralization, although the distribution of barite cementation and hematite leaching are also clearly spatially related to faults. The barite mineralization is not restricted to areas of hematite leaching, although generally, barite and hematite leaching show an intimate spatial relationship. At the level of the Alderley Horst

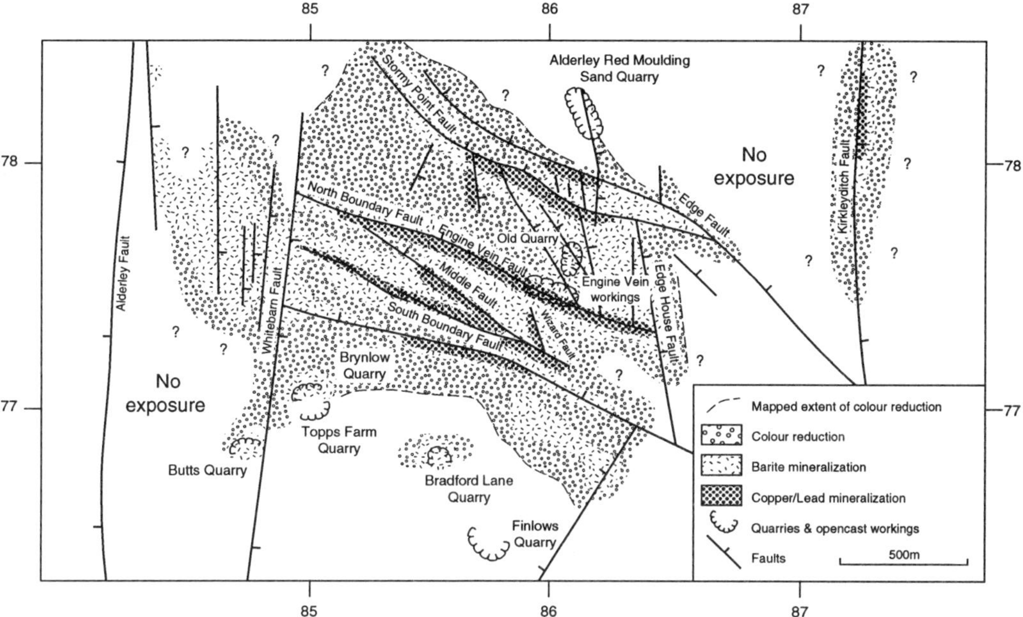

Fig. 13. Map showing the distribution and extent of mineralization around the crest of the Alderley Horst based on exposures at the surface and underground in mine workings. Boundaries are approximate, dashed where uncertain and unlined where speculative. Areas between Stormy Point and the Engine vein Fault and north of Brynlow Quarry are poorly exposed. Continuity of colour reduction here is inferred from adjacent outcrops. The distribution of Pb–Cu mineralization is based on the present distribution of sulphides and copper carbonates and inferred from the extent of cavernous mine workings.

structure exposed at outcrop, barite cementation and hematite reduction extend over many tens of metres away from the faults.

The mineralization is developed at the top of the Sherwood Sandstone Group in a tilted fault block structure. Evidence of the extent of mineralization at structurally higher levels is provided in the mine workings further south down dip on the Alderley Horst. As the sediments dip to the southwest at 15°, successively stratigraphically higher sediments are encountered to the south. The most intensely mineralized units of the Helsby Sandstone formation occur close to the base of the MMG and beneath the many local seals provided by mudstone horizons in the Engine Vein Conglomerates, Wood Mine Conglomerates, West Mine Sandstones and Brynlow Conglomerates. Vertical exposure in the mines indicates that the intensity of mineralization decreases at structurally lower levels on the WNW–ESE faults, thus precluding them as conduits for the vertical migration of reducing, mineralizing fluids. This is most readily documented in the Engine Vein Mine, where in the Hough Level in the Wilmslow Sandstone Formation virtually all mineralization and colour leaching of the sandstones are absent despite the presence of faults of significant

displacement at that depth. Only minor amounts of barite cementation and colour reduction are present in these lower levels, 50 m from the present surface. Projecting the base of the lowest formation of the MMG above the Alderley Horst indicates that the Engine Vein Fault at this level is ~180 m from the base of the regional seal that provided a cap for the fault block structure, giving a maximum thickness for the extent of mineralization.

Discussion

Reduction of SSG sandstones

In all onshore UK Permo-Triassic basins the SSG is a red bed sequence in which the reddening is the result of the early precipitation of authigenic iron oxides (Burley 1984). With time, these age to red coloured hematite (Walker 1967). As a consequence of early diagenesis and shallow burial, the SSG was thus completely reddened under oxidative conditions. The extensive hematite reduction in the SSG at the crest of the Alderley Horst indicates that the horst has been the focal point of migrating reducing fluids. Reduction of other red bed sandstone sequences has also been attributed to the migration of reducing fluids, which are commonly inferred or documented to be associated with hydrocarbons (e.g. Glennie *et al.* 1978; Elmore *et al.* 1989; Muchez *et al.* 1992; Lee and Bethke 1994). The reaction can be summarized:

$$CH_2O + 2Fe_2O_3 \rightleftharpoons CO_2 + H_2O + 4FeO$$

$$\updownarrow$$

$$HCO_3^- + H^+$$

The reduced iron can be either transported as solute or be incorporated in an authigenic mineral local to the site of reduction, such as ferroan carbonate (bicarbonate is a reaction product), or in the presence of dissolved sulphate, iron sulphide. Hydrocarbons are recorded at Alderley Edge, from Doc Mine, Stormy Point (Mohr 1964; Thompson 1970; Braithwaite 1994) and Engine vein (Braithwaite 1983; Thompson 1991), and ferroan carbonate is a late, post-hematite reduction cement, indicating that such a reaction could explain the mineralization and diagenetic assemblage.

Alternatively, in the absence of hydrocarbons, organic acids in the migrating fluid may have contributed to the reduction of hematite and precipitation of carbonates (Surdam *et al.* 1989):

$$2H_2O + CH_3COOH + 8Fe^{3+} \rightarrow$$
$$8Fe^{2+} + 2CO_2 + 8H^+$$

In either case, the volume of sandstone which has been reduced provides an indication of the scale of fluid movement around the Alderley Horst. On the basis of hematite reduction exposed at outcrop and in the mines, the mineralization developed over an area of 2.5 km by 1.8 km, and through an average thickness of sandstone of 100 m, representing a mineralized volume of 0.45 km^3, the scale of a moderate-sized hydrocarbon reservoir. The distribution of the hematite reduction indicates that the reducing fluid was channellized along the highly permeable horizons in the Helsby Sandstone Formation and was ponded at the crest of the Alderley Horst.

Fluid types and sources

The nature and distribution of the authigenic mineral assemblage, sulphide mineralization and hematite reduction indicate that at least two distinct fluids were involved in their precipitation, and that the sulphide mineralization probably formed as a result of fluid mixing.

Some constraints can be placed on the nature of these fluids and their sources. One fluid was a sulphate-rich chloride brine that was relatively oxidizing. According to the fluid inclusion data available from barite, this was a moderately to highly saline chloride brine (to 19 wt%NaCl equivalents) of a complex ionic nature, probably containing NaCl–CaCl$_2$–MgCl$_2$–SO$_4^{2-}$. This brine was derived from evaporitic units in either the Mercia Mudstone Group or Manchester Marl sediments as indicated by the high salinities required. Such an inference is consistent with the sulphur isotope data reported from barite cements at Alderley Edge by Naylor *et al.* (1989) which indicate an origin from marine evaporites.

The second fluid had to be capable of reducing hematite and sulphate, and either carrying base metals from an external source or leaching them on route through sandstones of the SSG. To be able to produce widespread hematite reduction, the reducing fluid was associated with either organic acids or hydrocarbons (Machel 1995). There is insufficient data to further constrain the characteristics of this fluid, although it was most probably derived from mudstones rich in organic matter. This suggests a Carboniferous source, such as marine Namurian or Westphalian mudstones present beneath the Permo-Triassic sequence in many parts of northern Cheshire.

Timing and conditions of mineralization

The almost complete absence of deformation affecting the mineralization indicates that barite and sulphide cementation took place after faulting. Petrographic textures further indicate that barite and sulphide precipitation also post-date quartz cementation. The mineralization therefore took place during considerable burial, either in the late Mesozoic or early Tertiary. Even though the burial history of the Cheshire Basin is poorly constrained, apatite fission track analysis (AFTA) indicates that the Helsby Sandstones at Alderley Edge attained maximum temperatures of 80–100°C in the early Tertiary, some 65 Ma ago (Lewis *et al.* 1992). Given a normal average palaeogeothermal gradient of 35°C km^{-1} and an ambient temperature of 10°C, the maximum burial depth for the Helsby Sandstone at Alderley Edge is some 2.5 km. Estimates of maximum burial based on the projection of stratigraphic thicknesses from eastern England across to the Cheshire Basin support attainment of such maximum burial depths (Burley 1987).

Fluid inclusion homogenization temperature (T_h) data from late calcites at Alderley Edge indicate a minimum trapping temperatures of 60–80°C, uncorrected for the effect of pressure on T_h at the time of entrapment (Naylor *et al.* 1989). The presence of two phase fluid inclusions in barite cements reported here also indicate a minimum temperature of 50–60°C. This temperature minimum is further supported by the low iron content of the sphalerite in the sulphide assemblage, which also provides a minimum temperature of precipitation of 60°C (Ixer and Vaughan 1982).

Although not entirely unequivocal, the available data are consistent with the barite–sulphide mineralization forming between 60 and 100°C during either late Mesozoic burial or Tertiary uplift at depths of approximately 2 km as a result of the mixing of two basinal fluids, one a saline and sulphate-rich chloride brine, and the other a reducing fluid associated with either organic acids or hydrocarbons.

Fluid migration pathways

The primary ore deposit, comprising barite and sulphide mineralization, together with the associated hematite reduction, is demonstrably spatially fault-related. What is more difficult to ascertain is whether the faults have provided a preferential vertical conduit for fluid migration, have facilitated cross-formational fluid flow or have acted as a barrier to fluid migration through the sandstones, or a combination of these. At least two fluid migration pathways are suggested, schematically illustrated in Fig. 14; one for the sulphate brine and one for the reducing fluid. Saline, sulphate-rich brines may have been sourced either via cross-formational flow from the downthrown MMG sediments on either side of the Alderley Horst or from the Manchester Marls in the same downthrown blocks. This latter source requires a vertical migration of some 700 m. An even greater vertical migration distance, in the region of 800 m, is required for the reducing fluid if it was sourced from Carboniferous marine mudstones beneath the Alderley Horst. An alternative

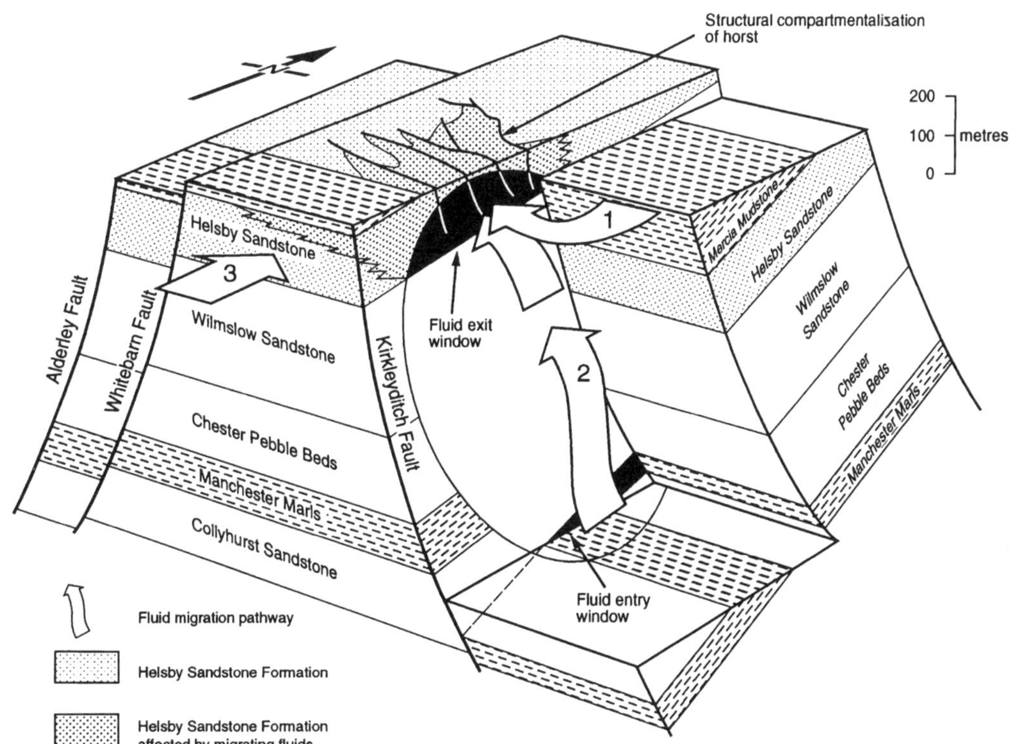

Fig. 14. Conceptual model of the probable fluid migration pathways responsible for the mineralization at Alderley Edge. Migration pathway 1 is for cross-formational flow of a saline, sulphate-rich chloride brine expelled from the MMG in the Kirkleyditch Fault Block. Pathway 2 is for overpressure valving of fluids from deeper stratigraphic horizons in the Kirkleyditch Fault Block. Example of a sulphate-rich chloride brine sourced from the Mancheter Marls is shown, but a similar vertical migration route is possible for a reducing fluid expelled from the underlying Carboniferous. Pathway 3 represents a regional flow of reducing fluid through the uppermost sandstones of the Sherwood Sandstone Group, sourced from the Carboniferous juxtaposed across the Red Rock Fault to the south. In all cases the fluids are trapped and ponded on the crest of the Alderley Horst. Fluid entry window and fluid exit window concepts based on Knipe (1993). Vertical exaggeration ×2.

possibility for the reducing fluid involves up-dip migration onto the crest of the Alderley Horst from the south, where the Sherwood Sandstone is fault juxtaposed against Carboniferous sediments across the Red Rock fault near Macclesfield some 6 km away.

None of the cements associated with faults show evidence of post-precipitation deformation, such as brecciation or cross-cutting vein relationships. The mineralization is thus post-faulting, an interpretation supported by the common observation that the highest concentrations of sulphide and barite cementation are developed against faults. Seismic pumping, in which faulting drives fluid movement (Sibson *et al.* 1975), cannot therefore be the mechanism responsible for fluid migration, although it could explain the movement of oxidative sulphate-rich chloride brines onto the structure from either the MMG or Manchester Marls early in the burial history of the Alderley Horst.

As the mineralization essentially overprints the faulting, and if faults are part of the fluid migration story, seismic valving, in which fluid overpressure within an aquifer juxtaposed against the fault drives fluid movement (Sibson 1981), is a more likely mechanism. It is possible that fluids were channellized by faults outside of the immediate area of the Alderley Horst and gained access to the structure via the horst-bounding faults. However, the Alderley Fault is not known to be mineralized, and mineralization on the crest of the structure appears to die out to the west of the Whitebarn Fault. The fluid migration pathway from the downthrown Cheshire Basin MMG sequence to the west implied by Naylor *et al.* (1989, their fig. 10b) cannot therefore be the vertical conduit for the migration of the reducing fluid. The Kirkleyditch Fault zone, however, is extensively mineralized (Warrington 1980) and has been mined in the past (Carlon 1979). Fluids probably either moved up or across the Kirkleyditch Fault as a consequence of overpressure valving and migrated laterally into juxtaposed permeable sandstone layers at the crest of the horst. Such cross-formational flow was baffled by the numerous subordinate WNW–ESE faults which cross the horst resulting in compartmentalization of the mineralizing fluids within fault-defined blocks (Fig. 14). The concentration of sulphide mineralization in the footwall crests of individual structural compartments defined by the WNW–ESE faults on the Alderley Horst is the result of this on-structure fluid flow (see Fig. 14). Redistribution of the fluids and the resultant WNW–ENE fault-related mineralization may have taken place in response to changes in pore fluid pressure controlled by the Kirkleyditch Fault, when the subsidiary WNW–ESE faults could act as local valves and allow cross-formational fluid flow. However, the WNW–ESE faults did not contribute to the vertical fluid migration because the intensity of hematite reduction and sulphide mineralization decreases downwards along them.

The influence of sub-seismic faults

CSBs and cataclastic slip zones (CSZs, Fowles and Burley 1993) are associated with all the faults that cut the SSG on the Alderley Horst. The spatial association of CSBs with faults of metre scale throw suggests that they result from brittle deformation of high porosity sandstones during fault displacement. The data presented here support the observation that CSBs are a common component of faulted high porosity sandstones (Aydin and Johnson, 1983; Fowles & Burley 1994; Knott 1993). Similar features are recognized in SSG sediments of the EISB (Edwards & Williams 1995; Beach *et al.* this volume).

However, the abundance of cataclastic deformation is not obviously related to the amount of displacement on related fault surfaces. On simple, straight faults, the abundance of CSBs is similar on faults with 10 m-scale displacement to those with 100 m-scale displacement. Even though the damage zone width is difficult to measure in these sandstones, it is also not obviously related to the amount of displacement on related faults. From qualitative observations the intensity of CSB development and width of damage zones appears related to the presence of bends, splays and breaches in faults rather than amount of displacement (cf. Knott, 1994).

The reduction in grain size, sorting and porosity within CSBs is likely to significantly reduce their permeability compared to the undeformed sandstone host. As CSBs are zones of reduced permeability they will be baffles to fluid flow. After deformation, fluid movement along or across the CSBs will be impeded because of their greatly reduced permeability. This inference is supported by the concentration of barite and sulphide cements in the footwalls of CSBs/CSZs. However, CSBs are, by their definition, thin features often restricted to a few millimetres in thickness and of finite length. Given high pore fluid pressures, single CSBs are likely to be leaky and allow both cross-CSB flow and tortuous flow around the ends of CSBs. Leakage of fluids through CSBs over geological time provides an explanation for the occurrences of barite and sulphide cement fabrics that appear to be concentrated on both sides of CSBs.

Despite the sand-dominated nature of the Sherwood Sandstone sequence at Alderley Edge, many of the WNW–ESE mineralized faults have thin shale smears along the fault plane. Barite and sulphide mineralization are most concentrated against these shale smears. Even though thin (a few centimetres in thickness), these shale smears are clearly major barriers to lateral fluid flow across the mineralized faults and are a major factor in the local concentration of cementation .

Implications for porosity modification in tilted fault block structures in the EISB

As the Alderley Horst is analogous to tilted fault block structures in the EISB, observations and interpretations drawn from this study have implications relating to porosity modification in structural highs that are hydrocarbon and fluid traps in the EISB. Porosity modification and thus fluid flow within reservoir sandstones of the EISB must be influenced by the distribution of permeability contrasts produced by faulting and fault-related cementation.

Damage zones caused by the presence of CSBs and associated sub-seismic scale faults are permeability baffles and as such will contribute to the compartmentalization of hydrocarbon reservoirs. Such fault compartmentalisation is likely to be beneath the limit of seismic resolution, but can be predicted from analogue studies such as reported here. The WNW–ESE faults on the Alderley Horst with displacements of <10 m for example, develop damage zones 2 m in width characterized by extensive CSB development, fault-related cementation and resultant porosity reduction. Such small scale faulting in hydrocarbon reservoirs on structural highs in the EISB is likely to have similar damage zones and may be associated with fault-related diagenesis.

Fault-related diagenesis at Alderley Edge records fluid flow in a structural high at the margin of the Cheshire Basin where a sulphate-rich chloride brine mixed with a reducing fluid, probably associated with hydrocarbons. The authigenic mineral assemblage present at the crest of fault block structures is dependent upon the composition of the pore fluids that have access to the fluid entry window onto the fault that define the structures and mix with fluids present on-structure (Fig. 14). Given the widespread development of evaporites in the MMG of the EISB, the occurrence of several large hydrocarbon accumulations (see Fig. 1) and the common presence of H_2S associated with hydrocarbons (Calder 4500 ppm, Asland 250 ppm, Lennox 1300 ppm and Douglas 1500 ppm; Haig et al., this volume; Yaliz, this volume; Blow and Hardman, this volume) then fault-related sulphide mineralization is highly probable in some of the fault-bounded structural highs within the basin. Its development should be restricted to fault blocks

where a sulphate-rich chloride brine has mixed with either a reducing brine associated with hydrocarbons or H_2S.

The burial diagenetic assemblage in hydrocarbon fields of the EISB includes quartz, ferroan dolomite–ankerite, illite, pyrite, anhydrite and ferroan calcite cements (Burley 1984; Bushell 1986; Stuart & Cowan 1991; Greenwood & Habesch, this volume). Some of these cements are demonstrably fault-related in the 110/2-1 and 110/3a-A3 wells (see Cowan et al. 1993) and must contribute to a reduction in porosity and permeability of the fault zones.

Conclusions

- Hematite reduction, barite and ferroan calcite cementation and sulphide mineralization developed in SSG sandstones at the crest of the Alderley Horst are spatially related to some of the faults that define and cross-cut the horst.
- The mineralization forms an integral part of the burial diagenetic sequence and occurs as pore-filling cements that are concentrated against faults and are unaffected by brecciation or cross-cutting veining, indicating formation after fault displacement.
- The mineralization does not appear to be significantly developed west of the Whitebarn Fault, and its abundance decreases with depth on the WNW–ESE faults. The Alderley Fault which defines the western margin of the horst and the WNW–ESE faults with which the mineralization is spatially associated, thus appear to be precluded as vertical migration pathways for the mineralizing fluids.
- The distribution of cementation and mineralization is consistent with the movement of a reducing fluid either vertically up or across the Kirkleyditch Fault, with cross-formational flow onto the crest of the structure. There it mixed with an oxidizing sulphate-rich chloride brine and was baffled against pre-existing permeability barriers.
- It is the WNW–ESE oriented faults within the Alderley Horst that created fluid compartments in which mineralization was preferentially localized and against which barite and sulphide cements were concentrated.
- Small scale deformation structures below the limit of seismic resolution (such as shale smears and CSBs) are a major influence on the concentration of mineralization against the WNW–ESE oriented faults.
- Clustering of CSBs around faults with displacements of as little as a metre produces damage zones up to 4 m in width. There is no obvious relationship between the displacement on faults

(range 10–100 m scale) and the widths of the damage zones. Rather, the width and deformation intensity of damage zones appear to be related to the presence of bends, splays and breaches in faults.

This research was undertaken as part of a PhD study by JR when both authors were at the University of Manchester. It was funded by BG Exploration & Production whose logistical and financial support is gratefully acknowledged. Thanks are also extended to the National Trust and its officers for kindly granting permission to conduct research on their land and take samples. Access to the mines of Alderley Edge was generously facilitated by the Derbyshire Caving Club. In particular, our sincere thanks are extended to Doug Kidd who repeatedly gave of his time and experience in guiding us around the mines; without him, sampling underground would not have been possible. The daily grind of drilling in Wood Mine was eased by the help of Simon Edwards and Sebastian Lewis. Early versions of the manuscript were significantly improved by thoughtful and constructive reviews by David Thompson, Rob Cook, Simon Guscott, Harry Shaw and Stephen Habesch. We are especially indebited to David Thompson who freely shared with us some of his experience and knowledge of the Sherwood Sandstones at Alderley Edge.

References

AYDIN, A. & JOHNSON, A. M. 1983. Analysis of faulting in porous sandstones. *Pure and Applied Geophysics*, **116**, 913–930.

BEACH, A., BROWN, J. L., WELBON, A. I., McCALLUM, J. E., BROCKBANK, P. & KNOTT, S. Characteristics of fault zones in sandstones from NW England: application to fault transmissibility. *This volume*.

BETHKE, C. M. 1986. Hydrologic constraints on the genesis of the Upper Mississippi Valley Mineral District from Illinois Basin Brines. *Economic Geology*, **81**, 233–249.

—— & MARSHAK, S. 1990. Brine migrations across North America; the Plate Tectonics of groundwater. *Annual Review Earth & Planetary Science Letters*, **18**, 287–315.

BLOW, R. A. & HARDMAN, M. Calder Field appraisal well 110/79-8, East Irish Sea Basin. *This volume*.

BRAITHWAITE, R. S. W. 1983. Minerals of the Derbyshire Orefield. *The Mineralogical Record*, **14**, 15–24.

—— 1994. Mineralogy of the Alderley Edge–Mottram St. Andrew Area, Cheshire, England. *Journal of the Russell Society*, **5**, 91–102.

BURLEY, S. D. 1984. Patterns of diagenesis in the Sherwood Sandstone Group (Triassic), United Kingdom. *Clay Minerals*, **19**, 403–440.

—— 1987. *Diagenetic modelling in the Triassic Sherwood Sandstone Group of England and its offshore equivalents, United Kingdom Continental Shelf*. PhD Thesis, Hull University.

——, MULLIS, J. & MATTER, A. 1989. Timing diagenesis in the Tartan Reservoir (UK North Sea): constraints from combined cathodoluminescence microscopy and fluid inclusion studies. *Marine Petroleum Geology*, **1989**, 98–120.

BUSHELL, T. P. 1986. Reservoir geology of the Morecambe Field, *In*: BROOKS, J., GOFF, J. C., & VAN HOORN, B. (eds) *Habitat of Palaeozoic Gas in Northwest Europe*, Special Publication of the Geological Society of London, **23**, 189–207.

CARLON, C. J. 1979. *The Alderley Edge Mines*. Sherratt, 144 pp.

—— 1981. The Gallantry Bank Copper Mine, Bickerton, Cheshire, with a review of mining in Triassic Rocks of the Cheshire–Shropshire Basin. *British Mining*, **16**, 1–50.

—— & THOMPSON, D. B. 1981. Aspects of the mineralization within the Cheshire Basin. *Journal of the Geological Society, London*, **138**, 220.

CATHLES, L. M. & SMITH, A. T. 1983. Thermal constraints on the formation of Mississippi Valley-Type Lead–Zinc deposits and their implications for episodic basin dewatering and deposit genesis. *Economic Geology*, **78**, 983–1002.

CHADWICK, R. A. Fault analysis of the Cheshire Basin, NW England. *This volume*.

CLARKE, A. M. 1980. *Mineralization in the Permo-Triassic Rocks of Britain with special reference to the Cheshire Basin*. PhD Thesis, Imperial College London.

COLTER, V. 1978. Exploration for gas in the Irish Sea. *Geologie en Mijnbouwe*, **57**, 503–516.

COWAN, G., OTTESEN, C. & STUART, I. A. 1993. The use of dipmeter logs in the structural interpretation and palaeocurrent analysis of Morecambe Fields, East Irish Sea Basin. *In*: PARKER, J. R. (ed.) *Petroleum Geology of Northwest Europe: Proceedings of the 4th Conference*. The Geological Society, London, 867– 882.

CRAWFORD, M. L. 1981. Phase equilibria in aqueous fluid inclusions. *In*: HOLLISTER, L. S. & CRAWFORD, M. L. (eds) *Fluid Inclusions: Applications to Petrology*. Mineral Association of Canada Short Course Handbook **6**, 75–100.

DUNHAM, K. C. 1970. mineralization by deep formation waters: a review. *Transactions of the Institute of Mining and Metallurgy*, **B1970**, 127–138.

EDWARDS, H. E. & WILLIAMS, G. D. 1995. Reservoir characterisation and structural compartmentalisation of the Sherwood Sandstone Group of Liverpool Bay and adjacent onshore areas: tectono-diagenetic. The petroleum geology of the Irish Sea and adjacent areas: Meeting Abstracts volume.

ELMORE, R. D., McCOLLUM, R. & ENGEL, M. H. 1989. Evidence for a relationship between hydrocarbon migration and diagenetic magnetic minerals - implications for petroleum exploration. *Bulletin of the Association of Petroleum Geochemical Explorationists*, **5**, 1–17.

EVANS, D. J., REES, J. G. & HOLLOWAY, S. 1993. The Permian to Jurassic stratigraphy and structural evolution of the Central Cheshire Basin. *Journal of the Geological Society, London*, **150**, 857–870.

FOWLES, J. & BURLEY, S. D. 1994. Textural and permeability characteristics of faulted, high porosity sandstones. *Marine and Petroleum Geology*, **11**, 608–623.

GLENNIE, K. W., MUDD, G. C. & NAGTEGAAL, P. J. C. 1978. Depositional environment and diagenesis of Permian Rotliegendes sandstones in Leman Bank and Sole Pit areas of the UK, southern North Sea. *Journal of the Geological Society, London*, **135**, 25–34.

GREENWOOD, P. J. & HABESCH. S. M. Diagenesis of the Sherwood Sandstone Group in the southern East Irish Sea Basin (Blocks 110/13, 110/14 and 110/15): constraints from preliminary isotopic and fluid inclusion studies. *This volume.*

HAIG, D. B., PICKERING, S. C. & PROBERT, R. The Lennox oil and gas field. *This volume.*

HOLMES, I., CHAMBERS, A. D., IXER, R. A., TURNER, P. & VAUGHAN, D. J. 1983. Diagenetic processes and the mineralization in the Triassic of Central England. *Mineral Deposits*, **16**, 365–377.

IXER, R. A. & VAUGHAN, D. J. 1982. The primary ore mineralogy of the Alderley Edge Ore deposit, Cheshire. *Mineralogical Magazine*, **46**, 485–492.

JACKSON, D. I. & MULHOLLAND, P. 1993 Tectonic and stratigraphic aspects of the East Irish Sea Basin and adjacent areas: contrasts in their post–Carboniferous structural styles. *In*: PARKER, J. R. (ed.) *Petroleum Geology of Northwest Europe: Proceedings of the 4th Conference*, The Geological Society, London, 791–808.

——, ——, JONES, S. M. & WARRINGTON, G. 1987. The geological framework of the East Irish Sea Basin. *In*: BROOKS, J. & GLENNIE, K. (eds) *Petroleum Geology of North Western Europe*. Graham & Trotman, 191–203.

JACKSON, S. A. & BEALES, F. W. 1967. An aspect of sedimentary basin evolution: the concentration of MVT ores during late stages of diagenesis. *Bulletin of Canadian Petroleum Geology*, **15**, 383–433.

KNIPE, R. J. 1993. The influence of fault zone processes on fluid flow and diagenesis. *In*: HORBURY, E. D. & ROBINSON, A. G. (eds) *Diagenesis and Basin Development*. Bulletin of the American Association Petroleum Geologists, Studies in Geology, **36**, 137–154.

——, COWAN, G. & BALENDRAN V. S. 1993. The tectonic history of the East Irish Sea Basin with reference to the Morecambe fields. *In*: PARKER, J. R. (ed.) Petroleum *Geology of Northwest Europe: Proceedings of the 4th Conference*. The Geological Society, London, 857–866.

KNOTT, S. D. 1994. Fault zone thickness versus displacement in Permo-Triassic sandstones of NW England. *Journal of the Geological Society, London*, **151**, 17–25.

LAND, L. S. 1987. The major ion chemistry of saline brines in sedimentary basins, *In*: BANAVAR, J. R., KOPLIK, J. & WINKLER, K. W. (eds) Physics and chemistry of porous media II: Ridgefield, Connetticut. *American Institute of Physics Conference Proceedings*, **154**, 160–170.

LEE, M. K. & BETHKE, C. M. 1994. Groundwater flow, late cementation and petroleum accumulation in the Permian Lyons Sandstone and its role in oil accumulation, Denver Basin. Bulletin of the American Association of Petoleum Geologists, **78**, 217–237.

LEWIS, C. L. E., GREEN, P. F., CARTER, A. & HURFORD, A. J. 1992. Elevated K/T palaeotemperatures throughout Northern England: three kilometres of Tertiary exhumation? *Earth and Planetary Science Letters*, **112**, 131–145.

MACHEL, H. G. 1995. Magnetic mineral assemblages and magnetic contrasts in diagenetic environments - with implications for studies of palaeomagnetism, hydrocarbon migration and exploration. *In*: TURNER, P. & TURNER, A. (eds) *Palaeomagnetic Applications in Hydrocarbon Exploration and Production*. Geological Society, London, Special Publication, **98**, 9–29.

MANNING, D. 1990. The ore deposits of Alderley Edge. *Amateur Geologist*, **18**, 31–35.

MEADOWS, N. S. & BEACH, A. 1993. Structural and climatic controls on facies distribution in a mixed fluvial and aeolian reservoir: the Triassic Sherwood Sandstone in the Irish Sea. *In*: NORTH, C. P. & PROSSER, D. J. (eds) *Characterisation of Fluvial and Aeolian Reservoirs*, Geological Society, London, Special Publication, **73**, 247–264.

METCALFE, R., ROCHELLE, C. A., SAVAGE, D., HOOKER, P. J. & GALE, I. N. 1993. Hydrochemical modelling of red bed diagenesis and heavy metal mineralization. *In*: PARNELL, J. (ed.) *Proceedings of Geofluids '93 Conference*, 276–279.

MONTACER, M., DISNAR, J. R., ORCEVAL, J. J. & TRICHET, J. 1987. Relationship between Zn–Pb ore and oil accumulation processes: Example of the Bou Grine Deposit (Tunisia). *Organic Geochemistry*, **13**, 423–431

MOHR, P. A. 1964. On the copper mineralized sandstones of Alderley Edge, England and Chercher, Ethiopia, and the problem of their genesis: an essay on Red Bed copper deposits. Contributions to Geophysics at the faculty of Science, Haile Selasse University, Addis Ababa. Series A., 4.

MUCHEZ, P. H., VIAENE, W., & DUSAR, M. 1992. Diagenetic control on secondary porosity in flood plain deposits: an example of the Lower Triassic of northeastern Belgium, *Sedimentary Geology*, **78**, 285–298.

NAYLOR, H., TURNER, P., VAUGHAN, D. J., BOYCE, A. J. & FALLICK, A. E. 1989. Genetic studies of red bed mineralization in the Triassic of the Cheshire Basin, northwest England. *Journal of the Geological Society, London*, **146**, 685–699.

PRICE, P. E., KYLE, J. R. & WESSEL, G. R. 1983. Salt dome related zinc-lead deposits. *In*: KISVARSANYI, G. (ed.) *Proceedings, International Conference on Mississippi Valley Type Lead–Zinc Deposits*, 558–571.

QUING, H. & MOUNTJOY, E. W. 1994. Formation of coarsely crystalline, hydrothermal dolomite reservoirs in the Presqu'ile Barrier, Western Canada sedimentary basin. *Bulletin of the American Association of Petroleum Geologists*, **78**, 55–77.

ROWE, J. P. 1994. *Evolution of Palaeofluid Flow in the East Irish Sea Basin and its Margins*. PhD thesis, University of Manchester.

——, BURLEY, S. D., GAWTHORPE, R. G., COWAN, G. & HARDMAN, M. 1993. Palaeo-fluid glow in the East Irish Sea Basin and its margins. *In*: PARNELL, J. (ed.) *Proceedings of Geofluids '93 Conference*, 358–362.

SIBSON, R. H. 1981. Fluid flow accompanying faulting: field evidence and models. *In*: SIMPSON, D. W. & RICHARDS, P. G. (eds) *Earthquake Prediction: An International Review*. American Geophysics Union, Maurice Ewing Series, **4**, 145–172.

——, MOORE, J. M. & RANKIN, A. H. 1975. Seismic pumping – a hydrothermal fluid transport mechanism. *Journal of the Geological Society, London*, **131**, 653–659.

STRONG, G. E. & PEARCE, J. M. 1995. Carbonate spheroids in Permo-Triassic sandstones of the Sellafield area, Cumbria. *Proceedings of the Yorkshire Geological Society*, **50**, 209–211.

STUART, I. A. 1993. The geology of the North Morecambe Gas Field, East Irish Sea Basin. *In*: PARKER, J. R. (ed.) *Petroleum Geology of Northwest Europe: Proceedings of the 4th Conference*. The Geological Society, London, 883–895.

—— & COWAN, G. 1991. The South Morecambe Field, Blocks 110/2a, 110/3a, 110/8a, UK East Irish Sea. *In*: ABBOT, I. L. (ed.) *United Kingdom Oil and Gas Fields, 25 Years Commemorative Volume*. Geological Society Memoir, **14**, 527–541.

SURDAM, R. C., CROSSEY, L. J., HAGEN, E. S. & HEASLER, H. P. 1989. Organic–inorganic interactions and sandstone diagenesis. *Bulletin of the American Association of Petroleum Geologists*, **73**, 1–23.

SVERJENSKI, D. A. 1984. Oil field brines as ore-forming solutions. *Economic Geology*, **79**, 23–37

THOMPSON, D. B. 1970. Itineraries X and XI, Alderley Edge. *In*: EAGAR, R. M. C. & BROADHURST, F. M. (eds) *The Area around Manchester*. Geologists Association Guide **7**, 39–51.

—— 1985. Field excursion guide to the Permo-Triassic of Cheshire and the adjacent basins. Poro-Perm Geochem, Chester, 172 pp.

—— 1991. Itineraries VIII and IX, Alderley Edge. *In*: EAGAR, R. M. C. & BROADHURST, F. M. (eds) *Geology of the Manchester Area*. Geologists Association Guide No. **7**, 57–74.

WALKER, T. R. 1967. Formation of red beds in modern and ancient deserts. *Geological Society of America Bulletin*, **78**, 353–368.

WALTER, L. M., STUEBER, A. M. & HUSTON, T. D. 1990. Br–Cl–Na systematics in Illinois basin fluids: constraints on fluid origin and evolution. *Geology*, **18**, 315–318

WARRINGTON, G. 1965. The metalliferous mining district of Alderley Edge, Cheshire. *Mercian Geologist*, **1**, 111–129.

—— 1980. The Alderley Edge mining district. *Amateur Geologist*, **8**, 4–13.

—— & THOMPSON, D. B. 1971. The Triassic Rocks of Alderley Edge. *Mercian Geologist*, **4**, 69–72.

——, AUDLEY-CHARLES, M. G., ELLIOT, R. E. *ET AL*. (1980) *A Correlation of Triassic Rocks in the British Isles*. Geological Society, London, Special Report, **13**, 78pp.

YALIZ, A. The Douglas oil field. *This volume*.

Diagenesis of the Sherwood Sandstone Group in the southern East Irish Sea Basin (Blocks 110/13, 110/14 and 110/15): constraints from preliminary isotopic and fluid inclusion studies

P. J. GREENWOOD[1,2] & S. M. HABESCH[1,3]

[1] *Geochem Group Limited, Chester Street, Chester CH4 8RD, UK*

[2] *Present address: NAM BV, Postbus 28000, 9400HH, Assen, The Netherlands.*

[3] *Present address: Advanced Geochemical Systems Ltd, 40 Springfield Drive, Crewe, CW2 6RA, UK.*

Abstract: Diagenesis in reservoir sandstones from the southern East Irish Sea Basin has resulted in the precipitation of multiple generations of quartz and carbonate cements. Other cements are volumetrically minor and do not affect sandstone reservoir quality, which overall retains high porosities and permeabilities. Stable isotope data suggest that precipitation of caliche calcite cement during early diagenesis occurred at low temperature (*c.* 30°C) from meteoric porewaters (−3‰ SMOW). By contrast $\delta^{34}S$ data for evaporitic gypsum/anhydrite from the Ormskirk Sandstone Formation and Mercia Mudstone Group suggest that locally, an early marine sulphate source was available following deposition. Late burial anhydrite and barytes cements were derived through dissolution and reprecipitation of the early sulphates, according to stable isotope data. Fluid inclusion data for burial diagenetic cements record elevated temperatures (>100°C) for authigenic quartz, carbonate and sulphate minerals, which precipitated from concentrated brines. Stable isotopes indicate that porewaters enriched in ^{18}O evolved during burial and late carbonates incorporated increasing quantities of carbon derived from organic maturation at depth. An episode of hydrocarbon migration during burial, predating the influx of hydrocarbons currently found, is preserved as bitumen in the sandstones. Further cementation postdated this episode, including late calcite cements postdating the migration of the current hydrocarbon charge. Petrographic data indicate that the elevated temperatures recorded in burial cements were not achieved entirely through depth of burial but probably reflect periods of elevated heatflow, as a result of the circulation of hot fluids through the reservoir sandstones.

Hydrocarbon exploration in the southern part of the East Irish Sea Basin (EISB) has led to the discovery of several important accumulations of gas and oil within the reservoir sandstones of the Triassic Ormskirk Sandstone Formation. The fields discovered to date in the study area are illustrated in Fig. 1, comprising the gas fields Hamilton, Hamilton North and Lambda, plus the Douglas oil field, in block 110/13; and the Lennox Field containing oil and gas in block 110/15 and extending into 110/14. In this paper the diagenetic history of the Ormskirk Sandstone Formation in this area is discussed in detail, together with some aspects of diagenesis in the overlying Mercia Mudstone Group. A number of previous studies in the EISB have included data on diagenesis in the reservoir sandstones. These have mainly concentrated on the Morecambe Field to the north of the study area in blocks 110/2a, 110/3a and 11018a (e.g. Colter & Ebbern 1978; Bushell 1986; Stuart & Cowan 1991). Much of the work has dealt with authigenic illite in the sandstones (e.g. Macchi *et al.* 1990), which occurs below a mappable datum (the 'Top Platy Illite Surface'), interpreted as representing a hydrocarbon–water palaeocontact. Other authors have studied diagenesis of the Sherwood Sandstone Group both in the EISB and onshore (Burley 1984; Knox *et al.* 1984; Strong 1993). Hardman *et al.* (1993) discussed diagenetic models in the context of the burial history of the sequence, utilizing petroleum geochemistry and fluid inclusion data from two EISB wells.

Stratigraphy and facies

The cored sequences in the study wells comprise the Ormskirk Sandstone Formation of the Sherwood Sandstone Group, the upper part of which forms the reservoir below the Mercia Mudstone Group seal (Jackson *et al.* this volume). In addition sidewall cores provided samples of the lower part of the Mercia Mudstone Group. The sedimentology of the Ormskirk Sandstone Formation has been described in detail by previous authors (e.g. Cowan 1993; Meadows & Beach 1993). The major depositional systems recognized from core studies are fluvial, aeolian and wet sandflat systems, composed of a range of facies

From Meadows, N. S., Trueblood, S. P., Hardman, M. & Cowan, G. (eds), 1997, *Petroleum Geology of the Irish Sea and Adjacent Areas,* Geological Society Special Publication No. 124, pp. 353–371.

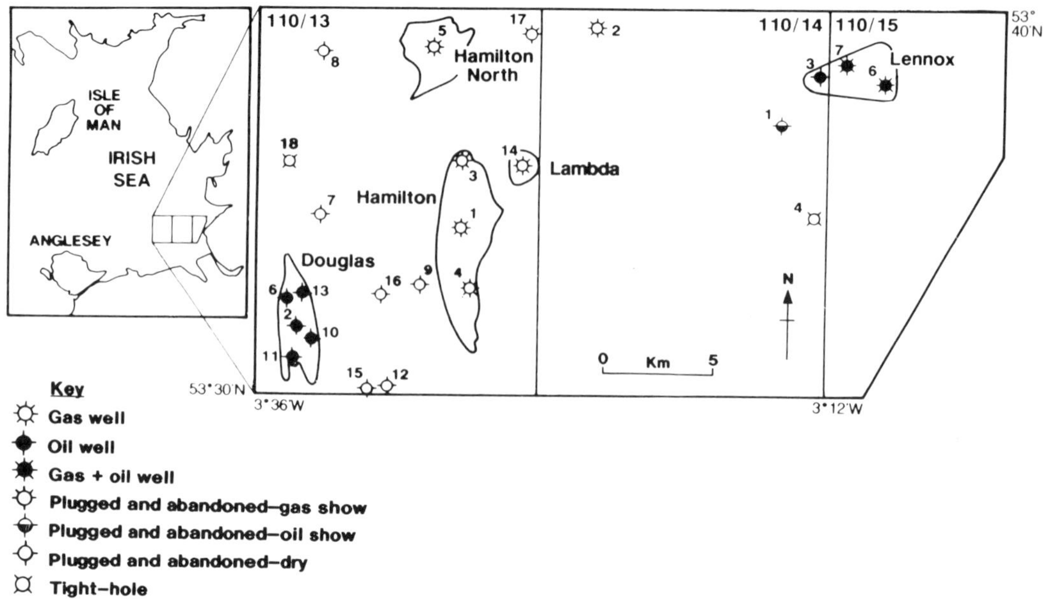

Fig. 1. Map of the study area displaying the location of wells drilled and fields discovered, to date.

types and consistent with deposition in a semi-arid continental environment.

Materials and methods

Data for this study has come from analysis of core material from five wells (110/13-13, 110/13-14, 110/14-1, 110/15-6Z and 110/15-7), plus sidewall cores from two further wells where no core material was available (110/13-15 and 110/13-16). A total of 105 thin sections were analysed from the cored intervals of the wells and quantified by modal analysis (200 points per sample). The thin sections were blue resin impregnated and stained for K-feldspar and carbonates. Modal data from the sidewall core samples is not included due to sample damage limiting the reliability of the data.

Scanning electron microscopy (SEM) analyses of 35 samples were undertaken on a Philips 515 SEM equipped with an EDAX 9100 microanalyser, providing detailed information on authigenic mineral textures and relationships.

X-ray diffraction (XRD) analyses of 35 whole rock and <2 μm clay fractions were performed on a Philips PW1710 diffractometer, to confirm the detrital and authigenic sandstone mineralogy.

Carbonate and quartz cements in 17 of the sandstones were further investigated under cathodoluminescence (CL), using a Technosyn 8200 Mk 2 Cold Cathodoluminescence system.

Isotope analyses were undertaken in order to constrain the pore fluid compositions and solute sources of the major cements during burial diagenesis. All isotope analyses were carried out at the Scottish Universities Research and Reactor Centre (SURRC) using standard procedures. All carbonate and sulphate isotope analyses were performed on powdered whole rock samples, following thin section and XRD screening to determine that the cement was monominerallic. Carbonate cements were analysed for stable carbon and oxygen isotopes ($\delta^{18}O$ and $\delta^{13}C$) in 52 samples (30 dolomite and 22 calcite) from six wells, of which 42 were core samples and 10 sidewall cores. CO_2 was extracted from calcite by reaction with phosphoric acid at 25°C (McCrea 1950) and at 100°C for dolomites. The CO_2 gas was purified prior to analysis in a VG Isogas SIRA 10 mass spectrometer. $\delta^{18}O$ and $\delta^{13}C$ values are presented as parts per mil (‰) relative to the PDB and SMOW standards as appropriate. Precision for monominerallic carbonate stable isotope analyses ($\delta^{18}O$ and $\delta^{13}C$) is ±0.2‰ (1σ). In addition the radiogenic strontium isotope signatures of 17 of the carbonates (10 dolomite and seven calcite) in five wells were determined. Strontium was separated using standard cation-exchange chromatography and isotope dilution, and $^{87}Sr/^{86}Sr$ ratios were measured on a VG Isomass 54E mass spectrometer, precision for $^{87}Sr/^{86}Sr$ in carbonates is ±0.001. Sulphur stable isotopes ($\delta^{34}S$) in anhydrite and barytes cements were determined for 13 samples, from seven wells. Sulphate samples were prepared using the method of Coleman and Moore (1978), the results are presented as $\delta^{34}S$ values in parts per mil (‰) relative to the CDT standard. Precision for sulphate $\delta^{34}S$ analyses is ± 0.2‰ (1σ). The oxygen stable isotope composition of quartz overgrowth cements in nine samples from three wells was measured using the technique described by Brint *et al.* (1991), after Milliken *et al.* (1981). The method involves HF dissolution and $\delta^{18}O$ analysis of two aliquots of purified quartz sample: (i) a bulk sample of quartz grains plus overgrowths; and (ii) a sample of detrital quartz grains after HF leaching to remove the overgrowths. A mass balance calculation is then performed to derive the overgrowth $\delta^{18}O$ value from the difference, using the modal percentage of authigenic quartz in the bulk sample. The precision for quartz $\delta^{18}O$ analyses is ± 0.2‰ (1σ) and Brint *et al.* (1991) estimate that the results are accurate to within ± 0.2‰, taking the modal analysis error into account.

Fluid inclusion microthermometry was carried out on 16 samples from five wells, prepared as double polished wafers and analysed using a Fluid Inc. adapted USGS gas-flow heating-freezing system, mounted on a Leitz Metallux 3 petrological microscope. The stage was calibrated using a distilled water ice bath and fluid inclusions of known pure composition (CO_2 –56.6°C and H_2O + 374.1°C) enclosed synthetically in quartz. Analytical precision is ± 0.1°C at –56.6°C and 0°C rising to ± 1°C at +374.1°C. Measurements were made on primary biphase inclusions, hosted within various mineral cements, which did not appear to have undergone post-entrapment changes. The criteria used to assess the suitability of the fluid inclusions for analysis was: (i) to relate inclusions to probable growth zones in the crystals, following the recommendations of Goldstein & Reynolds (1994), as distinct from planar arrays of secondary inclusions; (ii) the inclusions were of similar size and consistent liquid : vapour ratios; and (iii) irregular shaped inclusions were discounted to avoid the possibility of necking down. Analyses were made of the homogenization temperature (T_h), assumed to represent the minimum trapping temperature (T_t) of the inclusion, and the last ice melting temperature (T_m), representing the depression of freezing point from which the entrapped fluid salinity was derived. The latter is expressed as weight percent NaCl equivalents (wt% NaCl) in the absence of any reliable data on the entrapped fluid chemistry. No pressure corrections were applied to the T_h values, which thus represent the minimum trapping temperature and may be erroneously low, particularly for petroleum-bearing inclusions (Goldstein & Reynolds 1994). However, if the aqueous or petroleum inclusions are saturated with respect to methane then T_h is approximately equal to T_t, and even small amounts of dissolved methane will significantly reduce the difference between T_h and T_t (Burley *et al.* 1989).

Sandstone petrography

Detrital mineralogy

The sandstones from the cored Ormskirk Sandstone Formation consist mainly of quartz arenites, subfeldspathic arenites and sublithic arenites, with rare feldspathic and lithic arenites (Fig. 2), using a classification scheme modified from McBride (1963) and Dott (1964). The detrital minerals are dominated by quartz, K-feldspar, and lithic clasts, with minor plagioclase, mica and detrital clays, plus accessory tourmaline, zirzon and apatite. There are no significant variations in sandstone composition across the study area. The different facies within each well display limited compositional variation, with aeolian deposits generally the most quartzose and fluvial deposits incorporating the greatest lithic component, due to the presence of mudstone intraclasts deposited in the channels.

Authigenic mineralogy and diagenesis

The main cements precipitated in the sandstones are carbonates and quartz overgrowths. Other authigenic minerals are minor and it is noteworthy that there are no volumetrically significant authigenic clays in the reservoir sandstones, in contrast

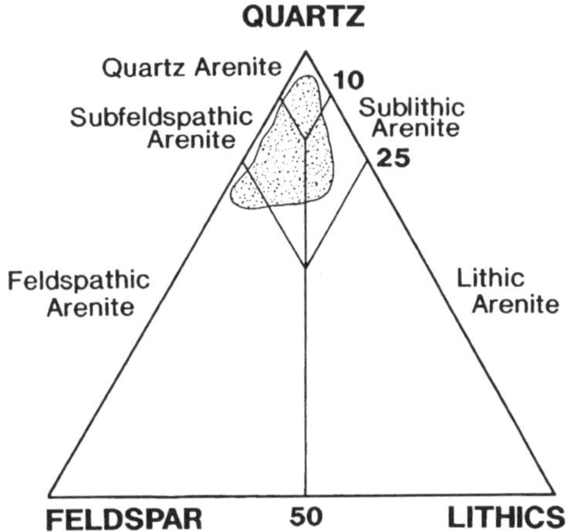

Fig. 2. Sandstone classification diagram, modified from McBride (1963) and Dott (1964). The stippled area represents the composition of >95% of the Ormskirk Sandstone Formation samples from the cored study wells.

to the Morecambe and Millom Fields to the north in which the reservoir sandstones are significantly illite cemented (e.g. Cowan & Bradney, this volume). Representative cements are illustrated in Fig. 3 and the diagenetic sequence established for the area is presented in Fig. 4 representing a sequence established from textural evidence from all the study wells. Thus all the cements are not necessarily present in every sample or every well. The overlapping range bars in Fig. 4 do not necessarily imply that the processes were contemporaneous, but authigenic textures indicate that some overlap occurred between certain cements and/or dissolution events. It is evident that in this part of the EISB the diagenetic history has involved multiple episodes of precipitation of the main cements during burial.

Near-surface diagenesis. Evidence is preserved in the reservoir sandstones to indicate early diagenetic modification following sandstone deposition. The processes are typical of red-bed style diagenesis in a semi-arid setting, as reviewed by Burley *et al.* (1985). Infiltrated detrital clays from percolating groundwaters are preserved as clay cutans around grains. The clays have an illitic composition at the present day but are likely to have been more smectite-rich upon deposition. Dissolution of labile detrital grains, such as feldspars, lithic fragments and ferromagnesian grains, occurred during burial and it is likely that dissolution was in part early, as is common in continental red beds (Walker *et al.* 1978). Ions released following grain dissolution were available for cement precipitation. Ferric iron from alteration of iron-bearing grains and smectitic clays coated grain surfaces and subsequently altered to haematite (Burley *et al.* 1985). Diagenetically stable K-feldspar overgrowths are common on many grains or surrounding secondary

pores, reflecting the reprecipitation of K^+, Si^{4+} and Al^{3+} released during feldspar dissolution (Waugh 1978; McKeever 1992). In addition minor quartz overgrowths may have formed at this stage of diagenesis.

Early non-ferroan calcite cements (Calcite I in Fig. 4), in the form of micronodules c. 300–500 μm in diameter composed of a radial arrangement of crystals (Fig. 3A), and as micrite cementing sedimentary intraclasts (Fig. 3C) are interpreted as caliche precipitates, although the cements may have recrystallized during burial. Locally dolomite mimics these textures and may reflect replacement of original calcite. The micronodules occur in all facies types and are probably largely *in situ*, whereas caliche-cemented intraclasts occur in fluvial channels and represent material reworked from the surrounding floodplain. The micronodular caliche cements display complex concentric zonation of alternating bright and dull luminescence in CL images (Fig. 3B), indicating variations in crystal chemistry between the zones. Strong & Pearce (1995) described similar structures, from Permo-Triassic sandstones in Cumbria, as carbonate spheroids c. 200–400 μm in diameter composed of radial-fibrous calcite or dolomite. These authors interpreted the spheroids to be early diagenetic, but whether precipitation was inorganic or biologically mediated could not be determined.

Early evaporitic gypsum/anhydrite form nodules on a millimetre to sub-centimetre scale composed of felted laths (Fig. 3D). These nodules have been identified in both the Ormskirk Sandstone Formation and the overlying Mercia Mudstone Group.

Burial diagenesis. During burial of the reservoir sandstones quartz overgrowths precipitated, probably over an extended period of time or in several discrete episodes. CL analyses record faint

---

Fig. 3. Representative photomicrographs of diagenetic textures in the reservoir sandstones, scale bars in microns. (**A**) Micronodule of caliche calcite (Calcite 1) is surrounded by zoned, non-ferroan to ferroan calcite (Calcite 2). The ferroan calcite locally postdates quartz overgrowths (e.g. arrow). 3451.0 feet, 110/13-14. (**B**) CL of calcite displays concentric bright and dull zonation in caliche nodules (1) and dark to bright to dull sequence in zoned cement (2), the dark and bright zones correspond to non-ferroan calcite and the dull zone corresponds to ferroan calcite. 3403.3 feet, 110/13-14. (**C**) Sedimentary intraclast in fluvial channel sandstone, cemented by caliche calcite (1) and surrounded by zoned, non-ferroan to ferroan calcite (2). 3344.0 feet, 110/13-14. (**D**) Evaporitic anhydrite nodule (A) composed of felted laths, partly replaced by later ferroan dolomite rhombs (D). 3497.4 feet, 110/14-1. (**E**) CL image showing detrital quartz grains with brown and violet luminescence contrasting with their mainly non-luminescent overgrowth cements, the overgrowths locally display a thin brown zone of luminescence (arrow). Later calcite cement with bright to dull orange luminescence, corresponds to zoned non-ferroan to ferroan calcite in (A). 3451.0 feet, 110/13-14. (**F**) a euhedral ferroan dolomite rhomb (D) occludes primary porosity and infills a secondary pore defined by remnants of K-feldspar (arrowed). Bitumen (B) coats the dolomite, but later anhydrite (A) encloses both dolomite and bitumen. 3538.0 feet, 110/13-13. (**G**) SEM image showing thick bitumen (B) lining pores and surrounding euhedral quartz crystals (Q) of varying size, and ferroan dolomite (D). 3501.6 feet, 110/14-1. (**H**) Late stage non-ferroan calcite (Calcite 3) encloses quartz overgrowths (Q), rhombs of ferroan dolomite (D) and grain coating bitumen (arrows). 3518.0 feet, 110/15-7.

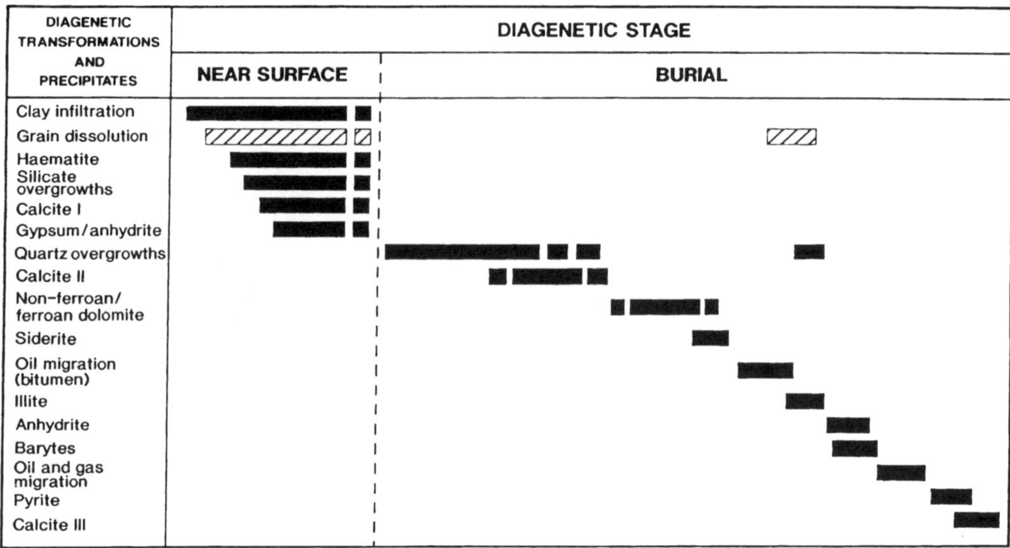

Fig. 4. General diagenetic sequence for the Ormskirk Sandstone Formation in the study wells. Note that overlapping range bars do not imply that processes were contemporaneous.

luminescent zonation (Fig. 3E) probably reflecting growth from fluids of varying composition and temperature.

Calcite cements, zoned from non-ferroan to ferroan in composition (Calcite II in Fig. 4), nucleate around the caliche calcite (Fig. 3A and B). The non-ferroan calcite generally predates much of the quartz cement, whereas ferroan calcite post-dates the overgrowths (Fig. 3E), indicating that timing of the two cements overlapped.

Dolomite cements comprise subhedral to euhedral rhombs up to c. 0.5 mm in size, staining shows the crystals are either zoned, generally from a non-ferroan to ferroan chemistry, or are entirely ferroan in composition (Fig. 3F), and postdate quartz and calcite cements. The dolomite is locally replacive of feldspar grains and mudstone intraclasts and may display a saddle crystal morphology indicative of elevated precipitation temperatures (Radke & Mathis 1980). Rarely siderite is associated with the dolomite cements. The $Fe^{2+}_{(aq)}$ for the ferroan carbonate cements may have been sourced from the replaced detrital components or from haematite, which has been reduced in many reservoir sections during burial resulting in loss of reddening (Stuart & Cowan 1991). Surdam *et al.* (1993) related hydrocarbon migration to the bleaching of red sandstones, with subsequent incorporation of the reduced iron into ferroan cements.

An episode of oil migration is recorded during burial as residual hydrocarbons (bitumen) coating the quartz, calcite and dolomite cements and detrital grains (Fig. 3F and G). There is insufficient evidence to indicate whether oil migration occurred in a single episode, or in several pulses as suggested by Bushell (1986) for the Morceambe Field. Further labile grain dissolution occurred during burial, represented by minor authigenic illite in secondary pores resulting from localized K-feldspar corrosion. Timing of this event is uncertain but may be tentatively correlated with source rock maturation generating corrosive fluids at depth (Schmidt & McDonald 1979; Surdam *et al.* 1989).

Late stage sulphate cements are present in the form of poikilotopic anhydrite (Fig. 3F) and barytes crystals, up to several millimetres in size. These sulphates are minor in terms of porosity occlusion and occur both within the hydrocarbon and water-bearing legs of the reservoirs, although they tend to be more abundant in the former. This suggests the sulphate cements predate migration of the oil and gas currently found in the reservoirs.

The latest cements recorded in the sandstones are pyrite and non-ferroan calcite (Fig. 3H; Calcite III in Fig. 4), which are restricted to the water-bearing leg of some reservoirs. The textural relationships and distribution of these cements indicates that precipitation postdated current hydrocarbon emplacement. Figure 5 illustrates the distribution of major cements and bitumen in a well from the Lennox Field. Non-ferroan calcite is the only cement which displays clear relationships with depth, occurring predominantly below the OWC.

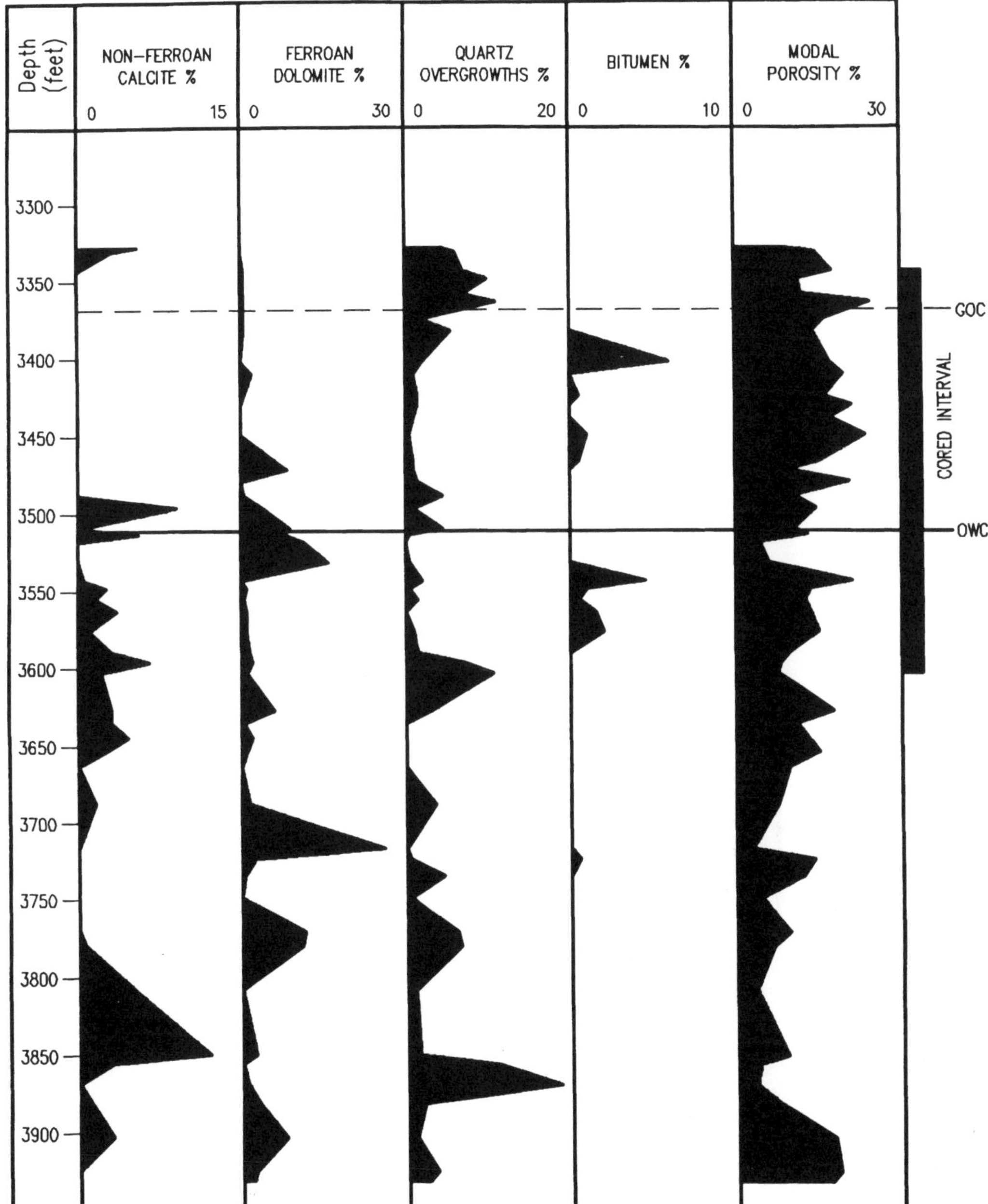

Fig. 5. Down-well distribution of major cements, bitumen and porosity over the reservoir interval in a representative well from the Lennox Field. Data is based on modal analyses of core and sidewall core samples.

Sandstone reservoir quality

The effects of diagenesis on sandstone poroperm characteristics are limited in the studied sequence and the only cements which are sufficiently abundant to affect reservoir quality are quartz and carbonates. Figure 6 illustrates the distribution of these cements in the three main depositional systems, based on modal analysis data. There is a clear facies control on the abundance of quartz

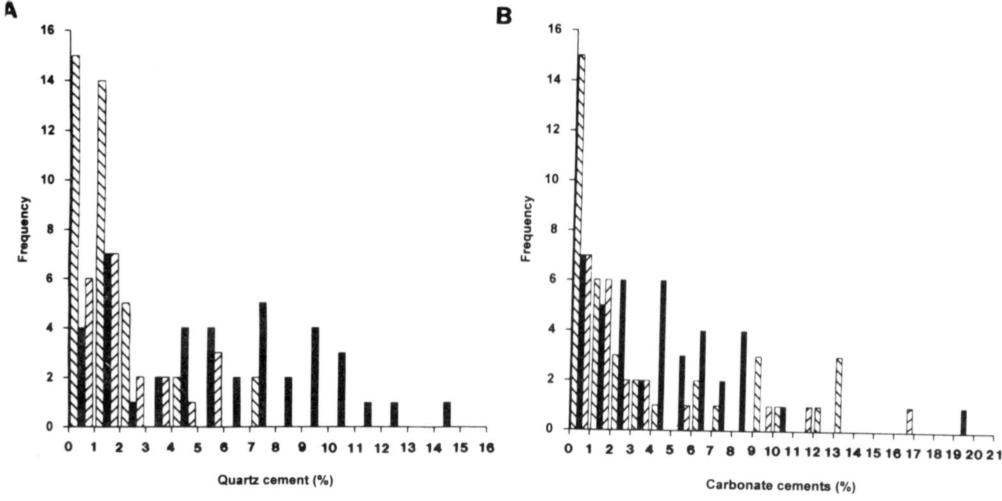

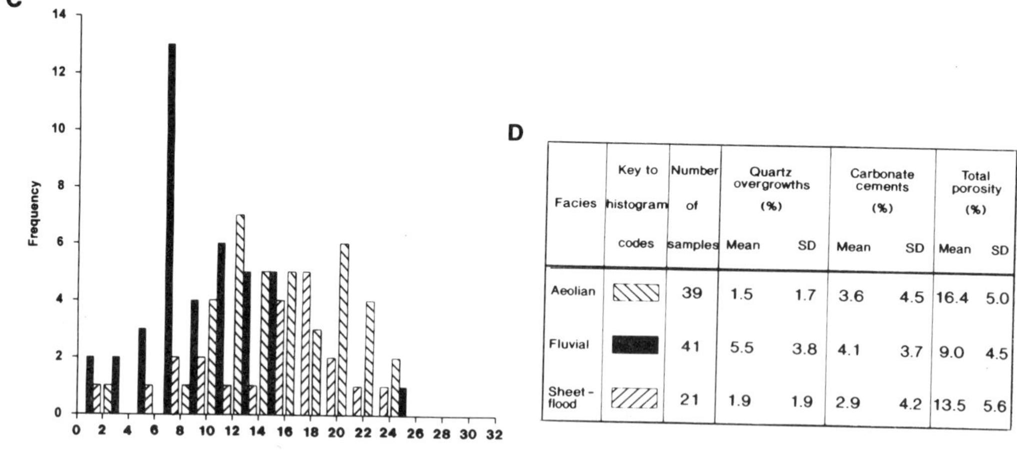

Fig. 6. Distribution of major cements and modal porosity in the three main depositional systems, based on modal analysis of 101 thin sections from five wells; (**A**) quartz overgrowths; (**B**) total carbonate cement; (**C**) modal porosity; (**D**) summary statistics for the cements and porosity.

cement (Fig. 6A), which is sparse in aeolian and sheetflood samples (means of 1.5 and 1.9%, respectively) and well developed in most fluvial sandstones (mean of 5.5%). Carbonate cements show a similar distribution (Fig. 6B), but the distinction between facies is less clear, as some aeolian sandstones are locally significantly carbonate cemented. The distribution of modal porosity in the three systems is shown in Fig. 6C, displaying the generally well preserved porosity in most samples. The influence of quartz and carbonate cements can be discerned, overprinting

depositional textures and resulting in the preferential loss of porosity in fluvial sandstones (mean of 9.0%) compared to the aeolian and sheetflood deposits (means of 16.4% and 13.5%, respectively).

Isotope data

Quartz cements

The oxygen isotope composition of quartz overgrowth cements has been determined in samples from wells 110/14-1, 110/15-6Z and 110/15-7

(Table 1), all the samples are from fluvial facies. The data is based upon analysis of two aliquots of purified quartz sample (with and without intact overgrowths), and calculation of the overgrowth value by mass balance, as described above. The $\delta^{18}O$ (SMOW) values of the quartz overgrowths range from +13.4 to +22.1‰, with a mean of +18.3‰ (SD = 2.9‰). The wide range of values probably reflects the errors inherent in the analytical method, mainly related to errors in modal analysis of the aliquots. There is also the possibility that multiple generations of quartz precipitated from fluids of variable composition and temperature, resulting in a range of $\delta^{18}O$ values. We are not aware of any published $\delta^{18}O$ data for quartz cements in the EISB, for comparison with the preliminary data presented here. However, the presently available values are typical of those reported for authigenic quartz in other reservoirs (see Brint *et al.* 1991). We regard the mean value of +18.3‰ as the most reliable current estimate of the $\delta^{18}O$ value for the overgrowths, as significant errors (i.e. ± 2‰) are associated with individual values (Brint *et al.* 1991). Considering these uncertainties no clear differences between the wells are evident from the data in Table 1.

Carbonate cements

Carbon and oxygen data. The oxygen and carbon stable isotope data for the calcite and dolomite cements are presented in Fig. 7. The distribution of data for calcite (Fig. 7A) reflects the three generations of calcite cement identified petrographically (i.e. I, II and III in Fig. 4), no systematic relationship to particular wells or facies is evident in the data. The caliche calcite samples (I) have the most positive $\delta^{18}O$ (–6.0 to –10.2‰ PDB) and $\delta^{13}C$ (–3.3 to –7.1‰) compositions and are similar to values for Triassic calcrete reported by Naylor *et al.* (1989*a*). For the burial calcites (II) the values trend towards lighter $\delta^{18}O$ (–9.2 to –17.3‰ PDB) and

$\delta^{13}C$ (–4.6 to –17.6‰) compositions reflecting the analysis of bulk samples with variable mixtures of caliche calcite (I), and zoned burial cements (II) precipitated at higher temperatures. The late calcite (III), postdating current hydrocarbons, has a discrete distribution with a narrow range of $\delta^{18}O$ (–11.0 to –11.5‰ PDB) and a wide range of very low $\delta^{13}C$ between –16.6 and –30.4‰ (PDB), derived from a thermally altered organic carbon source (Irwin *et al.* 1977).

The dolomite stable isotope data (Fig. 7B) display a wide range of $\delta^{18}O$ values, between –4.0 and –15.4‰ (PDB), possibly reflecting precipitation from fluids of variable composition under different temperature environments. $\delta^{13}C$ values range from –1.7 to –11.7‰ (PDB). No clear trends are observed with respect to the isotope composition and specific dolomite textures. There is a positive correlation between carbon and oxygen

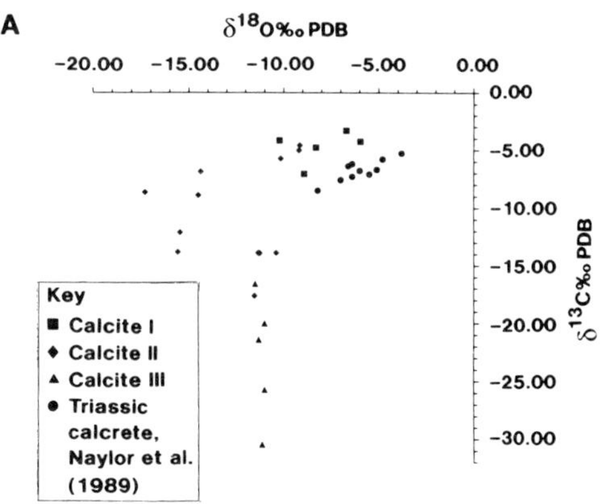

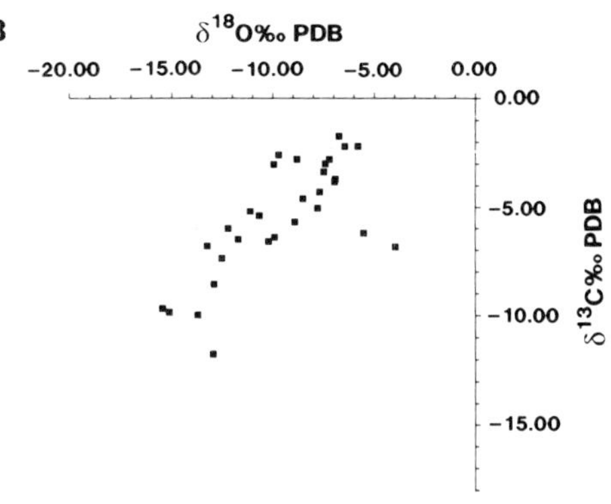

Fig. 7. Cross-plots of $\delta^{18}O$ (PDB) vs $\delta^{13}C$ (PDB) for (**A**) calcite cements; and (**B**) dolomite cements, in the study wells.

Table 1. *Oxygen isotope data for quartz overgrowth cements*

Well	Depth (ft MD)	$\delta^{18}O$ quartz (‰ SMOW)
110/14-1	3510.6	+17.2
110/14-1	3519.5	+22.1
110/14-1	3522.5	+21.7
110/15-6Z	4678.0	+18.7
110/15-6Z	4709.0	+20.3
110/15-6Z	4937.0	+16.8
110/15-7	3348.0	+19.2
110/15-7	3363.0	+13.4
110/15-7	3595.95	+15.1

isotope values, with the most negative $\delta^{13}C$ incorporated in dolomite with the lowest $\delta^{18}O$ values. There is no evident relationship between the stable isotope data and depth, well location or depositional facies.

Strontium isotope data. The calcite and dolomite cements have variable $^{87}Sr/^{86}Sr$ compositions with means of 0.7110 (SD = 0.0012) for calcite and 0.7129 (SD = 0.0029) for dolomite (Fig. 8). The data display a wide range and are consistently more radiogenic than contemporary Triassic seawater at 0.7078 (Koepnick *et al.* 1990), suggesting incorporation of varying amounts of ^{87}Sr from silicate grain dissolution, assuming an average $^{87}Sr/^{86}Sr$ of 0.720 (Burke *et al.* 1982). The values do not correlate with different cement generations in either calcite or dolomite, and no clear differences between individual wells or depositional facies can be determined.

Sulphate cements

The sulphur stable isotope compositions have been analysed for four early evaporitic anhydrite

samples from wells 110/14-1, 110/13-13, 110/13-15 and 110/13-16, six late poikilotopic anhydrites from wells 110/14-1, 110/15-6Z and 110/15-7, plus three samples of late poikilotopic barytes cement from 110/13-14 (Fig. 9). The early evaporitic anhydrite has $\delta^{34}S$ (CDT) values between +15.6 and +19.6‰, with similar values for samples from the Mercia Mudstone Group and Ormskirk Sandstone Formation. These values overlap with the late poikilotopic anhydrite which ranges from +14.8 to +17.0‰. There is no apparent difference between the $\delta^{34}S$ compositions of late anhydrite in the three wells studied. In contrast the data from late poikilotopic barytes from three depths in well 110/13-14 are lower than the anhydrite with a narrow range of $\delta^{34}S$ (CDT) between +13.1 and +13.6‰.

Fluid inclusion data

Fluid inclusion microthermometric data for quartz, carbonate and sulphate cements have been measured on samples from wells 110/13-13, 110/13-14, 110/14-1, 110/15-6Z and 110/15-7. The T_h and T_m values obtained for the cements are

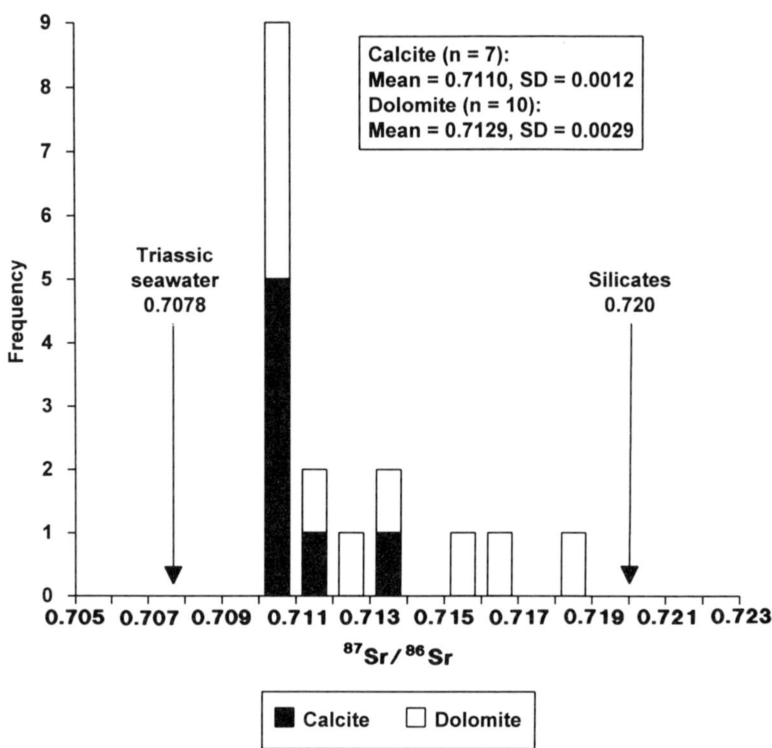

Fig. 8. Distribution of strontium isotope data for selected calcite and dolomite cements. Values for Early Triassic seawater (0.7078; Koepnick *et al.* 1990) and silicate grains (average 0.720; Burke *et al.* 1982) are shown for comparison.

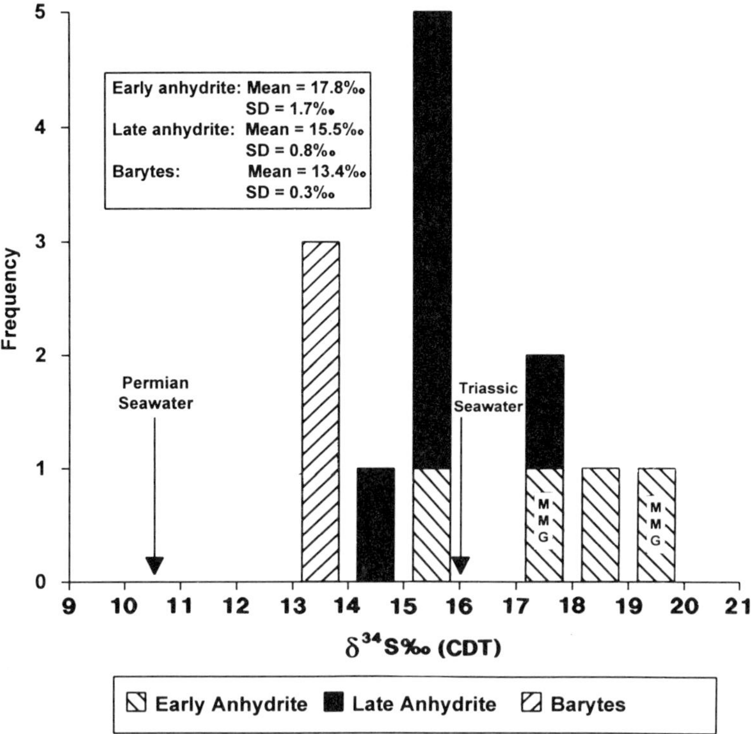

Fig. 9. Distribution of δ^{34}S (CDT) compositions for sulphate cements. Values labelled MMG were recovered from sidewall cores in the Mercia Mudstone Group, all other data is from the Ormskirk Sandstone Formation. Triassic (+16.0 ± 1.5‰) and Permian (+10.5 ± 1.0‰) seawater values from Claypool *et al.* (1980) are shown for comparison.

displayed in Fig. 10. This is the first published fluid inclusion data in this part of the EISB. Hardman *et al.* (1993) published fluid inclusion analyses from two unspecified EISB wells, but no other data is available for comparison. The data should therefore be regarded as a preliminary attempt to characterize the T_h and T_m ranges of the cements and further data is desirable to support that presented here. The number of measurements acquired during this study was limited because biphase inclusions were not common in any of the cements. The inclusions are small, all <10 µm and frequently <5 µm, so that phase changes were often difficult to record accurately, particularly during T_m measurements, and this also limited the amount of data collected. However, the T_h values are less likely to be reset in small inclusions and are more likely to record true growth temperature (Osborne & Haszeldine 1993). Some of the cements display a wide range in the T_h data which may reflect measurement of inclusions from separate assemblages, related to multiple growth events of the cement at different temperatures.

The analysed inclusions hosted in quartz overgrowths in three of the study wells (110/13-13, 110/13-14, and 110/15-7) range from 2 to 10 µm and are located both along the grain–overgrowth boundary and isolated within the overgrowths. The inclusions are aqueous biphase types, they have consistent liquid : vapour ratios of ≈10 : 1, and are considered to be primary in origin and probably associated with growth zones. No hydrocarbon inclusions were observed under ultraviolet (UV) light microscopy. The T_h values for these inclusions range from 89.6 to 158.1°C with 75% of the values between 130 and 150°C (Fig. 10A). T_m data for the quartz cements are sparse, as described above, with values between −14.8 and −17.4°C (Fig. 10D), corresponding to salinities of 18.5–20.5 wt% NaCl. No relationship between inclusion location (grain boundary or isolated) and T_h or T_m values is evident.

Calcite cements display two distributions of T_h data representing biphase aqueous and hydro-carbon-bearing inclusions (Fig. 10B). No fluid inclusion data were obtained from the early non-ferroan calcite (Calcite I) and all the data therefore relates to the later burial cements. The analysed

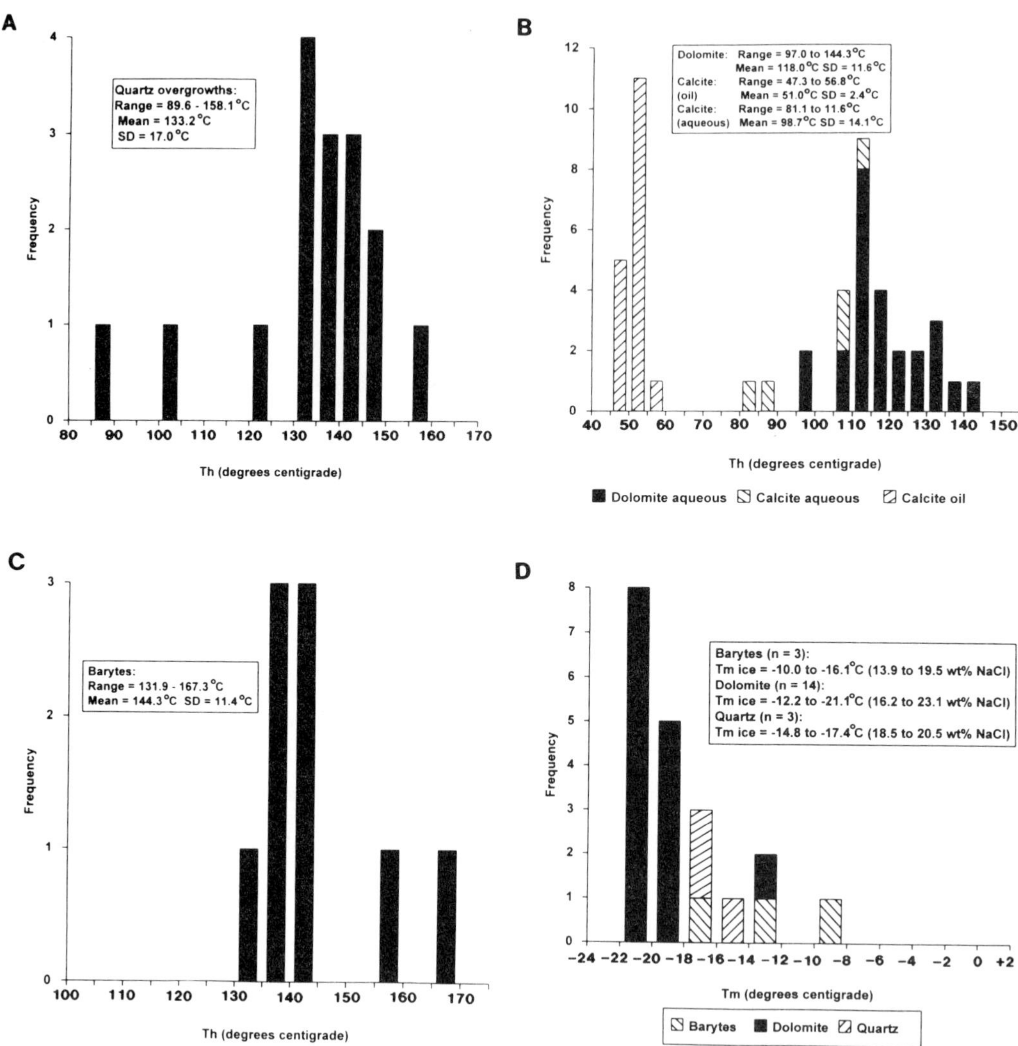

Fig. 10. (**A**) Homogenization temperatures for aqueous inclusions in quartz overgrowths; (**B**) homogenization temperatures for oil inclusions in calcite (III) cement, plus aqueous inclusions in calcite (II) and dolomite cements; (**C**) homogenization temperatures for aqueous inclusions in barytes cement; (**D**) last ice melting temperatures for aqueous inclusions in different cements.

aqueous inclusions are hosted in zoned burial cements (Calcite II) in well 110/13-14 and are between 3 and 5 µm in size with consistent liquid : vapour ratios of ≈10 : 1. The T_h data were measured on isolated inclusions which were not located along visible cleavage planes and T_h values range from 81.1 to 111.6°C. No reliable T_m data were obtained from these calcite cements, again due to difficulties recording phase changes in the small aqueous inclusions. The analysed hydrocarbon inclusions are hosted in the late calcite (Calcite III), restricted to the water leg of the

reservoir in well 110/15-7. The inclusions are between 3 and 10 µm in size with consistent liquid : vapour ratios of ≈10 : 1. Calcite III T_h measurements were obtained from isolated inclusions, away from visible cleavage planes and temperatures range from 47.3 to 56.8°C.

Aqueous biphase inclusions hosted in dolomite cements in four of the wells (110/13-14, 110/14-1, 110/15-6Z and 110/15-7) range from 2 to 10 µm in size and have consistent liquid : vapour ratios of ≈10 : 1. T_h data were obtained from isolated inclusions, located away from visible cleavage planes and display a temperature range between 97.0 and 144.3°C (Fig. 10B), T_m values between −12.2 and −21.1°C correspond to salinities of 16.2–23.1 wt% NaCl (Fig. 10D). No hydrocarbon inclusions were observed in dolomite cements under UV light microscopy.

Barytes cement from well 110/13-14 hosts aqueous biphase inclusions ranging from 2 to 6 μm in size, with consistent liquid : vapour ratios of ≈10 : 1. Measurements were made on isolated inclusions within the host cement which displays elevated T_h values of 131.9–167.3°C, with >75% of values between 130 and 145°C (Fig. 10C). T_m data for the barytes are similar to quartz and dolomite, with values between 10.0 and 16.1°C, corresponding to salinities of 13.9–19.5 wt% NaCl (Fig. 10D). No hydrocarbon inclusions were observed in barytes under UV light microscopy.

Overall the fluid inclusion data presented here are similar to that presented by Hardman et al. (1993) who reported T_h values for quartz, carbonate and anhydrite cements within the range 70 to 140°C, from two unspecified wells in the EISB.

Data interpretation

The combination of data from petrography, isotope geochemistry and fluid inclusion microthermometry allows some constraints to be placed on the diagenetic conditions during burial of the Ormskirk Sandstone in this part of the EISB.

Near-surface diagenesis

The caliche carbonate cements (Calcite I) precipitated during early diagenesis may have preserved the isotope signature of the early groundwaters, providing the cements were not recrystallized during burial. The non-ferroan calcites which preserve caliche fabrics display the more positive $\delta^{13}C$ and $\delta^{18}O$ compositions (Fig. 7), compared to the burial cements, suggesting that recrystallization has not resulted in isotopic re-equilibration during burial. An estimate of either the oxygen isotope composition of the groundwaters or the temperature of calcite precipitation is required to interpret the data. Naylor et al. (1989a) studied Triassic calcrete from the Inner Moray Firth and obtained a mean $\delta^{18}O$ (PDB) of −6.1‰ for the micrite. These authors estimated that Triassic continental meteoric waters may have had a $\delta^{18}O$ composition of approximately −3‰ (SMOW), which is relatively $\delta^{18}O$-enriched for meteoric water reflecting the likely effect of evaporation in a semi-arid climate. The most positive $\delta^{18}O$ value of −6.0‰ (PDB; +24.6‰ SMOW) obtained in the present study represents a calichified floodplain sample and has a similar oxygen isotope composition to the calcrete

of Naylor et al. (1989a). Assuming this value represents the least altered caliche cement, then line A in Fig. 11A shows that a precipitation temperature of c. 30°C is obtained from a water of −3‰ (SMOW). This corresponds to the Triassic surface temperature derived by Naylor et al. (1989a) from their data. The $\delta^{13}C$ composition of caliche cements would have been dominantly sourced from atmospheric and soil CO_2, with composition dependent upon climate and vegetation type. The $\delta^{13}C$ values of −7.1 to −3.3‰ (PDB) are slightly more positive than those obtained for Triassic calcretes by Naylor et al. (1989a; −8.5 to −5.3‰ PDB) or Purvis & Wright (1991; −9.03 to −7.9‰ PDB), and may indicate a carbon source dominated by soil CO_2 containing a significant contribution from atmospheric CO_2 (c. −6.5‰; Cerling 1991), due to diffusional mixing with the atmosphere (cf. Cerling et al. 1989).

Minor early evaporitic gypsum/anhydrite identified in the Ormskirk Sandstone Formation and the overlying Mercia Mudstone Group has $\delta^{34}S$ values between +15.7 and +19.6‰ (CDT), with a mean of +17.8‰ (SD = 1.7‰). These values are similar to those presented by Thompson & Meadows (this volume) for evaporitic anhydrite in well 110/8a-5 (+18.6 to +21.4‰) and correspond closely to published values for Triassic marine sulphate (+16.0 ±1.5‰; Claypool et al. 1980). Naylor et al. (1989b) interpreted anhydrite $\delta^{34}S$ values of +18.4 to +20.8‰, from the Mercia Mudstone Group in the Cheshire Basin, as having a Triassic marine origin. The identification of early evaporitic anhydrite with a marine sulphate signature in the study wells has important implications for depositional and palaeogeographic models of the EISB, as discussed by Thompson & Meadows (this volume).

Burial diagenesis

Quartz overgrowth cements are one of the most significant burial cements and are preferentially concentrated in the fluvial channel sandstones (Fig. 6A). The affinity of quartz cements for fluvial facies in the EISB has also been reported in previous studies. Meadows & Beach (1993) suggested that quartz may preferentially cement the fluvial sandstones, due to their lower textural maturity, in comparison with aeolian sequences. An alternative explanation involves early evaporite cement in aeolian sandstones, preventing significant quartz cementation, which has been removed to leave porous, uncompacted aeolian sands (Stuart & Cowan 1991; Cowan 1993). There is insufficient evidence at present to verify either theory. However, qualitative evidence suggests that fluvial channel sands with reworked aeolian grain types

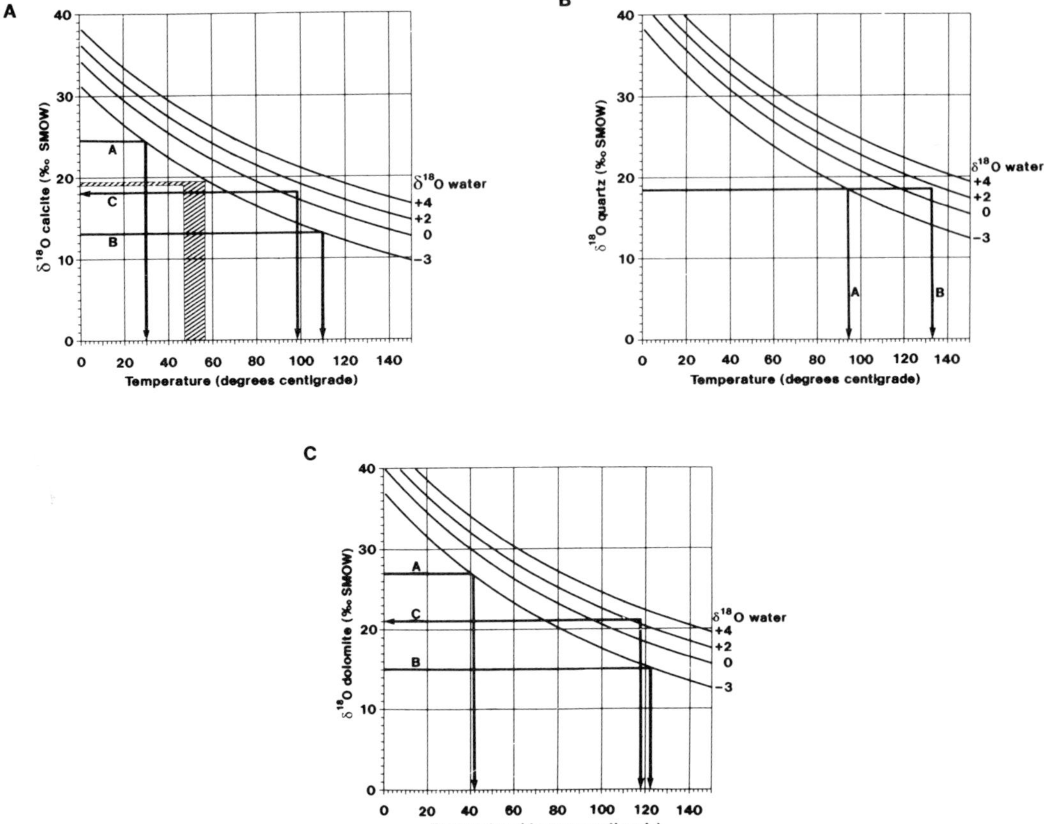

Fig. 11. Mineral $\delta^{18}O$ vs temperature and $\delta^{18}O$ porewater. (**A**) Calcite cements: line A represents the precipitation temperature from the most positive $\delta^{18}O$ value (+24.6‰ SMOW) for Calcite I, assumed to represent the least altered caliche calcite, and line B shows the maximum precipitation temperature for Calcite II from the lowest $\delta^{18}O$ value of +14.9‰ (SMOW), both assuming meteoric porewater; line C represents the porewater composition of +0.8‰ (SMOW) derived from the mean $\delta^{18}O$ of +18.0‰ (SMOW) and mean T_h of c. 99°C for Calcite II. The shaded area represents the $\delta^{18}O$ and T_h temperature ranges for late calcite (III) cement. (**B**) Quartz overgrowths: line A shows the precipitation temperature for mean $\delta^{18}O$ (+18.3‰) assuming meteoric porewater, and line B illustrates a porewater composition of c. +2‰ SMOW derived from the mean T_h of 133°C. (**C**) Dolomite cements: lines A and B show the precipitation temperature range derived from the $\delta^{18}O$ range assuming meteoric porewater, and line C illustrates the water composition of c. +2‰ SMOW derived from the mean T_h of 118°C for the mean $\delta^{18}O$ of +21.0‰. Curves representing different porewater isotope values were calculated using the following isotope fractionation equations:

$$1000\,\ln\alpha_{\text{calcite-water}} = 2.78 \times 10^6 \times T^{-2} - 2.89 \text{ (Friedman \& O'Neil 1977);}$$
$$1000\,\ln\alpha_{\text{quartz-water}} = 3.34 \times 10^6\, T^{-2} - 3.31 \text{ (Matsuhisa } et\ al.\text{1979); and}$$
$$1000\,\ln\alpha_{\text{domomite-water}} = 3.14 \times 10^6\, T^{-2} - 2.0 \text{ (Land 1983).}$$

are frequently quartz cemented (e.g. Fig. 3E), thus an explanation based purely on textural variations is unlikely.

Some constraints on the conditions during precipitation of quartz cement may be inferred from the fluid inclusion and isotope data. Line A in Fig. 11B shows that the mean quartz $\delta^{18}O$ value of +18.3‰ (SMOW) corresponds to a precipitation temperature of c. 95°C, assuming a porewater composition of −3‰ for Triassic meteoric waters. This temperature is lower than most of the fluid inclusion T_h values from authigenic quartz (Fig. 10A), and an ^{18}O-enriched formation water of c. +2‰ (SMOW) is required to explain a mean T_h value of 133°C for the mean quartz $\delta^{18}O$ composition of +18.3‰ SMOW (line B in Fig. 11B). T_m data for quartz indicate highly saline formation water with a salinity of over 5 times that of seawater (Fig. 10D), consistent with quartz precipitation from an evolved basinal brine. It must be stressed that such interpretations are at this stage speculative, due to the small database and uncertainties regarding the isotope data for quartz, in addition to the probability that quartz formed over a time interval representing variable porewater composition and temperature.

Zoned non-ferroan to ferroan calcite (Calcite II) partly postdates the quartz overgrowths. This calcite has a variable $\delta^{18}O$ signature potentially reflecting a range of precipitation temperatures and/or porewater compositions. Line B in Fig. 11A illustrates the maximum temperature of $c.$ 110°C, derived from the lowest $\delta^{18}O$ value of +14.9‰ (SMOW), assuming precipitation from Triassic meteoric water (–3‰ SMOW). Using the mean $\delta^{18}O$ (SMOW) value of +18.0‰ for Calcite II and the mean T_h of $c.$ 99°C, a porewater composition of +0.8‰ (SMOW) is derived for calcite precipitation (line C in Fig. 11A). This is more consistent with the quartz data, which suggests burial cements precipitated from fluids enriched in ^{18}O compared to depositional waters. Carbon isotope data for calcite suggests the later cements incorporate increasing quantities of more negative $\delta^{13}C$, derived from thermally altered organic sources during burial (Irwin et $al.$ 1977).

Dolomite cements which postdate quartz and calcite also display variable $\delta^{18}O$ signatures, suggesting precipitation over an interval encompassing a range of temperatures and/or porewater isotope compositions. The positive correlation between $\delta^{18}O$ and $\delta^{13}C$ in the dolomites (Fig. 7B) is similar to trends reported by Naylor et $al.$ (1989a) for Triassic calcrete and Spötl & Wright (1992) for Triassic groundwater dolocretes. Such trends can reflect evaporation of porewaters which result in an increase in $\delta^{18}O$ and $\delta^{13}C$ in precipitated cements (Naylor et $al.$ 1989a), although it seems unlikely that this process applies to burial cements. Lines A and B in Fig. 11C show that the range of dolomite $\delta^{18}O$ data give a temperature range of $c.$ 42–122°C with water of –3‰ (SMOW). Dolomite T_h data indicate a mean minimum trapping temperature of $c.$ 118°C (Fig. 10B) which are consistent with isotopically enriched water compositions, around +2‰ (SMOW) for the mean $\delta^{18}O$ of +21.0‰ SMOW (line C in Fig. 11C). T_m values for dolomite indicate highly saline brines, over 6 times seawater salinity during dolomite precipitation, similar to quartz overgrowth T_m data, and consistent with precipitation from basinal brines (Fig. 10D). The $\delta^{13}C$ data for dolomite indicate that the later cements with more negative $\delta^{18}O$ signatures incorporated greater proportions of ^{12}C-enriched carbon, originating from organic maturation, during burial. Strontium isotope data for the carbonates,

particularly dolomite, display a wide range and are consistent with an originally continental meteoric water, incorporating strontium from a range of sources (Spötl & Wright 1992).

The late poikilotopic anhydrite cements have $\delta^{34}S$ (CDT) compositions which overlap with the early evaporitic anhydrite and the mean value of +15.5‰ is close to published Triassic seawater values (Claypool et $al.$ 1980). This suggests that dissolution and reprecipitation of the early anhydrite sourced the later poikilotopic cement. The dissolved sulphate may have been derived from the Ormskirk Sandstone Formation or the overlying Mercia Mudstone Group. The poikilotopic barytes has a lower $\delta^{34}S$ signature (mean = +13.4‰) than the anhydrite. Dissolution of anhydrite produces aqueous sulphate 1.5‰ more depleted than the anhydrite, followed by barytes precipitation without significant isotopic fractionation (Naylor et $al.$ 1989b). It is therefore possible to source the barytes through dissolution of poikilotopic anhydrite, the alternative explanation being mixing of sulphate sources such as Triassic and Permian marine evaporites (Fig. 9). Barytes is not common in the EISB study wells, and where present in 110/13-14 it is concentrated along coarser, more permeable, laminae indicating that fluid flux was an important control on precipitation. In the Cheshire Basin Naylor et $al.$ (1989b) studied barytes mineralization and proposed that the mineralizing fluids were chloride-rich brines (17 wt% NaCl) at temperatures around 130–150°C, based upon fluid inclusion data and burial history considerations. Fluid inclusions from barytes in this study indicate very similar conditions with formation water salinities of 13.9 to 19.5 wt% NaCl and temperatures of 132–167°C.

The latest cements in the samples examined comprise non-ferroan calcite (Calcite III) and pyrite, restricted to the water leg of some reservoirs (Fig. 5). The calcite has a very low $\delta^{13}C$ signature (Fig. 7A), consistent with a dominantly thermally mature organic carbon source. The calcite hosts biphase hydrocarbon plus gas inclusions, also indicating that oil migration and calcite precipitation were related. T_h data are significantly lower than those for the aqueous inclusions in earlier cements, and may indicate meteoric water compositions in conjunction with the $\delta^{18}O$ data, illustrated by the shaded area in Fig. 11A. This implies a change in porewater composition compared to the basinal brines present during burial, possibly related to Tertiary uplift of the sequence and influx of meteoric waters. This interpretation assumes that the inclusions are methane saturated and that fluid inclusion T_h is approximately equal to T_t. However, if pressure corrections were applied they may result in a significant addition to the minimum trapping temperature and thus alter the isotope data interpretation, depending upon trapping pressure and hydrocarbon composition of the trapped fluid (Goldstein & Reynolds 1994).

The association of non-ferroan calcite, with very low $\delta^{13}C$, and pyrite may be explained by a reaction involving anaerobic oxidation of methane through reaction with mineral oxidants, such as sulphate, during thermochemical sulphate reduction as described by Machel (1987) (e.g. $CH_4 + SO_4^{2-} \rightarrow HCO_3^- + HS^- + H_2O$). In this reaction the isotopic composition of the produced bicarbonate will be similar to the reactant methane, enabling the precipitation of low $\delta^{13}C$ calcite and providing a source of sulphide for pyrite cement.

Discussion

Diagenesis in the study area is characterized by a range of cements, precipitated during early diagenesis and later burial of the sequence. The stable isotope data for calcite and dolomite display a wide range of $\delta^{18}O$ compositions and appear to reflect the multiple episodes of carbonate precipitation. This suggests that, despite some petrographic evidence for recrystallization of caliche cements, their isotope signatures have not re-equilibrated during burial.

In contrast to the potentially wide range of precipitation temperatures preserved by the stable isotope data, precipitation temperatures for quartz, carbonate and sulphate cements derived from fluid inclusions predominantly record elevated temperatures, mainly between 110 and 150°C. This may partly reflect the increased likelihood of forming biphase inclusions in minerals precipitated at higher temperature (Goldstein & Reynolds 1994), and assuming the T_h values are reliable indicates that significant cementation did occur at temperatures above 100°C. The fluid inclusion T_h data display a similar range to those from the Ormskirk Sandstone Formation elsewhere in the EISB (Hardman *et al.* 1993), and these authors suggested that higher temperature inclusions (100–130°C) may have re-equilibrated in response to continued burial or higher temperatures following precipitation of the host cement. In the present study the fluid inclusions were carefully analysed to try and exclude those which may have been reset, using the criteria described above, but the possibility of recording burial rather than precipitation temperatures cannot be completely discounted (cf. Osborne & Haszeldine 1993). If the T_h data are assumed to represent minimum trapping temperature then some interesting questions relating to the burial history are raised. No burial history modelling has been attempted in this study, but Hardman *et al.* (1993) present a model and discuss

the associated uncertainties regarding heatflow and inversion episodes in the basin. Assuming that the burial depths and geothermal data presented by Hardman *et al.* (1993; their fig. 11) are taken as a reasonable estimate for the present study area some problems are apparent. In order to achieve temperatures of 100°C at an average geothermal gradient of 30°C km^{-1}, burial depths of >10 000 feet are required. Such depths are significantly greater than those modelled for the upper Sherwood Sandstone Group at maximum burial during the late Cretaceous (*c.* 4500 feet). Thus either deeper burial or elevated heatflow is required to explain the fluid inclusion data. Compactional fabrics in the sandstones are not compatible with deep burial, as the grains are generally in point and long contact, and pressure solution fabrics are rare (Fig. 3). Apatite fission track data suggest an early Tertiary palaeothermal episode of 110°C or greater in the EISB (Hardman *et al.* 1993; Green *et al.* this volume), and recent studies indicate that it is unlikely that inversion exceeded 1–2 km during the Tertiary (Green *et al.* this volume; Cope this volume).

Overall the fluid inclusion data are best explained by pulsed migration of hot fluids through the reservoir sandstones, possibly focused through discrete conduits beneath aquicludes. Petrographic evidence from this study shows that barytes, a late stage high temperature cement (>130°C), is concentrated within more permeable laminae and is thus influenced by fluid flux. The T_h data provide additional evidence that diagenesis relates to discrete pulses of fluid, rather than progressive burial and increasing temperature, as quartz T_h (mean = 133°C) exceeds dolomite (mean = 118°C), but texturally dolomite postdates the quartz cement. Stable isotope and fluid inclusion T_m data indicate that the later cements precipitated from evolved highly saline fluids, compatible with influxes of deep burial brine.

Conclusions

The reservoir sandstones in the southern part of the EISB retain good reservoir quality, reflecting the general lack of abundant cements or strong compaction. The only significant cements in the sandstones are quartz overgrowths and carbonates, other cements are minor and it is significant that authigenic clays are rare in this part of the basin. The aeolian and sandflat facies have the highest porosities compared to fluvial sandstones, reflecting both primary depositional textures and the distribution of quartz and carbonate cements, which are more abundant in the fluvial deposits.

In spite of limited cement volume, the diagenetic sequences are often complex. Early diagenesis resulted in minor modification of the sandstones, in particular some calcite cements displaying caliche fabrics appear to retain an isotopic signature which has suffered little or no modification during burial. The $\delta^{18}O$ signature is consistent with precipitation from meteoric porewaters (*c.* $-3‰$ SMOW) at temperatures of around 30°C. Evaporitic gypsum/ anhydrite cements from the Ormskirk Sandstone Formation and Mercia Mudstone Group have $\delta^{34}S$ compositions close to Triassic seawater sulphate, suggesting at least limited marine porewater incursion during Mercia Mudstone Group and/or upper Sherwood Sandstone Group deposition.

The burial diagenetic cements in the sandstones record elevated temperatures during precipitation, with fluid inclusion T_h for authigenic quartz, carbonate and sulphate minerals mostly between 100 and 150°C. T_m data from the inclusions indicate the fluids were concentrated brines, typically 5–6 times seawater salinity. The $\delta^{18}O$ compositions of the quartz and carbonates are variable and may reflect multiple generations forming over a range of temperatures and/or porewater compositions, with isotopically evolved fluids (*c.* $+2‰$ SMOW) at depth. The later carbonate cements incorporated increasing quantities of carbon derived from organic sources, consistent with organic maturation during burial. The presence of bitumen in the reservoirs is testimony to an episode of oil migration postdating the quartz and carbonate cements but predating late sulphate, calcite and the current hydrocarbon accumulations. The $\delta^{34}S$ signatures of late stage anhydrite and barytes cements suggest that dissolution and reprecipitation of early gypsum/anhydrite sourced the late sulphate.

The only cements postdating current hydrocarbons are calcite and pyrite, restricted to the water-bearing leg of some reservoirs, and thus postdating uplift and emplacement of current hydrocarbons. The calcite traps petroleum inclusions and has low $\delta^{13}C$ consistent with a solely organic carbon source, possibly involving a methane oxidation reaction linked to thermochemical sulphate reduction. The $\delta^{18}O$ data and fluid inclusion T_h measurements for this calcite indicate a change towards lower $\delta^{18}O$ porewaters, possibly related to Tertiary uplift and an influx of meteoric waters.

The data obtained from this study suggests that during burial the reservoir sandstones were subjected to elevated temperatures, significantly in excess of 100°C. These temperatures are unlikely to have been achieved wholly as a result of burial depth, based on petrographic considerations and published data. It is more likely that periods of elevated heatflow, possibly due to the circulation of hot fluids through the sandstones, resulted in the precipitation of at least some of the burial cements.

The authors gratefully acknowledge Hamilton Oil Company Limited (now BHP Petroleum Limited) and their partners for permission to publish this data. BG E&P Limited allowed us to publish data from well 110/14-1. Tony Fallick and the staff at SURRC carried out the isotope analyses and provided useful assistance in its interpretation. The manuscript was significantly improved by the comments of the referees, Jim Marshall and John Rowe, whose help we gratefully acknowledge. The Geochem Group are thanked for providing typing and drafting facilities during the preparation of the manuscript.

References

BRINT, J. F., HAMILTON, P. J., HASZELDINE, R. S., FALLICK, A. E. & BROWN, S. 1991. Oxygen isotopic analysis of diagenetic quartz overgrowths from the Brent Sands: a comparison of two preparation methods. *Journal of Sedimentary Petrology*, **61**, 527–533.

BURKE, W. H., DENISON, R. E., HETHERINGTON, E. A., KOEPNICK, R. B., NELSON, H. F. & OTTO, J. B. 1982. Variation of seawater $^{87}Sr/^{86}Sr$ throughout Phanerozoic time. *Geology*, **10**, 516–519.

BURLEY, S. D. 1984. Patterns of diagenesis in the Sherwood Sandstone Group (Triassic), United Kingdom. *Clay Minerals*, **19**, 403–440.

——, KANTOROWICZ, J. D. & WAUGH, B. 1985. Clastic diagenesis. *In*: BRENCHLEY, P. J. & WILLIAMS, B. P. J. (eds) *Sedimentology: Recent Developments and Applied Aspects*. Geological Society, London, Special Publication, **18**, 198–226.

——, MULLIS, J. & MATTER, A. 1989. Timing diagenesis in the Tartan Reservoir (UK North Sea): constraints from combined cathodoluminescence microscopy and fluid inclusion studies. *Marine and Petroleum Geology*, **6**, 98–120.

BUSHELL, T. P. 1986. Reservoir geology of the Morecambe Field. *In*: BROOKS, J., GOFF, J. C. & VAN HOORN, B. (eds) Habitat of Palaeozoic Gas in N.W. Europe. Geological Society, London, Special Publication, **23**, 189–207.

CERLING, T. E. 1991. Carbon dioxide in the atmosphere: evidence from Cenozoic and Mesozoic paleosols. *American Journal of Science*, **291**, 377–400.

——, QUADE, J., WANG, Y. & BOWMAN, J. R. 1989. Carbon isotopes in soils and palaeosols as ecology and palaeoecology indicators. *Nature*, **341**, 138–139.

CLAYPOOL, G. E., HOLSER, W. T., KAPLAN, I. R., SAKAI, H. & ZAK, I. 1980. The age curves of sulphur and oxygen isotopes in marine sulfate and their mutual interpretation. *Chemical Geology*, **28**, 199–260.

COLEMAN, M. L. & MOORE, M. P. 1978. Direct reduction

of sulfates to sulfur dioxide for isotopic analysis. *Analytical Chemistry*, **50**, 1594–1595.

COLTER, V. S. & EBBERN, J. 1978. The petrography and reservoir properties of some Triassic sandstones of the Northern Irish Sea Basin. *Journal of the Geological Society, London*, **135**, 57–62.

COPE, J. C. W. 1997. The Mesozoic and Tertiary history of the Irish Sea. *This volume*.

COWAN, G. 1993. Identification and significance of aeolian deposits within the dominantly fluvial Sherwood Sandstone Group of the East Irish Sea Basin UK. *In*: NORTH, C. P. & PROSSER, D. J. (eds) *Characterization of Fluvial and Aeolian Reservoirs*, Geological Society, London, Special Publication, **73**, 231–245.

——& BRADNEY, J. 1997. Regional diagenetic controls on reservoir properties in the Millom accumulation: implications for field development. *This volume*.

DOTT, R. L. 1964. Wacke, greywacke and matrix – what approach to immature sandstone classification? *Journal of Sedimentary Petrology*, **34**, 625–632.

FRIEDMAN, I. & O'NEIL, J. R. 1977. Compilation of stable isotope fractionation factors of geochemical interest. *In*: FLEISCHER, M. (ed.) *Data of Geochemistry*, sixth edition. United States Geological Survey Professional Paper 440-KK.

GOLDSTEIN, R. H. & REYNOLDS, T. J. 1994. *Systematics of Fluid Inclusions in Diagenetic Minerals*. Society for Sedimentary Geology, Short Course **31**, 199 pp.

GREEN, P. F., DUDDY, I. R. & BRAY, R. J. 1997. Variation in thermal history styles around the Irish Sea and adjacent areas: implications for hydrocarbon occurrence and tectonic evolution. *This volume*.

HARDMAN, M., BUCHANAN, J., HERRINGTON, P. & CARR, A. 1993. Geochemical modelling of the East Irish Sea Basin: its influence on predicting hydrocarbon type and quality. *In*: PARKER, J. R. (ed.) *Petroleum Geology of Northwest Europe: Proceedings of the 4th Conference*. The Geological Society, London, 809–821.

IRWIN, H., CURTIS, C. & COLEMAN, M. 1977. Isotopic evidence for source of diagenetic carbonates formed during burial of organic-rich sediments. *Nature*, **269**, 209–213.

JACKSON, D. I., JOHNSON, H. & SMITH, N. J. P. 1997. Stratigraphical relationships and a revised lithostratigraphical nomenclature for the Carboniferous, Permian and Triassic rocks of the offshore East Irish Sea Basin. *This volume*.

KNOX, R. W., O'B., BURGESS, W. G., WILSON, K. S. & BATH, A. H. 1984. Diagenetic influences on reservoir properties of the Sherwood Sandstone (Triassic) in the Marchwood geothermal borehole, Southampton, UK. *Clay Minerals*, **19**, 441–456.

KOEPNICK, R. B., DENISON, R. E., BURKE, W. H., HETHERINGTON, E. A. & DAHL, D. A. 1990. Construction of the Triassic and Jurassic portion of the Phanerozoic curve of seawater $^{87}Sr/^{86}Sr$. *Chemical Geology* (Isotope Geoscience Section), **80**, 327–349.

LAND, L. S. 1983. The application of stable isotopes to studies of the origin of dolomite and to problems of diagenesis of clastic sediments. *In*: *Stable Isotopes in Sedimentary Geology*. Society of Economic Paleontologists and Mineralogists, Short Course **10**.

MACCHI, L., CURTIS, C. D., LEVISON, A., WOODWARD, K. & HUGHES, C. R. 1990. Chemistry, morphology, and distribution of illites from Morecambe Gas Field, Irish Sea, offshore United Kingdom. *The American Association of Petroleum Geologists Bulletin*, **74**, 296–308.

MACHEL, H. G. 1987. Some aspects of diagenetic sulphate-hydrocarbon redox reactions. *In*: MARSHALL, J. D. (ed.) *Diagenesis of Sedimentary Sequences*. Geological Society, London, Special Publication, **36**, 15–28.

MATSUHISA, Y., GOLDSMITH, J. R. & CLAYTON, R. N. 1979. Oxygen isotopic fractionation in the system quartz–albite–anorthite–water. *Geochimica et Cosmochimica Acta*, **43**, 1131–1140.

MCBRIDE, E. F. 1963. A classification of common sandstones. *Journal of Sedimentary Petrology*, **33**, 664–669.

MCCREA, J. M. 1950. On the isotope chemistry of carbonates and a palaeo-temperature scale. *Journal of Chemical Physics*, **18**, 849–857.

MCKEEVER, P. J. 1992. Petrography and diagenesis of the Permo-Triassic of Scotland. *In*: PARNELL, J. (ed.) *Basins on the Atlantic Seaboard: Petroleum Geology, Sedimentology and Basin Evolution*. Geological Society, London, Special Publication, **62**, 71–96.

MEADOWS, N. S. & BEACH, A. 1993. Controls on reservoir quality in the Triassic Sherwood Sandstone of the Irish Sea. *In*: PARKER, J. R. (ed.) *Petroleum Geology of Northwest Europe: proceedings of the 4th Conference*. The Geological Society, London, 823–833

MILLIKEN, K. L., LAND, L. S. & LOUCKS, R. G. 1981. History of burial diagenesis determined from isotope geochemistry, Frio Formation, Brazoria County, Texas. *The American Association of Petroleum Geologists Bulletin*, **65**, 1397–1413.

NAYLOR, H., TURNER, P., VAUGHAN, D. J. & FALLICK, A. E. 1989*a*. The Cherty Rock, Elgin: A petrographic and isotopic study of a Permo-Triassic calcrete. *Geological Journal*, **24**, 205–221.

——, ——, ——, BOYCE, A. J. & FALLICK, A. E. 1989*b*. Genetic studies of red bed mineralization in the Triassic of the Chesire Basin, northwest England. *Journal of the Geological Society, London*, **146**, 685–699.

OSBORNE, M. & HASZELDINE, S. 1993. Evidence for resetting of fluid inclusion temperatures from quartz cements in oilfields. *Marine and Petroleum Geology*, **10**, 271–278.

PURVIS, K. & WRIGHT, V. P. 1991. Calcretes related to phreatophytic vegetation from the Middle Triassic Otter Sandstone of South West England. *Sedimentology*, **38**, 539–551.

RADKE, B. M. & MATHIS, R. L. 1980. On the formation and occurrence of saddle dolomite. *Journal of Sedimentary Petrology*, **50**, 1149–1168.

SCHMIDT, V. & MCDONALD, D. A. 1979. The role of secondary porosity in the course of sandstone diagenesis. *In*: SCHOLLE, P. A. & SCHLUGER, P. R. (eds) *Aspects of Diagenesis*. Society of Economic

Paleontologists and Mineralogists, Special Publication, **26**, 175–207.

SPÖTL, C. & WRIGHT, V. P. 1992. Groundwater dolocretes from the Upper Triassic of the Paris Basin, France: a case study of an arid, continental diagenetic facies. *Sedimentology*, **39**, 1119–1136.

STRONG, G. E. 1993. Diagenesis of Triassic Sherwood Sandstone Group rocks, Preston, Lancashire, UK: a possible evaporitic cement precursor to secondary porosity? *In*: NORTH, C. P. & PROSSER, D. J. (eds) *Characterization of Fluvial and Aeolian Reservoirs*. Geological Society, London, Special Publication, **73**, 279–289.

—— & PEARCE, J. M. 1995. Carbonate spheroids in Permo-Triassic sandstones of the Sellafield area, Cumbria. *Proceedings of the Yorkshire Geological Society*, **50**, 209–211.

STUART, I. A. & COWAN, G. 1991. The South Morecambe Field, Blocks 110/2a, 110/3a, 110/8a, UK. *In*: ABBOTS, I. L. (ed.) *United Kingdom Oil and Gas Fields: 25 Years Commemorative Volume*. Geological Society, London, Memoir, **14**, 527–541.

SURDAM, R. C., CROSSEY, L. J., HAGEN, E. S. & HEASLER, H. P. 1989. Organic–inorganic interactions and sandstone diagenesis. *American Association of Petroleum Geologists Bulletin*, **73**, 1–23.

——, JIAO, Z. S. & MACGOWAN, D. B. 1993. Redox reactions involving hydrocarbons and mineral oxidants: a mechanism for significant porosity enhancement in sandstones. *American Association of Petroleum Geologists Bulletin*, **77**, 1509–1518.

THOMPSON, J. & MEADOWS, N. S. 1997. Clastic sabkhas and diachroneity at the top of the Sherwood Sandstone Group: East Irish Sea. *This volume*.

WALKER, T. R., WAUGH, B. & CRONE, A. J. 1978. Diagenesis in first-cycle desert alluvium of Cenozoic age, southwestern United States and northwestern Mexico. *Geological Society of America Bulletin*, **89**, 19–32.

WAUGH, B. 1978. Authigenic K-feldspar in British Permo-Triassic sandstones. *Journal of the Geological Society, London*, **135**, 51–56.

Regional diagenetic controls on reservoir properties in the Millom accumulation: implications for field development

GREIG COWAN & JOANNA BRADNEY

BG Exploration and Production Ltd, 100 Thames Valley Park Drive, Reading RG6 1PT, UK. (email: cowan.g@bgep.co.uk)

Abstract: The Millom accumulation comprises two dip-closed culminations situated down-flank of the 1 TCF North Morecambe Field. The depositional environments encountered in the Millom area are similar to those in the North Morecambe area and they share a common depositional layering. However, porosity and permeability are low compared to the Morecambe Fields. Although the presence of permeability reducing platy illite throughout the reservoir can explain the reduced permeabilities in the area, it does not account for the reduced porosity. The abundance of mineral cements (quartz and carbonates) increases rapidly away from the crest of the North Morecambe structure, down-dip towards the Millom area. The migration history of fluids outwards from the deeper parts of the basin is preserved by the cement distributions observed in the Morecambe–Millom area. Porosity in North Morecambe is preserved by an early hydrocarbon charge. The Millom structure had not formed during early hydrocarbon migration and was exposed to mineralizing fluids moving from the Keys Basin to the north. Both structures were filled with the present gas charge during late Cretaceous/early Tertiary gas migration. Poor reservoir properties, especially K_v, renders Millom unsuitable for development by horizontal wells.

The Millom accumulation, named after the Cumbrian village, was discovered in 1983 by well 113/26-1. Two additional wells were drilled between 1987 and 1993 which proved the structure to be three-way dip closed with two culminations joined by a saddle. It is located between 9 and 18 km northwest of North Morecambe and covers an area of over 30 sq km (Fig. 1). The field contains dry gas with a methane content of 88%, 10% N_2 and traces of CO_2. The Millom accumulation is operated by BG E&P Ltd with Purbeck Exploration having 18% equity in block 113/27a.

Although the discovery well encountered a gas column in excess of 500 ft in the Triassic Ormskirk Sandstone Fm, production test rates were poor, reaching a maximum 14.6 MMSCFD. Preliminary petrographic studies suggested that the structure suffered from the effects of permeability reducing platy illite and that porosity was reduced by quartz and carbonate cementation.

This paper presents a regional evaluation of reservoir quality over the Keys Basin–Keys Fault rollover region of the East Irish Sea Basin (EISB), and proposes a model for platy illite distribution in the EISB which explains the low porosity and permeability encountered in the Millom Field.

Database

A total of 647 feet of core was available from the three Millom wells as well as over 5000 ft of core

from four of the development wells in the nearby North Morecambe Field (Cowan 1996). In addition a full suite of wireline logs and measured core porosity and permeability data were available for all cored wells in the area. This paper is based upon the interpretation of point count (modal analysis) data from thin sections. Each sample was impregnated with blue epoxy and stained for carbonates (potassium ferricyanide and alizarin red) and K feldspar (sodium cobaltinitrite). Samples were selected to cover the full suite of lithofacies types, porosity–permeability range and reservoir layers. These were compared with log-derived 0.5 foot spacing porosity estimates and 1 foot spacing core data. Thin sections were sampled at roughly 20 foot spacing and a total of 330 thin sections were described and counted (300 points). Secondary electron imaging was carried out on selected samples to identify the clay mineral morphologies present. Clay-fraction X-ray diffraction was carried out on 12 samples to characterize the clay mineral species present.

Reservoir properties

Reservoir properties are generally excellent within the Morecambe area of the EISB, with arithmetic average gross porosities and permeabilities up to 15% and 200 mD, respectively. In production wells, flow into the well bore comes dominantly from thin layers (circa 10 ft) with porosities in

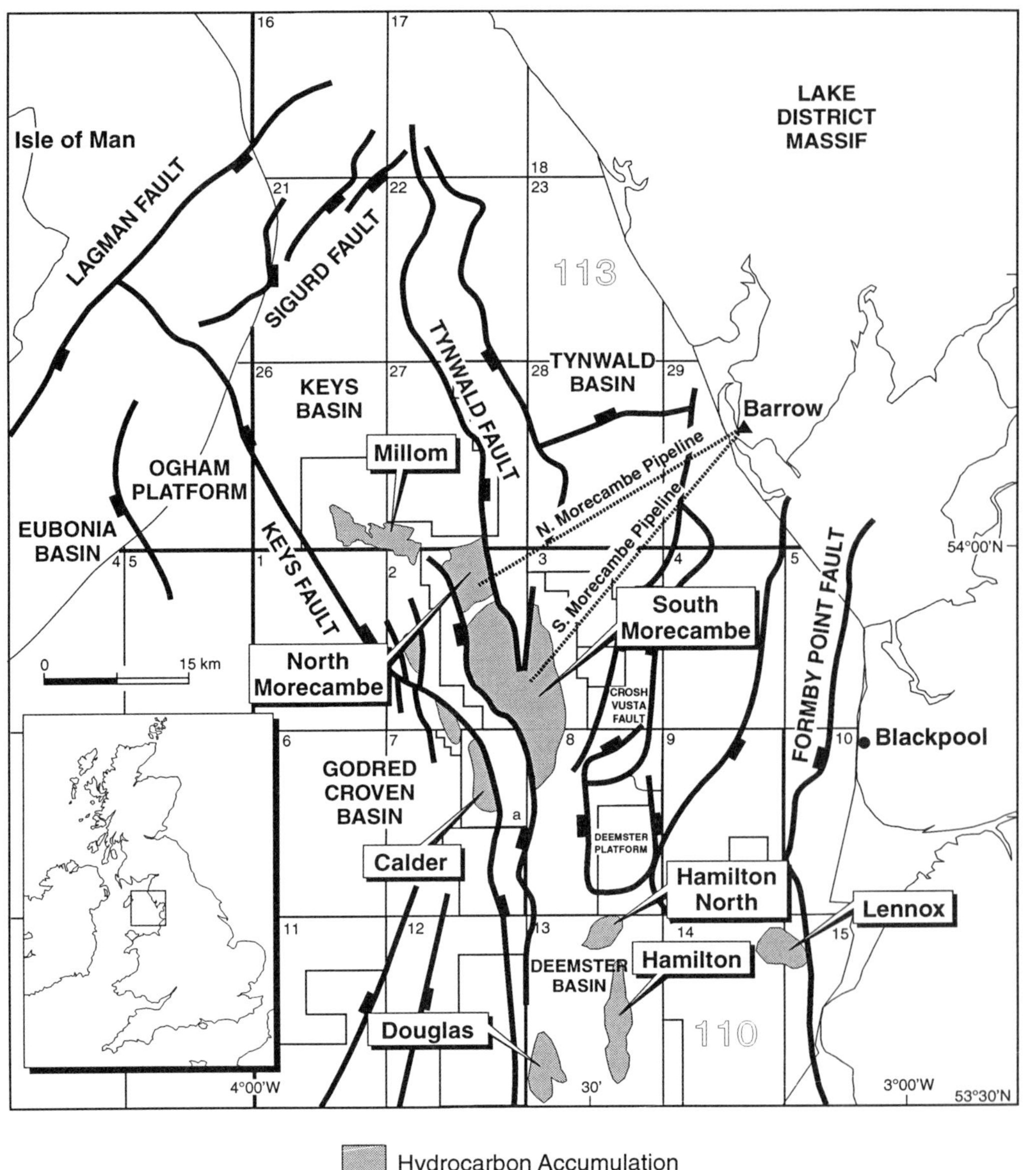

Fig. 1. Location map showing the position of Millom in relation to the major structural elements.

excess of 20% and permeabilities in the darcy range (Cowan 1993, Meadows and Beach 1993). Table 1 shows representative porosity and permeability data for the Ormskirk Sandstone in the Millom Field compared with the nearby North and South Morecambe Fields. Permeability is in parts reduced by the presence of authigenic platy illite in the North and South Morecambe Fields, hence the illite-free and illite-affected values shown in Table

1 (Bushell 1986, Woodward and Curtis 1987). Colter and Ebbern (1978) first suggested that the top of the platy illite affected zone marked a palaeo-hydrocarbon/water contact, below which illite precipitated. Over the Millom Field, the full section of the Ormskirk Sandstone Fm is affected by platy illite. The arithmetic average permeabilities of the Millom area are low for the basin, (0.2 mD) and the maximum values reach less than 100 mD. In the

Table 1. *Porosity and permeability data for the Morecambe Millom area (gross arithmetic averages from wells)*

	Millom	North Morecambe	South Morecambe
Illite free			
Porosity (%)	—	8–12	9–15
Permeability (mD)	—	5–180	30–1000
Illite affected			
Porosity (%)	6–11	9–15	12–17
Permeability (mD)	0.05–1.0	0.02–2.0	0.1–100

illite-free zones of North Morecambe permeabilities of up to 10 D are recorded, but even in the illite-affected zones of North Morecambe permeabilities can reach 500 mD (Fig. 2). Although the presence of illite can, in part, explain the low permeabilities in the field, it cannot explain the low porosity values encountered.

Depositional controls of reservoir properties

The Ormskirk Sandstones Fm varies in thickness from 650 to 850 ft in the Millom–Morecambe area, and was deposited in a series of fluvial and aeolian episodes within a semi arid setting. It overlies the dominantly fluvial St Bees Sandstone. Four major facies associations are recognized, each comprising a characteristic suite of lithofacies types (Table 2). These are: low sinuosity fluvial, siliciclastic sabkha[1], playa/channel abandonment and aeolian (dune and sandsheet) associations. Over the Morecambe and Millom area the distributions of the associations suggest the Ormskirk Sandstone Gp was deposited in a compound wetting-upwards megacycle some 800 ft thick. Meadows and Beach (1993) and Thompson & Meadows (this volume) have suggested that a rising and falling water table controlled the deposition of fluvial, sheetflood (*sic*) and aeolian sedimentation in the EISB. Herries and Cowan (this volume) have identified drying upwards patterns 10–20 ft thick and used these to construct lithostratigraphic correlations which extend over tens of kilometers. Cowan (1993) used pressure data from South Morecambe to verify similar lithostratigraphic layering over a 2–3 km spacing. A robust and detailed lithostratigraphic layering of facies associations can therefore be constructed over the study area (Fig. 2). However, over the southern part of the basin around blocks 110/13, 14 and 15 (Jackson *et al.* this volume,

Thompson and Meadows, this volume) a major aeolian system is developed at the top of the Ormskirk Sst Fm. This layering is significant for diagenetic studies since the primary porosity is very sensitive to the abundance of aeolian facies in any well. Thus, over the Millom–Morecambe area, cores taken near the base of the Ormskirk Sandstone Fm, where aeolian facies are more abundant, have higher primary porosities than the shallower zones, where sabkha and fluvial facies dominate. The opposite relationship holds in the southern part of the basin. It is therefore necessary to compare the diagenesis of similar lithofacies types on a layer by layer basis. Since the lithofacies distributions are similar, and can be correlated with confidence over the Millom–Morecambe area, the observed porosity decrease over the Millom area cannot be ascribed to lithofacies changes. Figure 3 shows how visible porosity is reduced by increased cementation in each facies association in the Millom area.

Diagenetic controls of reservoir properties

Various aspects of the diagenesis of EISB sediments have been described and illustrated by Colter & Ebbern (1978), Bushell (1986) Woodward & Curtis (1987), Stuart & Cowan (1991) and Meadows & Beach (1993), whilst Burley (1984) has described general aspects of diagenesis of Triassic reservoir sandstones. Stuart (1993) described the diagenesis of the North Morecambe field and identified tentatively a porosity decrease away from the crest of this structure which he attributed to progressive hydrocarbon filling of the structure, preserving porosity by inhibiting diagenetic mineral growth. Diagenetic fabrics have been well illustrated by the above authors (e.g. Stuart & Cowan 1991, fig. 8a–d, p. 534).

In this study, diagenesis has been simplified by considering quartz and carbonate cement for their effects on porosity, since over 90% of the pore blocking cement comprise quartz and dolomite

---

[1]Sensu Fyberger *et al.* 1993. Replaces the term 'sheetflood', see Herries and Cowan, this volume.

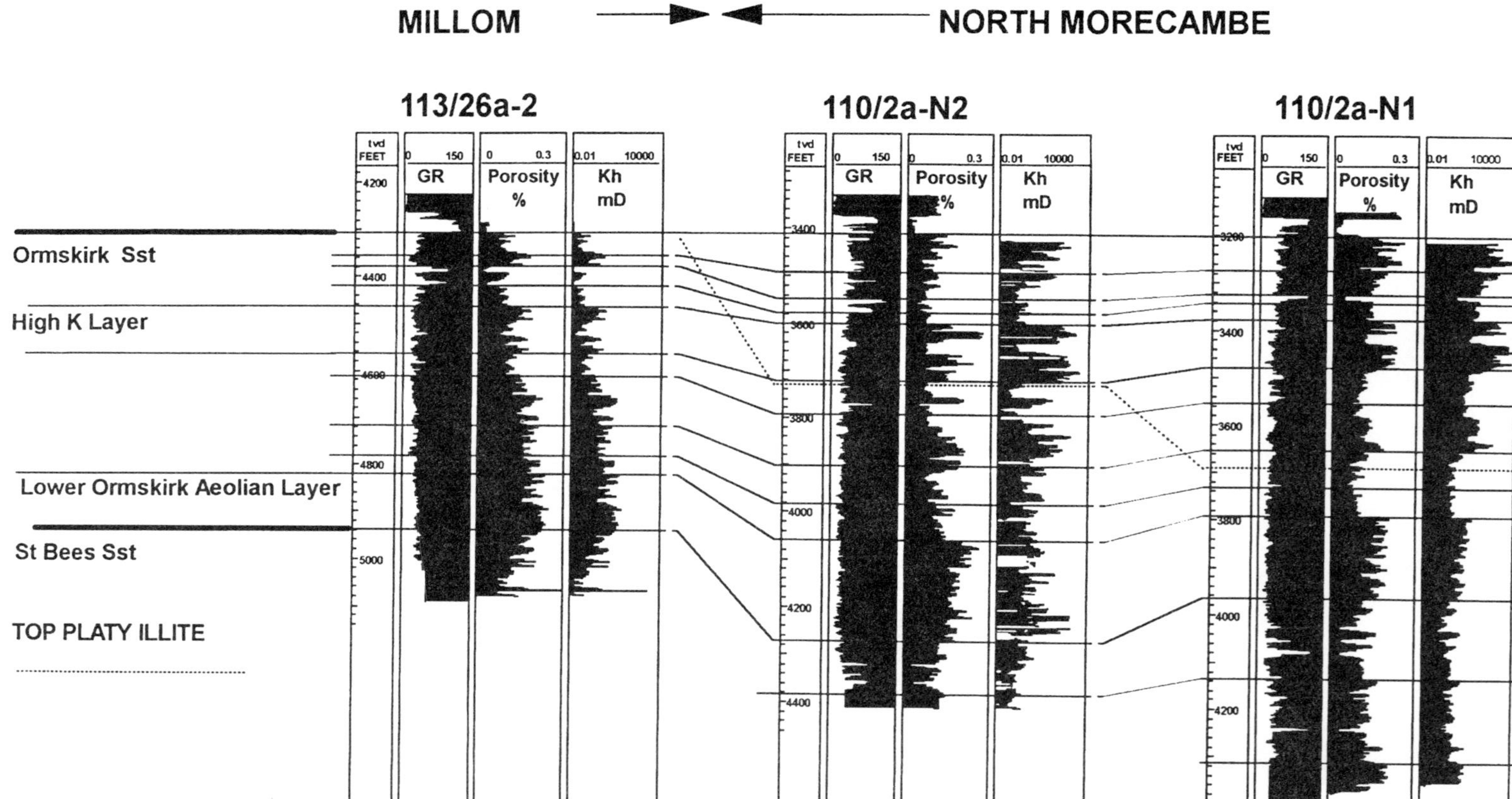

Fig. 2. Gamma-ray, porosity and permeability profiles through Millom and North Morecambe wells.

Table 2. *Lithofacies associations*

Major association	Sub associations	Lithofacies types
Fluvial	Ephemeral channels	Trough cross-laminated sst, massive sst, abandonment fines.
	Stacked channels	Planar cross-laminated sst, pebble-clast conglomerates, abandonment fines.
Aeolian	Aeolian dune	Avalanche and grainfall cross-laminated sst. Wind ripple laminated sst.
	Aeolian sandsheet	Wind ripple laminated sst.
Siliciclastic sabkha*	Non evaporitic sabkha	Wavy laminated argillaceous sst.
Non reservoir fines	Playa	Shales showing desiccation cracks, sand injection structures.
	Abandonment shales	Thin shales associated with Channel tops

* Sheetflood of Stuart and Cowan (1991) and Bushell (1986).

(both ferroan and non-ferroan). Evaporite cements (gypsum, anhydrite, barite, halite) only locally form abundant cement modes (up to 20%). The presence of platy illite is the principal factor controlling permeability (Colter & Ebbern 1978; Bushell 1986), and this overprints but does not destroy the primary depositional controls on permeability (Cowan 1996).

Detrital mineralogy

Most of the sandstones from the Millom area are quartz arenites, with subordinate sub arkoses and sublithic-arenites (McBride 1963) the most significant detrital components (after quartz) being K feldspar (0–7%) and rock fragments (1.3–7.7%), dominantly intraformational mudstone clasts. K feldspar is more abundant in North and South Morecambe (up to 10%) than Millom. Etched and dissolved feldspars are common in the EISB, and the fact that feldspar is absent from the down-dip well 113/26a-2 but present in correlatable layers in up-dip wells to west and east of this well suggests that the present day distribution of feldspar is strongly controlled by dissolution. Mica is common as an accessory but is not abundant (0–3%). The heavy mineral suite comprises rutile, tourmaline and zircon.

Authigenic mineralogy

Quartz and dolomite cements dominate the authigenic phases present, but anhydrite, gypsum, haematite, pyrite, anatase, calcite, kaolinite, siderite and halite are present in limited amounts. Illite is the most significant clay mineral. The evaporitic minerals can reach significant abundances (up to 20%) in limited numbers of samples, whilst halite

can locally form poikilotopic cements where the halites of the overlying Mercia Mudstone Group have been downfaulted against the Ormskirk Sandstone Fm. However these minor cements show no systematic relationships with the lithostratigraphy or lithofacies associations. Point counted estimates of the abundance of illite are not considered accurate since the width of illite plates (>30 μm) is similar to the thickness of the section. Illite abundance is therefore overestimated at the expense of porosity by point counting.

Quartz

Quartz cements exist as syntaxial overgrowths of which at least two phases can be seen, one which predates, and one which postdates illite. It is normally not possible to differentiate between the two phases optically but the relationship with illite can be seen clearly in SEM images. Early quartz cements postdate non-ferroan dolomite but pre-date ferroan dolomite (Stuart and Cowan 1991, Fig. 8a, p. 354). Quartz can be abundant, forming up to 25% of the mode, but average values are usually less than 10%.

Carbonates

Dolomite cements form poikilotopic micronodules which clearly predate platy illite and most of the quartz cementation. These nodules are ubiquitously overgrown by rhombic ferroan dolomite which postdates the illite and quartz cementation. Carbonate cements can reach over 40% of mode. In North and South Morecambe early non-ferroan dolomite cements are relatively more abundant than the later ferroan dolomite cements. In the Millom area the opposite relationship holds, with the later,

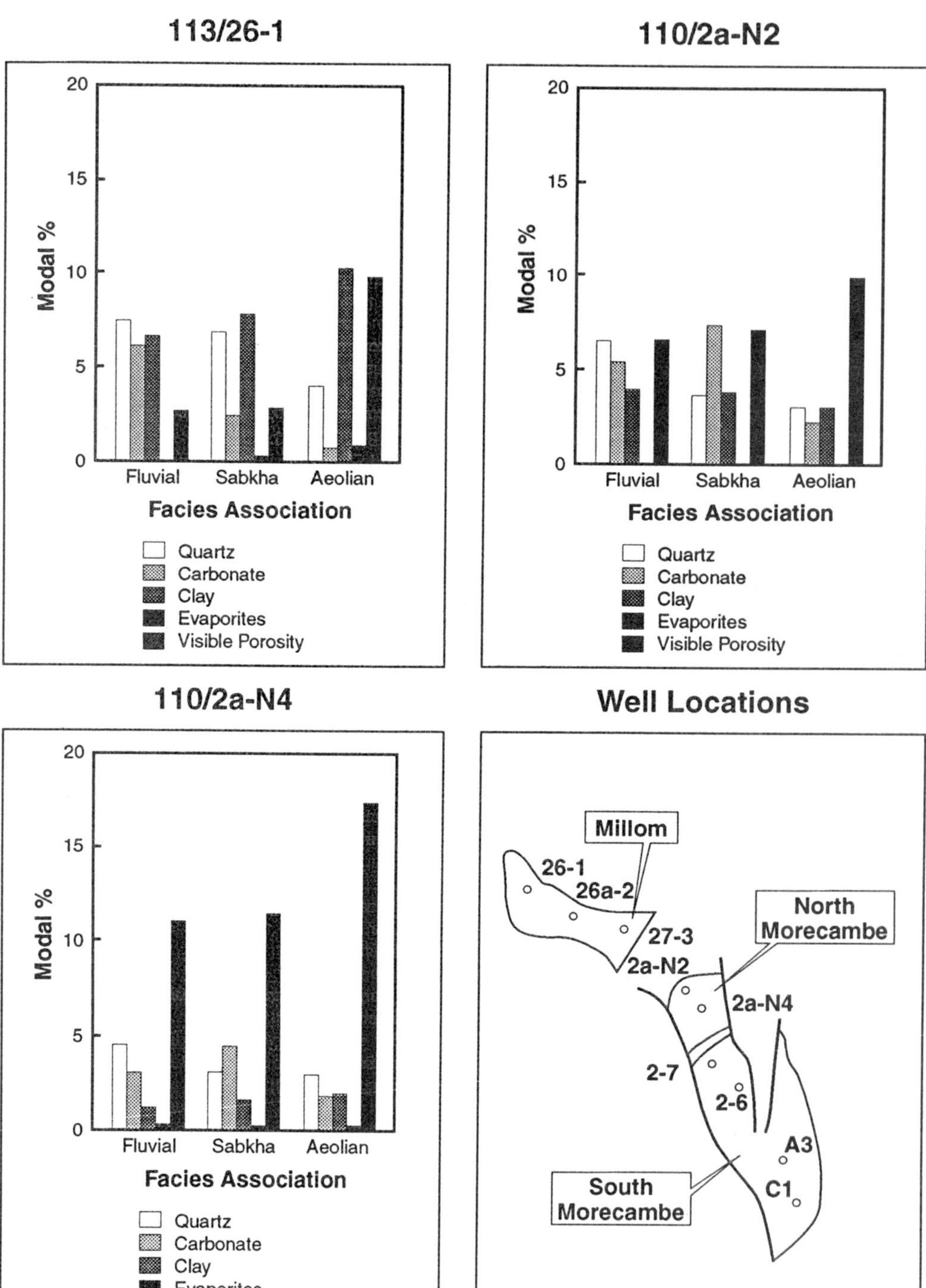

Fig. 3. Cement and porosity distributions in Millom and Morecambe displayed by reservoir facies. Non-reservoir playa and abandonment facies are omitted. A strong relationship exists between the cementation and facies, especially aeolian, and it can be seen that the Millom wells have lost porosity through enhanced cementation. For 113/26-1, $n = 12$; 110/2a-N2, $n = 72$; 110/2a-N4, $n = 42$.

Early Jurassic, 180 my

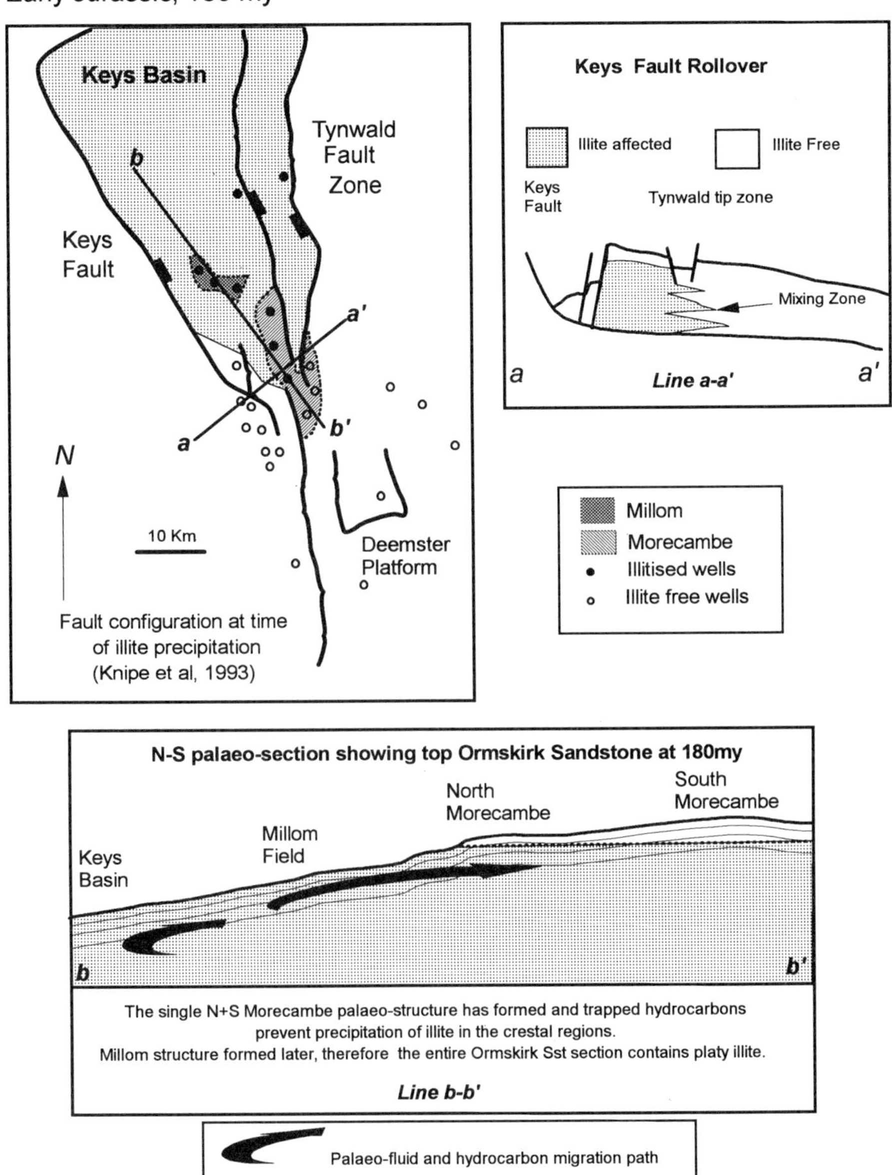

Fig. 4. Distribution of platy illite in the East Irish Sea Basin. Platy illite is confined to the Keys Basin and the broad ridge which forms the Morecambe structures. No platy illite has been encountered east or south of the dip-closed eastern flank of South Morecambe.

ferroan dolomite cements being more abundant. Early diagenetic ferroan dolomite nodules tend to be concentrated near the bases of the major channel-facies sandstones. These possibly represent aggradation on channel base lags rich in reworked caliche deposits.

Platy illite

Much of the diagenetic work carried out in the EISB has been concerned with the description and delineation of the layer of platy illite, which is believed to have precipitated below a palaeo-

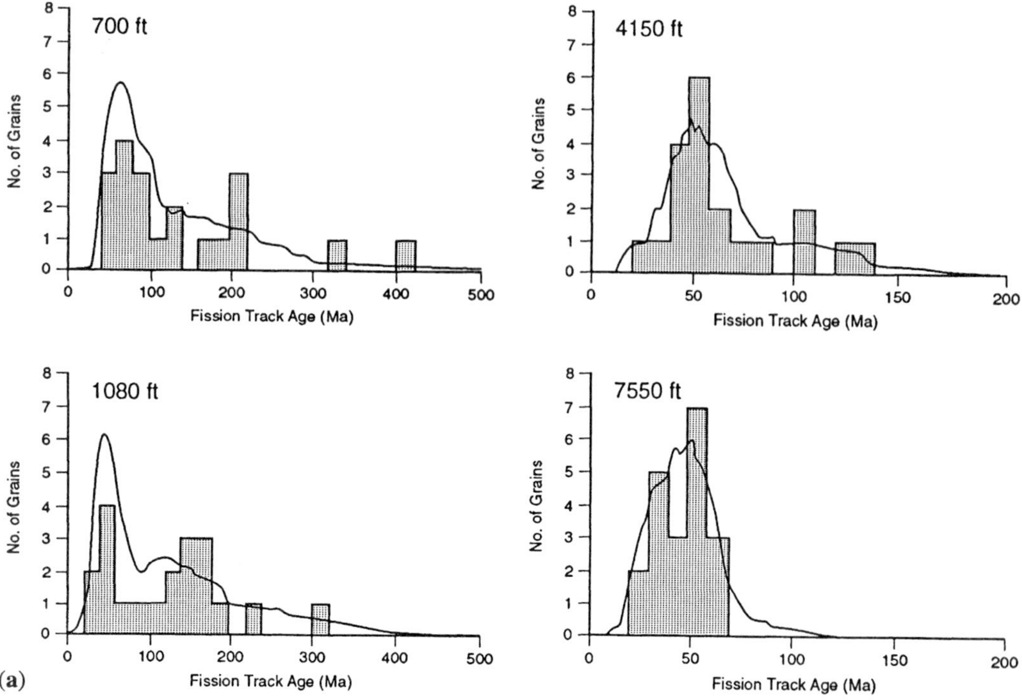

Fig. 5. (a) Apatite fission track single grain age histograms from well 113/26-1. Apatite fission track distribution analysis shows maximum palaeotemperatures were reached 55–60 Ma.

hydrocarbon water contact (Bushell; 1986, Woodward & Curtis 1987). Detailed study using transmission electron microscopy (TEM), X-ray diffraction (XRD) and microprobe analysis has shown that the illite morphologies observed in the Morecambe wells (tangential, platy and fibrous) have a similar phengitic composition and, therefore, probably precipitated in equilibrium with the pore fluids rather than via a temperature controlled replacement reaction (Woodward & Curtis 1987). K–Ar dating of the illite indicates that it precipitated *c.* 180 Ma, after the initial growth of the Morecambe structure (Bushell 1986).

Distribution of platy illite

Figure 4 shows the distribution of platy illite in the EISB, superimposed on the fault pattern which is likely to have been in existence at the time of illite precipitation, *c.* 180 Ma (Knipe *et al.* 1993). Much of this data comes from North and South Morecambe where over 50 wells have been drilled, as well as the 60 exploration wells drilled in the basin. No illite was encountered on the eastern flanks of South Morecambe, but all of the development wells reached TD above the South Morecambe free water level. At the time of drilling

it was assumed that these wells had not been drilled deep enough to encounter the top platy illite surface, and platy illite was present in the aquifer beneath the field. The palaeo-hydrocarbon contact model implied that the eastward-dipping top Ormskirk Sst Fm surface should intersect the top platy illite surface to the east of South Morecambe. An exploration well drilled to the east of south Morecambe in 1987 found no platy illite throughout the complete Permo-Triassic section. No platy illite has been encountered by exploration wells to the south or to the west of the South Morecambe Field, and the oil and gas fields in block 110/13 are also free of 'Morecambe type' platy illite (Yaliz, this volume; Greenwood & Habesch, this volume). Thus platy illite distribution is limited to an area between the Keys and Tynwald fault zones. The limited distribution gives strong clues to its likely origin. It is proposed that K+ rich fluids (from feldspar dissolution) and hydrocarbons migrating out of the deeply buried Keys Basin kitchen area (Fig. 4) were channelled along the Ormskirk Sst–Mercia Mudstone boundary between the Keys and Tynwald Fault zones, i.e. along a broad rollover of the Keys Fault. Upon reaching the Tynwald fault tip zone, these fluids mixed with aquifer waters of the less deeply buried Tynwald Basin, reducing the

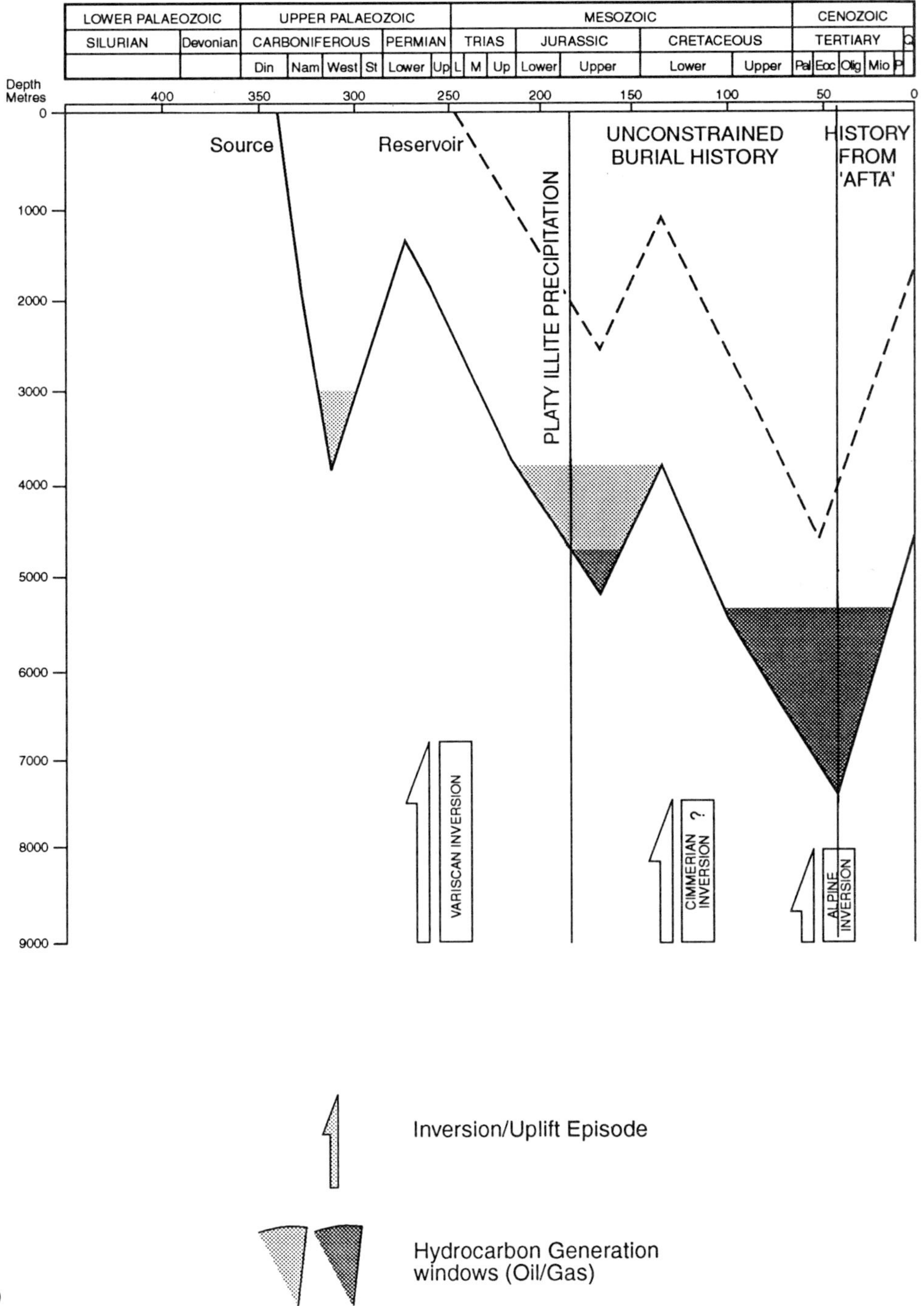

(b)

Fig. 5. (b) Burial history model of the Keys Basin area, where at the present day the Top Sherwood Sandstone is buried to depths in excess of 10 000 ft.

saturation of K$^+$ ions, and halting the precipitation of illite to the east of the zone of mixing.

Paragenesis

The basin has suffered from multiple episodes of uplift and burial (Hardman *et al.* 1993; Lewis *et al.* 1993; Green *et al.* 1993). Multi-phased cementation episodes and multiple phases of hydrocarbon generation have also been documented (Hardman *et al.* 1993; Stuart & Cowan 1991; Woodward & Curtis 1987; Greenwood & Habesch, this volume), Apatite fission track analysis of Sherwood Sandstone Group samples from the North Morecambe Field and Millom Well 113/26-1 (Fig. 5) show that apatite fission tracks were re-set between 55–65 Ma by temperatures in excess of 110°C and, therefore, that significant inversion and cooling occurred during the Tertiary (the present-day North Morecambe reservoir temperature is 32°C). Post-Triassic inversion was estimated to be around 1300–1700 m, (4000–5000 ft) by Colter & Barr (1975), based upon spore colouration, vitrinite reflectance of the underlying Carboniferous and shale velocities in the Mercia Mudstone. Jackson *et al.* (1987) estimated similar amounts of uplift by using Mercia Mudstone shale velocities. Lewis and co-workers (1993) have postulated that up to 3 km (10 000 ft) of uplift is required to explain the results of apatite fission track analysis (Lewis *et al.* 1993). There is strong evidence for at least two phases of hydrocarbon migration, the first, indicated by the platy illite palaeohydrocarbon-contact, dated

180 Ma, and the second as a consequence of the late Cretaceous/Tertiary heating/burial event (Stuart & Cowan 1991; Hardman *et al.* 1993; Armstrong *et al.*, this volume). This multiple hydrocarbon migration history creates a complex diagenetic history, the details of which have yet to be fully resolved and are outside the scope of the present study. A simplified paragenetic sequence for the Morecambe Millom area is shown in Fig. 6. Detailed descriptions of diagenetic pathways can be found in Burley (1984) and Stuart & Cowan, (1991). In examining the diagenetic history of a number of structures it is to be expected that the diagenetic histories of each structure will differ to a greater or lesser degree, however in the Millom–Morecambe area, and indeed throughout the EISB (Greenwood & Habesch, this volume), the diagenesis of the Ormskirk Sandstone Fm appears to be broadly similar, and multiple hydrocarbon fills appear to be the norm rather than the exception. The early diagenetic history of the Ormskirk Sst Gp shows features comparable to 'red-bed' type diagenesis described by Walker *et al.* (1978) and Burley (1984), namely eluviated clay rims being reddened by iron oxide cements liberated by the dissolution of ferromagnesian grains in an oxidizing geochemical environment. This red colouration has been removed from the top 800 ft or so of the hydrocarbon bearing structures by reducing fluids, probably associated with hydrocarbon fill. Significantly, this reduction appears to have had a limited effect in the Millom area since much of the recovered core appears red, with visible effects of

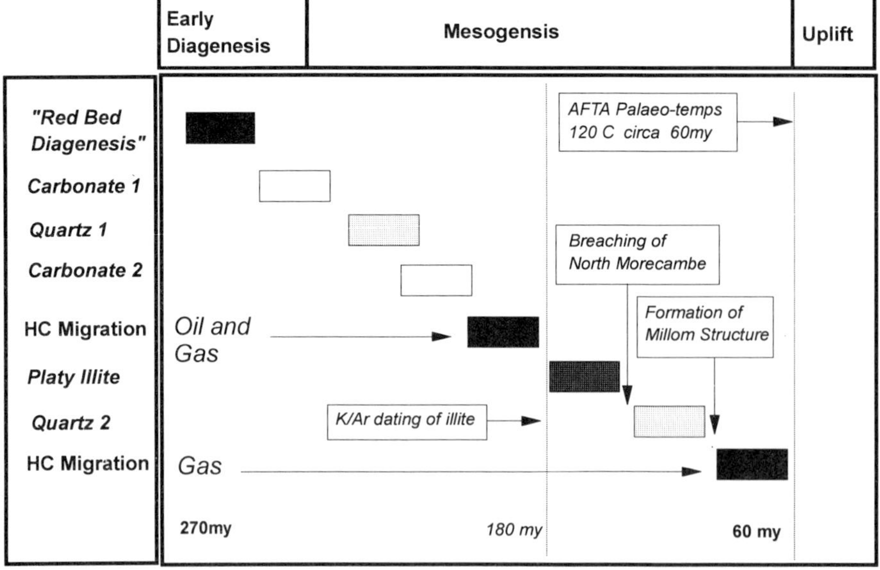

Fig. 6. Simplified Paragenetic sequence of the Morecambe Millom area.

reduction limited to thin bands and spots. In North and South Morecambe (Woodward & Curtis 1987) and in the Douglas Field (Yaliz, this volume) there is evidence of at least two discrete hydrocarbon migration phases separated by a phase of reservoir breaching and quartz cementation. Bitumen staining is common throughout the basin, (Stuart & Cowan 1991; Hardman *et al.* 1993; Greenwood & Habesch, this volume), and quartz overgrowths, which postdate this staining, provide unequivocal evidence of reservoir breaching. In order to allow precipitation of quartz, aqueous solutions must be able to circulate or pass through the reservoir section. Unfortunately, gas does not leave such a handy tell-tale behind, so breaching must be inferred by petrographic criteria. In South Morecambe this is manifested by an increase in post-illite quartz cements in the illite-free sections of the reservoir (Stuart & Cowan 1991). Evidence is less conclusive in North Morecambe; Stuart (1993) rejected a multiple fill model in favour of a single hydrocarbon charge. With a greater database available (Cowan 1996) proposed a multiple fill model for North Morecambe citing the presence of bitumen staining as evidence of an earlier hydrocarbon charge.

The fact that cementation occurred both before and after the first hydrocarbon charge of the reservoirs is fundamental to the understanding of cement distributions in the basin. Whether this is a reflection of a continuous background of precipitation or discrete episodes of cementation is unclear at the moment. Hardman *et al.* (1993) showed fluid inclusion evidence which supports two phases of cement precipitation in the south of the basin, however no fluid inclusion data has been published for the northern part of the basin. The evidence from the Keys Basin–Keys Fault rollover area suggests that if an early (Jurassic) hydrocarbon charge is not present in any particular structure, porosity will be reduced by quartz and late-stage ferroan-dolomite cementation.

Implications for Millom diagenesis

Since the Millom accumulation is located down-flank of the North and South Morecambe palaeo-high, it is within the migration pathway from the deeply buried Keys basin kitchen area to the Morecambe Field rollover. The structural style of Millom is quite different to that of the Morecambe Fields. Millom is *dip-closed* closed on three sides whilst the Morecambe Fields are both fault closed on three sides. Within the Ormskirk Sandstone Fm, Millom and North Morecambe show differing porosity depth trends. There is a progressive decrease in porosity upwards towards the Mercia Mudstone Gp within the Millom wells, but the

Morecambe crestal well 110/2a-N1 shows an upwards increase in porosity towards the top of the Orsmkirk Sandstone, whilst the core logs (Fig. 2) show similar distributions of facies associations. It is proposed that this upwards decrease in porosity reflects mineral precipitation along the Mercia Mudstone Gp/Ormskirk Sandstone Fm boundary. Mineralizing fluids flowing upwards out of the Keys Basin kitchen area were restricted from lateral flow by the Keys and Tynwald Fault zones and channeled through the Morecambe structures. The abundance of cements is greater in Millom because the structure formed late during basin evolution, probably as a consequence of Alpine compression and reverse movement along the Keys Fault during basin inversion (Knipe *et al.* 1993). The Millom area was therefore exposed to migrating fluids for most of the basin history. The increased abundance of post-quartz ferroan-dolomite in the Millom field supports this hypothesis. This model suggests that Ormskirk Sandstone reservoir properties will be uniformly poor in the Keys Basin unless early (pre-Jurassic) structures can be identified. It also suggests that cemented zones will be laterally extensive since the fluids sourcing them were migrating parallel to the depositional layering. This has significant implications for development scenarios which will be discussed below. It is

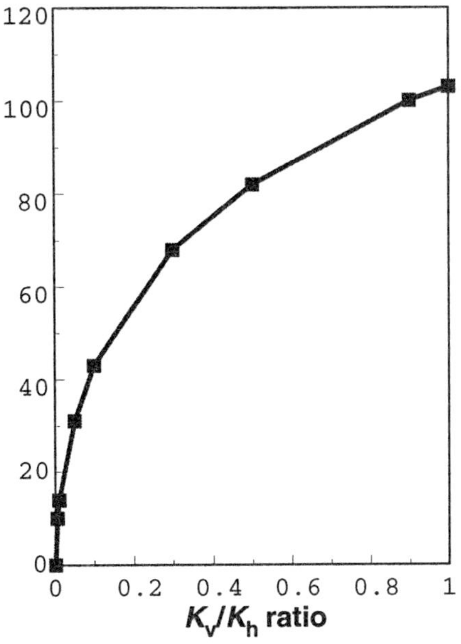

Fig. 7. Horizontal well deliverability as a function of K_v/K_h ratio. Rapid decrease in deliverability occurs below 0.2 K_v/K_h ratio.

obviously difficult to identify bedding-parallel cemented zones using a petrographic database with a 20 ft spacing, but the model predicts that mineral precipitation will enhance the vertical permeability heterogeneity inherent within the depositional system.

Diagenesis and development scenarios – does it make a difference?

Millom is a tight gas field which covers a large area. Under conventional scenarios developing such a field would require a large number of vertical wells, from at least two platforms to achieve efficient deliverability and drainage of the reservoir. Ostensibly, Millom is an ideal candidate for development by horizontal wells, which can show spectacular increases in productivity. How long high productivity is maintained is a function of a number of factors, the most important of which are reservoir architecture, including faulting, and the nature of the aquifer. Since the Millom reservoir is layered, and is largely unfaulted, uncertainty exists over the ability of a horizontal well to effectively drain each reservoir layer. In addition, the fact that porosity and permeability increase with depth, and that the aquifer is around 4000 ft thick, gives poorly sited horizontal wells a high probability of early water breakthrough. Figure 7 shows the predicted productivity of a Millom well as a function of field wide K_v/K_h ratio. It shows that at low K_v/K_h ratios, horizontal wells suffer severe productivity restriction. At low K_v, horizontal wells are unable able to drain layers above and below the producing zone and deliverability cannot be sustained. The absolute value of K_v in the Millom field increases with depth as a consequence of the diagenetic porosity/depth trend and an increase in abundance of aeolian sediment with depth. The lowest absolute values occur about 100 ft from the top of the reservoir since a number of field wide playa deposits are present which form effective barriers to vertical flow. However, cemented zones are also probably laterally extensive and will tend to exaggerate the primary depositional permeability

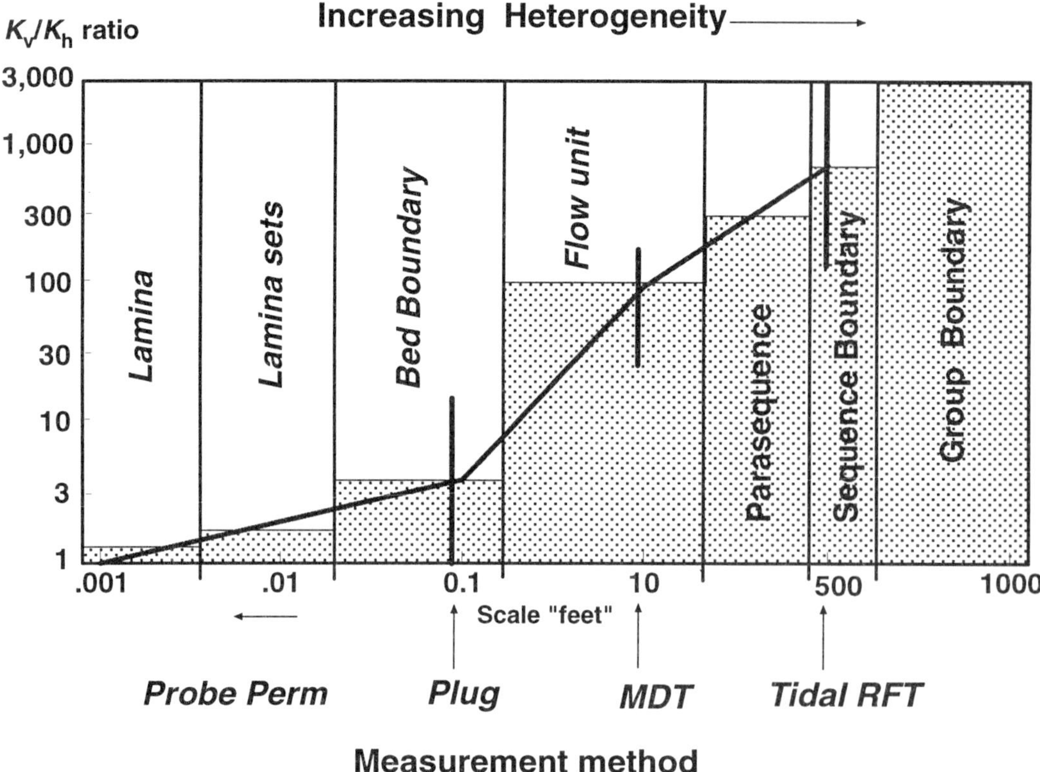

Fig. 8. K_v/K_h ratio as a function of scale. Measured values are shown with the solid line. The K_v/K_h ratio does not vary smoothly with scale. The stippled area represents likely step changes in K_v values as a function of the geological architecture.

heterogeneity. Figure 8 shows measured K_v/K_h values for the EISB estimated at a number of scales: plug data measuring on a >0.3 m scale. Schlumberger's Modular Dynamic Tester (MDT) in interference mode, measuring downhole on a 2 m scale (Ayan *et al.* 1995), and using the RFT response to tidal loading (Wannell & Morrison 1990) on a scale of hundreds of metres. It is tempting to draw a straight line through these data to give a log–log 'relationship'. However the K_v/K_h scale dependency does not vary smoothly but is a function of the geological architecture. Probe permeametery can help define the macro scales of heterogeneity, but the interlayer transmissibilities control K_v on a larger (10s of feet) scale. Estimates of K_v/K_h by upscaling probe permeametry data from the North Morecambe area (Thomas 1995) show that at a vertical thickness of over 15 ft the K_v/K_h ratio of illite-affected sediments varies from 0.012 to 0.018.

Development data from the South Morecambe Field has shown that the illite-free reservoir sections behave like a tank, i.e. gas depletes uniformly and pressure equilibrates quickly. However, sensitivity studies using a single phase simulation model have shown that the Millom reservoir, despite having similar primary depositional char-acteristics to South Morecambe, will behave as a heterogeneous layered system over the range of uncertainty in K_v. This modelling suggests each layer will deplete differentially and will therefore need to be drained separately. This contrast in reservoir behaviour is solely a consequence of the different diagenetic history of the Millom reservoir.

Conclusions

The Millom structure was formed late in the history of the East Irish Sea Basin. It is located between the North Morecambe Field and the Keys Basin kitchen area and as a consequence has suffered enhanced porosity reduction by precipitation of mineral cements, and permeability reduction by authigenic illitization. It is predicted that the field will be behave as a heterogeneous layered system during production and will not be suitable for development by horizontal wells alone.

We would like to thank the directors of British Gas Exploration and Production and Purbeck Exploration for permission to publish this paper. Reservoir modelling was carried out by Julian Slater and Jon Story. Jeremy Greenwood of Geochem has provided helpful discussion and with others, carried out the modal analysis.

References

ARMSTRONG, J. P., SMITH, J., D'ELIA, V. A. A. & TRUEBLOOD, S. P. 1997. The occurrence and correlation of oils and Namurian source rocks in the Liverpool Bay–North Wales area. *This volume.*

AYAN, C., COLLEY. N., COWAN. G. *ET AL.* 1994. Measuring permeability anisotropy, the latest approach. Unrefereed article, *Oilfield Review.*

BURLEY, S. D. 1984. Patterns of diagenesis of the Sherwood Sandstone Group (Triassic), U.K. *Clay Minerals*, **19**, 403–440.

BUSHELL, T. P. 1986. Reservoir geology of the Morecambe Field. *In*: BROOKS, J., GOFF, J. C. & VAN HOORN, B. (eds) *Habitat of Palaeozoic Gas in Northwest Europe.* Geological Society, London, Special Publication, **23**, 189–207.

COLTER, V. S. & BARR, K. W. 1975. Recent developments in the geology of the Irish Sea and Cheshire Basins. *In*: (WOODLAND, A. W. (ed.) *Petroleum and the Continental Shelf of North-West Europe*, Vol. 1. Applied Science, London, 61–75.

—— & EBBERN, J. 1978. The petrography and reservoir properties of some Triassic sandstones of the Northern Irish Sea Basin. *Journal of the Geological Society, London*, **135**, 57–62.

—— & —— 1979. SEM studies of Triassic. reservoir sandstones from the Morecambe Field, Irish Sea, UK. *Scanning Electron Microscopy*, **1**, 531–538.

COWAN, G. 1993. Identification and significance of aeolian deposits within the dominantly fluvial Sherwood Sandstone Group of the East Irish Sea Basin UK. *In*: NORTH, C. P. & PROSSER, J. (eds) *Characterization of fluvial and aeolian reservoirs.* Geological Society, London, Special Publication, **73**, 231–245.

—— 1996. The development of the North Morecambe Gas Field. *Petroleum Geoscience*, **2**, 43–52.

FRBERGER, S., AL-SARI, M. & CLISHAM, T. J. 1983. Eolian dune, interdune, sandsheet and siliciclastic sabkha sediments of an offshore prograding sand sea, Dharahn area Saudi Arabia. *American Association of Petroleum Geologists Bulletin*, **67**, 208–312.

GREEN, P. F. 1986. On the thermo-tectonic evolution of Northern England: evidence from fission track analysis. *Geological Magazine*, **123**, 493–506.

——, DUDDY, I. R. & BRAY R. J. 1993. Early Tertiary heating in Northwest England: fluids or burial (or both). *In*: PARNELL, J. *et al.* (eds) *Geofluids '93*, extended abstracts, 119–24

GREENWOOD, J. & HABESCH, S. 1997. Diagenesis of the Sherwood Sandstone Group in the Southern EISB (Blocks 110/13, 110/14 and 110/15): constraints from preliminary isotopic and fluid inclusion studies. *This volume.*

HARDMAN. M., BUCHANAN, J., HERRINGTON, P. & CARR, A. D. 1993. Geochemical modelling of the East Irish Sea Basin: its influence on predicting hydrocarbon type and quality. *In*: PARKER, J. R. (ed.) *Petroleum Geology of Northwest Europe: Proceedings of the 4th Conference.* Geological Society, London, 857–866.

HERRIES, R. D. & COWAN G. 1997. Challenging the 'sheetflood' myth: the role of water-table-controlled sabkha deposits in redefining the depositional model of Ormskirk Sandstone Formation (Lower Triassic), East Irish Sea Basin. *This volume.*

JACKSON, D. I., MULLHOLLAND, P., JONES, S. M. & WARRINGTON, G. 1987. The geological framework of the East Irish Sea Basin. *In*: BROOKS, J. & GLENNIE, K. (eds) *Petroleum Geology of North West Europe.* Graham & Trotman, London, 191–203.

—, JOHNSON, H. & SMITH N. J. P. 1997. A lithostratigraphic nomenclature for the Carboniferous–Triassic Rocks of the East Irish Sea Basin. *This volume.*

KNIPE, R., COWAN, G. & BALENDRAN, V. S. 1993. The tectonic history of the East Irish Sea basin with special reference to the Morecambe Fields. *In*: PARKER, J. R. (ed.) *Petroleum Geology of Northwest Europe: Proceedings of the 4th Conference.* Geological Society, London, 857–866.

LEWIS, C. L. E., GREEN, P. F., CARTER, A. & HURFORD, A. J. 1992. Elevated K/T palaeotemperatures throughout Northern England: three kilometres of Tertiary exhumation. *Earth and Planetary Science Letters,* **112**, 131–145.

MCBRIDE, E. F. 1963. A classification of common sandstones. *Journal of Sedimentary Petrology,* **33**, 664–669.

MEADOWS, N. S. & BEACH, A. 1993. Controls on reservoir quality in the Triassic Sherwood Sandstone of the East Irish Sea Basin. *In*: PARKER, J. R. (ed.) *Petroleum Geology of Northwest Europe: Proceedings of the 4th conference.* Geological Society, London, 823–833.

STUART, I. A. 1993. The geology of the North Morecambe gas field. *In*: PARKER, J. R. (ed.) *Petroleum geology of Northwest Europe: Proceedings of the 4th Conference.* Geological Society, London, 883–895.

—— & COWAN, G. 1991. The South Morecambe Field. *In*: ABBOTS, I. L. (ed.) *UK Oil and Gas Fields 25 Years Commemorative Volume.* Geological Society, London, Memoir, **14**, 527–541.

THOMPSON, J. & MEADOWS, N. S. 1997. Clastic sabkhas and diachroneity at the top of the Sherwood Sandstone Group, East Irish Sea Basin. *This volume.*

WALKER, T., WAUGH R. B. & CRONE, R. J. 1978. Diagenesis of first cycle desert alluvium of Cenozoic age. Southwestern United States and northwestern Mexico. *Bulletin of the Geological Society of America,* **89**, 19–32.

WANNELL, M, J. & MORISSON, S. J. 1990. Vertical permeability measurement in new reservoirs using tidal pressure changes. SPE paper 20532, 65th SPE Annual Technical Conference, New Orleans, Septemeber 1990.

WOODWARD, K. & CURTIS, C. D. 1987. Predictive modelling of the distribution of production constraining illites – Morecambe Gas Field, Irish Sea, Offshore UK. *In*: BROOKS, J. & GLENNIE, K. (eds) *Petroleum Geology of North West Europe.* Graham & Trotman, London, 205–215.

YALIZ, I. 1997. The Hamilton Field. *This volume.*

Calder Field appraisal well 110/7a-8, East Irish Sea Basin

R. A. BLOW & M. HARDMAN

*BG Exploration and Production Limited, 100 Thames Valley Park,
Reading RG6 1PT, UK*

Abstract: The Calder Field, Block 110/7a, is located approximately 5 km southwest of the South Morecambe Field in 100 feet of water. The field was discovered by well 110/7a-3 in 1982; this was the first well in the East Irish Sea Basin (EISB) to encounter sour gas (H_2S 4500 ppm) and live oil. Well 110/7a-4 was drilled to appraise the discovery in 1983; this well produced sour gas at a combined flow rate in excess of 25 MMSCFD from the Sherwood Sandstone Group reservoir. Whilst the two wells proved the presence of gas within the Sherwood Sandstone Group they also indicated a significant velocity variation within the Mercia Mudstone Group overburden. This resulted in uncertainty in depth conversion to top reservoir and also in reserve estimates based upon the depth maps. In an attempt to narrow the reserve uncertainty an integrated geological and geophysical re-evaluation of the field was carried out. This involved detailed subdivision of the Mercia Mudstone Group based on mudstone/halite stratigraphy together with intra-Mercia Mudstone Group seismic interpretation. This approach generated a velocity and depth model successfully tested by appraisal well 110/7a-8 which was drilled during 1994. Implicit in the field model are the occurrences on seismic section of a dipping 'flat spot' associated with the gas/oil contact and also pseudo-faulting at top reservoir level, both generated by lateral velocity variations in the near-surface section.

The Calder Field is located in the East Irish Sea Basin approximately 50 km west of the Lancashire coast (Fig. 1) and 5 km southwest of the South Morecambe Field. Structurally the field comprises a simple NNW–SSE trending tilted fault block. The reservoir is the Triassic Ormskirk Sandstone Formation of the Sherwood Sandstone Group sealed by the halite and shales of the Stanah and Fylde Halite Members of the Mercia Mudstone Group.

History

The Calder Field lies within block 110/7a; this block is licensed to BG Exploration and Production Ltd. (100%), a wholly owned subsidiary of BG plc. The licence P099 was awarded in 1970 and was formerly held 100% by Hydrocarbons Great Britain, also a wholly owned subsidiary of BG.

The Calder Field was discovered by well 110/7-3 which was drilled in 1982 to test a seismically defined closure at top Sherwood Sandstone level. This well encountered the top of the Sherwood Sandstone Group at 2521 (−2398) ft and penetrated 447 ft (gross) of hydrocarbon bearing section; a gas–oil contact was identified at −2825 ft and an oil–water contact at −2845 ft.

The planned test programme for well 110/7-3 was aborted due to the unexpected detection of high concentrations (4500 ppm) of hydrogen sulphide in the gas, for which the rig was not equipped. Well 110/7-3 was subsequently plugged and abandoned as a gas discovery. This was the first well in the East Irish sea to encounter both 'sour gas' and oil. Prior to curtailing the test programme the well flowed at rates up to 8.43 MMSCFD through a 48/64″ choke.

Well 110/7a-4, an appraisal well to the 110/7-3 discovery, was drilled during 1983. The primary objective of 110/7a-4 was to obtain further gas/oil samples for analysis and to complete a thorough programme of DSTs. Well 110/7a-4 encountered 233 ft (gross) of hydrocarbon-bearing Ormskirk Sandstone Formation. The three subsequent DSTs run on the well produced a cummulative 23.2 MMSCFD with 8910 ppm of hydrogen sulphide. The oil rim encountered on 110/7-3 was sampled via RFT and the gas–oil and oil–water contacts were confirmed.

Following the completion of well 110/7a-4 the recoverable gas reserves (proven + probable) of the Calder Field were calculated to be 120 BCF with approximately 5 MMBBL of oil recoverable.

1992–1993 seismic interpretation

In order to refine the understanding of the Calder structure, a 0.5 km dip line seismic grid was recorded in 1992. The time structure map at Top Sherwood Sandstone level (Fig. 2) confirmed the main elements of the previous time mapping but the improved seismic resolution also suggested increased minor north–south faulting downdip.

From Meadows, N. S., Trueblood, S. P., Hardman, M. & Cowan, G. (eds), 1997, *Petroleum Geology of the Irish Sea and Adjacent Areas*, Geological Society Special Publication No. 124, pp. 387–397.

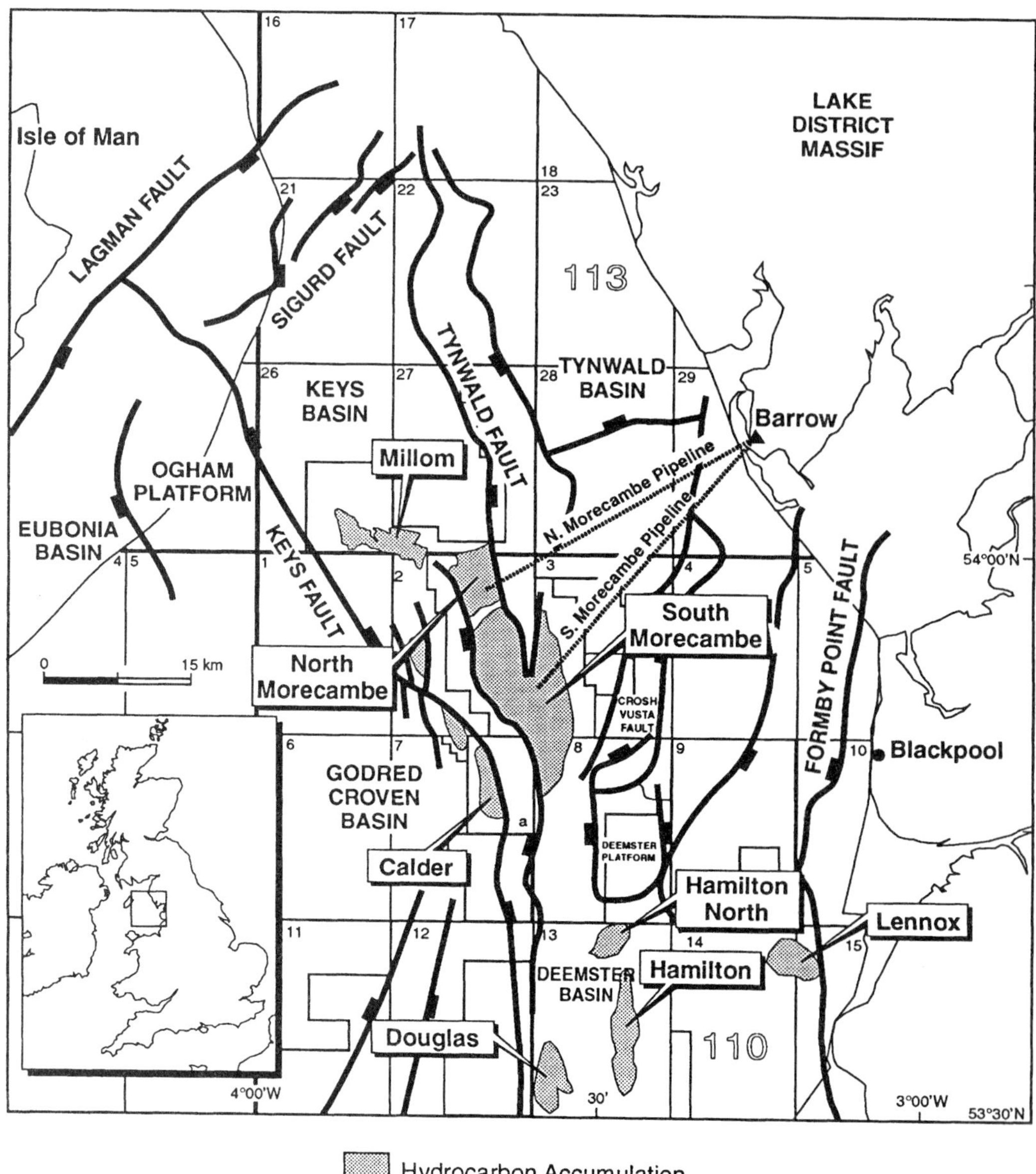

Fig. 1. Major tectonic elements of the EISB and location of Calder Field.

Additionally, amplitude brightening of the Top Sherwood event over the structure was apparent on some lines as was a high amplitude dipping 'pink event' immediately underneath and discordant to the bright Sherwood reflector (Fig. 3).

The two-way time (TWT) data for both wells 110/7-3 and 110/7a-4 are summarized in Table 1; these show a significant decrease in the average velocity down to the Top Sherwood event for well 110/7-3 (10140 ft s^{-1}) compared with well 110/7a-4 (10842 ft s^{-1}) and, associated with this, a significant increase in TWT to the gas–oil contact (0.550 s compared with 0.516 s). Figure 2 indicates the variation in areal extent of the Calder accumulation associated with this variation in TWT to the contact. Clearly an understanding of the cause of the velocity variations is a prerequisite to the accurate conversion of the time structure of Fig. 2 to

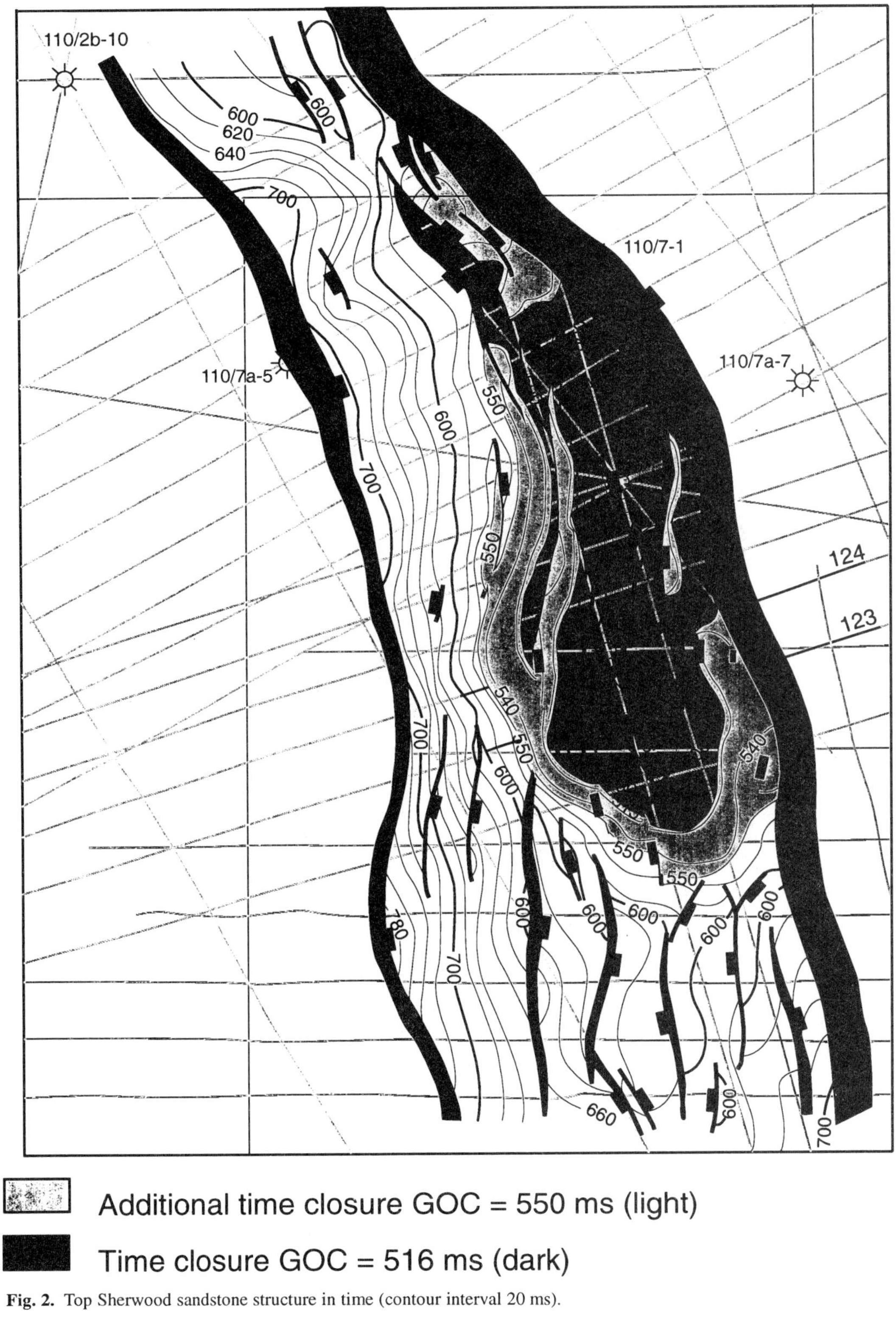

Fig. 2. Top Sherwood sandstone structure in time (contour interval 20 ms).

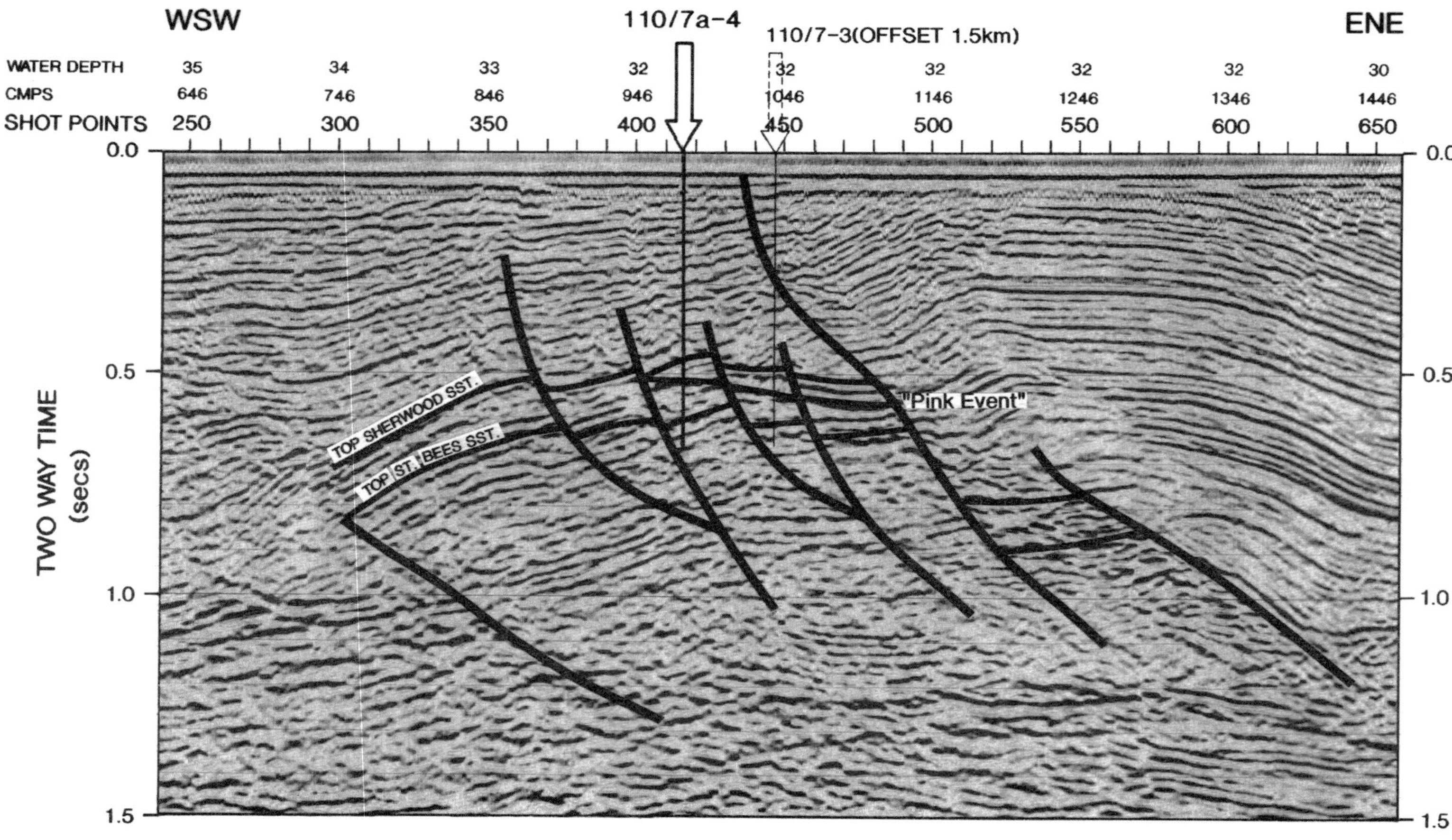

Fig. 3. Line BG923-124, showing well locations and high amplitude discordant 'pink event'.

Table 1. *Time–depth summary for wells 110/7-3,4*

	110/7-3				110/7-4			
	Z'	$\Delta Z'$	TWT (s)	V_{AV} (ft s^{-1})	Z'	$\Delta Z'$	TWT (s)	V_{AV} (ft s^{-1})
Sea bed	92				113			
		233				82		
Base Quaternary	325				195			
Top Sherwood Sandstone	2398		0.473	10140	2613		0.482	10842
GOC	2826		0.550		2826		0.516	

depth and thus to an accurate estimation of gas reserves.

It was recognized that these velocity variations could be associated with either the overlying Quaternary (1) or Mercia Mudstone (2) sections:

(1) Quaternary

Significant lateral changes in Quaternary cover associated with channeling were in part supported by the Quaternary thicknesses identified, albeit tenuously, in the two wells (Table 1). There was however no convincing evidence for channeling in well 110/7-3 from the available site survey data. A limited reprocessing exercise of the 1992 vintage seismic dataset was thus undertaken in order to enhance the near-surface section, including the use of a first break refraction statics routine normally associated with land data processing. Reprocessed sections (Fig. 4) did not substantiate the presence of Quaternary channels over the structure of sufficient magnitude to account for the observed average velocity variations. However refractor velocities calculated from the statics analysis indicated a marked change across the Keys Fault; footwall refractor velocities varied in the range 11500–14750 ft s^{-1} whilst hanging wall velocities varied little from a mean of 8850 ft s^{-1}.

(2) Mercia Mudstone

It was anticipated that significant changes in the Mercia Mudstone overburden would necessarily be associated with lithology variations. It was recognized on the higher resolution 1992 seismic data that well 110/7a-4 was unambiguously located on the upthrown side of the Keys Fault,whilst the shallow section of well-3 was located in the hanging wall (Fig. 3). It was thus possible that the low velocity of this well was simply associated with this younger hanging wall sequence. Given that gross Mercia Mudstone velocities can be as much determined by percentage halite within the unit as

age of mudstone and palaeoburial depths then a model which assumed low velocities in the hanging wall implied that this sequence represented an essentially halite-poor mudstone.

Mercia Mudstone stratigraphy

Encouraged by the results of the refraction statics noted above, it was decided to pursue the low velocity hanging wall model by detailed subdivision of the Mercia Mudstone encountered in the 110/7-3, 110/7a-4 and adjacent offset wells. The combined geological–geophysical approach adopted involved a detailed iterative integration of seismic and well data. The lack of log and detailed check shot data in all wells above the $9^{5}/_{8}''$ casing point (typically 2000 ft) introduced an uncertainty in this subdivision, including an ambiguity in the location of the Keys Fault in well 110/7-3. However well 110/7a-7, targeted at a Keys Fault terrace (Fig. 5) had locally encountered the thickest section of Mercia Mudstone and provided the most complete logged interval and thus control.

Based upon the available well penetrations the Mercia Mudstone Group was subdivided into a series of halite/shale cycles (Fig. 5). These were correlated both locally on the terraces associated with the Keys Fault system and more regionally into the North and South Morecambe fields. These cycles have now formally been given Member status within the stratigraphic hierarchy of the Mercia Mudstone Group (Jackson, this volume).

Application of the halite/mudstone stratigraphy thus predicted that a part of the upper section of Mercia Mudstone Group present in the hanging wall of the Keys Fault would be comparable in both lithological content and velocity to the upper sections of the Mercia Mudstone Group (Elswick Mudstone Formation and Warton Halite Formation) penetrated by wells 110/7a-5 and -7. The Elswick Mudstone Formation, was therefore predicted to overlie the eastern flank of the Calder Field and based upon the refraction statics where it

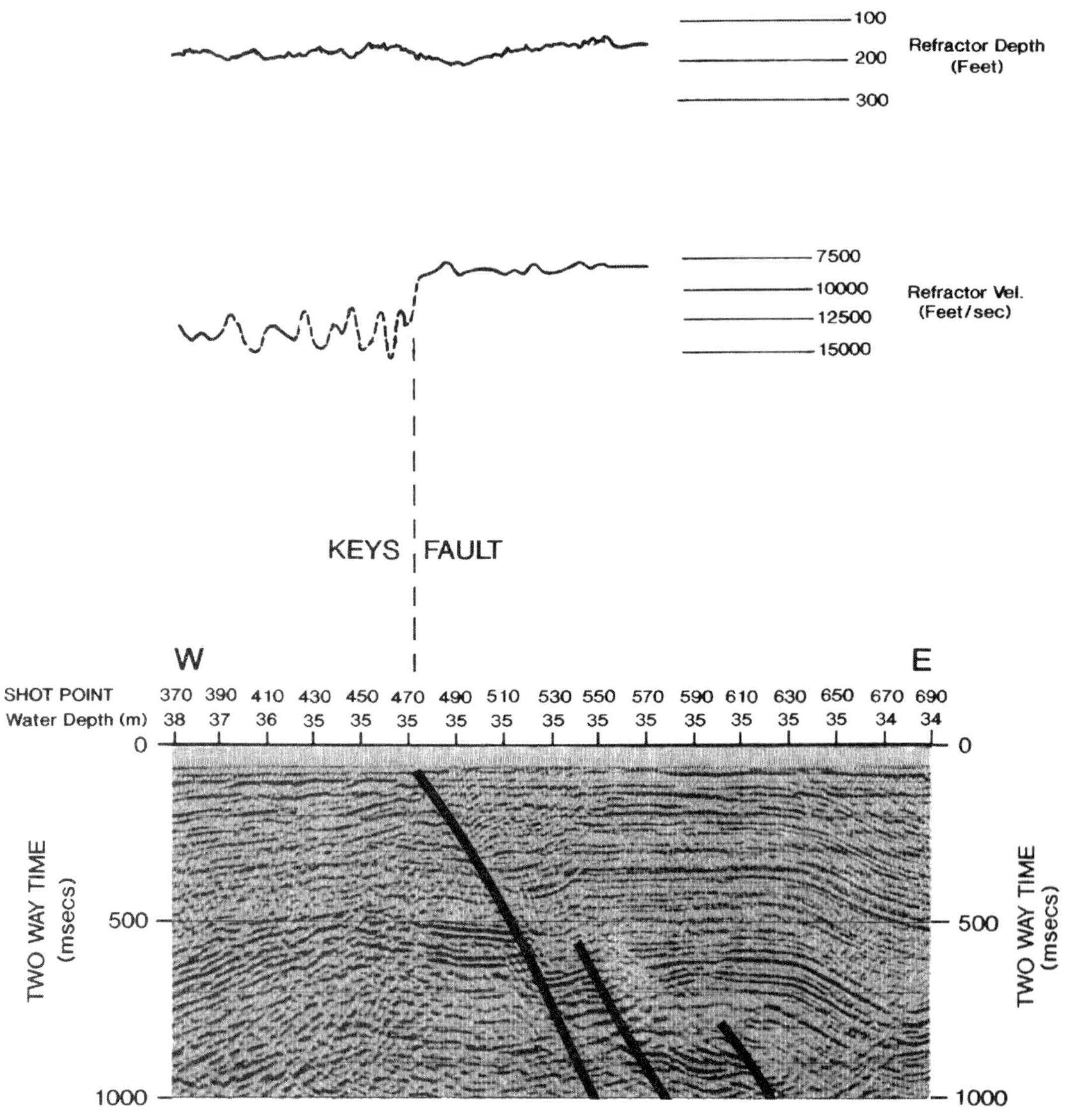

Fig. 4. Line BG923-123 reprocessed, showing results of first break refraction statics analysis.

outcropped at sea bed to have a velocity in the range 8500–9500 ft s^{-1}.

As discussed earlier this predicted interval velocity is considerably slower than the velocity encountered in well 110/7a-4 which penetrated a section comprising the Preesall Halite Formation to Stanah Member. Unfortunately this attempt to predict the lithology and velocity of the hanging wall section was hampered by a lack of logs and samples in the shallow section of well 110/7-3. As shown on Fig. 5, the section immediately above the

proposed Elswick Mudstone Formation was unknown, having not been penetrated by any well in the area. As a consequence the stratigraphic model still carried an element of uncertainty in terms of predicting lithology; the velocity of this unknown section was believed however from refraction statics to be comparable with the underlying Elswick Mudstone, suggesting that this unknown unit was the upper part of a significantly thicker Elswick Mudstone development than hitherto encountered.

Seismic modelling and depth conversion

Utilizing the detailed Mercia Mudstone stratigraphy, various velocity models were developed to generate synthetic sections in an attempt to accurately depth convert the Calder structure. The depth section of Fig. 6a essentially divides the Mercia Mudstone into an upper low velocity mudstone unit, a massive Warton Halite Formation, and a lower mudstone/halite unit (with a basal high velocity sequence equivalent to the Stanah member); additionally the velocity of the Sherwood Sandstone is reduced in the gas leg. The resultant synthetic section (Fig. 6b) indicates structurally that the crestal Sherwood event is depressed in time under the Keys Fault and that the apparent downdip faulting at this level is caused by the overlying lateral velocity variations across the fault. The amplitude brightening due to the gas fill is observed as is the seismic response of the gas contact which displays a discordant dip across the structure, due essentially to the low velocity upper mudstone unit.

The similarity of the significant features to emerge from the seismic modelling (Fig. 6b) to those observed on seismic section (Fig. 3), particularly the association of the discordant dipping event with the gas/oil 'flat spot', provided substantiation of a velocity model based on a subdivision of the Mercia Mudstone.

The final layer-cake velocity model used to depth convert the time structure is indicated in Fig. 7 together with the resultant depth map. Comparison with the previous mapping indicates that the field was potentially significantly larger than previously thought, both longer and with a shallower crest underneath the bounding fault.

Well 110/7a-8

Based on the new depth mapping, the potential reserves of the Calder field now carried significant uncertainty, with a P50 of 220 BCF some 100 BCF larger than the old proven plus probable reserves. Significantly, it was also suggested that much of the downdip faulting present at top Sherwood Sandstone level could be an artefact of the overburden velocity distribution. It was thus decided to drill a further appraisal well in an attempt to substantiate the velocity model and thus confirm the additional reserves and structural implications. Well 110/7a-8 was located some 1.5 km to the ESE of 110/7a-4 (Fig. 8). The location of the well was chosen such that:

(1) It tested the Calder structure at a location which was well within closure based upon the new depth conversion (predicted gas column

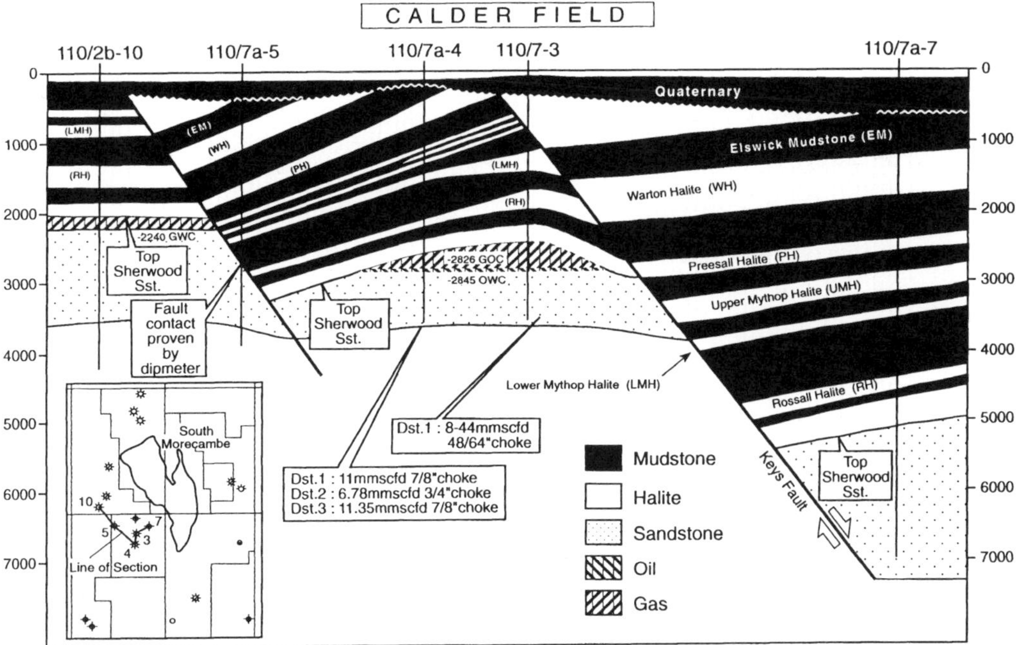

Fig. 5. Well correlation across Keys Fault Terraces.

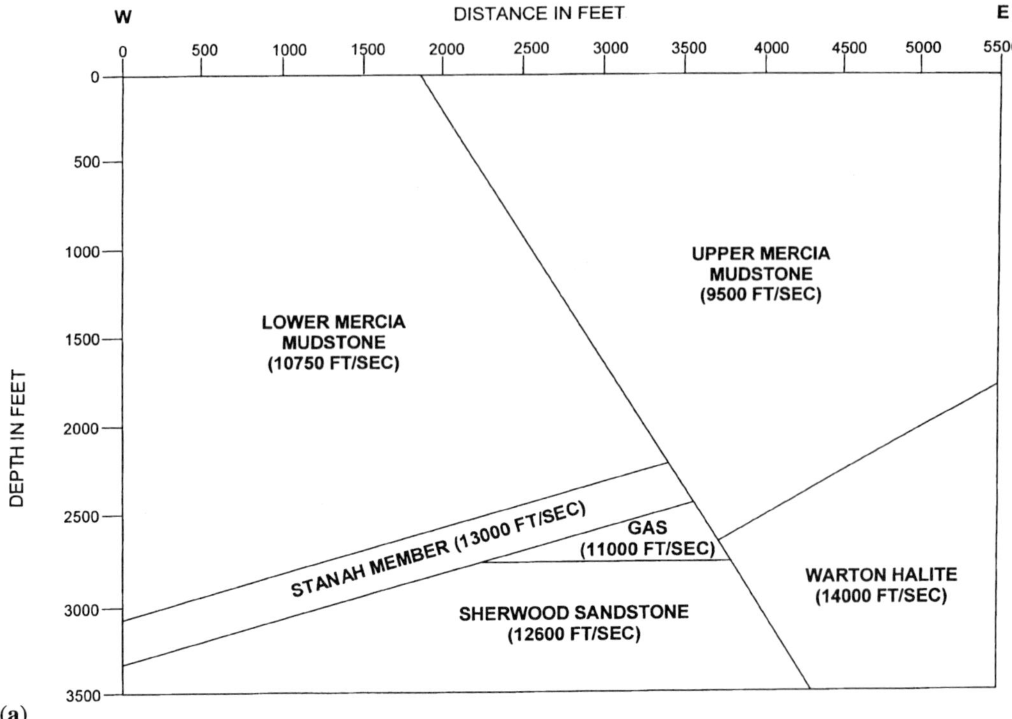

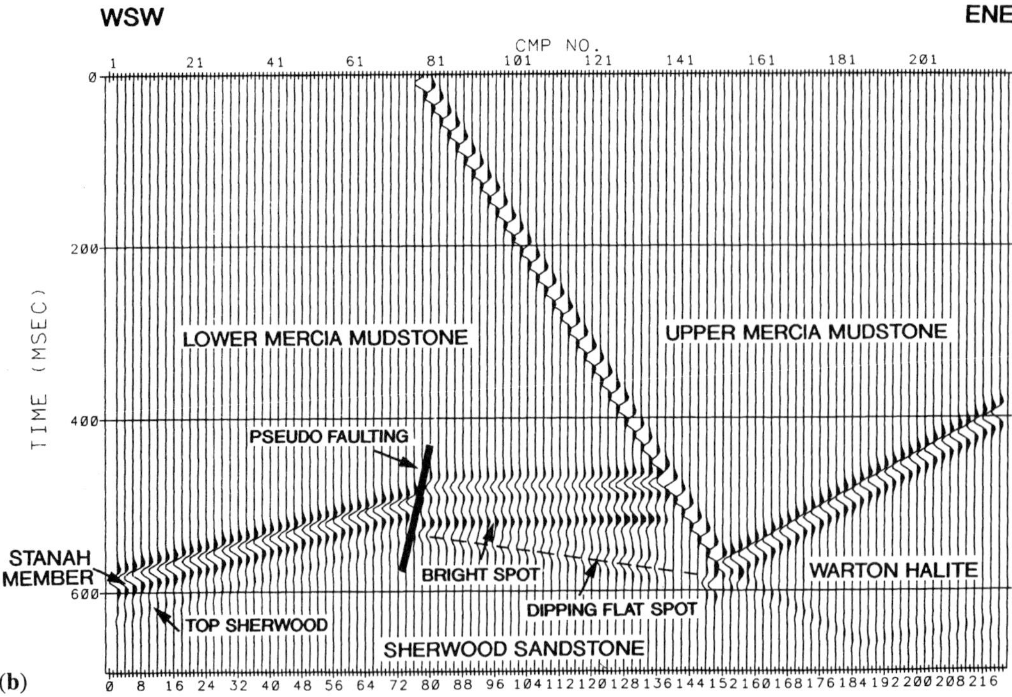

Fig. 6. (a) Velocity–depth model across Keys Fault. (b) Synthetic seismic section generated from (a) (image ray tracing, zero phase 30 Hz Ricker wavelet).

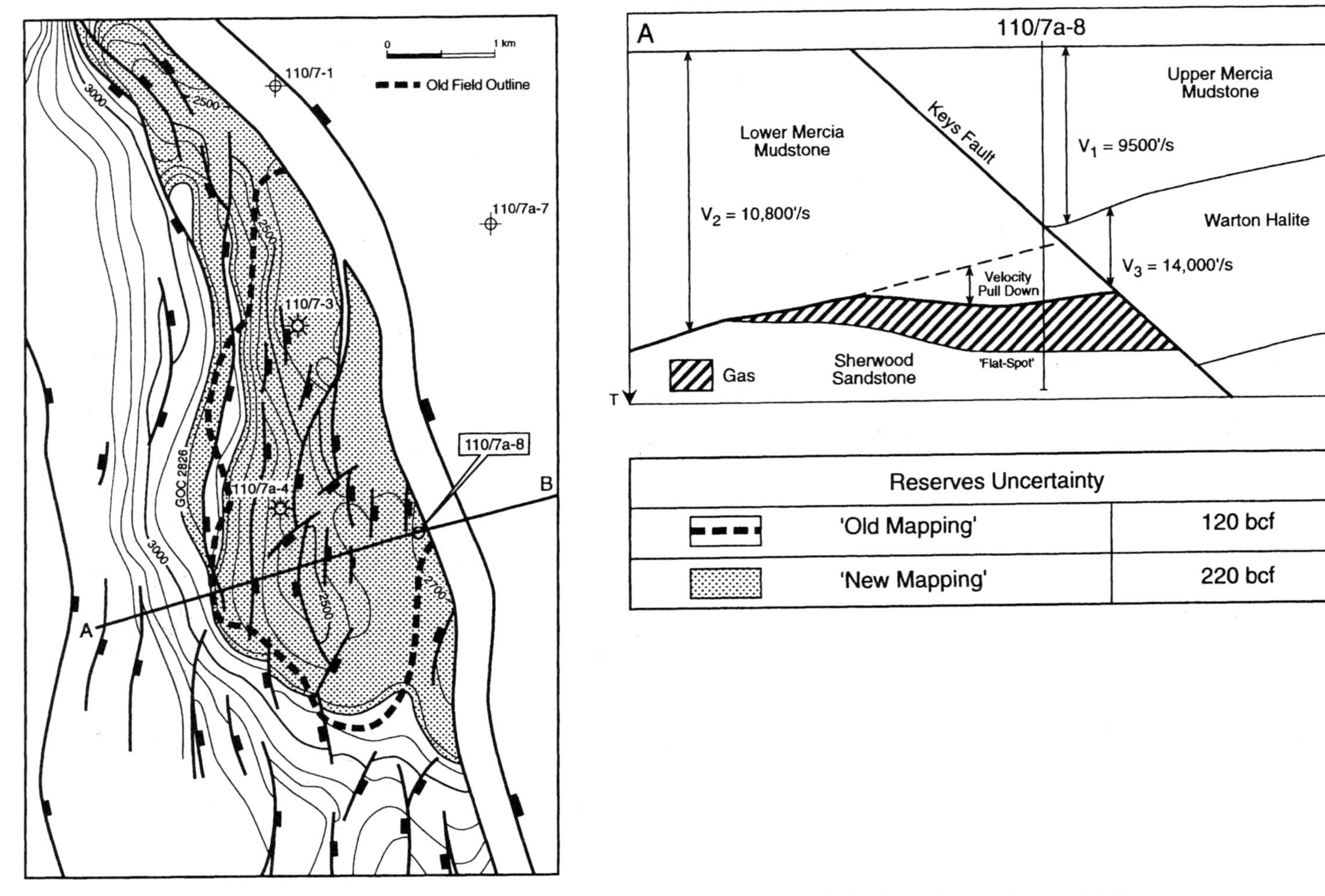

Fig. 7. Velocity–depth model for final depth conversion and resultant Top Sherwood Sandstone structure in depth map (Contour interval 100 ft).

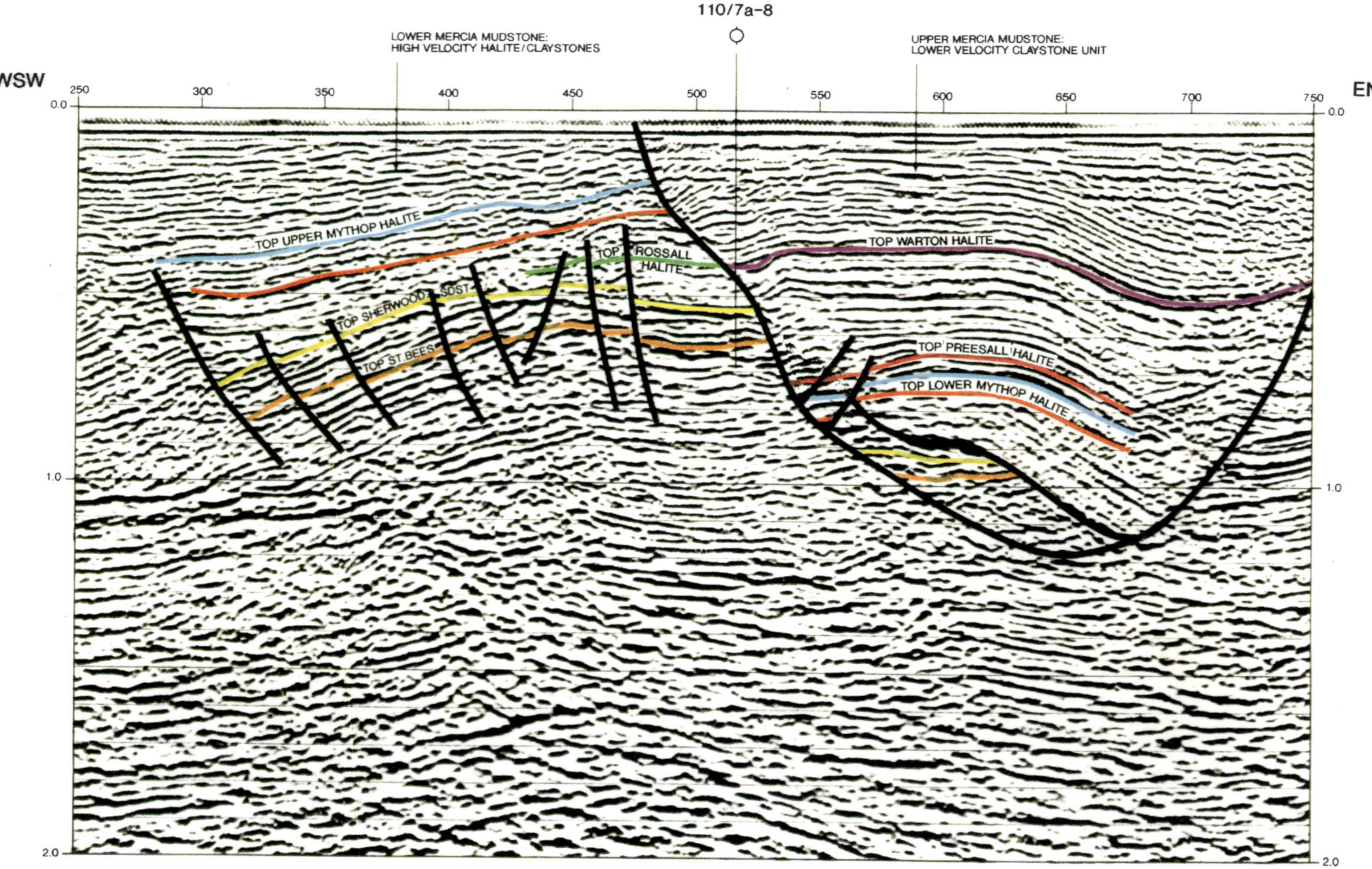

Fig. 8. Line BG923-123 showing location of 110/7a-8.

>200 ft). Based upon the previous depth conversion however this well would be spudded at a high risk location and would penetrate only a minor (<20 ft) gas column.

(2) The well would penetrate, sample and log the greatest development of a presumed thick Elswick Mudstone Formation to validate the lithological and velocity predictions inherrent in the model. The logging programme was thus designed to provide continuous log data from the 30″ casing point at approximately 360 ft ss, in marked contrast to previous BG wells in the EISB where logs typically were run only below 2000 ft.

The well was drilled in 1994 and encountered a hydrocarbon column of 174 ft, with the Sherwood Sandstone Group at 2671 ft TVDSS as compared with a pre-drill prognosis of 2624 ft TVDSS. The shallow 'unknown' section penetrated by the well was confirmed lithologically as the upper part of a thick Elswick Mudstone Formation which was comprised entirely of red brown claystone and siltstones, characterized by a low interval velocity of 9960 ft s^{-1} (cf. 9500 ft s^{-1} used in modelling) The well flowed gas at a rate of 16.2 MMSCFD from a 56/64″ choke with 1900 ppm hydrogen sulphide.

Conclusions

It is considered that well 110/7a-8 successfully validated a velocity–depth model based on a detailed subdivision of the Mercia Mudstone Group. This subdivision, based on well and seismic data, is considered critical to accurate depth conversion of hydrocarbon-bearing structures in structurally complex areas of the EISB, in particular those in footwall locations of major faults. Logging of EISB wells to as shallow a depth as possible is necessary to enable the Mercia Mudstone stratigraphy to be defined; this is now routine practice within BG following the 110/7a-8 well.

Additionally, the seismic 'flat spot' associated with the gas contact of a footwall accumulation may be expected to exhibit significant time relief where a major velocity contrast in the overlying Mercia Mudstone section is developed across the bounding fault. Apparent down-dip faulting at top reservoir level may simply result from such overlying velocity variations.

We would like to thank the directors of BG Exploration and Production for permission to publish this paper. The contribution of colleagues at BG involved in the Calder Field appraisal, particularly Jon Crawford, is gratefully acknowledged, and Geoff Collins is thanked for his work on the first break refraction study.

Reference

JACKSON, D. I., JOHNSON, H. & SMITH, N. J. P. Stratigraphical relationships and a revised lithostratigraphical nomenclature for the Carboniferous, Permian and Triassic rocks of the offshore East Irish Sea Basin. *This volume.*

The Douglas Oil Field

A. M. YALIZ

BHP Petroleum, Devonshire House, Piccadilly, London W1X 6AQ

Abstract: The Douglas Field, with an estimated STOIIP of 225 MMBBL and on stream in February 1996, is the first oil field to be developed in the offshore East Irish Sea basin. The field structure consists of three tilted fault blocks formed during extensional faulting in Triassic–early Jurassic times, and later readjusted by contractional movements during Tertiary inversion. The oil is trapped in the Triassic Helsby Sandstone Formation, which comprises moderate- to high-permeability aeolian and fluvial sandstones. The reservoir depth is shallow (2200 ft) with a maximum oil column of 330 ft. The sandstones can be divided into several laterally extensive units based on vertical facies variations. Although reservoir quality is principally controlled by primary depositional processes, a complex diagenetic history is recognized, with various phases of silica and carbonate cements. Authigenic clay minerals are not important, but bitumen is extensive in limited areas of the field, causing significant permeability reduction. The hydrocarbon filling history of the field was complex, with the occurrence of at least two phases of oil generation and migration. The Douglas Field contains a relatively 'dead' oil with a low GOR, thus 11 production and seven water injection wells are planned. Although the crude is classified as sweet it contains low levels of mercaptans and H_2S, which will be removed during processing offshore. First oil is planned for December 1995 with peak oil production reaching 40000 BPD.

The Douglas Field is situated in Block 110/13 of the East Irish Sea, about 40 km west of Liverpool in 110 ft of water (Fig. 1). Block 110/13 was originally awarded to Gulf Oil in 1965, but was relinquished in 1974 without any wells being drilled. It was reawarded to the P710 Licence Group, operated by Hamilton, in 1989 as part of the 11th UK Offshore Licensing Round.

The field was discovered by well 110/13-2 in 1990. This well was drilled on a sparse grid of speculative seismic data shot in 1986–1987. The Hamilton and Hamilton North gas fields were also discovered using the same lines. In late 1990, additional proprietary 2D seismic data were acquired. The field development is now based on a 3D survey, acquired in 1992, covering the entire Block 110/13, and four appraisal wells.

After the discovery of the Hamilton Gas Field with the first well in the block, the second well, 110/13-2 at the Douglas location, was prognosed to have a 700 ft gas column. However, the reservoir was 230 ft lower than prognosed and an oil-bearing section was encountered in the well. In this well the oil-water contact is at 2508 ft TVDSS giving an oil column of 106 ft with an average reservoir porosity of 19%. On testing the well flowed at a rate of 1800 BPD of 44° API oil. The original reservoir pressure is 1125 PSI at 2240 ft TVDSS and the reservoir temperature is just over 30°C. The oil is relatively 'dead' , generating only 170 SCF of solution gas per barrel of oil. Although low in total sulphur, the crude contains low levels of mercaptans and

hydrogen sulphide, H_2S. Mercaptans are sulphur compounds containing the sulfhydryl group (e.g. methyl mercaptan, CH_3SH). They are not harmful but they are characterized by a very strong and unpleasant odour. For this reason, they will be

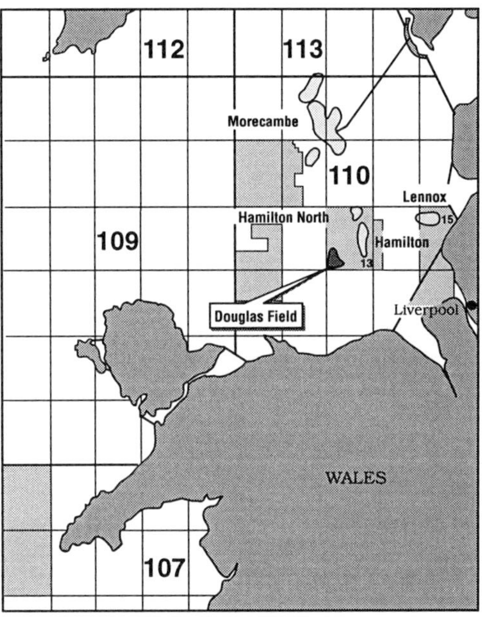

Fig. 1. Douglas Field location map.

From Meadows, N. S., Trueblood, S. P., Hardman, M. & Cowan, G. (eds), 1997, *Petroleum Geology of the Irish Sea and Adjacent Areas*, Geological Society Special Publication No. 124, pp. 399–416.

removed during processing along with hydrogen sulphide.

Further drilling between 1991 and 1993 confirmed the extent of the field. The field is currently understood to be a structural trap consisting of three tilted fault blocks elongated north–south and dipping to the west, about 5 km long and 2 km wide. The reservoir is the Triassic Sherwood Sandstone Group and is characterized by excellent permeability sandstones deposited by aeolian and fluvial processes. The individual sandstone packages are laterally correlatable over long distances. The characteristics of this reservoir along with the presence of a low energy aquifer require that recovery be assisted by water injection.

Annex B approval of the field development was given in 1993. Pre-drilling of 11 development wells was completed by April 1995 using a 12-slot subsea template and first oil was produced in February 1996. The remaining development wells were drilled in 1996. The expected peak oil production rate is 40 000 BPD and the expected field life is up to 15 years. A subsea pipeline transports crude from the Douglas complex to an offshore loading unit. The Douglas complex is also designed to handle the processing and export of hydrocarbons from the Hamilton (gas), Hamilton North (gas) and Lennox (oil and gas) fields.

Stratigraphy

A generalized stratigraphy for Block 110/13 is shown in Fig. 2. The Carboniferous Namurian Hollywell Shale Formation is the oldest formation penetrated in the area and is believed to constitute the source rock for the Douglas oil (Armstrong *et al.*, this volume). This formation is predominantly clastic with deep water shales deposited via a broad fluvio-deltaic regime that existed across north and east England during the Namurian. The overlying Permian consists of over 2400 ft of desert sandstones (Collyhurst Sandstone Formation) and 300–400 ft of red mudstones (Manchester Marl Formation).

The Triassic is divided into the Sherwood Sandstone Group and the Mercia Mudstone Group. The former consists of sandstone, siltstone and mudstones deposited in a semi-arid environment, whilst the latter is composed of shales and thick halite units deposited in lakes subject to periodic

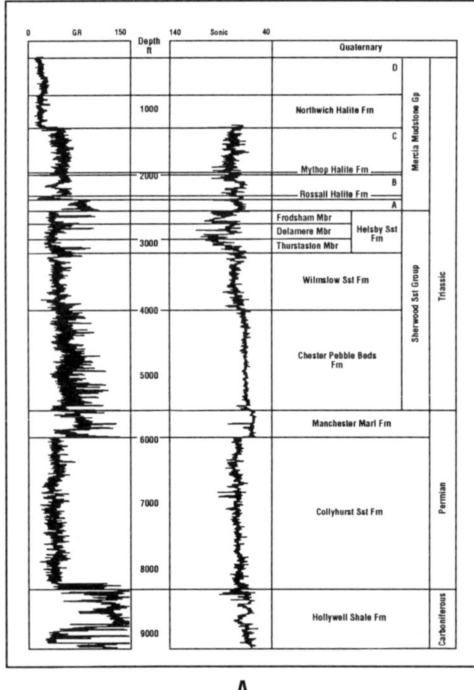

A

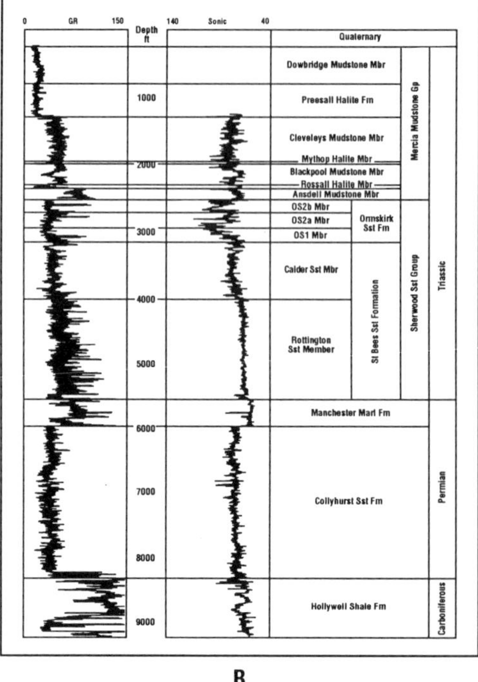

B

Fig. 2. Generalized stratigraphy of the Douglas Field based on Well 110/13-1, showing both the old (A) and the revised (B) versions.

flooding and desiccation, and forms a laterally extensive and competent seal over the Douglas Field.

The Sherwood Sandstone Group in NW England has been variously sub-divided by several authors (Thompson 1970; Smith *et al.* 1974; Colter & Barr 1975; Warrington *et al.* 1980; Jackson *et al.* 1987, this volume). Studies of available data in Blocks 110/13 and 110/15 indicate that the Douglas reservoir units can be correlated with those found on the Wirral and in the Cheshire basin. The lowest unit is the Chester Pebble Beds Formation, which was penetrated only in well 110/13-1 in the Hamilton Field. This is overlain by the Wilmslow Sandstone Formation (Bunter Sandstone equivalent), a sequence of predominantly fine to medium grained sandy braided river deposits. The overlying Helsby Sandstone Formation is the principal reservoir target in the East Irish Sea basin. It consists of fluvial and aeolian sandstones of variable grain size. Many of the fluvial sandstones have a high proportion of reworked aeolian grains and there are rapid thickness variations of fluvial and aeolian facies. The Helsby Sandstone Formation was divided by Thompson (1970) into three members; Thurstaston, Delamere and Frodsham. These can be easily recognized on the Wirral and in all the wells drilled in Block 110/13. They all exhibit very distinct gamma ray and sonic log characteristics. The Thurstaston Member includes a mixed sequence of fluvial and aeolian sediments. The Delamere Member is predominantly fluvial with occasional playa lake claystones. The overlying Frodsham Member represents predominantly aeolian conditions, although a sequence of fluvial facies is present in the middle of the unit.

The Mercia Mudstone Group consists of a cyclic sequence of alternating sandy mudstones and halites. The Rossall and Mythop halites are each less than 50 ft thick and the Northwich Halite has a thickness of around 500 ft.

The stratigraphic nomenclature described above has been used by BHP Petroleum and its partners until the present time. Within this publication a revised lithostratigraphic nomenclature has been proposed by Jackson *et al.* (1997), this has been accepted and will be used in the future (Fig. 2).

Seismic interpretation

Three principal seismic marker horizons can be identified from the 3D seismic data; Top Sherwood Sandstone Group, Top Collyhurst Sandstone Formation and Top Carboniferous. The Top Sherwood Sandstone Group pick is the most prominent and consistent seismic marker over the whole area. The main factors controlling seismic data quality over the Douglas Field are the variation of structural

complexity, diffractions and other fault noise and sea-floor conditions. The top reservoir map was depth-converted using a velocity versus seismic two-way-time function combined with a V_0 value that varies spatially depending on the geology of the overburden. Latest development drilling has shown that the most critical factor in reliable depth conversion is a good understanding of seismic velocities in the near surface formations.

Field structure

The structural history of the East Irish Sea Basin has been discussed in considerable detail elsewhere (Bushell 1986; Green 1986; Jackson *et al.* 1987; Stuart & Cowan 1991; Jackson & Mulholland 1993). Structural interpretation of seismic data in Block 110/13 supports the view that post-Variscan tectonics (extensional and contractional episodes with associated reactivation of some of the major faults) were responsible for basin evolution and trap-forming events in the area (Fig. 3). Extension during the Early Permian in the East Irish Sea Basin led to half-graben structures with N–S faults controlling facies and thickness variations. All faults of the Permian fault framework of the area were active

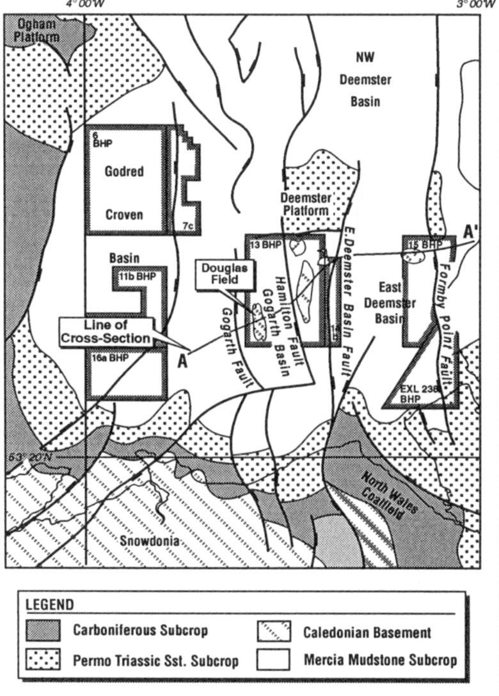

Fig. 3. Generalized structural elements of the East Irish Sea Basin (modified after Jackson *et al.* 1987).

during the Triassic. As a result of continuing fault activity during the Early Triassic the Sherwood Sandstone Group shows a marked increase in thickness towards the Gogarth fault (Fig. 4). The overall thickness of this group in the Douglas area is around 3000 ft compared with 5000 ft in the East Deemster basin (the Lennox area). Relatively more vigorous rifting and extension took place during the Late Triassic. The Mercia Mudstone Group of this age shows development of intermediate-scale structures and locally severe block-fault rotation. Due to erosion effects no Jurassic is seen, and extension during this period is based on data from outside the area (Jackson *et al.* 1987). Post-Jurassic history of the Douglas area is also difficult to constrain as the Triassic is immediately overlain by Quaternary drift. However, various authors have shown that the East Irish Sea Basin experienced a major phase of inversion and uplift during early Palaeogene (Lewis *et al.* 1992; Hardman *et al.* 1993; Williams & Eaton 1993; Cope 1994). Fault-controlled structures formed during Triassic–Jurassic rifting were modified during Early Tertiary inversion and new structures were generated.

The Douglas structure is principally controlled by the Gogarth fault, which is a shallowly dipping (0–45°) and N–S oriented feature lying to the west of Block 110/13 (Fig. 4). A deep master detachment fault extending into Carboniferous shales gives rise to the Gogarth fault ramp–flat listric system beneath the Douglas Field. The Douglas trap was formed within the fault footwalls of three tilted fault blocks during a major phase of extensional faulting in Triassic–early Jurassic times. The topology of the Gogarth fault plane has played a major role in the development of the trap during extension and its readjustment during later inversion. Contractional movements along the Gogarth ramp–flat listric fault during the Tertiary uplift caused the Douglas structure to move westwards to its present day position, and gave rise to a small net extension at the Gogarth fault. Inversion increases towards the west leading to elevation of the Douglas structure near the crest of an anticlinal feature. This anticlinal structure and a crestal collapse graben to the east of the Douglas Field correspond with the ramp–flat changes in the master fault. The Hamilton Field lies on the eastern side of the crestal collapse graben within an antithetic fault province.

The structural trap of the Douglas Field consists of three seismically defined tilted fault blocks which are elongated north-south and dip at approximately 15° to the west (Figs 5 and 6). The pre-drilling results have shown that the boundary faults are planar dipping at approximately 45° to the east. Seismic data indicate that cross-fault transfer zones are not common at reservoir level, but a few widely spaced examples can be mapped from high resolution aeromagnetic data. The structure as a whole is fault bounded to the east, and dip closed to the north, west and south. The culmination of the structure is at 2150 ft in the central fault panel. The top and across-fault seals are provided by the overlying Mercia Mudstone Group. The oil–water contact in the western block is 20 ft deeper than that in the central and eastern blocks. Formation pressure data from wireline logs show

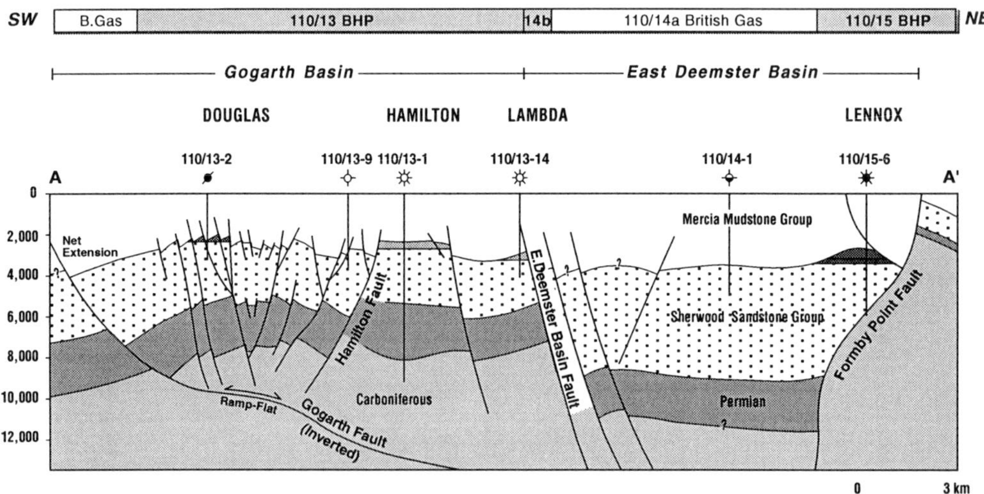

Fig. 4. East–west structural cross-section through Blocks 110/13, 110/14 and 110/15 in the East Irish Sea Basin .

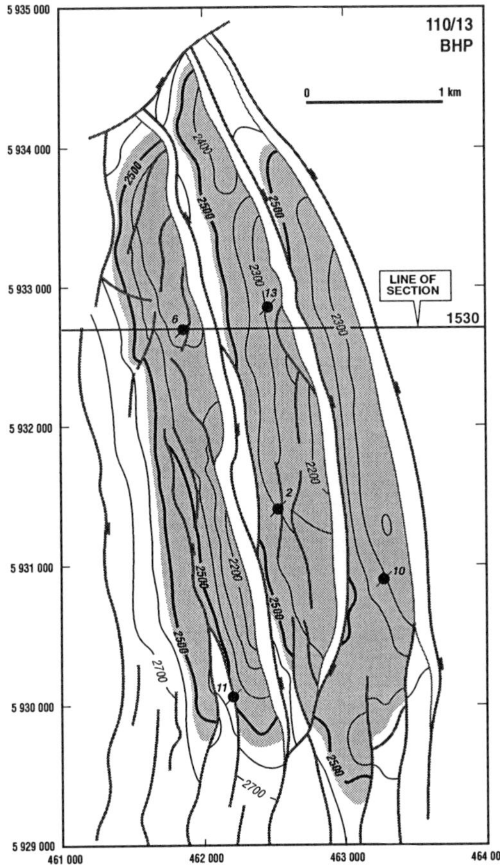

Fig. 5. Top reservoir depth structure map of the Douglas Field.

that all Block 110/13 fields have a common aquifer (Fig. 7). This suggests that in the aquifer the main faults are probably not sealing.

Hydrocarbon filling history

Various proprietary studies carried out by the Geochem Group (see References) on a suite of oil and gas samples from the Douglas and Lennox Fields indicate that the sequence of events leading to the hydrocarbon filling in these fields was complex but essentially identical (Fig. 8). The filling history for the Douglas Field reservoir can be explained as follows:

1. Early oil, and perhaps gas, generation and migration into anticlinal structures in Block 110/13 during Early Jurassic. This oil appears to have been heavily altered by deasphalting and/or water-washing processes resulting in

heavy bitumen formation. Due to subsequent breaching and faulting all early hydrocarbons were lost.

2. A second phase of oil generation from the same Carboniferous source area occurred during the Late Jurassic, leading to the current accumulation. Both the Douglas and Lennox fields contain oil of similar origin, which was derived from source rocks in the lower part of the oil window. The source facies for the crude contains largely algal organic matter deposited in a moderately anoxic and perhaps slightly restricted (but marine) environment. The oil was probably of medium to light gravity, probably sweet and was unlikely to be waxy.

3. The waxiness of the oil was probably caused by partial biodegradation during the Early Cretaceous, resulting in the severe alteration of the light ends to C_{12} but only slight degradation of the heavy ends beyond approximately C_{22}.

4. During further burial in the Cretaceous a very light condensate (effectively restricted to the gasolines lighter than C_8) entered the reservoir. The crude produced from Douglas is a blend of the earlier crude oil (paragraph 2) and this condensate.

5. The most soluble hydrocarbons (methane and the light aromatics) were removed, probably by water-washing, resulting in the low gas–oil ratio for the Douglas field. It would appear that the biodegradation had ceased by this time.

6. The source rock then entered the dry gas zone producing a charge consisting of nitrogen, carbon dioxide, hydrogen sulphide, mercaptans and probably some methane. This latest phase is presumed to be a geologically recent event, in which case the migration of these gases into the Douglas Field could still be occurring.

7. The Douglas oil appears to be quite different from the oil tested in the British well 110/8a-5, located 15 km north of the field, indicating an entirely different source facies.

Reservoir facies

An excellent review of facies distribution in the Irish Sea basin was provided by Meadows & Beach (1993). The main reservoir section was fully cored in the first Douglas appraisal well 110/13-6 (Fig. 9). Detailed facies analysis of cores and wireline logs has revealed the presence of six major facies types; (i) aeolian dune, (ii) aeolian sandsheet, (iii) 'sheetflood', (iv) fluvial channel, (v) fluvial abandonment, and (vi) playa lake-floodplain.

Aeolian dune sandstones (Facies A1) are medium-grained and well sorted, with subrounded to well-rounded grains. Preserved set thicknesses (1–7 ft) suggest accumulation as relatively small

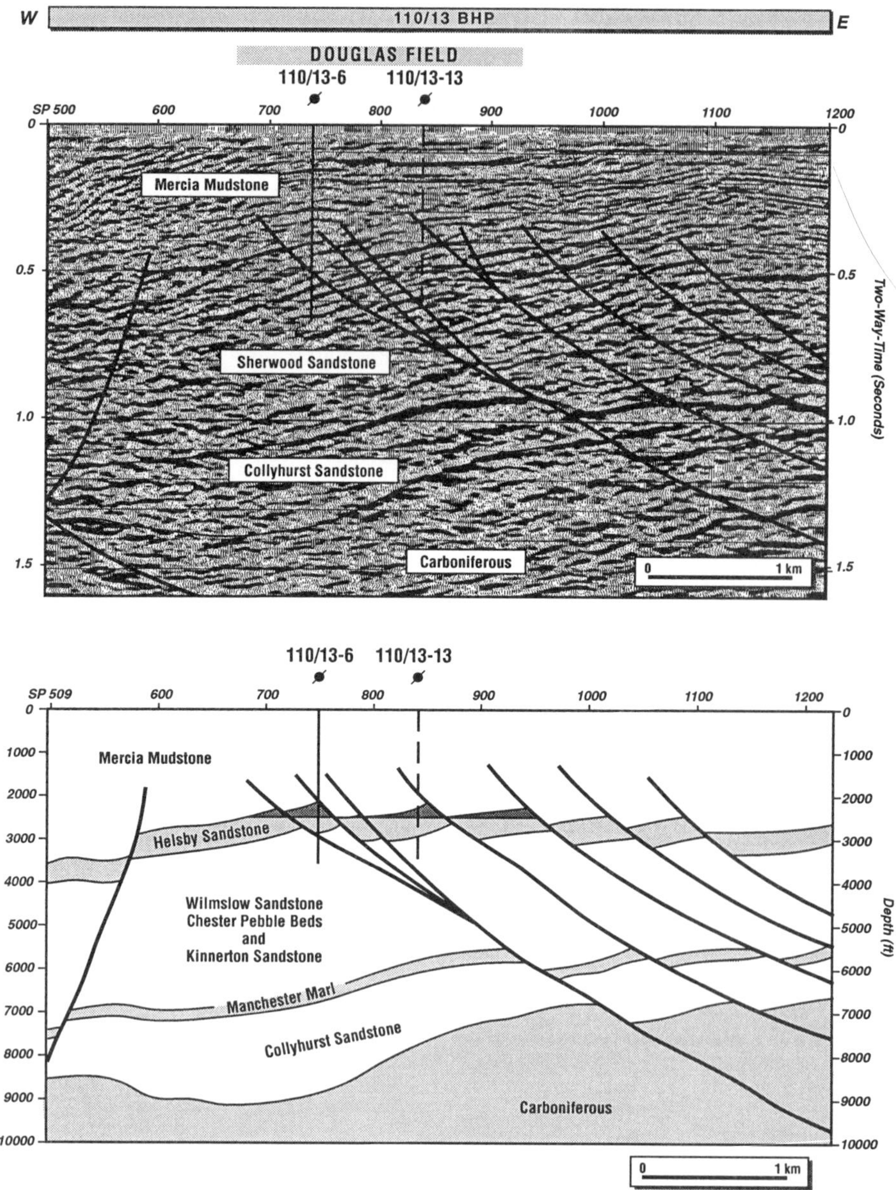

Fig. 6. Structural interpretation of an E–W seismic line across the Douglas Field.

dunes. They occasionally occur as isolated thin sands marked by a single set of cross-stratification suggesting that the volumes of sand available for aeolian reworking were limited. Thicker bodies of dunes with two or three sets of high angle, planar cross-stratification indicate the development and migration of several dunes in response to somewhat greater sand supply.

Aeolian sandsheet facies (A2) are characterized by single grain to millimetre scale horizontal and low angle, planar cross-stratification and are found interbedded with aeolian dune sands as finely laminated sequences. Their thickness range is 0.5–1.5 ft. This indicates that aeolian sandsheet conditions were episodically developed on a low relief sandflat immediately above the water table.

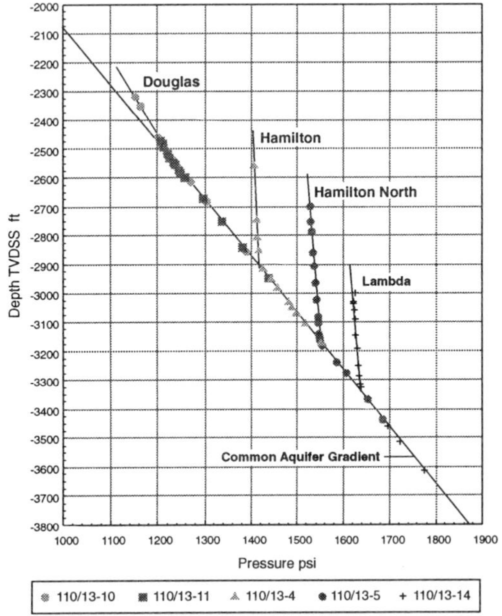

Fig. 7. Formation pressure data in Block 110/13.

'Sheetflood' sands (Facies S1) are very fine to fine-grained, poor to moderately sorted and comprise subangular to well-rounded grains. The individual beds range from 0.25 to 1.5 ft in thick-ness and are characterized by an irregular flat lamination style. They typically occur as 5–10 ft thick units which can be correlated over several kilometres within Block 110/13. They are almost always associated with the aeolian dune and aeolian sandsheet deposits and often contain high concentrations (up to 15%) of dolomite and anhydrite. Therefore, as proposed by Herries & Cowan (this volume), these facies could be described more appropriately as aeolian sabkha, deposited in response to rises in the water table.

Fluvial channel sandstones (Facies F1) were deposited within active and abandoning channels. They form stacked bodies of poorly sorted sandstones which grade from fine to coarse-grained. Sedimentary structures include low to moderate tabular cross-stratification as well as flat laminations. Preserved set thicknesses range 0.35–4 ft. A stacked sequence is commonly made of 10–11 channels and has an average thickness of about 15 ft. The sandstones are usually characterized by abundant small intraclasts, which are dolomite cemented green siltstones and pink very fine-grained sandstone fragments.

Fluvial abandonment sediments (Facies F2) are characterised by siltstone dominated units, some 0.4–10 ft thick. They form the upper section of broadly fining upward sequences of Facies F1.

Playa lake and floodplain deposits (Facies P1) are not common but where present they form laterally extensive grey-black siltstone units, generally occurring in the middle of fluvial channel

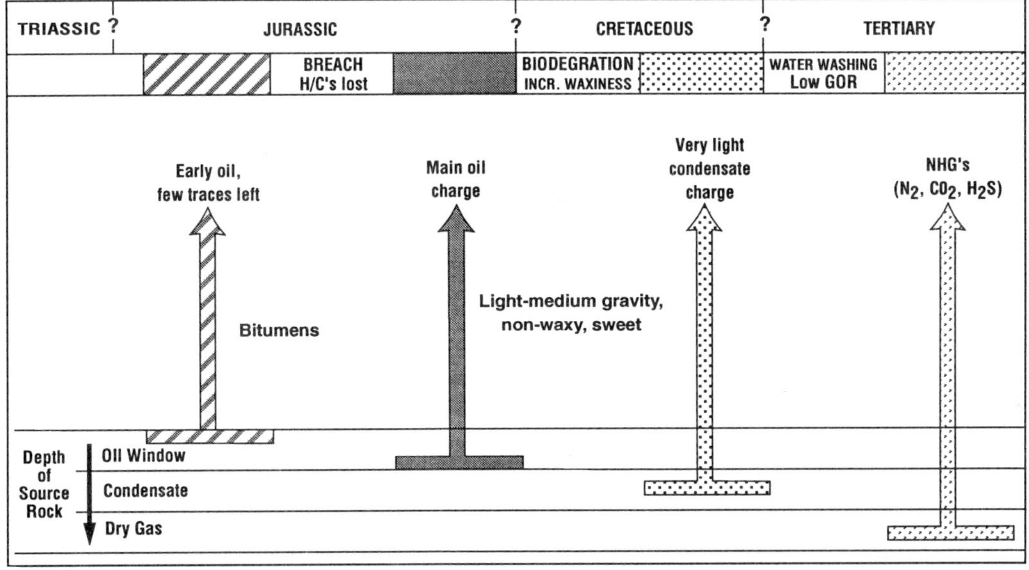

Fig. 8. Hydrocarbon filling history of the Douglas Field.

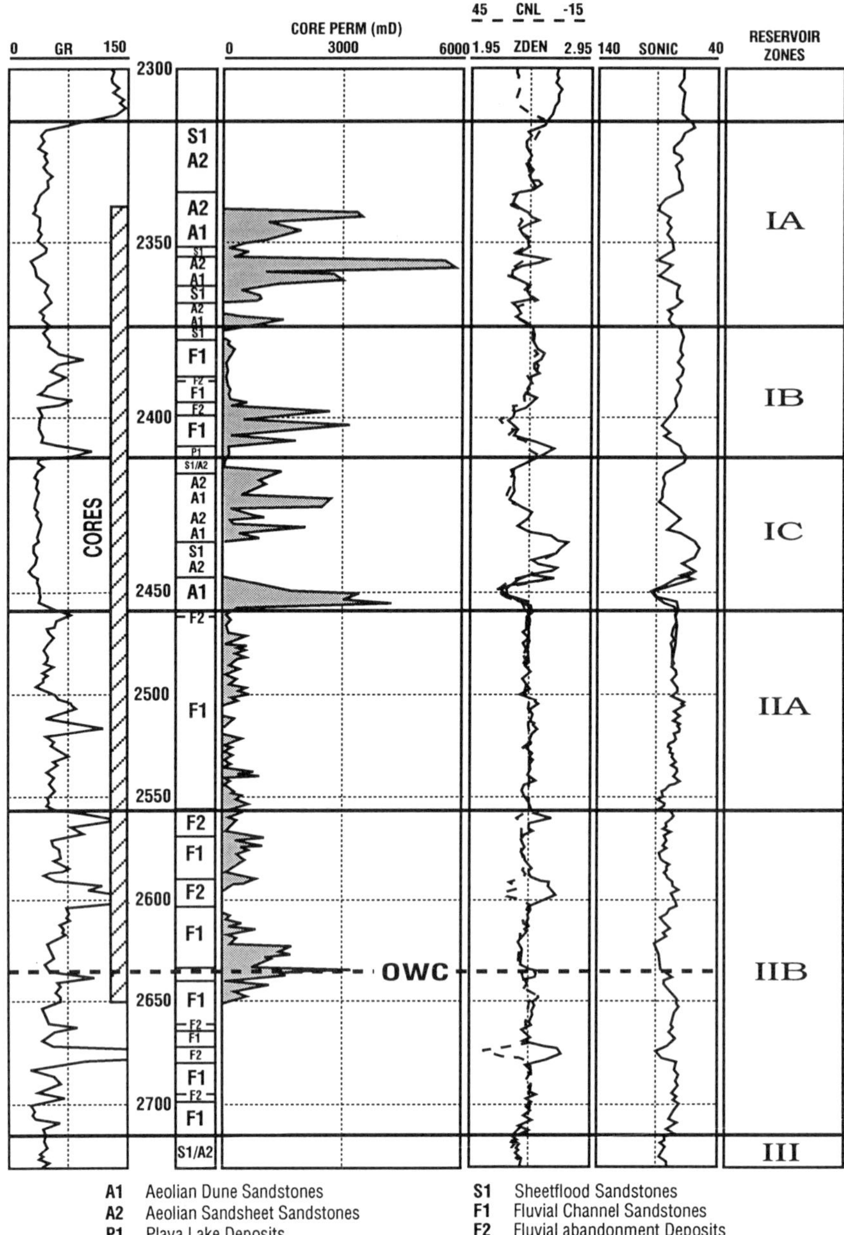

Fig. 9. Facies distribution and reservoir zonation in the Douglas Field (Well 110/13-6).

sequences. They are considered to form reservoir barriers.

Reservoir zonation

The vertical distribution of sedimentary lithofacies produces four distinct reservoir zones which can be easily identified by wireline logs and core analysis data (Figs 10 and 11). Zones I, II and III are the Frodsham, Delamere and Thurstaston members of The Helsby Sandstone Formation, respectively, and Zone IV is the Wilmslow Sandstone Formation.

Zone I is further sub-divided into three parts. The uppermost Zone IA with a fairly uniform thickness

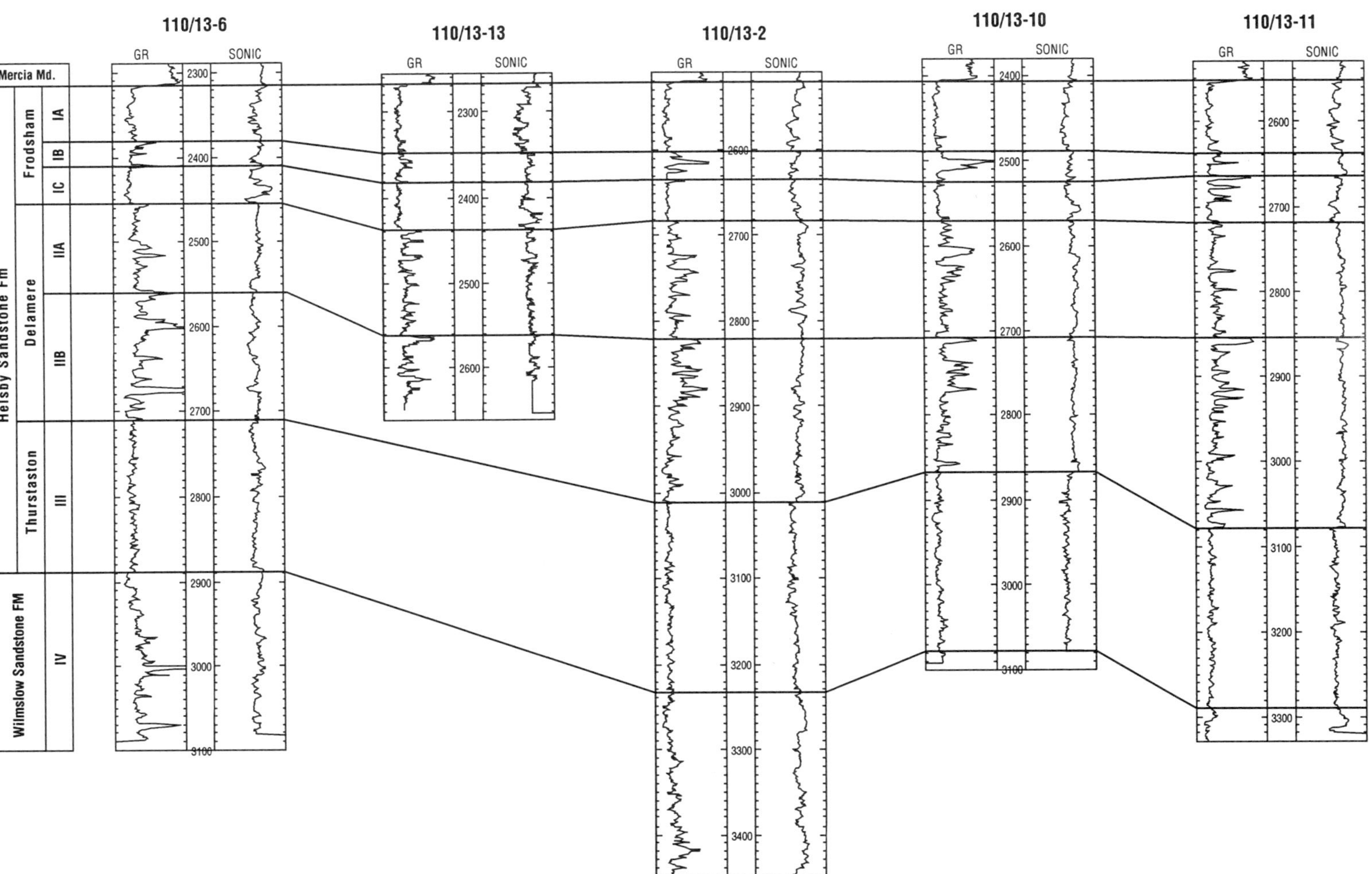

Fig. 10. Reservoir zone correlation in the Douglas Field.

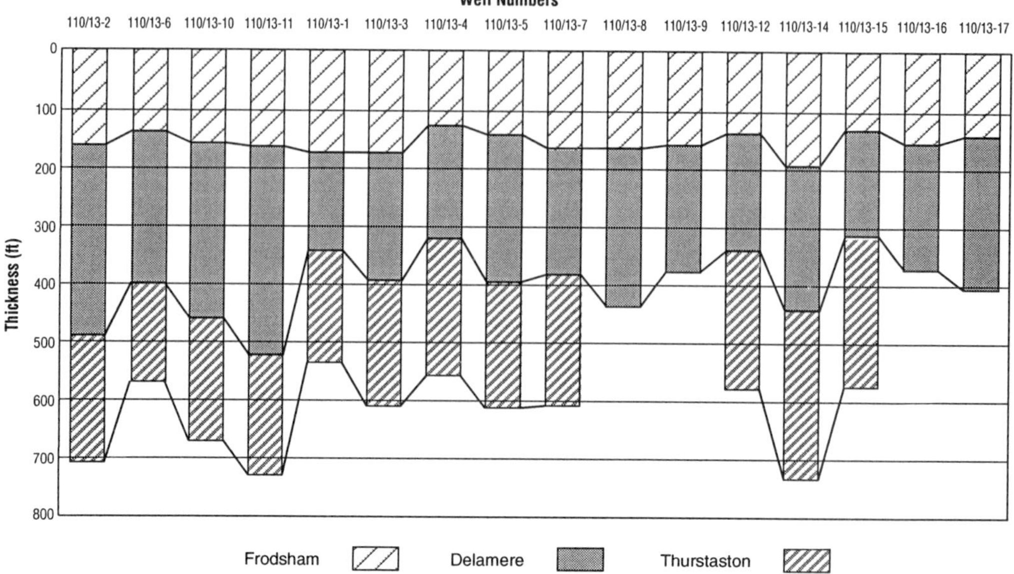

Fig. 11. Helsby sandstone thickness variation in Block 110/13.

of around 80 ft consists of several units of aeolian dunes and aeolian sandsheets alternating with 'sheetflood' sands. It is the most significant layer, containing almost half of the field's reserves. Permeability in the aeolian dune facies is often found to be several darcies, while porosities have a range of 20–30%. Zone IB, with an average thickness of 35 ft, is a low quality reservoir consisting of thinly interbedded fluvial channel and fluvial abandonment deposits. Zone IC, with high permeability aeolian sandstones, is similar to Zone IA, but its thickness averages only 45 ft.

Zone II consists exclusively of fluvial deposits. Its thickness increases from 250 ft in the east to 350 ft in the west, towards the Gogarth fault. For the purpose of calculating field volumetrics it has been subdivided into Zones IIA and IIB, which are separated by a thin but laterally extensive fluvial abandonment deposit.

Zone III has a fairly uniform thickness of around 200 ft and includes a variety of aeolian dune, sandsheet and 'sheetflood' facies. Although the reservoir quality is excellent, because of the geometry of the field structure very little oil is trapped in Zone III.

Zone IV consists of fluvial sandstones and is mapped principally as an aquifer layer.

The reservoir characteristics of each zone based on core porosity and permeability data are illustrated in Fig. 12. A fairly good log–linear relationship exists between the porosity and permeability values in all zones except Zone II. However, the slope and the intercept of the regression lines show slight differences between zones. The 'shotgun effect' exhibited by the Zone II data is thought to be an artefact of numerous small shale clasts it typically contains. A summary of the reservoir quality of the zones is given in Table 1.

Highest porosity and permeability values were measured in the aeolian dune and sandsheet facies in Zones IA and IC. However, the 'sheetflood' facies within the same zones have porosity and permeability values ranging from poor to excellent. The reservoir quality of Zone IB and Zone II is only fair to poor as their main constituents, fluvial, fluvial abandonment and playa-lake facies, are generally tight.

Table 1. *Summary of reservoir quality in the Douglas Field*

Zone	Av. porosity (%)	Core Perm. (mD)	Reservoir quality
IA	19	100–10000	excellent
IB	14	10–1000	fair
IC	18	100–10000	excellent
II	13	10–1000	fair
III	18	50–3000	good

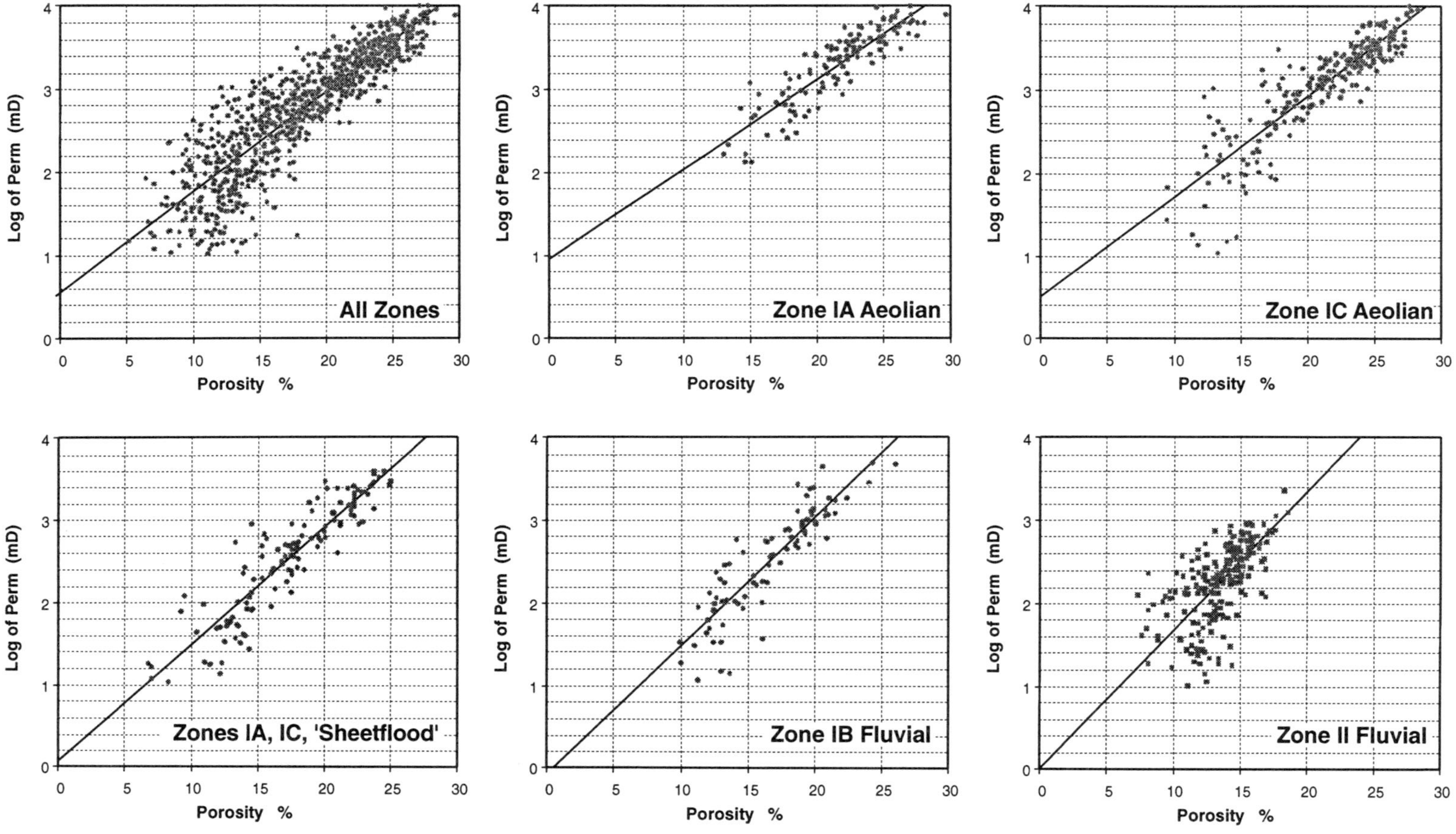

Fig. 12. Core porosity and permeability data in the Douglas Field.

Diagenetic history

The Douglas reservoir sandstones are classified as sublithic or subfeldspathic arenites, with 5–15% feldspar and up to 2% clay. Textural maturity is broadly related to facies type with aeolian sands being the most mature.

The authigenic minerals in order of decreasing abundance are carbonates (ferroan and non-ferroan calcite and dolomite), quartz, gypsum, anhydrite, bitumen, feldspar, pyrite, illite, chlorite and halite. Quartz cementation is greater within the fluvial facies, while non-ferroan calcite is largely confined to samples from below the oil-water contact. Illite cementation is practically non-existent in the Douglas Field.

Early diagenesis was largely controlled by the effects of weakly alkaline meteoric pore waters. This resulted in deposition of clays on grain surfaces, dissolution of lithic grains and feldspars, haematite staining, and cementation by quartz, caliche carbonates and evaporitic sulphates. Intermediate diagenesis produced dolomite cement that is frequently enclosed by bitumen, which suggests that a phase of hydrocarbon charging post-dated this cementation. The later stages of diagenesis produced locally intense gypsum and anhydrite cements, perhaps related to the expulsion of pore fluids from the overlying evaporitic Mercia Mudstone Group. A final phase of quartz overgrowths was also developed at this time. Finally, the present day oil was emplaced followed by non-ferroan calcite cementation below this in the water leg.

Bitumen evolution and distribution in reservoir

The nature and distribution of bitumen in this field have been studied in detail because of its probable negative effects on the reservoir performance. The bitumen has been found to be concentrated along a 30–100 ft thick field-wide layer, which extends both above and below the present oil-water contact and dips 1–2° towards the north–northwest. This bitumen zone does not appear to be related to either lithofacies or reservoir quality. However, within the zone (and when present in small quantities elsewhere) the bitumen is preferentially preserved along the coarser-grained laminae.

Highest concentrations of bitumen (10–20% gross rock volume) are found in the south-eastern and central parts of the field (Fig. 13). This decreases to less than 3% northwards. The bitumen layer cuts across all reservoir layers and the present oil–water contact showing no relationship to their

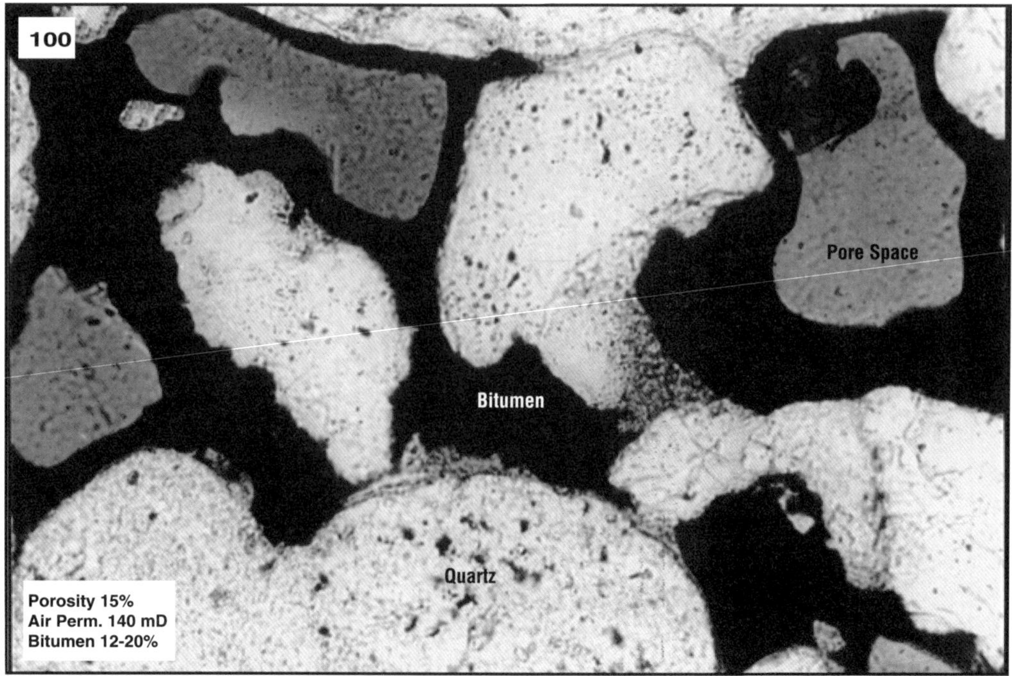

Fig. 13. Photomicrograph of bitumen-filled pore space (Well 110/13-10).

structural orientation. This suggests that its development was related to a palaeo-oil–water contact. An alternative origin is suggested from the fact that the reservoir interval above the bitumen layer is almost entirely bitumen-free. This could be interpreted to be due to the presence of a palaeo-gas cap, with the bitumen layer forming along the gas–oil contact.

Textural mineral relationships recorded in most Douglas wells suggest an earlier origin and bitumen is not related to the present oil charge. Bitumen post-dates early burial cements but late anhydrite and quartz post-date bitumen. Geochemical analysis (sulphur, nickel and vanadium contents and carbon isotope compositions) indicates a common source for all the bitumens seen in the field. The composition of the residual bitumen is compatible with the hypothesis that deasphalting would have involved the injection of gas into the oil-bearing reservoir causing the precipitation of asphaltenes as semi-solid platelets within the pore space (Lomando 1992). Therefore, although the mechanism of bitumen formation is still open to conjecture, the Douglas residual bitumen was probably derived from a pre-existing crude oil, most likely by deasphalting and/or water washing along a palaeo-oil–water contact.

The bitumen volumes in each well have been estimated by thin section point-counting using sidewall and conventional core samples. With the latter type of samples being available only over limited intervals, sidewall core samples proved to be the best source of data for the evaluation of the bitumen. Attempts to evaluate bitumen percentage from wireline logs have not been very successful, except when the bitumen is present below the oil–water contact. Such zones are characterized by higher than expected oil saturation values and the bitumen concentrations can be roughly estimated from resistivity porosity calculations.

High levels of reservoir bitumen (10–20% of gross rock volume) obviously have a detrimental effect on the permeability. For example, measurements on conventional core plugs have shown that the permeability in the aeolian sands in Zone IC may be drastically reduced, typically from 2–3 D to less than 100 mD, with the corresponding effective porosity values being reduced from around 25% to less than 15%. Studies on core samples indicated that very little bitumen was removed in the cleaning process using hexane as the cleaning agent. Therefore, in the reservoir simulation model, the bitumen layer in the southeastern area of the field is mapped as a partial barrier with greatly reduced permeabilities. It is difficult to predict at present how much effect this layer will have on the injection pressure support. At the start of field production, tracers will be used to help understand the effects of

the bitumen layer on the reservoir connectivity between specific injector and producer wells.

Data from the recently drilled development wells have generally supported the above interpretation of the bitumen layer. However, the recent data indicate that the bitumen layer was offset by movements along fault planes, probably during their reactivation by a Tertiary inversion event. This gives rise to slight differences in its dip within each fault block.

Reservoir fluid parameters

The Douglas field contains a low sulphur, 44° API, black oil, which is essentially a 'dead' oil with a formation volume factor of 1.075 RB/STB. The bubble point at reservoir temperature is only 285 PSIA. Total process GOR is 170 SCF/STB.

Reservoir fluid composition is summarized in Table 2. The bulk of the fluid is made up of components heavier than decanes (38%). Average methane, ethane and propane concentrations are 2.6, 4.6 and 6.1 mol %, respectively. The non-hydrocarbon components include nitrogen (1%), hydrogen sulphide (0.5%), mercaptans (0.5%) and carbon dioxide (0.1%).

Measurements on water samples indicate that the formation water is highly saline, with a total dissolved solids content of 270 000 mg^{-1} and a measured resistivity of 0.056 ohm-m at 60°F.

Field development

The platform location and the positions of the development wells on the top reservoir structure map are shown in Fig. 14. As a result of numerous

Table 2. *Reservoir fluid composition in the Douglas Field*

	mol % (110/13-13)
Methane	2.59
Ethane	4.57
Propane	6.05
i-Butane	2.80
n-Butane	6.33
i-Pentane	3.66
n-Pentane	4.61
Hexanes	7.30
Haptanes	8.48
Octanes	8.14
Nonanes	4.97
Decanes +	38.34
Nitrogen	1.01
Hydrogen Sulphide	0.47
Mercaptans	0.55
Carbon Dioxide	0.13

                                    A. M. YALIZ

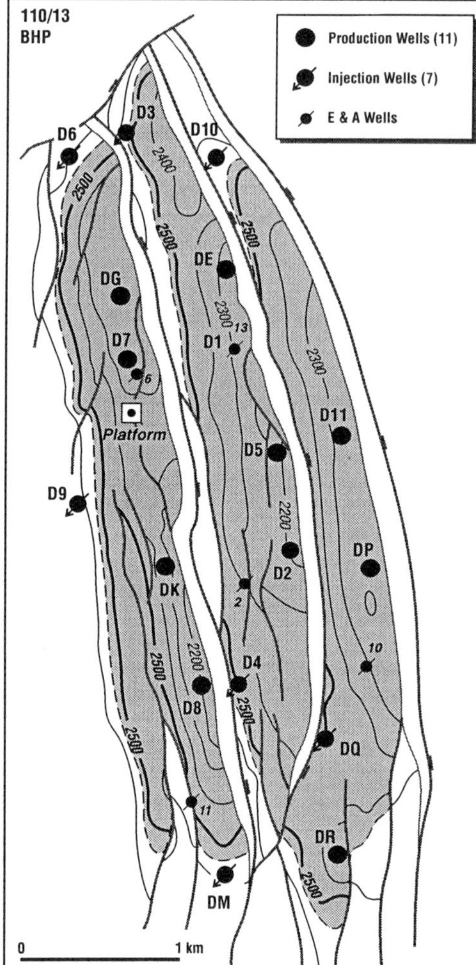

Fig. 14. Development well locations on the top reservoir depth structure map.

full-field model studies the development plan is based on 11 producing wells and seven water injectors. Typical well trajectories for both producers and injectors are illustrated in Fig. 15. Long step-out wells have S-shaped paths with an extended horizontal section in the Mercia Mudstone. Reservoir model studies were based on a STOIIP of 225 MMBBL (best estimate Annex B volume) with 51% of the oil being in Zone IA, 13% in Zone IB, 16% in Zone IC and 20% in Zone II.

The wells are designed to penetrate the reservoirs parallel to the major fault planes. The producers are placed near the crest of each tilted fault block, approximately 75 m away from the fault plane while the water injectors are drilled down dip, just below the oil–water contact. By placing the pro-

ducers close to faults the recovery of the 'attic' oil beneath the fault planes, which dip at around 45° to east, is greatly improved during the waterflood. The platform location has been chosen to maximize the capability of drilling such wells throughout the field. The maximum step-out is around 3 km. However, there still remains some uncertainty as to sealing capability of individual fault planes, which cannot be resolved until production commences.

In both production and injection wells the maximum reservoir interval, the entire sections of Zone I and Zone IIA, are perforated. This procedure is thought to be beneficial as it will maximize deliverability and injectivity, and allows the rapid provision of energy into the reservoir system via water injection. It is planned to use electrical submersible pumps (ESP) for artificial lift in all production wells, placed along their nearly horizontal sections and as close to the top reservoir as possible.

The first phase of development took place between September 1994 and April 1995, when 11 development wells were pre-drilled. The second phase of development was completed in January 1997. Of the 11 wells drilled in Phase I, six were completed as producers and five as water injectors (Fig. 14). There were a number of geological differences to the Annex B predictions. For instance, in most cases the depth to the top reservoir structure was found to be higher than expected (up to 80 ft). Within the bitumen layer log-derived effective porosities were over-estimated because of the uncertainties in the fluid density and matrix density values. Significantly, it was found that prediction of the major fault positions from the 3D seismic data proved to be difficult. Well 110/13-D2 is a typical producer and illustrates this point. It was located at the crest of the central tilted fault block, 75 m from the bounding fault (Fig. 16). As with all Douglas producers the well path was designed to be parallel to the fault plane in the reservoir section. Although the well was physically drilled within the very tightly specified target tolerance, an engineering success, it crossed the main fault into the eastern fault block while drilling in Zone I. Approximately 400 ft section of Mercia Mudstone was drilled before the well re-entered the Central panel by crossing the fault a second time in Zone II. The well was terminated at the base of Zone III. Several important conclusions can be drawn from the well results:

1. It is technically feasible to drill long step-out wells with extended horizontal sections and small target tolerances relatively quickly.
2. The actual fault position in the Central block was found to be approximately 75 m to the west

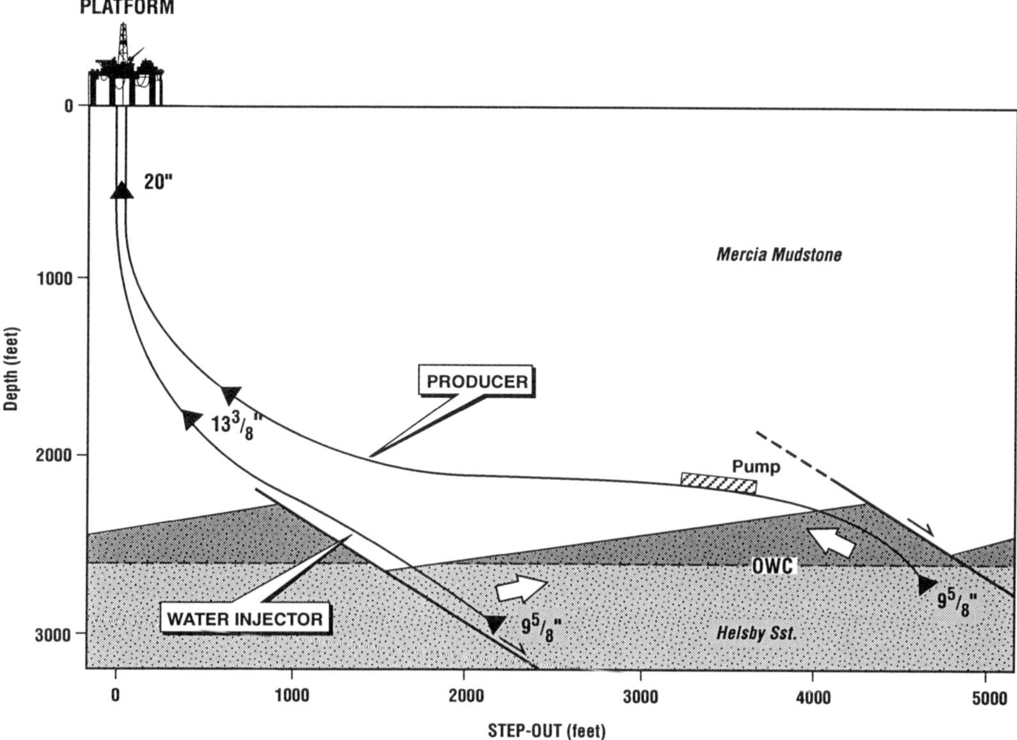

Fig. 15. Typical well trajectories for a producer and a water injector.

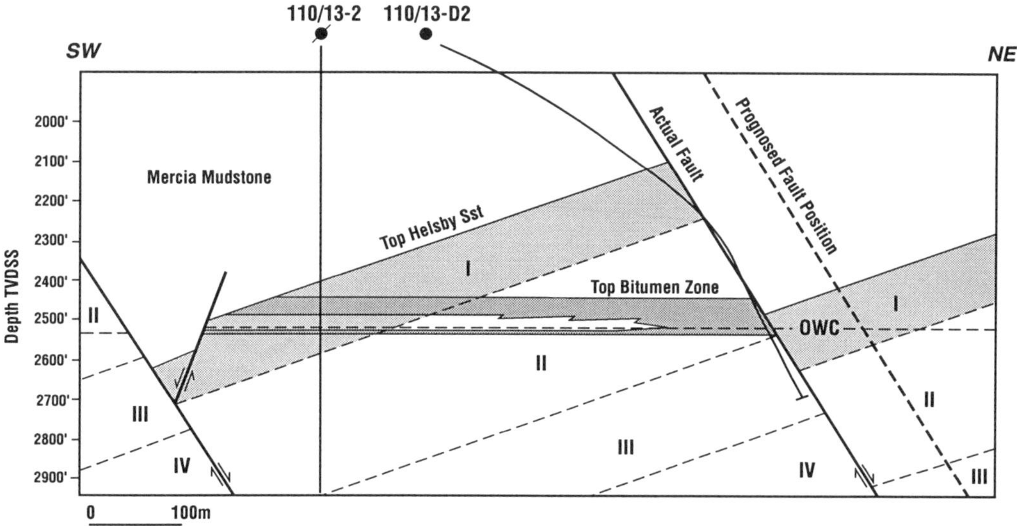

Fig. 16. Geological interpretation through the well paths of 110/13-2 discovery well and 110/13-D2 producer.

of the location predicted from the 3D-seismic data, and has been subsequently confirmed by other wells drilled in the same panel. The cause of this is thought to be related to the migration problems in seismic processing. As a result, the seismic data has now been reprocessed using pre-stack depth migration techniques.

3. There were no circulation losses while drilling through the fault plane. This suggests that where the reservoir sands are faulted against the Mercia Mudstone there are no open fractures associated with the faulting. However, in some of the later wells significant circulation losses were experienced along fault planes with sand-to-sand contact indicating the presence of large open fractures.

4. Contrary to earlier predictions, no deterioration of the reservoir quality was observed near the fault. Log analysis indicated excellent porosities. This was supported by the drill-stem testing of a limited interval in Zone I, which flowed at an extremely good rate of 3700 BPD with a PI of 20 BPD/PSI.

5. The bitumen zone encountered in 110/13-D2 is correlatable to that in the nearby discovery well 110/13-2. Both zones occur at the same structural elevation within the oil column and show similar thicknesses and bitumen concentrations. The bitumen zones were identified from core samples including sidewall cores.

The Douglas Field forms part of the BHP Petroleum's Liverpool Bay integrated development scheme as illustrated in Fig. 17. First oil was produced in February 1996 and additional wells were drilled during 1996 and early 1997. Drilling in the Hamilton North Field was completed in October 1996 and the drilling in the Lennox Field continues at the present time, Phase I having been completed in 1996. Hamilton Field wells will also be drilled in 1996. The Douglas and Lennox oils are being exported via a pipeline to an offshore storage barge in nearby Block 110/12 and shuttle tankers are used to ferry the crude. The Hamilton and Hamilton North gas goes via a 34 km, 20 inch export line to the BHP Petroleum's Point of Ayr terminal. After further processing and sulphur removal at the terminal the gas is delivered to

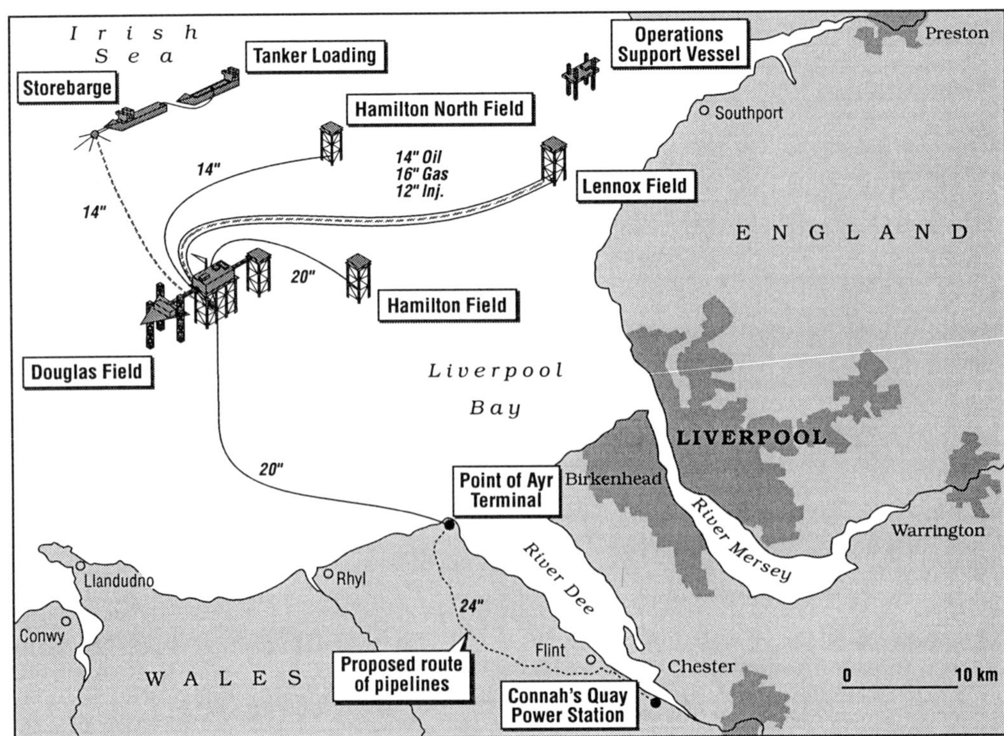

Fig. 17. Schematic map showing the Liverpool Bay Integrated Development scheme.

PowerGen's power station at Connahs Quay. The Lennox gas is initially being returned from Douglas for reinjection into the Lennox gas cap.

This paper is published by permission of BHP Petroleum, the operator of Blocks 110/13 and 110/15, and the Liverpool Bay Group participants, Lasmo (ULX) Limited, Monument Liverpool Bay Limited and PowerGen North Sea Limited. Thanks are also due to C. J. Bryan and S. Catterall, BHP personnel, whose work contributed significantly to the contents of this paper. Contributions of N. S. Meadows, S. M. Habesch and N. J. L. Bailey, the Geochem Group, to the understanding of the sedimentology and the geochemistry of the field, and of H. E. Edwards, Simon Petroleum Technology Ltd, to the understanding of the structural history of the field, are gratefully acknowledged.

References

ARMSTRONG, J. P., SMITH, J., D'ELIA, V. A. A. & TRUEBLOOD, S. P. The occurrence and correlation of oils and Namurian source rocks in the Liverpool Bay–North Wales area. *This volume.*

BUSHELL, T. P. 1986. Reservoir geology of the Morecambe Field. *In*: BROOKS, J., GOFF, J. C. & VAN HOORN, B. (eds) *Habitat of Palaeozoic Gas in N.W. Europe.* Geological Society, London, Special Publication, **23**, 189–208.

COPE, J. C. W. 1994. A latest Cretaceous hotspot and the southeasterly tilt of Britain. *Journal of the Geological Society, London*, **151**, 905–908.

COLTER, V. S. & BARR, K. W. 1975. Recent developments in the geology of the Irish Sea and Cheshire basins. *In*: WOODLAND, A. W. (ed.) *Petroleum and the Continental Shelf of North West Europe.* Applied Science, 61–75.

GEOCHEM GROUP, 1991. Geochemical evaluation of the residual bitumen in the Helsby Sandstone Formation reservoirs of wells 110/13-1 and 110/13-6, Morecambe Bay. Report Nos, 5130, 5205 & 4884. Proprietary report.

—— 1993. Investigation into the origin of mercaptans in the Douglas and Lennox fields, Liverpool Bay, Report No. 93/8085/01/01. Proprietary report.

—— 1995. A petrological and geochemical study of the bitumen-stained Sherwood Sandstone reservoir lithologies in various Douglas Field wells (Block 110/13), Report No. 95/9535/004/01. Proprietary report.

—— 1995. A calibration study of residual bitumen and remnant porosity in the Sherwood Sandstone reservoir in well 110/13-10 using combined core and petrographic analyses, Report No. 95/9622/004/01. Proprietary report.

GREEN, P. F. 1986. On the thermotectonic evolution of Northern England: evidence from fission track analysis. *Geological Magazine*, **123**, 493–506.

HARDMAN, M., BUCHANAN, J., HERRINGTON, P. & CARR, A. 1993. Geochemical modelling of the East Irish Sea Basin: its influence on predicting hydrocarbon type and quality. *In*: PARKER, J. R. (ed.) *Petroleum Geology of North West Europe: Proceedings of the 4th Conference.* Geological Society, London, 809–821.

HERRIES, R. D. & COWAN, G. 1997. Challenging the 'sheetflood' myth: the role of water-table-controlled sabkha deposits in redefining the depositional model for the Ormskirk Sandstone Formation (Lower Triassic), East Irish Sea Basin. *This volume.*

JACKSON, D. I., MULHOLLAND, P. 1993. Tectonic and stratigraphic aspects of the East Irish Sea basin and adjacent areas: contrasts in their post-Carboniferous structural styles. *In*: PARKER, J. R. (ed.) *Petroleum Geology of North West Europe: Proceedings of the 4th Conference.* Geological Society, London, 791–808.

——, ——, JONES, S. M. & WARRINGTON, G. 1987. The geological framework of the East Irish Sea basin. *In*: BROOKS, J. & GLENNIE, K. (eds) *Petroleum Geology of North West Europe.* Graham and Trotman, London, 191–203.

——, JOHNSON, H. & SMITH, N. J. P. 1997. Stratigraphical relationships and a revised lithostratigraphical nomenclature for the Carboniferous, Permian and Triassic rocks of the offshore East Irish Sea Basin. *This volume.*

LEWIS, C. L. E., GREEN, P. F., CARTER, A. & HURFORD, A. J. 1992. Elevated K/T palaeotemperatures throughout Northwest England: three kilometres of Tertiary erosion? *Earth and Planetary Science Letters*, **112**, 131–145.

LOMANDO, A. J. 1992. The influence of solid reservoir bitumen on reservoir quality. *Bulletin of the American Association of Petroleum Geologists*, **76**, 1137–1152.

MEADOWS, N. S. & BEACH, A. 1993. Structural and climatic controls on facies distribution in a mixed fluvial and aeolian reservoir: the Triassic Sherwood sandstone in the Irish Sea. *In*: NORTH, C. P. & PROSSER, D. J. (eds) *Characterization of Fluvial and Aeolian Reservoirs.* Geological Society, Special Publication, **73**, 247–264.

SIMON PETROLEUM TECHNOLOGY LTD, 1994. A structural study of the Douglas Field and adjacent areas, Block 110/13, East Irish Sea Basin. Report No. 7489/IIb. Proprietary report.

SMITH, D. B., BRUNSTROM, R. G. W., MANNING, P. I., SIMPSON, S. & SHOTTON, F. W. 1974. A correlation of Permian rocks in the British Isles. Geological Society, London, Special Report, **5**.

STUART, I. A. & COWAN, G. 1991. Morecambe Gas Field, Blocks 110/2a, 110/3a, 110/8a, U.K. *In*: ABBOTTS, I. L. (ed.) *United Kingdom Oil and Gas Fields: 25 years Commemorative Volume.* Geological Society, London, Memoir, **14**, 527–541.

THOMPSON, D. B. 1970. The stratigraphy of the so-called Keuper Sandstone Formation (Scythian–?Anisian) in the Permo-Triassic Cheshire basin. *Quarterly Journal of the Geological Society, London*, **126**, 151–181.

WARRINGTON, G., AUDLEY-CHARLES, M. G., ELLIOT, R. E.

ET AL. 1980. A correlation of the Triassic rocks in the British Isles. Geological Society, London, Special Report, **13**.

WILLIAMS, G. D. & EATON, G. P. 1993. Stratigraphic and structural analysis of the Late Palaeozoic–Mesozoic of NE Wales and Liverpool Bay: implications for hydrocarbon prospectivity. *Journal of the Geological Society, London*, **150**, 489–499.

Appendix: Douglas Field data summary

Trap

Type	Structural
Depth to crest	2140 ft TVSS
Hydrocarbon contacts	2515–2535 ft TVSS
Oil column	375 ft max

Main pay zone

Formation	Sherwood Sandstone
Age	Triassic
Net/gross ratio	0.85–0.95
Cut-off for N/G	10 mD
Porosity average	18% Zone I, 14% Zone II
Permeability average (air)	1000 mD Zone I, 250 mD Zone II
Productivity index	5–30 BPD/PSI

Hydrocarbons

Oil gravity	44° API
Oil type	Low sulphur
Bubble point	285 psi
Gas/oil ratio	170 SCF/BBL
Formation volume factor	1.075 RB/STB

Formation water

Salinity	270 000 ppm NaCl equivalent
Resistivity	0.030 ohm-m at 30°C

Reservoir conditions

Temperature	30°C
Pressure	1125 PSI at 2240 ft TVSS
Pressure gradient in reservoir	0.35 PSI/FT

Field size

Area	6.5 km^2
Gross rock volume	235 000 ac-ft
STOIIP	225 MMBBL
Drive mechanism	water injection

Production

Start-up date	February 1996
Development scheme	Central 24-slot wellhead tower, oil exported via pipeline to offshore storage unit for export by tanker.
Number/type of wells	5 exploration/appraisal 11 producers 7 water injectors
Production rate	40 000 BPD expected peak rate

The Lennox oil and gas Field

D. B. HAIG, S. C. PICKERING & R. PROBERT

*BHP Petroleum Limited, Devonshire House, Mayfair Place, Piccadilly,
London W1 6AQ, UK*

Abstract: The Lennox oil and gas Field is located in Block 110/15 of the East Irish Sea, approximately 10 km west of Southport, Lancashire. The field was discovered in 1992 with well 110/15-6 drilled by the P791 BHP group. The Lennox structure is a rollover anticline in Permo-Triassic sediments formed in the hanging wall of the Formby Point Fault. The reservoir comprises the Triassic Sherwood Sandstone Group, and is characterized by excellent permeability aeolian and fluvial sandstones. The structure is sealed by the overlying shales and evaporites of the Mercia Mudstone Group. To date four wells have been drilled on the Lennox Field and have delineated a free gas cap in communication with a constant thickness oil rim. The hydrocarbons have been sourced from Namurian Holywell shales and reflect complex migration into the Lennox structure. Reservoir studies indicate likely hydrocarbons-in-place of 463 BCF of gas and 218 MMBO of oil. Development approval for the Lennox Field was granted in 1993 with production due to commence in late 1995 from a series of horizontal oil producers. Gas produced during the early years will be recycled and re-injected into Lennox via a crestal gas well to help maximize the oil production. During the later years, production will switch to sales gas. Lennox will be developed as an integral part of BHP's Liverpool Bay Development.

The Lennox oil and gas Field is located in blocks 110/15 and 110/14 on the southeastern edge of the East Irish Sea Basin (EISB) only 4 km from the Lancashire coast (Fig. 1). BHP Petroleum Limited operate Block 110/15, including the Lennox Field, on behalf of the P791 group of companies comprising: BHP Petroleum Great Britain PLC, 46.100%; Lasmo (ULX) Limited, 25.000%; Monument Exploration and Production Limited, 20.000%; and PowerGen (North Sea) Limited, 8.900%

The field was discovered by the 110/15-6 well and has been subsequently appraised by the drilling of a further three wells, namely, 110/15-6Z, 110/15-7 and the 110/14-3 well drilled by BG which confirmed the extension of the Lennox Field into the neighbouring Block 110/14. Development approval for the Lennox Field was granted in 1993 and production is due to commence in November 1995.

Structural setting

The Lennox Field, in block 110/15, is located on the southeastern edge of the EISB within the East Deemster sub basin (Fig. 2). The block is bisected by a major north–south basin bounding fault, the Formby Point Fault, forming two distinct structural provinces.

- An eastern area where Carboniferous and Permo-Triassic rocks have not undergone extensive burial and where the Triassic Mercia Mudstone Group is thin or absent.
- A western area where deep burial has occurred followed by Alpine inversion. The Mercia Mudstone Group is thickly developed and preserved in this area.

The Lennox structure (Figs 3 & 4), is a faulted, roll-over anticline on the hanging wall of the Formby Point Fault. The structure has been formed in response to uplift of a ramp of the Formby Point Fault directly underlying the Lennox Field followed by subsequent collapse. The field is partially bounded to the north and to the south by east–west trending grabens which have formed in response to reactivation of underlying Caledonian-trending structures. A geoseismic line across the field illustrates the increase in the degree of faulting on the eastern side of the structure compared to the relatively 'quiet' western area. The crest of the structure at top reservoir level is at 2500 ft TVDSS with the gas–oil contact at 3257 ft TVDSS and the oil–water contact at 3400 ft TVDSS.

Reservoir stratigraphy

The hydrocarbons of the Lennox Field are reservoired entirely within sandstones of the Triassic Sherwood Sandstone Group and sealed by the overlying shales and evaporites of the Triassic Mercia Mudstone Group. A three-fold lithostratigraphic subdivision of the uppermost sandstones of

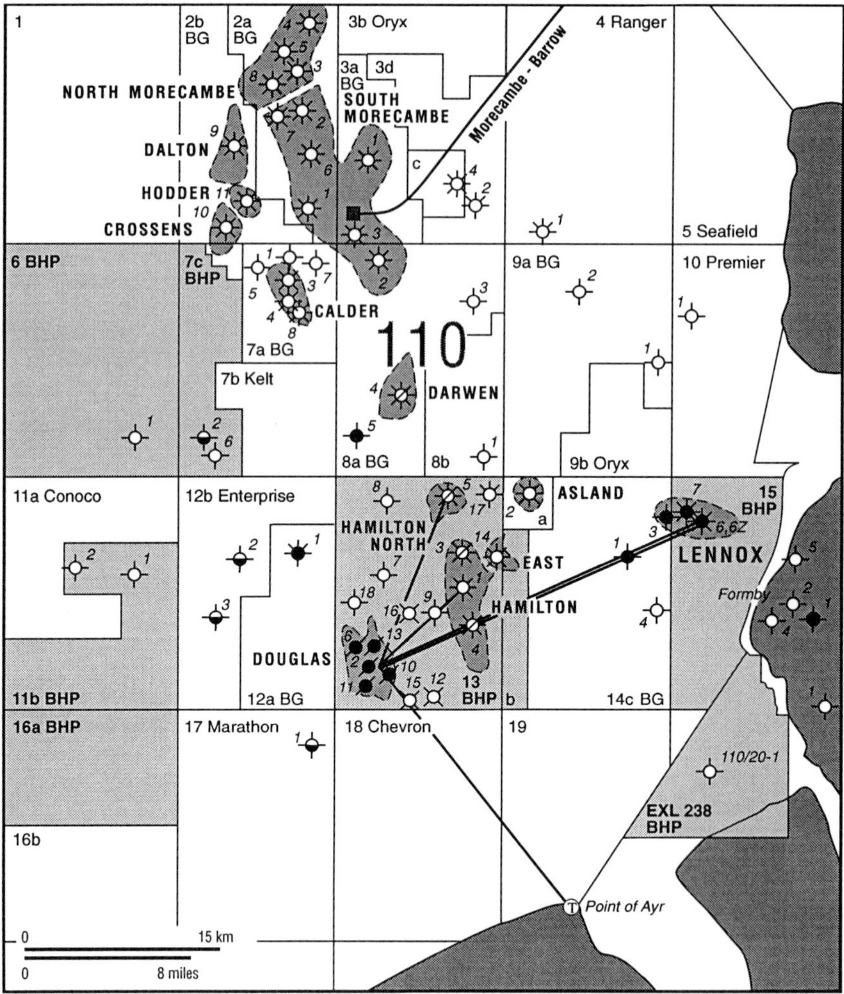

Fig 1. East Irish Sea Basin showing location of Lennox Field and other hydrocarbon discoveries.

the Sherwood Sandstone Group, i.e. the Ormskirk Sandstone Formation, has been applied based upon a combination of Liverpool, North Wirral and Cheshire Basin nomenclature (Fig. 5).

- The Thurstaston Member, which immediately overlies the St Bees Sandstone Formation, is a sequence of predominantly aeolian dune and sheetflood sandstones with minor fluvial channels.
- The Delamere Member overlies the Thurstaston and comprises predominantly fluvial channel and sheetflood sandstones.
- The Frodsham Member is the uppermost reservoir unit of the Lennox Field and comprises

predominantly aeolian dune sandstones with rare playa lake deposits.

For the purpose of field volumetric calculations, the reservoir sandstones have been subdivided into four zones (I–IV), following detailed facies analysis of core and wireline data. This facies analysis will be discussed later.

Exploration history

Block 110/15 was awarded to the P791 group of companies in the 12th offshore Licensing Round in 1991. A 2D seismic survey was carried out in early

1992 and the first exploration well, 110/15-6, was spudded in June 1992

Well 110/15-6

The Lennox discovery well, 110/15-6, was drilled to a location near to the crest of the Lennox structure encountering the top Sherwood Sandstone Group at 2617 ft (2514 FT SS). The reservoir section comprised 896 ft of Ormskirk Sandstone Formation with the underlying St Bees Sandstone Formation encountered at 3513 ft (3409 FT SS) (Fig. 6). The Ormskirk Sandstone reservoir contained 743 ft of gas bearing sandstones overlying 143 ft of oil bearing sandstones with a gas–oil contact at 3257 FT SS and an oil-water contact at 3400 FT SS. Overall reservoir quality was good with average zonal net to gross ratios varying from 96% to 98% and average zonal porosities varying from 13% to 19.6%. One drill-stem test, carried out over a 30 ft section of the oil bearing interval, flowed 870 BOPD through a 20/64" choke with a flowing wellhead pressure of 680 PSIG and a gas to oil ratio of 300–370 SCF/BBL. Wellsite chemical analysis of the crude recovered indicated an oil gravity of 45° API with a total gas to oil ratio of 480 SCF/BBL. Analysis of the gas recovered with the oil yielded H_2S levels of 2350 ppm and

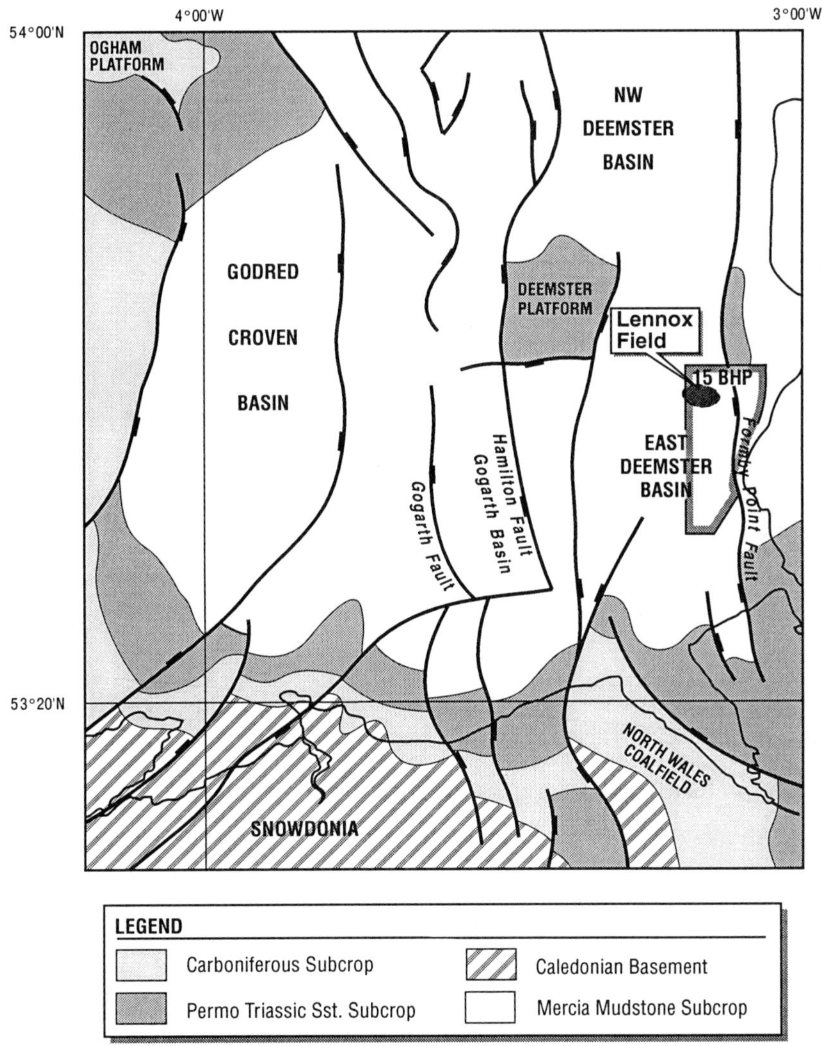

Fig 2. East Irish Sea Basin (south) structural elements and pre-Quaternary subcrop.

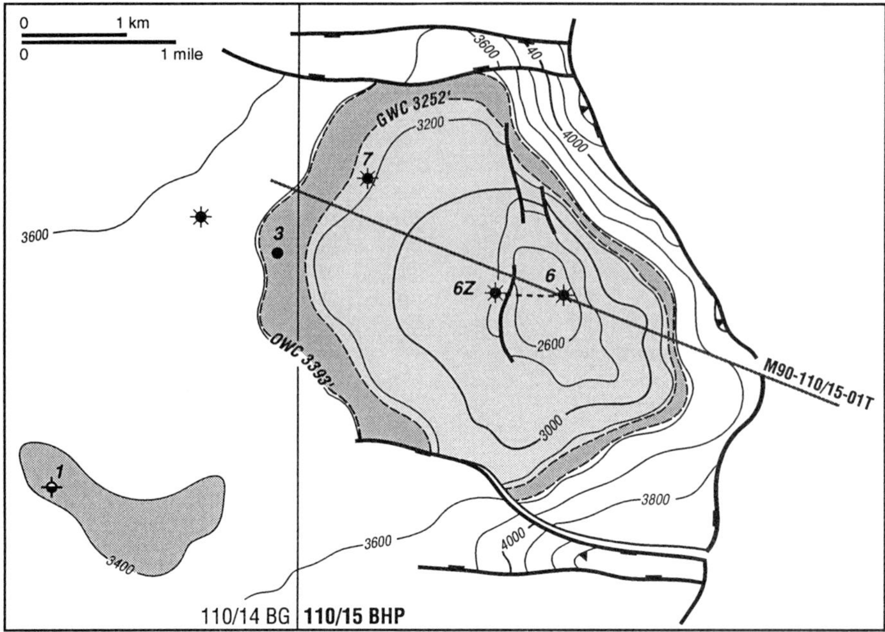

Fig. 3. Lennox Field – top Ormskirk Sandstone depth structure.

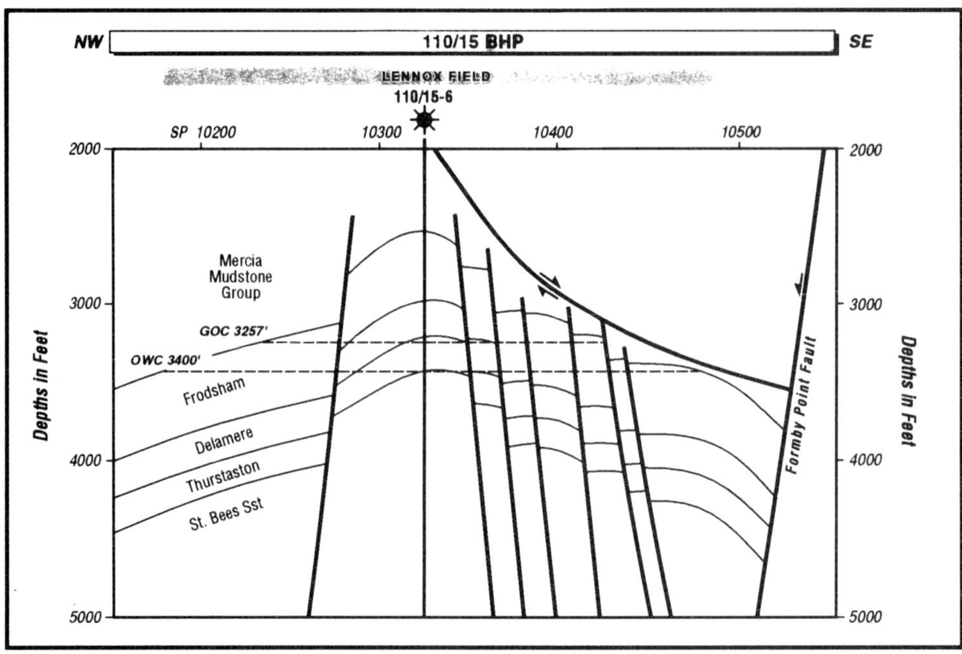

Fig. 4. Block 110/15, Lennox Field, Geoseismic line M90-110/15-01T showing heavily faulted eastern area and relatively 'quiet' western area.

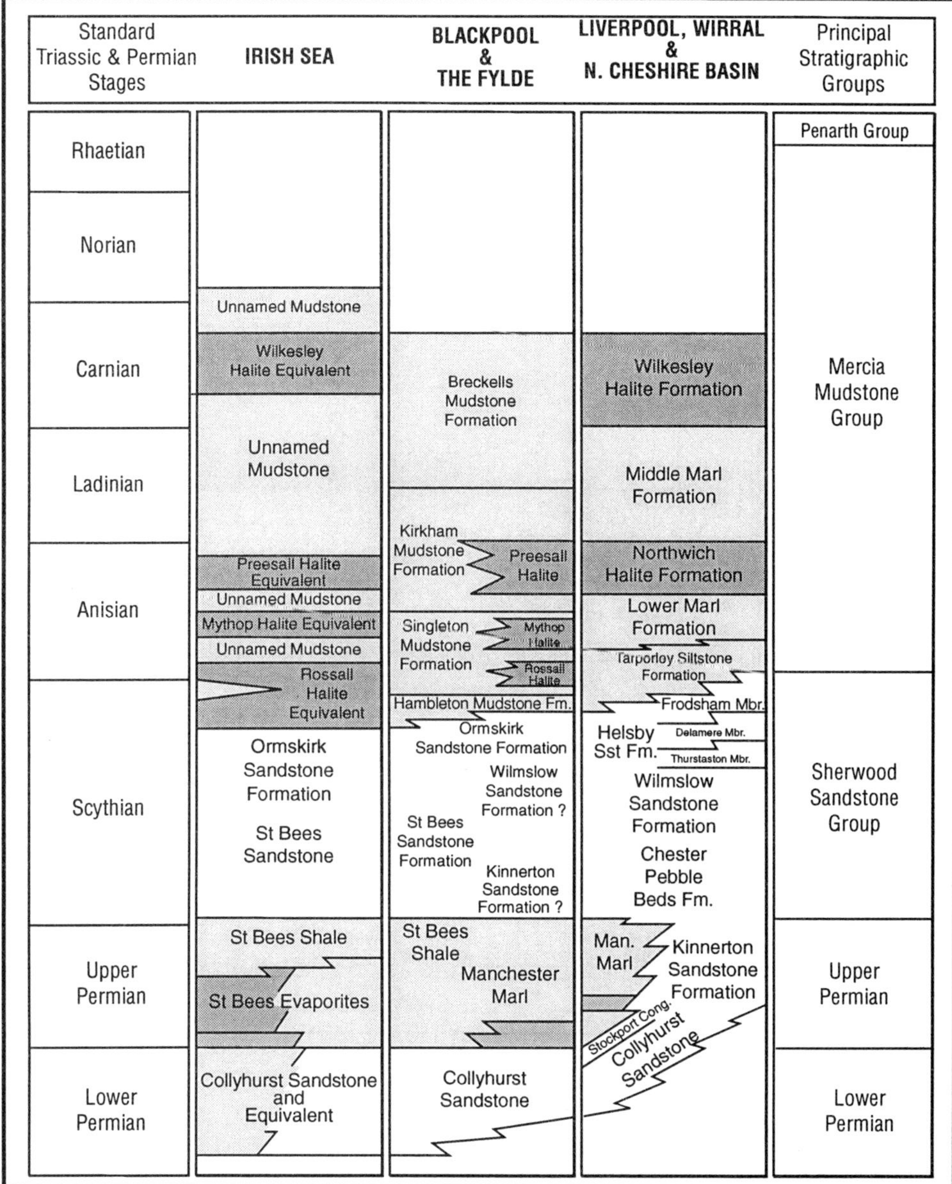

Fig. 5. East Irish Sea/Cheshire Basins – Permian and Triassic stratigraphy.

150 ppm of RSH mercaptans. The RSH mercaptan level within the crude oil was measured at 480 ppm sulphur by weight.

Well 110/15-6Z

The deviated appraisal well 110/15-6Z was designed to test and confirm the continuity of reservoir quality sandstones encountered in the discovery well and the presence of hydrocarbons in the western fault panel of the Lennox structure. The well was sidetracked to a location approximately 600 m to the west of the discovery well and established a gross hydrocarbon column of 1070 ft (573 ft TVT) extending from the top Ormskirk Sandstone Formation at 3874 ft (2820 FT SS) to

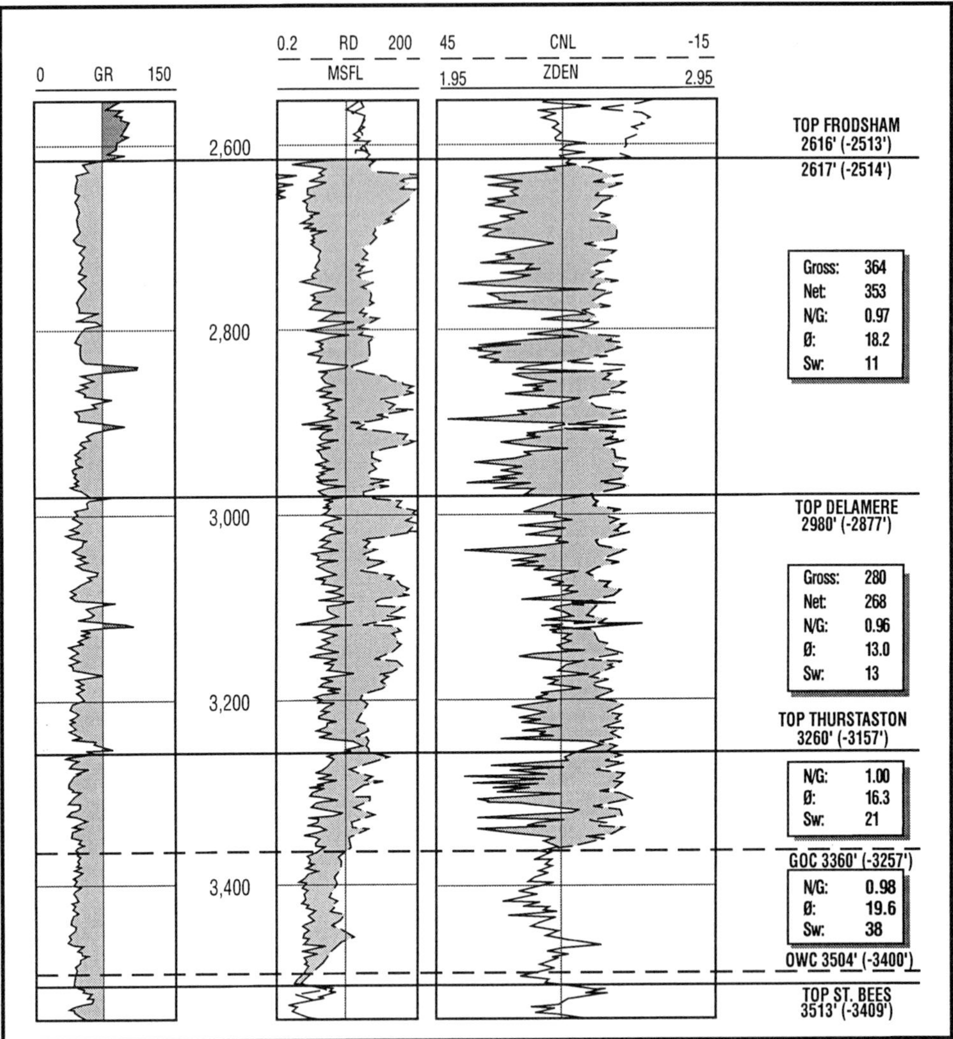

Fig. 6. Well 110/15-6 Ormskirk Sandstone reservoir section.

4944 ft (3393 FT SS) (Fig. 7). The hydrocarbon column consisted of 730 ft (834 ft TVT) of gas bearing sandstones down to a gas–water contact at 4604 ft (3252 FT SS), overlying 340 ft (141 ft TVT) of oil–bearing sandstones down to an oil–water contact of 4944 ft (3393 FT SS). The well was cored extensively through the oil leg. The reservoir qualities found in the discovery well were confirmed by well 110/15-6Z with average zonal net to gross ratios of 60–97% and average zonal porosities of 13.2%–16.7%. The oil-bearing section was confined to the predominantly fluvially dominated sandstones of the Delamere Member resulting in lower than average reservoir properties.

Well 110/15-7

The second appraisal well, 110/15-7 was drilled to test the reservoir and hydrocarbon continuity and the extent of the field in the western panel. The well encountered 45 ft TVT of gas-bearing Ormskirk Sandstone from 3323 ft (3212 FT SS) down to a gas–oil contact of 3368 ft (3257 FT SS), and 143 ft TVT of oil-bearing sandstones down to an oil–water contact of 3511 ft (3400 FT SS). The hydrocarbon column was confined to the predominantly aeolian Frodsham Member of the Ormskirk Sandstone Formation with an overall net to gross ratio of 100% and an average porosity of

19.3%. The well was extensively cored throughout the hydrocarbon-bearing section and a detailed facies analysis of the well was carried out using both core and wireline information (Fig. 8). This analysis subdivided the well into four main facies types: fluvial channel sandstones, sheetflood sandstones, aeolian dune sandstones and playa lake deposits. Based upon this subdivision and upon porosity/permeability relationships the Ormskirk Sandstone Formation reservoir was divided in to 18 layers for incorporating into reservoir simulation modelling. These were grouped in to four main zones used for mapping and reserve determinations.

Reservoir properties

Core data

A plot of core porosity versus permeability for well 110/15-6Z, demonstrates the importance of facies-derived porosity/permeability relationships in the Lennox Field (Fig. 9). Aeolian dune sandstones and sheetflood sandstones have the better reservoir properties and the shallower poroperm gradients, whereas the fluvial channel sandstones and the playa lake related deposits have poorer reservoir properties and steeper poroperm gradients. Detailed facies description is required to allow accurate

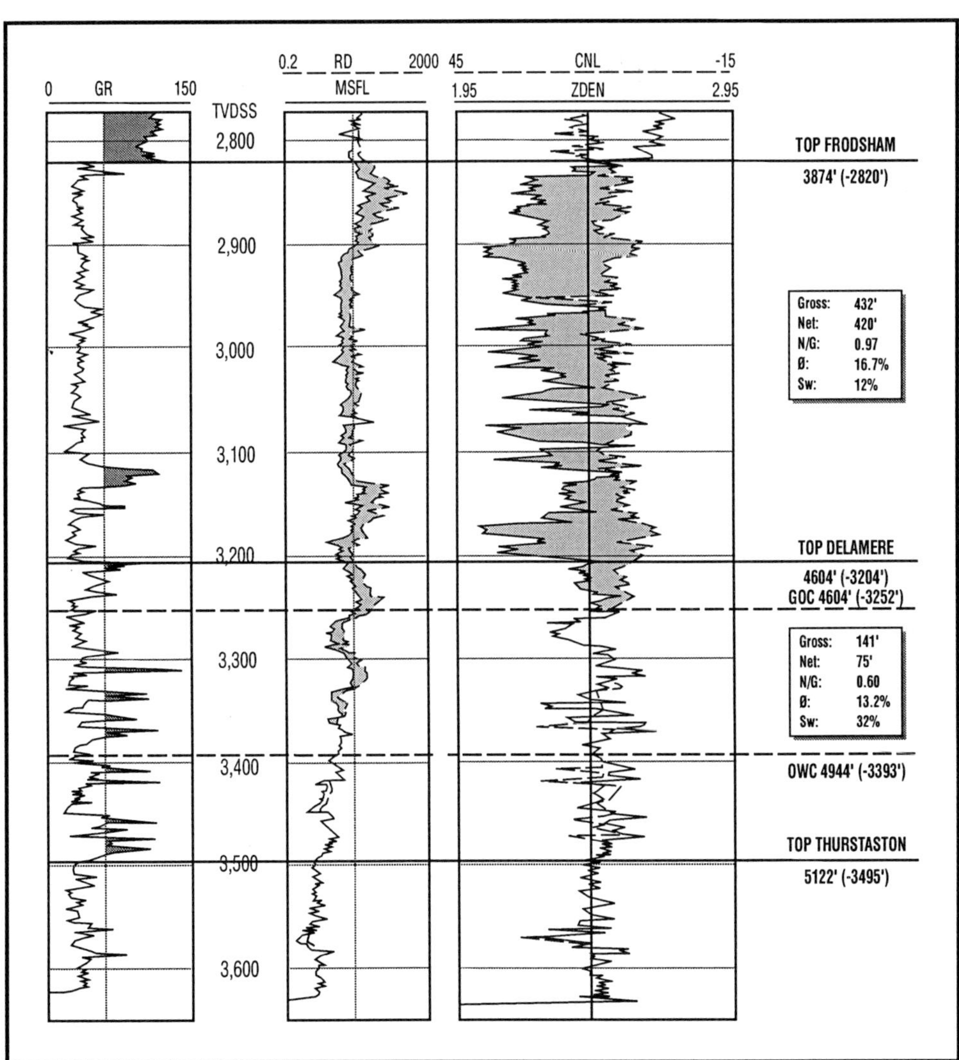

Fig. 7. Well 110/15-6Z Ormskirk Sandstone reservoir section.

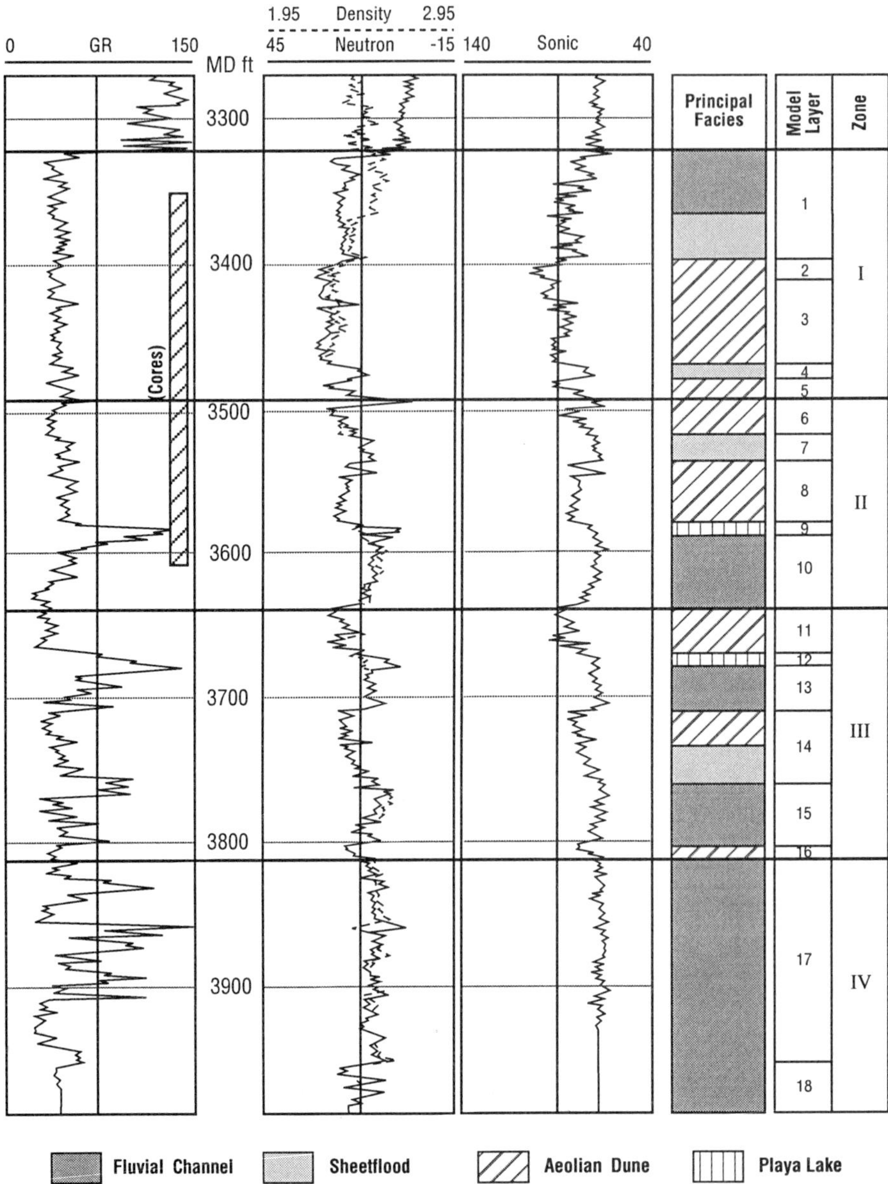

Fig. 8. Well 110/15-7 Ormskirk Sandstone reservoir section showing results of detailed facies analysis and reservoir model zones.

modelling of the behaviour of the reservoir when production from the Lennox Field commences.

Facies analyses

Detailed core and wireline analyses has identified the principal facies present within the Lennox reservoir resulting in recognition of four main reservoir zones.

- Zone I comprises predominantly aeolian facies overlain by a sequence of fluvial channel sandstones interpreted as reflecting an overall raising of the water table in the Lennox area towards the end of Ormskirk deposition. This zone corres-

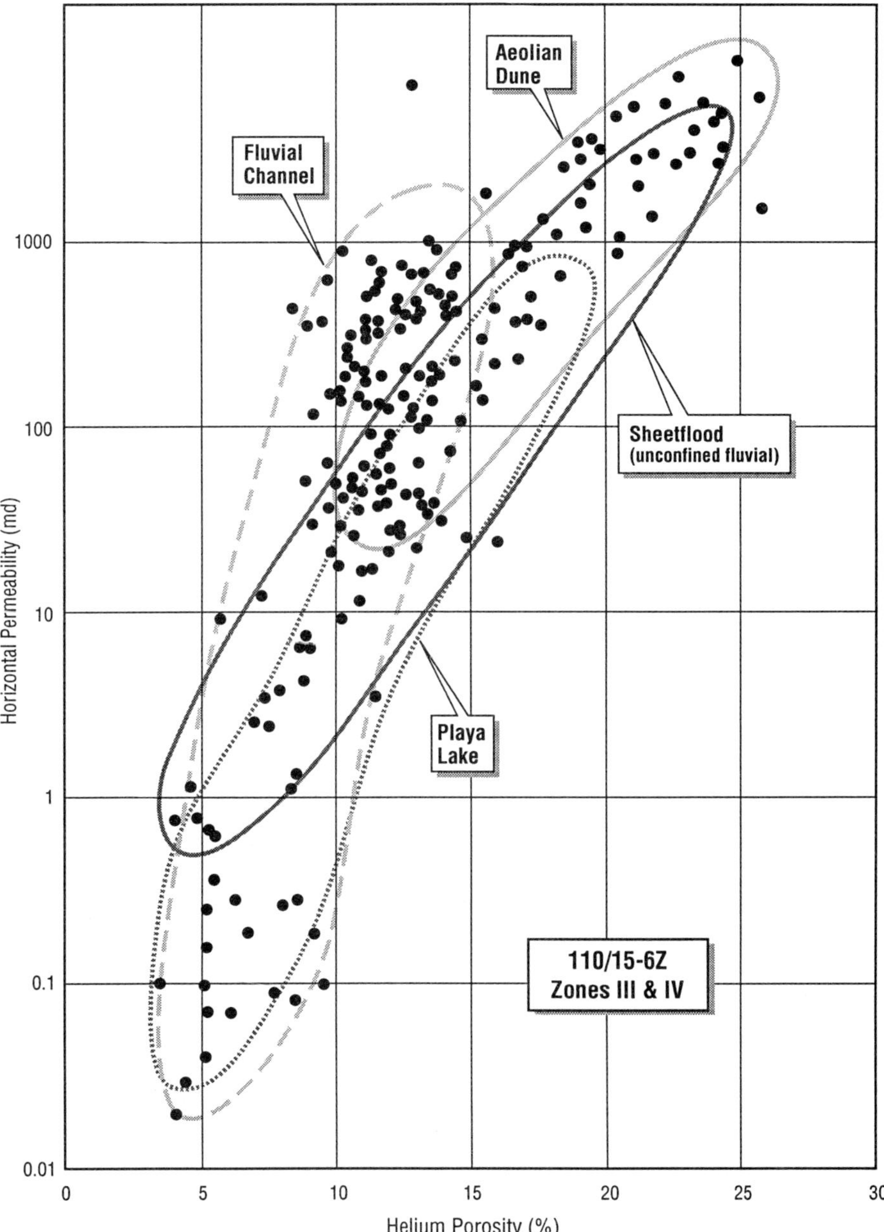

Fig. 9. Core porosity versus permeability data for well 110/15-6Z showing separate relationships for the main facies types.

ponds to the upper section of the Frodsham Member.
- Zone II comprises a fluvial sequence at the base overlain by predominantly aeolian facies. This zone corresponds to the lower section of the Frodsham Member.

- Zone III contains a mixed assemblage of fluvial channel deposits, aeolian dune sandstones and aeolian sheetflood deposits corresponding to the Delamere Member.
- Zone IV predominantly comprises a stacked sequence of fluvial channel deposits and is

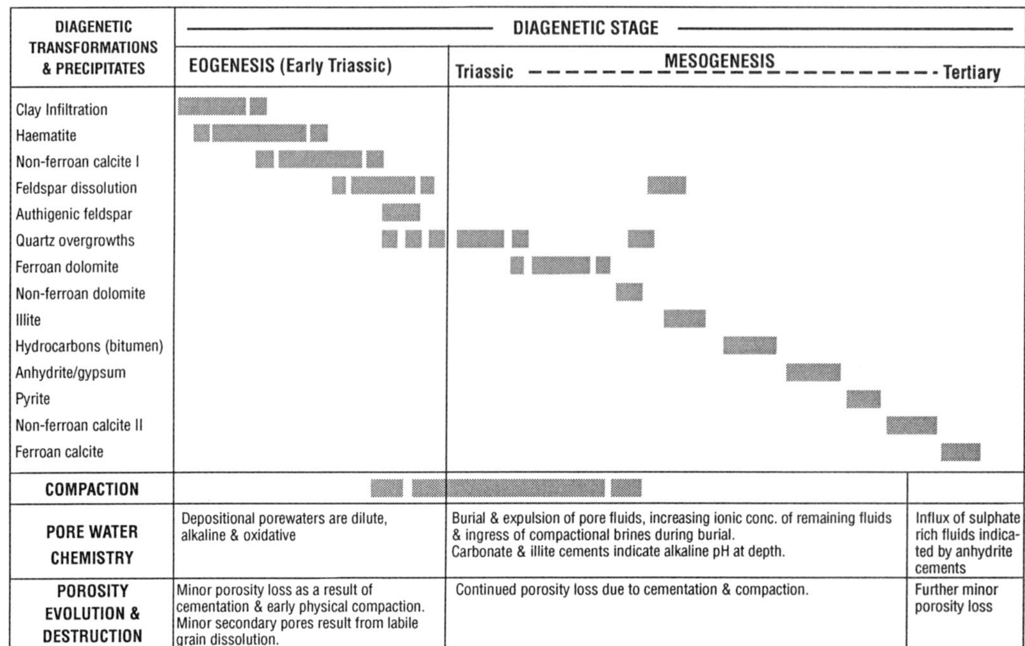

Fig. 10. Diagenetic history of the Ormskirk Sandstone reservoir in the Lennox Field. Note the late stage phase of bitumen emplacement.

Fig. 11. Well 110/15-7 core photomicrograph of aeolian sand facies. Note the pore rimming bitumen throughout, anhydrite cement (upper centre), and minor quartz overgrowths (lower left).

wholly contained within the Thurstaston Member.

These zones have subsequently been subdivided into 18 discrete layers for input to reservoir simulation modelling (Fig. 8).

Diagenesis

The Sherwood Sandstone Group reservoir sandstones of the Lennox Field exhibit a typical diagenetic sequence for the southern part of the East Irish Sea Basin (Fig. 10). During the eogensis phase (Early Triassic), the depositional porewaters were dilute, alkaline and oxidative resulting in early precipitation of non-ferroan calcite and minor feldspar dissolution. Early compaction resulted in the onset of minor quartz overgrowths. During the mesogenesis phase (Triassic to Tertiary), burial and expulsion of porefluids led to an increasing ionic concentration of the remaining fluids. The subsequent precipitation of carbonate and illite cements indicates an alkaline pH levels for these fluids. Porosity loss due to compaction and cementation is only minor indicating that palaeo burial depths have not been excessive. A phase of bitumen emplacement during the mesogenesis phase is indicative of an early phase of oil generation and migration. The bitumen is present as a pore-filling and pore-rimming phase which resulted in porosity and permeability reduction even in sandstones with excellent reservoir qualities (Fig. 11). The bitumen is often concentrated in thin layers marking-out the previously higher permeability bands in the cores (Fig. 12). A late stage influx of sulphate-rich fluids, post-bitumen emplacement, resulting in further porosity loss, is indicated by the presence of anhydrite cements.

Seismic interpretation

Seismic database

Block 110/15 is covered by a grid of good quality 2D seismic data (Fig. 13) comprising 586 km within the block and 311 km across the area of the Lennox Field. The data are a mixture of BHP 1992 proprietary data, traded data and speculative data. All the traded and speculative data were reprocessed to the same specifications as the BHP 1992 data. The quality of the proprietary data is good down to the Top Permian level and poorer below. The traded and speculative data are of slightly poorer quality due to their lower fold of coverage.

Horizon identification

The relationship between the synthetic seismogram and VSP, to the well results and their seismic cor-

relation is demonstrated for well 110/15-6Z (Fig. 14). The seismic character and time ties between the synthetic and the actual seismic data can be seen to be very good for the Sherwood Sandstone Group. The principal and most consistent seismic marker for the detailed interpretation is the Top Sherwood Sandstone Group, represented on the seismic by a strong black peak produced by a sudden decrease in acoustic impedance passing from the higher velocity Mercia Mudstone Group into the relatively lower velocity Sherwood Sandstone Group.

Hydrocarbon generation and migration

Oil to source rock correlation

A sterane fingerprint analysis of the crude oils from the Douglas, Lennox and Formby oilfields

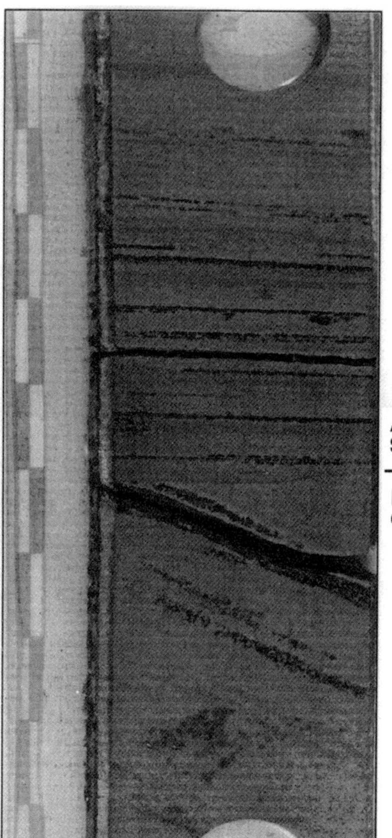

Fig. 12. Well 110/15-7 core photograph (3358′–3359′). Note black bitumen-stained laminae preferentially located in layers of originally higher permeabilities.

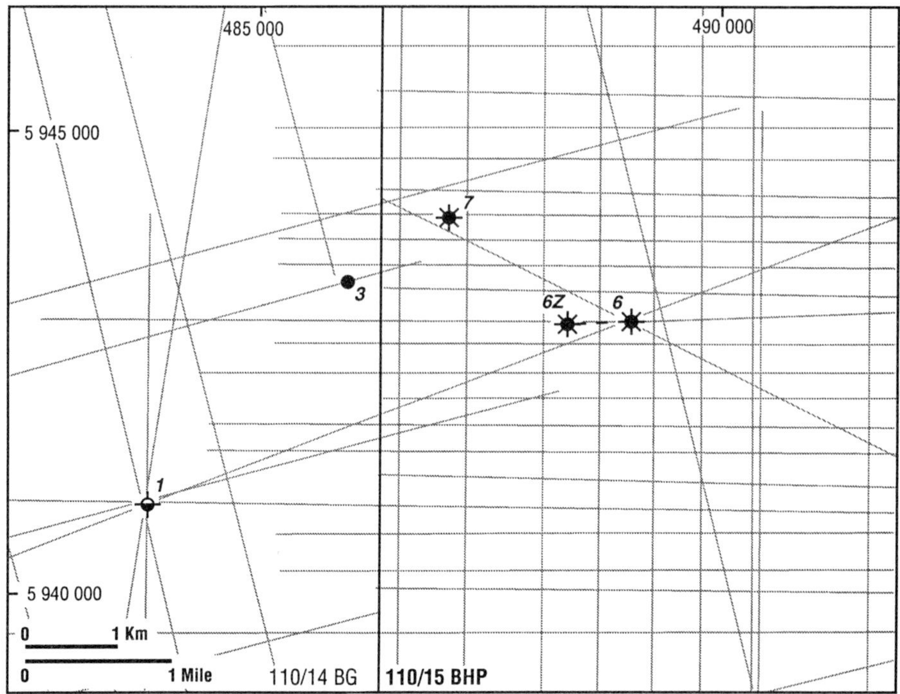

Fig. 13. Lennox Field – seismic database.

indicates that the oils in these three fields have all been generated from the same source rock (Fig. 15). Armstrong *et al.*, this volume) have identified the Namurian Holywell Shale as the source for both the gas and the oil reservoired in the Lennox Field. A map of vitrinite reflectance for the East Irish Sea Basin indicates that the Holywell Shale within the East Deemster Basin is mature for gas generation with vitrinite reflectance values of $R_o > 2.0\%$. On the footwall of the Formby Point Fault, however, values of $R_o \sim 1.0\%$ indicate that the Holywell Shale is currently mature for oil generation (Fig. 16).

Migration of hydrocarbons

The gas and oil reservoired within the Lennox Field, although sourced from the same stratigraphic source interval, are considered to have been sourced from different localities. The gas has been sourced from the deeply buried East Deemster Basin with migration up the Formby Point Fault. The oil is considered to have been sourced from the footwall of the Formby Point Fault with migration across the fault into the Lennox structure (Fig. 17). Current mapping of the Lennox structure indicates that the field may be full to spill with oil spilling across the fault and reservoiring in the footwall of the Formby Point Fault, e.g. the Formby Field.

A detailed picture of the hydrocarbon generation history for the southern EISB has been built up (Fig. 18) with slightly differing generating histories for the Douglas, Hamilton and Lennox Fields.

- Early oil generation during Jurassic times is thought to have occurred throughout the EISB with migration into the early formed structures. Cimmerian uplift and erosion is the likely cause of the breaching of these structures, resulting in the release of the mobile hydrocarbons and the bitumen staining seen today.
- Further deposition and burial during the Cretaceous resulted in renewed oil generation in Lancashire and in the Douglas area. This was followed by the main phase of gas generation during the latest Cretaceous in the more deeply buried sub basins of the EISB.
- The Tertiary was marked by a significant period of uplift, erosion and non-deposition across the entire EISB. This has not resulted in the breaching of the hydrocarbon-bearing structures.

Estimates of the amount of uplift undergone in the EISB vary between 1 and 3 km however the

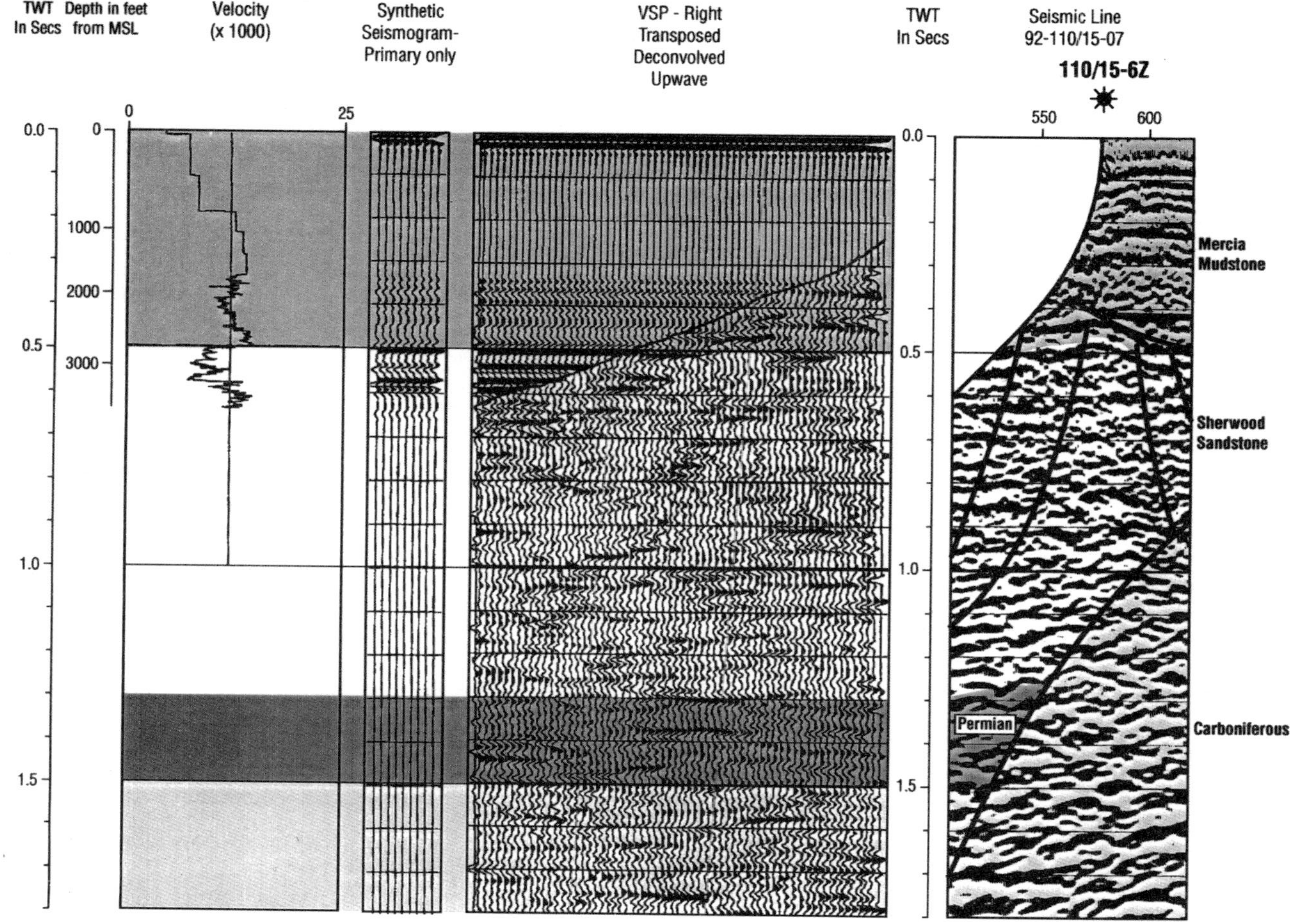

Fig. 14. Lennox Field, well 110/15-6Z synthetic seismogram and VSP.

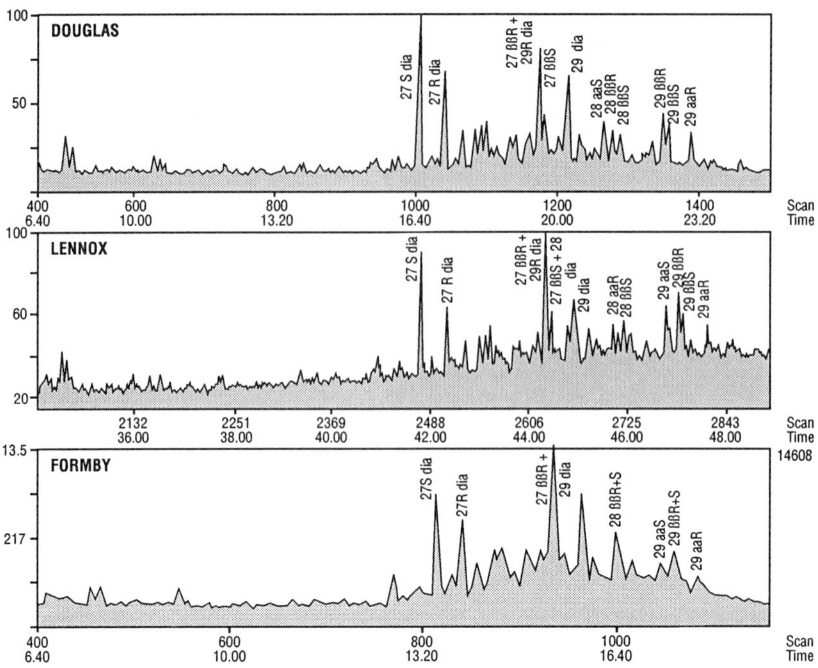

Fig. 15. Sterane fingerprint analysis for the Douglas, Lennox and Formby oil Fields.

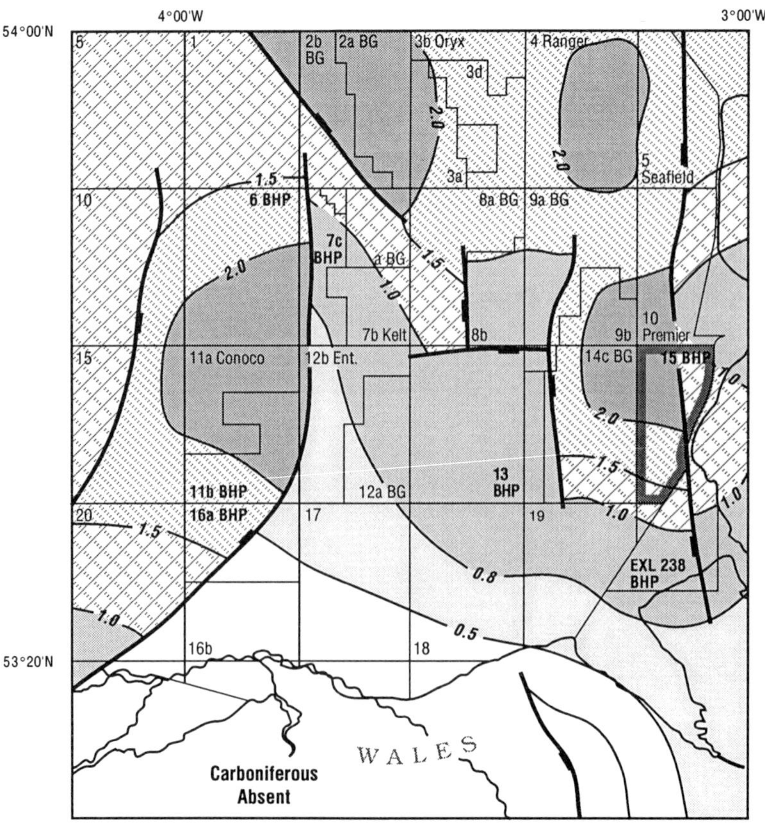

Fig. 16. East Irish Sea Basin (south), Holywell shale vitrinite reflectance map.

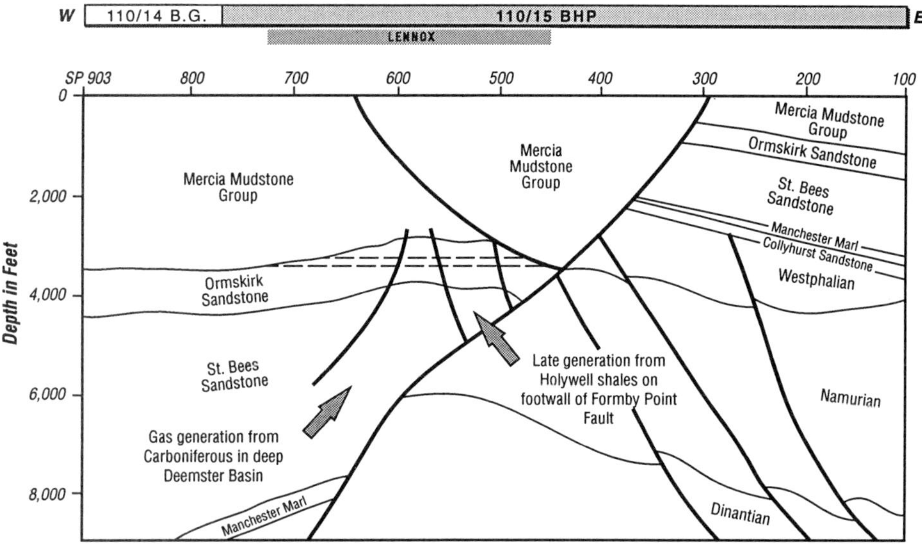

Fig. 17. Geoseismic line through the Lennox Field demonstrating migration models for oil and gas.

lower estimates are considered most likely due to the absence of (a) large scale halokinesis within the overlying Mercia Mudstone, and (b) major compaction-related diagenesis within the Ormskirk Sandstone. Both of these effects would be expected with larger amounts of burial.

Field data

RFT data

A plot of the RFT pressure data for the 110/15-6, 6Z and 7 wells (Fig. 19) demonstrates a well defined gas leg with a pressure gradient of

	TECTONICS	DOUGLAS	HAMILTON & HAMILTON NORTH FIELDS	LENNOX FIELD
TERT.	UPLIFT (~1km)	TRAP KEEPS INTEGRITY	TRAP KEEPS INTEGRITY	TRAP KEEPS INTEGRITY
		NHG GENERATION FROM DEEP BURIED CARB. LIME.		
CRETACEOUS	HEAT PULSE	LIGHT CONDENSATE GENERATION	GAS GENERATION IN EAST DEEMSTER	LIGHT CONDENSATE GENERATION IN LANCASHIRE
	BASIN SAG DEPOSITION	? PARTIAL BIO-DEGRADATION ?		
		RENEWED OIL GENERATION	RENEWED OIL GENERATION	RENEWED OIL GENERATION IN LANCASHIRE
JURASSIC	CIMMERIAN UPLIFT & EROSION	BREACH OF STRUCTURE	BREACH OF STRUCTURE(S)	BREACH OF STRUCTURE
	BASIN SAG DEPOSITION	BITUMEN STAINING	BITUMEN STAINING	BITUMEN STAINING
	RIFT BASIN DEPOSITION	EARLY OIL GENERATION	EARLY OIL GENERATION	EARLY OIL GENERATION
TRIASSIC	DEPOSITION OF MERCIA MUDSTONE GROUP	SEAL	SEAL	SEAL
	DEPOSITION OF SHERWOOD SANDSTONE GROUP	RESERVOIR	RESERVOIR	RESERVOIR

Fig. 18. East Irish Sea Basin (south), hydrocarbon generation history. (NHG = non hydrocarbon gases.)

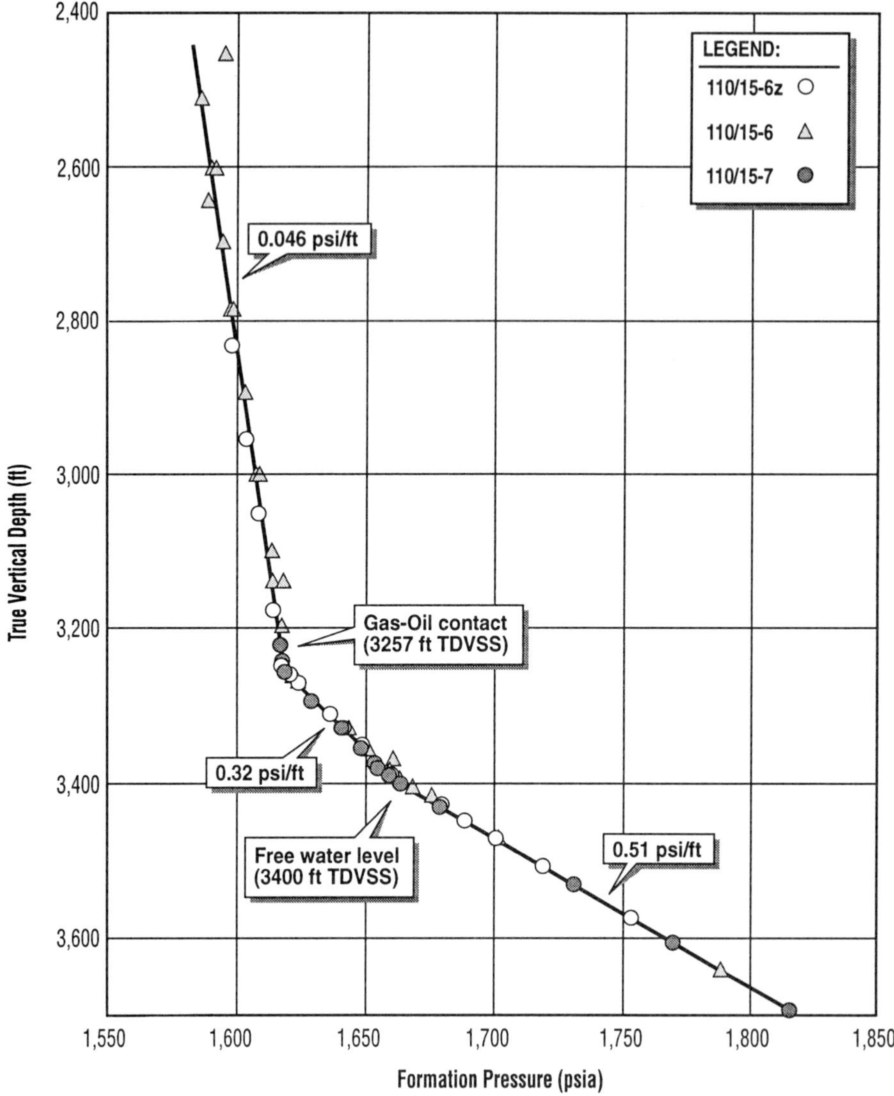

Fig. 19. Lennox Field repeat formation tester pressure data, wells 110/15-6, 6Z and 7.

Table 1. *Lennox Field hydrocarbon-in-place estimates*

	West	*Central*	*East*	*Total*
Gas-in-place (BCF)	206	125	132	463
Oil-in-place (MMBO)	123	18	77	218

Reservoir zone	*Gas-in-place (BCF)*	*Oil-in-place (MMBO)*
Zone I	258	85
Zone II	113	51
Zone III	68	55
Zone IV	24	27

0.046 psi/ft down to a gas–oil contact of 3257 ft TVDSS, an oil leg with a pressure gradient of 0.32 psi/ft down to a free water level of 3400 ft TVDSS, and an aquifer gradient of 0.51 psi/ft below this.

Hydrocarbons-in-place

The hydrocarbon-in-place estimates for the Lennox Field are presented in Table 1. Approximately 56% of the oil and 44% of the gas are located within the relatively unfaulted western area of the field. The more severely faulted eastern area contains 35% of the oil and 29% of the gas. Subdivision of the reserves by reservoir zone indicates that, as expected, the vast majority if the oil and gas is reservoired within Zone I.

Fluid analysis

Pressurized separator oil and gas samples and bottom hole samples were obtained from the oil leg during drill stem tests on wells 110/15-6 and 110/15-7. Pressurized gas and condensate samples from the gas leg were sampled at surface during drill stem testing on well 110/15-6 and wireline formation tester samples were obtained in wells 110/15-6Z and 110/15-7. Extensive laboratory analyses have been.performed on these samples.

The gas leg of the Lennox Field contains a dry gas with estimated condensate to gas ratios of 3.0 STB/MMSCF. The measured PVT parameters for the gas column, as obtained from the wireline formation tester sample from well 110/15-7 are summarized in Table 2.

The Lennox Field oil rim contains a light oil (45° API) with a moderate shrinkage ($B_o = 1.3$ RB/STB) and GOR (620 SCF/BBL). Saturation pressure measurements of both the bottom hole samples and surface recombination samples suggest the reservoir oil is fully saturated. Measured differential liberation data indicate that the oil from well 110/15-7 is marginally heavier than well 110/15-6 with a slightly lower solution GOR and

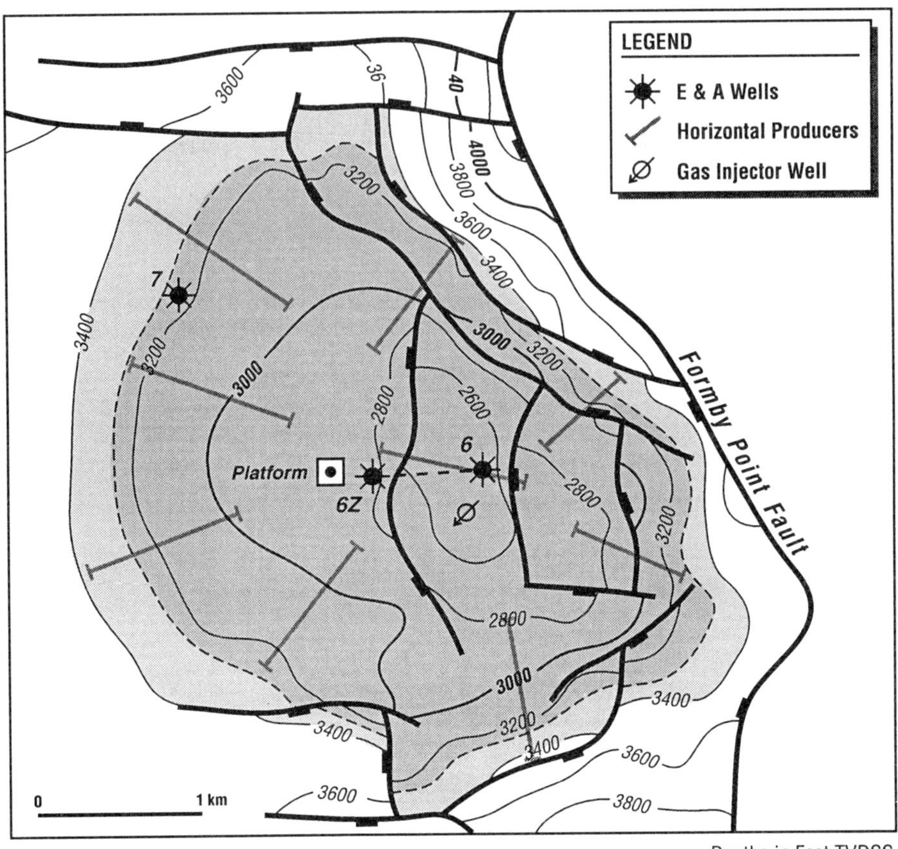

Fig. 20. Lennox Field development well locations.

Table 2. *Lennox Field Reservoir fluid composition – gas cap*

Compound	(Mol %)
Nitrogen	12.263
Hydrogen sulphide	0.040
Mercaptans	0.004
Carbon dioxide	0.120
Methane	77.479
Ethane	5.247
Propane	2.499
Isobutane	0.410
n-Butane	0.959
Isopentane	0.240
n-Pentane	0.250
Hexanes	0.210
Heptanes	0.190
Octanes	0.080
Nonanes	0.010
Decanes[+]	0.000

Table 3. *Lennox Field Reservoir fluid composition – oil leg*

Initial pressure	1620 PSIA (at datum of 3257 ft TVDSS)
Reservoir temperature	95° F
B_o	1.30 RB/STB
Bubble point	1620 PSIA
Oil compressibility	11×10^{-6} PSI^{-1}
Oil viscosity	0.5 CP (at reservoir conditions)
GOR	620 (SCF/BBL)
Oil gravity	45° API

lower shrinkage. The results of the PVT analyses of the Lennox oil are summarized in Table 3.

Field development and production

Development

The Lennox Field will be developed by nine horizontal oil producer wells drilled radially from a central platform location (Fig. 20). These wells will be drilled to penetrate the oil leg of the reservoir at a level approximately 50 ft above the oil–water contact and 100 ft below the gas–oil contact (Fig. 21). A typical producer well will have a horizontal step-out from the platform of over 7500 ft with a horizontal section through the reservoir of approximately 3500 ft. Solution gas will be separated and recycled at the Douglas central processing complex and reinjected into the Lennox gas cap via a

centrally located vertical gas well. This will maintain reservoir pressure within the gas cap, minimize the extent of water breakthrough and allow oil production to be maximized.

The Lennox oil and gas Field is to be developed as an integral part of the Liverpool Bay Integrated Development scheme linking the Hamilton and Hamilton North gas fields and the Douglas oil Field in Block 110/13 with the Lennox Field (Fig. 22). A not-normally manned platform at Lennox will be connected to the Central Processing Platform located at Douglas via a 14″ oil pipeline, a 16″ gas pipeline and a 12″ gas injection pipeline. Oil produced from Lennox will be commingled with Douglas crude, processed and then transmitted via a 14″ pipeline to a storage tanker to await loading onto a shuttle tanker for shipment to a refinery.

Production

First oil production from Lennox is expected to be in November 1995 with a peak production rate of over 30 000 STB/D (Fig. 23). The initial gas produced in solution with the oil will be recycled and reinjected into the Lennox gas cap via a crestally located gas injector well. When oil production starts to decline and water breakthrough

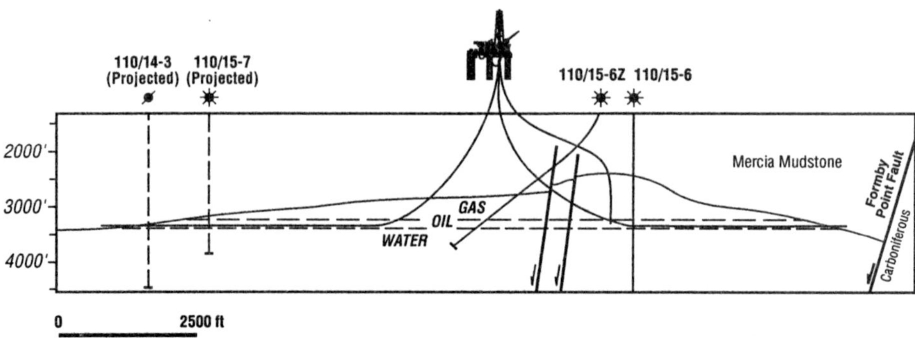

Fig. 21. Lennox Field true scale east–west cross-section showing typical oil-producing well track.

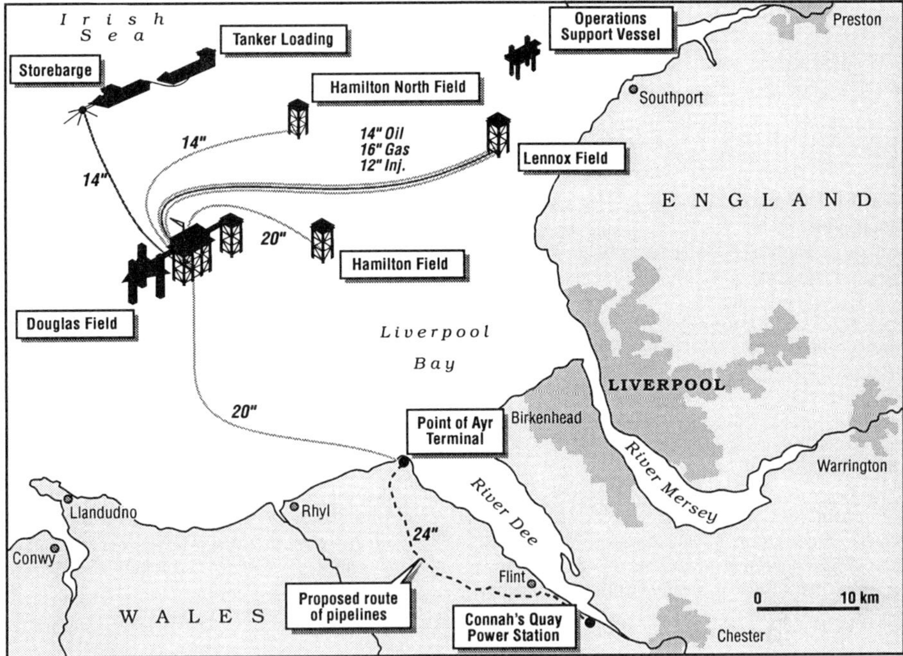

Fig. 22. Liverpool Bay integrated development scheme.

increases, this gas injector will be switched to gas production with gas being transmitted via the 12" injector pipeline to the Douglas Processing platform, commingled with gas from the Hamilton and Hamilton North gas fields, and transmitted via a 20" pipeline to BHP's Point of Ayr gas terminal. From there the gas will be piped to the PowerGen operated power station at Connah's Quay. First

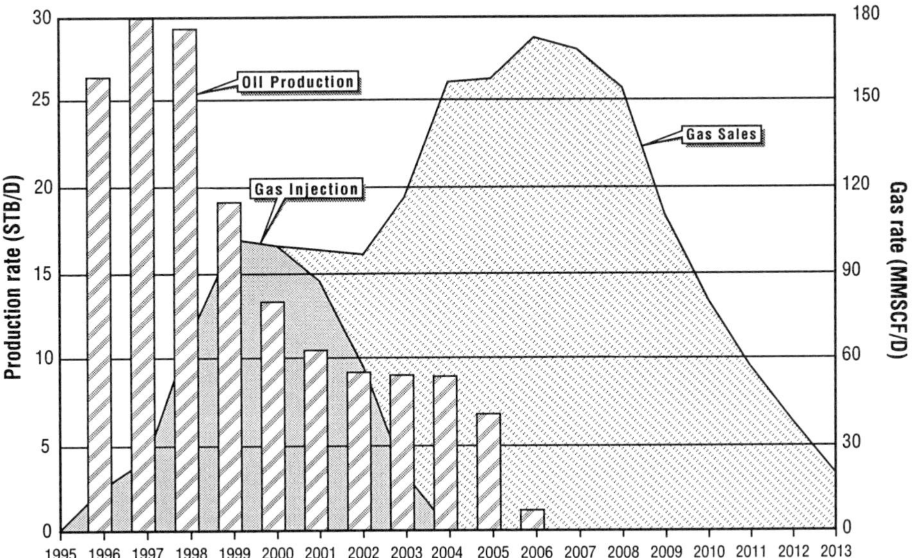

Fig. 23. Lennox Field most likely case production profiles.

sales gas production from the Lennox Field is anticipated in the year 2000 when the Lennox gas is expected to extend the production plateau from the Hamilton and Hamilton North Fields.

The authors would like to express their thanks to the P791 group of companies for their permission to publish this article.

References

ARMSTRONG, J. P., SMITH, J., D'ELIA, V. A. A., & TRUEBLOOD, S. P. 1997. The occurrence and correlation of oils and Namurian source rocks in the Liverpool Bay–North Wales area. *This volume.*

BARNES, R. P., AMBROSE, K., HOLLIDAY, D. W., & JONES, N. S. 1994. Lithostratigraphical subdivision of the Triassic Sherwood Sandstone Group in West Cumbria. *Proceedings of the Yorkshire Geological Society,* **50**, 51–60.

BRITISH GEOLOGICAL SURVEY 1997. A revised stratigraphic nomenclature for the East Irish Sea basin UKOOA/BGS.

HAMILTON OIL COMPANY, 1993. Lennox Field development plan. Vols 1 & 2.

HARDMAN, M., BUCHANAN, J., HERRINGTON, P. & CARR, A. 1993. Geochemical modelling of the East Irish Sea Basin: its influence on predicting hydrocarbon type and quality. *In*: PARKER, J. R. (ed.) *Petroleum Geology of Northwest Europe: Proceedings of the 4th Conference,* Vol. 2. Geological Society, London, 809–821.

JACKSON, D. I. & MULHOLLAND, P. 1993. Tectonic and stratigraphic aspects of the East Irish Sea Basin and adjacent areas: contrasts in their post-Carboniferous styles. *In*: PARKER, J. R. (ed.) *Petroleum Geology of Northwest Europe: Proceedings of the 4th Conference,* Vol. 2. Geological Society, London, 791–808.

MEADOWS, N. S. & BEACH, A. 1993. Controls on reservoir quality in the Triassic Sherwood Sandstone of the Irish Sea. *In*: PARKER, J. R. (ed.) *Petroleum Geology of Northwest Europe: Proceedings of the 4th Conference,* Vol. 2. Geological Society, London, 823–833.

Index

Some figures are also grouped under **cross-sections, stratigraphic sections** and **wireline logs**.

Aalian, 54–5
Abbey Mills, 197
Acadian, 63, 66–71, 125–31, 153, 297–9
   see also Caledonian reactivation
acritarchs, 33, 45, 137, 239
aeolian sandstones, 3, 172–3, 353–6, 373–6
   lithostratigraphy, 20–1, 24–7
   Liverpool Bay, 384, 403–16, 422–6
   Northern Ireland, 100
   Ormskirk Sandstone Formation, 75, 237–8, 242–50,
     253–60, 320
   sabkha, 260–274
AFT, 3–7, 219–26, 380–1
   EISB, 73–93, 343–4, 355, 362
   North Channel, 49–59, 104, 108, 124–5
Albian, 52, 56–7
Alderley Fault, 301–3, 326–7, 330
Alderley Horst, 4, 179, 325–51
algae, 145, 247, 262–4, 281–3
allocyclic control, 253–74
amplitude anomaly, 185–94, 388–91
Anderton, 179
Anglesey, 13, 49, 51–7, 83, 146, 168
anhydrite, Easton-1, 277–294
   Kimmeridge, 56
   Ormskirk Sandstone Formation, 6, 14–8, 23, 242–3
   South ESIB, 354–9, 365–9, 376, 410, 426–7
Annaghmore, 93
Ansdell Mudstone Member, 12–17, 19, 400
Ansian, 239–57
   see also Mercia Mudstone Group
Antrim basalts, 47, 50–1, 102–3, 115–23, 129–31, 213,
   216
Appleby Group, 11–14, 24–7
   see also Collyhurst, Penrith, Locharbriggs Sandstone
     Formations, Tormentil Volcanics
Archerbeck, 278
Auchencairn Bay, 213–4, 217
authigenic minerals, 213, 354–62, 377–8, 410

bacteria, sulphate reducing, 6, 229–35
Bajocian, 54–5, 57
Bakevillia Sea Basin, 11, 23, 25–6, 128
Bala Fault, 297–9, 318
Bala–Llanelidan Fault Zone, 163, 173–8
Ballavarkish, 136, 146–7
Ballyalton, 100
Ballycastle, 41, 100
Ballymacilroy, 38, 41–2, 116–7, 123–4, 131
Ballynumullen-1, 95
Ballytober, 95–6, 113–4, 116, 124, 128
Ballytober Fault, 120, 127–8
barite cement, 220, 326–50, 353–69, 376
Barrowmouth Mudstone Formation, 12, 21–6
   see also Manchester Marls and St Bees Evaporites
     Formations

Basal Breccia, 27, 281
Basal Units, Cumbrian Coast Group, 12, 24–6
basin orientations, 61, 97, 114–6, 126–31, 214, 297–312
Bathonian, 54–5
Beacon Lodge Sandstone Member, 328, 331
'beef', 117–9, 127–8
Belah Scar, 318–22
Belfast Harbour Evaporite Formation, 100–3
Belfast Harbour-1, 100–2, 117, 293
Berger curve, 292–4
Berw Basin, 13, 34–5, 51–4
Bewcastle Beds, 277, 280–1, 283, 289–91
Bickerton-Bulkley Hill-East Delamere Faults, 169, 175,
   180–1
biodegradation, 203–11
biomarker ratios, Hoylwell shales, 206–211
Bisat Group, 12, 12–14, 24–5, 27–30
bitumen, 145, 173, 325, 385
   analysed, 203, 206–7
   Cheshire Basin, 173
   North Solway Fault, 5, 215, 222–7
   South EISB, 358–9, 405, 410–11, 413–6, 426–34
Blackcraig, 217, 219–22, 227
Blackpool Mudstone Member, 12–19, 400
Blacon-1, 164, 173
Blakenhall, 163–4, 164, 166–9, 173
Block 110/8, 185
Block-111, 93
Blocks 110/13+14+15, 353–69, 380–5, 401–4, 417–36
Blocks 110/13, 110/14, 110/15, 353–69
Blocks-110, 51
Blue Anchor Formation, 12–13, 15, 36–7, 40–2, 166
Bodysgallen Hall, 198
Bogside Limestone, 280–6
Bold Formation, 24–5
Bollin Mudstone, 166, 168, 170–3, 179–80
Bolonian, 56–7
Boots Green Fault, 171, 180–1
Boots Green-1, 163–4
Border Groups, 277–94
Bowsey Wood, 164, 173, 176–8
bravoite, 326, 340
Bray Fault, 35–6, 54
Breckells Mudstone Formation, 15, 421
Bridgemere Fault, 298–305
brine, 37, 52, 217–22, 325–51
   Larne, 219
   Liverpool Bay, 410, 427–31
   reflector, 186–93
   Sherwood Sandstone Group, 353–74
Brockelschiefer, 4, 23, 123
Brockram facies, 22, 26–7, 102, 104
Brook House Fault, 300–3
Brooks Mill Mudstone, 15, 166, 170–2, 180
Brylow quarry, 329
Bucklow Fault, 169, 175, 181
Budleigh Salterton Pebble Beds, 23